Peter Stavroulakis (Ed.)

Third Generation Mobile Telecommunication Systems

Engineering ONLINE LIBRARY

http://www.springer.de/engine/

Springer-Verlag Berlin Heidelberg GmbH

Peter Stavroulakis (Ed.)

Third Generation Mobile Telecommunication Systems

UMTS and IMT - 2000

With 313 Figures

Springer

Professor Peter Stavroulakis
Telecommuncation System Institute (TSI)
37 Iroon Polytehniou str.
73133 Chania, Crete
Greece
e-mail: peter@tsinet.gr

Cataloging-in-Publication data applied for

Die Deutsche Bibliothek - CIP-Einheitsaufnahme
Third generation mobile telecommunication systems: UMTS and IMT-2000 / Peter Stavroulakis (ed)
Berlin; Heidelberg; NewYork; Barcelona; Hong Kong; London; Milano; Paris; Singapore; Tokyo:
Springer, 2001
(Engineering online library)

ISBN 978-3-540-67850-2 ISBN 978-3-642-56919-7 (eBook)
DOI 10.1007/978-3-642-56919-7

Originally published by Springer-Verlag Berlin Heidelberg New York in 2001

Typesetting: Camera ready copies from editor
Cover-Design: medio Technologies AG, Berlin
Printed on acid-free paper SPIN 10778061 62/3020 Rw 5 4 3 2 1 0

To my wife Nina whose love and
patience helps me continue when
work seems insurmountable

Acknowledgement

The editor is indebted to the staff of Telecommunications Systems Institute for their help and support throughout the preparation of this book. He is also especially appreciative of the work of his assistant H. Sandalidis who devoted countless hours in the collection and correction of the individual contributions. This book would not have been possible without the collaboration of the individual contributors to whom I'll owe a lifelong obligation.

List of Contents

1 Introduction

One hundred years ago, the notion of transmitting information without the use of wires must have seemed like magic. In 1896, the first patent for wireless communication was granted to Marchese Guglielmo Marconi. Since then the field of wireless communications which includes cellular systems has taken various forms of development. It basically evolved through three Eras. The Pioneer Era over the period of 1860-1921, the Precellular Era over 1921-1980 and the Cellular Era after 1980 and beyond. The first generation cellular era started with the Analog Systems and evolved in the digital domain utilizing Time Division Multiple Access (TDMA) and Code Division Multiple Access (CDMA), thus comprising the Second Generation Mobile Systems.

The first generation RF cellular communications systems deployed in the early to mid 1980's had air interfaces comprised of analog technology. Among them were AMPS (Advanced Mobile Phone System), NMT (Nordic Mobile Telephone), and TACS (Total Access Communications System). These were designed for use in a specific geographic area and not intended to be deployed in other areas. There was not much commonality beyond using the same air interface technology and same modulation. The air interface technology was Frequency Division Multiple Access (FDMA) and the modulation was analog FM, but with different deviations and channel spacings. The frequency bands, air interface protocols, number of channels, and data rates were different. In general, these systems provided local and national coverage.

As subscriber growth increased, capacity issues began to arise. A need for second generation systems was evident. In addition, some areas desired communication systems across countries as in Europe. The deployment of GSM (Global System for Mobile communications) in 1992 marked the beginning of second generation systems. NADC (North American Digital Cellular) and PDC (Personal Digital Cellular) followed in North America and Japan, respectively. For these second generation systems TDMA and FDMA air interface technologies were combined and used to help provide additional capacity. Again the frequency bands, air interface protocols, numbers of channels, data rates and, in this case, modulation were also different. For example, GSM used GMSK (Gaussian Minimum Shift Keying) and NADC used DQPSK (Differential Quadrature Phase Shift Keying) for the modulation type. These systems provided digital speech, privacy, and additional capacity.

Around 1995, enhanced second generation systems were introduced, some times referred to as second generation plus. These systems primarily centered around up-banding the existing systems to the 2 GHz PCS frequencies as well as the deployment of **cdmaOne**™ (IS-95 CDMA) systems at both the cellular and

PCS frequencies. **cdmaOne™** systems introduced a completely different air interface technology to RF cellular communications. It used code division multiple access, which promised even more capacity over the TDMA systems. With the increasing desire for systems with data transmission services these enhanced second generation systems offered additional services such as short messaging service, and improved the digital speech. Unlike the first generation systems that were designed for a specific geographic area, these second and enhanced second generation systems were intended to be deployed internationally.

Currently under development are third generation systems (3G). IMT -2000 is intended to be a global standard offering higher bit-rate services comparable to current wireline services. Services such as multimedia (video), internet, and e-mail are among those anticipated to fuel the subscriber growth for 3G systems.

The International Telecommunications Union (ITU) is an international standards body which is part of the United Nations. IMT-2000 is the ITU's vision of global wireless access in the 21st century. IMT stands for International Mobile Telecommunications; IMT-2000 is the new name for Future Public Land Mobile Telecommunications System (FPLMTS). FPLMTS was targeted at developing mobile telecommunications systems to be used "Anywhere-Anyplace" around the year 2000 and operate at approximately 2000 MHz. Hence, the name IMT-2000. The main data rates are: 1) up to 144 Kbits/second, mobile (vehicular) data rate; 2) up to 384 Kbits/second, portable (pedestrian) data rate; and 3) up to 2 Mbits/second, fixed data rate.

During the World Administrative Radio Conference in 1992 the frequency bands 1885-2025 MHz and 2110-2200 MHz were identified for use by countries worldwide intending to implement IMT-2000. It further identified bands 1980-2010 and 2170-2200 for the satellite component. The satellite component enables the deployment of services where there is little or no existing fixed infrastracture, such as in developing countries. In most countries this frequency spectrum has been reserved, however in the US much of it has been used for PCS (1850-1990 and 1930-1990 MHz).

One of the more popular technologies being proposed for IMT-2000 is wideband CDMA. Two versions of this technology appear to be the leading contenders for global standardization. While they differ in the details of their implemantation, they both promise wireless voice at significantly higher capacity and lower cost than can be provided by currently available systems. One version of this technology was proposed by ARIB (Association of Radio Industries and Businesses), Japan's regional standardization body. It is known as W-CDMA.

Another version of Wideband CDMA technology was proposed by the US's TIA. It is known as cdma2000. This is an advanced derivative, with a 5 MHz spreading bandwidth, of the TIA/EIA-95-B CDMA system that is currently deployed around the world. It is based on the IS-41 core network infrastracture.

In this book, major aspects of 3G systems are presented which cover coding, modulation, antennas and propagation, channel characterization and modeling, system architecture, traffic characterization and management, capacity as well as the evolving standard of 3G. In chapter 9, W-CDMA is examined from the point of view of power control, timing acquisition and network synchronization. Finally, the subject of services is covered in chapter 10.

References

1. R. Prasad and T. Ojanpera, Tutorial " Third Generation Wireless Personal and Multimedia Communications".
2. The 1999 3G Wireless R+D Symposium, Hewlett Packard.

2 CODING

2.1 Introduction

Coding plays a major role in the efficient transmission and improved error performance of the third generation mobile systems, as well as any future wireless system. This chapter will cover some major techniques in coding that have been adapted in the general framework of modern mobile systems. More specifically it will cover a partial response signaling with trellis coded modulation, and a source and channel coding technique as well as a multi-code CDMA and a source combining technique based on collaborative coding multiple access (CCMA).

In section 2.2 a blind maximum likelihood channel estimation algorithm is developed for quadrature partial response-trellis coded modulated (QPR-TCM) signals propagating through a Rician fading environment. A hidden Markov model (HMM) formulation of the problem is introduced and the Baum-Welch parameter estimation algorithm is modified to provide computationally efficient solution to the resulting optimization problem. Performance analysis of the proposed method is carried out through the evaluation of bit-error probability upper bound for Rician fading channels. Also, some illustrative simulations are presented.

Section 2.3 considers a link adaptation technique to select the optimum combination of source and channel coding for a particular link quality in 3rd generation mobile systems environment.

In section 2.4 a source combining technique based on collaborative coding multiple access (CCMA) is applied to UMTS which complies with the universal terrestrial radio access (UTRA). The basic concept of CCMA technique based on trellis coding together with a simple 2-user uniquely decodable scheme is presented. The use of CCMA as a source combiner for multimedia applications is presented and its performance over a selective fading channel based on the UTRA system is recorded.

In section 2.5 a low-powered multi-code correlator architecture is introduced for multi-code despreading. The suggested architecture can de-spread the input with several PN codes concurrently and both hardware overhead and power consumption are reduced significantly as the code number is two. According to the simulation results, about 24% and 46% power reductions are achieved compared with previous architectures, and the power efficiency is about 1 GOP/Watt. Some preliminary results are also discussed for multi-code correlator, including both hardware and software approaches. In fact such an architecture also can be used for code acquisition or code tracking.

2.2 Blind Equalization of Quadrature Partial Response-Trellis Coded Modulated Signals

Osman N. Ucan and Hakan A. Cirpan

2.2.1 Introduction

The ever increasing demand for reliable transmission of high quality voice and digital data in mobile communication systems has generated a great interest in developing both bandwidth and power efficient modulation techniques. Ungerboeck [1] has integrated multi-level modulation schemes and coding into a single system which is referred to as trellis coded modulation (TCM), to improve error performance in high data rate communication over bandlimited channels. To further increase the bandwidth efficiency of conventional modulation techniques, a controlled amount of intersymbol interference, known as partial response signalling (PRS) is introduced [2],[3]. In order to combine attractive features of PRS and TCM, quadrature partial response-trellis coded modulation (QPR-TCM)[4],[5] scheme is proposed. These signals have many desirable properties such as spectral efficiency and resistance to fading conditions.

The most frequently assumed model for a transmission system is the additive white Gaussian noise (AWGN). Unfortunately, in many cases digital transmission is also subject to fading due to multi-path reflections. In mobile satellite communication, multi-path can be modelled by Rician statistics. Hence, estimation of Rician fading channel parameters is of considerable practical interest and worth further investigation.

The goal of the present work is to introduce a general maximum likelihood (ML) blind method for estimating the Rician channel excited by QPR-TCM signals. We concentrate on ML since these estimation algorithms perform satisfactorily when only a short data record is available [6]. However, implementation of the method of ML is fairly involved for the blind problem at hand. We therefore need to adopt an iterative approach to maximize the likelihood.

The finite memory of the QPR-TCM along with the independent identicial distribution and finite alphabet structure of the input data allows us to model the systems response as a Markov chain [7] . However, the Markovchain is hidden since it can only be inferred from the noisy observations. We should therefore consider the hidden Markov model formulation of the problem and employ the computationally efficient Baum-Welch algorithm in order to find ML estimates of the distribution parameters of the hidden Markov chain.

The Baum-Welch algorithm was originally developed for the purpose of finding the ML estimates of the distribution parameters of a Markov chain probabilistic function and has been widely employed to adjust the model parameters of HMM's in the context of speech processing. It has also been applied to communication and source separation problems for estimating linear and non-linear parameters [8].

2.2.2 Motivation for QPR-TCM System

There is a growing need for reliable transmission of high quality voice and digital data in satellite-based land mobile communication systems. These systems which will be part of an emerging all digital network, are both power and bandwidth limitation, one can employ bandwidth efficient modulation techniques such as those that have been developed over the past several years for terrestrial communication systems.

In the past, coding and modulation were treated as separate operations with regard to overall system design. In particular, most earlier works on coded digital communication systems independently optimized: 1) conventional (block or convolutional) coding with maximized minimum Hamming distance and 2) conventional modulation with maximally separated signals.

In 1982, Ungerboeck [1] has introduced an effective modulation technique for band-limited channels called 'trellis coded modulation (TCM). TCM employ redundant nonbinary modulation in combination with a finite-state encoder which governs the selection of modulation signals to generate coded signal sequences. The detection errors caused by the channel noise is lowered by increasing the distance between channel signals. The essential concept of TCM is to use signal-set expansion to provide redundancy for coding, and to design coding and signal-mapping functions jointly so as to maximize directly the free distance (minimum Euclidean distance) between coded signal sequences.

At the receiver, the Viterbi decoder operates directly on the received channel signal instead of the corresponding binary sequences. Then the code design problem turns out to be searching the encoders with good Euclidean distance properties. At the same information rate, important coding gains are obtained without bandwidth expansion.

In most transmission channels, bandwidth is at premium, and an important attribute of any good digital signalling scheme is its ability to make efficient use of bandwidth. In high data rate communication systems, intersymbol interference (ISI) arises whenever the effects of one transmitted signals are not allowed to die away completely before transmission of the next. Partial response signalling (PRS) [2],[3] also known as correlative coding techniques, introduces ISI in a controlled way to achieve high data rate with fewer signal levels and hence with better error rate performance. Recently, PRS has been suggested for digital subscriber loops in order to reduce crosstalk interference. PRS designs are based on the premise that since the intersymbol interference is known, its effect can be removed. One of the merits of PRS is that the controlled intersymbol interference can be used to shape the system spectrum, for instance to place nulls in the frequency response. PRS consists of a nonrecursive digital filter in cascade with an analog filter. The digital filter, which represents the PRS polynomial,

$$c(D) = \sum_{n=0}^{m} c_n D^n \tag{1}$$

and has the periodic frequency response

$$C(f) = c(D)\Big|_{D=e^{-j2\pi fT}} \tag{2}$$

where $\{c_n\}$ represents nonzero sample values and (m+1) is the smallest number of contigous samples that span all the nonzero samples. The analog filter G(f) satisfies Nyquist-I criterion,

$$\sum_{k=-\infty}^{\infty} G(f - k/T) = T. \tag{3}$$

In general, it is the Nyquist-I filter G(f) which controls the spectral properties at high frequencies, while the system polynomial c(D) controls the low-frequency response. The PRS systems, widely used in practice, are duobinary (1+D), modified duobinary ($1-D^2$), and dicode (1-D). In duobinary signalling (1+D), two successive binary input pulses are added. Thus duobinary signalling introduces intersymbol interference (ISI), but it is controlled such a way that the interference comes only from the intermediate preceeding symbol. Note that the operation of adding two successive symbols has provided an effective spectral shaping to the transmitted signal. In modified duobinary signalling ($1-D^2$), binary pulses spaced by two sample periods are subtracted and it is suitable for single side band modulation. In dicode signalling (1-D), two successive binary pulses are subtracted.

The need for increase of data rates in band-limited channels leads to use a consideration of digital communication systems in which both amplitude and phase modulation are used. In phase shift keying (PSK), the in-phase and the quadrature components are not independent. Their values are constrained in order to produce a constant envelope signal, which is a fundamental characteristic of PSK. In quadrature amplitude modulation (QAM), this constraint is removed so that the quadrature channels may be independent. QAM offers the advantage of less required power than PSK for a given probability of error and alphabet size M, but at the cost of increased equipment complexity and a sensitivity to possible non-linearities in the channel.

In this chapter, to improve both bandwidth efficiency and error performance, partial response signalling (PRS) and trellis coded modulation (TCM) schemes are examined together and denoted as 'Quadrature Partial Response-Trellis Coded Modulation (QPR-TCM)' for quadrature amplitude modulation (QAM).

2.2.3 System Model

The block diagram of the QPR-TCM scheme is shown in Fig.2.2.1. At the ith epoch, the n-bit information vector **w**(i)=$(w_1(i), w_2(i),..., w_n(i))^T$ is input to QPR-TCM system that includes precoder, followed by n/n+1 convolutional encoder, a block interleaver, (1+D) MQAM mappers which correspond to PRS scheme in modulated signal space. The output of this system at the ith epoch is denoted as **g**(**w**(i)). The information sequence is precoded with an appropriate precoder to compensate catostrophic nature of the PRS structure, is then encoded by a rate n/n+1 convolutional encoder resulting in a n+1 binary output sequence. This sequence is then block interleaved to break up burst errors caused by amplitude fades of duration greater than one symbol time. The current and the previous interleaved sequences are then applied to the MQAM mappers to assign complex valued signals from the set $\{s_o, s_2, ...s_{M+1}\}$ where M=2^{n+1} (Fig 2.2.2). Finally, the resulting complex valued signals are then combined to generate QPR-TCM signals.

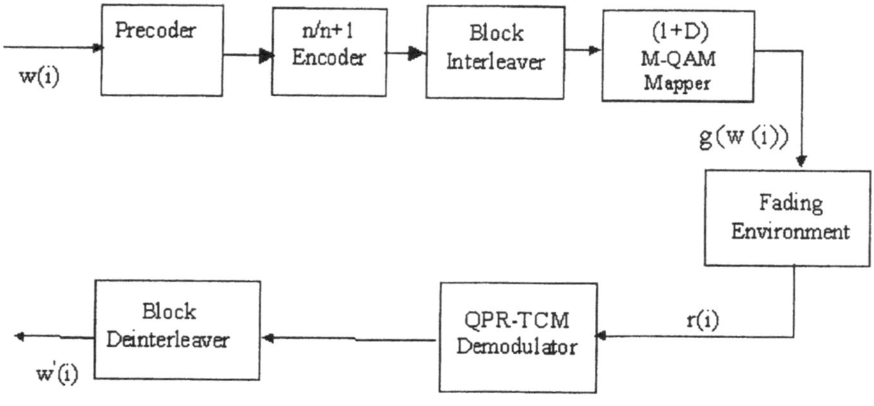

Fig. 2.2.1. Block Diagram of QPR-TCM System

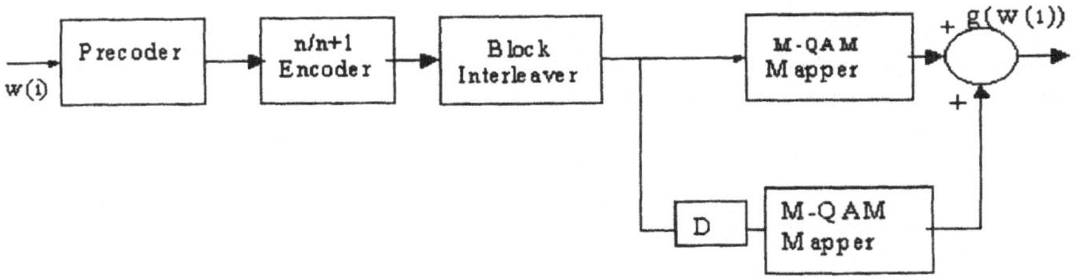

Fig. 2.2.2 QPR-TCM Encoder Structure

If we consider QPR-TCM signals over a Rician fading channel, the received signal at the signalling instant has the form,

$$r(i) = \rho(i)x(i) + v(i). \tag{4}$$

In order to provide a compact representation, we write (4) as

$$r(i) = \rho(i)\mathbf{g}(\mathbf{w}(i)) + v(i) \tag{5}$$

where x(i)=**g**(**w**(i)). The following assumptions are imposed on received signal model (5):

AS1: v(i) is a zero mean, i.i.d., additive Gaussian noise sequence with variance .

AS2: w(i) is a zero mean, i.i.d., binary, equiprobable input information sequence to the QPR-TCM.

AS3: ρ (i) represents a normalized fading magnitude having Rician probability density function (p.d.f.) for the signalling interval as,

$$P(\rho) = \begin{cases} 2\rho(1+K)\exp\left\{-K-\rho^2(1+K)\right\} I_0(2\rho\sqrt{K(1+K)}) & \rho \ge 0 \\ 0 & \rho < 0 \end{cases} \tag{6}$$

where $I_o(x)$ is the zero-order modified Bessel function of the first kind and the parameter K represents the ratio of the power in the direct or line-of-sight (LOS) plus specular components to that in the diffuse component. If the shadowing is severe, K approaches to zero. In the case of no fading K goes to infinity.

The basic problem dealt with here is the estimation of the Rician channel parameters from a set of observations. Due to its asymptotic efficiency, we focus on the ML approach for the problem at hand. Based on the assumption

AS1 and the parametric model (5), the maximum likelihood metric for w is proportional to the conditional probability density function of the received data sequence of length N (given *w*) and is of the form

$$m_j(r,w) = \sum_{i=1}^{N} \left| r(i) - \rho_j g(\mathbf{w}(i)) \right|^2 \tag{7}$$

where $w = (w(1), w(2), ..., w(N))$. Since the transmitted data sequence is not available at the receiver, the ML metric to be maximized can be obtained after averaging $m_j(g,w)$ over all possible transmitted sequences w, $m_j(r) = \sum_{w} m_j(r,w)$. Then the ML estimator of $\theta = [\rho_j, \sigma_\nu^2]$ is the one that maximizes average ML metric $\hat{\theta} = \arg\max_{\theta}[m_j(\rho)]$. Here slowly varying fading channels are assumed and ρ remains constant over the observation interval. The metric for w can be obtained by averaging (3) over all possible values of w(i),

$$m_j(r) = \sum_{w} \sum_{i=1}^{N} \left| r(i) - \rho_j g(\mathbf{w}(i)) \right|^2 \tag{8}$$

Since the direct minimization of (8) requires the evaluation of (7) over all possible transmitted sequences, it is complexity increases exponentially with the data length N. Fortunately, a general iterative approach to compute ML estimates known as the Baum-Welch algorithm [7], [9], [10] can be used to significantly reduce the computational burden of an exhaustive search.

2.2.4 Hidden Markov Model Framework

The finite memory QPR-TCM system along with the i.i.d. and finite alphabet input allow us to model the system response as a Markov chain. Since only noisy observations of the Markov chain are available due to Rician fading, the underlying Markov chain can not be observed directly. Therefore, a HMM framework is applicable [8].

In the sequel, we introduce the HMM formulation of a blind equalization problem. An QPR-TCM is described by a finite length state vector $\phi(i) = [\mathbf{w}(i), \mathbf{w}(i-1), ..., \mathbf{w}(i-\nu-k)]^T$. Since each element of w(i) takes on two possible values {-1,1}, the overall finite state machine has $P = 2^{n(\nu+k)}$ states where k is the memory of the appropriate precoder and ν the memory of the convolutional encoder.

In order to complete the HMM description, the following parameters of the HMM need to be defined:

- The transition probability of the underlying Markov chain or equivalently the transition matrix

$$a_{k,l} = \Pr(\phi(i+1) = \phi_l / \phi(i) = \phi_k) \tag{9}$$

or equivalently the transition matrix $A = (a_{k,l})$, $1 \le k,l \le P$, where $a_{k,l} \ge 0$ and $\sum a_{k,l} = 1$.

- The initial probability vector

$$\mathbf{p}=(p_k),\ p_k = \Pr(\phi(0) = \phi_k), 1 \le k \le P. \tag{10}$$

- Finally, the p.d.f. of the observations given the state needs to be determined, i.e.,

$$\mathbf{f}=(f_k),\ f_k(r(i)) = f(r(i)/\phi(i) = \phi_k) \qquad 1 \le k \le P. \tag{11}$$

Then the HMM is completely specified by the triplet φ=(A,**p**,**f**). Due to the Gaussian nature of the noise v(n) and AS4, the p.d.f. of the observations can be written as

$$f_k(r(i)) = \frac{1}{\sigma_v \sqrt{2\pi}} \exp\left\{ -\frac{1}{2\sigma_v} |r(i) - \eta_k|^2 \right\} \tag{12}$$

where $\eta_k = \rho_j g_s(\phi_k) = \rho_j g(\mathbf{w}(i))$. Here i belongs to i^{th} observation interval. Since A and p are predetermined for the current model, the only model parameters left to be estimated are the parameters of **f**. We therefore modify the Baum-Welch algorithm to estimate the parameters of **f** or equivalently $\theta \equiv [\rho_j, \sigma_v^2]^T$ directly.

2.2.5 Baum-Welch Algorithm

The Baum-Welch algorithm is an iterative ML estimation procedure, which was originally developed for estimating the parameters of a probabilistic functions of a Markov chain [7]. It is based on the maximization of an auxiliary function related to the Kullback-Leibler information measure instead of the likelihood function. It has been found applications in a diversity of parameter estimation problem, especially where there are multiple parameters and highly non-linear likelihood functions.

The Kullback-Leibler information measure is defined as,

$$KL(\theta_1,\theta_2) = E_{\theta_1} \log\{\frac{f_{\theta_1}(r)}{f_{\theta 2}(r)}\} = E_{\theta_1}\{\log f_{\theta_1}(r)\} - E_{\theta_1}\{\log f_{\theta_2}(r)\} \quad (13)$$

where $f_\theta(r)$ is the family of distribution identical by the unknown parameter of the available data. Kullback-Leibler gives some measure of discrepancy between $f_{\theta_1}(r)$ and $f_{\theta_2}(r)$.

Minimization of $KL(\theta_1,\theta_2)$ with respect to θ_2 is equivalent to maximizing the second term of Equation (13), since the term $E_{\theta_1}\{\log f_{\theta_1}(r)\}$ is fixed. The auxiliary function used in Baum-Welch algorithm is therefore defined as an approximation to the second term of Equation (13) as,

$$\Theta(\theta_1,\theta_2) \equiv \sum_{\phi} f_{\theta_1}(r,\phi)\log(f_{\theta_2}(r,\phi)) \quad (14)$$

where $\phi = \{\phi(i)\}_{i=1}^{N}$ takes all possible state sequences. The following theorem which forms the basis for the Baum-Welch algorithm, explains the reason for the choice of $\Theta(\theta_1,\theta_2)$ in Equation (14).

Theorem: $\Theta(\theta,\theta') \geq \Theta(\theta,\theta)$ implies $f_{\theta'}(r) \geq f_\theta(r)$.

The proof of the theorem is given by Baum et'al [7]. The iterative Baum-Welch algorithm for the evaluation of the estimates starts with an initial guess $\theta^{(0)}$. At the $(t+1)^{th}$ iteration, it takes estimate $\hat{\theta}^{(t)}$ from the $(t)^{th}$ iteration as an initial value and reestimates $\hat{\theta}^{(t+1)}$ by maximizing $\Theta(\hat{\theta}^{(t)},\theta')$ over θ'

$$\hat{\theta}^{(i+1)} = \arg\min_{\theta'} \Theta(\hat{\theta}^{(i)},\theta'). \quad (15)$$

This procedure is repeated untill the parameters converge to a stable point. It can be shown that the explicit form of the auxilary function at the jth observation interval has the form

$$\Theta(\hat{\theta}^{(t)},\theta'') = C + \sum_{i=1}^{N}\sum_{m=1}^{P} \gamma_m^{(t)}(i)\left\{\frac{1}{\sigma_v'^2}\left|r(i) - \rho_j' g_s(\phi_m)\right|^2\right\}. \quad (16)$$

where $\gamma_m^{(t)}(i) = f_{\theta'}(r, \phi(i) = \phi_k)$ is the joint likelihood of the data r and being at state ϕ_k at time i. Using the definitions of forward $\alpha_k(i)$ and backward $\beta_k(i)$ variables, which are the probabilities of the partial observation sequences, given the state ϕ_k at time i, $\gamma_k(i)$ can be recursively obtained as [8],[9]

$$\gamma_k^{(t)}(i) = \alpha_k^{(t)}(i)\beta_k^{(t)}(i). \tag{17}$$

The iterative estimation formulas can be obtained by setting the gradient of the auxilary function to zero $\nabla_{\rho_i'}\Theta = 0$, $\nabla_{\sigma_v'^2}\Theta = 0$. Then,

$$\nabla_{\rho_i'}\Theta = \frac{2}{\sigma_v'^2}\sum_{i=1}^{N}\sum_{m=1}^{P}\gamma_m^{(t)}(i)\Re\left\{\left[r(i) - \rho_j' g_s(\phi_m)\right]g^*(\phi_m)\right\} \tag{18}$$

$$\nabla_{\sigma_v'^2}\Theta = \frac{N}{2\sigma_v'^2} + \frac{N}{2\sigma_v'^4}\sum_{i=1}^{N}\sum_{m=1}^{P}\gamma_m^{(t)}(i)\Re\left\{\left[r(i) - \rho_j' g_s(\phi_m)\right]^2\right\} \tag{19}$$

Based on these results, the proposed algorithm is summarized in Table 2.2.1

Table 2.2.1: HMM based ML estimation algorithm.

Initialization: Set the parameters to some initial value $\theta^{(0)} = (\rho_j^{(0)}, \sigma^{2(0)})$.
Step 1: Compute the forward and backward variables from (10) to obtain $\gamma_m^{(t)}(i)$.
Step 2: Obtain $\hat{\rho}_j^{(t+1)}$ by solving the set of linear equations provided by (18).
Step 3: Solve $\nabla_{\sigma_v'^2}\Theta = 0$ from (19) to obtain $\sigma_v'^{2(t+1)}$.
Step 4: Repeat Step 1-Step 3 until $\left\|\theta'^{(t+1)} - \theta'^{(t)}\right\| \leq \varepsilon$, where ε is a predefined tolerance parameter.
Step 5: Employ the Viterbi algorithm on $\gamma^{(t)}$ to recover the transmitted symbols.

2.2.6 Performance Analysis

The primary advantage of TCM over modulation schemes employing traditional error correction coding is its ability to achieve increased power efficiency without the customary expansion of bandwidth introduced by the coding process. Thus any channel that is both power-limited and bandwidth limited would be ideally suited to TCM. One such application is the mobile satellite channel, where bandwidth is constrained by the desire to accommodate a large number of users in a given transmission bandwidth and power is constrained by the flux density limitation of the satellite's radiated transmission and the physical size of the mobile's antenna.

In addition to the usual additive thermal noise background, the mobile satellite channel is also impaired by Doppler frequency shift due to vehicle motion, the potential of a nonlinear channel, voice delay, multipath fading and shadowing. For most mobile satellite channels, fading can be assumed to be modelled by a Rician amplitude probability distribution as given at Equation (6).

An upper bound on the average bit error probability of TCM schemes for fading channel is obtained by Union bound as [11]

$$P_b \leq \frac{1}{n} \frac{\partial T(\overline{W}, L, I)}{\partial I}\bigg|_{I=1, L=2^{-n}} \tag{20}$$

where T(.) is the modified transfer function of the error state diagram. $\overline{W}$ is as,

$$\overline{W^{\lambda\rho^2}} = \frac{1+K}{1+K+\lambda\tau} e^{\frac{-K\lambda\tau}{1+K+\lambda\tau}} \tag{21}$$

where the probability density function (p.d.f) of ρ is given in Equation (6).

In fading channel, evaluation of an upper bound depends on decoding metric, the presence or absence of channel state information (CSI) and type of the detection used. In the present chapter, coherent detection and CSI with additive Maximum Likelihood metric is assumed.

For the case of coherent detection with ideal CSI and a Gaussian decoding metric, it has been shown that at sufficiently high Eb/No the bit error probability is approximately equal to [11],

$$P_b \approx \prod_{n=1}^{L} \frac{e^{-KL}}{\left(\frac{E_b}{N_o}\right)^L \frac{|x_n - \hat{x}_n|^2}{(4(1+K))^L}} \tag{22}$$

where L is the length of the shortest event path by the non-zero pairwise distances between symbols along its branches and those along the current path of the trellis diagram. $|x_n - \hat{x}_n|^2$ represents pairwise Euclidean distance between corresponding symbols on the pair of paths being compared.

2.2.7 Examples

In this section, we investigate the performance of 6QPR-TCM system [4] over fading mobile satellite channel. The block diagram of 6QPR-TCM is shown in Fig. 2.2.3 which consists of a unit memory precoder (k=1) two-state encoder (ν=1) followed by (1+D) PRS. The i.i.d. binary data sequence is modulated with 6QPR-TCM and then transmitted through fading channel. The random variable ρ representing the normalized fading amplitude at each signalling interval were generated from the pdf [4] with K=0.

The suggested algorithm based on the HMM, takes the noisy observations and estimates the channel parameters (Fig 2.2.4). Then it uses these estimates to recover transmitted data sequences with Viterbi algoritm which makes decisions at every 10 signalling interval. HMM parameters of 6QPR-TCM are predetermined as follows,

$$A = \begin{bmatrix} \frac{1}{2} & 0 & \frac{1}{2} & 0 & 0 & 0 \\ \frac{1}{2} & 0 & \frac{1}{2} & 0 & 0 & 0 \\ 0 & \frac{1}{4} & 0 & \frac{1}{2} & 0 & \frac{1}{4} \\ 0 & \frac{1}{4} & 0 & \frac{1}{2} & 0 & \frac{1}{4} \\ 0 & 0 & \frac{1}{2} & 0 & \frac{1}{2} & 0 \\ 0 & 0 & \frac{1}{2} & 0 & \frac{1}{2} & 0 \end{bmatrix}$$

$$P = \left\lfloor \frac{1}{8} \quad \frac{1}{8} \quad \frac{2}{8} \quad \frac{2}{8} \quad \frac{1}{8} \quad \frac{1}{8} \right\rfloor .$$

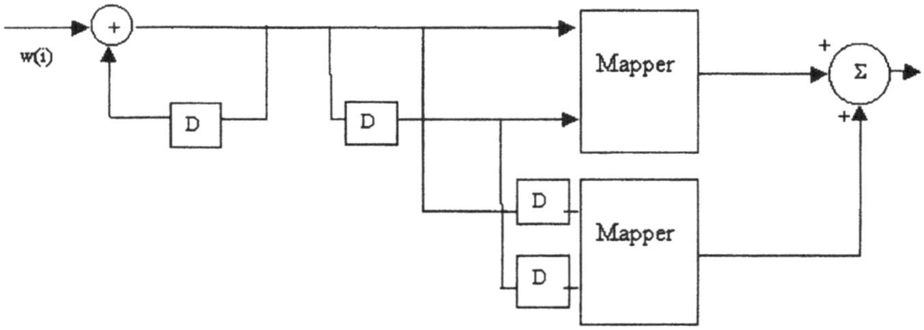

Fig 2.2.3 Block Diagram of 6QPR-TCM System

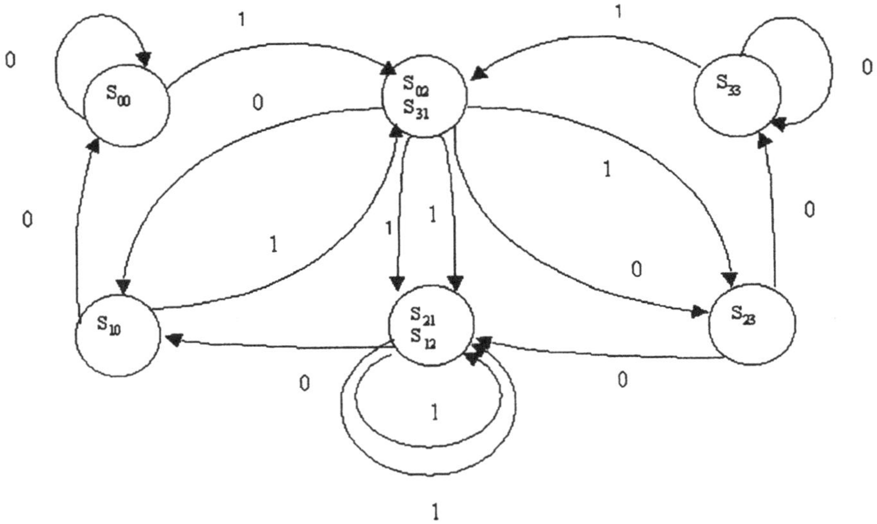

Fig 2.2.4 6QPR-TCM HMM state transition diagram

The proposed method is tested accros a range of signal to noise ratio points and the number of bit errors are recorded in each trial. The number of bit errors obtained from simulations are compared with analytical upper bounds derived as in [4].

In AWGN case, by using trellis structure in Fig 2.2.5, 6QPR-TCM system has the modified transfer function of the form,

$$T(W,L,I) = \frac{\dfrac{8W^8W^4W^8aL^4I^2}{(1-0.25a.aL^2I^2)(0.5-0.5W^{16})-0.25a.aI^2} + 0.5W^4W^{20}a^2}{1-aL^2I^2} \qquad (23)$$

where $a = (W^4 + W^{20})$. In the presence of fading, the bit error probability upper bound for the case of ideal state information is obtained by simply replacing W^λ by $\overline{W}^{\lambda\rho^2}$ given in Equation (21). 6QPR-TCM is upper bounded by replacing (23) into (20) and compared to the classical 4QAM-TCM correspondence. It is shown that both analytical and simulation results of the proposed sheme has better performance than classical systems for SNR values greater than 10 dB (Fig 2.2.6).

To demonstrate the validity of these interesting results, another analytical approach has been applied. In 6QPR-TCM, the length of the shortest error path is L=3 and the product of the branch distances for the shortest error event path, normalized to the average bit energy, Es=4, is computed as,

$$\prod_{n=1}^{L}\left|x_n - \hat{x}_n\right|^2 = (\frac{8}{4})(\frac{4}{4})(\frac{4}{4}) = 2$$

Thus (22) becomes,

$$P_b \approx \frac{64}{2}\frac{(1+K)^3 e^{-3K}}{(\frac{E_b}{N_o})^3} \qquad (24)$$

Comparing (24) and 4QAM-TCM [4],[5], it is concluded that at high SNR ratios, 6QPR-TCM performs better than 4QAM-TCM if $\frac{E_b}{N_o} \geq 16(1+K)e^{-K}$. Thus one important observation is that, after a certain value of the signal to noise ratio, the TCM scheme combined with partial response coding performs better than 4QAM-TCM on fading channel.

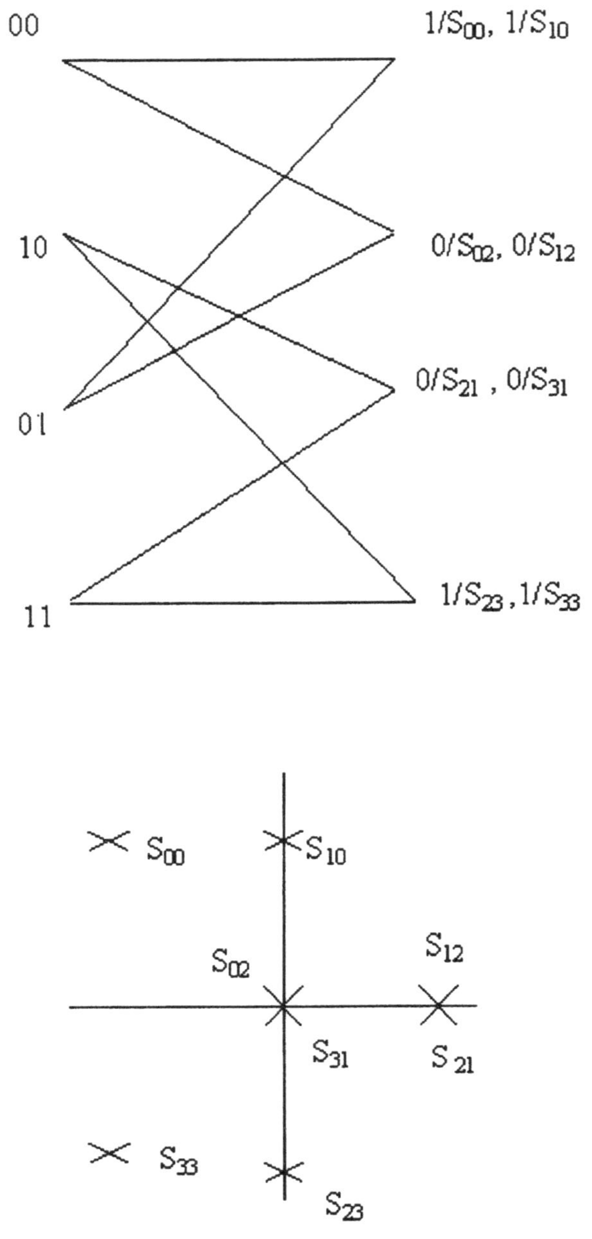

Fig 2.2.5 6QPR-TCM Trellis structure and its signal constellation

Therefore, the combined schemes improve both error performance and bandwidth efficiency. Especially for Rayleigh fading which corresponds to severe fading, 6QPR-TCM performs better than the 4QAM-TCM for all signal to noise values greater than 10 dB. This can be best explained by the following facts:

i) the 6QPR-TCM achieves diversities

ii) the product of squared branch distances become larger than those achievable with conventional trellis coded 4QAM having same number of states .

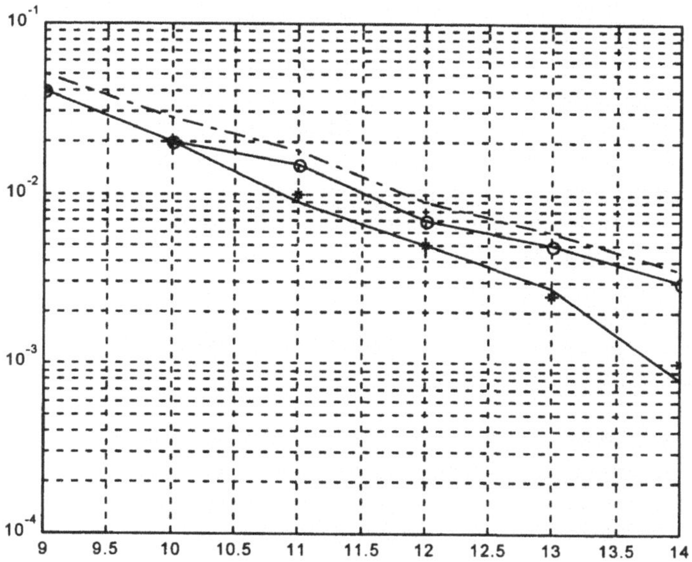

Fig 2.2.6 Performance Analysis

-.-. Simulation results for 4QAM-TCM

0 0 Analytical upper bound for 4QAM-TCM

__ Simulation results for 6QPR-TCM

** Analytical upper bound 6QPR-TCM

Thus it is demonstrated that combining the TCM with a PRS provides an additional degree of freedom for designing a code to meet the optimization criteria on a fading channel. It is obvious that QPR-TCM systems, appear to be one of the best choice, both in terms of spectral efficiency and bit-error-rate for finite values of K (Rice/Rayleigh channel).

In this chapter, a stochastic ML blind fading channel estimation algorithm is proposed and it`s performance is evaluated based on the derivation of bit error probability upper bound. The interesting result obtained from this work is that, the bit error probability upper bounds of the proposed 6QPR-TCM is better than the related 4QAM-TCM schemes on fading channel for high SNR values.

Hence, fading environments could benefit from the proposed QPR-TCM schemes and ML channel estimation procedures.

References

1. G. Ungerboeck, " Channel coding with multilevel/phase signals ", *IEEE Trans. on Information Th.*, Vol. 28, pp. 56-67, January, 1982.

2. P. Kabal and S. Pasupathy, " Partial Response Signalling", *IEEE Trans. on Commun.*, Vol. 23, pp. 921-934, September, 1975.

3. A. Grami and S. Pasupathy, " Pulse shape, excess bandwidth, and timing error sensitivity in PRS systems", *IEEE Trans. On Commun.* Vol. 35 , pp. 475-480, April, 1987.

4. Osman N. Ucan , "Performance Analysis of Quadrature Partial-Response Trellis Coded Modulation (QPR-TCM)", *Dissertation Thesis, Istanbul Technical University*, October, 1994.

5. E. Panayirci, U. Aygolu, and O. N. Ucan, "Error performance analysis of quadrature partial response trellis coded modulation (QPR-TCM) in fading mobile satellite channels ", *IEEE Trans. on Commun.*, Vol. 43, pp. 1653-1662, February/March/April ,1995.

6. M. Erkurt and J. G. Proakis, "Joint data detection and channel estimation for rapidly fading channels ", Proc. *IEEE Globecom*, pp. 910-914, 1992.

7. L. E. Baum, T. Petrie, G. Soules and N. Weiss, "A maximization technique occurring in the statistical analysis of probabilistic functions of Markov chains", *The Annals of Mathematical Statistics*, Vol. 41, pp. 164-171, January, 1970.

8. L.R. Rabiner, "A tutorial on hidden Markov models and selected applications in speech recognition", Proc. *IEEE*, Vol. 77, pp.257-274, February, 1989.

9. K. Kaleh and R. Vallet, "Joint parameter estimation and symbol detection for linear or nonlinear unknown channels", *IEEE Trans. on Commun.*, Vol. 42, pp. 2406-2413, July, 1994.

10. H. A. Cirpan and M. K. Tsatsanis, "Maximum likelihood blind channel estimation in the presence of doppler shifts", *IEEE Transactions on Signal Processing,* Vol. 47, No. 6, pp. 1559-1569, June, 1999.

11. D. Divsalar and M.K. Simon, ''Trellis coded modulation for 4800-9600 bits/s transmission over a fading mobile satellite channel'', *IEEE J. Select. Areas in Commun.*, Vol. SAC-5, pp.162-175, February, 1987.

2.3 Range Extension in Mobile Communications Based on Link Adaptation

John Dunlop and Juan Pons

2.3.1
Introduction

Adaptive channel coding schemes, where a transmission code is chosen from a set of possible codes of varying robustness depending on the quality of the channel, have been widely reported in the literature [1,2], indicating that considerable improvements in capacity and coverage can be achieved.

Link adaptation is a technique which has been considered for 3rd generation systems. However, in recent times, considerable interest has been shown in the migration of link adaptation techniques to 2nd generation systems and the potential increase in performance which may be achieved. This paper addresses the application of gross rate link adaptation to range extension and presents a control mechanism which may be employed for control of both gross rate and net rate link adaptation mechanisms in enhanced 2nd generation and 3rd generation systems.

2.3.2
Voice Coding In Gsm

Voice coding is accomplished in GSM by modelling the actual speech production process, using a technique known as *analysis-by-synthesis* predictive coding. This is illustrated in Fig 2.3.1, in which the speech production mechanism is shown.

This mechanism can be represented as a source of excitation (the lungs), with a given pitch and loudness, which is filtered by the characteristics of the vocal tract (teeth, tongue, pharynx etc.). The filtering operation can be emulated by a series of digital filters with specific (time varying) parameters. Analysis-by-synthesis voice coders are based on the principle of minimising the mean squared error between the actual speech waveform and the synthesised speech by adjusting the excitation and the filter characteristics of the synthesiser. A schematic diagram of this technique is shown in Fig. 2.3.2 in which $\hat{v}(n)$ represents samples of the synthesised speech.

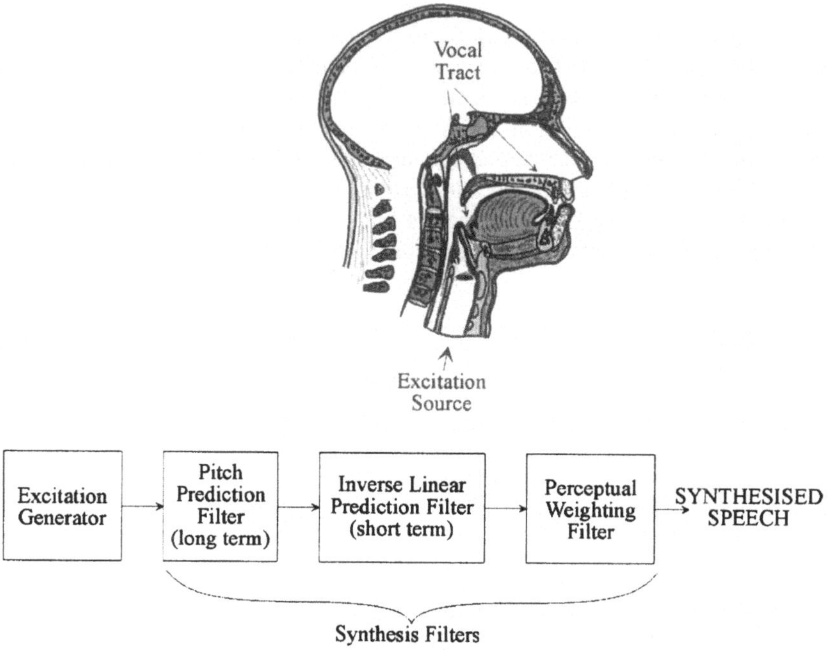

Fig 2.3.1 Human speech production mechanism and model

2.3.2.1
The GSM Full Rate Coder

In the GSM full rate voice coder the excitation is modelled by regular pulse excitation (RPE-LTP). The synthesis filter parameters of Fig 2.3.2 calculated outside the optimisation loop from blocks of speech of duration 20 ms. The excitation sequence for this filter is calculated by dividing the speech blocks into 5 ms sub-blocks of N samples and synthesising the N speech samples by a sequence of M pulses spaced by an interval D where $M = N$ div D (N div D represents an integer division). The quantised filter parameters and excitation are then transmitted to the receiver (with channel coding).

The pulse positions are given by $m_i^{(k)} = k + iD$ where $k = 0, \ldots, D - 1$ is the position of the first pulse (or intial phase) and $i = 0, \ldots, M\text{-}1$ and the pulse amplitudes $= \beta_i^{(k)}$. The RPE algorithm requires solving a set of M simultaneous equations D times (for GSM $M = 13$, $D = 4$). The position of the first pulse and the amplitudes $\beta_i^{(k)}$ (all quantised) are transmitted and used at the receiver to reproduce the excitation sequence which is passed through the synthesis filters to reproduce the synthesised samples $\hat{v}(n)$ which are an accurate representation of the original speech samples $v(n)$.

The full rate voice coder produces a block of 260 bits every 20 ms which is equivalent to a net rate of 13 kb/s. The 260 bits produced by the voice coder are sorted into 3 classes (according to function and importance). Channel coding (for error protection) is added which increases the number of bits per 20 ms to 456. The most important bits are 50 class Ia bits (these describe speech sub block amplitudes, filter coefficients and LTP parameters) which receive highest coding. Next are 132 class Ib bits, which describe the excitation sequence and further LTP parameters and receive less robust coding. The remaining 78 (class II bits) receive no channel coding. This produces a coder gross rate of 22.8 kb/s.

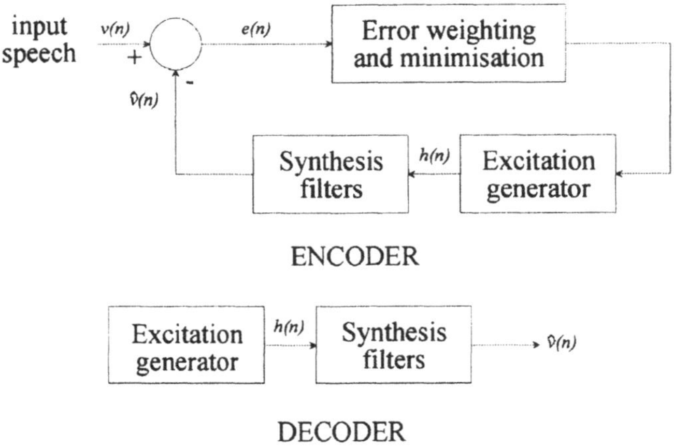

Fig. 2.3.2 Analysis-by-synthesis voice coding

2.3.2.2 The GSM Half rate coder

When the net coder rate, in a regular pulse excited voice coder, drops below about 9.6 kb/s a significant deterioration in synthesised voice quality occurs. The GSM half rate coder has a net rate of 5.6 kb/s and uses an excitation source based upon a codebook of Gaussian sequences. It belongs to a class of coders known as codebook excited linear predictive (CELP) coders. The excitation codebook contains L codewords of length N samples which correspond to an excitation frame of the speech. The excitation sequence for a speech sub-block, which produces minimum error is then chosen from the codebook after a scaling process. The address of this sequence and the scaling factors are then transmitted to the receiver, together with the appropriate filter coefficients. The particular codebook used in the GSM half rate coder is based on highly structured algebraic codes which results in a form of coder known as a Vector Sum Excited Liner Predictive Coder (VSELP).

The VSELP technique uses special innovation codebooks having an algebraic structure which has several advantages in terms of storage, search complexity, and robustness. The receiver is equipped with the same codebook and the synthesised speech is produced by choosing the appropriate sequence and filter coefficients for each speech sub-block.

The half rate coder produces an output of 112 bits every 20 ms which is equivalent to a net rate of 5.6 kb/s. This is transformed to 228 bits by channel coding giving a gross bit rate is of 11.4 kb/s. The half rate coder shares alternate slots on a full rate traffic channel and is interleaved over 4 frames (a full rate channel is interleaved over 8 frames).

2.3.2.3
The GSM Enhanced Full Rate Coder

The main difference between this and the full rate coder is that the excitation is based on the CELP principal rather than the regular pulse excitation principle. The EFR coder actually produces 244 bits per 20 ms (fewer bits than the FR codec). Parity and repetition bits are inserted to bring this up to 260 bits. Channel coding is then applied, as in the full rate case, to bring the total number of bits per 20 ms to 456.

2.3.3
Link Adaptation

Link adaptation consists of selecting, dynamically, a transport mode defined by the vocoder net rate and the channel coding scheme. This leads to a variable net rate and/or gross rate link adaptation and the possible requirement to vary the allocated resources dynamically during transmission. Typical response characteristics which become candidates for link adaptation are shown in Fig. 2.3.3.

This article concentrates, in particular, on the potential *range extension* which may be gained by application of gross rate link adaptation. The process is illustrated with reference to the GSM VSELP half rate coder, although the process is also applicable to UTRA and may be regarded as a sub-set of the general application of adaptive multirate coders to 2nd and 3rd generation mobile communication systems. In all cases link adaptation requires the existence of a link adaptation algorithm and a means of triggering such an algorithm from measurements of received signal quality.

2.3.4
Gross Rate Link Adaptation

Gross rate link adaptation assumes that the change in channel coding rate is sufficient to require a change in transmission resources and thus must be closely linked with the resource allocation controller which acts as arbitrator in a dynamic system. Referring to fig. 2.3.31, the most obvious example of gross rate link adaptation is a switch between half rate and full rate coders. In the case of GSM a half rate channel requires a transmission rate of 11.4 kb/s, based on a vector sum excited linear predictive coder (VSELP), and a full rate channel requires a transmission rate of 22.8 kb/s based on a regular pulse excitation linear predictive coder (RPE/LPC).

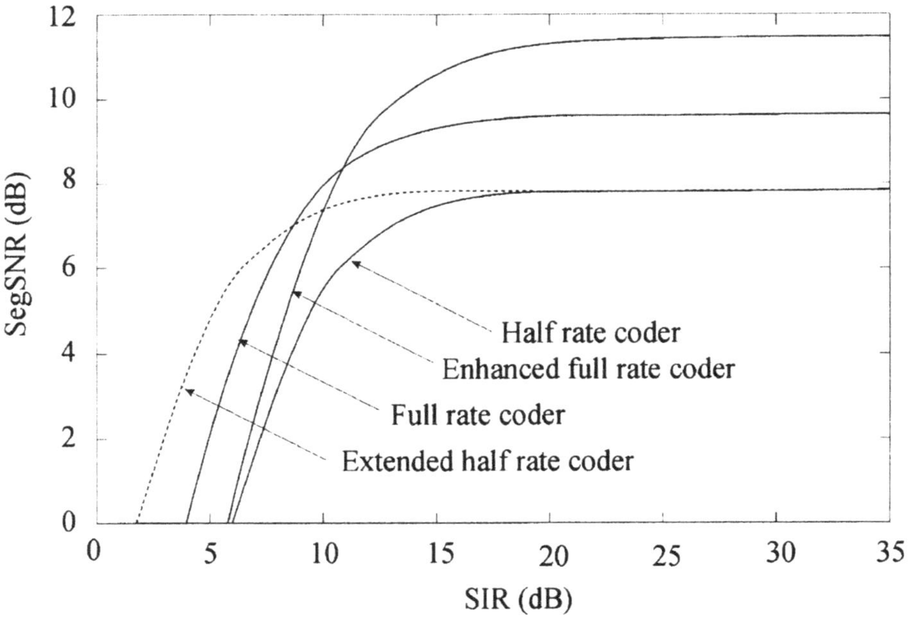

Fig. 2.3.3 Link adaptation candidates

The object of the exercise in range extension is to provide "acceptable voice quality" when the received carrier to interference ratio drops below the normally accepted minimum value. It will be noted from Fig. 2.3.3 that a switch between half rate and full rate coder will not provide a significant improvement in voice quality. However a significant improvement can be achieved if the resource required by the full rate coder is allocated instead to an extended half rate coder. The extended half rate coder is essentially a half rate voice coder with additional channel coding.

Fig. 2.3.4 shows the typical response of the recently adopted GSM adaptive multirate (AMR) coder. It will be noted from this figure that there is considerable similarity between the AMR full rate characteristic and the extended half rate characteristic shown in Fig. 2.3.1. In the GSM example shown in Fig. 2.3.3, the extended half rate coder is based on the coding of the 112 bit block of the GSM half rate voice codec into 456 bits, and then transmitting this over a full rate speech traffic channel. The channel coding scheme employed for the full rate channel for the purpose of this paper is as follows:

The 112 bit half rate codec output block consists of 95 class I bits and 17 class II bits. The 22 most significant class I bits are used to calculate a 3-bit CRC, as defined in the standard.

- The 95 class I bits, plus the 3 CRC bits, plus 6 tail bits are encoded with a convolutional code of rate 1/4, producing 416 bits.
- The 17 class II bits, plus 3 tail bits are encoded by a 1/2 rate convolutional code, producing 40 bits.

The overall frame length is 4×[95+3+6]+2×[17+3] = 456 bits. These bits are interleaved over 8 bursts as defined in the GSM recommendations for the full rate channel coding. This scheme is termed a 'robust full rate speech traffic channel', TCH/RFS and is shown in Fig. 2.3.5.

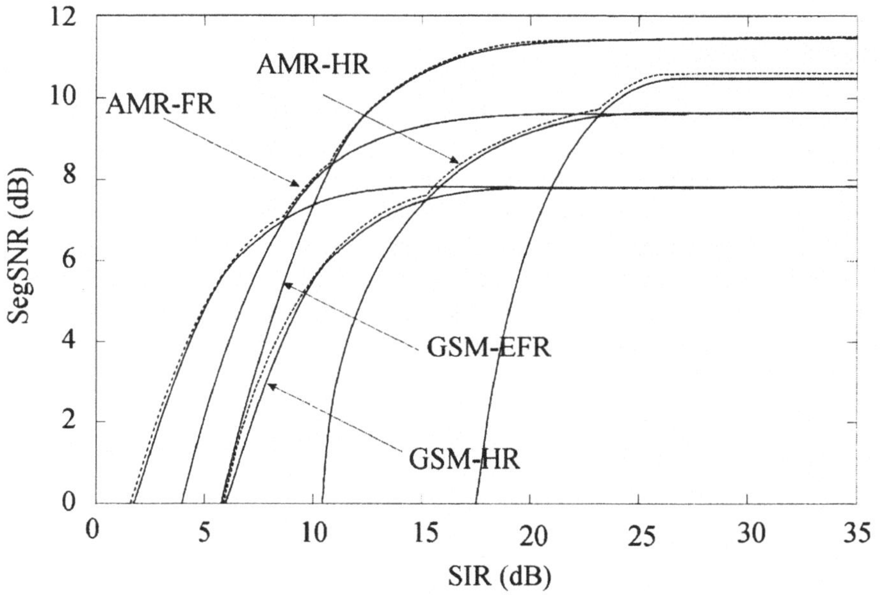

Fig. 2.3.4 GSM adaptive multirate (AMR) coder

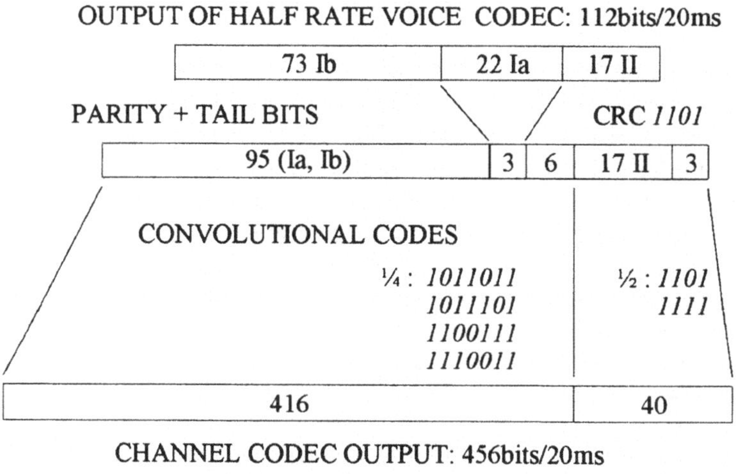

Fig. 2.3.5 Channel coding scheme

In the case of this paper channel quality is based on received SIR which is used to select the optimum mode. The triggering algorithm is based on a hysteresis window as shown in Fig. 2.3.6, the thresholds being defined according to the characteristics of the combined voice and channel coding schemes.

2.3.5 System Simulation

The system performance evaluation described in this paper is based on 3 simulation tools. Bit level simulations were performed using the COSSAP® simulation tool. The receiver module includes a 16-state Soft Viterbi Equaliser and a Soft Viterbi Convolutional Decoder which uses the knowledge of tail bits to improve error correction. Two channel propagation models have been considered: Typical Urban with speed 50 km/h (TU50), and Rural Area with speed 250 km/h (RA250). Co-channel interference plus thermal noise is modelled, the cochannel interference being produced by a single strong interferer whose signal is uncorrelated with the reference signal. This follows the reference testbed reported in [4].

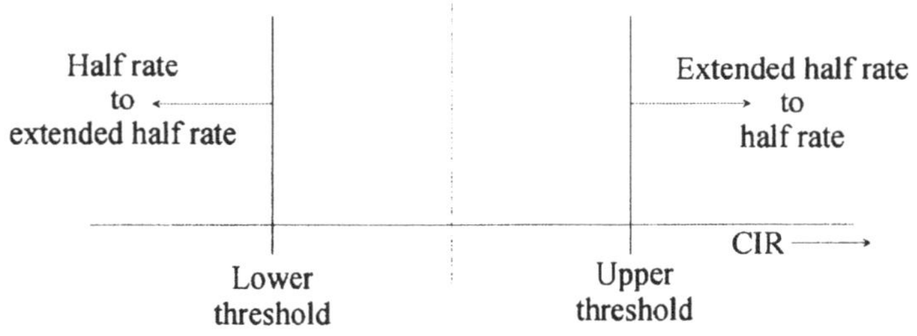

Fig. 2.3.6 Triggering thresholds.

A Montecarlo simulator was implemented to model interference in a mobile radio cellular environment which allows the probability distribution of the signal to interference ratio (SIR) to be obtained. This distribution is then used to extract the outage probability, defined as the probability of having a SIR value lower than a threshold which ensures an acceptable link quality. An acceptable performance target is assumed to be an outage probability < 5% for 99% of the calls.

Finally an event-driven simulator was implemented to model the dynamic behaviour of SIR. This simulator produces statistics on usage of modes, outage probability, and final perceived quality. The Montecarlo and Event Driven simulators are based on the following assumptions concerning cellular structure and radio environment:

1. When DTX is applied, voice activity has a negative exponential distribution with a mean talkspurt duration of 1.3 seconds and a mean silence period of 1.7 seconds.

2. The frequency band is 1800 MHz (DCS). Modified Hata-Okumura models are used to predict pathloss.

3. Shadowing has a log normal distribution with standard deviation 6dB, and a decorrelation distance of 20 m.

4. Thermal noise power is -120 dBm.

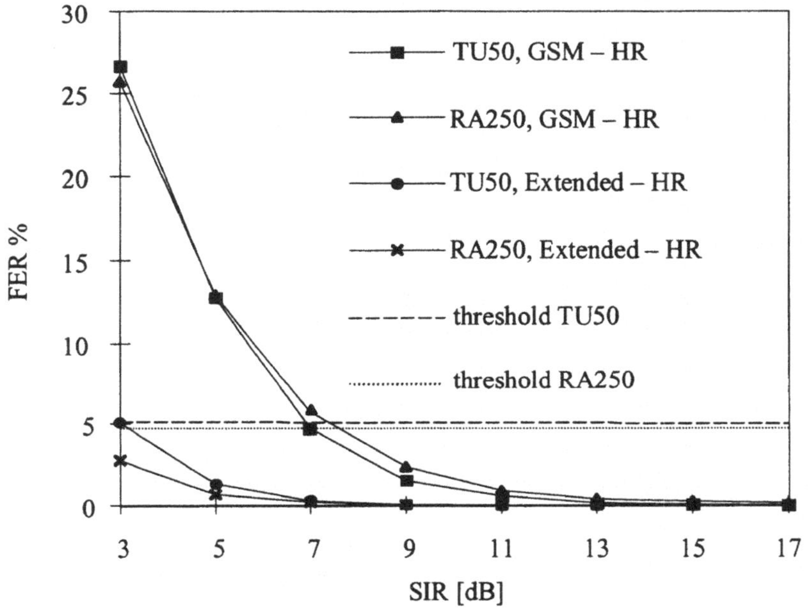

Fig. 2.3.7 Frame erasure rate characteristic

5. Perfect power control is assumed, each mobile attempting a target power level at the receiver of -87 dBm subject to a maximum transmit power of 37 dBm. The values are set according to [4] for mobile type 3 terminals

2.3.6 Simulation Results

The GSM recommendations [4] specify quality thresholds for frame erasure rate (FER), residual bit error rate for class Ib bits (RBER Ib) and residual bit error rate for class I bits (RBER II). Fig. 2.3.7, 2.3.8 and 2.3.9 show the results obtained for the half rate coder (GSM – HR) and the extended half rate coder (Extended – HR) as a function of SIR. It should be noted that to ensure that all thresholds are met GSM – HR requires SIR > 9.5 dB for TU50 and SIR > 10.5 dB for RA250. To ensure that all threholds are met in the case of the Extended – HR a SIR > 4.5 dB is required for TU50 and a SIR > 4 dB is require d for RA250 . The Extended – HR characteristic thus provides a margin of approximately 5 dB in SIR which may be used to achieve an extension in range (for an acceptable audio quality).

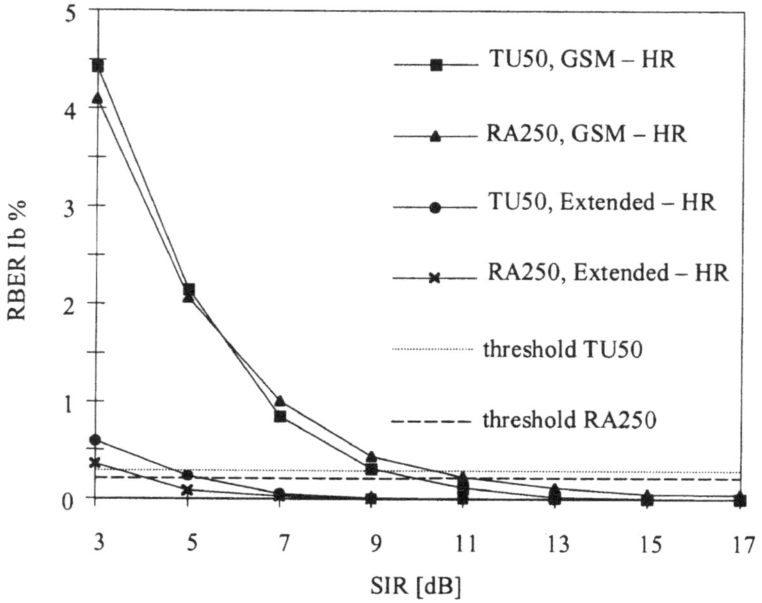

Fig. 2.3.8 Residual bit error rate for class_Ib bits (RBER_Ib)

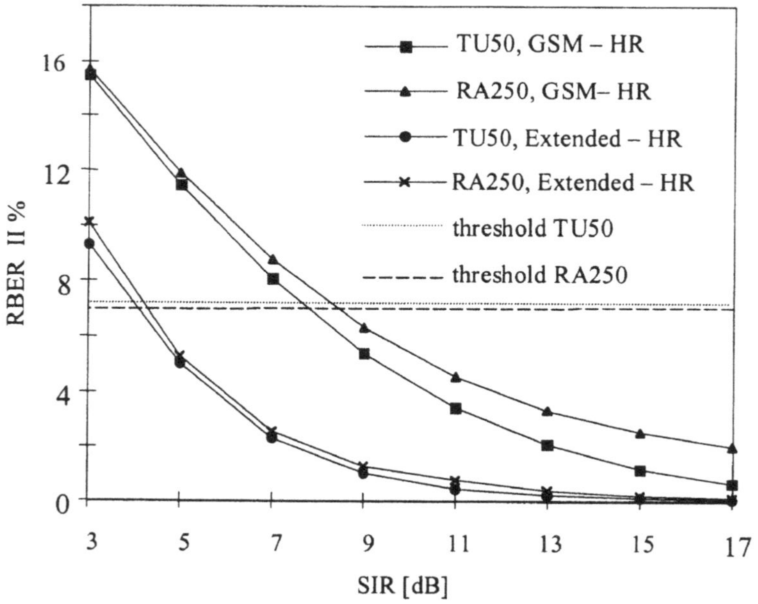

Fig. 2.3. 9 Residual bit error rate for class_II bits (RBER_II)

2.3.7 Link Adaptation Algorithm

The link adaptation algorithm is configured in order to achieve the best possible speech quality while using the least resources. Fig. 2.3.7 illustrates that SIR values > 14 dB provide a residual bit error rate for class_Ib bits (RBER_Ib) which is effectively zero for the GSM-HR coder. Thus 14 dB is used as the link adaptation SIR threshold with a hysteresis a window size of 2 dB to reduce ping-pong effects.

Table 2.3.1 Link Adaptation Performance

EXAMPLE	A1	A2	B	C1	C2
Usage, Ex – HR (%)	9.3	15.9	8.0	6.7	1.5
Usage GSM – HR (%)	90.8	84.1	92.0	93.3	98.5
Ex – HR error (%)	22.5	20.8	27.2	56.2	76.5
GSM – HR error (%)	2.4	4.1	2.5	2.3	0.7
Outage Ex – HR (%)	7.8	6.4	8.4	13.9	14.2
Outage GSM – HR (%)	0.1	0.1	0.5	2.0	1.5
Outage (overall)	0.8	1.1	1.1	2.8	1.7
outage without LA	3.7	6.0	3.6	5.1	2.0
$\frac{\textit{outage with LA}}{\textit{outage without LA}}$	0.21	0.22	0.30	0.54	0.85

The performance obtained is shown in Table 1, in which:

A1 = URBAN, NO DTX, NO FH, one active interferer
A2 = URBAN, NO DTX, NO FH, two active interferers
B = URBAN, DTX, two active interferes
C1 =URBAN, FH, full load.
C2 = URBAN, FH, DTX , full load.

With 2 interferers the final probability of outage is 1.14% when link adaptation is applied, compared to 6.07% without link adaptation. Extended – HR channels are used 16% of the time. The mode error for the GSM – HR represents the time extra resources were provided when they were not needed. This time is around a 2% for A1 and a 4% for A2. The Extended – HR error is an indication of incorrect use of GSM – HR and indicates the period the quality could have been improved had the Extended - HR coder been used. The Extended – HR error for A1 indicates that quality was poorer than it could have been for 22% of the time. The error corresponds mainly to the time required for the mode prediction algorithm to respond to a change in link conditions.

Mode error does not significantly affect the final probability of outage, since the probability of outage for GSM – HR channels is only 0.13%. Extended – HR channels bear most of the outages, the outage probability for this mode being 6.4%. Outage probability is considerably reduced when link adaptation is implemented. DTX modifies the SIR dynamics, and therefore the performance of the link adaptation algorithm. When According to [1], the best option at the beginning of a new talkspurt is to keep the last active mode.

2.3.8 Range Extension For Rural Cells

Fig. 2.3.10 shows the cdf of SIR for a rural area scenario with cell radii of 10 km and 14 km resulting from the application of gross rate link adaptation (DTX is implemented in each case). Gross rate link adaptation has two effects, it increases overall interference, since the Extended – HR requires extra resources, but it also increases robustness which allows operation at lower SIR values.

In the 10 km cell radius case, the outage probability requirement is met for the GSM – HR at a 10.5 dB SIR threshold, with a narrow margin. The threshold for the Extended – HR is 4.5 dB, which means that a cell radius of 14 km could be achieved (an increase in area of 1.96).

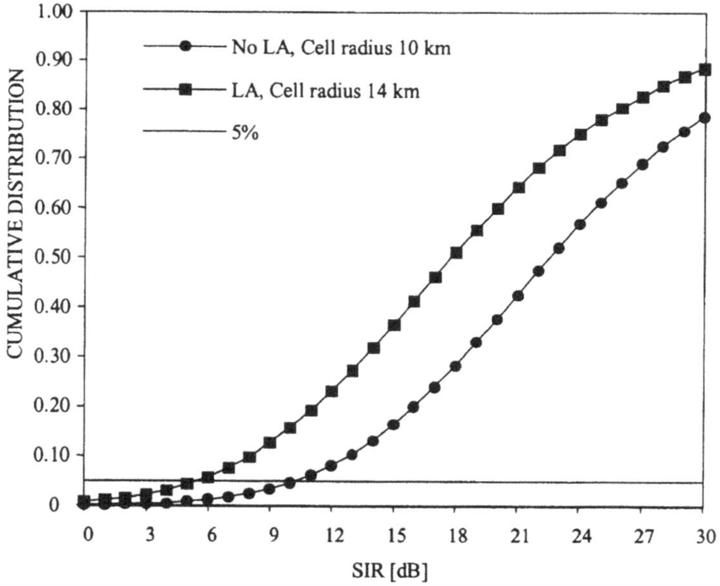

Figure 2.3.10 Cumulative distribution of SIR (rural environment)

2.3.9 Range Extension For Urban Cells

Fig. 2.3.11 shows the cdf of SIR for an urban environment with shadowing standard deviations of 6 dB and 8.5 dB, when LA is applied. The 6 dB curve meets the outage probability requirement for a 9.5 dB SIR threshold with a narrow margin. With link adaptation, a shadowing standard deviation of 8.5 dB could be accommodated with an outage probability of 4.3%, the percentage of mode 2channel usage being 20%. Without LA a standard deviation of 8.5 dB would lead to an outage probability of 16%.

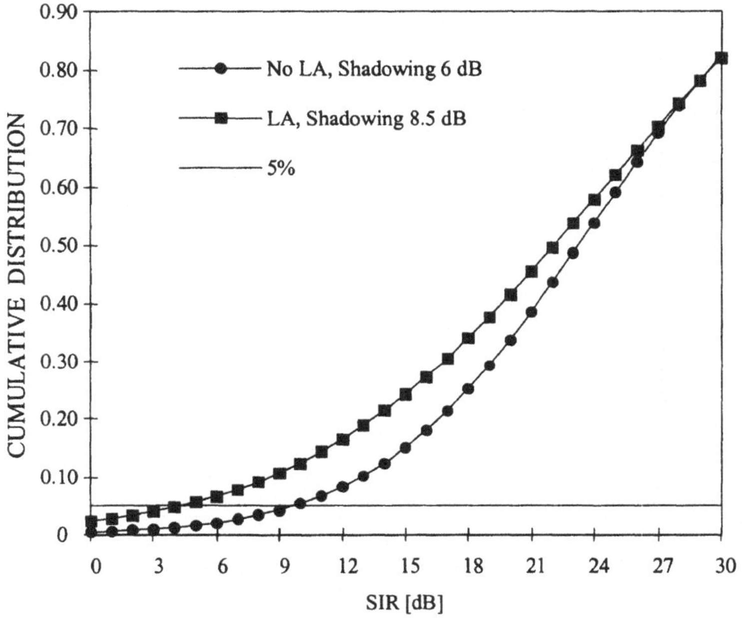

Fig. 2.3.11 Cumulative distribution of SIR (urban environment)

The almost doubling of shadowing variance which can be accommodated with LA thus allows a significant increase of coverage, and a reduction of the effect of specific propagation problems such as in-building penetration or corner effects. It should be noted that a further trade off between quality and coverage is possible.

2.3.10 Measurement Of Link Quality

The results presented in the previous sections are based on a knowledge of the received SIR. Alternative methods of assessing channel quality have also been considered. One readily available measure is to compare the received training sequence with the known training sequence and to predict the BER for the channel from this measurement. The error between the predicted BER and the actual BER is then a measure of the validity of this approach.

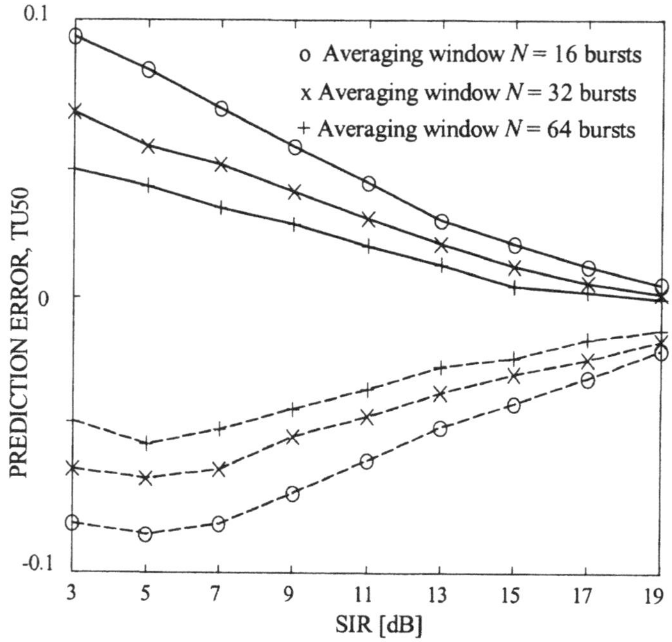

Fig. 2.3.12 BER prediction errors derived from the training sequence

The prediction error is plotted as a function of SIR for several averaging window sizes for the TU50 scenario in Fig 2.3.12. The prediction error is defined as

Prediction error = (actual BER – predicted BER)

In this figure N represents the averaging window in terms of bursts. The curves of Fig. 2.3.12 show the limits of prediction error for 90% of mobiles (e.g. 90% of mobiles will have a prediction error between +4% and –3% when SIR = 3dB and N = 64), this error margin decreases as the SIR increases. Thus such an estimate of channel quality, with its relatively small prediction error, could be incorporated in the link adaptation algorithm in a straightforward manner.

2.3.11 Conclusions

This article has illustrated the effects of applying gross rate link adaptation to 2nd and 3rd generation mobile systems. The gains to be achieved by the application of these concepts have been investigated by simulation. A triggering algorithm based on an hysteresis window has been proposed which has been tuned to achieve the optimum quality. It has been shown that a large increase in coverage plus a general enhancement of perceived quality can be expected. In addition a method of estimating channel quality based on error in the training sequence has been proposed and statistics for prediction error have been derived. These results show the practicality of incorporating range extension techniques in existing 2nd generation systems and future 3rd generation systems.

REFERENCES

1. Dunlop J, Irvine J and Cosimini P "Estimation of the performance of Link Adaptation in Mobile Radio" Proc. IEEE VTC-95, Chicago 1995, Vol. 1, pp 326-330.
2. Irvine J, Cosimini P, Dunlop J "Implementation considerations for Gross Rate Link Adaptation" in Proc. IEEE VTC-96, Atlanta 1996, Vol. 3, pp 1766-1770.
3. GSM recommendations. GSM 05.03: Channel coding.
4. GSM recommendations. GSM 05.05: Radio transmission and reception.
5. Honkasalo Z, Wang L, Silverntoinen M and Kajala M "A Cell Range Extension Technique for GSM/DCS1800 using Half-rate Codec over Full-rate Channel" Proc. IEEE VTC-95, Chicago 1995, pp 893-898.
6. Olofsson H, "Interference diversity as a means for increased capacity in GSM" Proc. IEEE EPMCC'95, Bologna, pp 97-102.
7. Zorzi M, "Minimum duration outages in Markov channels" Proc. IEEE ICUPC'97, San Diego, Vol. 2, pp 583-587.

2.4 Application of CCMA Codes to Future Mobile Radio Communications

Ming Ma, Nader Zein, Phillip Benachour , and Bahram Hon

2.4.1 Introduction

The advent of the Third Generation of mobile systems is making a new range of services available (for example mobile multimedia) which require a larger bandwidth and more flexibility in the use of the spectrum. The universal mobile telecommunication system (UMTS) has attracted growing interest since its adoption by the European Telecommunications Standards Institute (ETSI). The European effort to define a third generation mobile communication system has resulted in a consensus decision in January 1998 on the universal terrestrial radio access (UTRA) concept [1]. Code division multiple access (CDMA) has been chosen by the ETSI as a multiple access scheme for UMTS. For the UTRA two multiple access schemes have been adopted: wide band CDMA (WCDMA) in the frequency division multiplex (FDD) mode for the paired frequency bands of UMTS and time division CDMA (TD-CDMA) in the time division duplex (TDD) mode for the unpaired frequency bands [2].

In this chapter, a source combining technique based on collaborative coding multiple access (CCMA) is applied to UMTS which complies with the UTRA FDD access. CCMA and CDMA techniques allow efficient simultaneous communications by several users sharing a common channel without sub-division in time and frequency like time division multiple access (TDMA) and frequency division multiple access (FDMA) [3, 4]. In addition, CCMA allows a substantial increase in the number of users that can access the system simultaneously, leading to a higher combined information rate and hence a potentially more efficient system. As well as providing the multiple access function, these schemes can incorporate a certain degree of error protection against noise. Our interest in CCMA coding techniques is in their multiple access function which is exploited to achieve higher throughput as a source combiner for multimedia applications. Furthermore, a novel trellis decoding technique for the CCMA codeword is employed and is shown to give improved error correction capability. The UMTS standard is specified to support various services which in turn require the support for multi-rate and variable level of protection for the transmitted information. CCMA allows the encoder to be designed so as to increase the error protection level of one source at the expense of reduced protection level of the other sources.

This property of the CCMA matches very well with the aforementioned requirement of the UMTS system.

The outline of the chapter is as follows: The basic concept of CCMA techniques together with a simple 2-user uniquely decodable scheme are presented in section 2.4.2. Section 2.4.3 introduces the trellis design procedure for CCMA schemes based on Shannon product of trellises; in particular, trellis constructions and representations for the 2-user scheme for different block lengths N are illustrated. Further, a proposed modification of such a code that will increase its error correction capability is investigated. Finally, the use of CCMA codes as a source combiner for multimedia applications is presented and its performance over a selective fading channel based on the UTRA system is recorded.

2.4.2 Collaborative Coding Multiple Access

In a situation where the bandwidth is a restricted resource such as radio frequency bands, it is necessary to study efficient ways of sharing it between as many users as possible. Furthermore, it is of considerable importance to use a simple and effective multiple access coding technique capable of error control. CCMA is an attractive proposition since it allows a substantial increase in the number of users that can access the channel simultaneously leading to a higher combined information rate. CCMA techniques exist which lie between the two extreme cases of TDMA and CDMA and offer in certain circumstances the possibility of rate sums higher than unity with modest synchronisation requirements [3]. There are two main approaches for the CCMA code design for the discrete adder channel. The first one focuses on achieving the bounds promised by multiple access information theory where all the users are active [3-17]; the second approach aims at code construction for T active users out of M multiple access system where the primary goal of code construction is not to achieve channel capacity [18].

Code constructions for CCMA schemes are restricted since the composite code resulting from the individual users code combinations have to be uniquely decodable. A composite code is said to be uniquely decodable if it can decode each of the component codes uniquely and deliver the corresponding sink information reliably to their intended destinations. Various block codes have been designed to meet the unique decodability criteria . It was found that the best rate sum would be achieved if N is kept to a minimum [10]. The rate sum decreases with increase in N tending to unity. Code constructions in this instance are based on the multiple access information theory (MAIT) approach which began with a coding theorem developed by Ahlswede [19] and Liao [20]. Codes that belong to this category are the ones given by [9-16]. The search for codes in this case is complicated by the fact that at least one of the component codes must be non-linear in order to achieve a rate point near the boundary of the capacity region of the MA adder channel. A similar approach based on achieving channel capacity asymptotically as the number of users (M) goes to infinity is also described. Here, each user gets two codewords and the overall rate sum is M/N (bits/channel use). Code construction belonging to this class of codes are given by [5,6,8,17].

The original model of such a scheme was proposed by Kasami and Lin [10] and represents a uniquely decodable code pair of block length N=2, as is shown in Table 2.4.1.

Table 2.4.1 : 2-user Block Code

$User_2$\ $User_1$	00	11
00	00	11
01	01	12
10	10	21

User one has two code words C_1=(00,11) and User two has three code words C_2=(00,01,10). The individual rates for User one and User two are R_1=0.5 and R_2=0.792 respectively. The composite coding scheme, shown in Table 2.4.1, has a total rate sum $R_T=R_1+R_2$=1.292 (bits/channel use).
The rate of a component code is expressed as

$$R_i = \frac{\log_2 W_i}{N} \quad \text{(bits/channel use)} \tag{1}$$

where W_i is the number of distinct codewords in component code C_i and N is the block length. The rate sum $R_T(M)$ of an M-user code (C_1, C_2, ..., C_M) is defined as

$$R_T(M) = R_1 + R_2 + ... + R_M \quad \text{(bits/channel use)} \tag{2}$$

The simple coding scheme above can be extended to length N, where C_1 is the two N-tuples (000... 0) and (111... 1), and C_2 is the N-tuples (000... 0) and all the other N-tuples except the all one vector. The omission of the all one vector from C_2 is made in order to maintain unique decodability. It is clear that User one code is a repetition code that has one message symbol which is repeated N times. The total rate sum of a 2-user scheme based on this construction can be seen to decrease with increase in N tending to unity. Table 2.4.2 below shows the different rate sums associated with composite codewords for N=2, 3, 4, 5, 6 and 7.

Table 2.4.2 : Rate Sum Versus Code Length

N	R_1	R_2	R_T
2	0.5	0.792	1.292
3	0.33	0.935	1.265
4	0.25	0.976	1.226
5	0.2	0.991	1.191
6	0.166	0.996	1.162
7	0.143	0.998	1.141

The reduction in the rate sum is affected mainly by User one component code rate. Since the number component codes for User one is two for all values of N, it's rate R_1 is 1/N . In the User two case, the total number of component codes is 2^N-1 and can be seen to increase with increasing values of N and, as a result, the rate R_2 will also increase.

2.4.3 Trellis Design of CCMA Scheme

Collaborative coding and decoding techniques have been studied and tested for MA baseband transmission [21-22]. However, practical applications of this type of MA are restricted due to the poor synchronisation properties and lack of efficient soft maximum likelihood trellis decoding techniques [3]. Although a number of different trellis decoding techniques have been proposed [15,23], the general solution of this problem remains unsolved due to the non-linear structure of the overall coding scheme. To date, trellis design for the MA adder channel has recently been investigated and implemented for synchronous and quasi-synchronous users [24-26]. The technique is based on the concept of the Shannon product of trellises [27-28] and is applicable to MA coding schemes with M ($M \geq 2$) users. Results have shown that a performance improvement can be achieved when comparing trellis and conventional hard decision decoding. Trellis diagrams for some known 2-user schemes have been constructed; in addition, a DC-free 2-user binary code together with its trellis structure was introduced and investigated. It was shown that such a technique achieves about 2.2 dB energy gain for the overall coding scheme in comparison with the conventional hard decision case.

2.4.3.1 The Shannon Product of Trellises

Shannon[29] describes the product of two "channels" corresponding to a situation where both channels are used at each unit of time. Based on this method, Sidorenko et al presented trellis constructions for block and convolutional codes [27]. The well known trellises, such as syndrome trellis [30] and Viterbi trellis [31], can also be presented as the SP of trellises. The construction technique is outlined as follows: A trellis T is a directed graph. Nodes of the graph are decomposed into a union of disjoint subsets that are called levels. The levels are numerated by 0,1, ... m. The t-th level consists of N_t nodes $S_1^{(t)}, S_2^{(t)}, \ldots, S_{N_t}^{(t)}$. Two nodes $S_i^{(t-1)}$ and $S_j^{(t)}$ of adjacent levels can be connected by (one or more) directed branches $b_l(S_i^{(t-1)}, S_j^{(t)})$ labelled by h-tuple binary vector $W_l(S_i^{(t-1)}, S_j^{(t)})$ of code symbols, and by vector $u_l(S_j^{(t-1)}, S_j^{(t)})$ of information symbols [27]. A path is a sequence of branches connecting the nodes of levels 0,1, ... m. Each path has a correspondent sequence of code branch labels w and of information branch labels u. Let a pair (x,c) be the information sequence x and corresponding codeword $c \in C$. A trellis is called a code trellis if for each pair (x,c), $c \in C$, there exists a path in the trellis T labelled by (u,v)=(x,c) and vice versa.

Given trellises T and T' with m+1 levels and h symbols. The SP T*T' of these trellises is a trellis that at level t consists of $N_t N_t'$ nodes numbered by pairs $[S_i^{(t)}, S_i^{'(t)}]$, i=1,...,N_t, j=1,...,N_t'. Two nodes $[S_i^{(t-1)}, S_i^{'(t-1)}]$ and $[S_i^{(t)}, S_i^{'(t)}]$ of adjacent levels are connected by a branch labelled by code symbols:

$$W_1[S_i^{(t-1)}, S_j^{(t)}] + W_1^{'}[S_i^{'(t-1)}, S_j^{'(t)}] \quad (3)$$

and by information symbols:

$$u_1[S_i^{(t-1)}, S_j^{(t)}], u_1^{'}[S_i^{'(t-1)}, S_j^{'(t)}] \quad (4)$$

if and only if trellis T has branch $b_1[S_i^{(t-1)}, S_j^{(t)}]$ and trellis T' has branch $b_1^{'}[S_i^{'(t-1)}, S_j^{'(t)}]$. Fig,. 2.4.1 shows an example of this method.

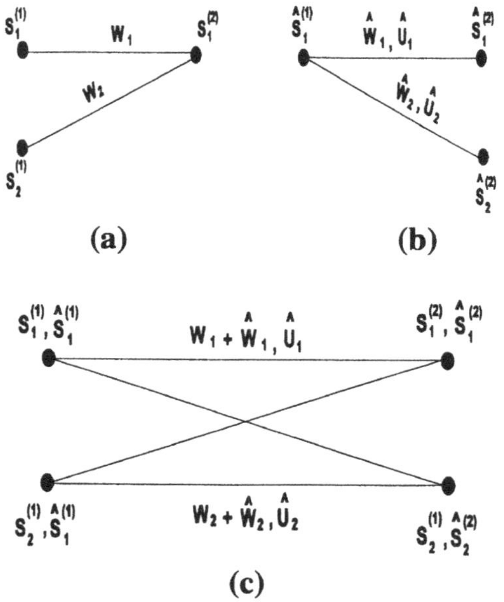

Fig. 2.4.1: Trellis diagram (**c**) is the Shannon product of trellises (**a**) and (**b**)

2.4.3.2
Trellis Design Procedure for Binary Adder Channel

Consider the M-user binary adder channel, where i-th user is employing (n,k_i,d_i) code C_i, i=1,2,...M. We define an overall code $C = \sum_{i=1} C_i$ as the set of all possible sums $C = \sum_{i=1}^{M} C_i$, where c_i is a codeword from C_i, $c_i \in C_i$, and addition represents algebraic sum.

Let T_i, i=1,2,...M, be a BCJR trellis of C_i, with the following state and branch profiles:

$$N^i = [N_0^1, N_1^i, ..., N_n^i] \tag{5}$$

$$B^i = [B_1^i, B_2^i, ..., B_n^i] \tag{6}$$

where the N_t^i, t=0,1,2,...n, i=1,2,...M, represent the number of states at t-th level of T_i, and the B_l^i, l=1,2,...n, represent the number of branches at the l-th depth of T_i. It is known that such a trellis can be obtained for any linear block code [31,32] and although for non-linear block codes the general solution is still unknown, it is always possible to design a "trivial" trellis with parallel paths. Let also two states $S_j^{(t-1)}$ and $S_m^{(t)}$ of the adjacent levels (t-1) and t of T_i be connected by one (or more) directed branch $\omega_t^i[S_j^{(t-1)}, S_m^{(t)}]$, labelled as x_t^i/c_t^i, where x_t^i and c_t^i represent input and output symbols in C_i.

The overall trellis, T, of C can be obtained as a SP [27] of component trellises $T=T_1*T_2*...*T_M$, with the following state and branch profiles:

$$N(t) = \left[\prod_{i=1}^{M} N_0^1, \prod_{i=1}^{M} N_1^i, \prod_{i=1}^{M} N_n^i\right] \tag{7}$$

$$B(t) = \left[\prod_{i=1}^{M} B_1^i, \prod_{i=1}^{M} B_n^i\right] \tag{8}$$

In this trellis two states of adjacent levels are connected by a branch

$$\omega_{\Psi} = \omega_t^1[S_j^{t-1}, S_m^t] + \omega_t^2[S_o^{t-1}, S_q^t] + ... + \omega_t^M[S_x^{t-1}, S_y^t] \tag{9}$$

where $j, o, ..., x = N_{t-1}$ and $m, q, ..., y = N_t$ and if and only if the trellis T_i has a branch $\omega_t^i[S_j^{(t-1)}, S_m^{(t)}]$ (addition in 3.7 represents algebraic sum). At the t-th depth of the trellis, branches are labelled as $x_t^1, x_t^2, ... x_t^M/c_t$, where for x_t^i; i=1,2,...M, t=1,2,... n, represent the t-th information symbol in the i-th code C_i, and for c_t; t=1,2,...n, represent the t-th encoded symbol in the overall code, C.

Example:

As an example we use a 2-user coding scheme designed by [9] where both users employ block codes of length n=4 and User one and User two have component codes

C_1=(0000,0001,0010)

C_2=(0000,0100,1000,1100,0011,0111,1011,1111)

The information rate of the overall code is R=1.146 (bits/channel use). C_1 is a ternary (2,1,1) code with matrix G_1=[0 1] and symbol mapping

"0"⇔ (00)

"1"⇔ (01)

"2"⇔ (10)

Code C_2 is a (4,3,1) binary code with parameters (n=4, k=3, d=1) with the following generator matrix:

$$G_2 = \begin{bmatrix} 1 & 0 & 0 & 0 \\ 0 & 1 & 0 & 0 \\ 0 & 0 & 1 & 1 \end{bmatrix}$$

Trellises, T_1 and T_2, of these codes and their SP can be designed using any of the known techniques proposed by [31,32] and are shown in Fig. 2.4.2 and 2.4.3 respectively. In these figures, branch labels include only encoded symbols from the component and composite codes resulting from C_1 and C_2. The node and branch profiles can be seen to have the following parameters

$$N^1(t) = [1,2,2,2,1] \qquad N^2(t) = [1,2,2,2,1]$$

$$B^1(t) = [1,2,3,2] \qquad B^2(t) = [2,4,4,2]$$

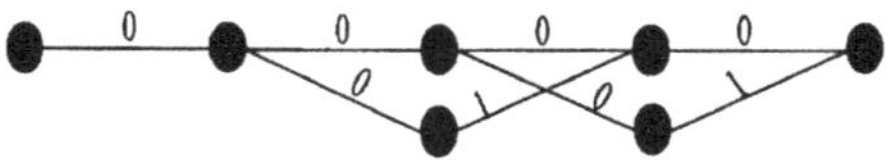

(a)

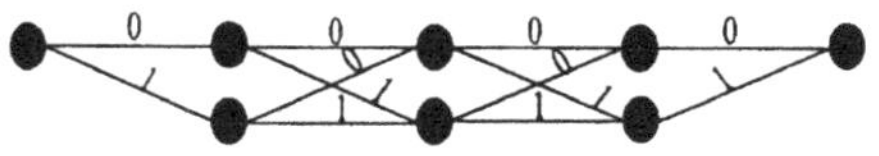

(b)

Fig. 2.4.2: Trellis diagram of component codes (**a**) User one and (**b**)User two

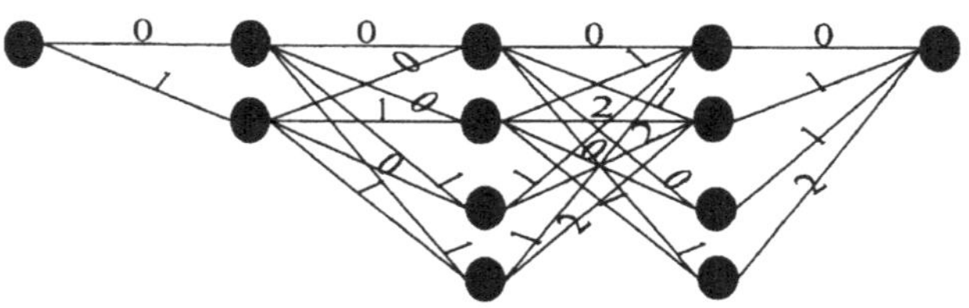

Fig. 2.4.3: Shannon Product of the 2-user scheme

2.4.3.3
Trellis Design of CCMA Block Codes

In this section, trellis representation of the 2-user scheme described in section 2.4.3.1 is investigated. The composite trellis structure for such a scheme can be obtained using the SP of trellises technique introduced in 2.4.3.1 where each user has a trellis structure from which an overall trellis diagram can be constructed. Fig. 2.4.4 and 2.4.5 represent trellis structures for the component codes of User one and User two and the overall 2-user scheme of the simplest and most efficient case.

Fig. 2.4.4: Trellis Diagram for Component Codes with length N=2

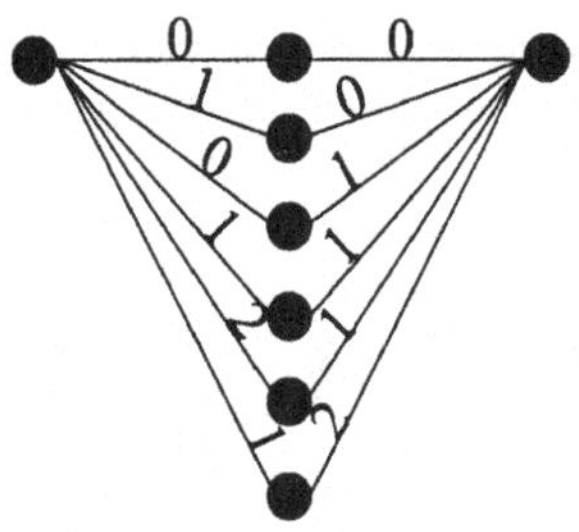

Fig. 2.4.5: Trellis Diagram for 2-user Scheme with length N=2

The total number of composite code words is six since $C_1=\{(00),(11)\}$ and $C_2=\{(00),(01),(10)\}$ and can be calculated using the expression in equation 10 as follows

$$CW_T = CW_1 \times CW_2 \times \ldots \times CW_i \tag{10}$$

where CW_T is the total number of composite code words and i=1, 2, …, M where M is the number of users in the system.

In section 2.4.3.2, equations 5 and 6 characterised expressions for the node and branch profiles of a trellis. In order to assess the complexity of the overall structure of this code, the node and branch profiles of the constituent and composite codes are depicted in Table 2.4.3 for block lengths N=2,3,4 and 5. The profiles shown represent the minimal trellis designs for the component and composite codes. It will be indicated further on that User two trellis codes can have different structures from which one is selected to characterise a general case for N>=3.

Table 2.4.3 : Node and Branch Profiles of Component and Composite codes

N	$N^1(t)$	$B^1(t)$	$N^2(t)$	$B^2(t)$	$N^T(t)$	$B^T(t)$
2	[1,2,1]	[2,2]	[1,3,1]	[3,3]	[1,6,1]	[6,6]
3	[1,2,2,1]	[2,2,2]	[1,2,4,1]	[2,7,4]	[1,4,8,1]	[4,14,8]
4	[1,2,2,2,1]	[2,2,2,2]	[1,2,5,2,1]	[2,8,9,2]	[1,4,10,4,1]	[4,16,18,4]
5	[1,2,2,2,2,1]	[2,2,2,2,2]	[1,2,5,2,2,1]	[2,8,9,4,2]	[1,4,10,4,4,1]	[4,16,18,18,4]

It can also be seen from table 2.4.3 that User two has equal node and branch numbers for a particular block length N and has ratio L=2 where L is defined as the ratio of block length N to the number of nodes or branches. This is based on the fact that User one has only two component codes being the N-tuples (000… 0) and (111… 1) for different lengths N.

$$L = \frac{N}{N_0 \text{ of Nodes}} = \frac{N}{N_0 \text{ of Branches}} \tag{11}$$

Since User two has only two code words assigned for all values of N, a general case for such a code can be constructed and is shown in Fig. 2.4.6.

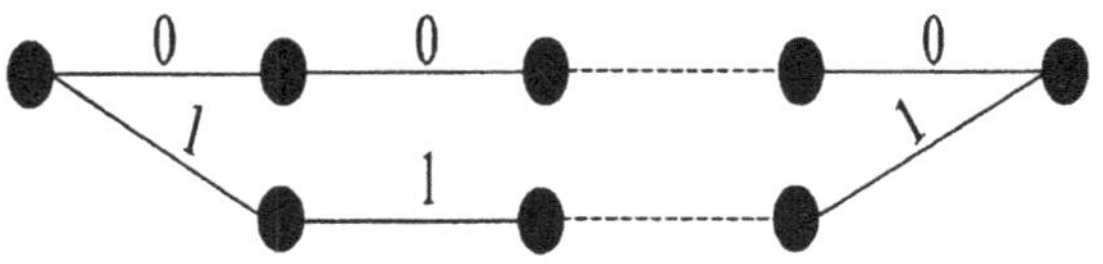

Fig.2.4.6: Trellis Diagram for User one Code with length N>2

This is not the case with User two node and branch profiles since the component codes are made up of the N-tuples (000…) and all other N-tuples except the all one vector.

The size of User two code book increases by a factor of (2N + 1) for each increment of N by one and, since such a code is non-linear, the trellis diagram for different values of N can have distinct appearances. As a result of it's non linearity, such a code will have a non-homogeneous structure and consequently the overall trellis structure resulting from the SP will be a non-homogeneous trellis. By definition, a trellis code is non-homogeneous if the number of branches emanating from each node is not the same. For instance a trellis diagram with block length N=3 can be seen to have one of the following two structures

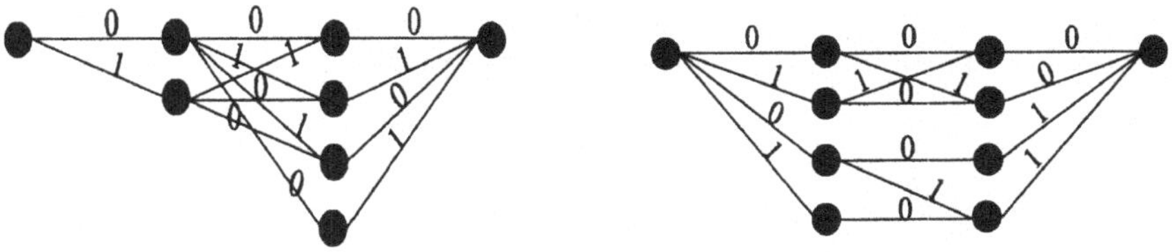

Fig. 2.4.7: Trellis Diagrams for User two Code with length N=3

Fig. 2.4.7 a has a minimal structure but is not unique in the sense that the complexity issue is not so significant since the length of the code, the number of nodes and branches are relatively low in both cases. In the case of N=4 (Fig. 2.4.8), User two code can be seen to have at least four structures. Similar observations can be made regarding the complexity issue as the N=3 case and on a closer inspection of the two sets of trellis structures reveals that there are certain similarities mainly between trellises in Fig. 2.4.7b and 2.4.8c; the main difference being an increase in the number of nodes to five at depth three when N=4.

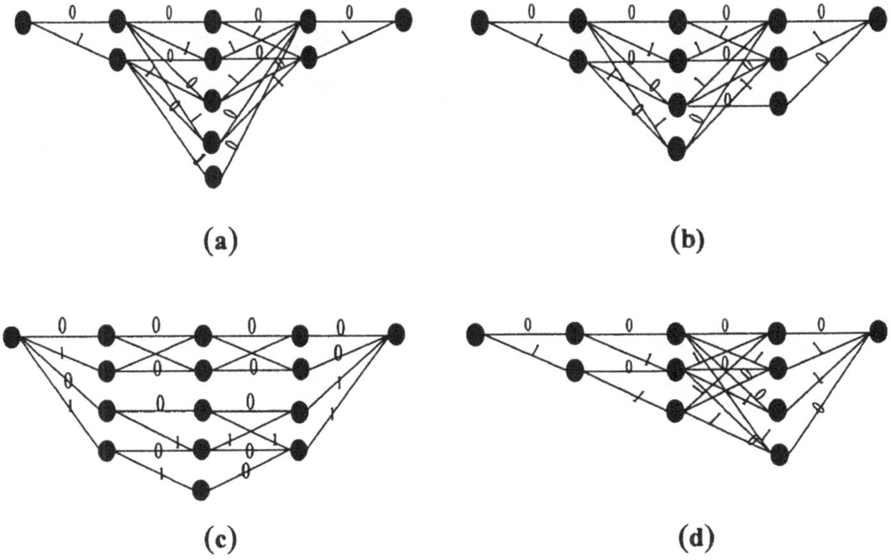

Fig2.4.8: Trellis Diagrams for User two Code with length N=4

If the length of the code increases to five, one of the possible trellis structures, Fig. 2.4.9, would resemble the ones in Fig. 2.4.7b and 2.4.8c for the N=3 and N=4 examples. Such a situation becomes prominent for larger values of N and a comprehensive structure for User two code can be obtained by increasing the depth of the trellis with N whilst maintaining an equal number of nodes (Fig. 2.4.10). As a result, the 2-user scheme can be presented as depicted in Fig. 2.4.11.

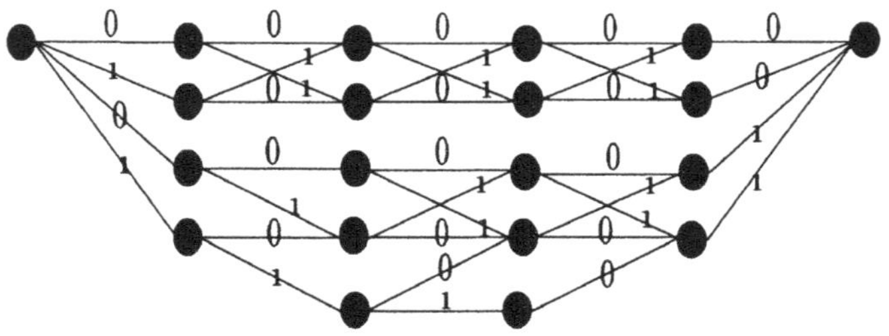

Fig2.4.9: Trellis Diagrams for User two code with length N=5

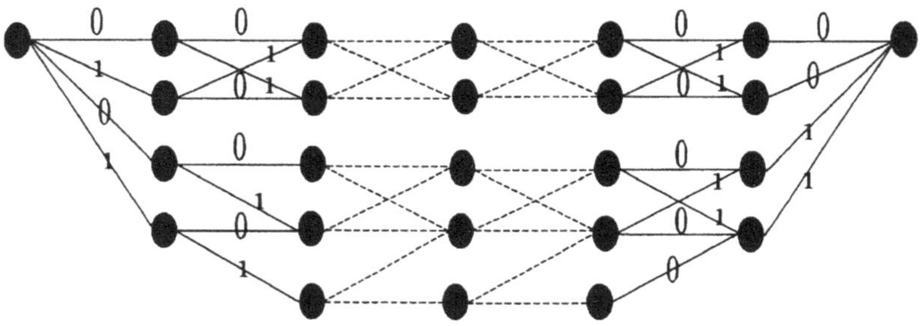

Fig. 2.4.10: General structure for User two Code (N>=3)

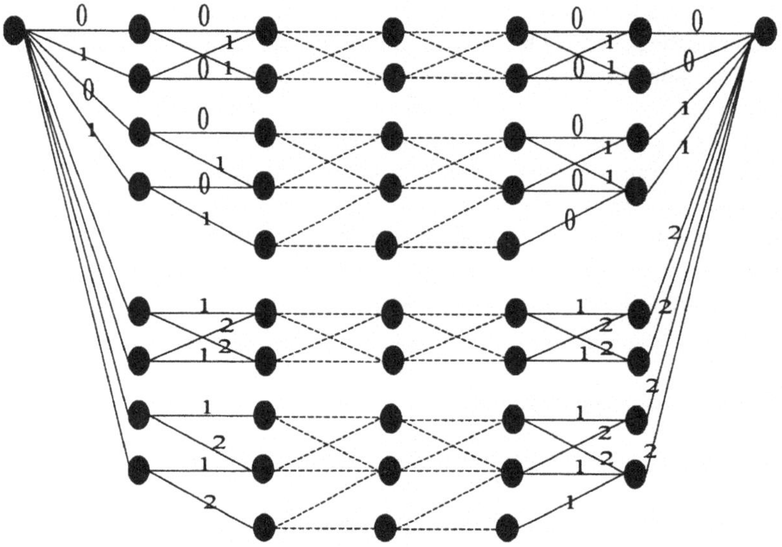

Fig2.4.11: Trellis Diagram for 2-user Scheme with length N>=3

2.4.3.4
An Improved Error Correction Scheme

It is noted in the coding scheme above that since the number of component codes for User two is 2^N-1 it's minimum distance d_{min2} will always be one regardless of the size of N whilst d_{min1} for User one increases and is equal to N. It was proved by [5] that the minimum L-distance, $d_{min(M)}$ of an M-user collaborative code is the smallest value of $d_L(Z,Z')$ over all distinct Z and Z' where Z and Z' are defined as N-symbol composite code words where

$$Z = Z_1 + Z_2 + \ldots + Z_M \tag{12}$$

and

$$Z' = Z'_1 + Z'_2 + \ldots + Z'_M \tag{13}$$

It was further established by the same authors that $d_{min(M)}$ can also be determined in terms of the minimum distance of the component codes as

$$d_{min(M)} = \min\left(d_{min1}, d_{min2}, \ldots, d_{minM}\right) \tag{14}$$

where d_{min1} and d_{min2} are the minimum distances of User one and User two respectively.

This can be seen to be a disadvantage which limits the error correction capability of the composite scheme. One way of improving the error correction is to increase d_{min2} whilst maintaining d_{min1} constant. The price paid for such an increase in d_{min2} and, subsequently $d_{min(M)}$, is the reduction in the overall code rate of the multi-user system since the ith-user's rate is proportional to the number of code words CW_i . Furthermore, it can be shown that it is possible to achieve a $d_{min2} = 2$ for User two component code with an overall code rate sum equalling time sharing. The advantage in this case is the added error protection provided in the system. As an example, we treat the 2-user case with length N=3 where the original coding scheme has the following composite codes

Table 2.4.4 : 2-user Block Code

$User_2$\User	000	111
000	000	111
001	001	111
010	010	121
011	011	122
100	100	211
101	101	212
110	110	221

The 2-user scheme shown in table 2.4.4 has $d_{min1} = 3$ and $d_{min2} = 1$ for users one and two respectively and a code rate sum of 1.265 (bits/channel use). A modification for such a table in order to improve the error correction capability would be to choose User two component code with increased d_{min2} . Table 2.4.5 shows such a modification where $d_{min1} = 3$ and $d_{min2} = 2$ and the composite scheme maintains the unique decodability criteria.

Table 2.4.5 : 2-user Block Code

$User_2/User_1$	000	111
000	000	111
011	011	122
101	101	212
110	110	221

In this case, User two has four component codes which corresponds to the minimum number of code words required to achieve the time share limit. The number of code words for User two has decreased by a factor of $2^N/2^{N+1}$ to $2^N/2$. This can be shown to be true for all values of N where d_{min2} =2. Furthermore, it has been found that for odd values of N, User two code will maintain the all-zero and all-one codes whereas, for even values of N, they may change but such a change will not effect d_{min1}. The change in User one component codes is a consequence of User two acquiring the all-one code as part of it's code book. This can be demonstrated by examining the modified trellis structures for lengths N=3 and 4 in Fig. 2.4.12.

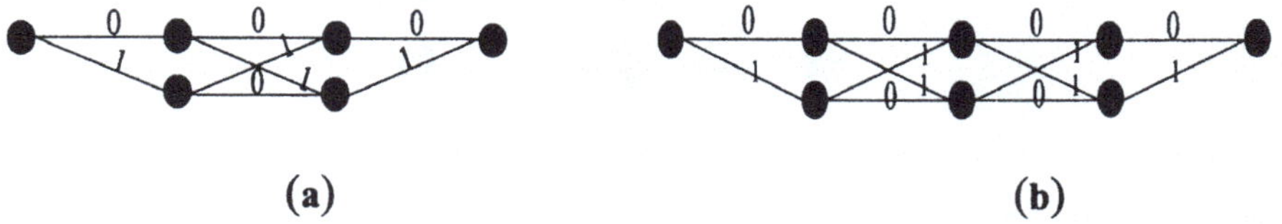

Fig.2.4.12: Trellis diagram of modified User two code for (**a**) N=3 and (**b**) N=4

Inspection of Fig. 2.4.12b shows that the trellis representation for N=4 is designed in such a way that the all-one code is part of it's code book. In order to maintain unique decodability, User one cannot have as part of it's code the all-one or the all-zero codes. Further examination show that User two trellis codes can be generalised for all values of N (Fig. 2.4.14). Furthermore, User one code can also be generalised separately for odd and even N, (Fig. 2.4.13), from which a overall trellis structure is attained as shown in Fig. 2.4.15 and 2.4.16 for odd and even N respectively.

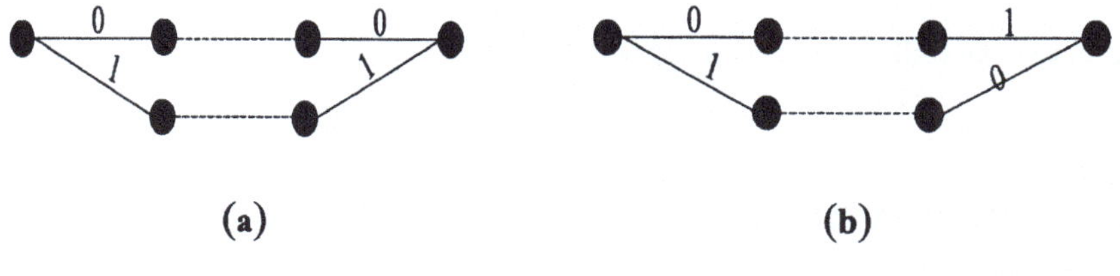

Fig.2.4. 13: Trellis structure User one code for (a) odd and (b) even N

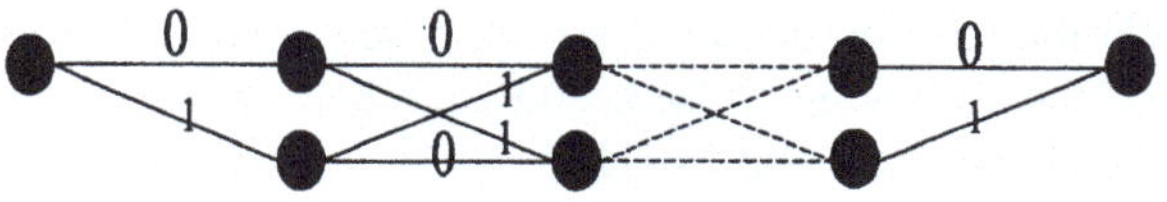

Fig. 2.4.14: Trellis structure User two code for length N

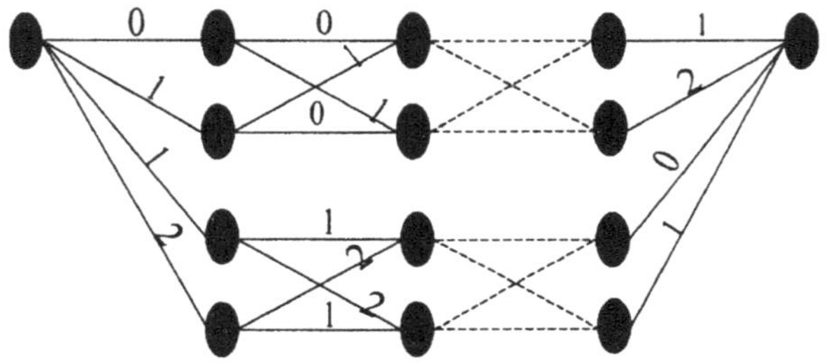

Fig2.4.15: Overall structure with even N

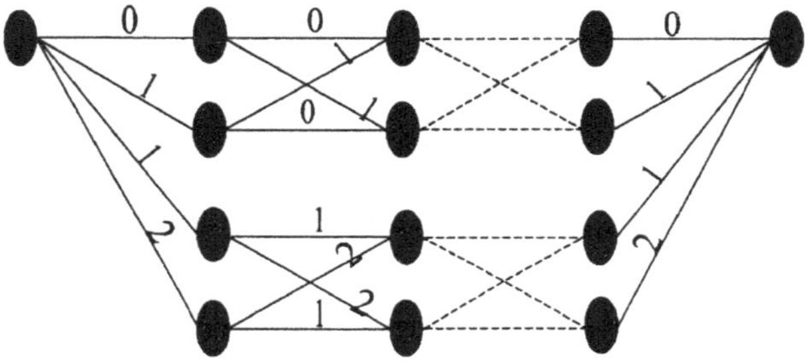

Fig. 2.4.16: Overall structure with odd N

2.4.4 Source combiner based on CCMA schemes

In this section we introduce the application of CCMA schemes as a source combiner and we present techniques in which the CCMA codewords are conveniently mapped to a complex signal that is then spread prior to its transmission. For this purpose we use 2-user CCMA coding examples of block length n =4 as a source combiner. In order to illustrate the different rate and protection levels that can be obtained with the new source combining technique we make use of the following CCMA coding schemes, which are based on the schemes proposed by Kasami and Lin [10].

The first CCMA scheme uses the following codes for user one and two;

$C_1 = (0000, 1111)$

C_2 = (0000, 0001, 0010, 0011, 0100, 0101, 0110, 0111, 1000, 1001, 1010, 1011, 1100, 1101, 1110),
the second CCMA scheme uses the following codes;
C_1 = (0000, 0011, 1100, 1111)
C_2 = (0000, 0001, 0010, 0100, 0101, 0110, 1000, 1001, 1010)
The rate of each code is obtained using eq. (1). For the first CCMA scheme, the rate of C_1 is $R_1 = \log_2 2/4 = 0.25$, the rate of C_2 is $R_2 = \log_2 15/4 = 0.976$, the overall rate sum is $R_T = R_1 + R_2 = 1.226$, which is better than time-sharing. The minimum distance of these codes are $d_{min1} = 4$, $d_{min2} = 1$,respectively . For CCMA scheme 2, the rate of C_1 is $R_1 = \log_2 4/4 = 0.5$, the rate of C_2 is $R_2 = \log_2 9/4 = 0.792$. The overall rate sum is $R_T = R_1 + R_2 = 1.292$. The minimum distance is, $d_{min1} = 2$, $d_{min2} = 1$, which is different to that of the first scheme. We code source one using C_1 and source two using C_2. Because the information is generated from same terminal the synchronization problem of CCMA is eliminated. C_1 and C_2 have different minimum distances, so two sources will have different levels of protection. The output of source combiner $z^{(r)}$ is non-binary (assuming Q-ary number, here Q = 3), we use the following phase modulation to map the Q-ary output of the CCMA encoder to a more convenient complex form as follows:

$$b^{(r)} = e^{i\frac{2\pi}{Q}z^{(r)}} \tag{15}$$

This is then spread using one of the orthogonal spreading codes proposed for the UTRA system. For simplicity we use a simple Walsh function as the spreading code. The system model which we used in the simulation is shown in Fig. 2.4.17.

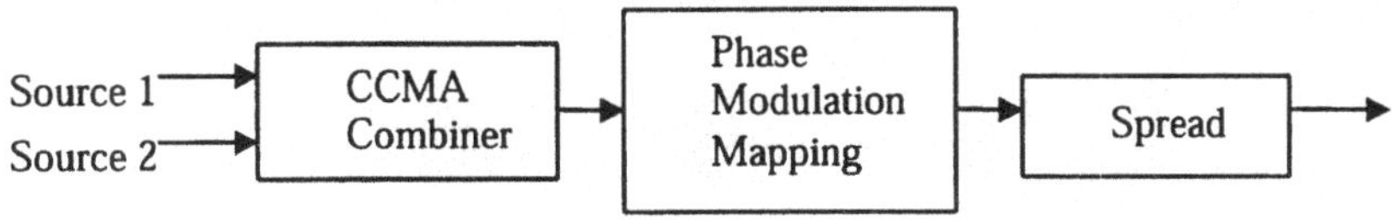

Fig. 2.4.17: system model using source combiner (transmitter side)

2.4.4.1
Trellis structures of CCMA codes

The trellises for each of the coding schemes is constructed which is then used for soft decision decoding.

Starting with the first coding schemes, the component trellises of source one, C_1, and source two, C_2, are T_1 and T_2. These are obtained using the Shannon product technique described in section 2.4.3. The constructed trellises for these codes are shown in Fig. 2.4.18 and 2.4.19. The overall trellis is shown in Fig. 2.4.20. As it follows from these figures, the component trellises have the following state and branch profiles:

$$N^1(t) = [1,2,2,2,1]; \qquad N^2(t) = [1,4,5,4,1];$$

$$B^1(t) = [2,2,2,2]; \qquad B^2(t) = [4,8,9,4]$$

Following the procedure outlined in 3.1, the overall trellis will have the following state and branch profiles:

$$N(t) = [1,8,10,8,1];$$

$$B(t) = [8,16,18,8]$$

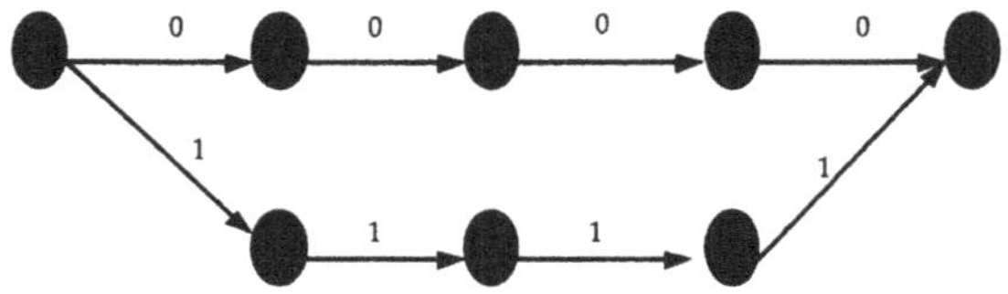

Fig.2.4.18: Trellis Diagram for Source 1 code

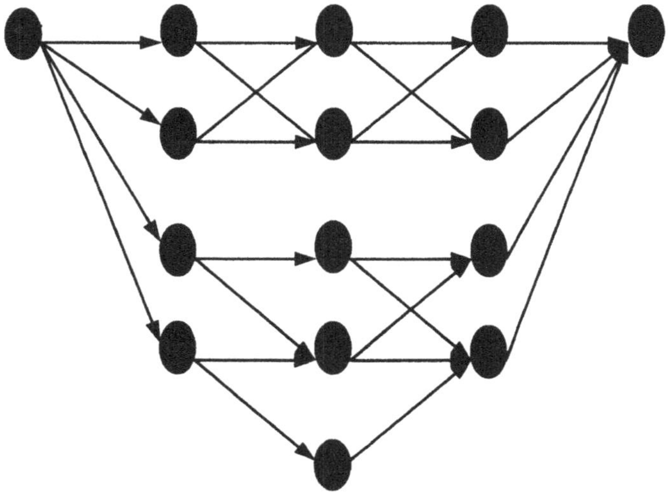

Fig. 2.4.19: Trellis Diagram for Source 2 code

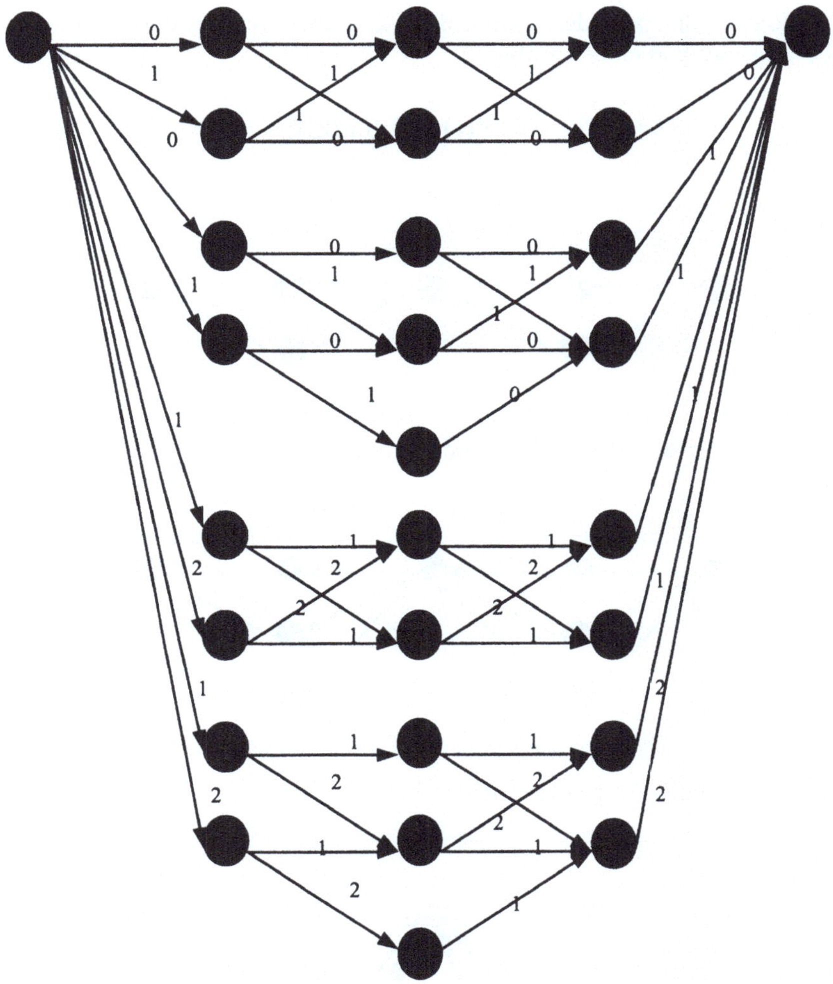

Fig.2.4.20: Overall Trellis Diagram for Source 1 and Source 2 codes

Similarly, the component trellises of source one and two, T_1 and T_2, of the coding scheme 2 are constructed and are shown in Fig. 2.4.21 and 2.4.22, and the overall trellis is shown in Fig. 2.4.23. As it follows from these figures, the component trellises have the following state and branch profiles:

$$N^1(t) = [1,2,1,2,1]; \qquad N^2(t) = [1,3,1,3,1];$$
$$B^1(t) = [2,2,2,2]; \qquad B^2(t) = [3,3,3,3]$$

Following the procedure outlined in 2.1, the overall trellis will have the following state and branch profiles:

$$N(t) = [1,6,1,6,1];$$

$$B(t) = [6,6,6,6]$$

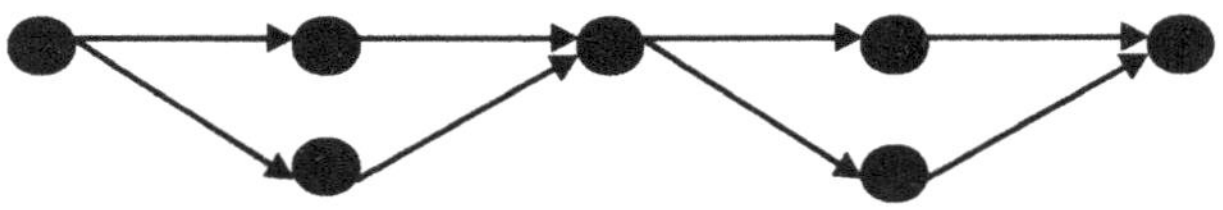

Fig. 2.4.21: Trellis Diagram for Source 1 code

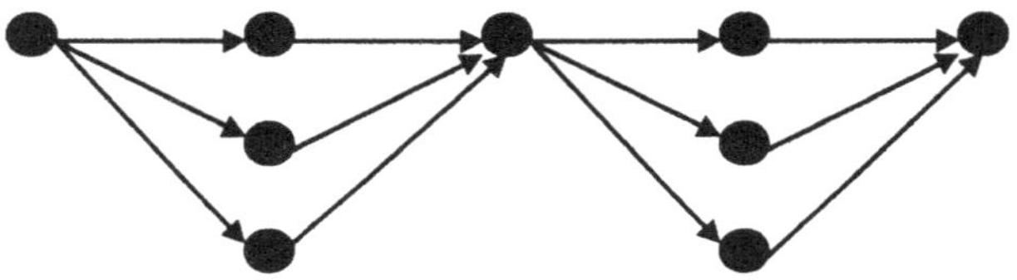

Fig. 2.4.22: Trellis Diagram for Source 2 code

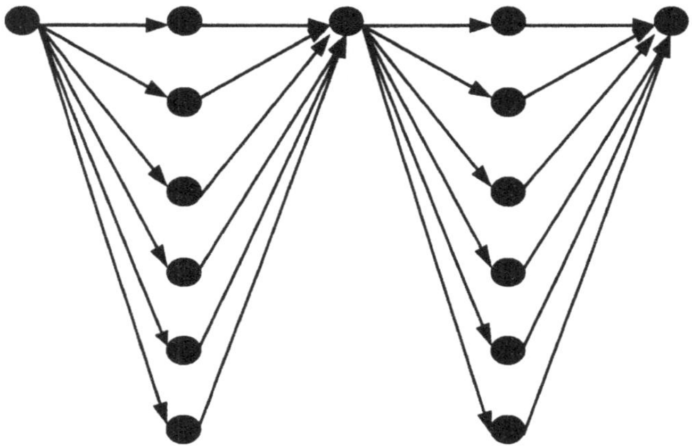

Fig. 2.4.23: Overall Trellis Diagram for Source 1 and Source 2 codes

2.4.4.2
Performance of the source combiner schemes

The system shown in Figure 17 is simulated for the two coding schemes described in section 2.4.4.1 over a frequency selective channel. The parameters of the system which are used in the simulation are as follows. Walsh function are used as the spreading code with spreading factor of 4 and chip rate 4.096 Mcps. The symbol rate for both CCMA schemes is 1024 ksps. This makes the overall information rate for each scheme as follows: the information bit rate of Scheme one source one is 256kbps, and for source 2 it is 999.424 kbps; for the second scheme, source one the information bit rate is 512 kbps, and for source two it is 811.008 kbps. Three path Rayleigh channel model, with maximum Doppler frequency is 50 Hz, is used with the delay profile, h[3] = {1, 0.7, 0.4}. The spreading code is Walsh function (length = 4). Figure 24 and 25 gives the simulated BER results of source 1 and 2 under the proposed system with scheme 1 and scheme 2, respectively. It observed from this figures that source one provides a much better error protection level comparing with source two due to the larger minimum distances. For scheme 1, a BER of 10^{-4} is obtained at E_b/N_0= 15dB for source one, while the BER of source two is about 10^{-3}. For scheme 2 the BER of source one is about $4.9x10^{-4}$, while the BER of source two is about $1.2x10^{-3}$. The error protection level improvement of scheme 1 is better than that of scheme 2 because the minimum distance difference for scheme 1 (d_{min1} =4, d_{min2} = 1) is bigger than that of scheme 2 (d_{min1} = 2, d_{min2} = 1). Fig. 2.4.26 to 2.4.29 show the simulation results for source one and two using hard decision and trellis decoding techniques for the two different schemes. For scheme 1, source one of the trellis decoding gives about 1.8 dB improvement in comparison with hard decision decoding for CCMA at BER of 10^{-3}, while the improvement of source two is not significant because of the small minimum distance of its code; for scheme 2, source one of the trellis decoding gives about 1 dB improvement which is less than that of scheme 1. This is also due to the minimum distance difference of scheme 1 which is bigger than scheme 2. The results of Fig. 2.4.24 to 2.4.29 are obtained using conventional receiver.

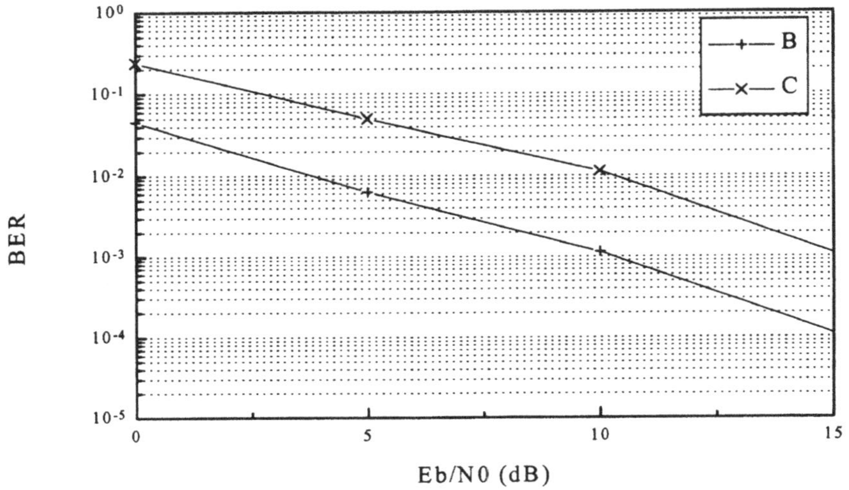

Fig. 2.4.24: BER performance of Source one (B) and source two (C) for scheme 1.

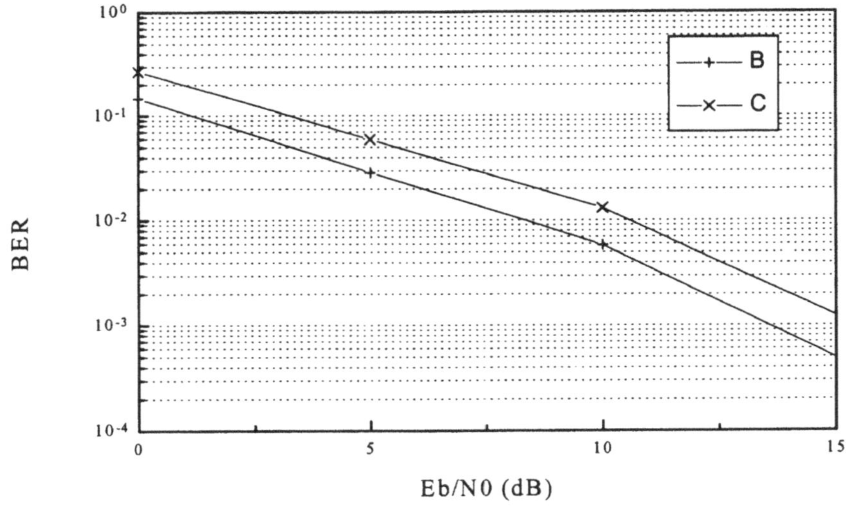

Fig. 2.4.25: BER performance of Source one (B) and source two (C) for scheme 2.

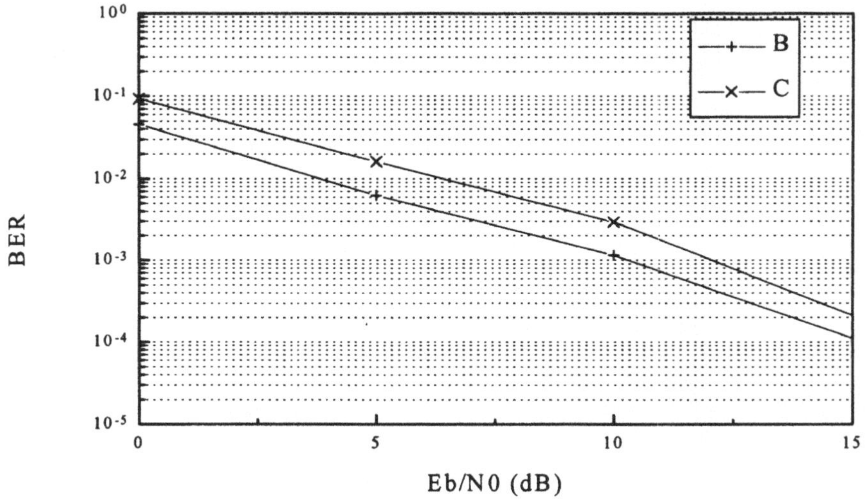

Fig.2.4.26: BER of Source one using trellis (B) and hard decision (C) decoding for scheme 1.

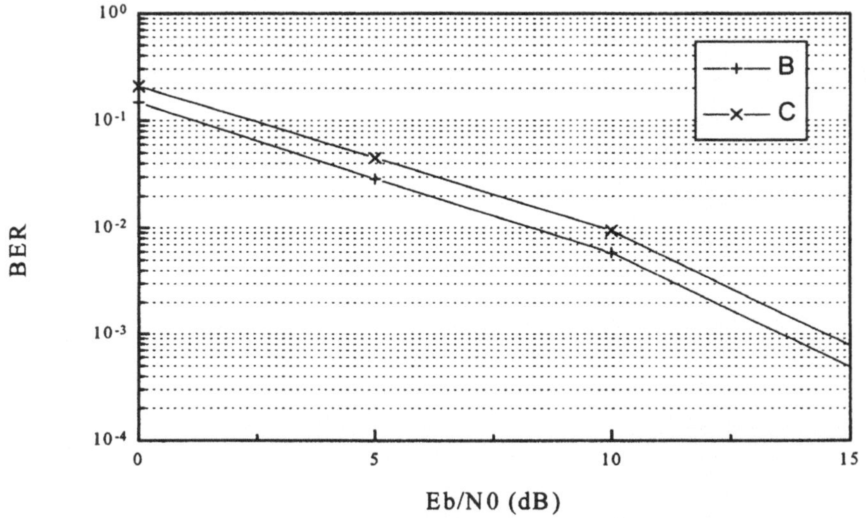

Fig. 2.4.27: BER of Source one using trellis (B) and hard decision (C) decoding for scheme 2.

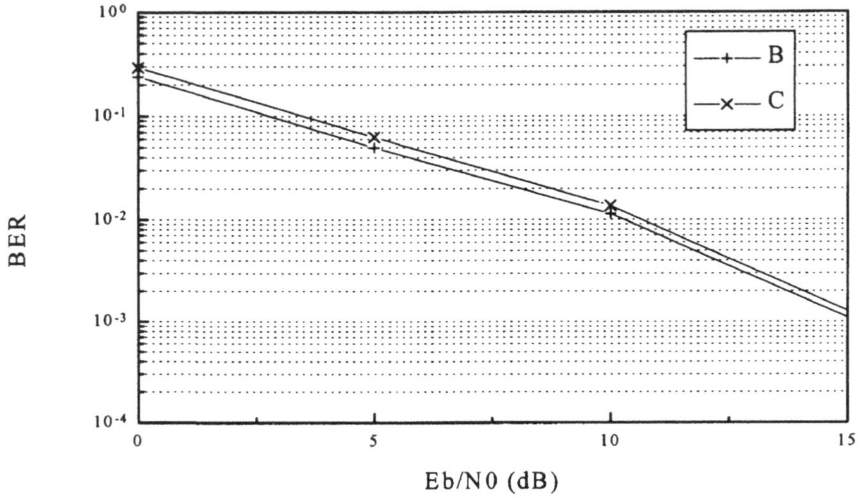

Fig. 2.4.28: BER of Source two using trellis (B) and hard decision (C) decoding for scheme 1.

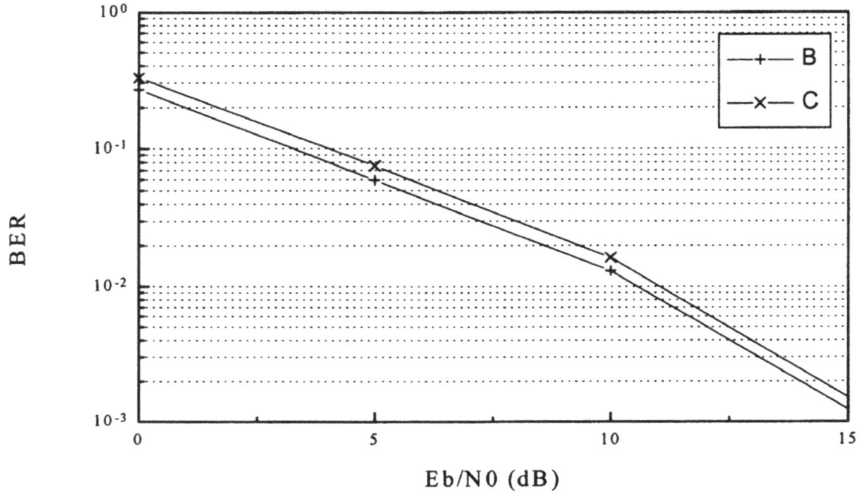

Fig. 2.4.29: BER of Source two using trellis (B) and hard decision (C) decoding for scheme

REFERENCES

1. P. Anderson, "FRAMES contribution to UMTS", Proceedings of *FRAMES Workshop 1999*, Jan. 1999, Delft, the Netherlands, pp. 2-8.
2. K. Kalliojarvi and E. Nikula, "Physical Layer of UTRA FDD Mode - WCDMA", Proceedings of *FRAMES Workshop 1999*, Jan. 1999, Delft, the Netherlands, pp. 9-13.
3. P. G. Farrell,"Survey of channel coding for multi-user systems" *in SKWIRZYNSKI, J.K.* (Ed.): 'New concept in multi-user communication'(Sijthoff and Noordhoff, 1981), pp. 133-159.
4. R. Gallager, "A perspective on multiaccess channels", *IEEE Trans.*, 1985, IT-31, (2), pp. 124-142.
5. S. C. Chang and E. J. Weldon, Jr., "Coding for T-user multiple access channel," *IEEE Trans. Inform. Theory,* vol. IT-25, no. 6, pp. 684-691, Nov. 1979.
6. S. C. Chang, "Further results on coding for T-user multiple access channels, " *IEEE Trans. Inform. Theory,* vol. IT-30, no. 2, pp.411-415, Mar. 1984.
7. P.A.B. M. Coebergh van den Braak and H. C. A. van Tilborg, "A class of uniquely decodable code pairs for the tw-access binary adder channel," *IEEE Trans. Inform. Theory,* vol. IT-31, no.1, pp. 3-9, Jan. 1985.
8. T. J. Ferguson , "Generalised T-user codes for multiple access channels," *IEEE Trans. Inform. Theory,* vol IT-28, no. 5, pp. 775-778, Sept. 1982
9. T. Kasami, S. Lin, V. Wei, and S. Yamamura, "Further results on coding for a multiple access channel, " in Topics Information Theory. I. Csiszar and P. Elias, Eds. New York: North-Holland Publishing Company, pp. 369-394, 1975.
10. T.Kasami and S.Lin, "Coding for a multiple access channel", *IEEE Trans. Inform. Theory.* Vol.. IT-22, No.2, pp. 129-137, March 1976.
11. T. Kasami, S. Lin, V. Wei, and S. Yamamura, " Graph theoretic approaches to the code construction for the two-user multiple access binary adder channel," *IEEE Trans. Inform. Theory,* vol. IT-29, pp.114-130, 1983.
12. G. H. Khachatrian , " On the construction of codes for noiseless synchronised 2-user channels," *Prob. Contr. Inform. Theory,* vol. 11, pp.319-324, 1982.
13. G. H. Khachatrian , " Decoding for a noiseless adder channel with 2-users," *Prob. Peredach. Inform.* vol. 19, no. 2,pp.8-13, (pp. 159-163 in English translation), Apr.-June 1983.
14. G. H. Khachatrian , "δ-decodable code constructions for a T-user adder channel" *Prob. Peredach. Inform.* vol. 24, no. 2, pp. 94-99, (pp. 159-163 in English translation), Apr.-June 1988.
15. S. Lin, V. K. Wei, "Non-homogeneous trellis codes for the quasi-synchronous multiple access binary adder channel with two users", *IEEE Trans. Inform. Theory,* vol. IT.32 pp.787-796, Nov. 1986.
16. E. J. Weldon, Jr., " Coding for a multiple access channel," *Inform. Contr,* vol. 36, no. 3, pp.256-274, Mar. 1978.
17. J. H. Wilson, "Error correcting codes for a T-user binary adder channel", *IEEE Trans. Inform. Theory,* vol. IT.34, no. 34, pp.888-890, Jul. 1988.
18. P.Mathys, ''A class of codes for T active users out of N, for a multiple access communication system,'' *IEEE Trans. Inform. Theory,* vol. 36 pp.1206-1219, Nov.1990.
19. R. Ahlswede, "Multi-way communications channels," in *Proc. Of the 2nd Int. symp. Inform. Theory, Tsahkadsor, Armenian S.S.R,* pp. 23-52, 1971.

20. H. Liao, "A coding theorem for multiple access communications," *in the Int.Symp. Inform. Theory Asilomar, 1972.*
21. F.H. Ali and B. Honary, ''Low complexity soft decision decoding technique for T-user collaborative coding multiple access channels'', *Electronic Letters, 27,* (13), pp.1167-1169, 1991.
22. F.H. Ali and B. Honary, ''Collaborative coding and decoding techniques for multiple access channel'', *IEE Proceedings on Communications,* Vol. 141, No. 2, pp.56-62, April 1994.
23. V. da Rocha, G. Markarian and B. Honary, "Trellis decoding of block codes for the 2-user binary adder channel", *Proceedings of ICC'93, Geneva,* vol. 2/3, pp. 821-826.
24. G.Markarian, B.Honary, P.Benachour, ''Trellis decoding techniques for the binary adder channel with M users'', *IEE Proceedings on Communications,* vol.144, 1997.
25. P. Benachour. G. Markarian, B. Honary, " Trellis decoding techniques in an M-choose-T multiple access adder channel ", *in Proceedings of the 4th Int. Symp. On Communication Theory and Applications, Ambleside, UK,* pp 84-85, July 1997.
26. G. Markarian, P. Benachour and B. Honary, "Novel decoding technique for the synchronous and quasi-synchronous multiple access adder channel " *in Proceedings of the 6th International Conference On Cryptography and Coding, Cirencester, UK,* pp.307-316, December 1997.
27. V.Sidorenko, G.Markarian, B.Honary, ''Minimal trellis design for linear codes based on the Shannon product '', *IEEE Trans. Inform. Theory,* vol. 42 pp.2048-2053, Nov.1996.
28. B. Honary, G. Markarian, " Trellis decoding of block codes : A practical approach ", *Kluwer Academic Publishers,* 1997.
29. C. E. Shannon, "The zero capacity of a noisy channel", *IRE Trans. on Infrom. Theory,* vol.2, pp.s8-s19, September 1956.
30. L. R. Bahl, J. Coke, F. Jelinek and J. Raviv, "Optimal decoding of linear codes for minimising symbol error rate", *IEEE Trans. on Inform. Theory,* vol. IT-20, 1974, pp. 284-287.
31. R. J. McEliece, "The Vitrebi decoding complexity of linear block codes", *Proc. Of the IEEE Int. Symp. On Inform. Theory,* Trondheim, Norway, p. 341, 1994.
32. J. K. Wolf, "Efficient maximum likelihood decoding of linear block codes using a trellis", *IEEE Trans. on Inform. Theory,* vol.24, pp. 76-80, 1978.

2.5 Low-Power Multi-Code Correlator for IMT-2000

Chung-Wei Ku and Fu-Yen Kuo
Liang-Gee Chen and Chi-Kuang Chen

2.5.1 Introduction

Wireless communications become more and more important in terms of technology trends. In recent years, the demand of data transmission over mobile radio channels increased very fast. Although GSM standard becomes very popular and begins to finalize the data service specifications [8], most of the next generation proposals for mobile communications are based on Code Division Multiple Access (CDMA) technologies, such as ETSI UMTS or TR45.5 cdma2000 [9][10]. In order to support future multimedia transmission, wide bandwidth and high data rate are both expected.

For modern CDMA systems, either different orthogonal codes or the same code with different shifts are used by both mobile station and base station. One example is multi-code DS-CDMA. That is, one user uses several codes or channels to transmit his data [11][12]. Such a scheme is suggested in ETSI UMTS and TR45.5 cdma2000 for supporting high data rate services [9][10]; the above two proposals are active candidates for ITU IMT-2000 standard. On the other hand, in terms of code acquisition, we generally have to correlate the input with one PN code with different shifts. As a result, it is quite useful to design a hardware for concurrent de-spreading with multi-code.

In CDMA systems, receiver has to de-spread the received signals back into original input symbols by correlating input data with PN sequence. Most of the implementations utilize chip matched filter or correlator to achieve the above purpose. In terms of general CDMA receiver design, for the sake of trade-off between flexibility and cost, digital signal processor is preferred for symbol-rate processing; other processing, either for chip-rate or high-speed process, might be handled by other ASIC-based engines. For example, Viterbi engine is designed for convolutional decoding, and correlator array is designed for chip-rate de-spreading. The above approach is efficient for hardware, and becomes a starting approach for future software defined radio [16].

In fact, in terms of power consumption or real-time speed requirement, de-spreader is one of the most important blocks in RAKE receiver. Some previous researchers try to increase the throughput by adding two PN sequences as the FIR coefficients [4][5]. No multipliers are needed actually because the FIR coefficients are all powers of two. In addition, since many FIR coefficients become zeros, hardware cost can be reduced furthermore. For the results in [4][5], the two PN sequences in fact come from the same code with different shifts. Such a shift is equal to 1 chip in [4] but half of spreading factor in [5]. Another innovative idea to reduce power consumption utilizes sign-magnitude data format and two accumulators to correlate positive and negative partial sums respectively [3][7]. In order to satisfy the speed and power requirements, carry-save adder and gated-clock technique are suggested in this work. Some researcher utilized the properties of over-sampled input and dedicated circuit design to reduce power consumption furthermore [6].

In this chapter, a 1 GOP/Watt low-powered architecture combining the above ideas is proposed for concurrent dual-code de-spreading at first. The suggested architecture can de-spread the input signal with two PN sequences at the same time but both power consumption and hardware overhead are reduced significantly. Similar idea is also extended to the design of multi-code correlator and some preliminary simulation results are discussed. Generally speaking, there is no restrictions of code inputs for the proposed architecture. In fact, the proposed architecture can be regarded as a general representation of previous works [3][4][5].

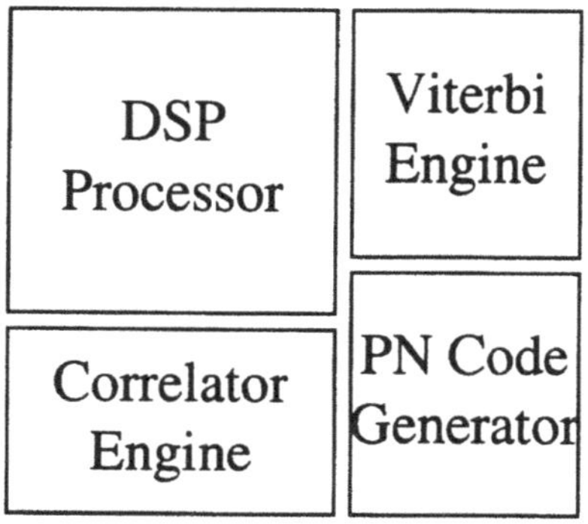

Fig. 2.5.1. General CDMA baseband receiver architecture.

2.5.2 CDMA Baseband Receiver Architecture

Most of the implementation of CDMA baseband receiver can be divided into several parts: the digital signal processor core, Viterbi engine for soft-decision convolutional decoding or demodulation (this part might be included in the DSP core for some modern DSP processors), correlator array (probably for code acquisition, code tracking, channel estimation, and the frond-end of RAKE combining), and the PN code generator. In addition to the DSP core, other engines are designed for chip-rate processing or mass computation of symbols. A block diagram of the above idea is shown in Fig. 2.5.1. The correlator array can be implemented in many ways, such as integrate-and-dump correlators, or chip matched filter approach in FIR form. Generally speaking, correlator array is the most computation-intensive part for the entire baseband receiver, therefore how to reduce the power computation of this part might be critical issues for mobile stations. Some previous researchers already proposed some innovative ideas to solve this problem. We firstly discuss the techniques they proposed in this section then introduce the idea we proposed.

2.5.2.1 Pipelined Differential Chip Matched Filter

Liu proposed a pipelined differential matched filter for code acquisition [4]. Suppose the N-tapped FIR filter output at time T-1 is described as

$$f_c(T\text{-}1) = a_N x_0 + a_{N\text{-}1} x_1 + \ldots. + a_2 x_{T\text{-}2} + a_1 x_{T\text{-}1.}$$

Similarly, the output at time T can be found as

$$f_c(T) = a_N x1 + a_{N\text{-}1} x2 + \ldots. + a_2 x_{T\text{-}1} + a_1 x_{T.}$$

We define the differential output $f_D(T)$ as

$$f_D(T) = [c_{N+1} x_0 + c_N x_1 + c_{N\text{-}1} x_2 + \ldots. + c_2 x_{T\text{-}1} + c_1 x_T] + f_D(T\text{-}1),$$

which is equal to

$$[\text{-}a_N x_0 + (a_N \text{ - } a_{N\text{-}1}) x_1 + (a_{N\text{-}1} \text{ - } a_{N\text{-}2}) x_2 + \ldots. + (a_2 \text{ -} a_1) x_{T\text{-}1} + a_1 x_T] + f_c(T\text{-}1).$$

In fact, $f_D(T)$ is identical to $f_c(T)$ if:

$$c_1 = a_1,\ c_i = a_i \text{ -} a_{i\text{-}1} \text{ for i} = 2 \text{ to } N, \text{ and } c_{N+1} = \text{-}a_N.$$

As a result, since all of the c_i's are either 0, 2, or -2, some hardware cost can be reduced. Generally speaking, half of the coefficients will be zeros; and others are either 2 or -2. In addition, multiplied by 2 or -2 can be easily implemented by simple hard-wired shift operation; no actual multipliers are needed for the pipelined differential matched filter. The block diagram of such an idea is shown in Fig. 2.5.2. The differential matched filter receives the input signal and generates the output in a differential form. The output is accumulated by an accumulator to obtain the correct correlation values clock by clock.

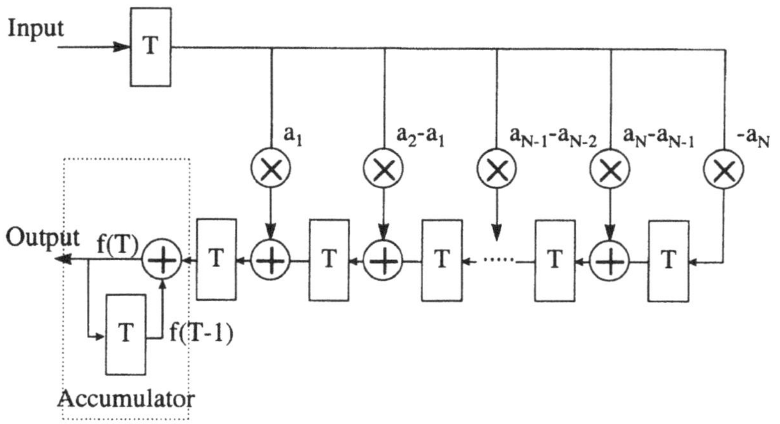

Fig. 2.5.2. Block diagram of pipelined differential matched filter.

2.5.2.2
Low Complexity Matched Filter

The idea in section 2.5.2.1 utilizes the property about one PN code with one chip shift. The difference between one PN code and its shift includes lots of zeros therefore the number of multipliers (in fact, XOR gates for 2's complement representation) can be saved. Similar idea named low complexity matched filter is suggested in [5]. Fig. 2.5.3 displays the idea about low complexity matched filter. The bold line in the figure denotes the beginning of the PN sequences. If we would like to use chip matched filter for code acquisition, the proposed idea is to get the difference of two codes, which comes from the same PN code with different delays. In fact, such an idea is similar to the pipelined differential matched filter, but the delay in this design is one half of spreading factor. As a result, the matched filter will obtain two peaks which is located in the peak of the original PN code but with opposite signs. For code acquisition application, we just have to find the location with the most correlation amplitude. However, for channel estimation cases, the sign ambiguity might be a problem. The authors also suggested some ways to examine the maximal correlation values with other considerations to obtain the correct sign because the modulation scheme is OQPSK for wireless LAN. As a result, the hardware cost is reduced and the acquisition time is saved because the throughput of the matched filter is doubled in fact. However, overhead of control circuits for sign ambiguity decision and timing decision is unavoidable. In fact, such an idea also depends on the correlation properties of the PN code used. In the case of wireless LAN, the PN codes are fixed Barker code and the purpose of the code is for spectrum spreading. However, in the case of CDMA systems, the code is further used for multiple access; therefore, the PN code is not fixed anymore. In addition, the auto-correlation and cross-correlation for PN codes must be considered.

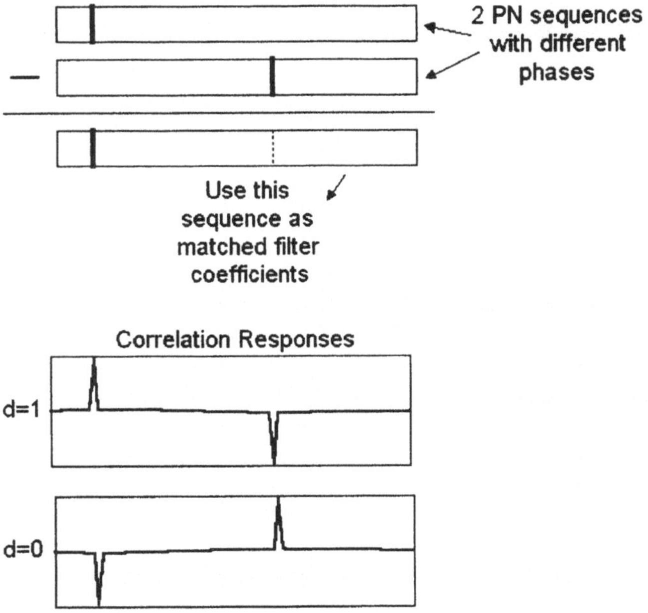

Fig. 2.5.3. The idea of low Complexity matched filter architecture.

2.5.3
Theory of Multi-Code De-Spreading

In previous section, some previous works utilized the property of single PN code is reviewed. For modern third generation technologies, Code Division Multiple Access (CDMA) seems to be the trend. For example, most of the proposals for ITU International Mobile Telecommunications in year 2000 (IMT-2000) are based on CDMA technologies. Two representative proposals are TIA cdma2000 [9], and ETSI UMTS Universal Terrestrial Radio Access (UTRA) proposal [10]. Both of the proposals are developed by the third generation partnership projects (3GPP and 3GPP2) which devoted to the migration from the second generation to the third generation. An uniformed standard of IMT-2000 is still under development. However, in order to support multimedia services and the integration of a variety of handover or roaming, some important issues, such as variable spreading factor supported by orthogonal variable spreading factor (OVSF), multi-code transmission, or multi-carrier option, can be found from those proposals. All of these try to support variable bit-rate transmission, or different quality of services under different environments. Basically speaking, variable spreading factor receiver can be easily implemented by programmable correlators. We just need to change the integration time for integrate-and-dump, therefore the spreading factor is changed. On the other hand, multi-code CDMA receiver also can be easily implemented by several independent correlators. Since multi-code CDMA is useful for future multimedia transmissions, a low-powered architecture for multi-code correlator will be discussed in next section.

2.5.3.1 Code Division Multiple Access

The basic idea of CDMA is spreading one user's data by a specific PN code. In terms of mathematical formula, we can express the transmitted base-band signal as

$$S_T = \sum_i b_i[j]C_i,$$

where $b_i[j]$ is a scalar in 1 or -1 format representing the *j*th bit of user *i*, and C_i is a column vector whose component represents the code sequence; the entries in C_i are also either 1's or -1's. Finally all the transmitted signals are summed together and transmitted. We neglect the effects of multi-path fading channel for simplification purpose. However, the results considering those effects should give similar concepts. On the other hand, the receiver de-spreads the data by correlating received signal with one specific PN code C_k as follows:

$$b'_k[j] = sign(C_k^T S_R).$$

Sign function outputs 1 if the input is positive and outputs -1 for negative input. Generally

$$S_R = S_T + n,$$

where *n* means the AWGN white noise from the environment. Therefore, we can get

$$\begin{aligned} b'_k[j] &= sign[C_k^T(S_T + n)] \\ &= sign[C_k^T(\sum_i b_i[j]C_i + n)] \\ &= sign[b_k[j]C_k^T C_k + \sum_{i \neq k}(b_i[j]C_k^T C_i + C_k^T n)]. \end{aligned}$$

For orthogonal PN codes,

$$C_k^T C_i \approx 0, \text{if } i \text{ is not equal to } k.$$

In addition, noise *n* is small compared with original signal and

$$C_k^T C_k = m, \text{ where } m \text{ is the spreading factor.}$$

Since $b_k[j]$ is either 1 or -1, we obtain the de-spread symbols as

$$b_k'[j] \approx sign(b_k[j]m)$$
$$= b_k[j].$$

In other words, in addition to the noise signal, the other user's data can be regarded as another source of interference for CDMA systems. Generally speaking, the cross-correlation between different PN codes should be good enough to eliminate the interference to other users. As a result, the capacity of the whole CDMA systems is guaranteed.

2.5.3.2 *Dual-Code De-Spreading*

We consider the situation about de-spreading two different PN codes concurrently at first. In order to de-spread two codes simultaneously, we re-design two new codes which comes from the sum and difference of the original two codes :

$$C_1 = (C_a + C_b)/2$$
$$C_2 = (C_a - C_b)/2.$$

C_a and C_b are two codes which probably are two orthogonal codes or come from the same code but with different shifts, depending on how we use the de-spreader. Since the entry of C_a and C_b is either 1 or -1, it is obvious that the entry of C_1 and C_2 is 1, -1, or 0. In addition, suppose $C_{1,i}$ represents the *i*th entry of vector C_1, we find

$$C_{1,i} = 0 \text{ if } C_{2,i} \neq 0, \text{ and}$$

$$C_{1,i} \neq 0 \text{ if } C_{2,i} = 0.$$

That is, the entries in C_1 and C_2 are exclusively non-zero. Furthermore, according to the property of PN codes, about half of the items in the two re-designed codes should be zeros. Without losing the generality, suppose the transmitted signal includes two users' data:

$$S = b_a[j]C_a + b_b[j]C_b.$$

To simplify the process, assume noise is zero or negligible. As a result, if we correlate this signal with one of the re-designed new codes as

$$b_1[j] = C_1^T S$$
$$= (C_a^T + C_b^T)S/2$$
$$= (b_a[j] + b_b[j])m/2.$$

Similarly,

$$
\begin{aligned}
b_2[j] &= C_2^T S \\
&= (C_a^T - C_b^T)S/2 \\
&= (b_a[j] - b_b[j])m/2 .
\end{aligned}
$$

Therefore, we can obtain the results by simply addition and subtraction of the correlation values as:

$$
\begin{aligned}
b_a[j] &= sign(b_1[j] + b_2[j]) \\
&= sign(b_a[j]m) \\
&= b_a[j] \\
b_b[j] &= sign(b_1[j] - b_2[j]) \\
&= sign(b_b[j]m) \\
&= b_b[j]
\end{aligned}
$$

The above implies we can use the two new codes to de-spread the input signal. The obtained correlation values can be further utilized to get the correct correlation values for original PN codes. However, since the total numbers of computations (1's and -1's) in C_1 and C_2 is equal to the spreading factor, that implies we just need extra addition/subtraction in addition to the original computations for single code de-spreading to accomplish the correlation for dual-code. That significantly reduces the operations compared with direct de-spreading by correlating the input with the original two PN codes.

2.5.3.3 An Example of Dual-Code De-Spreading

In order to explain the dual-code de-spreading method more clearly, we assume the two codes are Walsh codes, $[1\ -1\ 1\ -1]^T$ and $[1\ 1\ -1\ -1]^T$. Therefore, the new two codes can be easily obtained as $[1\ 0\ 0\ -1]^T$ and $[0\ -1\ 1\ 0]^T$. Now we try to correlate the input with these two codes, and suppose the received input is $[4\ -2\ 3\ -3]^T$. We found the correlation is

$$[1\ 0\ 0\ -1][4\ -2\ 3\ -3]^T=7,$$

$$[0\ -1\ 1\ 0]\ [4\ -2\ 3\ -3]^T=5.$$

In addition, the total operations for multiply and accumulate (MAC) are almost the same as for single code de-spreading (four multiply operations and two additions). The correct correlation values can be obtained with two more operations as

$$(7+5)=12 \text{ and } (7-5)=2.$$

We can check out and find those are the same as

$$[1\ -1\ 1\ -1][4\ -2\ 3\ -3]^T=12,$$

$$[1\ 1\ -1\ -1]\ [4\ -2\ 3\ -3]^T=2.$$

2.5.3.4 *Multi-Code De-Spreading*

In fact, the above idea also can be extended for multi-code de-spreading. We can regard the dual-code de-spreading as an special case for the following multi-code de-spreading but the code number M is fixed as two. If we try to use M binary codes C_1 .... C_M with spreading factor L for multi-code CDMA, where $C_m = (C_{m,1}....C_{m,L})$ with 0 or 1 elements. The correlated output with input x_n for each code is

$$d_m = \sum_{n=0}^{L-1} (-1)^{C_{m,n}} x_n .$$

Let $B_n=(C_{1,n}$ .... $C_{M,n})$ represents the bit vector which consists of the n'th element of M PN codes. Define P_i as a binary vector whose component is just equal to the binary representation of integer i (e.g. $P_0=(0,0,....,0)$, $P_1=(0,0,....1)$, $P_2= (0,0,....1,0)$, etc.). We also define P_{ij} is the j'th element of P_i. As a result, d_m can be written as

$$\begin{aligned} d_m &= \sum_{n=0,C_{m,n}=0}^{L-1} x_n - \sum_{n=0,C_{m,n}=1}^{L-1} x_n \\ &= \sum_{i=0,P_{im}=0}^{2^M-1} \sum_{n=0,B_n=P_i}^{L-1} x_n - \sum_{i=0,P_{im}=1}^{2^M-1} \sum_{n=0,B_n=P_i}^{L-1} x_n \\ &= \sum_{i=0,P_{im}=0}^{2^M-1} s_i - \sum_{i=0,P_{im}=1}^{2^M-1} s_i \end{aligned}$$

where $S_i = \sum_{n=0,B_n=P_i}^{L-1} x_n$. That implies we can obtain the correlation values for multi-code with some common parts.

Input	y_0	y_1	y_2	y_3	y_4	y_5	y_6	y_7
Code 1	1	1	0	0	1	0	1	0
Code 2	1	0	1	0	0	1	1	0
Code 3	0	1	1	1	0	0	0	0
Group No.	6	5	3	1	4	5	6	0
Reduced No.	1	2	3	1	3	2	1	0

$s_0 = y_7$
$s_1 = -y_0+y_3-y_6$
$s_2 = -y_1+y_5$
$s_3 = y_2-y_4$

$x_1 = s_0+s_1+s_2+s_3$
$x_2 = s_0+s_1-s_2-s_3$
$x_3 = s_0-s_1+s_2-s_3$

Fig. 2.5.4. Example for multi-code de-spreading (M=3, L=8).

In order to explain this idea, we give a simple example in Fig. 2.5.4. We put all the elements for all the codes as a matrix firstly. The group number is assigned to each column according to the binary representation of this integer, which also implies one kind of common terms for multi-code de-spreading. Furthermore, the group number can be reduced to one-half because binary pattern (110) and (001) can be shared; one uses addition and the other uses subtraction operation. As a result, we got four groups of common terms from this example. We can find the exact correlation values for these three codes via the addition/subtraction of the four common terms. Our conclusion is the group number is equal to 2^{M-1}; in addition, as the spreading factor increases, the saving of operations also increases therefore power consumption can be reduced. The final operation to get the exact correlation values is similar to a Hadamard-Walsh transform which can be easily implemented by simple adder/subtractor.

2.5.4 Architecture Considerations

2.5.4.1 Low-Power Architecture Design

Low power design becomes more and more important for modern VLSI design, especially for portable or mobile communications. Reduced power consumption can make the battery life longer; in other words, lighter weight or cheaper cost of the system can be achieved. For CMOS circuits, the most significant power dissipation comes from dynamic power consumption, $P_d = \alpha f C V^2$, where α means the transition probability; f is the clock frequency; C represents the effective load capacitance, and V is the supplied voltage. A representative introduction paper about low power CMOS VLSI design from different aspects of system design was given in [13]. According to the methods described in the above paper, in order to reduce power consumption, we can try to make some alternative designs from algorithm, architecture, circuit, or physical levels. As a result, we can reduce the effective load capacitance, or reduce supplied voltages, or reduce the operated clock frequency to decrease the consumed power. However, in order to support the required computation ability for real-time processing, parallel or pipelined design methodologies are utilized in many cases. Another method is to reduce the transition probability α. Obviously, we can reduce the data transition probability of correlators by using sign-magnitude representation because the input data need to multiply a code which is either 1 or –1. If we use sign-magnitude representation, we only need to multiply the sign bit of the input signal with the PN code. From previous works [5], it shows that sign-magnitude representation can reduce about 30% power consumption. Such an idea is proposed by *Infopad*, and we will discuss all these issues in the following sections.

2.5.4.2
2's Complement Representation

Firstly we consider the architecture for 2's complement representation. The general architecture of a single code correlator is shown in Fig. 2.5.5. The code input is in 0 or 1 format in real implementations. As the code input is zero, input is not changed and sent to the accumulator. As the code input is one, the input signal is inverted and the carry-in of the adder is set to one; as a result, the input of the accumulator is exactly the same as the negative input for 2's complement representation. The proposed dual-code correlator can be easily implemented as shown in Fig. 2.5.6. Since the two new codes are always exclusively non-zero at any time, only one adder is utilized and two registers are used for the storage of two partial correlation results. The correct correlation values are obtained by the final adder and subtractor as explained in section 3.2. In fact, they can be implemented in bit-serial manner if the spreading factor is larger than the adder/substractor word-length. Another important issue is that the dynamic range of the adder in Fig. 2.5.5 can be reduced by 1 bit because of the property of orthogonal codes; generally the non-zero parts of the two re-designed codes are nearly one half of the spreading factor. Therefore, the hardware cost can be further reduced. However, for modern VLSI technologies, generally power consumption issues are more important than area issues. So we suggest to use sign-magnitude representation to reduce power consumption with a little overhead of silicon area.

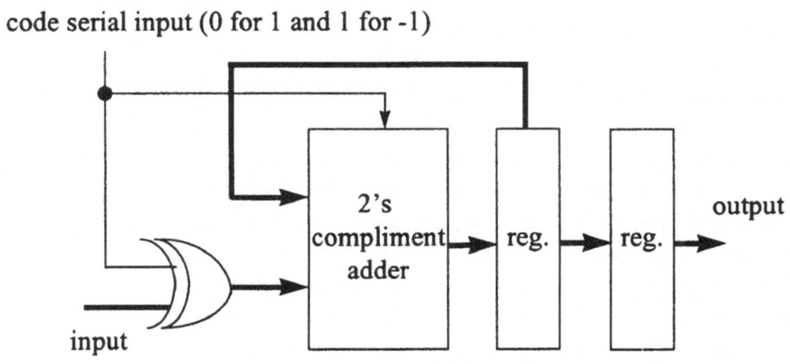

Fig.2.5.5. Block diagram of general 2's complement single code correlator.

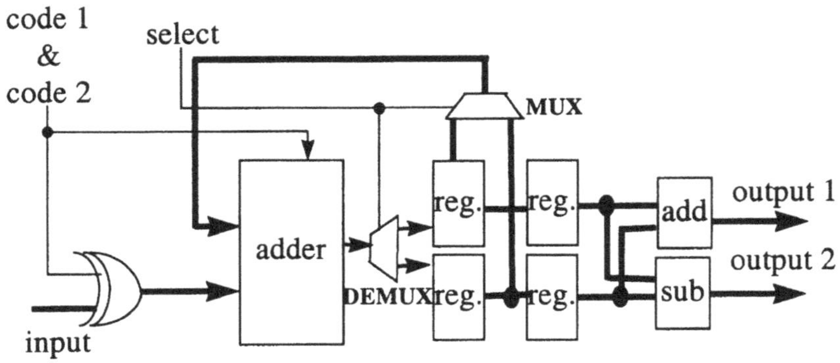

Fig. 2.5.6. Block diagram of proposed 2's complement dual codes correlator.

2.5.4.3
Sign-Magnitude Representation

For the sake of low-powered design, most of the signal processing systems favor the sign-magnitude representation because the transition probability of VLSI circuits is reduced [12]. In order to achieve the low-power purpose, *InfoPad* proposed an elegant architecture which can significantly reduce the power consumption of a correlator as shown in Fig. 2.5.7. The basic idea is to use two accumulators, one for positive and the other for negative correlation values [6]. Therefore, the transition probability for either positive or negative adder is further reduced. In addition, gated-clock scheme is adopted; that is, no actual MUX/DEMUX but gated clock to control the data flow of input signal. More power saving is achieved because only one path of the positive and negative accumulators is activated. That means only the active region with clock frequency *f* but others are with zero frequency. We also can use two such correlators to de-spread dual-code concurrently. However, we proposed some alternative solutions. Combing the proposed dual-code architecture with the two-accumulator idea, we can get two possible architectures which are depicted in Fig. 2.5.8 and 2.5.9. In Fig. 2.5.8, two adders are utilized; one for the accumulation of positive values, and the other for negative values. Each adder uses two registers to store the partial sums for the two new codes. The exact correlation results are obtained by final adders/subtractors. However, as shown in Fig. 2.5.9, we also can use just one adder to implement. In this way the area of one adder is saved with some overhead about MUX/DEMUX. Since gated clock technique is suggested to control the data flow of all registers, there is no actual MUX/DEMUX in fact as in [6]. As a result, the power consumption can be further reduced because one adder is eliminated and the gated-clock scheme makes unused module consume much less power. Similarly, the dynamic range of adders can be further reduced by 2 bits for architecture in Fig. 2.5.8 as explained in 2's complement case. Bit-serial adder/subtractor can also be used for the saving of hardware cost if the spreading factor is larger than the word-length of adder/substractor.

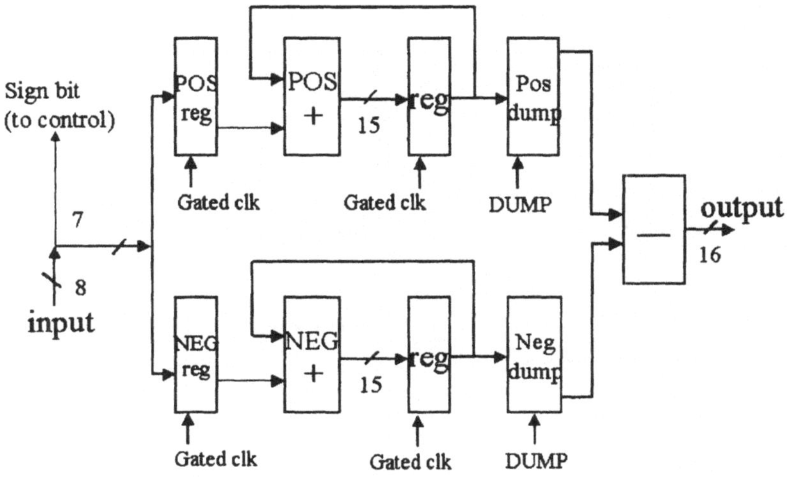

Fig. 2.5.7 Sign-magnitude approach of single code correlator proposed by *Infopad*.

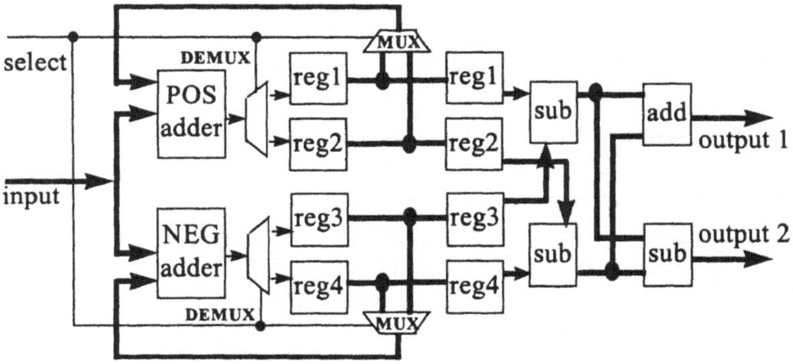

Fig. 2.5.8 Low powered architecture for sign-magnitude with two adders.

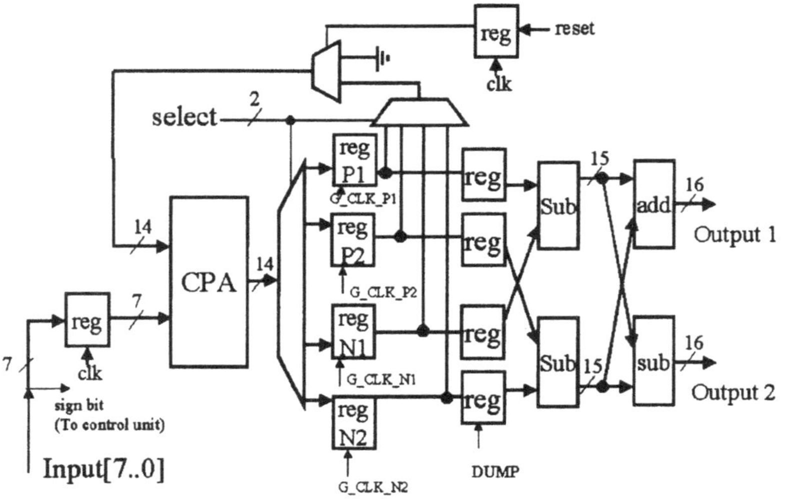

Fig. 2.5.9. Low powered architecture for sign-magnitude with single adder.

2.5.4.4
Chip Matched Filter Architecture

Of course, the proposed architecture can also be applied to the design of chip matched filter. The coefficients of the FIR can be either hard-wired or programmable depending on the implementation. Fig. 2.5.10 displays the difference between single code de-spreader and the proposed dual-code de-spreader for hard-wired implementation. Because the coefficients of the chip matched filter are either 1 or -1 and all pre-determined, no multipliers are needed as displayed in Fig. 2.5.10.a. We neglected the XOR gates in the figure because they can be directly implemented by hard-wired and these circuits are included in the adders in Fig. 2.5.10.b. Compared with using two filters in Fig. 2.5.10.a to de-spread two codes, only additional adder/subtractor and some pipeline registers are needed to generate the correct outputs in Fig. 2.5.10.b. Since the coefficients of the two re-designed codes are always exclusively non-zero at any time, only one new code uses the adder in each pipeline stage. If the length of the matched code (or spreading factor) is quite long, saving of the area is significant. Sign-magnitude representation also can be applied to reduce power consumption; However, we will not discuss the details because sign-magnitude approach is easy to be extended from the correlator in section 2.5.4.3. Bit-serial techniques can not be applied to the final adder/subtractor anymore due to the pipelined constraint. If programmable FIR hardware is used, the proposed scheme still can save power consumption because the total operations are reduced as explained in section 2.5.3.2. However, the support of extra pipeline registers is necessary for the storage of partial results; therefore some overhead of power consumption is possible.

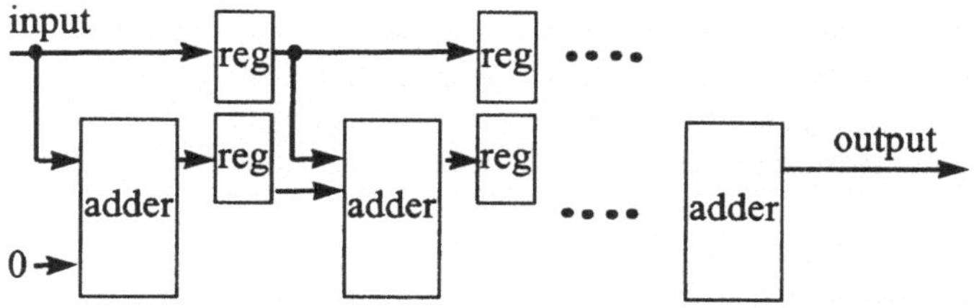

Fig. 2.5.10.a Single code hard-wired chip matched filter architecture.

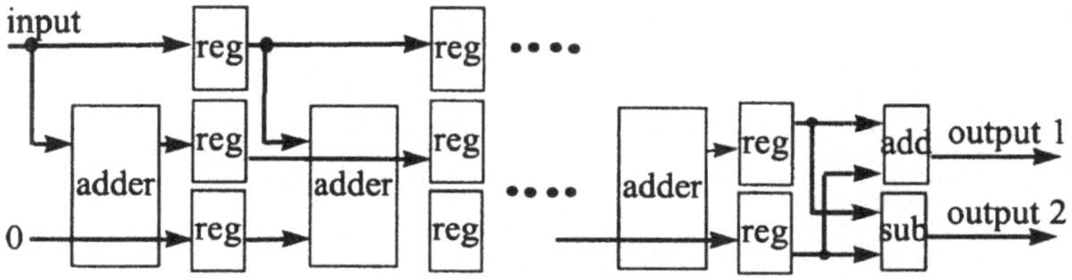

Fig. 2.5.10.b Block diagram for suggested hard-wired dual-code chip matched filter.

2.5.4.5 Multi-Code Correlator Architecture

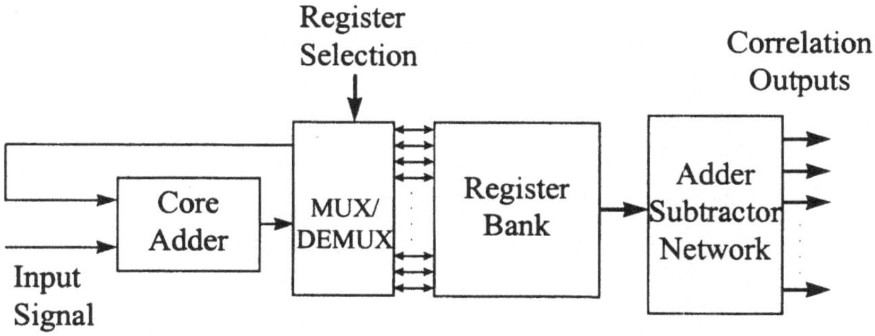

Fig. 2.5.11. Multi-code correlator architecture.

Similar architecture is applied to multi-code correlators. The architecture for multi-code correlator is displayed in Fig. 2.5.11. Since the calculations of s_i are all exclusive each other, all the operations can be achieved by a core adder similarly. The partial results are saved in a register bank which includes 2^{M-1} number of registers.. The MUX/DEMUX block delivers appropriate partial sum to the core adder, which generates the associated partial correlation with input signal. However, as suggested in dual-code correlator, gated-clock scheme can be used to reduce the hardware cost of MUX/DEMUX and power consumption simultaneously. The suggested architecture can be implemented in either 2's complement or sign-magnitude method. If sign-magnitude approach is used, power consumption can be reduced but the number of registers should be doubled into 2^M for both positive and negative correlation values as described before. The disadvantage of multi-code correlator is its hardware complexity grows very fast as the code number increases. Basically speaking, the group number is 2^{M-1}, which implies an exponential grow of register bank and MUX/DEMUX network. The trade-off between area and power will be demonstrated in later discussions.

2.5.5 Performance Evaluation

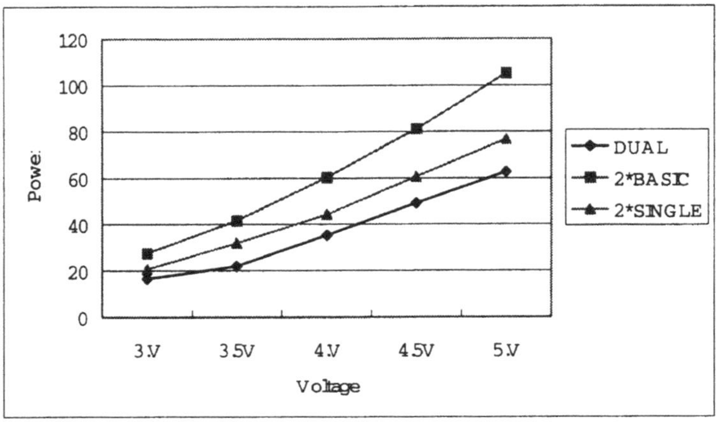

Fig. 2.5.12. The comparison of power consumption under different supplied voltages.

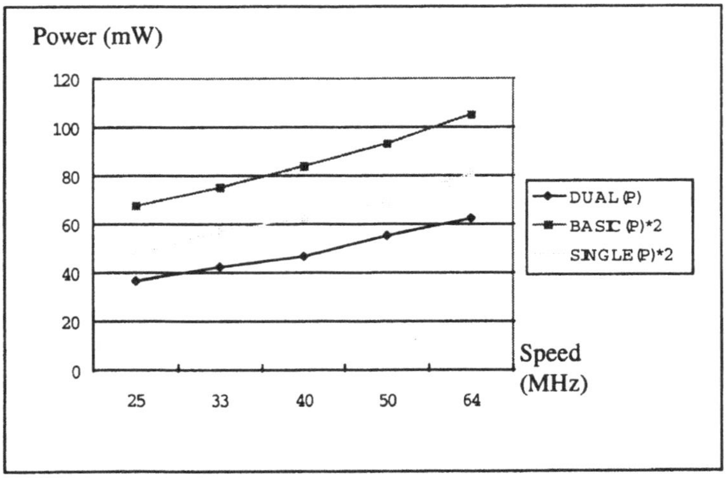

Fig. 2.5.13. The comparison of power consumption under different clock speed.

In order to verify and estimate the performance of the proposed dual-code de-spreader architecture, VERILOG hardware description language is used to simulate the results. In addition, cell-based design flow and TSMC 0.6 mm SPDM Compass library is used for synthesis. The power consumption is estimated by PowerMill EDA tool from the layout. In the following comparison, three kinds of architectures are simulated; the first one is 2's complement single-code correlator (BASIC) shown if Fig. 2.5.5, the second one is sign-magnitude single-code correlator (SINGLE) proposed in [6] and depicted in Fig. 2.5.7, and the last is the proposed dual-code correlator (DUAL) in Fig. 2.5.9. As explained previously, all the MUX/DEMUX are implemented via gated clock schemes; that is, no actual circuits to implement the MUX/DEMUX but gated clock control circuits to arrange the flow of data. The word-length of all adders are set to 16 bits. In addition, bit-serial adder/subtractor are not used in our simulation. That means some advantages of the proposed architecture are below-estimated in the simulation results. In order to estimate the power consumption, 40960 chips of signal coming from the sum of 256 spreading codes (Gold code with length 2^{18}-1 and random symbol input) is used as the input of correlator. The resolution of input chip is 8 bits and spreading factor is assumed to be 256. The results come from the average of 100 iterations. As expected, the proposed dual-code correlator can generate the correct input symbols of two codes concurrently. In Fig. 2.5.12, the performance of the above three architectures are compared with different supplied voltages. The values for both SINGLE and BASIC are multiplied by two to make a fair comparison. The clock rate is fixed as 64 MHz. Obviously the power consumption follows the square rule for voltages, and sign-magnitude implementation can save consumed power indeed. In addition, the proposed dual-code architecture can reduce power consumption furthermore as expected. Similarly another comparison is shown in Fig. 2.5.13, which displays the performance of the three architectures under different clock speed and the supplied voltage is set as 5V. The consumed power follows the linear rule as expected, and the proposed scheme consumes the least power. Some comparison data is listed in Table 1. The last two rows give the area and power ratios of the suggested dual-code correlator compared with the other two architectures. Let's look at the case for 25 MHz clock firstly. The area of a simple 2's complement correlator (BASIC) is the smallest. However, it consumes the largest power. If we use the architecture proposed in [6], the area is doubled but the power consumption is reduced about 29%. On the other hand, the proposed dual-code correlator can reduce consumed power furthermore, about 46% reduction compared with BASIC, and 24% reduction compared with SINGLE. Although the area overhead of DUAL is 68% compared with BASIC, it saves 20% hardware cost compared with SINGLE. The same conclusion can be obtained for 64 MHz clock; the power reductions become 41% and 23% respectively. However, area issues are not quite critical for current deep sub-micron VLSI technologies. It is obvious the suggested dual-codes architecture can reduce both hardware cost and power consumption significantly. If we use the data collected from 64 MHz clock rate, the power efficiency of the proposed architecture is as high as 1 GOP/Watt. The performance for chip matched filter should show similar results; however, the suggested dual-code scheme can give more savings of area and power because the number of registers is reduced significantly. In addition, some advantages of the proposed architecture are neglected in our simulation as explained.

In fact, we also made some preliminary investigations about the implementation results for multi-code correlator. All the simulation conditions are the same as that for dual-code correlator. Currently we concluded that the consumed power is optimal for dual-code correlator. If the code number is larger than 3, the power consumption becomes almost the same as that for sign-magnitude approach (SINGLE). In addition, the area increases very fast as code number increases; therefore the overhead for register banks and gated-clock control also increase, as we explained in previous sections. More research is under investigations for multi-code correlator implementation, including both hardware and software approaches.

Table 2.5.1. The comparison of area and power for three architectures.

	GATE COUNT	CORE AREA (MM^2)	POWER@25MHZ	POWER@64MHZ
DUAL (Dual-Code)	1657	0.545	36.6 mW	62.55 mW
SINGLE (Sign-Magnitude)	1034 x 2 = 2068	0.34 x 2 = 0.68	23.88 x 2 = 47.8 mW	40.55 x 2 = 81.1 mW
BASIC (2's Complement)	450 x 2 = 900	0.162 x 2 = 0.324	33.85 x 2 = 67.7 mW	52.5 x 2 = 105 mW
DUAL/SINGLE	0.8	0.8	0.76	0.77
DUAL/BASIC	1.84	1.68	0.54	0.59

2.5.6 Conclusion

In this chapter, a 1 GOP/Watt low-powered architecture is proposed to de-spread dual PN codes concurrently. Re-designing the two codes, both hardware overhead and power consumption are reduced significantly. The same idea can be applied to the design of chip matched filter and the improvement should be more obvious. In addition, a similar architecture is extended to the design of multi-code correlator. According to our simulation results, about 46% power reduction compared with 2's complement single code correlator, and 24% saving for sign-magnitude single code correlator in the case of dual-code correlator. The area overhead is limited and even smaller than previous sign-magnitude implementation. Obviously the benefit should be larger for the design of dual-code chip matched filter. On the other hand, currently the results for multi-code correlator is still under investigations. Especially in terms of software radio approach, more efforts might be put on flexible architectures in FPGA form or software implementations by DSP processors [15][16]. Combining the idea of correlator array proposed in [6], a flexible correlator array which can be used as a chip matched filter or parallel correlators is under development. Since multi-code correlator is useful for future multi-code CDMA systems, we believed the proposed architecture can be applied to the design of UMTS and IMT-2000 receiver, including mobile stations and base stations.

2.5.7 Acknowledgement

The authors would like to thank the editor of this book and the support of R.O.C. *Ministry of Economic Affairs* under the project No.3P12100 conducted by CCL/ITRI.

References

1. A. J. Viterbi, *CDMA - Principle of Spread Spectrum Communications,* Addison-Wesley, 1995.
2. M. Y. Rhee, *CDMA Cellular Mobile Communications and Network Security*, Prentice Hall, 1998.
3. S. Sheng, L. Lynn, J. Peroulas, K. Stone, L. O'Donnel, and R. Brodersen, "A Low-Power CMOS Chipset for Spread-Spectrum Communications", *Proceeding of IEEE International Solid-State Circuits Conference,* pp. 346-347, 1996.
4. K.-C. Liu, W.-C. Lin, and C.-K. Wang, "A Pipelined Digital Differential Matched Filter FPGA Implementation and VLSI Design", *Proceeding of IEEE Custom Integrated Circuits Conference*, pp. 75-78, 1996.
5. J. S. Wu, M. L. Liou, H. P. Ma, and T. D. Chiueh, "A 2.6-V, 44-MHz All-Digital QPSK Direct-Sequence Spread-Spectrum Transceiver IC", *IEEE Journal of Solid-State Circuits*, pp.1499-1510, Oct. 1997.
6. W. Namgoong and T. Meng, "Power Consumption of Parallel Spread Spectrum Correlator Architectures", *Proceeding of IEEE Custom Integrated Circuits Conference*, pp. 75-78, 1996.
7. S. Sheng and R. Brodersen, *Low-Power CMOS Wireless Communications - A Wideband CDMA System Design*, Kluwer Academic Publisher, 1998.
8. ETSI TS02.60 V5.0.0 - Digital Cellular Telecommunication System (phase 2+); General Packet Radio Service (GPRS); Service Description, 1997.
9. TR45.5 - The cdma2000 for ITU-R RTT Candidate Submission, Jun. 1998.
10. The ETSI UMTS Terrestrial Radio Access (UTRA) ITU-R RTT Candidate Submission, Jun. 1998.

11. T. Dohi, Y. Okumura, A. Higashi, K. Ohno, and F. Adachi, "Experiments on Coherent Multi-code DS-CDMA" *Proceeding of IEEE Vehicular Technology Conference*, pp. 889-893, 1996.
12. E. Dahlman and K. Jamal, "Wide-band Services in a DS-CDMA based FPLMTS System", *Proceeding of IEEE Vehicular Technology Conference*, pp. 1656-1660, 1996.
13. A. Chandrakasan and R. Brodersen, "Minimizing Power Consumption in Digital CMOS Circuits", *Proceeding of the IEEE*, vol. 83, no. 4, pp. 498-523, Apr. 1995.
14. A. Chandrakasan, A. Burstein, and R. Brodersen, "A Low-Power Chipset for a Portable Multimedia I/O Terminal", *IEEE Journal of Solid-State Circuits*, vol. 29, no. 12, pp. 1415-1428, Dec. 1994.
15. J. Mitola, "Technical Challenges in the Globalization of Software Radio", *IEEE Communications Magazine*, vol. 37, no. 2, pp. 84-89, Feb. 1999.
16. J. E. Gunn, K. S. Barron, and W. Rucaczyk, "A Low-Power DSP Core-Based Software Radio Architecture", *IEEE Journal on Selected Areas in Communications*, vol. 17, no. 4, pp. 574-590, Apr. 1999.
17. TS 25.213, 3rd Generation Partnership Project (3GPP) Technical Specification Group (TSG) RAN WG1; Spreading and Modulation, Apr. 1999.
18. TS 25.101, 3rd Generation Partnership Project (3GPP) Technical Specification Group (TSG) RAN WG4; UTRA (UE) FDD; Radio Transmission and Reception, Apr. 1999.
19. C.-W. Ku and F.-Y. Kuo, "Low-Powered Concurrent Dual-Code De-Spreader for UMTS or IMT-2000", *to appear in Proceeding of International Workshop on Mobile Communications, Crete Island*, Jun. 1999.
20. C.-W. Ku, F.-Y. Kuo, C.-K. Chen, and L.-G. Chen, "Low Powered Multi-Code Correlator Architecture for IMT-2000", *to appear in Proceeding of the 10th International Symposium on Personal, Indoor and Mobile Radio Communications, Osaka*, Sep. 1999.

3 MODULATION

3.1 Introduction

The evolution of the wireless networks especially mobiles, as well as the increasing demand for data transmission of higher and higher bit rates created the need for specially efficient modulation techniques.

In section 3.2 a Variable Minimum Shift Keying (VMSK) technique is presented. The spectrum analysis and the techniques of realizing the modulation is given. The required bandwidth in conjunction with the RF modulation is explained. Two encoding rules are given and the efficiency of the system is calculated. Analysis for the demodulation is presented and a special study for the low group delay filters implementation is made. Comparison with other modulation techniques shows the high spectral efficiency and the positive prospects of VMSK/2.

In section 3.3 the Hidden Markov Model (HMM) is used for modelling the physical layer of a typical Wireless-ATM indoor system. The physical layer is based on a multicarrier modulation scheme (OFDM) with QPSK modulation on each carrier. Due to the long coherence time of the indoor channel, long time consuming programs are needed for the realistic simulation of the physical layer. By using the HMM, the received bit sequences are generated with much less computational time and they describe, with a good approach, the error distribution of the real sequences. Detailed physical layer simulations have also been carried out since they are still needed to train the HMM and to obtain the necessary validation parameters.

3.2 Wireless Communications Using Spectrally Efficient VMSK/2 Modulation

H. R. Walker, J. C. Pliatsikas, C. S. Koukourlis and J. N. Sahalos

3.2.1 Introduction

Due to the increasing demand for data transmission, the need for spectrally efficient modulation techniques has become a necessity. Single Sideband Biphase Modulation claims to be one of the most spectrally efficient digital modulation techniques. In the past, Biphase Modulation was the well known "Manchester Encoding", which was used in Ethernet LANs without any high bandwidth efficiency. There exist other biphase coding methods that require less bandwidth without any significant increase in circuit complexity. These are the *phase modulated biphase codes* which are transmitted by a Single Sideband-Suppressed Carrier at RF frequencies. The above have an advantage over other methods in that they do not lose bit energy with increasing spectral efficiency (compression). Their bit error rates (BER) are shown to be obtainable at C/N ratios approaching Shannon's Limit.

The first of these codes was the Variable Phase Shift Keying (VPSK)[1,2], which has been used satisfactorily at bandwidth efficiencies as high as 15 bits/sec/Hz. VMSK (Very Minimum Shift Keying) [3] was introduced in 1997 to simplify the encoding/decoding chips and raise the spectral efficiency to as high as 60 bits/sec/Hz. Hardware for field use has been tested at 53 bits/sec/Hz. Some changes have been made in the above methods that merge them into a single one referred as VMSK/2, or the VMSK code divided by 2 [4,5].

In this chapter, the VMSK and VMSK/2 encoding methods will be presented along with a rough calculation of the occupied bandwidth. Also the modulation and the demodulation techniques will be presented. Moreover, a critical analysis of the filters used will be made. Finally, the BER of the system will be calculated.

Conventional modulation techniques, like Quadrature Phase Shift Keying (QPSK), use Non Return to Zero (NRZ) data presentation at the baseband. QPSK concentrates the modulated signal around the carrier (Fig.3.2.1) occupying at least the full Nyquist bandwidth [6]. In order to maintain high spectral efficiency the signal must be transmitted by using Single Side Band (SSB) transmission. This could call for filters that impose serious implementation difficulties along with phase and amplitude distortion to the signal.

SSB transmission is possible if data are pre-encoded by means of biphase encoding technique, like Miller Coding or Manchester. This action separates the occupied signal bandwidth at RF into two "isles" lying away from the carrier. Unfortunately, Miller and Manchester codes utilize excessive bandwidth, resulting in a reduced spectral efficiency compared to most NRZ encoding schemes.

Consider the case (worst) of having an AA Hex repeating NRZ data pattern (i.e. 101010...) of bit rate f_b. This pattern presents at the frequency domain a pair of dominant spectral lines (inside the main lobe) at the positions $\pm f_b/2$ (Fig.3.2.1a). For another pattern, e.g. CC Hex (11001100...), the spectral lines are presented at $\pm f_b/4$, i.e. closer to dc. Every individual pattern, or random data, will give dominant spectral lines lying at $f \leq |f_b/2|$. To transfer the desired data up to a bit rate f_b it is clear that

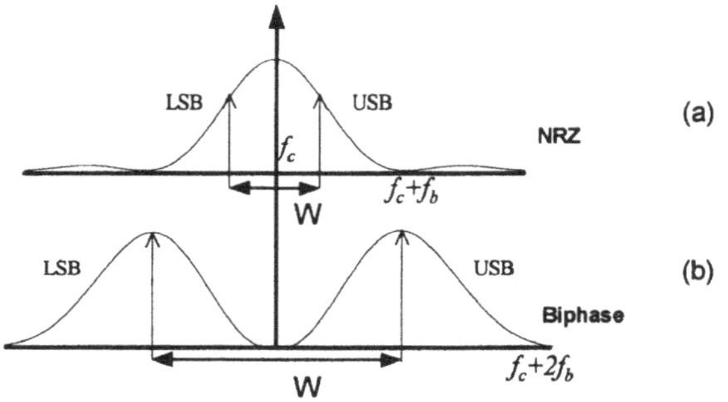

Fig. 3.2.1. RF Spectrum of an (a) NRZ and (b) biphase encoded signal. f_b = Bit Rate, f_c = Carrier Frequency W = minimum Bandwidth = Sampling Rate.

the whole spectrum from - $f_b/2$ to $f_b/2$ is required. If the NRZ data are modulated using BPSK, the signal requires around the carrier a minimum bandwidth f_b. This is the well known Nyquist bandwidth. At a first glance it seems that someone is trapped to use a minimum bandwidth of f_b. To reduce it, a higher order modulation scheme has to be used. In this case the multiple bit/symbol increases the spectral efficiency and thus reduces the sampling rate.

A non-conventional method which overcomes the Nyquist trap will be described here. In the past Nyquist's minimum bandwidth theorem has been "defeated" by A. Lender [7,8] who has shown that it is possible to transmit 2W symbols/sec with zero

InterSymbol Interference (ISI) using the theoretical minimum of W Hertz, without infinitely sharp filters. By introducing controlled interference, Lender canceled out ISI and achieved the ideal symbol-rate packing of 2 symbols/sec/Hz, an amount that had been considered unrealizable [7]. Returning to the non-conventional way of thinking, consider that the NRZ data are encoded prior to BPSK modulation. If an encoding scheme could be found that would concentrate the dominant spectral lines of any possible data pattern around a single line ($\pm f_b/2$ for VMSK/2 or $\pm f_b$ for VMSK), then the whole Nyquist bandwidth down to zero hertz would not be necessary. This could improve dramatically the spectral efficiency of the modulation.

3.2.2
Spectral Efficiency of VMSK (Rough Calculation)

VMSK code is shown in Fig. 3.2.2. In VMSK, the information is encoded at the falling edge or in the aperture of a clock like signal. The encoding algorithm is the following: Let us assume that a 13 times higher than the bit rate clock i.e. the incoming data bits are divided into 13 fractional parts. Logic 0 is encoded as a symbol having a transition from high to low at 6/13 of the original data bit duration and logic 1 as a symbol with a similar transition at 7/13 of the data bit duration. There is a transition back to high at the end of each bit period so that the symbol pattern resembles the x1 clock.

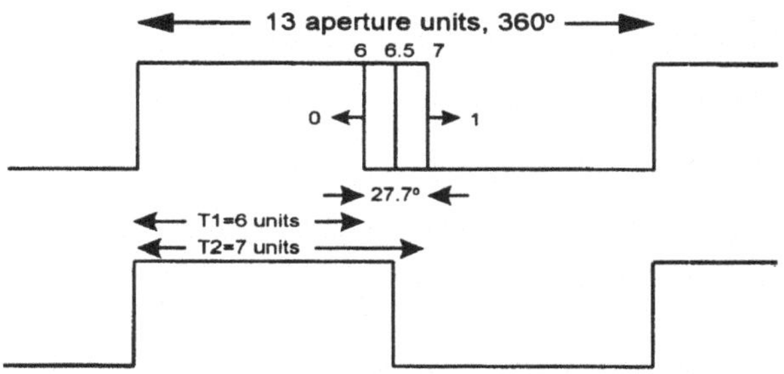

Fig. 3.2.2. The 6,7,(13) VMSK Aperture Code. Bottom trace is the clock waveform.

The other code presented here is VMSK/2. This is realized by inverting every other neighboring symbol of the VMSK encoded signal forcing the second part of a symbol to be at the same state (level) with the first part of the next, diminishing the transitions at the borders of the encoded symbols. This action halves the required spectrum and creates the " /2" factor in the name of the modulation (VMSK/2). The VMSK/2 waveform can be obtained directly from VMSK by driving the later in a D flip-flop. This is equivalent to a *divide by two* operation. The encoding algorithm after dividing by two becomes as follows: shift to logic 1 is presented by a pulse 14/13 of the data bit width (either level), shift to logic 0 by a 12/13 bit width (either level) and finally a pulse 13/13 bit width indicates that the previous logic state is preserved . Consecutive pulses are of opposite level. This rule is the *Encoding Rule 2* which schematically is presented in Fig. 3.2.3.

To indicate a repeating digital " one" pattern that is followed by a " zero", the symbol pattern is the following: a pulse 14/13 indicates the shift to 1, thereafter 13/13 pulses until the next 12/13 which indicates the shift to 0. A repeating " zero" pattern that is followed by " one" would be: a pulse 12/13 indicates the shift to 0, thereafter 13/13 pulses until the next 14/13 pulse, that indicates the shift to 1. Note that the zero crossings occur early or late relative to the x1 clock (indicated by dotted lines in Fig. 3.2.3), thus making the detection possible. The algorithm does not permit a 12/13 followed by a 12/13, or a 14/13 followed by a 14/13. Relative to the x1 clock, only a phase shift of $\pm\pi/26$ radians appears.

The use of various higher than the bit rate clocks in VMSK and VMSK/2 result to the different encoding schemes. The notation for both VMSK and VMSK/2 encoding rules make use of 3 numbers. The first two numbers in VMSK notation indicate the transition points of the signal, while the third one, in parenthesis, denotes the higher clock or the total number of the fractional parts. In VMSK/2, all the three numbers correspond to all the possible pulse durations that can occur. Using the above, the codes described earlier are 6,7,(13) for VMSK and 12,13,14 VMSK/2.

The waveform of the VMSK encoded data shown in Figs. 3.2.2 and 3.2.3 is very similar to the data clock. The data clock is obtained directly from it by using a form of PLL or high Q ringing coil. Utilizing this clock as a reference, the receiver needs only to determine whether the zero crossing occurs early or late. This baseband waveform is used directly in power line modems and DSL circuits at 1.544 Mb/s and higher [9].

In the following it will be shown that detection can be done by means of early / late measurements of the zero crossings relative to the clock. The above is presented by the dashed lines of Fig. 3.2.3. Also in Fig. 3.2.3 it can be seen that the phase shift is ± •/26 radians relative to the clock.

Using now a 13 times higher than the clock rate, the occupied by 6,7(13) VMSK bandwidth, shown in the example, can be calculated. The Fourier series of a periodic waveform with period T_0 is given by [10,11]:

$$\upsilon(t) = \sum_{n=0}^{\infty} a_n \cos(n\omega_0 t) + \sum_{n=1}^{\infty} b_n \sin(n\omega_0 t) \tag{1}$$

where $\omega_0=2\pi/T_o$ and a_o, a_n ,b_n are the Fourier coefficients given by: [11]

$$a_n = \begin{cases} \frac{1}{T_0}\int_a^{a+T_0} \upsilon(t)\,dt, & n = 0 \\ \frac{2}{T_0}\int_a^{a+T_0} \upsilon(t)\cos n\omega_0 t\,dt, & n \geq 1 \end{cases} \tag{2}$$

$$b_n = \frac{2}{T_0}\int_a^{a+T_0} \upsilon(t)\sin n\omega_0 t\,dt, \qquad n > 0$$

A periodic waveform taking values −1 and +1 with high (+1) pulse duration τ, starting at a and having period T_0 gives the following Fourier coefficients:

$$a_n = \begin{cases} \frac{1}{T_0}\int_a^{a+\tau} 1\,dt + \frac{1}{T_0}\int_{a+\tau}^{a+T_0} (-1)\,dt, & n = 0 \\ \frac{2}{T_0}\int_a^{a+\tau} \cos n\omega_0 t\,dt + \frac{2}{T_0}\int_{a+\tau}^{a+T_0} (-1)\cos n\omega_0 t\,dt, & n \geq 1 \end{cases} \tag{3}$$

$$b_n = \frac{2}{T_0}\int_a^{a+\tau} \sin n\omega_0 t\,dt + \frac{2}{T_0}\int_{a+\tau}^{a+T_0} (-1)\sin n\omega_0 t\,dt, \qquad n > 0$$

Preserving the main lobe, we keep the a_0, a_1 and b_1 Fourier coefficients. Referring to 6,7,(13) VMSK encoding rule, τ, the high portion of the encoded symbol, is equal to $T_b/2 \pm T_b/26$. It can also be stated that $\tau = T_b/2 + d(t)\cdot T_b/26$, with $d(t)$ the data stream taking values ± 1. If $d(t) = +1$, the high portion of the encoded signal is extended by $T_b/26$ while if $d(t) = -1$ it is reduced accordingly. Replacing τ and taking into account the first three terms we get:

$$\upsilon_{VMSK}(t) = \frac{2\,d(t)}{13} - \frac{2}{\pi}\sin\left(\frac{2\,d(t)\pi}{13}\right)\cos\left(\frac{2\pi t}{T_b}\right) + \frac{4}{\pi}\cos^2\left(\frac{d(t)\pi}{13}\right)\sin\left(\frac{2\pi t}{T_b}\right) \tag{4}$$

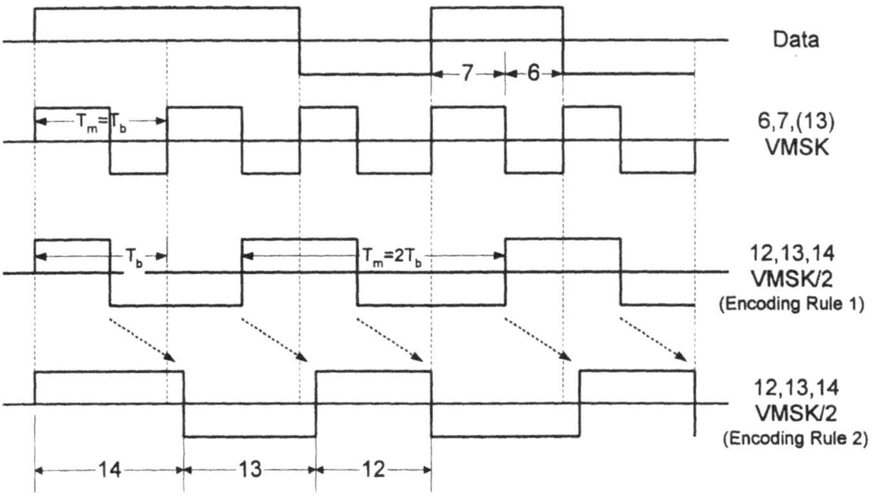

Fig. 3.2.3. A 6,7,(13) VMSK code, which when divided by 2 yields a 12,13,14 VMSK/2 code.

Transforming the trigonometric functions, the above equation yields to:

$$\upsilon_{VMSK}(t) = \frac{2\,d(t)}{13} + \frac{4}{\pi}\cos\left(\frac{d(t)\pi}{13}\right)\cos\left(\frac{2\pi t}{T_b} + \frac{13 + 2\,d(t)}{26}\pi\right) \tag{5}$$

The information term d(t) appears to be into the amplitude and the phase of the signal. It is noticed that the information could be recovered by the use either the amplitude or the phase term of the above. In our case the amplitude term is ignored and is discarded by the use of a limiter or a comparator. The phase term contains all the information and points at the necessary bandwidth. The occupied bandwidth of the encoded signal and the corresponding spectral efficiency Q can be calculated by taking the phase derivative of the signal. The derivation of the phase term gives the total frequency deviation which is:

$$2\Delta f_{VMSK} = \frac{1}{13}\frac{\mathrm{d}}{\mathrm{dt}}d(t) \tag{6}$$

$2\Delta f$ is the total frequency deviation, i.e. the occupied bandwidth, of the encoded signal.

In order to estimate the spectral efficiency Q, we compare it with a BPSK signal, which can be represented as:

$$\begin{aligned}\upsilon_{BPSK}(t) &= d(t)\cos\left(\omega_0 t\right) = \\ &= A\cos\left(\omega_0 t + d(t)\frac{\pi}{2}\right)\end{aligned} \tag{7}$$

This gives a total frequency deviation which is:

$$2\Delta f_{BPSK} = \frac{1}{2}\frac{\mathrm{d}}{\mathrm{dt}}d(t) \tag{8}$$

Comparing the occupied bandwidth of 6,7,(13) VMSK with that of BPSK we get an improvement of 6.5. BPSK is transmitted using Double Side Band transmission while 6,7,(13) VMSK Single Side Band transmission. The above doubles the RF spectral efficiency of 6,7,(13) VMSK, rising finally it to 13. Note that the spectral efficiency here cannot be expressed in terms of " N", which is the number of bits/symbol.

The total phase change in the alternation from logic 0 to logic 1 and vice versa for the 6,7(13) VMSK is 2•/13 radians or 27.7° as shown in Fig. 3.2.2. The phase change of (2•/13) results in a Δf of 1/13 of the bit rate f_b which corresponds to a spectral efficiency of 13 bits/sec/Hz. The 6,7(13) VMSK divides the data bit into 13 fractional parts, while the possible transitions occur at 6 and 7. The transitions, marked in Fig.3.2.2 as T1 and T2 give a bandwidth improvement which is expressed by Q. It is:

$$Q = \frac{\mathrm{T1}+\mathrm{T2}}{\mathrm{T1}-\mathrm{T2}} = \frac{13}{7-6} = 13 \tag{9}$$

Looking at the spectrum which is taken in the analyzer we can see that it does not show any deviation. Only a single spectral line appears (see Fig. 3.2.4). This spectral line however, does contain phase modulation in the form of timed phase reversals.

From Fig. 3.2.3 it would be expected that the spectrum for VMSK/2 should consist of a set of narrow band of frequencies grouped around one half the bit rate. This does not occur in practice. There is an unusual phenomenon which occurs when the phase change is less than •/4. When the phase change is more than •/4, the phase appears to be constant and the frequency varies. When the phase change is less than •/4, the frequency appears to be a constant and the phase varies. This is indeed the case in the spectrum plotted in Fig. 3.2.4 for an 8, 9,10 VMSK/2 code. The same occurs also in VMSK. The center line is at one half of the bit rate and shows no frequency deviation. The lower noise or " grass" is a collection of Fourier products at sub harmonics of the data rate plus a sinx/x ringing. The above are not a necessary parts of the signal and can be removed by filtering.

Looking at the modulation index β which is

$$\beta = \frac{\Delta f}{f_m} = \frac{\Delta\theta}{\pi} \tag{10}$$

we have that if β is equal or less than 0.2 the spectrum no longer shows any frequency deviation at all. The spectrum contains a single line with phase modulation that cannot be observed in the analyzer (see Fig. 3.2.4).

A waveform with double sideband modulation according to Carson's Rule requires a bandwidth which is:

$$BW = 2\ (f_m + \bullet f) \tag{11}$$

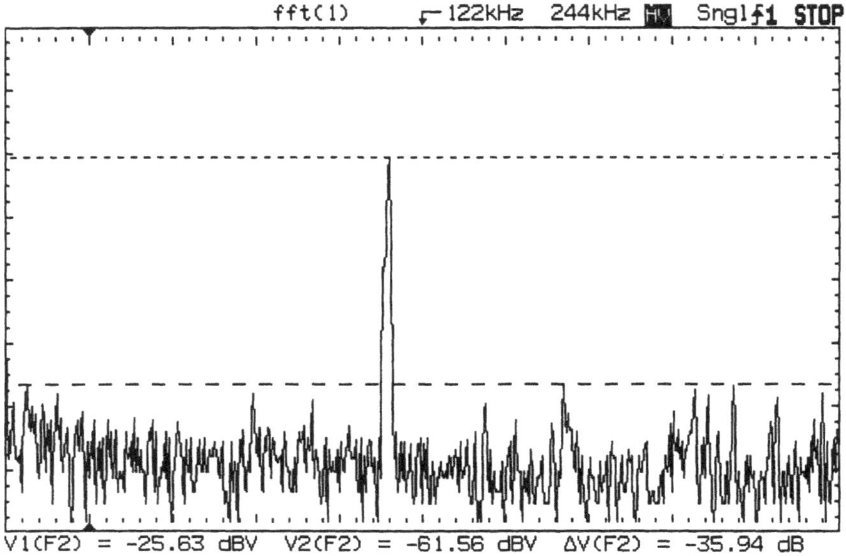

Fig. 3.2.4. Baseband spectrum of an 8,9,10 VMSK/2 code at a data rate of 180 Kb/s

If a single sideband modulation with suppressed carrier (SSB-SC) is used, the bandwidth is presumed to be only 2•*f*. That happens in VMSK and VMSK/2. In the example of 8,9,10 VMSK/2 it will be shown that the bandwidth is be 1/18 the data rate. It is noticed also that in practice •*f does not occur, but for calculation purposes, it can be assumed to exist.* Coming back in the spectrum of Fig. 3.2.4 we can say that the center line is the result of phase modulation where the modulation angle is less than •/4. The " grass" level has been determined empirically as $-20 \log_{10} [2(R^2) / \bullet]$ dB (volts), where R is the center number in the three number code. Thus for a 8,9,10 code, the level is –34.2 dB which matches with measurements. This value is valid for a spectrum analyzer with a 3 KHz IF filter. For other filter bandwidths (FBW) the following formula can be used:

$$P = -20 \log_{10} [6(R^2) / (FBW)\bullet]\ dB \tag{12}$$

FBW is in kHz.

3.2.3
Baseband Pulse Shaping

It is well known that when a square wave pulse passes through a filter, it is transformed to a sinx/x form, spreading to the nearby pulses and creating an intersymbol interference (ISI). VMSK and VMSK/2 can reduce the intersymbol interference. Their phase does not change randomly, according to the information data, but has an almost steady phase reversal every half bit (VMSK) or every bit (VMSK/2) period. This phase reversal also reverses the sinx/x pulses, resulting in an almost triangular pulse with zero crossings at half bit period for VMSK and every bit period for VMSK/2 (see Fig. 3.2.5).

Let us suppose that the available bandwidth is narrow and the pulses are extending in time to the neighboring pulse intervals. Each individual pulse can then be analyzed in two consecutive pulses of the same polarity – say positive. One of them is in the present symbol interval and one is into the next. The second creates a rather heavy ISI. The next symbol is also analyzed into two opposite pulses, i.e. negative consecutive ones. The first one is into the same interval as the previous second pulse. As they have opposite polarities (due to the coding effect of VMSK/2 where every other pulse is inverted) they eliminate each other. Looking at the sum of the sinx/x waveforms we can see that it becomes narrower with almost triangular $(\sin x/x)^2$ form with zero crossings every T_b. The information is taken at the time / phase varying peaks. Of course there is some residual intersymbol interference that is caused from the varying pulse width of the VMSK and VMSK/2 pulses. This ISI creates the "grass" that is shown in Fig. 3.2.4.The level of the grass is low enough so that the effect on the modulation angle is much less than 1% in practical systems. This grass can be removed by filtering and must be reduced to meet regulatory limits.

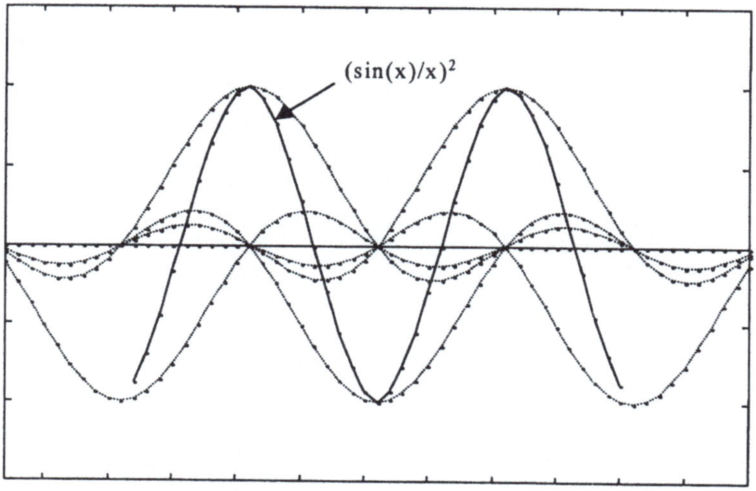

Fig. 3.2.5. The *(sinx/x)² - like* pulses resulting from summing phase alternating sinx/x pulses

3.2.4 RF Modulation

RF modulation is an important approach in wireless systems. If the waveform shown in Fig. 3.2.3 is used to modulate an FM or PM generator, the sidebands become Bessel products which decrease in an unusable level. It seems that FM/PM modulations are
not suitable for VMSK or VMSK/2. The proper choice is an AM or a BPSK type phase alternating modulation to generate the sidebands. The carrier can be suppressed in a balanced modulator and the undesired sideband can be removed by filtering.

If the sideband is a single frequency of constant amplitude, the receiving device is not able to recognize whether it is an AM, FM or PM signal. In most cases a carrier must be added. The phase of the carrier will determine whether the signal is detected as AM or FM/PM. The detector can use a product detector like that used for SSB-AM, or a phase detector such as an XOR gate. The above have been used satisfactorily to VPSK and are being used now to VMSK/2. The carrier can be restored from the sideband. [2].

It must be pointed out that is not necessary to generate the opposite sideband. In some applications this single sideband signal is being detected without recreating a carrier - all of the modulation energy is in the sideband. The carrier is a non-existent phantom.

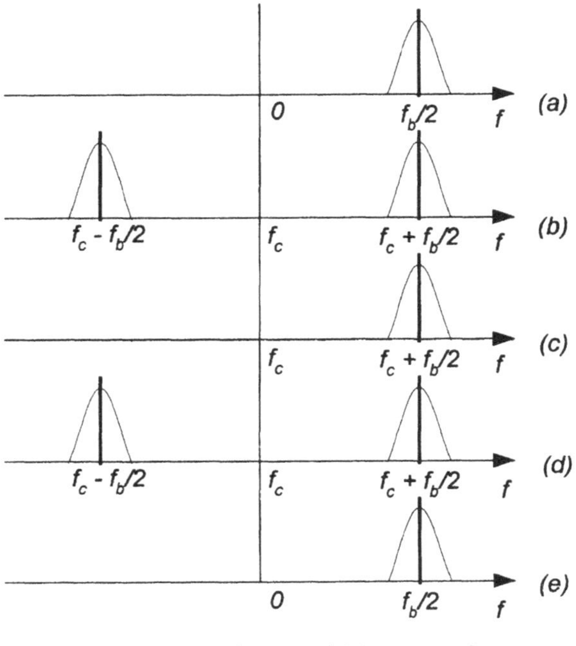

Fig. 3.2.6. Alternating Aperture Coded (VMSK/2) Modulation Spectrum.

Fig. 3.2.6 shows the spectrum in the different steps. Fig. 3.2.6(a) presents the spectrum at the baseband. Fig. 3.2.6(b) shows the double sideband modulated spectrum after passing through a balanced modulator. Fig. 3.2.6(c) shows the single sideband spectrum as transmitted. Fig. 3.2.6(d) shows the restored DSB signal at a coherent detector and Fig. 3.2.6(e) shows the restored baseband signal after detection. If a narrow band RF filter is used at 6(c), the transmitted signal is within a very narrow spectral band away from the carrier, having a much lower noise bandwidth than the full Nyquist bandwidth, which extends from 0 Hz to $f_b/2$. Only this narrow bandwidth of Fig, 3.2.6(c) need be transmitted and received. The modulation frequency is $f_m = ½ f_b$ (Bit Rate).

3.2.5 VMSK/2 Spectrum

3.2.5.1 VMSK/2 Encoding

In the following, the VMSK/2 encoding method will be presented analytically. This method, in contrast to other biphase ones, does not extend the required bandwidth but it compresses it [1,3,4,5,9]. It will be shown in more detail that transmitted bandwidth at RF (or IF) is concentrated around a pair of spectral lines at frequencies $f_c + f_b/2$ and $f_c - f_b/2$, where f_c is the carrier frequency and f_b the bit rate. One of the two frequency " isles" is selected by proper filtering, resulting in an SSB modulation scheme.

The encoding algorithm of VMSK is the following: It is assumed that of an 8 (or 16) times higher than the bit rate clock is used. The incoming data bits are divided into 8 fractional parts (Fig.3.2.7). Logic 0 is encoded as a symbol having a transition from high to low at 3.5/8 (7/16) of the original data bit duration and logic 1 as a symbol with a similar transition at 4.5/8 (9/16) of the data bit duration. There is a transition back to high at the end of each bit period so that the symbol pattern resembles the x1 clock. This rule is presented in Fig. 3.2.3 and Fig. 3.2.7 as " Encoding Rule 1" . Inverting the neighboring encoded symbols, the second part of a symbol is at the same state (level) with the first of the next, diminishing the transitions at the borders. The above divides the required spectrum by two and is the reason why the " /2" factor is incorporated in the name of the modulation (VMSK/2). The *divide by two* action is accomplished by adding a toggling flip flop section responding to the negative transitions only.

To indicate a repeating digital " one" pattern that is followed by a " zero" , the symbol pattern is the following: a pulse 9/8 the data bit width to shift to 1, thereafter 8/8 pulses until the next 7/8 pulse that indicates the shift to 0. A repeating " zero" pattern that is followed by an " one" would be: a pulse 7/8 data bit width to get to 0, thereafter 8/8 pulses until the next 9/8 pulse, indicating the shift to 1.

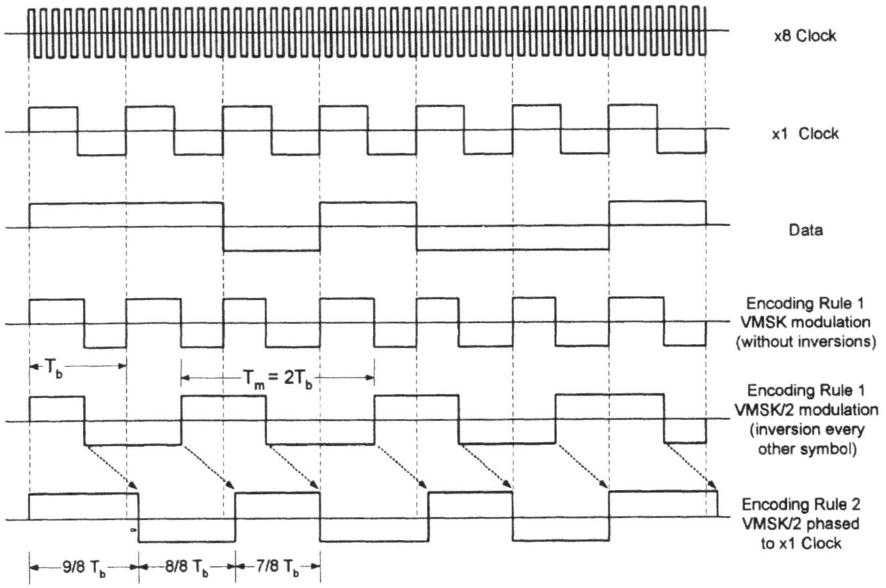

Fig. 3.2.7. Random incoming data stream and the corresponding encoded data stream. Notice that the 8 times higher clock is used.

Fig. 3.2.7 shows the data and clock relationship during the various stages of the modulation. The VMSK code has aperture transitions at the middle of each bit. The division by 2 causes the aperture crossings to alternate in polarity every other bit. Due to the above VMSK/2 is sometimes referred to as *alternating aperture coding*. When the divide by 2 code is phased properly with the clock, it can be seen that the crossings occur early or late with respect to the rising edge of the clock. A 9/8 time period will cause an early crossing to shift to a late crossing. An 8/8 period causes it to remain as a late crossing. A 7/8 period causes a shift from late back to early. In this example, an early crossing signifies the a zero and a late crossing the one. Relative to the x1 clock, only a phase shift of $\pm\pi/16$ appears.

The *shift* action of the encoding is a kind of differential encoding. From Fig.3.2.7, we can see that any inversion of the transmitted waveform will have no effect to the interpretation of the digital data. As the demodulation is based on the early / late interpretation of the incoming edges for a long string of similar logic bits the phase information could probably be lost. The problem is solved, as usually, by passing the data through a *pseudo-random* generator.

3.2.5.2 Signal Analysis of VMSK/2

In order to calculate the occupied by VMSK/2 spectrum, the encoded waveform is analyzed into three fundamental ones. These waveforms are shown in Fig.3.2.8. $\upsilon_o(t)$

contains the odd encoded symbols, $\upsilon_e(t)$ the even ones which will be inverted, and $\upsilon_m(t)$, a square-waveform with the fundamental frequency $\omega_m = \omega_b/2 = 1/2 \cdot 2\pi/T_b$.

Getting back to our earlier analysis and preserving the main lobe, we keep the a_0, a_1 and b_1 Fourier coefficients. Referring to Fig. 3.2.7 and the first encoding rule, we have that the high portion of the encoded symbol, is equal to $\tau = T_b/2 \pm T_b/16$ (3.5/8 to 4.5/8 the data bit period). It can also be stated that $\tau = T_b/2 + d(t) \cdot T_b/16$, with $d(t)$ the data stream taking values $\pm$ 1. If $d(t) = +1$, the high portion of the encoded signal is extended by $T_b/16$ and if $d(t) = -1$ is reduced accordingly. Replacing τ into Equation (3) and using Equation (1) we have:

$$\upsilon(t) = \frac{d(t)}{8} - \frac{2}{\pi}\sin\left(\frac{d(t)\pi}{8}\right)\cos\left(\frac{2\pi t}{T_b}\right) + \frac{4}{\pi}\cos^2\left(\frac{d(t)\pi}{16}\right)\sin\left(\frac{2\pi t}{T_b}\right) \quad (13)$$

Taking into account that the fundamental period T_m of the three waveforms that compose VMSK/2 is $T_m = 2T_b$, τ is expressed as $\tau = T_m/4 + d(t) \cdot T_m/32$, and Equation (13) becomes:

$$\begin{aligned}\upsilon_o(t) = {} & \frac{-8 + d_o(t)}{16} + \frac{2}{\pi}\cos\left(\frac{d_o(t)\pi}{16}\right)\cos(\omega_m t) \\ & + \frac{2}{\pi}\left(1 + \sin\left(\frac{d_o(t)\pi}{16}\right)\right)\sin(\omega_m t)\end{aligned} \quad (14)$$

where $d_o(t)$ is the odd data bits.

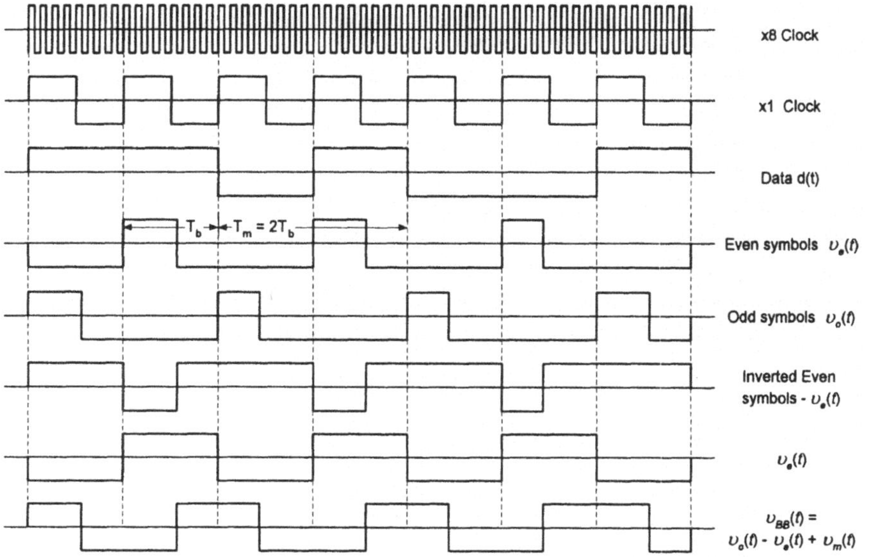

Fig. 3.2.8. Decomposition of VMSK/2 signal into fundamental components.

Equation (14) gives the odd waveform $\upsilon_o(t)$. Using the same procedure, the inverted even waveform $\upsilon_e(t)$ and the square wave pulse train $\upsilon_m(t)$ are expressed by:

$$\upsilon_e(t) = \frac{8-d_e(t)}{16} + \frac{2}{\pi}\cos\left(\frac{d_e(t)\pi}{16}\right)\cos(\omega_m t) + \frac{2}{\pi}\left(1+\sin\left(\frac{d_e(t)\pi}{16}\right)\right)\sin(\omega_m t) \tag{15}$$

$$\upsilon_m(t) = -\frac{4}{\pi}\sin(\omega_m t) \tag{16}$$

where $d_e(t)$ is the even data bits and ω_m is the angular frequency of the encoded symbol stream.

Adding $\upsilon_o(t)$and $\upsilon_e(t)$ it can be found the VMSK/2 baseband waveform. It is:

$$\upsilon_{VMSK/2}(t) = \frac{1}{8}\frac{d_o(t)-d_e(t)}{2} + \frac{4}{\pi}\cos\left(\frac{\pi}{16}\frac{d_o(t)-d_e(t)}{2}\right)\cos\left(\omega_m t - \frac{\pi}{16}\frac{d_o(t)+d_e(t)}{2}\right) \tag{17}$$

It is well known that the Fourier analysis of a waveform can start at any point in the waveform, but it is more convenient to start at the first rise, or at the center of the first half. When started arbitrarily, the transform has both sine and cosine terms. When started at the center it is an even function and there are only cosine terms. When started at the first rise, it is an odd function and there are only sine terms. This implies that the transform can be either sine or cosine series by merely changing the phase, or the start time of the analysis.

Digital modulation methods are usually analyzed in terms of pulses. A square wave pulse has a Fourier transform: Y = sinx/x, or a sinc function [10]. The sinx/x has a principal repetition rate equal to f_m and overshoot components at multiples of $1/T_b$. The pulse time T_b varies with the modulation pattern (see Fig. 3.2.3), hence the overshoot components will be in varying time/phase positions with regard to the bit period. These overshoot components plus the changing times above and below zero cause the " grass" seen below the center line of Fig. 3.2.4.

Using the Fourier transform we convert time / level changes v(t) to frequency and phase changes V(f). The corresponding inverse transform returns to the time domain:

$$v(t) = F^{1}\ V(f). \tag{18}$$

The expression of v(t) given earlier by (1), (2), (3) is:

$$v(t) = \sum_{n=-\infty}^{\infty} c_n e^{jn\omega_o t} \tag{19}$$

where $c_n = A\tau/T \cdot \sin(n\pi\tau/T)/n\pi\tau/T = A_{av} \cdot \sin x/x$ and $x = n\bullet(\tau/T)$, $A_{av} = A(\tau/T)$ $\Rightarrow A_{av}/n\bullet(t/T) = A/n\bullet$

3.2.6 Phase Multiplication and Modulation

In order to simplify and further manipulate the VMSK/2, we keep from Equation 17 only the last term where the information is buried in its phase. Thus we have:

$$\upsilon_{VMSK/2}(t) \approx \cos\left(\omega_m t - \frac{\pi}{16}\frac{d_o(t)+d_e(t)}{2}\right) \tag{20}$$

The term $\frac{d_o(t)+d_e(t)}{2} = -e(t)$ takes values −1, 0, +1, depending on the values of the input data, $d_o(t)$ and $d_e(t)$, which are ±1. Equation 20 yields:

$$\upsilon_{VMSK/2}(t) = \cos\left(\omega_m t + \frac{\pi}{16}e(t)\right) = \cos(\omega_m t \pm \Delta\theta) \tag{21}$$

where the phase change (deviation) $\pi/16$ e(t) has been replaced by $\Delta\theta$. Actually, the phase deviation is + $\Delta\theta$, -$\Delta\theta$ or zero.

The RF modulated VMSK/2, where the BPSK modulation is incorporated, can be stated as:

$$\upsilon_{SSB}(t) = \cos[(\omega_c + \omega_m)t \pm \Delta\theta] \tag{22}$$

where $\bullet_c$ is the carrier angular frequency. It is assumed that only the fundamental Fourier term is kept and is mixed with the carrier and finally is filtered to give the upper sideband. Actually, the mixing process of the encoded data waveform with the carrier takes place first and the filtering follows.

Multiplying the phase in the above equation by 3 as in the Armstrong PM generating method yields:

$$\upsilon_{SSB}(t) = \cos[3(\omega_c + \omega_m)t \pm 3\Delta\theta] \tag{23}$$

Notice that after the multiplication the phase change (deviation) becomes ± $3\Delta\theta$ or zero for a 7,8,9 code. It is noticed that the signal of Equation (23) is mixed with $\cos(2(\bullet_c + \bullet_m)t)$, obtained by passing the original signal through a high group delay filter and multiplying by 2. The mixer output is the $\cos[(\bullet_c + \bullet_m)t \pm 3\Delta\theta]$. This multiplication and mixing can be shared between the transmitter and receiver. The RF transmitter stages can multiply the phase, while the receiver local oscillator can restore the signal to $\cos[(\bullet_c + \bullet_m)t \pm 3\Delta\theta]$. Here the data rate ($f_m = 1/T_m$) remains unchanged, but the modulation phase changes to achieve that data rate are tripled. Multiplying phase in this manner we have a Q reduction without changing the narrow filter bandwidth or raising the grass level.

Let us consider an example of an 11,12,13 VMSK/2 with a 3KHz grass level -39 dB and an error angle • of •/24. Multiplying the phase by 3 we have an error angle of •/8 which is the equivalent of a 3,4,5 VMSK/2 code. The required narrow filter bandwidth remains unchanged and the grass level is observed unchanged. A 3,4,5 VMSK/2 has a grass level of -20 dB instead of -39 dB and is more difficult to be filtered off to meet FCC or other regulatory specifications. The error angle reduction means that much more noise can be tolerated before an error is caused. In this case it is 9.5 dB. This effect will be analyzed later.

3.2.7
VMSK/2 Demodulation

To detect the signal it is important to compare the single sideband that has been transmitted with a reference one. The first and most complicated method is to restore the carrier. For analysis purposes, the carrier can be assumed to be $\sin(\omega_o)t$ or $\cos(\omega_o)t$. If it is the $\cos(\omega_o)t$, it is considered to be along the I axis and the detected result is the I mode signal. If it is $\sin(\omega_o)t$ it is considered to be along the quadrature axis and the detected result is the Q mode signal.

When a carrier ω_c is restored and multiplied with the SSB signal which is $(\omega_c + \omega_m)t \pm \Delta\theta$, the product term $(\omega_c + \omega_m)t \pm \Delta\theta - \omega_c t = \omega_m t \pm \Delta\theta$ is the desired signal, which must then be detected (see Fig. 3.2.9) with an early / late circuit. The output of an XOR gate followed by a LPF and used as a demodulator or phase detector appears as a triangular instead of a sine waveform.

Using the In-phase reinserted carrier, $\cos(\omega_o)t$, the information appears at the peaks of the detected triangular waveform and must be recovered by the use of a peak detector (with the meaning given in this chapter) or a differentiator. If the carrier is shifted 90° the information appears at the crossings and must be detected with a zero crossing detector.

3.2.7.1
Quadrature Detection.

The sideband itself can be used as a reference, resulting in the outputs shown in Fig. 3.2.10 and 3.2.11 if the modulation is removed by means of a crystal with large group delay. No coherent carrier is necessary.

Using the reference at 90° out of phase with the signals average phase, the information data can be detected directly by sampling the mixer output. The output is shown in Fig. 3.2.10. This is not the preferred method because the noise shifts the amplitude and the R effect is minimized. Using the $\sin(\omega_c+\omega_m)t$, as a reference and multiplying it with the SSB signal we get:

$$\upsilon_{OUT}(t) = \cos[(\omega_c + \omega_m)t \pm \Delta\theta]\sin[(\omega_c + \omega_m)t] \tag{24}$$

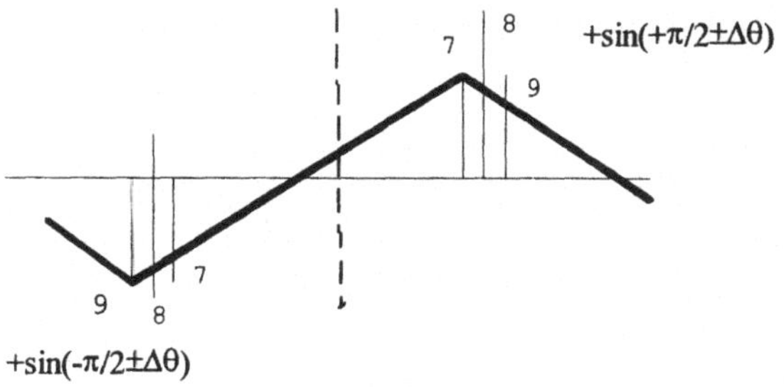

Fig. 3.2.9. Detected signal with re-inserted carrier.

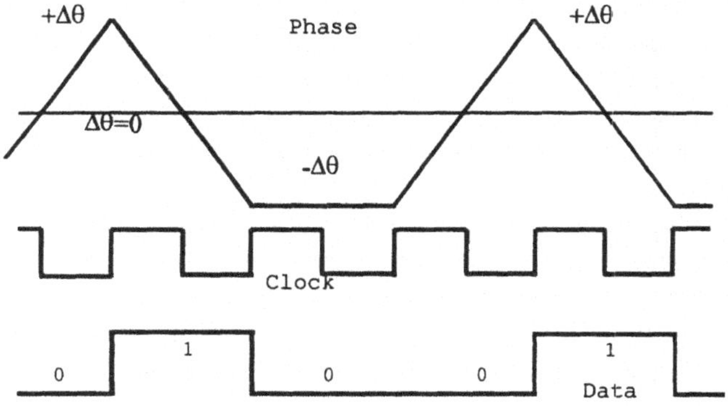

Fig. 3.2.10. Demodulation with recovered carrier (f_c+f_m) used as a frequency reference.

The output of the mixer makes use of a signal twice the recreated carrier frequency. For small •• it is sin (••) ≈ •• and:

$$\upsilon_{OUT}(t) = \sin(\pm\,\Delta\theta) = \pm\sin(\Delta\theta) \approx \pm\Delta\theta \tag{25}$$

Fig. 3.2.10 shows the output of a phase detector using the Q mode, that is the reference and signal are in quadrature. The information (ones and zeroes) are readable directly by sampling at the peaks or using a differentiator. The output is ± •• or zero.

3.2.7.2 XOR-based Detection.

When an XOR-based phase detector is used the carrier frequency is doubled. This is expected, because for each input pulse two output ones are produced. The gate is functioning as an In-Phase mixer. Filtering out the gate's output we get a waveform at $2f_m$.

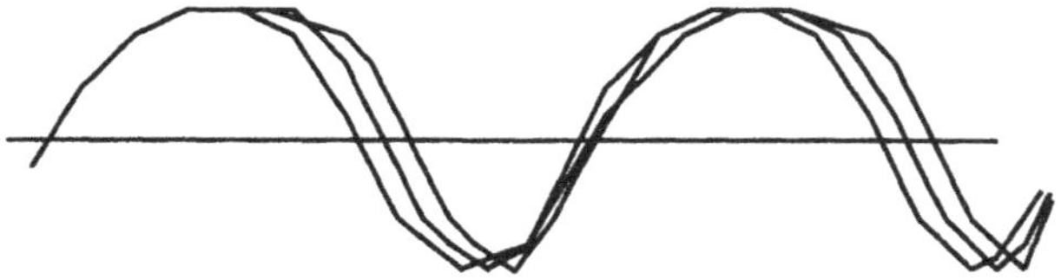

Fig. 3.2.11. Detection takes place at the zero crossings of the down slopes. Peak-to-peak is one bit period. Frequency = bit rate. The Reference is in phase with the sideband average. (I mode).

The information term appears from the early/late crossings of ½ cos(•$_b t$). In this case the tuning is adjusted to use the negative zero crossings. When it is applied to a divide by 2 flip flop the encoding pattern of Fig. 3.2.3 is restored. In the case where it passes through a bi-directional zero crossing detector, the VMSK encoding pattern of Fig. 3.2.2 is restored. Detection is by means the early/late zero crossings relative to a reference one.

The data clock is reset by the early crossings. The late crossings have no effect. Noise causes a low frequency phase shift in the early / late crossings. By resetting the clock at each early crossing, the clock is caused to shift following the noise effect, largely canceling it. This is a practical application of the *R* effect, where sampling occurs at a higher rate than the drift frequency so that only a small portion of the noise appears in each sample.

3.2.8 Filtering

An ideal or brick wall filter response is shown in Fig. 3.2.12. This filter has a phase shift from edge to edge of • radians over a frequency shift "Δf".

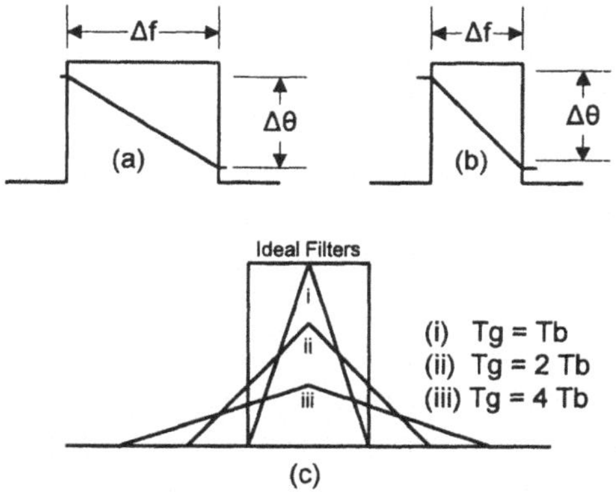

Fig. 3.2.12. The Ideal Filter.

The group delay of a filter can be obtained from Equation 9. It is:

$$T_g = [\bullet\bullet / 2\bullet \bullet f] \tag{26}$$

When the filter bandwidth is narrowed by reducing Δf as shown in Fig. 3.2.12b, the group delay rises. The effects of this narrowing can be seen in Fig. 3.2.12c. If the filter bandwidth is equal or greater than that of the sinx/x pulse, the maximum energy will pass. From the group delay equation it is evident that Δf and T_b are inversely related. For a narrow bandwidth the group delay is too large. The problem is that as the group delay increases the filter output decreases. The above is equated with a *slew rate*: The phase cannot slew fast enough if the group delay is too large. This topic is discussed by Taub and Schilling [10]. The group delay must not be confused with the *differential group delay* quoted for SAW filters. The differential group delay is the derivative of the group delay and is associated with the different delays of the individual frequency components of the waveform.

Fig. 3.2.13 shows the bit period vs. the relative phase output levels (vertical scale), with group delay exceeding the bit period. If the group delay is 5 times the bit period, the phase shift output is reduced to 20% of the peak. This is observed on the projection of the cross-point A at T_g and the slopped line of the phase shift.

To obtain the advantages of VMSK/2 modulation, conventional filters must be narrower, which in turn increases the group delay. Considering that BW $=1/T_g =$ B and $T_g = T_m$. We have B$T_g = 0.25$. This quantity is commonly used (e.g. in GMSK modulation).

Filters in cascade have the sum of the group delay, i.e., two filters with a group delay of 25 microseconds each have a cumulative group delay of 50 microseconds. The effect of this is that the system can cause a phase output that is less than the calculated values. Using a 4,5,6 code, the expected end to end phase shift is •/5. Excessive group delay may reduce this to •/6 or •/7. Having a $\Delta\theta$ equal to •, which is the case of an ideal filter and a bandwidth Δf equal to $f_b/Q = 2f_m/Q$, where Q is the middle number of the VMSK/2 notation (i.e. the spectral efficiency) the group delay is:

$$T_g = 1/2\pi \times \Delta\theta/\Delta f = QT_m/4 \tag{27}$$

Let $\Delta\theta = \pi/2$. Then $T_g = (\pi/2 / 2\bullet \bullet f) = 1/(4\bullet f)$. Since $\bullet f$ is reduced by Q the above equation becomes: $T_g = Q/(4f_b)$. This implies that a filter with a narrow bandwidth suitable for VMSK/2 has Q times more group delay to slew the phase of • in one bit period."

Ordinary filters (real or simulated), since they have long group delay cannot be used. Special filters, or oscillating ones of the locked or synchronous oscillators must be used. Filters with group delays exceeding the bit period can remove all of the modulation, passing only f_m as a single unmodulated frequency. This problem has lead to the failure of simulation programs such as SysView and MatLab. When the full Nyquist bandwidth is used there is no difficulty with the group delay of a normal filter. VMSK/2 however, uses only a small portion of the Nyquist bandwidth and the group delay rises for narrowband filters. In this case the modulation can be lost.

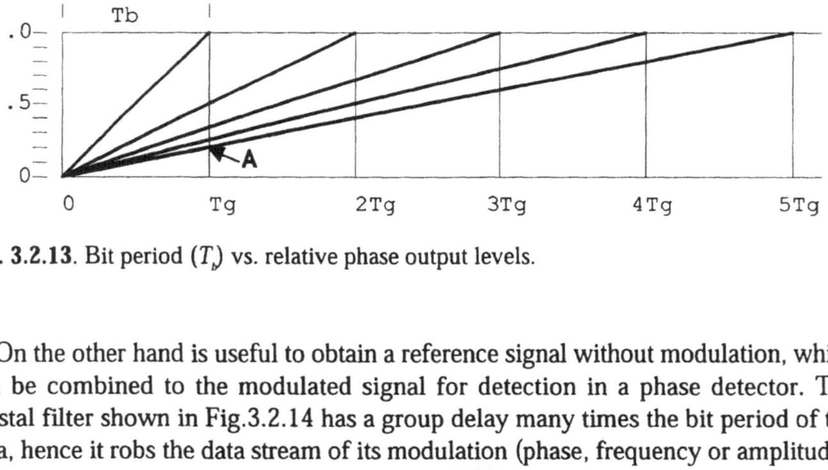

Fig. 3.2.13. Bit period (T_b) vs. relative phase output levels.

On the other hand is useful to obtain a reference signal without modulation, which can be combined to the modulated signal for detection in a phase detector. The crystal filter shown in Fig.3.2.14 has a group delay many times the bit period of the data, hence it robs the data stream of its modulation (phase, frequency or amplitude). Passing the original signal to one input of an XOR gate used as a phase detector we compare it with the reference and take a baseband output corresponding to the phase changes. Fig. 3.2.10 shows the output of such a detector.

The Ideal filter and the Nyquist filter have one thing in common. They comply with the accepted formula for Group Delay: $T_g = \bullet\bullet/2\bullet\bullet f$. The narrowest possible bandwidth is equal to the symbol rate. The best possible group delay is equal to $1/f_b$ and the phase change wall to wall is ±90°.

Thus, there is a trap as far as VMSK is concerned. Special low group delay filters are required to accommodate the narrow bandwidth of VMSK modulation.

There is not any other way at this time except of this trap at baseband. There is no also any known filter at baseband that can meet the low group delay requirements of VMSK. A filter having a Nyquist bandwidth be used.

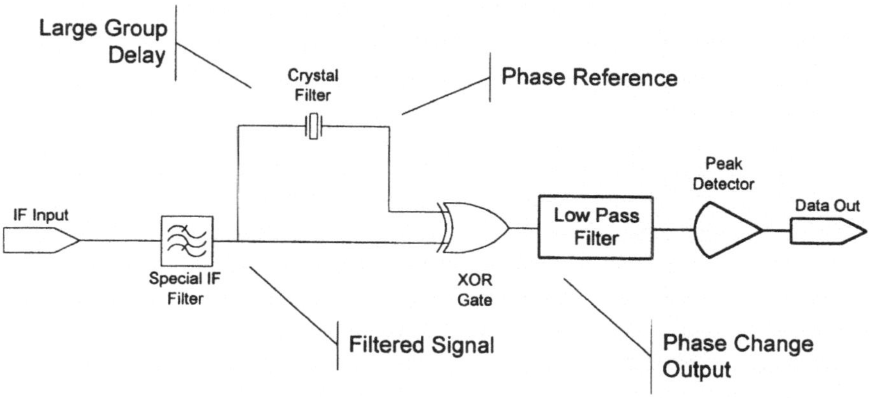

Fig. 3.2.14. Large group delay (crystal filter) output is used as a reference frequency at the demodulator.

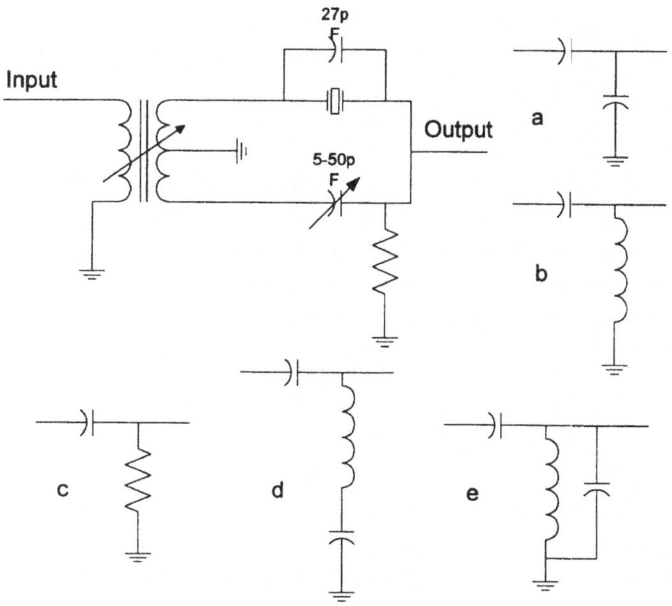

Fig. 3.2.15. Low group delay filter implementation with various loads.

At the RF there is a solution. Non-Nyquist Filtering (NNF) gives the opportunity to pass a very narrow bandwidth with the minimum group delay. NNF is a special filter that bypasses the bandwidth determining element. Two points must be kept in mind:

1. If the signal passes through the filtering element, it is subject to the group delay.
2. If the filtering element, which is oscillating or ringing at the signal frequency, has a large group delay, is used as a reference only, and the group delay can be defeated.

Fig. 3.2.15 shows a novel circuit that defeats the group delay problem. This circuit makes use of a single crystal in a bridge circuit that can be tuned to utilize it in several modes depending upon the values of the capacitors. By adjusting these values, the crystal can be made to respond in the parallel mode as shown in 'e', in which it offers a very high impedance at resonance so that very little of the signal passes through the crystal and most of it passes via the capacitor in the opposite bridge arm. This is represented by 'c' where the resistor has an almost infinite value. In this case the group delay of the crystal is ignored.

A second and preferred mode is that of using the crystal in the series mode as shown in 'd'. The nominal shunt capacity of the crystal is approximately 5 pf. If this is shunted by a 27 pf capacitor as shown, the total capacity in that arm is 32 pf. If the variable capacitor is then set to 32 pf, the capacitances balance in the bridge, creating a null, and only the series reactance of the crystal remains to off balance the bridge. The crystal has a very high Q and large group delay, so that it can not respond to the modulation, but behaves as a low value resistor as shown in 'c'. The signal passes through the crystal as if it did not exist at the resonant frequency, which is the single frequency representing the VMSK/2 modulation.

Off frequency, the crystal has a capacitive or inductive reactance as shown in 'a' or 'b'. These are large compared to the series resonant resistance of the crystal. Off resonance, the circuit tends to null the output. At resonance, the bridge reaches a maximum unbalance. It must be noted that this is not the case in Fig. 3.2.14. The behavior in this bridge circuit is very different from that in Fig. 3.2.14 where the modulation is removed by the group delay of the crystal. In the bridge circuit there is a negative feedback to provide cancellation.

This is an interesting concept. In practice, the filter can have a 3 dB bandwidth of 5 KHz at 24 MHz IF frequency. The data rate used in testing was 270 Kb/s. The spectral efficiency [bit rate/bandwidth] is 270/5 = 64 bits/sec/Hz. A 4,5,6 VMSK/2 code was used, with a nominal Δf of 26.6 KHz. Normally it would not be possible to pass this much deviation in such a narrow bandwidth without considerable distortion and loss of signal. *This is considered to be an indication that there is no frequency deviation but only a phase variation.*

The *grass* is reduced by approximately 20-25 dB on the filter shoulders. There is no visible indication that this filter introduces any group delay at all. This is not a very high degree of shoulder rejection, since a 2 pole crystal filter would do much better, but these filter stages can be cascaded to obtain much higher side attenuation. LC tuning is advisable in addition to the crystal to remove spurious frequencies. More recent developments in crystal manufacture have resulted in crystals with high fundamental frequencies ranging from 30 to 300 MHz as compared to the older upper limit of about 20 MHz. These crystals also have a high " rubber effect" enabling them to be tuned off by as much as 1000 ppm as compared to the 20 ppm for the older designs. They have an advantage in the circuits above in that they are almost free of spurs and allow a grass reduction of about 20 dB.

3.2.9 SNR Improvement

A data stream that is encoded with a 7,8,9 code is seen in Fig.3.2.16 as waveform " A" . When the phase is multiplied by 3, the distances away from the clock center are expanded by three times as seen in " B" . This is the equivalent of a 2,3,4 code. If the filter has excess group delay, the spread will decrease the approximate equivalent of a 4,5,6 code. There is still a net improvement in the signal delectability, since it can accept twice as much noise (phase drift) as the original. If the group delay is three times the bit period, the signal after filtering has the same spread with respect to the clock as before spreading and filtering.

Another advantage is obtained from the fact that at the receiver, narrow band IF filters are used that pass only a portion of the Nyquist baseband. When white noise passes through a narrowband filter, it changes from white Gaussian noise to a single frequency waveform, approximately that of the center frequency of the filter, with a slow moving envelop in frequency and level. R.E. Best [12] had studied this case giving the SNR improvement due to the narrowband filtering.

This additional improvement factor has been taken the name R factor. It is applicable to biphase modulation methods such as VMSK and VMSK/2, as well as wideband FM [10,12]. The improvement due to filtering is:

$$SNR_L = [P_S / P_{\cdot}]\ [B_I / 2B_L]\ E_b/\bullet \tag{28}$$

where B_I is the double sided RF bandwidth and B_L is the filter bandwidth.

The above equation from [12] contains two parts. The first half $[P_S / P_{\cdot}]$ is the noise reduction due to the spectral efficiency Q, which is 16 for 7,8,9 VMSK/2. The second half $[B_I / 2B_L]$ is the phase noise reduction factor R. The value SNR_L is the signal to noise ratio at the detector input (filter output), which is better than the input $E_b/\bullet$. For any one bit per symbol method $E_b/\bullet$ = C/N. The phase noise reduction factor R can be seen from Fig. 3.2.17.

In VMSK/2, there is a narrow bandwidth filter centered at ½ $f_b = f_m$ above zero Hz (at baseband). Any interfering signal must be within this filter where it cannot act directly. It must show up as a low beat or difference frequency starting at zero Hz. The sampling rate is f_b , so the signal is being sampled at a much higher rate than the interference beat. R is a phase noise reduction factor and is not the same with the noise improvement that is obtained by bandwidth reduction alone, as in Q.

When the sampling rate exceeds the low frequency beat from the narrow band filter, only a small portion of the noise appears in each sample.

The *R* effect applies only to wideband FM and Biphase modulation methods such as VPSK and VMSK. It does not apply to any NRZ line code method such as QAM or QPSK. A prerequisite is that the signal must be located away from zero Hz or the carrier as shown in Fig. 3.2.17.

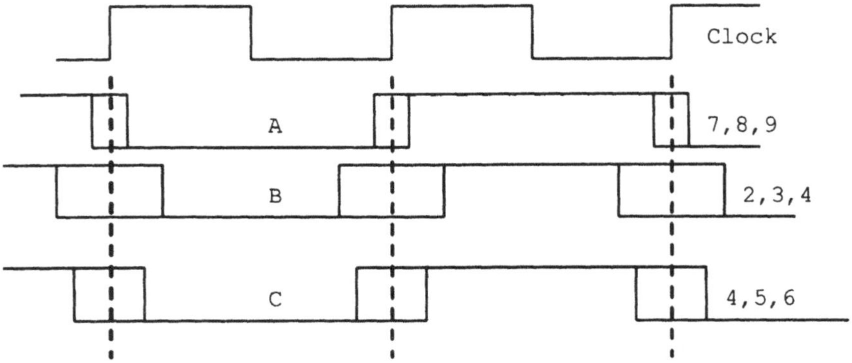

Fig. 3.2.16 The combined effects of phase multiplication and cumulative filter group delay.

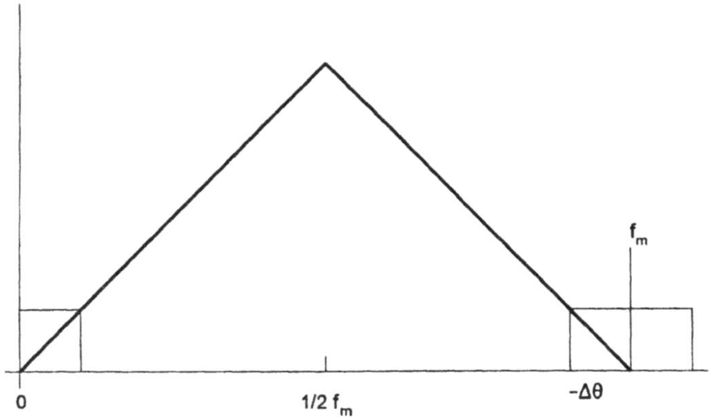

Fig. 3.2.17. The *R* Effect. Only the signal within the boxes is effected. Interference in the filter appears as a low frequency beat signal shown on left. Only a small portion of it appears in each sample.

Zero interference is predicted for the interfering frequency being exactly equal to the signal frequency. The interfering effect increases at the edge of the filter. This was analyzed by Best [12], Camies [13], and by Taub and Schilling [10] resulting in the relationship given for *R*. It is a reformulation or restatement of the FM interference analysis to be found in most texts where the interference at 0 Hz is said to be zero.

3.2.10 Universal Formula for the SNR

The "Universal Equation", [14], for the post detection SNR in order to analyze digital modulation methods is as follows:

$$\text{SNR} = NOPQR\,(E_b/\bullet) \tag{29}$$

where

- SNR or (S/N) is the post detection signal to noise ratio
- CNR or (C/N)= [N $(E_b/\bullet)$] is the (input) pre-filtered or *off the air* carrier to noise ratio.
- N = Bits/Symbol (in this case 1).
- $E_b/\bullet$ = Input signal bit energy to noise density = CNR when N =1.
- O = Other factors, such as 3/2 in FM (in this case 1).
- $P = \bullet^2$, where $\bullet$ = ½ of the modulation angle = error angle.
- Q = Sample Rate / Bandwidth (•f), the spectral efficiency in bits/sec/Hz.
- R = The phase noise improvement (reduction) factor = Nyquist BW/Filter BW.
- Q is also known as "Processing Gain" and is equal to $2\bullet/\bullet\bullet$ and to $2f_m / \bullet f = f_b / \bullet f$

The term $\bullet f / f_m$ is the well-known modulation index for FM. The letter sequence *NOPQR* is a memory aid.

The value of $O = 3/2$ is given in most textbooks for the universal equation and comes from the fact that the noise relationship area within the filter in Fig. 3.2.18 is not that of a uniform rectangular area, but a triangular area whose effective power is 1/3 that of a rectangular noise area. These effects (*R* and *O*) can be measured by means of a CW signal generator to represent an interfering signal. For VMSK and VMSK/2, *O* has been measured and found to behave somewhere between the rectangular noise effect and the triangular, so the value of $O = 3/2$ will be omitted from the formula and the value 1 will be substituted. In wideband FM, the *R* improvement is achieved by preemphasis and deemphasis networks in the audio path [10].

Substituting the above values, the post detection SNR from the Universal Equation is:

$$\text{SNR} = 2.44\ E_b/\bullet \text{ for VMSK and SNR} = 4.9\ E_b/\bullet \text{ for VMSK/2} \tag{30}$$

There is another generally accepted equation [15], other than the Universal Equation in the form:

$$\text{SNR} = \bullet^2\ [\text{ Bit Rate/ Filter BW}]\ E_b/\eta . \tag{31}$$

It can be seen that this is the same with as the universal equation if QR is replaced by the [Bit Rate/ Filter BW]. R applies only to SSB Biphase digital modulation, hence is not generally observed. Q = [Bit Rate/ Filter BW], replaces N (bits/symbol) in older texts.

VMSK/2, which is obtained by dividing the coded VMSK waveform by 2, is an obvious improvement over VMSK, but it should also be noted that this SNR value for both methods is independent on the spectral efficiency Q, which is not true for other digital modulation methods. The SNR is the same for 50 bits/sec/Hz as it is for 10 bits/sec/Hz. For BPSK, SNR = E_b/•. VMSK/2 is better than BPSK by 4.9/1, or 7 dB. Given a perfect theoretical filter, VMSK/2 would have a 3.8 dB C/N for a 10^{-6} BER.

When phase multiplication is used, the SNR becomes even better, approaching Shannon's lower limit of -1.4 dB for any code. With phase multiplication the 10^{-6} BER is theoretically obtainable near 0 dB C/N. There is a similarity then to the characteristics of MFSK where the use of an infinite number of frequencies over a broad bandwidth approaches Shannon's limit.

In theoretical terms, • is the modulation error angle, which is the ½ of the modulation angle. When the error angle is multiplied - for example by 3 - the value of • increases by 3 while QR remains unchanged. The equation then becomes:

$$\mathrm{SNR} = (3\bullet)^2 QR(E_b/\bullet). \qquad (32)$$

The above is an improvement in the value of SNR by 9.5 dB. C/N for VMSK/2 is the same as E_b/•. The C/N for any method cannot be less than Shannon's Limit (-1.4 dB), so the implication of the above equation is that C/N is approaching Shannon's limit.

3.2.11 Comparisons

3.2.11.1 FM or PM vs. VMSK

We assume that instead of the time domain method, a DDS synthesized signal is used to obtain FM, PM or FSK modulation. The 7,8,9 code produces a phase change of 22.5°, or a peak to peak change of 45°. Modulating for a phase change of 22.5° or the equivalent frequency change *results in Bessel products which are not the same as Fourier products in their effect.* In the above case the Fourier " grass" is replaced by Bessel products using FM or PM which would be necessary to detect the signal but do not exist using the AM time domain. The detected output level drops as the " grass" is lowered. In the FM/PM case the signal becomes a J_0 plus J_1 pattern in which the noise cannot exceed the J_1 levels. The carrier to noise ratio required for a 10^{-6} BER using a 7,8,9 code would be almost 30 dB. The difference in SNR relative to the time domain method gives a loss of 20 dB or more. Without filtering, the synthesized signal is a phase modulated signal. A narrow band filter has an integrating effect. Integrating PM at RF results in FM. The detected signal is therefor converted to FM which needs the Bessel products. Actually it is FSK. The three frequencies or phases resulting from the encoding have a phase integrating effect with time as shown in Fig. 3.2.10, so that only two represent a change. The center frequency is a " hold the last position" piece of information.

FM or PM require a broadband filter, not the very narrow band filter used with the VMSK/2 time domain method. The detected output level drops if the J_1 levels are reduced in the filter. This is not the case with VMSK/2 where the grass can be removed entirely without affecting the detected output level

3.2.11.2 Shannon's Limit

Shannon's limit [10] is always raised as an objection when high bandwidth efficiencies are proposed. For any bit-per-symbol method, Shannon's limit is 0 dB.

$$\frac{f_b}{W} = \log_2\left[1 + \frac{f_b}{W}\left(\frac{E_b}{\eta}\right)\right] \tag{33}$$

f_b is the data rate, W is the sampling rate, *fb/W* is the number of bits per symbol N, which is 1 for VMSK/2. The equation above is the equation for Shannon's limit [10], which gives the value of $E_b/$•.

VMSK/2 has been consistently demonstrated to operate at about 7 dB or less for $1\mathrm{x}10^{-6}$ BER with crude filtering. If a narrow transmitted bandwidth *W* is used in the Shannon's equation as a substitute for the sampling rate, i.e. f_b/W is substituted by *Q*, the required C/N would be approximately 64 dB for 10^{-6} BER, with $Q = 26$. This is obviously not the case.

VMSK and VMSK/2 are always 1 bit per symbol. *W* cannot be replaced by $1/Q$ (the narrow bandwidth) that would imply the data to be sampled $1/Q$ times per bit, which is an obvious violation of Nyquist's Sampling Theorem . When bandwidth is used in this formula it must be the RF Nyquist BW = Bit Rate.

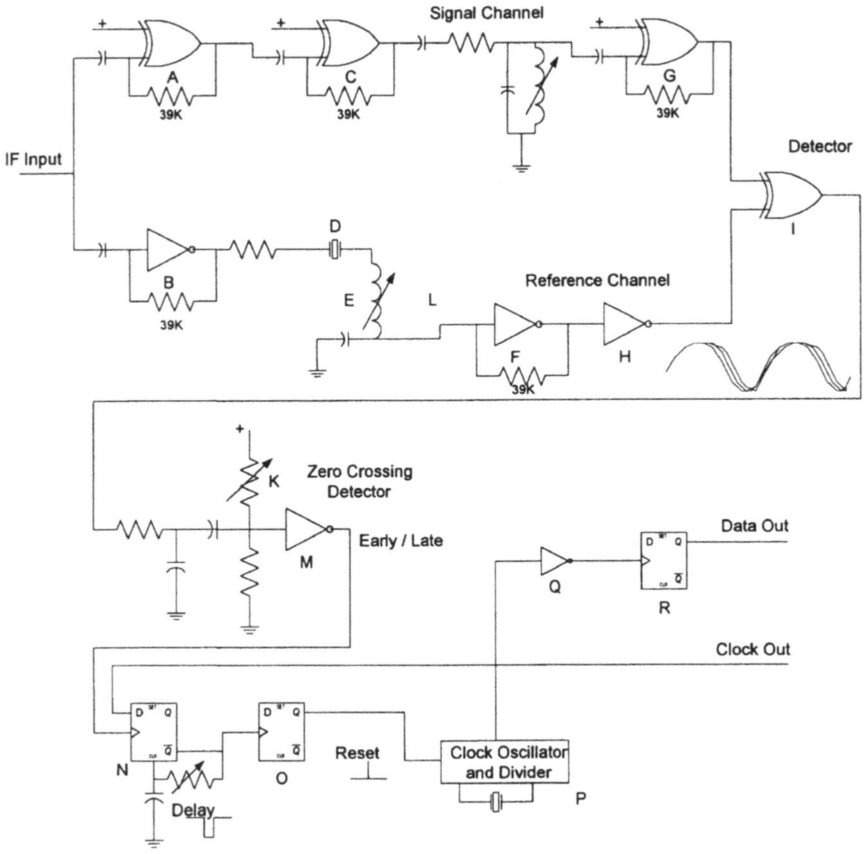

Fig. 3.2.18. Physical implementation of VMSK/2 demodulator. Due to the *R* effect, noise causes a low frequency phase drift. Resetting the clock each early period make it to follow the drift, hence to include only a small portion of the drift in each sample.

3.2.11.3
Detection and Clock Recovery.

The upper portion of Fig. 3.2.18 is the physical realization of the block diagram given in Fig. 3.2.14. The circuits are tuned to have the two paths in phase at the XOR phase detector " I", the output of which is shown in Fig. 3.2.11. When this is passed through the inverting zero crossing detector " M", a rectangular signal with edges occurring early or late with respect to the clock are obtained. The D flip flop R, converts these early / late changes to ones and zeroes.

Changes in a positive direction cause the one shot " N" to generate a negative pulse with delayed rising edge. This edge is timed to cause a narrow spike that resets the clocking oscillator " P" so that the D flip flop R samples exactly between the early / late occurrences. The flip flop " N" will respond only if the D input is held high by the counter " P" . Only early pulses will reset the clock counter.

3.2.11.4 Differential Coding and Error Correction.

Some modulation methods require the reinsertion of a carrier that can be ambiguous in phase. These methods require the use of differential coding ahead of the modulation. That is not the case with VMSK/2, so differential encoding of the data is not required or useful. It only degrades the system. On the other hand, the differential property is already self-included as has been stated previously in the *encoding rule* analysis paragraph. Convolutional error correcting codes such as " Viterbi" can be used while Reed Solomon is not particularly applicable.

3.2.12 Summary

VMSK/2 is one of the most bandwidth efficient modulation techniques known today. The information is " compressed" around a single frequency at half the bit rate, as seen in the spectrum analyzer. This single frequency line is the result of a phenomenon that occurs when the modulation index of a frequency shift keying method is reduced to less than 0.2 radians. There is no visible frequency shift and does not do any frequency shift appear to be necessary to detect the signal.

Since the signal is a single frequency there is no need to pass any signal other than that as long as the filtering has zero, or a very minimal, group delay.

The baseband signal is analyzed using Fourier Series and the mathematical presentation, shows that the information is encoded on the phase changes of the signal. Calculations based on the phase shift are made in order to estimate the occupied bandwidth and to illustrate the used modulation and demodulation circuitry. Also, special non-Nyquist filters with minimal group delay are used, enabling the VMSK/2 encoded signal to be transmitted and detected at bandwidth efficiencies greater than 60 bits/sec/Hz.

This spectral efficiency is being achieved without loss of carrier to noise ratio. In fact, the method has been shown to achieve the same SNR ratio with BPSK at several dB lower C/N. Also, in contrary to all known modulation methods which loose signal energy due to the compression (*P* in the Universal Equation), VMSK and VMSK/2 do not. The Signal to Noise ratio is kept constant, regardless the bandwidth efficiency.

An unusual factor called the *R*, or phase Noise Reduction factor, is applicable to this modulation method. This results in a significant signal to noise improvement. It is known that a filter only 1/6 as wide as would normally be called for is being used without signal degradation. The main speculation is that the minimum filter bandwidth is equal to 1/QR where *Q* is the spectral efficiency and *R* is the phase noise improvement factor.

It is seen that the method complies with the formula:

$$SNR= \bullet^{2}\ [\text{Bit Rate / Filter BW}]\ E_b/\bullet. \quad (34)$$

where [Bit Rate / Filter BW] is greater than *Q* bits/sec /Hz. Apparently [Bit Rate / Filter BW] = *QR*. Therefor a minimum filter bandwidth can be calculated to be 1/QR without signal distortion. The maximum obtainable spectral efficiency is equal to QR bits /sec /Hz, which in the case of a 4,5,6 code is 10 x 5 = 50 bits/sec/Hz. The obvious conclusion is that a spectral efficiency calculated on the basis of frequency deviation cannot be used. The much higher spectral efficiency related to filter bandwidth is valid in our case.

Also one of the advantages is that a coherent carrier is not required. All of the necessary information is contained around a single sideband spectral line which can be used as a detector reference. Transmitter and receiver IF frequencies can be off from one another by tens of parts per million since detecting the signal is not coherent carrier dependent when using the sideband itself as a reference.

The phase deviation can be multiplied at the transmitter as in the Armstrong PM method to decrease the noise sensitivity at the receiver and improve the bit error rate. In so doing, the C/N for a very low BER approaches 0 dB.

Finally as a conclusion one can say that no other known modulation method combines so high spectral efficiency with such low C/N values.

References

1. H. R. Walker, "High Speed Data Communications System Using Phase Shift Key Coding" U.S. Pats. 4,742,532 and 5,185,765. (Basic VPSK and " slip code" Patents.) U.S. Pat. 5,930,303 covers the aperture coding method and the phantom carrier signaling method (VMSK). Additional U.S. and Foreign patents are pending.
2. B. Stryzak and H.R. Walker, " Improve Data Transmission Using Single Sideband FM With Suppressed Carrier" , Microwaves and RF Magazine, Wireless Design Supplement, Nov. 1994.
3. H. R. Walker, " VPSK and VMSK Modulation Transmit Digital Audio and Video at 15 Bits/Sec/Hz" , IEEE Trans. on Broadcasting, Mar. 1997, pp. 96-103.
4. J. C. Pliatsikas, C. S. Koukourlis, J. N. Sahalos, H. R. Walker, " VMSK Modulation Boosts Wireless Communication Efficiencies" , Wireless Systems Design, Jan. 1998, pp. 35-43.
5. C. S. Koukourlis, J.C. Pliatsikas, J.N. Sahalos and H.R. Walker, " Spectrally Efficient Biphase Modulation" , Applied Microwave & Wireless magazine, May 1998, pp. 74- 81.
6. Martin S. Roden, " Analog and Digital Communication Systems" , Prentice Hall International, N.J., 3rd edition, 1991.
7. B. Sklar, " Digital Communications Fundamentals and Applications" Prentice Hall, Englewood Cliffs, New Jersey, 1988.
8. K. Feher, " Digital Communications: Microwave Applications" , Prentice Hall, Englewood Cliffs, New Jersey, 1981.
9. H.R.Walker, " High Data Rate Power Line Modem Uses VMSK Modulation" , Proceedings, Wireless Symposium, Santa Clara, Ca., Feb. 1997.

10. Taub and Schilling, "Principles of Communications Systems" McGraw Hill, 2nd edition, 1986.
11. Leon W. Couch II, "Modern Communication Systems, Principles and Applications", Prentice Hall International, N.J., 1995.
12. R.E. Best, "Phase Locked Loops", Mc Graw Hill, N.Y., 1984. (Analyzes the PLL as an FM detector, Includes analysis of the bandwidth efficiency (processing gain) Q and the R factor).
13. B.S. Camies, "Principles of Frequency Modulation", John Rider Publisher, Inc. N.Y., 1959.
14 H.R. Walker, "Universal Formula Analyzes All Modulation Methods", Applied Microwave and Wireless, July/Aug. 1997. This formula also appears in the "Encyclopedia of Electrical and Electronic Engineering" John Wiley Publishing, 1999 under "Modulation Analysis".
15 K. Feher, "Wireless Digital Communications", Prentice Hall, N.J., 1995.

3.3 Modelling of a Wireless ATM Physical Layer Based on an OFDM Modulation Scheme Using Hidden Markov Model

Christos V.Verikoukis, Juan J. Olmos

3.3.1 Introduction

The evolution of the wireline networks to support broadband services and the increasing success of the personal communication systems are the two major trends in the telecommunication world in the recent years. Thus, an increasing demand to connect mobile devices to the broadband wired network is expected in the future.

Asynchronous Transfer Mode (ATM), due to its inherent support of all kind of communications applications into various environments, and its ability to support a wide range of services with QoS guarantees, has been proposed by ITU to be the transfer mode of the future Broadband Integrated Services Digital Network (B-ISDN) [1].

To extend the capabilities of ATM over the wireless channel, the concept of Wireless ATM (W-ATM) was first proposed in 1994 [2]. The combination of wireless air interface and ATM can provide the freedom of movement for end users that wireless networks provide and the advantages that wired ATM can offer. Research within several projects [2]-[7] shows that W-ATM can be a solution for the future wireless indoor broadband communication systems.

The main components in a typical W-ATM system are (Fig. 3.3.1): mobile terminals (the end user equipment), access points (the base station of the cellular environment) and an ATM switch to support interconnection with the rest of the AMT network. The area is covered by a number of cells. In each cell the access point is connected with the wireline network. At a given time, a mobile is associated with the access point in one cell. The access points communicate with the mobiles, on one hand, and with the other access points (through the wired network infrastructure) on the other hand. Therefore, the access point provides the mobiles with access to wireline hosts and to mobiles located in other cells.

In Fig. 3.3.2 the protocol stack of a typical W-ATM system is presented. The wireless air interface consists of the Wireless Physical Layer (W-PHY) and Data Link Control (DLC) layer. This last contains a Medium Access Control (MAC) and a Logical Link Control (LLC) sublayers. The MAC sublayer is in charge of implementing the multiple access scheme of the W-ATM system, while the LLC sublayer performs a service class specific error correction scheme in order to guarantee high transmission quality.

The design of the wireless physical layer is one of the key issues in a W-ATM system. The objective is to build a suitable high speed radio physical layer to transmit ATM cells efficiently in an indoor radio environment taking into account parameters as the required performance, operational environment, range and mobility, frequency band and available technology for implementation.

One of the indoor radio channel characteristics is that the impulse response of the indoor channel is slowly changing in comparison with the bit rate of the system. Therefore a long time and high computational load is needed to simulate

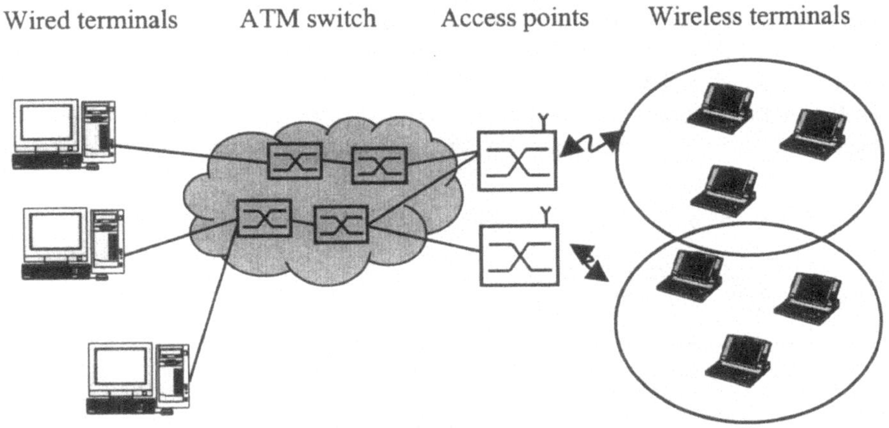

Fig.3.3.1 A typical Wireless ATM system

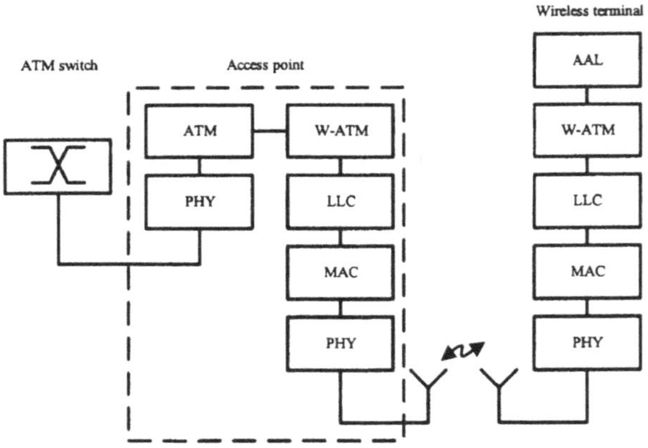

Fig.3.3.2 Protocol stack for integration of W-ATM terminals into a fixed ATM network

all the possible channel conditions of the W-ATM physical layer in an indoor environment. One method to overcome this problem is to model the physical layer by means of a Hidden Markov Model (HMM).

In what follows the modelling, using Hidden Markov Model [8], of a typical W-ATM physical layer with an OFDM modulation scheme is described. The HMM method reproduces the statistical error distribution, previously obtained by means of detailed simulations, of the real physical layer. Using HMM the

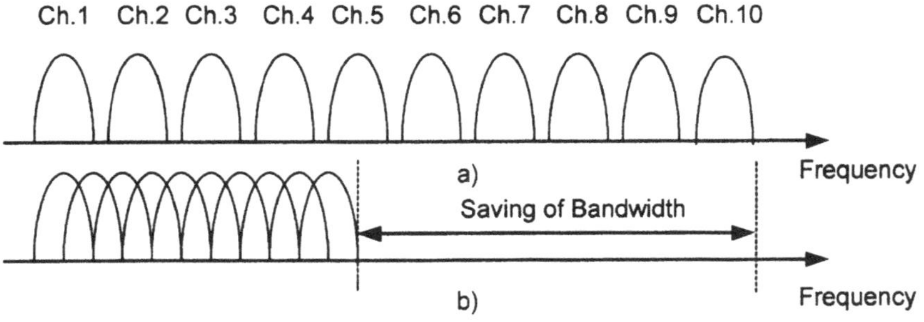

Fig. 3.3.3 a) Conventional multicarrier technique and b) Orthogonal multicarrier technique

computational load for obtaining the received bit sequences is reduced. This will allow for fast testing and comparison of several layer 2 protocols and error control schemes. The study is focused in the uplink (from mobile terminal to access point) since, due to the limited power of the mobile transmitter, it is the worst conditioned of the two links.

The description of the block diagram of the W-ATM physical layer is done in section 3.3.2. Some OFDM basics and a short OFDM history are also presented in the same section. A brief introduction of HMM theory, a description of the adopted HMM model as well as simulation results are given in section 3.3.3.

3.3.2
Physical Layer

3.3.2.1
OFDM Basics

After more than thirty years of research and development, Orthogonal Frequency Division Multiplexing (OFDM) has been widely implemented in high speed digital communications. Due to recent advances of digital signal processing (DSP) and very large scale integrated circuit (VLSI) technologies, the initial obstacles of OFDM implementation, such as massive complex computation and high speed memory do not exist anymore.

Currently, the use of Fast Fourier Transform (FFT) algorithms that eliminates the need for array sinusoidal generators make the implementation of the technology cost effective.

In an OFDM scheme, several modulated orthogonal carriers with overlapping spectrum are transmitted in parallel. Due to the frequency multiplexing, an OFDM system can convert a wide band frequency selective fading channel into a series of narrow band and frequency non selective fading subchannels. This is an efficient way of having several subchannels in a fixed bandwidth. The subcarriers are not

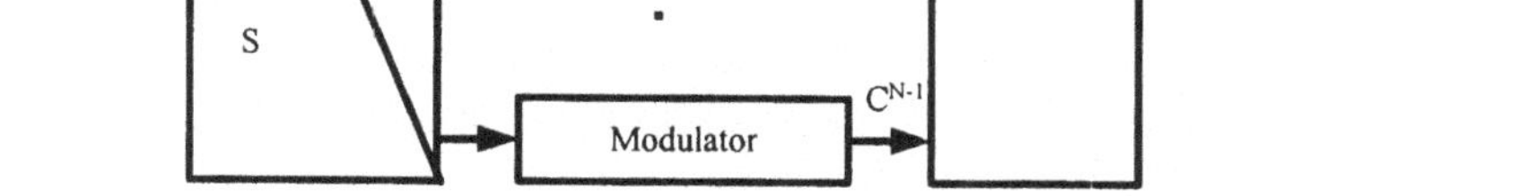

Fig. 3.3.4 Typical OFDM transmitter

separated by the distance corresponding to their modulation bandwidth. Instead, the spacing between the subcarriers is arranged in such a way that they become orthogonal.

A typical OFDM transmitter is shown in fig. 3.3.4. The parallel bit substreams are modulated, by an IFFT process, onto a set of evenly spaced carrier frequencies. In the receiver the bit streams are jointly demodulated using a FFT process.

One of the main problems in an OFDM system is the loose of the subcarrier orthogonality due to the Intersymbol Interference (ISI). The ISI is due to the multipath propagation of the radio waves, which gives rise to echoes of the signal that arrive at the receiving antenna with different delays. In order to avoid ISI, the symbol duration is extended by a cyclic prefix which absorbs the transient response of the channel between two consecutive modulation symbols.

Some important parameters in the design of an OFDM system are:

- Number of subcarriers for a given bandwidth and required bit rate
- Guard time, symbol duration
- Subcarrier spacing
- Modulation type per subcarrier

The use of an OFDM transmission scheme has the following advantages:

- OFDM is an efficient way to deal with multipath. For a given delay spread the implementation complexity is lower than that of a single carrier system with an equaliser.
- It is robust against narrowband interference, since such an interference affects only a small percentage of the subcarriers.
- It makes single frequency networks possible, which is especially attractive for broadcasting applications.

Comparing to single carrier modulation, OFDM:

- Is more sensitive to frequency offset and phase noise.
- The transmitted signals have a nonconstant power envelope. This means that any nonlinearity in the transmitter will cause in-band noise and spectral scatter causing degraded performance and adjacent channel interference.
- The peak power of an OFDM signal is in theory N times higher than the average power, where N is the number of subcarriers. In an OFDM system it is necessary to back off the transmit power amplifier which tends to reduce the power efficiency of the battery.

3.3.2.2
OFDM history

The history of OFDM has been presented in the literature [9], [10]. Following is a brief summary based on these works. The history of OFDM dates back to the mid 60s, when Chang [11] published his paper on the synthesis of bandlimited signals for multichannel transmission. He presents a principle for transmitting messages simultaneously through a linear bandlimited channel without interchannel interference (ICI) and intersymbol interference (ISI). Shortly after, Saltzberg performed an analysis of the performance [12], where he concluded that the strategy of designing an efficient parallel system should concentrate more on reducing crosstalk between adjacent channels than on perfecting the individual channels themselves, since the distortions due to crosstlalk tend to dominate. During these years OFDM was used in high frequency military systems as KINEPLEX, ANDEFT and KATHRYN.

A major contribution to OFDM was presented in 1971 by Weinstein and Elbert [13], who used to Discrete Fourier Transform (DFT) to perform baseband modulation and demodulation. To combat ISI and ICI they used both guard space between the symbols and raised-cosine windowing in the time domain.

Another important contribution was due to Peled and Ruiz in 1980 [14], who introduced the cyclic prefix, solving the orthogonality problem. Instead of using an empty guard space, they filled the guard space with a cyclic extension of the OFDM symbol. In the 1980's, OFDM had been studied for high-speed modems, digital mobile communications and high density recording [15]-[21].

In the 1990's, OFDM has been exploited for wideband data communications over mobile radio channels, high-bit-rate digital subscriber lines (HDSL; 1.6Mbps), asymmetric digital subscriber lines (ADSL;up to 6Mbps), very high-speed digital subscriber lines (VHDSL;100Mbps), digital television and HDTV terrestrial broadcasting [22]. Recently, it has been used in the European Digital Audio Broadcasting (DAB) standard [23] and in the W-ATM systems [4],[5].

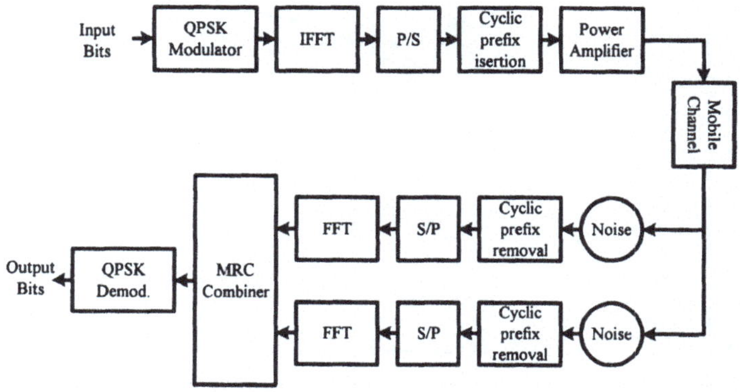

Fig. 3.3.5 Block diagram of the physical layer

3.3.2.3
Physical layer description

In order to study the W-ATM performances, we specify an OFDM physical layer that will provide a 20 Mb/s (uncoded) wireless link for ATM traffic in a bandwidth of 25 MHz (including the side lobes of the OFDM spectrum).

According to the world-wide available frequency bands, a carrier frequency of 5.2 GHz is a good selection. This physical layer, including the mobile channel, can be simulated in a computer to get an exact idea of the impact of the radio layer on the W-ATM system.

In the proposed OFDM system, 16 sub-carriers are used in parallel with QPSK modulation on each one. An Inverse Fast Fourrier Transformation (IFFT) and an oversampling factor of 4 is used to generate the time domain samples transmitted during one OFDM symbol. Since the mobile channel model that is used includes digital filtering, oversampling is necessary to avoid aliasing in the generated signal spectrum. To add the cyclic prefix, some samples at the end of each OFDM symbol are replicated to the beginning of the symbol. The use of a time domain raised cosine (roll-off=0.5) windowing of each OFDM symbol is necessary to reduce the adjacent channel interference. The raised cosine window lasts 15.6% of the OFDM symbol at the beginning and at the end of each symbol. Thus the total symbol is made up of the windowing time, a FFT period of 1.28 μs and a 160 ns guard time (cyclic period).

After modulation, the signal is up-converted to the RF channel band. Before being transmitted, the signal passes through a nonlinear power amplifier. The distortion of the amplifier can be modelled using the following expression for the

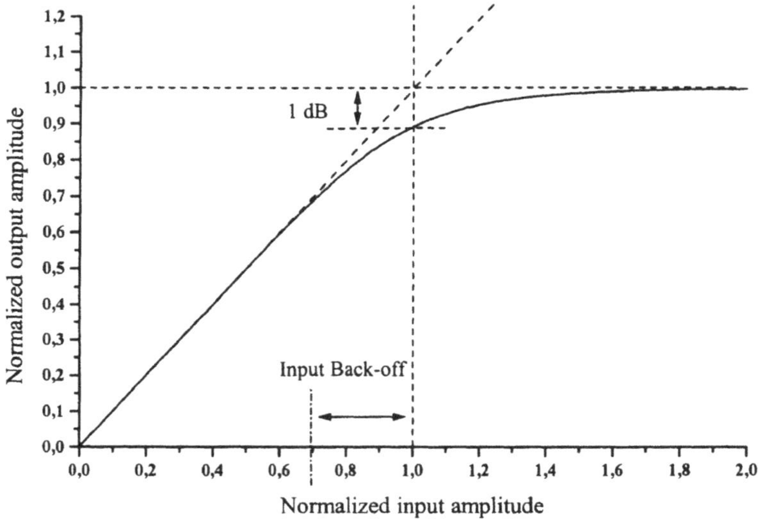

Fig.3.3. 6 Normalised output vs. normalised input of the amplifier

AM/AM conversion [24]:

$$g(A) = \frac{A}{\left(1 + A^{2p}\right)^{1/2p}}$$

Where A is the signal envelope at the input of the amplifier, g(A) the output envelope and p is equal to 3. In fig. 3.3.6 the normalised output envelope versus normalised input of the amplifier is shown.

Due to the non-linearity of the amplifier, the bit error rate of the link is increased. A back-off of the power amplifier, with an impact in the transmitter efficiency, is required to keep a reasonable BER.

At the receiver the signal is down-converted, it is digitised by the Analogue to Digital Converter (A/D), the cyclic extension is removed from the time domain samples and the frequency domain data symbols are recovered using FFT. In the receiving end, the QPSK signal is demodulated coherently and the bits are detected one by one using minimum distance criterion. Ideal time and frequency synchronism can be assumed in order to simplify simulations.

To estimate the transfer function of the channel, a 10 QPSK symbol pilot sequence is used. Since the indoor channel is slowly time variant, the pilot sequence, known a priori by the receiver, is transmitted only once every 1000 OFDM blocks (1 OFDM block=1 FFT period). In this way it is possible to implement a two antenna spatial diversity with two independent receivers whose

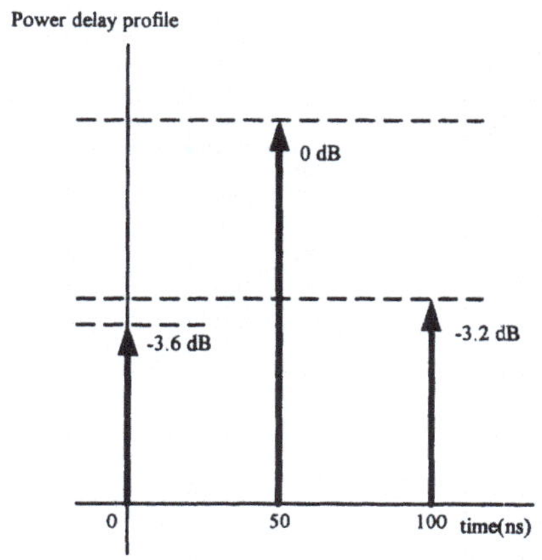

Fig3.3. 7 COST 207 pico-cellular channel model

outputs are combined using maximal ratio combining (MRC) [25].

The modulated signal will pass through a picocellular mobile radio channel modelled by three time-variant propagation paths. The power delay profile (fig. 3.3.7) is based on the COST 207 model [26], and it has a maximum propagation delay of 100ns, being the simulated velocity of the mobile terminal 1m/s (equivalent to 17 Hz Doppler spread). A FIR filter with the appropriate number of taps to model the channel impulse response, is used in the simulations.

Fig. 3.3.8 shows the mean BER curves for constant back-off of 3 dB and the

BER when pilot symbols are protected with an additional back-off of 9 dB. Notice that, despite that the additional back-of diminishes the signal to noise ratio of the pilots, the BER improves noticeably. The interpretation is that the use of a pilot sequence produces a correlation gain equal to the sequence length (10 in this case) that compensates for the signal to noise ratio lost with the back-off.

3.3.3 Hidden Markov Models

3.3.3.1 Introduction

The Hidden Markov Models (HMMs) are a popular and powerful tool for modelling stochastic random processes [8]. The basic theory was published in several papers by Baum and his colleagues in the late 1960s and early 1970s, but have become popular in the last several years. HMMs are general enough to approximate various statistical data, which explain their popularity in many applications such as speech recognition, control theory, biology, communications theory over channels with burst errors, queuing theory, etc.

In several works [27]-[33] channels with memory as well as the error sequences

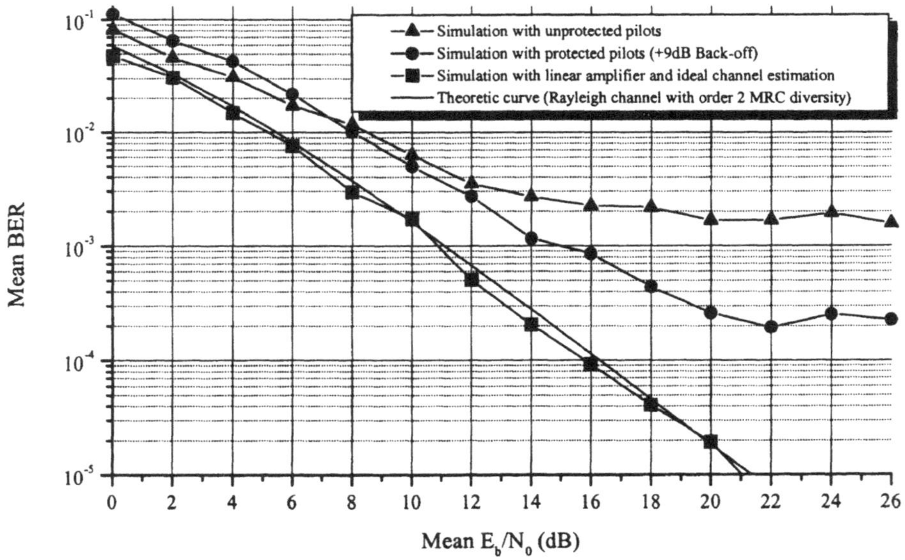

Fig.3.3. 8 Mean BER vs. mean E_b/N_0

in digital communications systems are modelled using HMMs.

A short introduction based on [34] to the Discrete Markov Process, which is a base of the HMMs, is presented in the following section.

3.3.3.2 Discrete Markov Processes

Consider a system which may be described at any time by one of N different states: S_1, S_2, ...S_N. At discrete time instants, the system undergoes a change of state according to a set of probabilities associated with the states. Assume that the time instants associated with the state changes are t=1,2,... and that the state at time t is q_t.

A full probabilistic description of such a kind of system requires to specify the current state at instant t and the previous states. For the special case of a discrete, first-order, Markov chain, this description is simply given by the current state and the previous one:

$$P[q_t = S_j | q_{t-1} = S_i, q_{t-2} = S_k, \dots] = P[q_t = S_j | q_{t-1} = S_i]$$

If it turns out that the transition probabilities are independent of time, then we have what is referred to as a homogeneous Markov chain. In this case a set of state transition probabilities, a_{ij}, can be defined:

$$a_{ij} = P[q_t = S_j | q_{t-1} = S_i] \quad 1 \le i, j \le N$$

with the state transition coefficients having the properties

$$\sum_{j=1}^{N} a_{ij} = 1$$

$$a_{ij} \ge 0$$

The above stochastic process could be called an observable Markov Model since the output of the process is the set of states at each instant of time, where each state corresponds to a physical, observable event.

3.3.3.3 Hidden Markov Models definition

In the above Markov model each state corresponds to an observable event. The HMM is an extension of the concept of the Markov model to include the case where the observation is a probabilistic function of the state rather than deterministic. In this case the resultant model is a double embedded stochastic process with an underlying state process, which is not observable. For this reason it is called "hidden", since it can only be observed through another set of stochastic processes that produce the sequence observation.

We denote q_t the current state at time t, $S=\{S_1, S_2, \dots, S_N\}$ the individual states, $V=\{v_1, v_2, \dots, v_D\}$ the individual symbols that can be observed, a_{ij} the probability of transition from state S_i to state S_j and $b_j(k)$ the probability of getting the output v_k when we are in state S_j.

A HMM can be fully defined by the following elements:

- **N**, the number of states in the model.
- **D**, the number of the different observation symbols per state. The observation symbols correspond to the physical output of the system being modelled.
- **A**=$\{a_{ij}\ 1\leq i, j \leq N\}$, the state transition probability matrix.
- **B**=$\{b_j(k)\ 1\leq j \leq N, 1\leq k\leq D\}$, the observation symbol probability matrix
- The initial state distribution vector $\pi=\{\pi_i\}$, where π_i is the probability for the initial state to be S_i,

Giving appropriate values to N, D, A, B and π, the HMM can be used as a generator to produce the observation sequence $O=\{O_1,O_2,...,O_T\}$ (where O_t is one of the symbols of V and T is the number of observations in the sequence) according to the following procedure:

1. Choose an initial state $q_1 = S_i$ according to the initial state distribution π.
2. Initialise the time $t = 1$.
3. Choose $O_t = v_k$ according to the symbol probability distribution at state S_i, that is, $b_i(k)$.
4. Transit to a new state $q_{t+1} = S_j$ according to the transition probability of state S_i, that is, a_{ij}.
5. Set $t = t+1$,

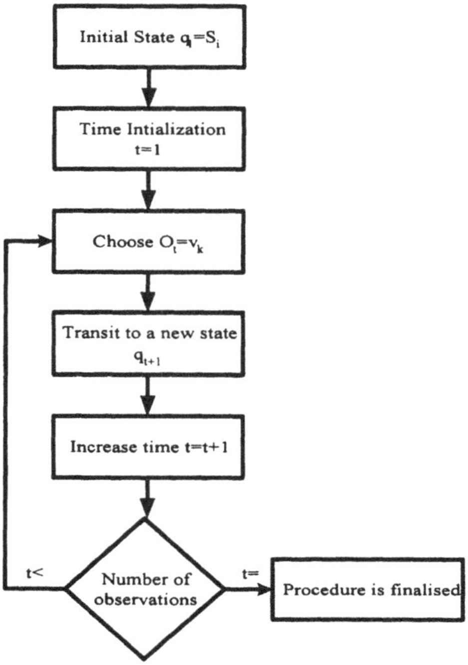

Fig. 3.3.9 Observation sequence generator flow diagram

6. if t<T go back to (3).
7. If t=T the procedure is finalised.

This procedure can be used as an observation generator, and also as a model of how a given observation sequence is generated by a so-designed HMM.

In shorthand notation, a given model can be summarised as:

$$\lambda=(A,B,\pi)$$

3.3.3.4 Description of the HMM adopted for physical layer modelling

The model adopted, divides the transmitted data into constant length blocks, where each one is characterised by the number of errors it contains. The number of errors in a block may be generated to match the statistical distribution of the number of errors in a block produced by a detailed simulation of the system that the model represents.

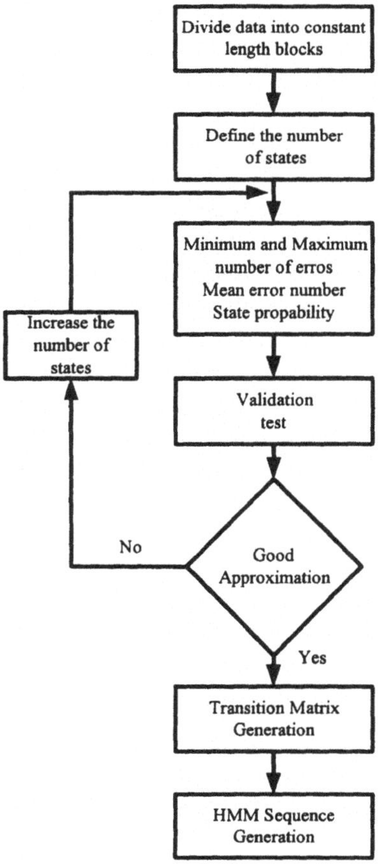

Fig. 3.3.10 HMM flow diagram

When a model is based on a Markov process, there is a correlation between consecutive blocks, and the number of errors within a transmitted block only depends on the number of errors of the previous block. Ideally, each state of the Markov model should represent a given number of errors within a block, but, in order to simplify the model, we will consider that each state represents a range of number of errors within the block. The range of number of errors assigned to every state should be as equiprobable as possible. For every E_b/N_0 value, in steps of 1 or 2 dB, we define the number of states, in this case 4, 8 or 16 states.

Choosing the correct number of states is a critical point to obtain an accurate model. A small number of states force the HMM to group together observations which may introduce a substantial variance and would be better modelled by separate states. On the other hand, if the model has too many states, it may not have enough distinct events to estimate the parameters of each state correctly.

Using detailed physical layer simulations the minimum and maximum number of errors is tentatively defined for every state. Then, the mean error number and the probability of the state are calculated. Having defined the number of states and its parameters, the probabilities of passing from one state to any one of the other states are calculated. This probabilities form the "transition matrix". This matrix is used as the input to the HMM program that generates a sequence of bits (with errors) statistically similar to the sequence used for training but in a much shorter time. The flow diagram of the model is shown in the Fig. 3.3.10.

For running the Markov model it is sufficient to generate random numbers distributed in accordance to the transition matrix. In this way the simulated system passes from one state to the other with the right probabilities. Depending on the actual state corresponding to each transmitted packet, a number of errors are placed on it according to the algorithm proposed in [35]. In [35] the errors are distributed inside the packet with uniform statistics. In our case, the errors in the real system appear surrounding the carrier where the fading is centered (Fig. 3.3.13). For a OFDM system, a better method for error distribution consists in obtaining, form the detailed simulations, the p.d.f. of the error distribution in the packets conditioned to the fact that the fading is centered in one of the 16 carriers. This is done for every carrier. Then, when running the model, the errors are distributed inside the packet according to one (randomly selected) of the stored p.d.f.'s.

To check the overall performance of the model, validation tests, consisting in analysing the errors introduced by HMM within the frame and comparing them with those of the real channel simulation, should be carried out. The validation is based on the comparison of the following three statistical parameters:

- The probability that a block of M bits contain exactly K errors
- The probability of having a run of T bits in error
- The probability of having error free intervals of length N bits

In case of obtaining large deviations, the number of states is increased and the above procedure is repeated.

3.3.3.5
HMM performance

Our goal is to find suitable values for the number of states N for a packet length of 512 bits where the error distribution of the HMM sequences approaches the error distribution of the real sequences. For this purpose detailed physical layer simulations were carried out, taking into account the parameters already described in section 3.3.2.

In Fig. 3.3.11 the probability of the number of errors in a block, for different numbers of states in the HMM, for a SNR value of 10dB is presented. It can be

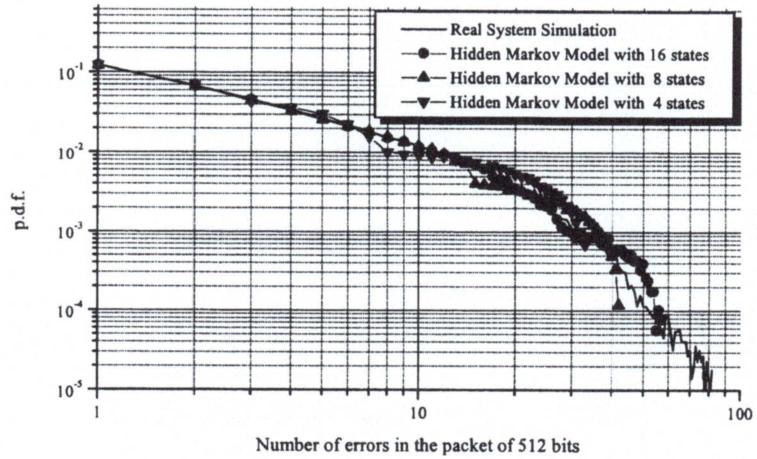

Fig.3.3. 11 p.d.f of number of errors in the packet of 512 bits for different number of states in the HMM for mean E_b/N_0 10dB

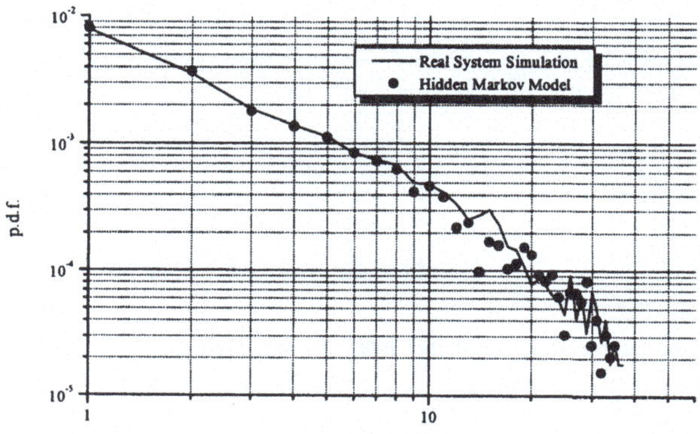

Fig.3.3. 12 p.d.f of number of errors in the packet of 512 bits for mean E_b/N_0 18dB

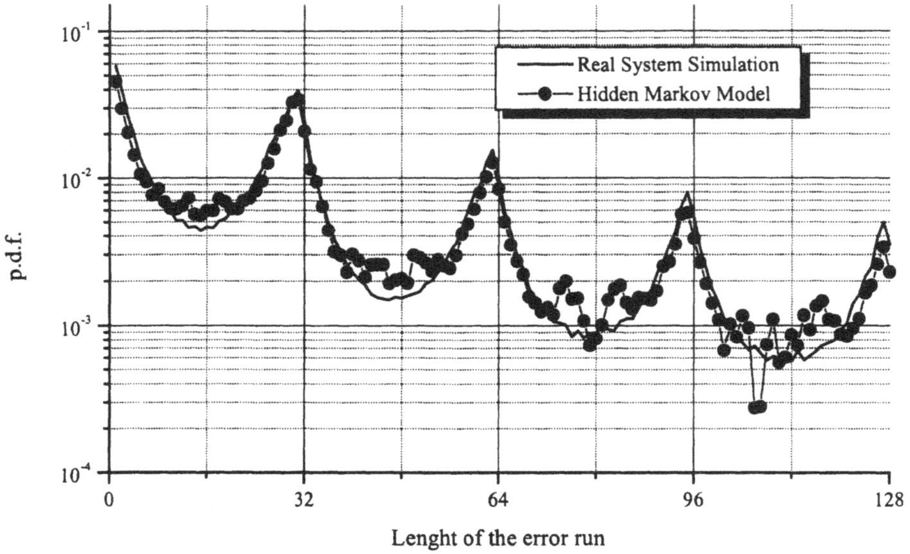

Fig.3.3. 13 p.d.f of the length of the error runs for mean E_b/N_0 10dB

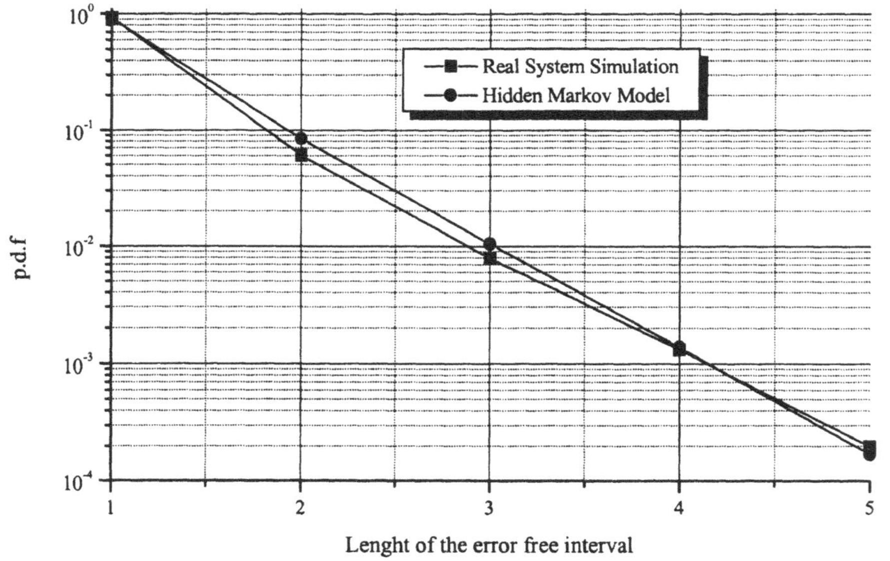

Fig3.3. 14 p.d.f of the length of the error free intervals for mean E_b/N_0 10dB

Table 3.3.1 State Transition Matrix A, in a block of 512bits, for mean E_b/N_0 10dB.

State	1	2	3	4	5	6	7	8
1	0.9214	0.0537	0.0150	0.0057	0.0031	0.0009	0.0002	0.0000
2	0.4838	0.2567	0.1340	0.0635	0.0456	0.0134	0.0025	0.0005
3	0.2639	0.2634	0.1907	0.1196	0.1121	0.0388	0.0097	0.0018
4	0.1439	0.2095	0.1902	0.1570	0.1754	0.0951	0.0253	0.0036
5	0.0781	0.1219	0.1522	0.1507	0.2411	0.1818	0.0662	0.0079
6	0.0297	0.0418	0.0656	0.0920	0.2163	0.3037	0.2113	0.0396
7	0.0089	0.0141	0.0213	0.0302	0.0888	0.2455	0.3952	0.1959
8	0.0017	0.0048	0.0046	0.0056	0.0143	0.0363	0.1737	0.7590

observed that using 8 states in the HMM gives better accuracy than using 4 states, but when the number of states is increased to 16 the performance improvement is not so significant. Thus, the use of 8 states is preferable because it runs faster.

Fig. 3.3.12 compares the p.d.f. of the number of errors per packet obtained with HMM and with the real system for a mean E_b/N_0 of 18dB. Fig. 3.3.13 allows to compare the statistics of the length of the error runs obtained by Markov with the real system for a mean E_b/N_0 of 10dB, finally in Fig. 3.3.14 the p.d.f. of the length of the error free intervals for a mean E_b/N_0 of 10dB is presented.

Once we obtain the optimum number of states for each SNR, state transition matrices as the one is presented in table 3.3.1 are produced. According to that matrix, the HMM sequences are produced. From the above results it can be seen that HMM using 8 or 16 states in a packet of 512 bits can describe the error distribution of the real sequences with good accuracy. Thus the HMM is a realistic procedure which reduces the complexity and the computational effort of the system in comparison with the bit by bit simulation of the real sequences.

References

1. D.Bertsekas and R.Gallager, *Data Networks*. Upper Saddle River, NJ: Prentice-Hall, second ed., 1992.

2. D.Raychaudhuri and N.D.Wilson, "ATM-Based transport architecture for multiservices wireless personal communication networks", *IEEE JSAC*, vol.12, pp.1401-1414, October 1994.

3. M.Umehira et al., "An ATM Wireless Access System for Tetherless Multimedia Services", *Proc. ICUPC'95.*

4. J.Mikkonen et all., "The Magic WAND-Functional Overview", *IEEE JSAC*, Vol. 16, no. 6 pp. 953-972, August 1998.

5. C. Giotti, J. Borowski "The AC006 MEDIAN Project-Overview and State of the Art" *ACTS Mobile Commun. Summit*, Granada, Spain, Nov. 1996.

6. J.Porter et all, "The ORL Radio ATM System Architecture and Implementation" Tech. Paper 1/1996, *ftp://ftp.orl.co.uk/pub/docs/ORL.*

7. B.Walke, D.Petras, and D.Plassmann, "Wireless ATM: Air Interface and Network Protocols of Mobile Broadband System", *IEEE Personal Commun. Mag.*,pp50-56, Aug. 1996.

8. Lawrence R. Rabiner, "A Tutorial on Hidden Markov Models and Selected Applications in Speech Recognition", *Proceedings of the IEEE*, Vol.77, No.2, pp. 257-286, February 1989.

9. W.Y.Zou, Y.Wu, "COFDM an Overview", *IEEE Trans on Bradcasting*, Vol. 41, No. 1, pp:1-8, March 1995.

10. J.Bingham, "Multicarrier modulation for data transmission: an idea whose time has come" *IEEE Communications magazine*, Vol 28, pp.5-14, May 1990.

11. R.W. Chang "Synthesis of band-limited signals for multichannel data transmission", *Bell System Tech. J.*,45 pp.1775-1796, Dec. 1966.

12. B.R. Saltzberg, "Performance of an efficient parallel data transmission system", *IEEE Trans. Commun.*, COM-15, Vol-5, pp.805-811, Dec. 1967

13. S.B.Weinstein and P.M.Elbert. "Data transmission by frequency-division multiplexing using the Discrete Fourrier Transformation", *IEEE Trans. Commun.*, COM-19, Vol.5, pp.628-634, Oct.1971

14. A.Peled and A.Ruiz "A Frequency Domain Data Transmission Using Reduced Computational Complexity Algorithms" *Proc. IEEE Inter. Conf. Acoust., Speech, Signal Processing*, pp.964-967, Denver, CO,1980.

15. B. Hirosaki "An orthogonally multiplexed QAM system using the discrete Fourrier transform", *IEEE Trans. Commun.*, vol. COM-29, pp.982-989, July 1981.

16. B. Hirosaki, S.B.Hasegawa and A. Sabato, "Advanced groupband data modem using orthogonally multiplexed QAM technique", *IEEE Trans. Commun.*, vol COM-34, pp.587-592, June 1986

17. B. Hirosaki et al., "A 19.2 kbits voiceband data modem based on orthogonally multiplexed QAM techniques", *Proc. of IEEE ICC'85*, pp.21.1.1-5, 1985

18. W.E. Keasler and D.L.Bitzer, "High speed modem suitable for operating with a switched network", *U.S. patent No. 4*,206,320, June 1980.

19. Telebit Corporation, "Comparative performance results for asymmetrical duplex, V.32 (extended), and multicarrier modems", *CCITT SG XVII, Contribution D56*, Sept. 1989

20. S.P.Bernand, "A low baud rate 9600 bps voice band modem employing frequency division multiplexing", *Workshop on Communications*, Commun. Res. Lab. McMaster Univ Hamilton, Ont., 1982

21. T.Kelly, "Digital modem packs data onto 40 bauds for 9600 bps data comms over voice lines", *Canadian Datasyst.*, vol.11, Apr.1979

22. R.V. Paiment, "Evolution of single carrier and multi-carrier modulation techniques for digital ATV terrestial broadcasting", *CRC report No. CRC-RP-004*, Ottawa, Dec. 1994.

23. Radio broadcasting systems; Digital Audio Broadcasting (DAB) to mobile, portable and fixed receivers, *ETS 300 401*, ETSI-European Telecommunications Standards Institute, Valbonne, France, Feb. 1995

24. C.Rapp, "Effects of HPA-Nonlinearity on a 4-DPSK/OFDM Signal for Digital Sound Broadcasting System", *Proc. of Second European Conference on Satellite Communications*, Liège, Belgium, October 22-24, 1991, pp.179-184.

25. J.G.Proakis, *"Digital Communications"*, McGraw-Hill Book Company, 1989

26. COST 207, Digital land mobile radio communications, Final report, *ECSC-EEC-EAEC*, Brussels, Luxembourg, ISBN 92-825-9946-9.

27. L.N.Kanal and A.R.K Sastry "Models for channels with memory and their applications to error control", *Proc of IEEE*, vol. 66, pp.724-744, no.7, July.

28. S.Sivaprakasam and K.S. Shanmugan, "An equivalent Markov model for burst errors in ditigal channels", *IEEE Trans. on Commun.*, vol 43, pp.282-286, April 1995.

29. S.Tsai, "Markov characterization of the HF channel", *IEEE Trans. on Commun.*, vol. COM-17, pp.24-32, Feb. 1969.

30. W.Turin, *"Performance Analysis of Digital Transmission Systmes"*, Computer Science Press, New York.

31. W.Turin, an M.M.Sondhi, " Modeling error sources in digital channels", *IEEE JSAC*, vol.11, No.3, pp.340-347, December 1993.

32. J.Garciá-Frías, P.M.Crespo, "Hidden Markov Models for Burst Error Characterization in Indoor Radio Channels".*IEEE Trans. on Vehicular Technology*, Vol. 46, No.4, pp. 1006-1020, November 1997

33. J.J.Olmos, P.Diaz, O.Sallent, F.Casadevall, I.Berberana, "Hidden Markov Models Used for Modelling the Physical Layers of Third Generation Radio Interfaces" *ACTS Summit 1997*, vol.1 pp.138-143, October 1997, Aalborg, Denmark.

34. L.Kleinrock, "*Queuing Systems*", John Wiley & Sons, 1974

35. C.M.Dearlove, "A Channel Error Model for the Simulation of Error Control Coding in a Digital Communications System", *RMTP/RB/L177, MEC025*, June 1991.

4 Antennas & Propagation

4.1 Introduction

Radio propagation plays a major role in the design of 3G Mobile Systems with acceptable quality for the offered service. The system requirements are the high capacity, the high data rate, the high operational frequency and the low transmitting power. In a mobile radio system where a single antenna is used in both the BS and the MS the channel between the input of the transmitter antenna and the output of the receiver antenna can be described by a channel impulse response. The mobile channel, however, is defined between the output of the transmitting antenna and the output of the receiver antenna. This knowledge is necessary for designing the mobile system to operate at a certain quality level.

In section 4.2 and 4.3 the benefits of the adaptive antennas are presented. In section 4.4 a propagation model is analysed which is used to derive an optimum procedure in cell planning and in section 4.5 a stochastic model is used for radio propagation which is based on the physics of electromagnetic wave propagation and takes into account the average power loss due to reflection of signals from surfaces. During the modeling process, the statistical properties of the received signal are computed and compared with experimental results.

4.2 Benefits of Adaptive Antennas in the Uplink of TD-CDMA Mobile Radio Systems

A. Papathanassiou, J.J. Blanz, M. Weckerle, and R. Schmalenberger

4.2.1 Introduction

In the early 1990ies the development of a novel air interface for 3G (Third Generation) mobile radio systems began. In this air interface the combination of the multiple access methods FDMA and TDMA known from GSM is supplemented by a CDMA component. Further, in the receivers the increased multiple access interference (MAI) typical of CDMA is mitigated by joint detection (JD), which eliminates intracell MAI [1]. The air interface concept is termed TD (Time Division)-CDMA and has been verified by extensive computer simulations [2] and by field tests performed with a TD-CDMA hardware demonstrator [3]. TD-CDMA forms part of the third generation mobile radio standard for UMTS adopted by ETSI, and is intensively studied within the standardization activities of the third generation partnership project (3GPP) [4], which support the works towards IMT-2000. Important pros of TD-CDMA are its backward compatibility with GSM, its suitability for both the duplexing scheme TDD (Time Division Duplex) to be applied in the unpaired bands and the duplexing scheme FDD (Frequency Division Duplex) to be applied in the paired bands provided for UMTS, and its high spectrum efficiency and capacity [2]. As another advantage of TD-CDMA, it easily lends itself to the utilization of adaptive antennas because, due to its TDMA component, the number of user signals to be simultaneously treated is much smaller than in CDMA systems which work without TDMA. As shown in [5], the benefits achievable by adaptive antennas are mainly based on the three physical effects of path separation, introduction of additional paths, and reuse of paths. In addition, a classification of multi-antenna configurations was presented in [5] and the types of diversity, namely directional diversity, field component diversity as well as spatial micro and macro diversity, achievable by each configuration was explained. In this paper, recently gained insights into the specific benefits achievable by adaptive antennas in TD-CDMA will be given. These benefits concern both channel estimation and data detection in TD-CDMA. Further, computer simulation results support the theoretical analysis and illustrate impressively the benefits of adaptive antennas in TD-CDMA mobile radio systems.

The paper is organized as follows: In Section 4.2.2, mobile radio propagation and channel modeling issues associated with systems using adaptive antennas are given. Section 4.2.3 deals with advanced channel estimation techniques developed for TD-CDMA. In Section 4.2.4 the benefits of adaptive antennas associated with the data detection are presented. Section 4.2.5 deals with the advantages of employing information about the intercell interference in the receivers of TD-CDMA mobile radio systems. In Section 4.2.6, the theoretical analysis of Sections 4.2.2 to 4.2.4 is supported by computer simulation results performed in the uplink of TD-CDMA. Finally, Section 4.2.7 summarizes the basic results and concludes the paper.

4.2.2 Mobile Radio Propagation

4.2.2.1 General

In a mobile radio system where a single antenna is used in both the BS and the MS, the channel between the input of the transmitter antenna and the output of the receiver antenna can be described by a channel impulse response, the lowpass equivalent of which is denoted by $\underline{h}(\tau, t)$. $\underline{h}(\tau, t)$ originates from a weighted superposition of the impulse responses of a selection of paths, through which part of the electromagnetic energy radiated by the transmitter reaches the receiver. The weighting and the selection of paths is determined by the employed antennas [5]. The channel impulse response $\underline{h}(\tau, t)$ is spread along the delay axis, i.e., delay dispersion occurs, and is time variant [6]. $\underline{h}(\tau, t)$ is related via Fourier transformation to a time variant transfer function $\underline{H}(f, t)$, which is frequency selective on account of the delay dispersion of $\underline{h}(\tau, t)$. $\underline{h}(\tau, t)$ and $\underline{H}(f, t)$ are lump manifestations of the active paths of the propagation environment as seen through the employed antennas [5]. In [5] differential directional channel impulse responses are defined for obtaining a theoretical description of the propagation environment between the transmitter and receiver locations. With the antenna characteristics and these differential directional channel impulse responses, the time variant impulse response $\underline{h}(\tau, t)$ between the input of the transmitter antenna and the output of the receiver antenna can be determined [5]. Knowledge of the properties of the mobile radio channel, i.e., ideally about the above mentioned differential channel impulse responses, is important in two different respects. On the one hand, such knowledge is required for designing mobile radio systems in an optimum way for the environments in which they have to work and for comparing different system concepts on a realistic basis. The information required for these purposes is usually obtained by channel sounders (CS) [7] in expensive measurement campaigns [8, 9, 10] and by evaluation of the measurement results, but also by implementing models which imitate the geometrical and electrical conditions of the considered scenario on the computer [11, 12]. On the other hand, when operating a mobile radio network, the valid channel properties have to be estimated continuously in order to be able to adapt the transmitter and/or receiver including their antennas to the channel and thus optimize transmission with the ultimate aim to achieve a large capacity. To solve this task, channel estimation devices are required which work in real time [13].

Usually, not the total knowledge on the channel, i.e., not the above introduced differential channel impulse responses, but rather a partial knowledge becomes available by channel measurements as well as by channel estimation devices [5]. Furthermore, carrying out channel measuring campaigns [8, 9, 10, 14] is rather expensive. Therefore, it is sometimes attractive to establish channel models on the computer to represent the behavior of real world mobile radio channels. The important research area of channel modeling is the subject of the next section.

4.2.2.2
Channel Modeling

Worldwide research activities have aimed at the modeling of the mobile radio channel including its directional characteristics [15, 16, 17, 18, 19, 20, 10]. These activities were accompanied and supported by channel sounding campaigns [8, 9, 10, 14], the results of which were useful for the parametrization of the established models. These established models can be divided into two general categories:

- models for propagation prediction, and
- models generating channel impulse responses which retain the properties of the mobile radio channel.

Propagation prediction models aim at the development of models which provide an accurate estimate of the mean received power or path loss (PL) based on geographical information about the propagation environment [15]. These models are very helpful for mobile radio service providers for planning their networks because they allow optimization of the cell coverage while minimizing interference effects [15]. Within this category three model classes can be identified: empirical, deterministic, and semi-deterministic models [15]. Empirical models consist of diagrams or equations for PL calculation which are obtained from statistical analysis of a large number of measurements. Such a model is the well-known Hata empirical model [21]. Deterministic models are established by two approaches. First, a PL equation is provided which is derived based on an idealized environment, retaining some basic features of the physical one in order to make a theoretical treatment possible [15]. Such models can be found in [22, 23]. Second, geographical information about the investigated site supplied by means of digital maps is exploited to predict the dominant propagation paths and their loss. Some of the models of this kind can be found in [24, 25]. Finally, semi-deterministic models result from an empirical modification of deterministic models to improve the agreement with measurements [15]. The most well-known model of this class is the COST231-Walfish-Ikegami semi-deterministic model [26].

The second category of channel models aim at the generation of channel impulse responses which retain the properties of the mobile radio channel including its directional characteristics [15, 5]. The models of this category can be divided into two classes:

- algorithmic channel models, and
- geometry-based channel models.

In algorithmic channel models, the mobile radio channels are represented by tapped delay line arrangements, see [5] for a detailed analysis. In geometry-based channel models, the model of the wave propagation in the real mobile radio channel is based on geographical, topographical and morphological data of the considered scenario [11, 16, 19, 10]. In the following section, the geometry-based channel models used in the paper, which include the directional characteristics of the mobile radio channel, are descibed.

4.2.2.3
The UKL 2 Directional Channel Models

In this section, the channel models used in the paper are described. These models are called UKL (University of KaisersLautern) 2 directional channel models and they are geometry-based channel models. The UKL 2 directional channel models are an evolution of the UKL directional channel models presented in [16, 12] and they are appropriate for investigating the behaviour of TD-CDMA using adaptive antennas at the BS receiver [27, 28].

In addition to the advantages offered by the UKL channel models, the UKL 2 directional channel models enable the modeling of the movement of a MS in a scenario and they considerably reduce the number of necessary scatterers for producing the channels. In the sequel, the main topics connected with the UKL 2 directional channel models are described.

The transmission channel between a MS and a BS with single antennas is considered in the following. The channel impulse response at a time instance t is created as a linear combination of the impulse responses of individual echoes, the number K_{e} of which is much greater than one [12, 29, 30]. Further, only those echoes are considered, which are produced by single reflection, single scattering, or single diffraction [16, 12]. Moreover, a line of sight connection between the MS and the BS is excluded. A time delay $\tau^{(k_{\mathrm{e}})}$, a Doppler shift $f_{\mathrm{D}}^{(k_{\mathrm{e}})}$, a zero phase $\vartheta^{(k_{\mathrm{e}})}$, a direction of departure $\varphi_{\mathrm{e}}^{(k_{\mathrm{e}})}$, a direction of arrival $\varphi_{\mathrm{e}}^{(k_{\mathrm{e}})}$, and

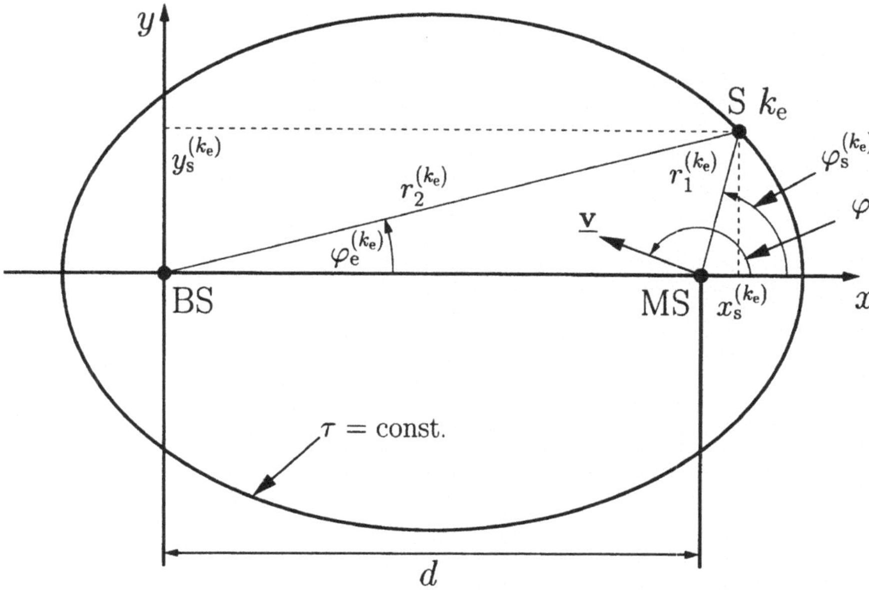

Fig. 4.2.1. Definition of model parameters for the UKL 2 directional channel models

an attenuation $\alpha^{(k_e)}$ are assigned to each echo k_e, $k_e = 1 \ldots K_e$ [12]. The above defined quantities can be directly determined by the position $(x_S^{(k_e)}, y_S^{(k_e)})$ of the scatterer (S) k_e, $k_e = 1 \ldots K_e$, relative to the position of the MS and the BS in the coordinate system of Fig. 4.2.1 [12]. In Fig. 4.2.1, d is the distance between the MS and the BS, $r_1^{(k_e)}$ is the distance between the MS and the scatterer k_e, $r_2^{(k_e)}$ is the distance between the scatterer k_e and the BS, $\underline{\mathbf{v}}$ is the velocity vector of the MS, and φ_V is the angle characterizing the direction of movement of the MS [12]. If the attenuation for each echo is given by [12]

$$a^{(k_e)} = \frac{\alpha_0}{\left(r_1^{(k_e)} r_2^{(k_e)}\right)^{\alpha/2}}, \qquad k_e = 1 \ldots K_e, \tag{1}$$

i.e., all K_e scattering processes create the constant attenuation α_0, then the directional channel impulse response density at a reference point (RP) of the BS is given by [12]

$$\vartheta(\tau, t, \varphi) = \sum_{k_e=1}^{K_e} \frac{a_0}{r_1^{(k_e)} r_2^{(k_e)}} \cdot e^{j\left(2\pi f_D^{(ke)} t + \vartheta^{(ke)}\right)} \cdot \delta\left(\tau - \tau^{(k_e)}\right) \cdot \delta\left(\varphi - \varphi_e^{(k_e)}\right). \tag{2}$$

In the case where sectorized antenna configurations or centro-symmetric array configurations are used at the BS and K users are simultaneously active, a

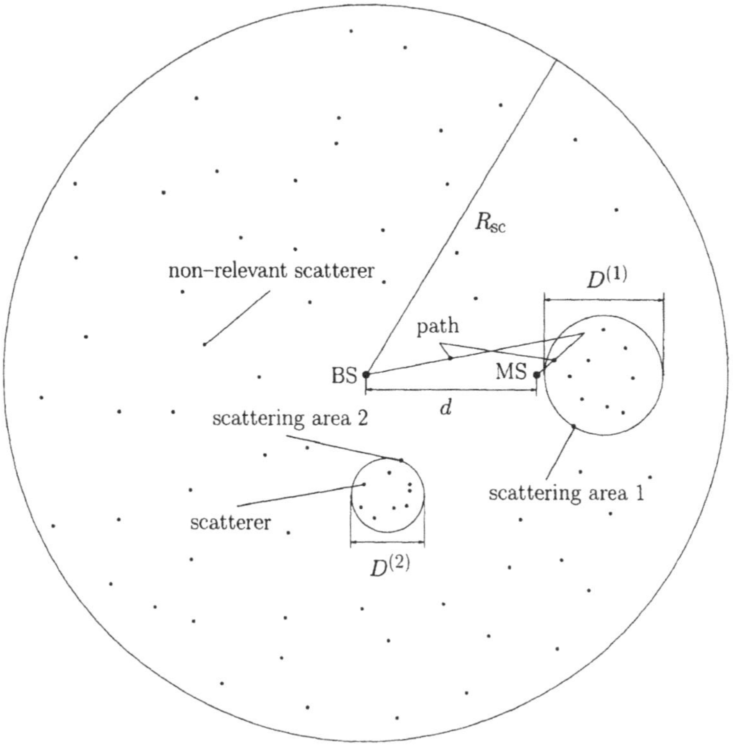

Fig. 4.2.2. Hierarchical division of the propagation environment into non-relevant scatterers, scattering areas, and scatterers

directional channel impulse response density $\vartheta^{(k,k_a)}(\tau, t, \varphi)$, for each user k, $k = 1 \ldots K$, and each antenna k_a, $k_a = 1 \ldots K_a$, can be defined by taking into account the antenna characteristics or the antenna positions relative to the RP of the BS, respectively [12].

In order to model propagation geometrically, a circle of radius R_{sc}, called scenario area and representing an instantaneous scenario, is covered with scatterers uniformly chosen in the (x, y) plane, see Fig. 4.2.2. The position of the scatterers within the scenario area is not changed. The BS is placed at the center of the scenario area and the position of each MS is randomly distributed within the scenario area. Then, the scenario area is hierarchically subdivided into:

1. Regions with scatterers which are not relevant for the link between the considered MS and the BS.

2. N_z circular scattering areas each of diameter D_{n_z}, $n_z = 1 \dots N_z$. The scattering areas may represent, e.g., urban or suburban buildings. The scattering area 1 with diameter D_1 is always placed around the considered MS.
3. $N_p{}^{(n_z)}$ scatterers which are defined by each circular scattering area n_z, $n_z = 1 \dots N_z$. These scatterers model, e.g., buildings from which scattered electromagnetic waves originate.

An example of the hierarchy of the UKL 2 directional channel models is illustrated in Fig. 4.2.2 with

$$N_z = 2, \quad N_p{}^{(1)} = 9, \quad N_p{}^{(2)} = 8. \tag{3}$$

The above mentioned model allows the movement of the MS and the scattering area 1 placed around the MS. As the MS takes another position within the scenario area, the scattering area 1 defines a new number of scatterers $N_p{}^{(1)}$, which are relevant for the link between the considered MS and the BS. If there are additional scattering areas within the scenario area which do not pertain to a certain user, these scattering areas do not change their position as the MS moves within the scenario area [27, 28]. Compared to the UKL channel models defined in [16, 12], the UKL 2 directional channel models

- allow the modeling of the movement of the MS by defining a scenario area and descriminating between non-relevant scatterers and relevant scatterers for a considered link, and
- do not contain the hierarchy level of circular scatterers with scattering points of the UKL channel models [16], thus reducing considerably the number of propagation paths, and consequently the simulation time for producing directional channel impulse responses, retaining simultaneously the accuracy of the UKL channel models concerning the characteristics of the produced directional channel impulse responses [27, 28].

In the rest of this section model parameters for three typical macrocellular propagation environments are proposed. These environments are the rural, the urban, and the dense urban propagation environment. The constellation of the scattering areas for these environments are shown in Fig. 4.2.3 and correspond to the model types rural, urban 1, and urban 2 of the UKL channel models [16, 12]. As already mentioned above in this section, the scattering area 1, which is always centered around the MS, follows the movement of the MS, whereas the position of the additional scattering area 2, if present, remains unchanged while the MS moves. The proposed parameters for the distances and the number of scatterers are shown in Table 4.2.1, where $\bar{N}_p^{(n_z)}$, $n_z = 1 \dots N_z$, is the mean number of scatterers of scattering area n_z. In the next section of this chapter, a novel approach for a unified treatment of the mobile radio propagation in macrocell environments is presented.

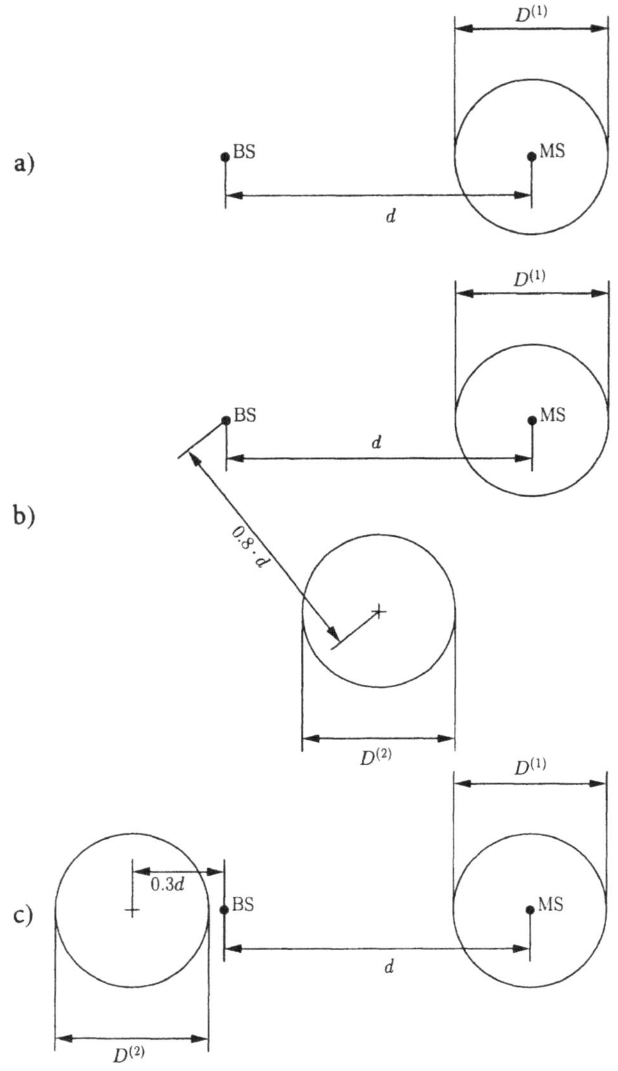

Fig. 4.2.3. Constellation of the scattering areas for the models: **a)** rural, **b)** urban, **c)** dense urban.

Table 4.2.1. Parameters for the models rural, urban und dense urban

	$\frac{R_{sc}}{km}$	N_z	$\frac{D^{(1)}}{km}$	$\frac{D^{(2)}}{km}$	N_P	$\bar{N}_P^{(1)}$	$\bar{N}_P^{(2)}$
rural	5	1	0.1	-	6000	10	-
urban	2	2	0.5	0.25	1500	20	10
dense urban	2	2	0.5	0.25	1500	20	10

4.2.2.4
A Novel Approach for a Unified Treatment of the Different Propagation Environments

4.2.2.4.1
General

The channel impulse responses are typical of the environment and the mobility scenarios of the MSs. For each mobile radio scenario which is determined by the topology and the morphology of the considered environment, the cell and cluster sizes, and the distribution of BSs and MSs, these channel impulse responses have typical properties. In mobile radio communications each element of the total discrete time channel impulse response $\underline{\mathbf{h}}$ in the equivalent lowpass domain is usually made up by partial waves leaving the transmitter and impinging at the receiver with a variety of directions at the receiver locations. As mentioned Section 4.2.2.1, the channel impulse responses can be expressed by time variant differential directional channel impulse responses in combination with the characteristics of the applied antennas. Further, the mobile radio propagation including directional information can be described by directional channel models, see Section 4.2.2.3, which are useful for evaluating systems with adaptive antennas. In this paper, in order to evaluate adaptive antenna concepts in the TD-CDMA air interface concept, we assume a general classification into large cell and small cell channel types, which are of special interest for mobile communications. In oder to define the large cell and the small cell channel type, it is assumed that the full solid angle domain 4π is discretized into $K_s >> 1$ differential elements of size $4\pi/K_s$. In general, a directional channel impulse response is characterized by $K_s W$ complex elements. In the sequel, the characteristics of the large cell and the small channel type are described.

4.2.2.4.2
Large Cell Channel Type

For the large cell channel type it is assumed that the multipath propagation arises due to ${K_d}^{(k)}$, $k = 1 \ldots K$, scattering areas for each of the K MSs. Each scattering area contains uniformly distributed local scatterers and is modeled by a circle of radius $\rho^{(k_d)}$, $k_d = 1 \ldots {K_d}^{(k)}$. The radius $\rho^{(k_d)}$ of each scattering area is assumed to be small compared to the distance $D^{(k_d)}$, $k_d = 1 \ldots {K_d}^{(k)}$, of its center to the BS site. Therefore, in large cells it can be assumed that signals coming from a MS impinge at the BS only from a limited number $K_d^{(k)} << K_s$ of different directions. In this case, a directional channel impulse response is determined by $K_d^{(k)} W$ complex elements. In Fig. 4.2.4 a scenario with a single MS is depicted where there are ${K_d}^{(k)} = 2$ scattering areas, i.e., two directional channel impulse responses arise at the BS receiver. The UKL 2

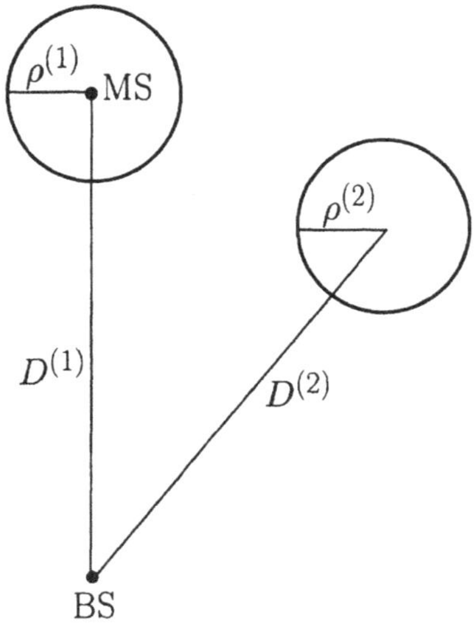

Fig. 4.2.4. Examplary scenario of a large-cell channel type

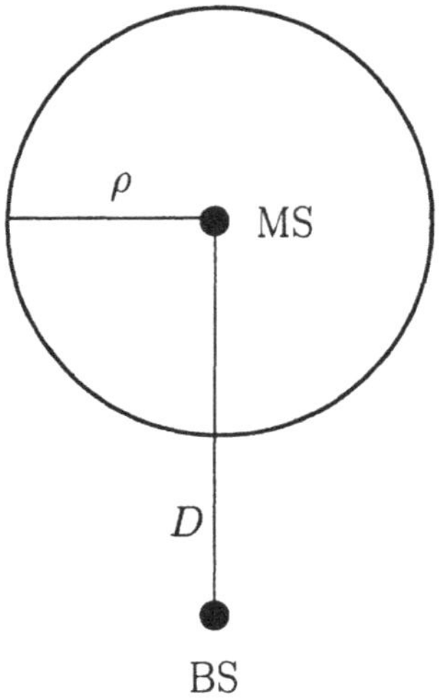

Fig. 4.2.5. Examplary scenario of a small-cell channel type

rural and the UKL 2 urban channel models, see Section 4.2.2.3, belong to this channel type and, as it will be presented in Section 4.2.3, they can be treated in the same manner in TD-CDMA mobile radio systems when channel estimation is performed.

4.2.2.4.3
Small Cells

In the small cell channel type it is assumed that the radius $\rho^{(k)}$ of each user k, $k = 1 \ldots K$, of the single scattering area, see Fig. 4.2.5, is in the order of the distance between the MS and the BS. Therefore, it is reasonable that each of the W elements of a channel impulse response is associated with only a single direction of incidence, see also [20] for a similar assumption. In this case, a directional channel impulse response is determined by only W complex elements. The UKL 2 dense urban channel model belongs to the small cell channel type, because of the directional characteristics of the produced channel impulse responses. As for the models of the large cell channel type, the models of the small channel type are treated in the same manner when channel estimation is performed in TD-CDMA, see Section 4.2.3.

4.2.3
Advanced Channel Estimation Techniques for TD-CDMA

4.2.3.1
Steiner Estimator

Prior to presenting the advances in channel estimation techniques for TD-CDMA, the channel estimation technique which is already state of the art in TD-CDMA is briefly discussed. This technique has been proposed by Steiner in [13] and has been verified by extensive field tests performed with a TD-CDMA hardware demonstrator [3]. Steiner developed a technique for jointly estimating the channel impulse responses between the links of each active user in the considered cell and a single antenna at the receiver. By taking advantage of the midamble section inserted between the two data sections of the burst transmitted by each user [1], the novelty of the Steiner estimator consists in the design process of the midamble training sequences known at the receiver. If the midamble codes of all users are derived from a single periodic basic code, the channel estimator can be implemented as a low cost cyclic correlator [13]. In general, the channel estimation in TD-CDMA for a single antenna element at the receiver can be described mathematically as follows. The received signal $\underline{\mathbf{e}}_{\mathrm{m}}$ of length L, which depends exclusively on the transmitted midamble section and not on the data sections, is given by

$$\underline{\mathbf{e}}_{\mathrm{m}} = \underline{\mathbf{G}}\,\underline{\mathbf{h}} + \underline{\mathbf{n}}_{\mathrm{m}}, \tag{4}$$

where $\underline{\mathbf{h}}$ is the total channel impulse response vector of length KW from all K users, the matrix $\underline{\mathbf{G}}$ contains the midamble training sequences of all K users, and $\underline{\mathbf{n}}_{\mathrm{m}}$ represents the additive intercell MAI vector valid for the channel estimation. It is noted that generally L is slighty larger than KW, which means that the system of equations in (4) is slightly overdetermined. From (4), according to the maximum-likelihood principle, the estimate of $\underline{\mathbf{h}}$ given by

$$\hat{\underline{\mathbf{h}}} = \left(\underline{\mathbf{G}}^{*\mathrm{T}}\underline{\mathbf{R}}_{\mathrm{t}}^{-1}\underline{\mathbf{G}}\right)^{-1}\underline{\mathbf{G}}^{*\mathrm{T}}\underline{\mathbf{R}}_{\mathrm{t}}^{-1}\underline{\mathbf{e}}_{\mathrm{m}} \tag{5}$$

follows [13], where $\underline{\mathbf{R}}_{\mathrm{t}}$ denotes the covariance matrix of the noise vector $\underline{\mathbf{n}}_{\mathrm{m}}$ in (4) [31]. In [31], it is shown that the Steiner estimator can be generalized to take into acount the use of multi-antenna configurations at the BS receiver. In this case, the total received vector $\underline{\mathbf{e}}_{\mathrm{m}}$ from all antennas can be written as

$$\underline{\mathbf{e}}_{\mathrm{m}} = \left(\mathbf{I}^{(K_{\mathrm{a}})} \otimes \underline{\mathbf{G}}\right)\,\underline{\mathbf{h}} + \underline{\mathbf{n}}_{\mathrm{m}}, \tag{6}$$

where now $\underline{\mathbf{h}}$ is the total channel impulse response vector of length $K_{\mathrm{a}}KW$ valid for all the links between each user and each antenna, $\underline{\mathbf{n}}_{\mathrm{m}}$ is the total received noise vector from all K_a antennas, and $\otimes$ denotes the Kronecker product [32]. The estimate of the total channel impulse response $\underline{\mathbf{h}}$, see (6), is now given by [31]

$$\hat{\underline{\mathbf{h}}} = \left[\mathbf{I}^{(K_a)} \otimes \left(\underline{\mathbf{G}}^{*\mathrm{T}}\underline{\mathbf{R}}_{\mathrm{t}}^{-1}\underline{\mathbf{G}}\right)^{-1} \underline{\mathbf{G}}^{*\mathrm{T}}\underline{\mathbf{R}}_{\mathrm{t}}^{-1}\right] \underline{\mathbf{e}}_{\mathrm{m}}, \tag{7}$$

which expresses the generalization of the Steiner estimator, see (5), when multi-antenna configurations are used at the BS receiver. In the subsequent parts of this section, two novel channel estimation techniques, one for the large cell and one for the small cell channel type, are presented, when centro-symmetric array configurations [31, 33] are employed at the BS receiver.

4.2.3.2
Large Cell Channel Type

As assumed in Section 4.2.2.4 for the large cell channel type, there are ${K_\mathrm{d}}^{(k)}$ DOAs per user k, $k = 1 \ldots K$, see Fig. 4.2.4. Therefore, a channel impulse response $\underline{\mathbf{h}}_{\mathrm{d}}^{(k)}$ valid for a user k, $k = 1 \ldots K$, is determined by ${K_\mathrm{d}}^{(k)}W$ complex elements. Following the considerations of [31, 33], we define a total directional channel impulse response

$$\underline{\mathbf{h}}_{\mathrm{d}} = \left[\underline{\mathbf{h}}_{\mathrm{d}}^{(1)\mathrm{T}} \underline{\mathbf{h}}_{\mathrm{d}}^{(2)\mathrm{T}} \ldots \underline{\mathbf{h}}_{\mathrm{d}}^{(K)\mathrm{T}}\right]^{\mathrm{T}}, \tag{8}$$

which has $K_\mathrm{d}W$ complex elements, where

$$K_\mathrm{d} = \sum_{k=1}^{K} {K_\mathrm{d}}^{(k)}. \tag{9}$$

If $\underline{\mathbf{A}}_{\mathrm{d,l}}$ contains the information about the DOAs of the directional channel impulse responses of all K users [31, 33], the total channel impulse response vector $\underline{\mathbf{h}}$ of (6) can be written in the following fashion, see also [31, 33]:

$$\underline{\mathbf{h}} = \underline{\mathbf{A}}_{\mathrm{d,l}}\, \underline{\mathbf{h}}_{\mathrm{d}}. \tag{10}$$

Further, the total received vector from all antennas now becomes

$$\underline{\mathbf{e}}_{\mathrm{m}} = \left(\mathbf{I}^{(K_a)} \otimes \underline{\mathbf{G}}\right)\, \underline{\mathbf{A}}_{\mathrm{d,l}}\, \underline{\mathbf{h}}_{\mathrm{d}} + \underline{\mathbf{n}}_{\mathrm{m}}. \tag{11}$$

In the large cell channel type, it can be assumed that

$$K_\mathrm{d} < K_\mathrm{a}K \tag{12}$$

holds, see also [34]. Therefore, we observe from (11) and (10) that the number of unknown complex elements contained in $\underline{\mathbf{h}}_{\mathrm{d}}$ has been reduced to $K_\mathrm{d}W$, see

(9), compared to the number $K_a KW$ of unknown complex elements of $\underline{\mathbf{h}}$ in (6). Since the number of equations in both (6) and (11) is the same, this reduction of the number of the unknown channel impulse reponse components leads to an improved quality of the channel estimates, see also [31, 33]. According to the maximum-likelihood principle, the channel impulse response estimate of $\underline{\mathbf{h}}_{\mathrm{d}}$ is obtained from (11) as

$$\hat{\underline{\mathbf{h}}}_{\mathrm{d}} = \left[\underline{\mathbf{A}}_{\mathrm{d,l}}^{*\mathrm{T}}(\mathbf{I}^{(K_a)} \otimes \underline{\mathbf{G}}^{*\mathrm{T}})\underline{\mathbf{R}}_{\mathrm{m}}^{-1}(\mathbf{I}^{(K_a)} \otimes \underline{\mathbf{G}})\underline{\mathbf{A}}_{\mathrm{d,l}}\right]^{-1} \underline{\mathbf{A}}_{\mathrm{d,l}}^{*\mathrm{T}}(\mathbf{I}^{(K_a)} \otimes \underline{\mathbf{G}}^{*\mathrm{T}})\underline{\mathbf{R}}_{\mathrm{m}}^{-1}\underline{\mathbf{e}}_{\mathrm{m}}, \quad (13)$$

see also (7). Here, $\underline{\mathbf{R}}_{\mathrm{m}}$ denotes the covariance matrix of the total noise vector $\underline{\mathbf{n}}_{\mathrm{m}}$ from all K_a array elements. In Section 4.2.6, the influence of this improved channel estimation on the overall system performance is demonstrated by simulation results for large cell channel types.

4.2.3.3
Small Cell Channel Type

In the small cell channel type, see Section 4.2.2.4, it is assumed that there is a single directional channel impulse response $\underline{\mathbf{h}}_{\mathrm{d}}^{(k)}$ for each user k, $k = 1 \dots K$. As in Section 4.2.3.2, see (8), we define a total directional channel impulse reponse

$$\underline{\mathbf{h}}_{\mathrm{d}} = \left[\underline{\mathbf{h}}_{\mathrm{d}}^{(1)\mathrm{T}} \underline{\mathbf{h}}_{\mathrm{d}}^{(2)\mathrm{T}} \dots \underline{\mathbf{h}}_{\mathrm{d}}^{(K)\mathrm{T}}\right]^{\mathrm{T}}, \quad (14)$$

which has KW complex elements. If $\underline{\mathbf{A}}_{\mathrm{d,s}}$ contains the information about the DOAs associated with each component of the directional channel impulse response of all K users, the total channel impulse response vector $\underline{\mathbf{h}}$ of (6) can be expressed as

$$\underline{\mathbf{h}} = \underline{\mathbf{A}}_{\mathrm{d,s}}\, \underline{\mathbf{h}}_{\mathrm{d}}, \quad (15)$$

where $\underline{\mathbf{h}}_{\mathrm{d}}$ is defined in (14). Then, the total receiver vector $\underline{\mathbf{e}}_{\mathrm{m}}$ from all antennas is expressed in this case as

$$\underline{\mathbf{e}}_{\mathrm{m}} = \left(\mathbf{I}^{(K_a)} \otimes \underline{\mathbf{G}}\right)\, \underline{\mathbf{A}}_{\mathrm{d,s}}\, \underline{\mathbf{h}}_{\mathrm{d}} + \underline{\mathbf{n}}_{\mathrm{m}}. \quad (16)$$

Since

$$KW < K_a KW \quad (17)$$

is always valid in the small cell channel type, the number of unknown channel impulse reponse components contained in $\underline{\mathbf{h}}_{\mathrm{d}}$ is reduced compared to the number of unknown channel impulse reponse components contained in $\underline{\mathbf{h}}$, see (15) and (17). Therefore, following the rationale of Section 4.2.3.2, this reduction leads to an improved quality of the channel impulse response estimates in the small cell channel type. Finally, the channel impulse response estimate of $\underline{\mathbf{h}}_{\mathrm{d}}$ is obtained from (16) according to the maximum-likelihood principle as

$$\underline{\hat{\mathbf{h}}}_{\mathrm{d}} = \left[\underline{\mathbf{A}}_{\mathrm{d,s}}^{*\mathrm{T}}(\mathbf{I}^{(K_{\mathrm{a}})} \otimes \underline{\mathbf{G}}^{*\mathrm{T}})\underline{\mathbf{R}}_{\mathrm{m}}^{-1}(\mathbf{I}^{(K_{\mathrm{a}})} \otimes \underline{\mathbf{G}})\underline{\mathbf{A}}_{\mathrm{d,s}}\right]^{-1} \underline{\mathbf{A}}_{\mathrm{d,s}}^{*\mathrm{T}}(\mathbf{I}^{(K_{\mathrm{a}})} \otimes \underline{\mathbf{G}}^{*\mathrm{T}})\underline{\mathbf{R}}_{\mathrm{m}}^{-1}\underline{\mathbf{e}}_{\mathrm{m}}. \quad (18)$$

In Section 4.2.6, the improvement of the system performance introduced by the improved channel estimation in small cell channel types is demonstrated by simulation results.

4.2.4 Joint Data Detection

4.2.4.1 ZF-BLE

The multiple access scheme and the details of the burst and frame structures applied in TD-CDMA are well known [31, 35]. With K the number of users simultaneously active within a cell in the same frequency band, each burst of user k, $k = 1 \ldots K$, consists of two data sections $\underline{\mathbf{d}}^{(k)}$ with N data symbols each, and a midamble inserted between these data sections, which allows channel estimation at the receivers [13] and, subsequently, data detection by the JD algorithm [1]. Prior to transmission, the data symbols are spectrally spread by the user specific CDMA codes $\underline{\mathbf{c}}^{(k)}$, $k = 1 \ldots K$, of length Q. It is assumed that K_{a} antennas are utilized at the BS and only one antenna at each MS. Consequently, the mobile radio channel can be characterized by KK_{a} channel impulse responses $\underline{\mathbf{h}}^{(k,k_{\mathrm{a}})}$, $k = 1 \ldots K$, $k_{\mathrm{a}} = 1 \ldots K_{\mathrm{a}}$, see Fig. 4.2.6. As shown for instance in [1], a system matrix $\underline{\mathbf{A}}$ can be established, which is determined by the transmitter and receiver filters and by the total vectors $\underline{\mathbf{c}}$ and $\underline{\mathbf{h}}$. $\underline{\mathbf{A}}$ is a $K_{\mathrm{a}}(NQ + W - 1) \times KN$ matrix in the uplink case. Setting out from $\underline{\mathbf{A}}$, from the total data vector $\underline{\mathbf{d}}$ and from the total intercell MAI vector $\underline{\mathbf{n}}$, we can represent

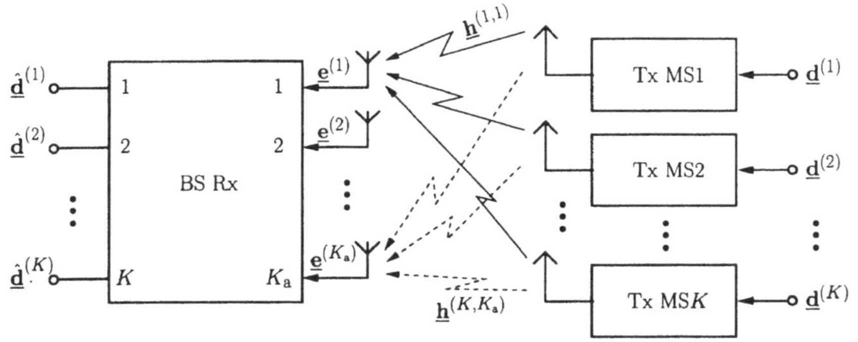

Fig. 4.2.6. System model for the uplink of TD-CDMA

the total received signal in the form of a total vector

$$\underline{\mathbf{e}} = \underline{\mathbf{A}}\,\underline{\mathbf{d}} + \underline{\mathbf{n}}. \tag{19}$$

In this equation, $\underline{\mathbf{e}}$ is known at the receiver. This is also approximately true for the system matrix $\underline{\mathbf{A}}$ with the accuracy of $\underline{\mathbf{A}}$ depending on the accuracy of the channel estimates, see Section 4.2.3. The vectors $\underline{\mathbf{d}}$ and $\underline{\mathbf{n}}$ are unknown. However, the covariance matrix

$$\underline{\mathbf{R}}_{\mathrm{n}} = \mathrm{E}\left\{\underline{\mathbf{n}}\,\underline{\mathbf{n}}^{*\mathrm{T}}\right\} \tag{20}$$

of $\underline{\mathbf{n}}$ shall be assumed to be known. The task of data detection consists in determining $\underline{\mathbf{d}}$ from (19). This equation can be considered as a linear system of equations for the KN unknown components of $\underline{\mathbf{d}}$. In order to enable a solution, this system of equations has to be determined or over-determined. Then, by applying the zero-forcing algorithm to perform JD [1, 34], a linear estimate

$$\hat{\underline{\mathbf{d}}} = \left(\underline{\mathbf{A}}^{*\mathrm{T}}\underline{\mathbf{R}}_{\mathrm{n}}^{-1}\underline{\mathbf{A}}\right)^{-1}\underline{\mathbf{A}}^{*\mathrm{T}}\underline{\mathbf{R}}_{\mathrm{n}}^{-1}\underline{\mathbf{e}} = \underline{\mathbf{M}}\,\underline{\mathbf{e}} \tag{21}$$

for $\underline{\mathbf{d}}$ can be obtained from (19), where

$$\underline{\mathbf{M}} = \left(\underline{\mathbf{A}}^{*\mathrm{T}}\underline{\mathbf{R}}_{\mathrm{n}}^{-1}\underline{\mathbf{A}}\right)^{-1}\underline{\mathbf{A}}^{*\mathrm{T}}\underline{\mathbf{R}}_{\mathrm{n}}^{-1}. \tag{22}$$

If the expression (19) for $\underline{\mathbf{e}}$ is substituted into (21), the equation

$$\hat{\underline{\mathbf{d}}} = \underline{\mathbf{d}} + \underline{\mathbf{M}}\,\underline{\mathbf{n}} = \underline{\mathbf{d}} + \underline{\mathbf{n}}_{\mathrm{out}} \tag{23}$$

is valid, where

$$\underline{\mathbf{n}}_{\mathrm{out}} = \underline{\mathbf{M}}\,\underline{\mathbf{n}}. \tag{24}$$

According to (24), the covariance matrix of the vector $\underline{\mathbf{n}}_{\mathrm{out}}$, under the assumption that the covariance matrix $\underline{\mathbf{R}}_{\mathrm{n}}$ of the input intercell MAI vector $\underline{\mathbf{n}}$ equals the identity matrix, is given by

$$\underline{\mathbf{R}}_{\mathrm{n,out}} = \mathrm{E}\left\{\underline{\mathbf{n}}_{\mathrm{out}}\,\underline{\mathbf{n}}_{\mathrm{out}}^{*\mathrm{T}}\right\} = \left(\underline{\mathbf{A}}^{*\mathrm{T}}\underline{\mathbf{A}}\right)^{-1}. \tag{25}$$

It can be observed from (25) that the matrix

$$\underline{\mathbf{N}} = \underline{\mathbf{A}}^{*\mathrm{T}}\underline{\mathbf{A}} \tag{26}$$

plays a fundamental role in the system performance.

In Fig. 4.2.7, the structure of a block $\underline{\mathbf{A}}^{(k,k_\mathrm{a})}$ is shown with the respective dimensions. Because of the structure of the system matrix $\underline{\mathbf{A}}$, the matrix $\underline{\mathbf{N}}$ of (26) is Hermitian and consists of K^2 blocks $\underline{\mathbf{N}}^{(k_1,k_2)}$, $k_1 = 1\ldots K$, $k_2 = 1\ldots K$,

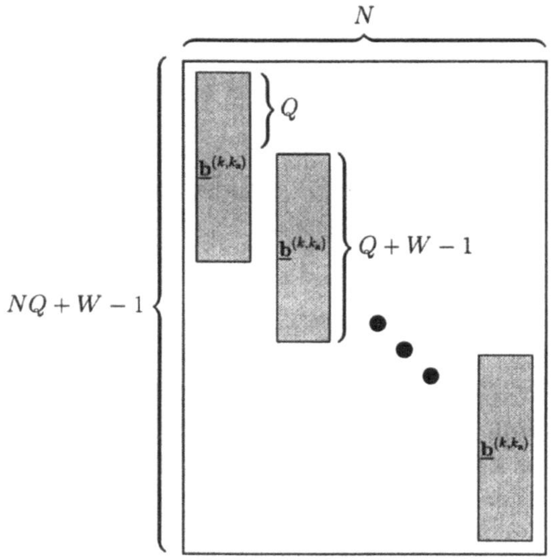

Fig. 4.2.7. Structure of a block $\underline{\mathbf{A}}^{(k,k_a)}$ of the system matrix $\underline{\mathbf{A}}$

of dimensions $N \times N$ each. The diagonal blocks $\underline{\mathbf{N}}^{(k,k)}$, $k = 1 \ldots K$, are Toeplitz and Hermitian matrices, whereas the off-diagonal blocks $\underline{\mathbf{N}}^{(k_1,k_2)}$, $k_1 \neq k_2$, are only Toeplitz. Moreover, since

$$NQ + W - 1 >> Q + W - 1 \tag{27}$$

in TD-CDMA, the blocks $\underline{\mathbf{N}}^{(k_1,k_2)}$, $k_1 = 1 \ldots K$, $k_2 = 1 \ldots K$, of (26) are rather sparse matrices. Now, we proceed with the explanation of the influence of the matrix $\underline{\mathbf{N}}$ on the system performance. If the elements of the off-diagonal blocks $\underline{\mathbf{N}}^{(k_1,k_2)}$, $k_1 \neq k_2$, take large values compared to the values of the elements of the diagonal blocks $\underline{\mathbf{N}}^{(k,k)}$, $k = 1 \ldots K$, i.e., if the structure of $\underline{\mathbf{N}}$ differs considerably from the block diagonal structure, then the diagonal elements of the inverse matrix $\underline{\mathbf{N}}^{-1}$ take also large values and the system performance is degraded. In contrast to this situation, if the elements of the off-diagonal blocks $\underline{\mathbf{N}}^{(k_1,k_2)}$, $k_1 \neq k_2$, take small values compared to the values of the elements of the diagonal blocks $\underline{\mathbf{N}}^{(k,k)}$, $k = 1 \ldots K$, then a structure closer to the block diagonal structure occurs for $\underline{\mathbf{N}}$, and, therefore, the system performs more favourably. In the following section, the conditions under which the system behaviour is favourable in TD-CDMA mobile radio systems with adaptive antennas is investigated.

4.2.4.2
Benefits of Adaptive Antennas

4.2.4.2.1
Antennas

In order to evaluate the benefits of adaptive antennas in TD-CDMA, two types of multiple array arrangements are considered in this paper, which belong to the class of micro structures according to the classification adopted in [5]. In this class, the K_a employed antenna elements are so closely spaced that each of them experiences the same impinging wavefronts, however, with different delays, which depend on the antenna locations and the directions of arrival (DOA) [5]. Two practical antenna configurations which belong to the class of micro structures are the sectorized antennas and the centro-symmetric array configurations. If the sectorized antennas are considered, the K_a antenna elements are virtually at the same location, however, each with a different characteristic covering a different spatial domain [5], see Fig. 4.2.8. It is assumed that each antenna element of the sectorized antenna configuration covers an ideal sector of $4\pi/K_a$. In the centro-symmetric array configurations, e.g., cross array configuration, it is assumed that each of the K_a antenna elements has an omnidirectional characteristic and that the antenna elements are so closely spaced that a wave impinging from DOA (Θ, Φ) arrives at each of the K_a anten-

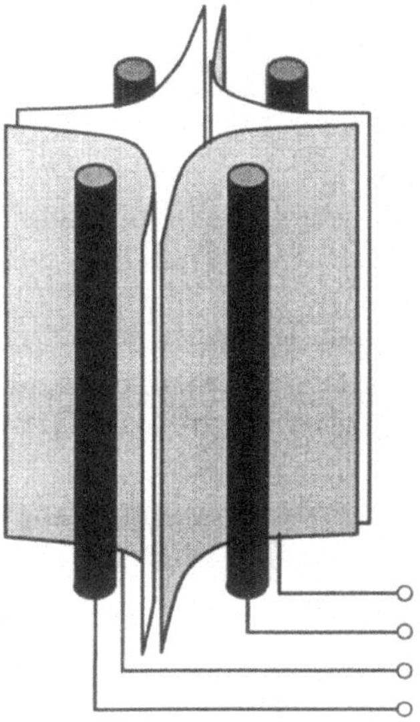

Fig. 4.2.8. Sectorized antenna configuration of $K_a = 4$ antennas

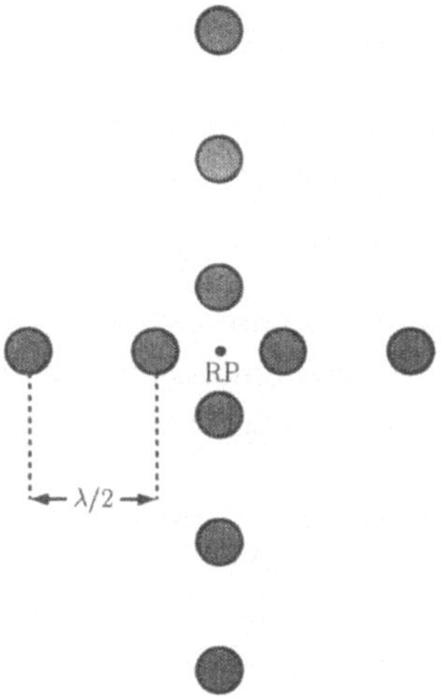

Fig. 4.2.9. Cross configuration with $K_a = 10$ antennas

nas with a different phase shift depending on the DOA (Θ, Φ) and the position of the antenna element relative to a reference point (RP) of the array configuration, see Fig. 4.2.9. In Sections 4.2.4.2.2 and 4.2.4.2.3, the benefits of adaptive antennas in TD-CDMA are investigated for these two configurations and for the two channel types introduced in Section 4.2.2.4, i.e., the large cell and the small cell channel type.

4.2.4.2.2
Sectorized Antenna Configurations

As explained in Section 4.2.4.2.1, each antenna of the sectorized configuration covers an ideal sector of $4\pi/K_a$, where K_a is the total number of antennas. In the large cell channel type, see Section 4.2.2.4, the considered spatial scenario of the impinging user signals has a direct influence on the system behaviour. If the user signals are spatially well separated, i.e., if the mean number of user signals received per antenna is rather small, then the structure of the system matrix $\underline{\mathbf{A}}$, see (19), and consequently of the matrix $\underline{\mathbf{N}}$, see (26), is closer to the block diagonal structure and, therefore, the diagonal elements of $\underline{\mathbf{N}}^{-1}$ take small values and the system behaves favourably. In the case where the user signals are spatially poorly separated — in the worst case, all user signals are received by the same antenna element of the sectorized configuration—, the structure of the matrix $\underline{\mathbf{N}}$ departs considerably from the block diagonal structure and the system behaves unfavourably. However, the TDMA component of TD-CDMA offers the possibility of channel assignment strategies which take into account the directional information of the user signals to be assigned to the different bursts of a TDMA frame, see [35] for a detailed analysis. Such channel assignment strategies lead to a more favourable spatial separation of the users simultaneously active within a TDMA burst. Therefore, a sectorized antenna configuration supported by a rather simple channel assignment strategy [35] will considerably enhance the system performance of TD-CDMA in the large cell channel type.

On the other hand, in the small cell channel type, where large angular spreads normally prevail, see Section 4.2.2.4, the utilization of sectorized antenna configurations normally leads to an unfavourable structure of the matrix $\underline{\mathbf{N}}$, see (26), since increased values of the elements of the off-diagonal blocks $\underline{\mathbf{N}}^{(k_1,k_2)}$, $k_1 \neq k_2$, are observed compared to the values of the elements of the diagonal blocks $\underline{\mathbf{N}}^{(k,k)}$, $k = 1 \ldots K$, see also Section 4.2.4.1. Therefore, in the case of this channel type the use of sectorized antennas is not suggested unless it is accompanied by sophisticated channel assignment strategies.

4.2.4.2.3
Centro-Symmetric Array Configurations

When centro-symmetric array configurations are used in the large cell channel type, see Section 4.2.2.4, there is a rather easy procedure for evaluating the benefits of adaptive antennas in TD-CDMA. Without loss of generality we assume that the number of DOAs $K_{\mathrm{d}}{}^{(k)}$ from user k, $k = 1 \ldots K$, equals one, see Section 4.2.2.4. In the worst case, where all user signals impinge from the same DOA, the centro-symmetric array configurations offer a gain factor of K_{a} compared to single antenna systems. In any other case of the spatial distribution of the user signals in the large cell channel type, an additional gain factor is observed, since the values of the elements of the off-diagonal blocks of $\underline{\mathbf{N}}$, see (26), take smaller values compared to the elements of the diagonal blocks. Certainly, a simple channel assignment strategy within the TDMA frame as the one presented in [35] would result in a more favourable system behaviour, if centro-symmetric array configurations are used.

Opposed to the behaviour of sectorized antenna configurations, centro-symmetric array configurations work satisfactorily also in the small cell channel type. Here, the elements of the off-diagonal blocks $\underline{\mathbf{N}}^{(k_1,k_2)}$, $k_1 \neq k_2$, are kept small, since it is not probable for all W DOAs of different users in the small cell channel type to be closely spatially situated. Moreover, good crosscorrelation properties of the CDMA codes $\underline{\mathbf{c}}^{(k)}$, $k = 1 \ldots K$, lead to small values of the elements of the off-diagonal blocks even in the case, when the DOAs of different users are closely spatially situated. In Section 4.2.6, the benefits of centro-symmetric array cofigurations in TD-CDMA are demonstrated by simulation results.

4.2.5
Consideration of the MAI Covariance Matrix $\underline{\mathbf{R}}_{\mathrm{n}}$

In Section 4.2.4.1, see (21) and (22), it is shown that the knowledge of the covariance matrix $\underline{\mathbf{R}}_{\mathrm{n}}$ of the intercell noise vector $\underline{\mathbf{n}}$, see (19), is required in order to get an estimate $\hat{\underline{\mathbf{d}}}$ of the unknown total data vector $\underline{\mathbf{d}}$. There, the covariance matrix $\underline{\mathbf{R}}_{\mathrm{n}}$ is assumed to be known at the BS receiver. In the case of uncorrelated white Gaussian noise at the different antenna elements of the BS receiver, the matrix $\underline{\mathbf{R}}_{\mathrm{n}}$ equals the identity matrix, and the matrix $\underline{\mathbf{M}}$ in (22) can be written in the form

$$\underline{\mathbf{M}} = \left(\underline{\mathbf{A}}^{*\mathrm{T}}\,\underline{\mathbf{A}}\right)^{-1}\underline{\mathbf{A}}^{*\mathrm{T}}. \tag{28}$$

However, in a TD-CDMA mobile radio system, the intercell MAI is band limited, which leads to time correlations between the components of the received intercell MAI vector $\underline{\mathbf{n}}$, see (19). Moreover, depending on the directional inhomogeneity of the intercell MAI and the geometry of the employed antenna configuration at the BS receiver, there are also spatial correlations between the components of the intercell MAI vectors received at the different antenna elements. Therefore, the covariance matrix $\underline{\mathbf{R}}_\mathrm{n}$ of (20) is no longer the identity matrix, and the more $\underline{\mathbf{R}}_\mathrm{n}$ differs from the identity matrix, the more important it becomes to consider this information in the JD process given by (21). If it is assumed that each single interfering wavefront impinging at the BS receiver originates from independent sources, then the time correlations of the intercell MAI are the same at each antenna element. In this case, the covariance matrix $\underline{\mathbf{R}}_\mathrm{n}$ of (20) can be expressed as the tensorial product of the matrix $\underline{\mathbf{R}}_\mathrm{s}$ which contains the spatial correlations and the matrix $\underline{\mathbf{R}}_\mathrm{t}$ which contains the time correlations of the intercell MAI vector $\underline{\mathbf{n}}$ of (19) [31, 32, 33]:

$$\underline{\mathbf{R}}_\mathrm{n} = \underline{\mathbf{R}}_\mathrm{s} \otimes \underline{\mathbf{R}}_\mathrm{t}. \tag{29}$$

Since we refer to a cellular mobile radio system which uses signals with the same spectral form, the time correlations of the intercell MAI are the same as the time correlations of the desired signals [33]. Therefore, the covariance matrix $\underline{\mathbf{R}}_\mathrm{t}$ of (29) is a-priori known at the receiver. Due to the unknown DOAs of the intercell MAI, the spatial correlations contained in the matrix $\underline{\mathbf{R}}_\mathrm{s}$ of (29) are not known at the receiver. Moreover, the spatial correlations change with time due to the movement of the intercell interferers. Therefore, if the matrix $\underline{\mathbf{R}}_\mathrm{s}$ is not known, it must be replaced by the identity matrix in (29), and a performance degradation should be expected. A performance improvement can be achieved by an online estimation $\hat{\underline{\mathbf{R}}}_\mathrm{s}$ of the spatial covariance matrix $\underline{\mathbf{R}}_\mathrm{s}$, with the estimate $\hat{\underline{\mathbf{R}}}_\mathrm{s}$ being permanently updated and considered in the total covariance matrix $\underline{\mathbf{R}}_\mathrm{n}$, see (29). The receiver structure with estimation of $\underline{\mathbf{R}}_\mathrm{n}$ is shown in Fig. 4.2.10. In a first step, the JD process is performed without consideration of $\underline{\mathbf{R}}_\mathrm{s}$. Then, the estimated total data vector $\hat{\underline{\mathbf{d}}}$ together with the approximately known system matrix $\underline{\mathbf{A}}$, with the accuracy of $\underline{\mathbf{A}}$ depending on the accuracy of the channel estimation, can be used for a reconstruction

$$\hat{\underline{\mathbf{e}}}_\mathrm{d} = \underline{\mathbf{A}}\,\hat{\underline{\mathbf{d}}} \tag{30}$$

of the noise-free received signal. An estimation $\hat{\underline{\mathbf{n}}}$ of $\underline{\mathbf{n}}$ can now be obtained by subtracting the reconstructed signal $\hat{\underline{\mathbf{e}}}_\mathrm{d}$ of (30) from the total received signal $\underline{\mathbf{e}}$ of (19):

$$\hat{\underline{\mathbf{n}}} = \underline{\mathbf{e}} - \hat{\underline{\mathbf{e}}}_\mathrm{d}. \tag{31}$$

With $\hat{\underline{\mathbf{n}}}$ from (31) and the interference power σ^2 an estimate

$$\hat{\underline{\mathbf{R}}}_\mathrm{s} = \frac{1}{\sigma^2} \sum_{i=1}^{NQ+W-1} \left[\hat{\underline{n}}_i^{(1)}\, \hat{\underline{n}}_i^{(2)} \ldots \hat{\underline{n}}_i^{(K_\mathrm{a})}\right]^\mathrm{T} \cdot \left[\hat{\underline{n}}_i^{(1)}\, \hat{\underline{n}}_i^{(2)} \ldots \hat{\underline{n}}_i^{(K_\mathrm{a})}\right]^* \tag{32}$$

of the spatial covariance matrix $\underline{\mathbf{R}}_\mathrm{s}$ can be obtained. In (32), $NQ+W-1$ is the number of samples of $\underline{\mathbf{n}}^{(k_\mathrm{a})}$, $k_\mathrm{a} = 1 \ldots K_\mathrm{a}$. In a second step, the received signal

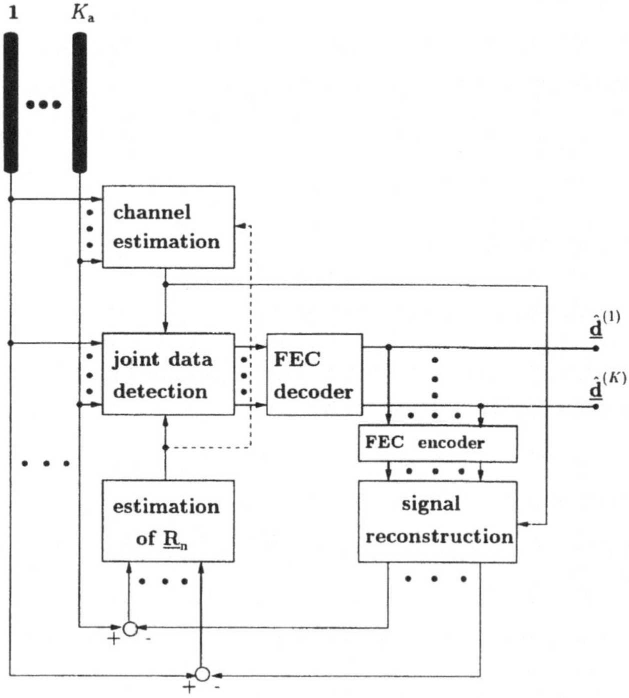

Fig. 4.2.10. Receiver structure with estimation of $\underline{\mathbf{R}}_n$

$\underline{\mathbf{e}}_d$ can be once more estimated, now considering the estimated matrix $\hat{\underline{\mathbf{R}}}_s$ and the a-priori known matrix $\underline{\mathbf{R}}_t$. This procedure can be repeated iteratively. It is noted that if the interference scenario does not change quickly compared to the duration of the estimation process of $\underline{\mathbf{R}}_s$, the estimated covariance matrix $\hat{\underline{\mathbf{R}}}_s$ can already be considered in the first step. Finally, for improving the channel estimation process described in Section 4.2.3, an estimation of $\underline{\mathbf{R}}_m$, see (13) and (18), can be obtained in the same way as described for data estimation in the present section.

4.2.6 Simulation Results

4.2.6.1 General

In this section, the benefits of adaptive antennas presented in the paper are demonstrated by simulations in the uplink of a TD-CDMA mobile radio system. The average coded bit error rate (BER) P_b is investigated as a function of the mean input SNR $\bar{\gamma}_{in}$ per net information bit with respect to a single receiver antenna by Monte Carlo simulations of data transmission in a single cell. The main parameters of the TD-CDMA mobile radio system used for the simulations are given in [2]. Further, it is noted that the real time implementation of the JD receiver can be already achieved with the hardware available today, see [3] for a detailed analysis.

4.2.6.2
Consideration of Data Detection

Figure 4.2.11 shows the benefits of adaptive antennas, when the channel impulse response vector $\underline{\mathbf{h}}$, see (10), is known at the JD receiver for the UKL 2 rural channel models, see Section 4.2.2.3. These channel models can be classified into the large cell channel type with

$$K_{\mathrm{d}}{}^{(k)} = 1, \quad k = 1 \ldots K, \tag{33}$$

see Section 4.2.2.4 and Fig. 4.2.4. The DOAs of the $K = 8$ simultaneously active users are assumed to be uniformly spatially distributed. The employed antenna configuration for $K_{\mathrm{a}} = 16$, see Fig. 4.2.11, is an eight-by-eight cross array configuration. The additive intercell MAI vector $\underline{\mathbf{n}}$, see (19), which originates from intercell interference in the interference-limited case, is assumed to be white and Gaussian distributed with zero mean, i.e., the link level system perfor-

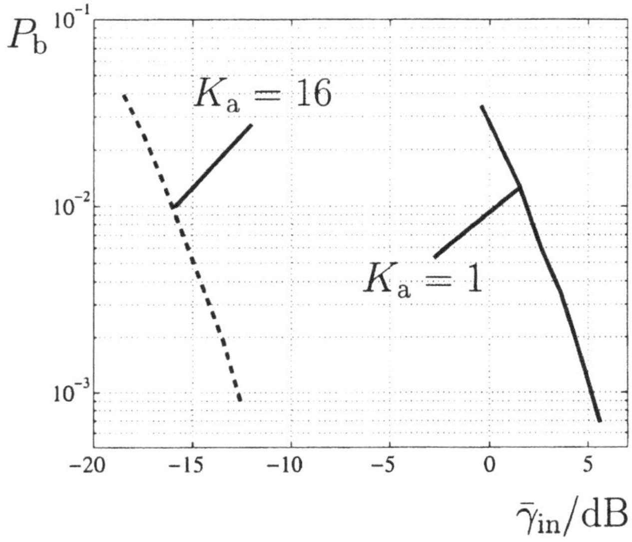

Fig. 4.2.11. P_b versus $\bar{\gamma}_{\mathrm{in}}$ for $K_{\mathrm{a}} = 16$ and $K_{\mathrm{a}} = 1$; propagation environment: UKL 2 rural; no channel estimation; mobile speed: 90 km/h

mance is evaluated. Under the above considerations, Fig. 4.2.11 shows that a cross array configuration of $K_a = 16$ antennas offers a gain of more than 17 dB at a coded BER $P_b = 10^{-3}$ compared to a single antenna system. In these simulations the channel estimation error is not considered, i.e., the channel impulse responses are assumed to be perfectly known at the BS receiver. Therefore, the results of Fig. 4.2.11 should be viewed at as a demonstration of the potential of adaptive antennas in TD-CDMA for system performance enhancement.

4.2.6.3
Consideration of Channel Estimation

In Fig. 4.2.12, the benefits of adaptive antennas are demonstrated, when channel estimation is additionally considered in the BS receiver, see Section 4.2.3. In this simulation scenario, the UKL 2 rural channel models are used, see Section 4.2.2.3, and a cross array configuration of $K_a = 16$ antennas is employed. The DOAs of the $K = 8$ simultaneously active users are assumed to be uniformly distributed. Moreover, the intercell MAI vector $\underline{\mathbf{n}}$, see (19), is assumed to be white and Gaussian distributed. The link level performance of an uplink receiver in TD-CDMA using the generalized Steiner estimator for more than a single antenna, see Section 4.2.3.1, is compared with the link level performance of a receiver that utilizes the directional information of the user signals as described in Section 4.2.3.2. for the large cell channel type. The receiver using the

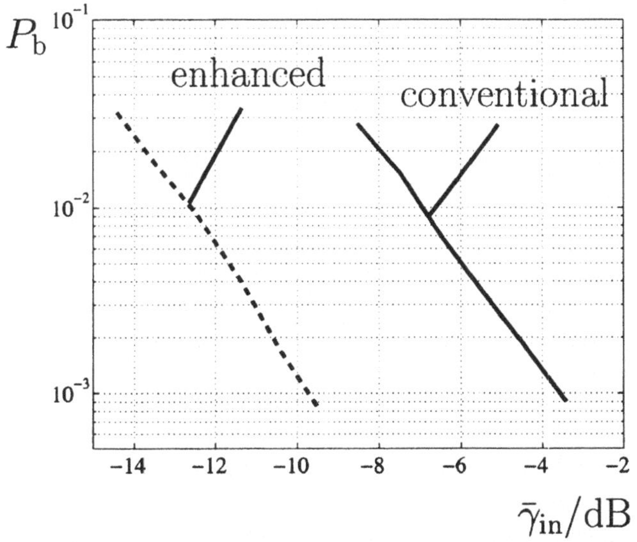

Fig. 4.2.12. P_b versus $\bar{\gamma}_{in}$ for eight-by-eight cross array configuration; propagation environment: UKL 2 rural; conventional and enhanced scheme; mobile speed: 90 km/h

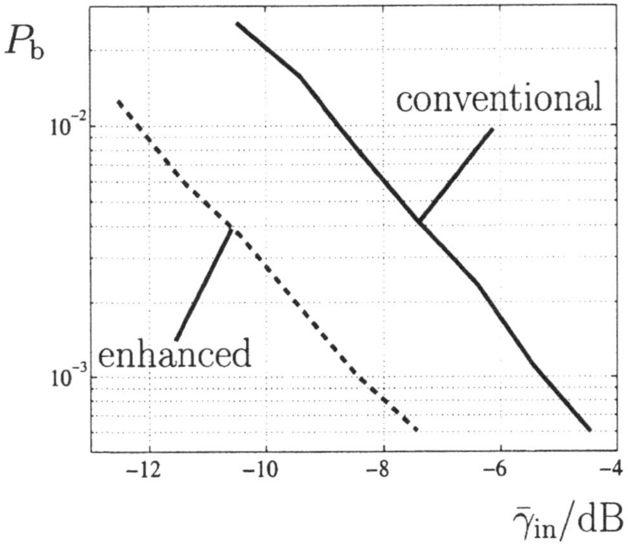

Fig. 4.2.13. P_b versus $\bar{\gamma}_{in}$ for eight-by-eight cross array configuration; propagation environment: UKL 2 dense urban; conventional and enhanced scheme; mobile speed: 30 km/h

generalized Steiner estimator is addressed as the conventional scheme since it does not utilize the directional characteristics of the user signals, whereas the receiver utilizing the directional information of the user signals is addressed as the enhanced scheme. Fig. 4.2.12 shows the potential of the concept described in Section 4.2.3.2 for large cell channel types, which results in a gain of more than 6 dB at a coded BER $P_b = 10^{-3}$ compared to the multi-antenna receiver which does not use the DOA information.

For the simulation results shown in Fig. 4.2.13, the channel model UKL 2 dense urban of Section 4.2.2.3 is used. This channel model can be classified into the small cell channel type introduced in Section 4.2.2.4 because of the characteristics of the channel impulse response of a user k, $k = 1 \ldots K$, the elements of which are associated with more or less a single direction of incidence. The other parameters of the simulation scenario are the same as the ones of the simulation scenario of Fig. 4.2.12. The curves in Fig. 4.2.13 show the potential of the enhanced scheme of Section 4.2.3.3 compared to the conventional receiver that uses the generalized Steiner estimator. The gain of more than 3 dB at a coded BER $P_b = 10^{-3}$ offered by the enhanced scheme demonstrates that channel estimation techniques using the directional characteristics of the user signals should be utilized even in propagation environments which already contain an increased diversity potential, like the one of the UKL 2 dense urban channel model. Certainly, the diversity of such propagation environments limits the gain achievable by adaptive antennas compared to the gain in the large cell channel types, see Fig. 4.2.12.

4.2.6.4
Consideration of Intercell MAI

In Figs. 4.2.11–4.2.13 the link level performance in a TD-CDMA mobile radio system is considered, i.e., the MAI covariance matrix $\underline{\mathbf{R}}_\mathrm{n}$, see (20) and (29), equals the identity matrix. In the curves of Fig. 4.2.14, a simple intercell interference scenario is considered in order to demonstrate the benefits of adaptive antennas in TD-CDMA, when the information about the covariance matrix $\underline{\mathbf{R}}_\mathrm{n}$ is used at the BS receiver, see Section 4.2.5 for a detailed theoretical analysis. Concerning the intracell scenario, measured channel impulse responses in an urban propagation environment have been used in order to produce the noise-free received signal $\underline{\mathbf{e}}_\mathrm{d}$, see (19) and Section 4.2.5. For the intercell interference scenario, it is assumed that the intercell MAI is composed of two components:

- A component which consists of a single interfering signal originating from one of the neighbouring cells. This interfering signal impinges on the BS from a discrete DOA of 90° and causes 80% of the total received intercell

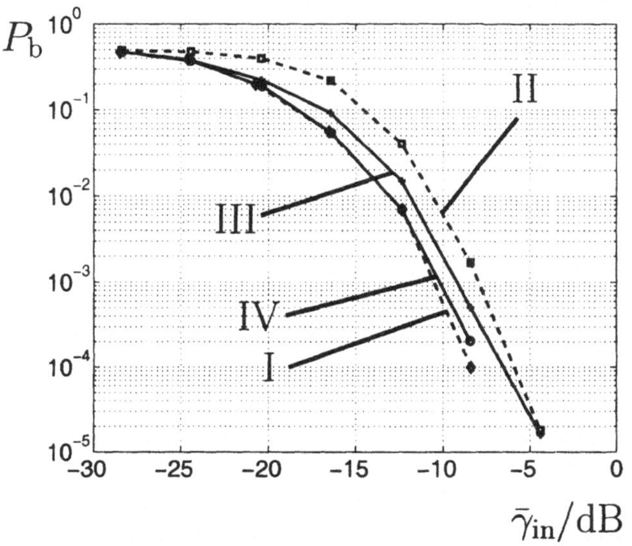

Fig. 4.2.14. P_b versus $\bar{\gamma}_\mathrm{in}$ for a uniform array configuration of $K_\mathrm{a} = 4$ antennas; measured channel impulse responses in an urban environment; different conditions of considering the intercell MAI covariance matrix in the JD process

noise power. The employed antenna configuration is a uniform rectangular array of $K_a = 4$ antennas.

- A component which is created by a very large number of interfering signals with a uniform distribution of their DOAs over the total range $[0, 2\pi]$ in azimuth. These interfering signals are pairwise uncorrelated, and they are also uncorrelated with the single interfering signal impinging from the discrete DOA of 90°. The second intercell interference component causes 20% of the total received intercell noise power.

According to the above assumptions, the total intercell MAI covariance matrix can be expressed by the tensorial product of (29), where the information about $\underline{\mathbf{R}}_t$ is assumed to be a-priori known at the BS receiver, see also Section 4.2.5. Fig. 4.2.14 shows the average coded BER P_b as a function of the mean input $\bar{\gamma}_{in}$ per net information bit for the below listed conditions concerning the spatial covariance matrix $\underline{\mathbf{R}}_s$, see (29), of the intercell MAI in the investigated intercell interference scenario:

- Condition I: $\underline{\mathbf{R}}_s$ is assumed to be perfectly known at the BS receiver.
- Condition II: $\underline{\mathbf{R}}_s$ is not known at the BS receiver and is replaced by the identity matrix $\mathbf{I}^{(K_a)}$ in the JD process.
- Condition III: $\underline{\mathbf{R}}_s$ is estimated and considered in the second step of the process described in Section 4.2.5.
- Condition IV: $\underline{\mathbf{R}}_s$ is estimated and is also considered in the first step of the process, see the analysis of Section 4.2.5.

From Fig. 4.2.14 the benefits of considering the intercell MAI covariance matrix in the JD process become obvious. The consideration of the spatial covariance matrix $\underline{\mathbf{R}}_s$, and consequently of the total intercell MAI covariance matrix $\underline{\mathbf{R}}_n$ since $\underline{\mathbf{R}}_t$ is a-priori known, offers a gain which varies between 1 dB and 3 dB at a coded BER $P_b = 10^{-3}$, see curves I, III, and IV of Fig. 4.2.14, compared to the receiver which does not use this information, see curve II. This variation of the achieved gain depends on the estimation quality of the spatial covariance matrix $\underline{\mathbf{R}}_s$. Fig. 4.2.14 shows that the performance remains approximately the same, when $\underline{\mathbf{R}}_s$ is estimated, but already considered in the first step of the estimation process presented in Section 4.2.5, compare the curves I and IV. If $\underline{\mathbf{R}}_s$ is estimated but only considered in the second step of the estimation process of Section 4.2.5, see curve III of Fig. 4.2.14, a performance degradation of about 1dB at a coded BER $P_b = 10^{-3}$ is observed compared to the case where $\underline{\mathbf{R}}_s$ is already considered in the first step of the estimation process of Section 1.6. In every case, though, the consideration of the intercell MAI covariance matrix $\underline{\mathbf{R}}_n$ in the JD process leads to an improved performance of TD-CDMA mobile radio systems with adaptive antennas.

4.2.7 Conclusions

In this paper, the benefits of adaptive antennas in the uplink of TD-CDMA mobile radio systems have been theoretically analyzed and demonstrated by computer simulation results. The directional inhomogeneity of the mobile radio channel is modeled by using directional channel models which explicitely incorporate the directional characteristics of the impinging signals at the BS receiver. It is shown that the exploitation of this directional inhomogeneity of the user signals in the considered cell leads to considerable performance improvements both for the joint data detection and the channel estimation in TD-CDMA compared to single antenna systems. Moreover, the consideration of the information about the statistical properties of the intercell interference in the joint detection receiver offers an additional benefit achievable by adaptive antennas in TD-CDMA mobile radio systems.

Acknowledgements

This paper is devoted to Prof. Dr.-Ing. habil. P.W. Baier on the occasion of the 25th anniversary of the Research Group for RF Communications. The authors thank the German Research Society (DFG) which by a research grant within "Schwerpunktprogramm Mobilkommunikation" contributed to the progress of this work. The invaluable support of the supercomputer staff at the Regionales Hochschulrechenzentrum Kaiserslautern (RHRK) is also highly acknowledged. Finally, the authors extend their thanks to their partners at Siemens AG for fruitful discussions.

References

[1] A. Klein and P. Baier, "Simultaneous cancellation of cross interference and ISI in CDMA mobile radio communications," Proc. IEEE 3rd International Symposium on Personal, Indoor and Mobile Radio Communications (PIMRC '92), Boston, pp. 118–122, 1992.

[2] J. Blanz, A. Klein, M. Naßhan, and A. Steil, "Performance of a cellular hybrid C/TDMA mobile radio system applying joint detection and coherent receiver antenna diversity," IEEE Journal on Selected Areas in Communications, vol. 12, pp. 568–579, 1994.

[3] J. Mayer, J. Schlee, and T. Weber, "Realtime feasibility of joint detection CDMA," Proc. 2nd European Personal Mobile Communications Conference (EPMCC'97), Bonn, pp. 245–252, 1997.

[4] 3GPP, "About the Third Generation Partnership Project (3GPP)", published in the Internet under the URL http://www.3gpp.org/ABOUT_3GPP/ 3gpp.htm, 1999.

[5] P. Baier, J. Blanz, and R. Schmalenberger, "Fundamentals of smart antennas for mobile radio applications," IEEE 8th International Symposium on Personal, Indoor and Mobile Radio Communications (PIMRC'97), S. Glisic and P. Leppänen, eds., Helsinki, pp. 345–376, Kluwer, 1997.

[6] J. Parsons, The mobile radio propagation channel. London: Pentech Press, 1992.

[7] T. Felhauer, P. Baier, W. König, and W. Mohr, "Optimized wideband system for unbiased mobile radio channel sounding with periodic spread spectrum signals," IEICE Transactions on Communications, vol. E76-B, pp. 1016–1029, 1993.
[8] P. Eggers, "TSUNAMI: Spatial radio spreading as seen by directive antennas," COST 231 TD(94) 119, Darmstadt, 1994.
[9] A. Klein, W. Mohr, R. Thomas, P. Weber, and B. Wirth, "Direction-of-arrival of partial waves in wideband mobile radio channels for intelligent antenna concepts," Proc. IEEE 46th Vehicular Technology Conference (VTC '96), Atlanta, pp. 849–853, 1996.
[10] U. Martin and M. Grigat, "A statistical simulation model for the directional mobile radio channel and its configuration," Proc. IEEE 4th International Symposium on Spread Spectrum Techniques & Applications (ISSSTA'96), Mainz, pp. 86–90, 1996.
[11] T. Kürner, D. Cichon, and W. Wiesbeck, "Concepts and results for 3D digital terrain based wave propagation models — an overview," IEEE Journal on Selected Areas in Communications, vol. 11, pp. 1002–1012, 1993.
[12] J. Blanz, Empfangsantennendiversität in CDMA-Mobilfunksystemen mit gemeinsamer Detection der Teilnehmersignale, series 10, no. 535, Fortschrittberichte VDI. Düsseldorf: VDI-Verlag, 1998.
[13] B. Steiner and P. Jung, "Optimum and suboptimum channel estimation for the uplink of CDMA mobile radio systems with joint detection," European Transactions on Telecommunications and Related Technologies, vol. 5, pp. 39–50, 1994.
[14] B. Fleury, M. Tschudin, R. Heddergott, D. Dahlhaus, and K. Pedersen, "Channel parameter estimation in mobile radio environments using the SAGE algorithm," IEEE Journal on Selected Areas in Communications, vol. 17, no. 3, pp. 434–450, 1999.
[15] B. Fleury and P. Leuthold, "Radiowave propagation in mobile communications: An overview of European research," IEEE Communications Magazine, vol. 34, pp. 70–81, 1996.
[16] J. Blanz, P. Baier, and P. Jung, "A flexibly configurable statistical channel model for mobile radio systems with directional diversity," Proc. ITG-Fachtagung Mobile Kommunikation, Neu-Ulm, pp. 93–100, 1995.
[17] J. Blanz, A. Klein, and W. Mohr, "Measurement-based parameter adaptation of wideband spatial mobile radio channel models," Proc. IEEE 4th International Symposium on Spread Spectrum Techniques & Applications (ISSSTA'96), Mainz, pp. 91–97, 1996.
[18] B. Fleury, D. Dahlhaus, R. Heddergott, and M. Tschudin, "Wideband angle of arrival estimation using the SAGE algorithm," Proc. IEEE 4th International Symposium on Spread Spectrum Techniques & Applications (ISSSTA '96), Mainz, pp. 79–85, 1996.
[19] J. Fuhl, Smart antennas for second and third generation mobile communications systems. Technical University of Vienna: Dissertation, 1997.
[20] A. Klein and W. Mohr, "A statistical wideband mobile radio channel model including the directions-of-arrival," Proc. IEEE 4th International Symposium on Spread Spectrum Techniques & Applications (ISSSTA'96), Mainz, pp. 102–106, 1996.
[21] M. Hata, "Empirical formula for propagation loss in land mobile radio services," IEEE Transactions on Vehicular Technology, vol. VT-29, pp. 317–325, 1980.
[22] J. Berg and Holmquist, "An fft multiple half-screen diffraction model," Proc. IEEE 44th Vehicular Technology Conference (VTC'94), Stockholm, pp. 195–199, 1994.
[23] M. Lebherz, W. Wiesbeck, and W. Krank, "A versatile wave propagation model for the VHF/UHF range considring three-dimensional terrain," IEEE Transactions on Communications, vol. COM-40, pp. 1121–1131, 1992.
[24] T. Kürner, D. Cichon, and W. Wiesbeck, "Concepts and results for 3D digital terrain-based wave propagation models: An overview," IEEE Transactions on Communications, vol. COM-II, pp. 1002–1012, 1993.

[25] U. Liebenow and P. Kuhlmann, "Theoretical investigations and wideband measurements on wave propagation in hilly terrain," Proc. IEEE 44th Vehicular Technology Conference (VTC'94), Stockholm, pp. 1803–1806, 1994.

[26] K. Loew, "Comparison of urban propagation models with CW measurements," Proc. IEEE 42nd Vehicular Technology Conference (VTC '92), Denver, pp. 936–942, 1992.

[27] R. Schmalenberger and A. Papathanassiou, "Zwei kompatible Kanalmodelle für Simulationen auf System- und Linklevel," Proc. ITG Diskussionssitzung Modellszenarien für Systeme mit intelligenten Antennen, Kaiserslautern, 1997.

[28] R. Schmalenberger and A. Papathanassiou, "Two compatible channel models for system and link level simulations of mobile radio systems," COST 259 TD (98) 055, Bradford, 1998.

[29] P. Höher, Kohärenter Empfang trelliscodierter PSK-Signale auf frequenzselektiven Mobilfunkkanälen — Entzerrung, Decodierung und Kanalparameterschätzung, series. 10, no. 147, Fortschrittberichte VDI. Düsseldorf: VDI-Verlag, 1990.

[30] P. Höher, "A statistical discrete-time model for the WSSUS multipath channel," IEEE Transactions on Vehicular Technology, vol. VT-35, pp. 461–468, 1992.

[31] J. Blanz, M. Haardt, A. Papathanassiou, I. Furió, and P. Jung, "Combined direction of arrival and channel estimation for time-slotted CDMA," Proc. IEEE 4th International Conference on Telecommunications (ICT'97), Melbourne, pp. 395–400, 1997.

[32] A. Graham, Kronecker products and matrix calculus with applications. Chichester: Ellis Horwood, 1981.

[33] A. Papathanassiou, M. Haardt, I. Furió, and J. Blanz, "Multi-user direction of arrival and channel estimation for time-slotted CDMA with joint detection," Proc. IEEE 13th International Conference on Digital Signal Processing (DSP'97), Santorini, pp. 375–378, 1997.

[34] A. Klein, Multi-user detection of CDMA-signals — algorithms and their application to cellular mobile radio, series. 10, no. 423, Fortschrittberichte VDI. Düsseldorf: VDI-Verlag, 1996.

[35] A. Papathanassiou, J. Blanz, M. Haardt, and B. P.W., "Spatial channel assignment considerations in a joint detection CDMA mobile radio system employing smart antennas," Proc. IEEE 5th International Conference on Telecommunications (ICT'98), Chalkidiki, pp. vol. II, 318–322, 1998.

4.3 Hybrid Adaptive Algorithm Based On Temporal Update And Spatial Spectrum Estimation For Adaptive Array Antenna

Kazunori WATANABE, Isamu YOSHII And Ryuji KOHNO

4.3.1 Introduction

In an adaptive array antenna using an temporal updating algorithm like LMS (Least Mean Square error), RLS (Recursive Least-Squares), it takes a long time to converge into the optimum antenna weights, because the algorithm requires updating antenna weights sample by sample in time domain. If channel situation is varying quickly in time domain, it may hardly track such time-variance.

On the other hand, in an adaptive array antenna based on spatial spectral estimation methods like DFT (Discrete Fourier Transform), MEM (Maximum Entropy Method) or MUSIC (MUltiple Signal Classification), it is possible to estimate arrival angles of received signals that is a direction of arrival (DOA), and SNRs by analyzing spectrum of arrival signals which are estimated with spatially sampled signals by each element. And we can derive the optimum weight coefficients by substituting them into the Wiener solution. It is important to obtain the optimum weight coefficients by one snapshot for the purpose of quick adaptation in a time-varying channel. In case of DFT array antenna, however, there are problems in the estimation accuracy of DOA which are restricted by the number of antenna elements. Recently, MUSIC and ESPRIT have been highlighted for its feature of high resolution method. However, as compared with DFT arrays, there are problems that requires larger complexity for calculating eigenvalues of the correlative matrix.

This chapter proposes and investigates an adaptive array antenna using hybrid or combined DFT and LMS algorithm in order to improve performance of both temporal updating algorithm and spatial spectrum estimating one maintaining both advantages such as quick adaptation and less complexity. In this study, we adopted the following algorithm. Firstly, we derive approximate best weight coefficients by estimating DOA with DFT algorithm based on spatially sampled signals by using antenna array. Secondly, the weight coefficients derived by DFT algorithm are set as initial coefficients and weight coefficients are updated by using LMS algorithm. This hybrid algorithm of DFT and LMS algorithm enables us to quickly converge better weight coefficients than either LMS algorithm or DFT one. Furthermore, it makes possible to catch up time-variance of a channel.

4.3.2 Dft Adaptive Array Antenna

4.3.2.1 Optimum Weight Coefficients

Let $y(t)$ be an array output and $r(t)$ be a reference signal. The error signal $e(t)$ is defined by

$$e(t) = y(t) - r(t) = \mathbf{WX} - r(t) \qquad (1)$$

By the evaluation function J, weight coefficients are calculated to minimize the mean square error $E[r^2(t)]$.

$$E[e^2(t)] = \mathbf{W}^{\mathrm{T}} f^3 \mathbf{W} - 2\mathbf{W}^{\mathrm{T}}\mathbf{S} + E[r^2(t)] \qquad (2)$$

where $\mathbf{S}$ is a correlative vector between antenna input signals and reference signal, i.e.

$$\mathbf{S} = E[\mathbf{X}^* r(t)] = E[\mathbf{X}_{\mathrm{d}}^* r(t)] = A_d R \mathbf{U}_{\mathrm{d}} \qquad (3)$$

where $\mathbf{A}_{\mathrm{d}}$ is an amplitude of desired signals, and R is an amplitude of reference signals. $\mathbf{U}_{\mathrm{d}}$ is a steering vector, which is represented by the phase difference between the desired signals at adjacent antenna elements. $\mathbf{U}_\mathbf{i}$ denotes a steering vector for interference signals. Desired signals are independent on interference signals. The reference signal correlates with only desired signals. $\boldsymbol{f}^3$ denotes a covariance matrix of input signals, and is expressed by $\mathbf{U}_\mathbf{d}$ and $\mathbf{U}_\mathbf{i}$ as follows,

$$\begin{aligned}\Phi &= E\left(\mathbf{X}_\mathbf{d}^*\mathbf{X}_\mathbf{d}^\mathrm{T}\right) + E\left(\mathbf{X}_\mathbf{i}^*\mathbf{X}_\mathbf{i}^\mathrm{T}\right) + E\left(\mathbf{X}_\mathbf{n}^*\mathbf{X}_\mathbf{n}^\mathrm{T}\right) \\ &= A_d^2\mathbf{U}_\mathbf{d}^*\mathbf{U}_\mathbf{d}^\mathrm{T} + A_i^2\mathbf{U}_\mathbf{i}^*\mathbf{U}_\mathbf{i}^\mathrm{T} + \sigma^2\mathbf{I}\end{aligned} \qquad (4)$$

where $\mathbf{A}_i$ is an amplitude of interference signals, σ^2 is a variance of noise, and $\mathbf{I}$ is a unit matrix. The evaluation function J becomes a quadratic function of these factors. The evaluation function J can be minimized if $\mathbf{W}$ is the optimum weight coefficient $\mathbf{W}_{opt}$, i.e.

$$\frac{\partial\left\{E\left[e^2(t)\right]\right\}}{\partial\mathbf{W}} = 0 \qquad (5)$$

By substituting (2) to (5), it is rewritten by

$$\frac{\partial\left\{E\left[e^2(t)\right]\right\}}{\partial\mathbf{W}} = 2\boldsymbol{f}^3\mathbf{W} - 2\mathbf{S} \qquad (6)$$

Therefore,

$$f^3\mathbf{W}_\mathbf{opt} = \mathbf{S} \qquad (7)$$

If $\boldsymbol{f}^3$ is positive definite, inverse matrix $\boldsymbol{f}^{3-1}$ exists. $\mathbf{W}_{opt}$ is represented by

$$\mathbf{W}_{opt} = \boldsymbol{f}^{3-1}\mathbf{S} \qquad (8)$$

By using (3) and (4), (8) is rewritten by

$$\mathbf{W}_\mathbf{opt} = SNR\left(SNR\mathbf{U}_\mathbf{d}^*\mathbf{U}_\mathbf{d}^\mathrm{T} + INR\mathbf{U}_\mathbf{i}^*\mathbf{U}_\mathbf{i}^\mathrm{T} + \mathbf{I}\right)^{-1}\mathbf{U}_\mathbf{d}^* \quad (9)$$

where SNR denotes desired signal-to-noise ratio, i.e. $SNR = A_d^2/\sigma^2$, and INR dose interfering signal-to-noise ratio, i.e., $INR = A_i^2/\sigma^2$

If SNR and DOA are correctly estimated, the optimum weight coefficients can be determined at one instant without making the temporal updating.

4.3.2.2
DFT Array Antenna

As an example of array antenna using a spatial spectrum estimation algorithm, DFT array antenna is applied. The flow-chart of DFT algorithm is shown in Fig. 4.3.1. DOA can be estimated by using DFT of spatially sampled signals with antenna array. However, accuracy of estimated DOA depends on the number of antenna elements. Increase of the number of elements results in larger scale of the system.

4.3.3
Lms Adaptive Array Antenna

LMS array antenna recursively derives the optimum weight coefficients temporal sample by sample in order to minimize mean square error (MSE) between reference signal and array outputs. This calculation is made under the maximum inclination method using a value in a moment of input signal by updating weight matrix $\mathbf{W}$, in case which DOA and signal powers are unchanged. At initial period, a reference signal is required for convergence.

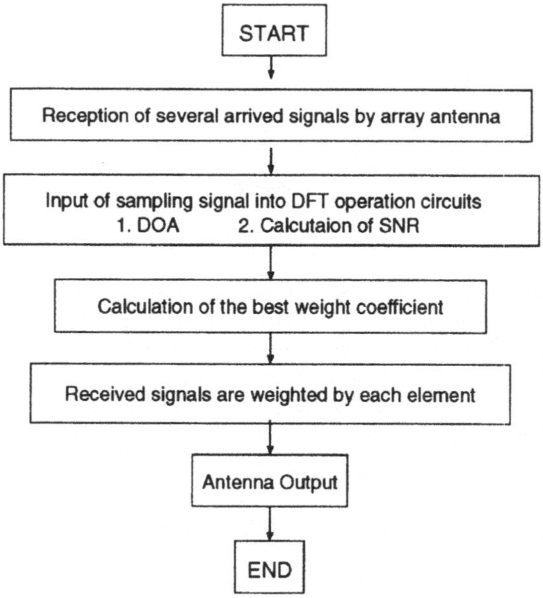

Fig. 4.3.1 Flow-chart of DFT array.

4.3.4 An Adaptive Array Using Combined Dft And Lms Algorithm

4.3.4.1 Advantages and Disadvantages of DFT and LMS Algorithm

In LMS algorithm, it takes a long time to converge into the optimum antenna weight, because the algorithm requires updating antenna weights sample by sample in time domain. Moreover, if channel situation is varing quickly in time domain, it may hardly track such time-variance. These are disadvantages of LMS algorithm. However, by using DFT algorithm, it is possible to obtain the optimum weight coefficients by one snapshot for the purpose of quick adaptation in a time-varing channel. This is advantage of DFT algorithm.

But DFT algorithm has problems in the estimation accuracy of DOA which are restricted by the number of antenna elements. This is disadvantage of DFT algorithm. If channel situation is stable in time domain, it is possible to converge into the optimum antenna weight by using LMS algorithm. This is advantage of LMS algorithm.

The above shows that DFT and LMS algorithm can help each other to catch up time-variance of a channel.

4.3.4.2 Proposed Hybrid Algorithm

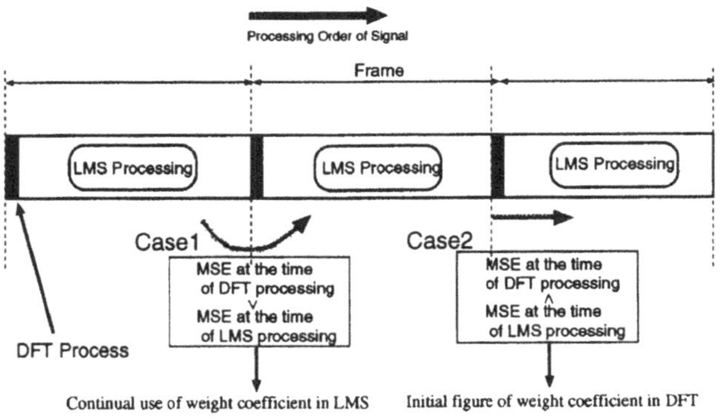

Fig. 4.3.2 Simulation Model.

First of all, an incoming signal wave received at each antenna element is sampled at one snapshot. Obtained set of sampled signal is called spatially sampled signals. A signal in one snapshot is processed in DFT to obtain sub-optimal weight coefficients. If the sub-optimal weight coefficients are employed as the initial values at temporal updating based on LMS algorithm for the purpose of improving convergence speed.

As shown Fig. 4.3.2, receiving signals are processed by the unit of a fixed length frame. In the beginning signal of a frame, an input signal is processed by both DFT and LMS algorithm. In this model (Fig. 4.3.2), it is assumed that all data are already known for simplicity.

The proposed algorithm is shown in Fig. 4.3.3.

(1) Compare process to check whether the input data is the beginning of frame
(2) If the input data is the beginning of frame, go to LMS algorithm block (4) and DFT algorithm block (5) at once
(3) If the input data isn't the beginning of frame, go to LMS algorithm block (7)
(4-a) Composition multiplied by weight coefficients to received signal
(4-b) Calculation of signal error
(4-c) Calculation of weight coefficients by successive update formula
(5-a) Getting Spectrum by DFT processing of receipt signal
(5-b) Discrimination of the desired wave and interference wave
(5-c) Calculation of weight coefficients by Wiener-Hoff solution
(6) After adoption of smaller weight coefficients in MSE (Fig.4.3.2), go to (8)
(7) After process of LMS algorithm, go to (8)
(8) final determination of weight coefficients

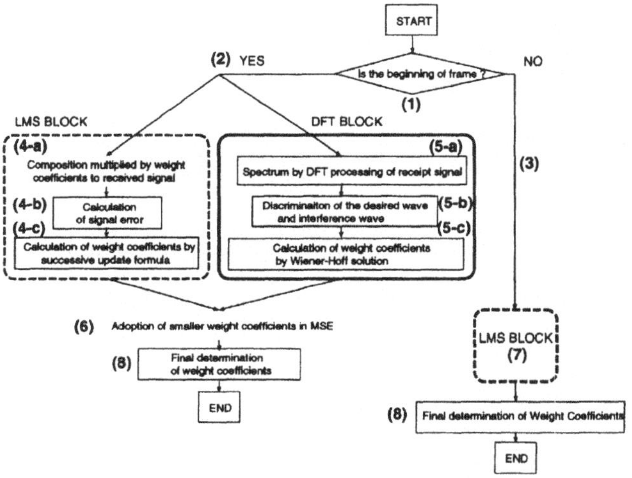

Fig. 4.3.3 Flow-Chart of Updating Algorithm.

4.3.4.3 Modified Algorithm

Under the proposed system, only beginning of frame is required for a possible DFT processing. Depending on the time-variance of a channel being static or dynamic, it is better to change the balance of DFT and LMS processing. As shown in Fig. 4.3.4, the ratio of LMS processing is higher in static situation, while that of DFT processing is higher in dynamic situation. In the most dynamic situation,

DFT processing will be always used throughout one frame instead of LMS processing.

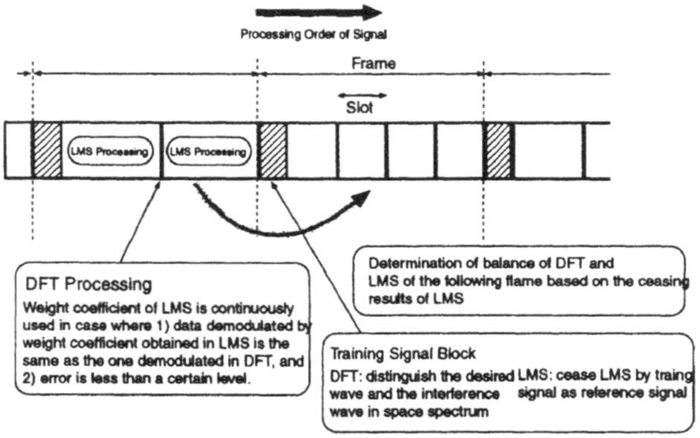

Fig. 4.3.4 Frame Model which enable to change the ratio of LMS and DFT processing

4.3.5 Computer Simulation

4.3.5.1 Channel Model

The transmitted signal, $s(t)$, can be written as

$$s(t) = A(t)\exp(j\omega_c t)\exp(j\phi(t))$$

where $A(t)$ is a parameter of the amplitude modulation, ω_c is a center frequency and $\phi(t)$ is a parameter of the phase modulation. The received signal $u(t)$ can be written as

$$u(t) = \sum_{i=1}^{m} R_i s(t-\tau_i) e^{j2\pi f_d \cos\theta_i} e^{j\alpha_i}$$
$$= \sum_{i=1}^{m} R_i s(t-\tau_i) e^{j\omega_c t} e^{j\phi(t-\tau_i)} e^{j2\pi f_d \cos\theta_i} e^{j\tilde{\alpha}_i}$$

where $\tilde{\alpha}_i = \alpha + \omega_c \tau_i$ and R_i is an amplitude of each arrival signal and τ_i is delay time and θ_i is the arrival angle for the direction of movement and α_i is phase and f_d is maximum value of a doppler frequency. The behavior of a fading for the received signal depends on the relationship between the change speed of $A(t)$ and $\phi(t)$, and $\tau(t)$.

4.3.5.2
Simulation Specification

Table 4.3.1: shows specification of simulations.

Carrier Frequency	800MHz
SNR	3.0dB
SIR	1.0dB
Numbers of arrival waves	2
Direction of arraival	D:0° I:30.0°
Frame Size	100 Symbols
Step Size	1/(2N) (N is number of lements)

4.3.5.3
Numerical Results

I ascertained the effectiveness of combining both DFT processing and LMS processing in the model based on fig. 4.3.2. Fig. 4.3.5,4.3.6 and 4.3.7 show convergency of MSE according to the number of elements. The more antenna elements, the smaller estimated errors are observed in DFT algorithm. It is noted that MSE at the beginning is smaller in case of 64 elements (Fig. 4.3.7) than 4 elements (Fig. 4.3.5). There is a problem that the more elements, the slower cease of errors are observed at the time of weight coefficients being updated by LMS. However, by using weight coefficients obtained by DFT processing at the beginning, such problem can be improved as the optimum value is already seen at the beginning of updating (Fig. 4.3.7).

It is confirmed that the more elements, the better coefficient value we can obtain in the proposed hybrid algorithm.

4.3.6
An Implementation Of A Digital Beam Forming Antenna Array System

We made a Digital Beam Forming (DBF) antenna array trial system. By using this system, it seems to enable us to evaluate updating algorithms that includes the proposed hybrid algorithm. The system structure is show in Fig. 4.3.8, and the implementation specification is show in Table 4.3.2.

Table 4.3.2: Implementation Specification

Carrier Frequency	2.5GHz-2.7GHz
Numbers of Antenna Elements	4
IF Frequency	128kHz
Sampling Frequency	512kHz-1024kHz

4.3.7 Conclusion

This chapter has proposed and investigated an adaptive array antenna using combined DFT and LMS algorithm. This algorithm can improve performance of both time-updating algorithm and spatial estimating one mainting both advantages such as quick adaptation, accurate estimation and less complexity.

4.3.8 Acknowledgment

About an implementation of a DBF antenna array system, I would like to express my gratitude to Prof. Hiroyuki Arai in Yokohama National University and the members of his laboratory.

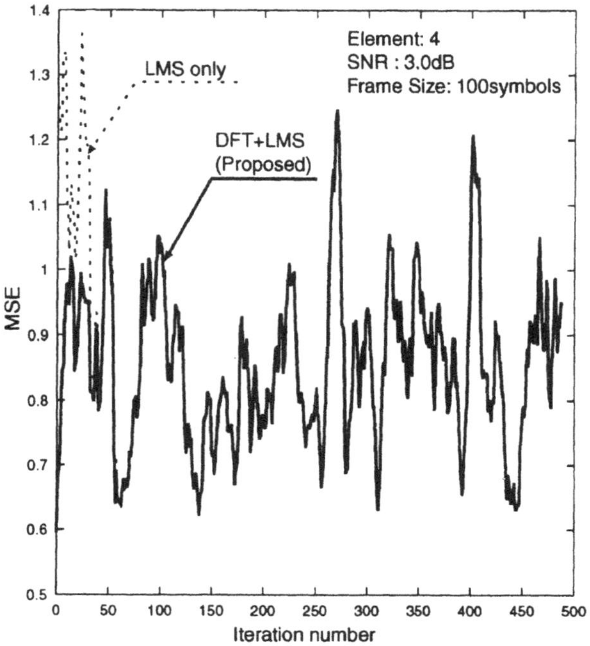

Fig.4.3.5 Convergency of MSE
(Antenna Element Number:4)

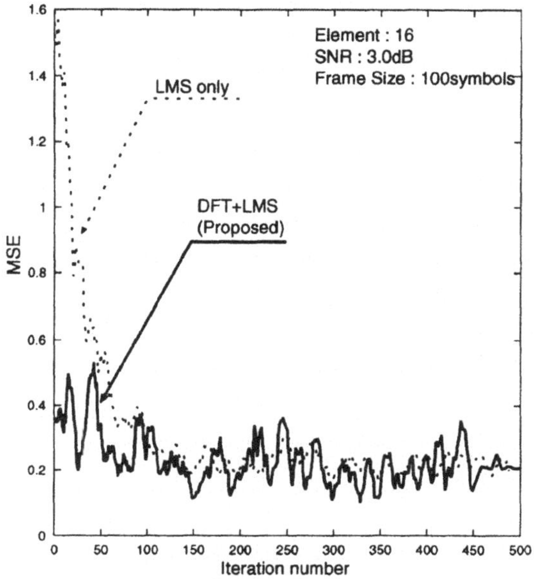

Fig. 4.3.6 Convergency of MSE
(Antenna Element Number:16)

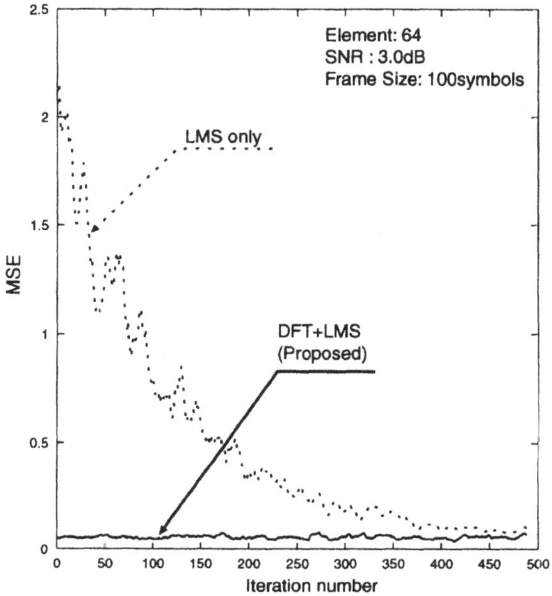

Fig. 4.3.7 Convergency of MSE
(Antenna Element Number:64)

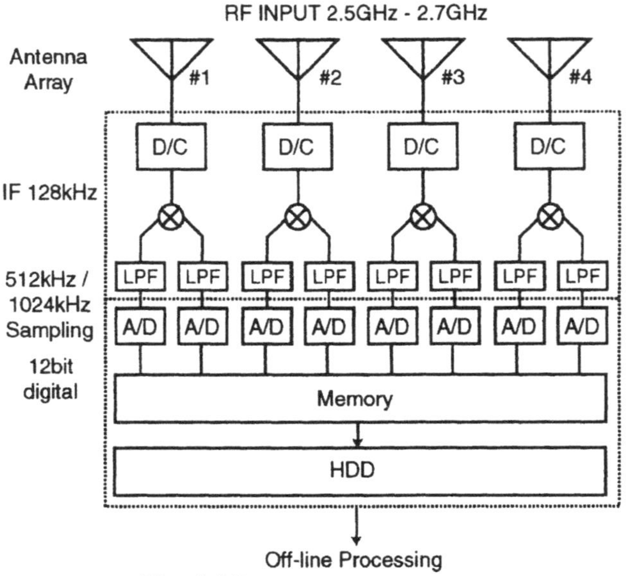

Fig. 4.3.8 System Structure of DBF.

References

[1] Chrysostomos L. Nikias / Athina P. Petropulu, *" HIGER-ORDER SPECTRA ANALYSIS"*, PRENTICE HOLL

[2] M.Nagatsuka, N.Ishii, R.Kohno and H.Imai, "Array Antenna Based on Spatial Spectrum Estimation Using Maximum Entropy Method", *IEICE Trans. Communication*, 5EB, E77-B, May 1994.

[3] N. Ishii and R. Kohno, "Spatially and Temporally Joint Transmitter - Receiver Using an Adaptive Array Antenna," *IEICE Trans. Communication*, Vol. EB-79, pp. 361 -- 367, March 1996.

[4] M. Nagatsuka, R. Kohno and H. Imai, "Optimal Receiver in Spatial and Temporal Domains Using Array Antenna," *International Symposium on Information Theory and Its Applications*, pp. 893- 898, Nov. 1994.

[5] Don H. Johnoson, Dan E. Dudgeon:, "*Array Signal Processing*" Prentice-Hall,1993

4.4 Optimisation of the cell-planning RF Design Parameters in complex geographic areas

S. Bouzouki, S. Kotsopoulos and G. Karagiannidi

4.4.1 Introduction

Radio propagation is one of the most important scientific areas of the telecommunication's sector that determines the quality of the corresponding pre-defined communication service. More precisely, for each provided communication service, the investigation of the received signal's quality in non-homogeneous geographical areas, plays a very important role in the overall system's performance and is extensively studied by the involved communication operators and the systems active manufacturers. It is noted that the signal's quality is studied in urban areas with many physical and technical obstacles, in rural areas and finally inside buildings.

Transmitting radio waves of the allocated frequency band for mobile telephony, have a great difficulty in propagating through various types of physical and technical obstacles [1]. The propagation procedure causes the appearance of various radios shadowing phenomena in large geographical regions that affect on the level of the received signal-to-noise ratio (S/N).

Moreover, the horizon reduces considerably the range of the transmitted signals due to the occurred large losses caused by the lower atmospheric layers. Conversely, reuse of frequencies beyond the horizon can be achieved without any co-channel interference. In this case, the frequency reuse criterion should be valid. This is the basic requirement for the operation of a cellular mobile radio communication system.

RF modelling is needed as an attempt to analyse the relation between the level of the received signal power and the corresponding level of the transmitted signal power [2]. The study of the propagation procedure remains a very complex scientific area, which is the subject of optimum modelling and experimental terrain measurements by service operators. In such case, 100% coverage and service quality cannot be guaranteed due to the occurred propagation problems and the cost the complete coverage.

The 3rd Generation Mobile Communication systems (G3G) opens a new era in the sector of the provided mobile services. The system's requirements are the high capacity, the high data rate, the high operational frequency and the low transmitting power. The cell planning scenario is one of the most important technical procedure of the operators, in order to achieve from one hand the desirable radio coverage and from the other hand the best cost effective provided communication services to the involved mobile subscribers. To be effectual, a cell planning procedure is important to have the knowledge on the effects of both the radio shadowing and multipath phenomena in different types of environment.

The decision to implement a cellular systems approach to high capacity mobile radio communication services, has set extremely demanding engineering targets for the operating involved consortia. In common with all terrestrial mobile radio transceivers, cellular equipment should operate in environments degraded by a variety man-made interference. Also, it is noted that spectrum efficiency is one of the most important factors in designing cellular mobile radio systems. It depends mainly upon the selected cluster scheme, the applied modulation-demodulation techniques and the set of technical parameters that affect the service quality [3]. In this case the conditional co-channel interference probability is a measure to control the interference's level helping the designers to re-adjust the system's operating parameters in order to keep this type of interference in low levels. For this reason, when attempting to increase the system's capacity by increasing the total number of radio channels, then the probability of co-channel interference sometimes is increased in high levels. In this case the transmission quality cannot be maintained at the specified level and new techniques should be applied in order to increase the system's capacity by having as a reference point the predicted conditional co-channel interference probability.

Moreover, the combination of the three solution's element (system's core solutions, network solution, mobile unit solution) is the key factor to provide increased capacity solution with the appropriate Quality of Service and the corresponding Grade of Service. As systems core solutions are the dynamic power control, the discontinuous transmission, the frequency hopping and the frequency reuse pattern. For network solutions are the cell splitting, the hierarchical cells, the dynamic channel allocation and the smart antennas. Finally, mobile unit solutions are the single and dual band users, the half rate, the full rate and the adaptive multi rate.

Section 4.4.2 states the radio coverage problem for terrestrial mobile communication systems. In section 4.4.3 a brief description of the mobile environment and the types of cells for terrestrial mobile communication systems, is given. In section 4.4.4 the system's cell planning procedure is specified. In section 4.4.5, the fundamentals of the applied RF modelling are described. In section 4.4.6 is described a hybrid prediction RF algorithm for the G3G Mobile Systems. Finally, section 4.4.7 presents the concluding remarks.

4.4.2 The radio-coverage problem

General, the term *coverage,* refers to the presence of sufficient desired signal power relative to the sum of undesired signal and noise powers to achieve the desired communication performance level or quality.

In the following, the *definitions of some terms pertinent to the subject of coverage are given:*

- *Service Area:* The operational area within which the radio communication system should meet some specified channel performance criteria. The area can be defined in many ways. For example, the service area might be defined as some radius around a site, as opposing corners of a rectangle centred about a site, as a political boundary, as a political boundary plus some buffer distance, or as an irregular polygon of connected line segments.
- *Channel Performance Criterion:* The channel performance criterion (CPC) is some minimum required level under faded conditions. For example, 5% message-error probability for a packet data radio system or circuit merit 3 (quality for voice systems). Such requirements in turn relate to the presence of some minimum desired signal power relative to the sum of interference and noise powers, C/(I+N).
- *Channel Performance Criterion:* The channel performance criterion (CPC) is some minimum required level under faded conditions. For example, 5% message-error probability for a packet data radio system or circuit merit 3 (quality for voice systems). Such requirements in turn relate to the presence of some minimum desired signal power relative to the sum of interference and noise powers, C/(I+N).
- *CPC Reliability:* reliability in the coverage refers to the probability that the required C/(I+N) exists for the region of interest. That region might be a contour or an area within some contour.

On the subject of contours, it is note that regulatory contours represent a specific criterion and prediction model used with a single height above average terrain (HAAT).

Reliability is what end users desire; probability is what radio system design engineers predict. Such predictions must allow some margin (excess) in probability to provide confidence that the desired reliability is in fact delivered and proof of performance tests can be passed. It is used the acronym coverage acceptance test plan (CATP) to refer to such tests.

Throughout the years, published literature on cellular system coverage design criteria has not been sufficient precise to preclude different interpretations. For example, V. H. MacDonald states:

" The AMPS transmission-quality objective was therefore quantified for design purposes as a requirement that this S/N value (18 dB) be exceeded in 90% of the area…". (V. H. MacDonald, [4])

A later statement by T. Brenig reads:

> "it is common practice to design for a carrier to co-channel interference ratio (C/I) of greater than 17 dB at the 90th percentile...". (T. Brening, [5])

For the newer digital cellular systems in GSM Recommendation 3.30 it is written

> " For car mobile traffic a usual measure is 90% area coverage per cell, taking into account the minimum signal-to-noise ration E_c/N_0 under multipath fading conditions...".

The GSM recommendation makes it clear that the area coverage being referred to be that of individual cells. The diversity benefit of handoff will then generally map the 90% spatial coverage to something approaching 100% temporal coverage. The designer could use a best-server distribution for design purposes, but then a coverage percentage much higher than 90% would be necessary. This is because 90% spatial coverage means that 10% of the call seconds involve substandard audio, a fraction that is far out of line with another common cellular specification: 2% blocking due to traffic overload.

Cell planning specifies substantially the Quality of Service (QoS) of a Network user. Mobile Communication Systems are based on better quality performance in a lot of ways. One of them is the determination of neighbor cells so that:

- The user is always being served by the base station that gives the best signal level quality.
- The maximum possible number of successful Handovers will take place.

In practice, there are up to six neighbors determined in mobile GSM 900 and 1800 systems. This means that for a Base station position planning, as long as for the frequency assigment there, we must have an actual radio-coverage area from any of the responsible antennas of a cell.

4.4.3 Mobile radio environment

The operating success of a cellular mobile system is relative with the surroundings of the base stations and the mobile users.

The terrain and the urban, suburban and rural environments are quite complex of the geometrical point of view because of the number of different object are involved: mountain, sea, hills, buildings, trees, lamp posts, furniture etc. Moreover, some of them are mobile objects such as cars and persons, all of which are involved in the radio propagation phenomenon, although each has a different influence. Because of that, without certain simplifications, radio propagation in such environments is impossible to simulate even approximate electromagnetic techniques.

At frequencies in the 900 MHz and above, the electromagnetic wavelength is small compared to the dimensions of buildings, which permits the use of ray-optical methods as a means for describing propagation in and around buildings.

Ray optics treats the electromagnetic waves as travelling along localized ray paths. For cellular mobile radio (CMR) and PCS (Personal Communication Systems) applications, the most important local features encountered by the rays are the ground and planar building elements, such as walls, floors and roofs.

It is important to make the separation of the existing environment in each type of cell (macrocell, microcell, picocell).

- The design of a wireless system with macrocell structure, employing high base station antennas and cell radii greater than several kilometers. The environment of a macrocell includes all the possible types such as rural, suburban and urban.
- The microcellular design employing low base station antennas and propagation over limited ranges (radii<1 km). For microcell prediction, it is recognised that different building environments and different propagation path in each environment will give different characteristics. First it is consider line-of-sight propagation along a street with the buildings assumed to form a canyon with uniform walls. For residential and commercial environments, the propagation is assumed to be over the roofs for low antennas by neglecting the cross streets and consider the buildings to be of uniform height and organised in uniformly spaced rows. In a high-rise environment is treated assuming the buildings to be so tall that propagation occurs down streets and around corners.
- The picocellular design employing propagation on the same floor as the base station or the same and other floors. That type of environment.

4.4.4 System cell planning procedures

Specifying a system's equipment and layout to achieve some coverage goal and proving to the customer that in fact that goal has been met are two separate but related activities. Despite its obvious importance, the subject of coverage verification has been little discussed in readily available technical literature.

The objectives of a cellular design are [6]:

1. The electrical coverage of the cell area must be guaranteed. The received signal must always be above a threshold level and holes in the coverage should be avoided. Coverage with a constant signal value in all the cell area are also desirable. The power coverage should be above the receiver sensitivity threshold in a maximum possible percentage of points of the cell, for example, greater than 90%. To avoid interfere with other cells, the signal strength of each BS outside its cell must be reduced.
2. The C/I ratio in a large percentage of the cell area (e.g., 90%, 95% of the points in the cell) must satisfy a minimum of quality in the received signal.
3. The grade of service (GOS), is enviable to provides, in each cell, enough channels to satisfy its traffic demand for new calls and for handoffs.

4. It must minimise the number of BSs so as to reduce the cost and the time of cellular deployment. The design approach must be flexible in order to locate the BS transmitters (antennas) at the available sites.
5. The radio system should be integrated with the land-fixed network. This usually means that the microcellular design should fit some requirements to connect correctly to a previously deployed land-fixed network.

4.4.5 The fundamentals of the applied RF modeling

One of the main aspects in the design of a wireless mobile system is to satisfy the traffic demand of the potential users with a minimum standard of quality in the received signal and with an acceptable percentage of blocked connections. The quality of the signal is mainly determined by the propagation conditions between the transmitters (desired and interferers) and the receiver. When the receiver moves along a trajectory, the local average signal suffers fluctuations known as slow fading or large-scale signal variations. These variations of the local average are a consequence of the attenuation of the signal with the distance between the transmitter and receiver and the presence of the obstacles in the environment. The rate of the slow fading variations is of the order of the size of the obstacles surrounding the receiver. Therefore, in rural environments the local average shows a smoother variation than in urban scenarios. The large-scale propagation models predict the local average signal level considering the influence of environmental obstacles.

According to the nature of the propagation model, they can be classified as:

- Empirical models
- Semi-empirical or semi-deterministic models
- Deterministic models

The empirical models are described by equations derived from statistical analysis of a large number of measurements. These methods are simple and do not require detailed information about the environment. They are also easy and fast to apply because the estimation is usually obtained from closed expressions. On the other hand, they cannot provide a very accurate estimation of the path loss.

Deterministic radio propagation prediction in areas as urban, mountain and indoor cases, is a highly complicated electromagnetic problem. The mathematical complexity of the electromagnetic coverage makes it impossible to predict radio propagation with a high degree of accuracy. The radio propagation depends on the characteristics of the environments such as buildings high, road width, type of the ground etc.

Rigorous techniques are difficult to use because of the electric size of the environmental obstacles. Several techniques have been used in the deterministic usually based on ray tracing electromagnetic methods: geometrical theory of diffraction (GTD), physical optics (PO) and, nor as frequently, rigorous methods such as integral equation (IE) methods or finite-difference time-domain (FDTD).

Semi-empirical or semi-deterministic models are based on the equations derived from the application of deterministic methods to generic urban or indoor models. Sometimes, the equations have been corrected experimentally in order to improve their agreement with the measurements. The resulting equations are functions of the characteristics of the areas surrounding the antennas and certain specific characteristics of the scenario. These methods require more detailed information about the environment than the empirical methods but not as much as the deterministic models. With the required data known, they are easy and fast to apply because, as with the empirical models, the results are obtained from closed expressions.

A typical block diagram of the prediction process is given in Fig. 4.4.1:

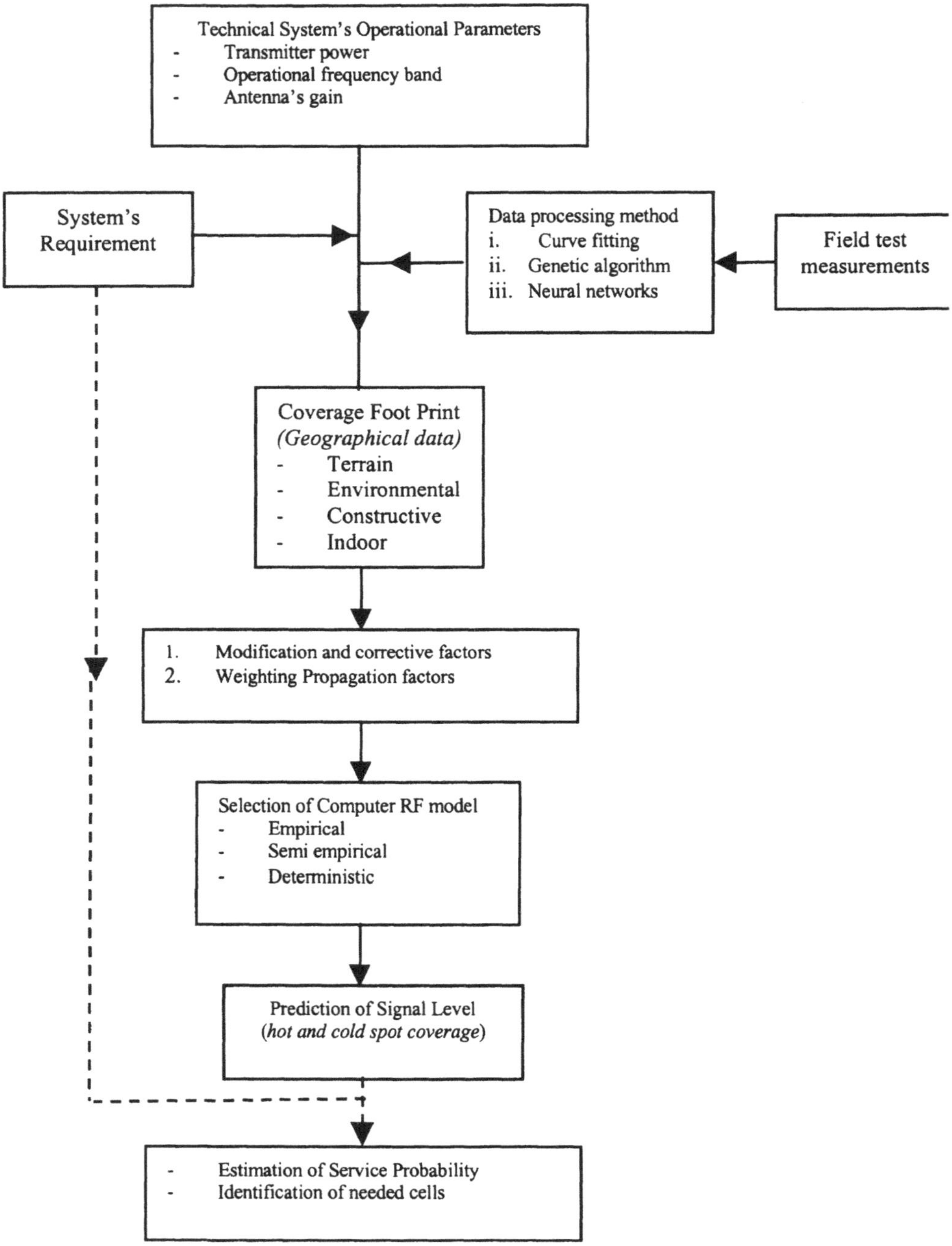

Fig. 4.4.1. Prediction process block diagram

The used propagation models, in mobile communications, can be classified according to the cellular plan scenario to which they are applied. So, the cell planning of a cellular system is consisted from:

✓ Macrocells
✓ Microcelss
✓ Picocells

The cell grading has to do with the size of the radio coverage area. Because of that there is a relationship between the three types of models and the types of cells for which they are suitable. Owing to their empirical nature, the empirical and semi-empirical models are suitable for macrocells with homogeneous characteristics. The semi-empirical models are also suitable for homogeneous microcells where the parameters considered by the method characterize the whole scenario well. The third type is suitable for microcells and picocells independently of their shape.

According to the existing literature, all the reported models are classified and given in the follow table:

Table 4.4.1. Existing propagation Models

Macrocell			Microcell	Indoor
Empirical	*Semi-empirical*	*Other Models*		*Empirical*
1. Allsebrook	1. Ikegami	1. Andersen[a]	1. Two-Ray	1. Long-distance path loss
2. Okumura-Hata	2. Walfish & Bertoni	2. Zhang[a]	2. Multiple-Ray	2. Attenuation Factor
3. COST 231-Hata	3. Xia & Bertoni	3. Saunder[b]	3. Multi-slit Waveguide Model	3. Keenan-Motley
4. Lee	4. COST 231 Walfish-Ikegami	4. Bonar[b]	4. Uni-Lund Model	4. multiple-Wall model[c]
5. Ibrahim & parsons				
6. McGeehan & Griffiths				
7. Atefi & parsons				
8. Sakagami-Kuboi				

[a] Model based on UTD for multiple diffraction
[b] PO-based expresion
[c] Multiople GTD-reflection on walls + UTD-diffraction at their corner+GTD-reflection from the ground

4.4.5.1
Experimental procedure

Measurements should be conducted in sites, which represent possible environments where PCS will operate. In order to characterise propagation in different directions relative to the streets for urban environment, selected test routes should be chosen during the measurements. The routes involved measurements in LOS (Line-of-sight) and NLOS (no Line- of-sight) paths, parallel and perpendicular streets and zig-zag and staircase streets. That kind of measurement campaigns have taken place in many different areas (urban, suburban etc) around the world.

The measurement set-up includes the mobile test equipment. One example of measurement devices that can be used is shown schematically in Fig. 4.4.2.

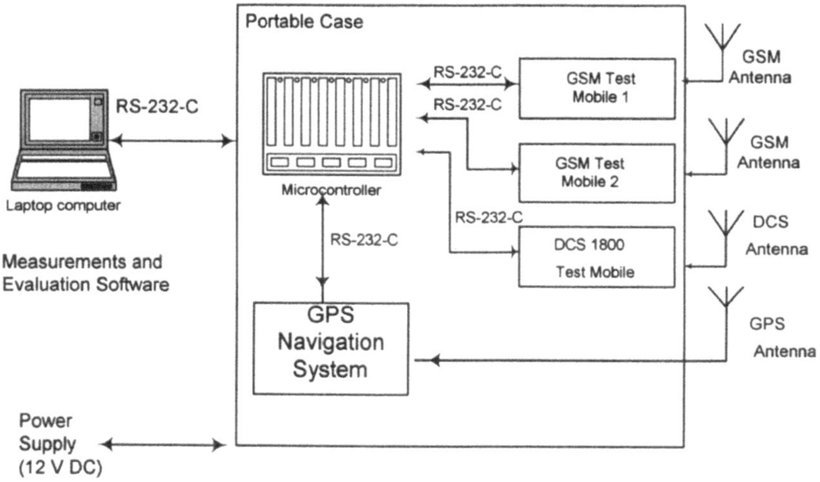

Fig. 4.4.2. Block diagram of the experimental system

The system uses the intelligence of the used radiotelephones. This means, that they automatically find the operating frequencies of the radio service. Some of the standard measurement functions of the system are the automatic call set up and hang up with separate phone number for each test mobile, variable call duration settings and idle time jointly adjustable for all mobiles (120 sec in our experiment). The most important function of the system is the recording of the peak receiving power. The micro-controller is responsible for the manipulation of both the test mobiles and the GPS (Global Positioning System) receiver. It allows data to be pre-processed before data are stored in an IBM-compatible laptop. The final software program enables us to observe at any time the signal fading on the computer monitor during the measurement procedure and to have a realistic representation of the followed path.

The used receiving omni-directional antenna with gain 1 dBi was located at the center of the car's roof at a 1.5 m height above the ground.

4.4.6 A hybrid prediction RF Algorithm for the 3rd Generation Mobile Systems

The aim of the existing prediction RF models is to study the behavior of the electromagnetic wave propagation in a specific RF service area. The need of a complex path loss, prediction algorithm has to deal with the adopting of the small and micro cell structures in the future cellular systems [7]. The small cell and low base stations in micro cell, building height, street width and the terrain elevation have profound effects on radio propagation. These RF models, in order to be applicable for a system's realistic cell planning, a set of necessary concessions have been considered for simplifying the geometry of the BTS-mobile RF link. In this case, the received signal's power is characterized by a caused mean error in the corresponding prediction values due to the above-considered concessions in link's geometry.

4.4.6.1 Basic model's consideration

The applied ETSI specifications to the RF modeling suggest the usage of the Hata [8] and COST 231-Hata Model for macro/small cell planning, and for micro cell planning the Walfish-Ikegami model [9]. The main disadvantage of these models is their statistical experimental dependence on the added constants that are derived from a measurement campaign in specific geographical area. This disadvantage is responsible for the different caused predict mean square error for the signals level in different environments (i.e. where the GSM 900/1800 systems are in operating mode).

One of the basic elements, which made a prediction model dependable, is the accuracy of the calculation of the received signal's power. The accuracy is achieved by taking into consideration the most realistic visualized technique for the representation of the environment that is wanted to be under radio coverage.

The answer for the most suitable RF propagation model will be the one which calculates the path loss in all types of environment as a result of signal reduction due to free space wavefront spreading as well as building and terrain influence. In the proposed algorithm all the basic system parameters, such as operational frequency, base and mobile station height are taken into consideration. Also, by using the terrain's elevation, the average building height and street width the radio coverage environments are specified.

For the G3G (Global 3rd Generation) systems the used path-loss prediction model is characterized by the next two statuses:

To be efficacious in the parameters which their influence can not be analytical taken into account in the total path-loss calculation (added constants).

To provide, the capability of achievement a procedure, in the cell planning, for finding the optimum BTS locations.

The above statuses are accomplished with the application for one hands the appropriate mathematical formula for the field strength calculation that depends on the environmental peculiarities. For the other hand the determination (recalculation) of the correction factors in dB which derived from the statistical processing of the local experimental measurements.

The needed for the calculation detailed knowledge of the natural terrain, is provided by a digital database which is derived from a Geographical Information System (GIS). The most significant feature concerning the feasibility of a complex propagation model, as the proposed one, is the application of the automatic ray-tracing algorithm.

4.4.6.2
Mathematical description

The received power on a territorial pixel is calculated from the transmitted ERP (Equivalent Radiated Power) (in dB) and subtracting the total losses L_T that evaluated by adding the four elements: the intermediate power L_M, the path loss L_P, the diffraction loss L_D and the reflection loss L_R. In this case the received power can be written as follows:

$$P_r = P_t + G_t + G_r - L_T \qquad (1)$$

Where: G_t is the transmitter antenna gain
G_r is the receiver antenna gain

And L_T the total loss, expressed by the following formula:

$$L_T = L_M + L_P + L_D + L_R \qquad (2)$$

The used L_M factor is derived from the following formula:

$$L_M = L_{en} + \frac{sd}{\sqrt{2}}\left[erf^{-1}\left(1 - 2\frac{percent\_\log}{100}\right) + 10\log\left(\frac{\ln\frac{percent\_Rayleigh}{100}}{\ln 0.5}\right)\right] \qquad (3)$$

Where:
L_{en} represents the environmental losses (or land cover correction factors [10],)
sd is the lognormal standard deviation (typical 6 to 10 dB range)

The term's *percent_log* and *percent_Rayleigh*, are correspond to the statistical desired radio coverage.

The factor L_M is taken into account regardless of the environmental type (rural, urban, suburban, open area). The L_{en} is a value that minimizes the overall mean square error that is the r.m.s. of the differences between predicted field values and measurements taken over a limited area.

L_P represents the path loss component expressed in logarithmic scale as a linear function of distance d:

$$L_P = C + 10\,\gamma\,\log d \qquad (4)$$

Where: d is the distance between base and mobile station

C is an intercept point generally taken as the path loss at a distance of 1 m or 1 km for outdoor macrocellular systems.

γ is the path loss exponent which indicate the rate at which the path loss increases with distance.

The following two cases are examined:

- *LOS (Line-of-Sight)*

With the term LOS is meant that the geographical area is flat with the restriction that the base–mobile station distance is less than the distance where the first Fresnel zone is obstructed by the different terrain elevation. The distance d_F, at which the first Fresnel zone becomes obstructed, is given by the following formula [11]:

$$d_F = \frac{1}{\lambda}\sqrt{\left(\Sigma^2 - \Delta^2\right)^2 - 2\left(\Sigma^2 + \Delta^2\right)\left(\frac{\lambda}{2}\right)^2 + \left(\frac{\lambda}{2}\right)^4} \qquad (5)$$

Where: $\Sigma = h_t + h_r$

$\Delta = h_t - h_r$

h_t, h_r = the base and mobile station heights respectivily

For the development of a propagation program tool from the research needs of our laboratory, the used formula for L_P loss is derived from Okumura's report [12] and organized for computational use from Hata and COST 231 [8] [9]. The involved parameters are the frequency *f*, the base station (BTS) height h_b, the mobile height h_m, and the distance *d* between BTS and mobile. Two formulae are used to estimate the L_P by considering the frequency range (150-1500 MHz and 1500-2000 MHz) [10].

$$L_P = \begin{cases} 69{,}55 + 26{,}16\log f - 13{,}82\log h_b + \left(44{,}9 - 6{,}55\log h_b\right)\log d & (150 - 1000MHz) \\ 46{,}3 + 33{,}9\log f - 13{,}8\log h_b + \left(44{,}9 - 6{,}55\log h_b\right)\log d + Cm & (1500 - 2000MHz) \end{cases} \qquad (6)$$

Where: *f* is the operational frequency (MHz)

h_b is the effective antenna height (m)

d is the distance (Km)

The h_b can be evaluated from the algorithm proposed in [12].

- *OLOS (Obstructed-Line-of-Sight)*

The case of OLOS is recognized in the following situations:

1. Non homogeneous elevation street canyon profile
2. Building discontinuity

In the first case of OLOS an additional loss factor is added to the LOS losses. This factor is denoted as L_{nh} and is estimated by the first Fresnel zone clearance. The first Fresnel zone radius r_F is calculated:

$$r_F = \sqrt{\frac{\lambda \cdot d_1 \cdot d_2}{d}} \qquad (7)$$

Where: λ is the wavelength
d_1 is distance between base station and the break point
d_2 is the distance between the break point and the mobile
d is the separation distance between base and mobile station

The attenuation a_z due to obstruction is calculated by the term $\Delta h/r_F$ from the curve given in Fig. 4.4.3. The parameter Δh is the difference between the r_F and the elevation of the break point.

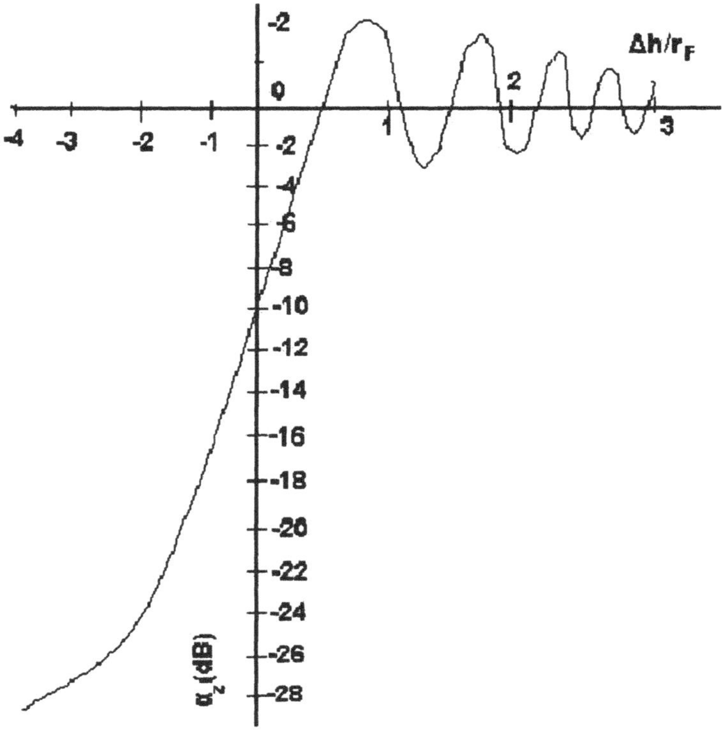

Fig.4.4.3. Attenuation due to obstructions as a function of normalized path clearence

In the second case of OLOS an additional loss factor is also added to the LOS losses. This factor is denoted as L_{open} and is derived by the following formula:

L_{open} = -[4,78(log$f$)^2-18,33log$f$+40,94] (8)

In urban and suburban environment, the diffraction losses L_D are calculated by the UTD-based approach for cellular mobile propagation for urban areas. Fig. 4.4.4 depicts the visualization of the propagation geometry in a small cell with the presence of buildings.

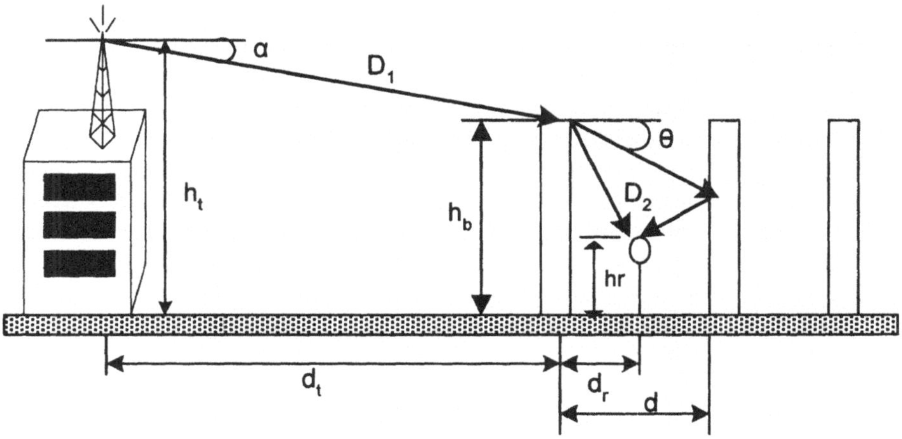

Fig. 4.4.4. Geometry of a small environment

That kind of losses are caused by the wave diffraction in local screen and are derived from the next formula [13]:

$$L_D = -20\log_{10}\left|D^{I}_{s,h}\right| + 10\log_{10} D_2(m) + 10\log_{10}\left[\frac{D_1(D_1+D_2)}{D^2}\right] \quad (9)$$

Where: $D^{I}_{s,h}$ is the UTD diffraction coefficient for a spherical field from the BTS antenna [14]

D_1, D_2 are clearly defined in Fig. 4.4.4.

$D = \sqrt{(d_t + d_r)^2 + (h_t - h_r)^2}$

s, h are denote the soft and hard boundaries, respectively

The loss L_R caused by the reflection can be described as [13]:

$$L_R = -20\log_{10}\left|1 + R_{H,V}\sqrt{\frac{D_2(D_1+D_2)}{r_2(D_1+r_2)}}\right| \cdot \frac{D^{II}_{s,h}}{D^{I}_{s,h}} e^{-jk(r_2-D_2)} \quad (10)$$

Where: $D^{II}_{s,h}$ is the UTD diffraction coefficient for a spherical field from the BTS antenna [14]

k is the wavenumber

$R_{H,V}$ is the reflection coefficient [14]

The above used UTD diffraction coefficients are valid only when the mobile is in the shadowing zone. This restriction leads to the condition dt> 600 m.

4.4.6.3
An application example

In this section, are described some case studies and are compared predicted and measured results for a cell site in an urban small-cell environment.

Propagation measurements were made in Patras area during a two-day period in autumn. The measurement signal was undertaken using fixed base station transmitters of two GSM providers and a DCS-1800 provider. For the radio coverage of the urban area of Patras, the both type of provider use a combination of macro-cellular and small-cellular structure. In the case of small-cellular structure the antenna is sited above the median but bellow the maximum height of the surrounding rooftops and the maximum cell range is typically less than 1-3 km. The Fig. 4.4.5 and 4.4.6 are shown the location of the BTS (Base Transmission Station) responsible for the radio coverage of the wide area of Patras.

Fig. 4.4.5. The east part of the wide area of Patras

Fig. 4.4.6. The west part of the wide area of Patras

All the transmitters are located on the roof of several buildings around the city except those that are installed in position BTS1. This location is the top of a hill with 160-m elevation and from there an antenna can almost illuminate all the area.

Because of the mountain on the south of the city, not to have any functional problems from the wave reflection in the mountain, none cell illuminate in that direction. To radio coverage the port area, the providers use BTS where are installed in locations across the port of Patras. The GSM providers follow this strategy and the most faraway BTS can be 10 km.

The street-plan of Patras is contain many squares and that means the urban environment contains a number of small open areas (the existing city structure is expected to influence the signal behaviour). Also as it can be noticed from Fig. 4.4.1 and 4.4.2, the block structure of the city has create very long streets which cross the city from the east up to west part. In combination with the heights of the surrounding buildings we have a kind of waveguide propagation.

The tests were taken place over the streets in order to have a better view about propagation:

1. in different directions relative to the streets
2. in different dimensions and type of building in the center of Patras.

To examine the dominant propagation mechanisms in that environment, the path we follow to take the measurements were in line-of-sight and non-line-of-sight case.

Each BS consisted of transmitter feeding an antenna with an EIRP (Equivalent Isotropic Radiative Power) given by the formula:

$$EIRP = Tx\ effect - Total\ Loss + Gain \qquad (11)$$

Where: *Tx effect* is the transmitter's output power
Total Loss is the total signal's attenuation in the coax cable in the transmitter site
Gain is the antenna's gain

The range of EIRP is from 45 dBm to 54 dBm. The antenna of each BTS was elevated from 20m to 42m. All the BTS illuminate a cluster of three cells and each cell has registered the most 4 frequencies. For the transmission of each frequency it's needed 5 to 10 Watt transmitting power.

Finally, some typical building characteristics are the following:

- Building heights: 21m to 30 m
- Street width: 6m to 10 m

The technical characteristics of the BTS 1 are the following:

- Mast height: 35m
- Transmitting frequency: 1856.6 MHz
- Receiving frequency: 1770.6 MHz
- EIRP: 54 dBm
- Position: -6° (elevation)
  286° (azimuth)

Fig. 4.4.7, shows the variation of the received local power versus the distance.

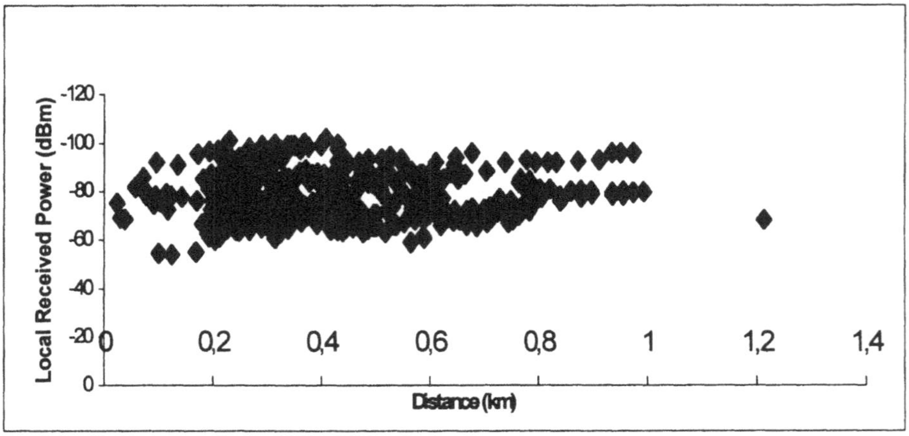

Fig. 4.4.7. The experimental results of the received power from cell C

The input data in the simulation program are:

✓ Technical parameters (i.e antenna gains, antenna heights, operating frequencies)
✓ Terrain profile in the testing area.

The digital database concerning the map of the Patras area, is based on GIS (Geographic Information System) technique. The used digital database includes all the necessary information (heights, coordinates, and type of ground) but not any information about technical characteristics such as building heights and road widths.

Fig. 4.4.8 shows the variation of the received power in various zig-zag street points.

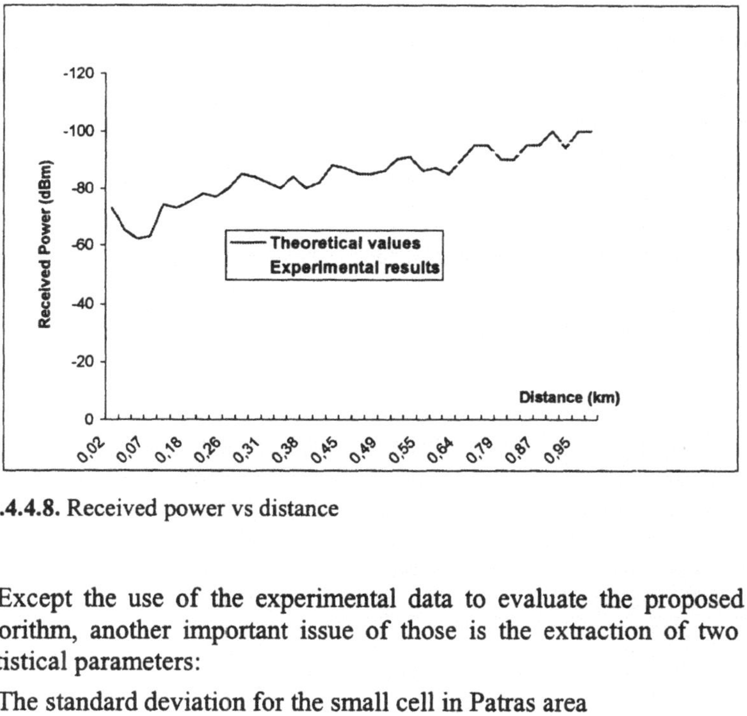

Fig.4.4.8. Received power vs distance

Except the use of the experimental data to evaluate the proposed prediction algorithm, another important issue of those is the extraction of two significant statistical parameters:

1. The standard deviation for the small cell in Patras area
2. The pdf of the signal distribution of the small cell in Patras area

The estimated pdf is very useful for the calculation of system's outage probability.

The standard deviation is calculated by the following formula:

$$\sigma = \sqrt{\frac{\sum_{i=1}^{N}\left(x_i - \bar{x}\right)^2}{N-1}} \qquad (12)$$

Where:

σ = standard deviation
N = the number of datum

x_i = the individual measurement

$\bar{x}$ = the mean of all measurements

For the examining small cell in the Patras area it is found that $\sigma = 9{,}846$.

In Fig. 4.4.9, the pdf of the ratio P_{local} to P_{mean} is shown. This analysis is based on the statistical processing of the obtained experimental results.

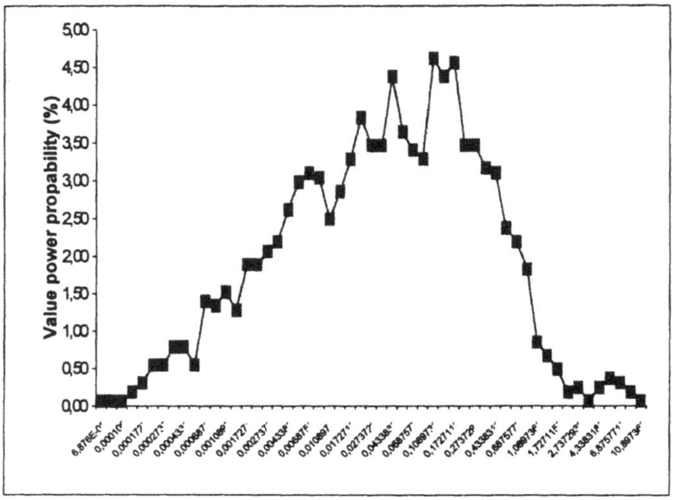

Fig.4.4.9. The pdf of the ratio P_{local}/P_{mean}

For the theoretical statistical analysis, it is attempted to approximate the data curve with the Nakagami distribution. The Nakagami distribution (or m-distribution), contains a set of sub-distributions (i.e. Rayleigh, etc) and provides an optimum fit to collected data in outdoor and indoor environments, in the frequency range from 800 MHz to 4 GHz [15, 16].

If r is a Nakagami variable, the pdf is given by the following formula:

$$f(r) = 2 \cdot \left(\frac{m}{\Omega}\right)^m \cdot \frac{\xi_k^{2 \cdot m - 1}}{\Gamma(m)} \cdot \exp\left(-\frac{m}{\Omega} \cdot r\right) \qquad (12)$$

Where:

ξ = received signal

m = an arbitrary fading parameter

Ω = average power

$\Gamma(m)$ = Gamma function

The Kolmogorov-Smirnov test is used for the best-fit test [16]. This test generates a reliability interval around the PDF of the received signal power normalized on its mean power. The best fit is created when the parameters m, Ω take the values of correspondingly. From the theoretical results of Nakagami pdf the best fit is when (m, Ω) is equal to (4, 14).

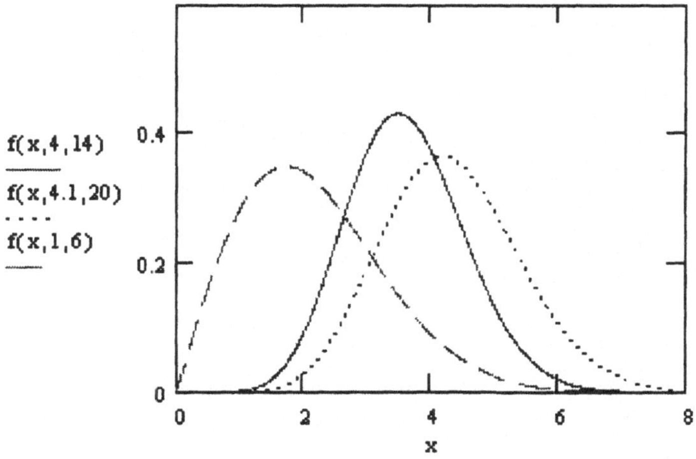

Fig. 4.4.10: The theoretical simulation of the pdf. For m=1 is the Rayleigh distribution.

Reliant on the theoretical results, the range of the receiving power is from –69 dBm to -110 dBm. Even if, the receiver's sensitivity of the mobile station is –102 dBm, according the manufacturer's specifications, the probability for unsuccessful connection is very high when the receiving power is less than –96 dBm. This is observed during the performed experimental procedure.

The theoretical model, which is used for predicting the propagation behavior of a signal in the band of 900 MHz, was based on UTD theory presented in [13].

In that case the total path losses is given by the formula:

$$L_t = L_0 + L_1 + L_{md} + L_r \qquad (13)$$

Where: L_0 are the free space losses

L_l are the losses due to the local numbering n+1 buildings in the vicinity of the mobile

L_{md} are the losses due to multiple diffraction caused by rows of buildings

L_r is the loss caused by the reflection of the diffracted electric fields from the wall of the building next to the mobile.

The examined BTS of the GSM provider is installed in the location BTS4, with EIRP= 48 dBm, f_c=947 MHz.

For two different rays from the BTS the results are given in Fig. 4.4.11 and 4.4.12.

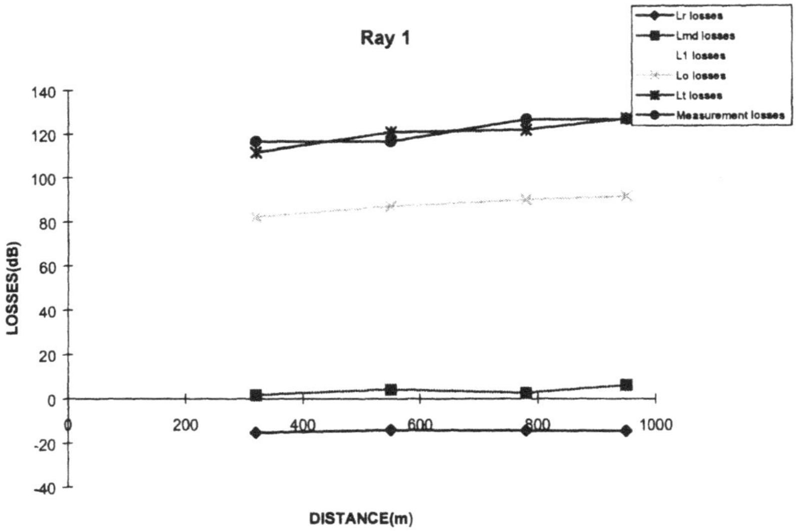

Fig. 4.4.11. The behavior of a GSM BTS (ray 1)

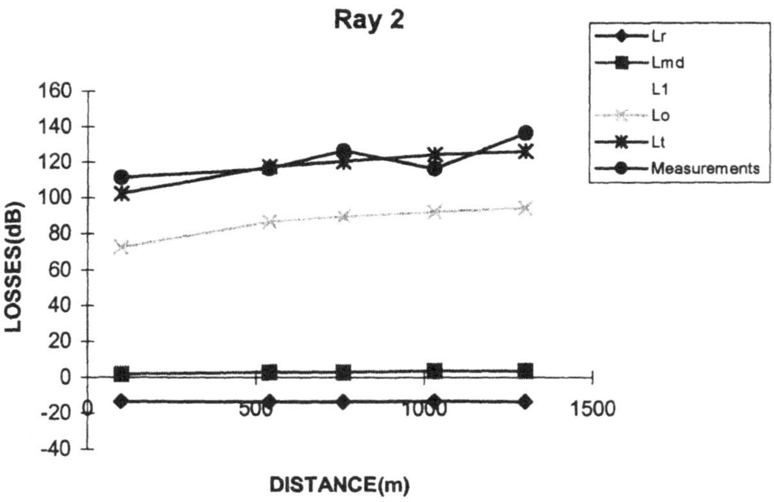

Fig. 4.4.12. The behavior of a GSM BTS (ray 2)

4.4.7 Conclusions

The significance of the role of the RF modeling in cell planning procedures has been pointed out. Moreover, a comparative overview of various propagation models has been presented. Results have been summarized and have also been used as a basis for a new proposed propagation model for calculating reference transmission loss at UHF frequencies as a function of involved geographic data and propagation phenomena. The peculiarities of in-door and out-door propagation parameters have been outlined in order to organize the core of our proposed RF model. The propagation approach proposed in this chapter is a contribution to this scientific area, helping the designers of mobile communication systems of the 2nd and 3rd generation to improve their efforts in order to satisfy the required system's Quality of Service and corresponding Grade of Service demands.

References

1. Kambezidis H and Kotsopoulos S, An Algorithm for the Computation of Diffraction of Electromagnetic Waves by Knife-Edge Obstacles. Technical Chronical-Scientific Journal, Vol. 4, No 3, 1984
2. Kotsopoulos S, Kambezidis H and Zevgolis D, Application of Efficient RF Modeling to Wireless Dispatch Systems, IEE 9th International Conference on Antenna and Propagation, Eindhoven, Netherlands, April 4-7, 1995
3. Kotsopoulos S, Palatos S and Kakavas M, Simulation of a mobile radio communication network based on a new flexible dynamic assignment algorithm, IASTED International Conference on Applied Modeling and Simulation, Lugano, Switzerland, June 20-22, 1994
4. A. Jagoda and M. De Villepin, Mobile Communications, Publishing Co: J. Wiley
5. W. Jakes, Microwave Mobile Communications, Publishing Co: IEEE
6. Garry Chess, Handbook of land-mobile radio system coverage, Artech House Publishers, BOSTON-LONDON
7. S. Bouzouki, G. Karagiannidis and S. Kotsopoulos, A Novel master-Slave Based Architecture for Wireless Pico-Cellular Communications, ComCon 7, Athens, Greece, 28 lune – 2 July 1999
8. Hata M., Empirical Formula for propagation loss in Land Mobile Radio Services, IEEE Tran. On Vehic. Vol. 29, No 3, August 1980.
9.COST 231, Urban transmission loss models for mobile radio in the 900- and 1800- MHz bands, Revision 2, TD(91) 73, Hange, September 1991
10. RMTP/RB/R105, ISSUE 1.0, APPENDIX 2, 20th November 1990
11. M. J. Feuersteun, T. Rappaport, H. Xia, path loss, Delay Spread and Outage models as Functions of Antenna Height for Microcellular System Design, IEEE Tran. On Vehiv, Vol. 43, No 3, August 1994
12. Okumura Y., Field strength and its vatiability in UHF and VHF land-mobile radio service, Rev. Elec. Commun. Lab., Vol. 16, 1968.
13. Zhang W., A Wide-band propagation model based in UTD for cellular mobile radio communications, IEEE tran. On Antenna and prop., Vol. 45, No 11, November 1997.
14. Kouyoumijian R. G., Pathak P. H., A uniform geometrical theory of diffraction for an edge in a conducting surface, Proc. IEEE Vol. 62, pp. 1448-1461, November 1974.
15. Al-Hussaini E., Effects of Nakagami fading on antijam performance requirements, Electronic Letters, ol. 24, No 4, 1988
16. Stuart A., ord J., Advanced Theory of Statistics, Vol. 1, New York, oxford, 1987.

4.5 Stochastic Models for Long-Term Multipath Fading Channels and their Statistical Properties

Charalambos D. Charalambous and Nickie Menemenlis

4.5.1 Introduction

A common hypothesis about the (power) path loss in wireless transmission of signals over large areas is the log-normal distribution. This hypothesis implies that when the path loss is measured in dB's it is normally distributed [1, 2]. This model does not account for variations in the propagation environment (channel) and any possible relative motion between the transmitter and the receiver. In general, the channel statistics describing the propagation environment change; therefore, data collected over different observation intervals at the receiver, may not fit the stationary hypothesis and thus, the log-normal distribution which assumes time-invariant statistics. In order to capture the dynamic changes of the propagation environment, there is a need for diversified models which describe these changes. In modeling these dynamic changes, one has to investigate the mechanism of electromagnetic wave propagation which is a distributed parameter phenomenon in space. At the microscopic level the distributed parameter nature of the channel can be captured by models which describe reflections, diffractions and scattering of the transmitted signals [1, 2, 3, 4]. Typically when an electromagnetic wave is injected into the channel, it interacts with the environment by means of reflections and scattering, when the wavelength is smaller and larger, respectively, than the encountered object, and by means of diffraction in presence of sharp objects.

These interactions give rise to the time-spread properties of the channel which suggest that a pulse, sent from a particular transmitter, it will manifest itself into a number of pulses arriving at the receiver at different times each experiencing amplitude and phase distortions. Thus, the received instantaneous signal consists of many multipath components, arriving from different directions having different attenuation factors, phase shifts and time delays, all of which are random [5].

In addition to the time-spread property, the channel experiences time variations. This is evidenced by repeating the pulse sounding experiment and observing that the received multipath signal experiences time variations in the number of paths, the arrival times of each path and amplitude and phase distortions [5].

The objective of this work is to introduce detail stochastic dynamical models which describe the time-spread and time variations of the impulse response of wireless channels over large areas. These stochastic models are introduced based on the physics of electromagnetic wave propagation, described at the microscopic level, and measurements (e.g. evidence and justification) reported in the literature [1, 2, 6, 7, 8]. Specifically, the dynamical model describing the power loss proposed in this note takes into account the (i) average power loss due to long distance propagation and (ii) power loss due to reflection of signals from surfaces. When expressed in dB's the dynamical model obtained from (i) and (ii) is described by an Ornstein-Uhlenbeck process which has a mean reverting structure. The mean reversion models the average power loss due to long distance propagation. This gives rise to power loss trajectories which are log-normally distributed. The modeling assumption of power loss accounts for the shadowing or long-term effects of signal fading for locations which have different multipath configurations [1, 2, 9].

Further, the randomness of the number of paths and the arrival times of the these paths suggest the use of point processes.

We examine the arrival times from their macroscopic point of view, as we try to capture the overall characteristics of the signal. We assume that the arrival times are completely random, and the birth or death of any particular occurrence is neither influenced by previous, nor does influence any subsequent occurrence. If in addition, we assume an orderliness of the phenomenon, indicating that the arrival times do not occur in batches, we can model them as the points of a non-homogeneous Poisson-counting process having a time-varying rate. In the past, it has been found that this model disagrees with experimental data [9]. We attribute these discrepancies to the fact that the arrival times were examined from the microscopic point of view in isolation of the remaining properties of the received signal. It is to the authors' belief that this model will behave reasonably well if one considers the overall properties of the signal. This is because inaccuracies at the microscopic level which considers the arrival times alone, are smoothed out at the macroscopic level, where one views the overall received signal. Unfortunately, to the authors' knowledge, this point has not yet been examined in the literature and will be examined here. The non-homogeneous Poisson counting process is a reasonable model because of the simplicity and it should form the core for subsequent generalizations. With this model at hand, the statistical properties of the received signal are examined by performing an analysis which may be considered as a generalization of the shot-noise analysis investigated by Rice [10] in the mid 1940's. The moment generating function of the received signal is calculated and exploited to derive a central limit theorem, the power delay profile, the Doppler spread, and second-order properties of the channel. It is important to note that an analogous analysis for short-term multipath fading channels (MFC) is also feasible [11].

During the modeling process we compute various statistical properties of the received signal. These computations illustrate the mathematical tractability of the models introduced and their consistency with previous experimental data and theoretical models.

In Section 4.5.2, we review the static multipath long-term power loss models which give rise to log-normal distributions.

In Section 4.5.3 we extend this static multipath long-term power loss model to its time-varying analog. It turns out that the dynamical behavior of power loss is best described by a mean-reverting Ornstein-Uhlenbeck process. First and second moments are derived therein.

Finally in Section 4.5.4 we introduce mathematical models for the arrival times of the individual paths as well as for the number of reflections or sub-paths. We compute various statistical properties for the three models discussed above.

4.5.2 Log-Normal Multipath Models: Time-Invariant

In this section, we introduce a static multipath long-term power loss model, for subsequent reference in deriving dynamical analogs.

4.5.2.1 *Long-Term Multi-path Model*

The long-term power loss model describes the total power loss when the transmitter receiver separation distance is large and the propagation environment is not heavily populated. For such channels, the standard power loss model consists of two parts: (i) the standard power loss, due to distance, which implies that the average power loss in dB's increases logarithmically with distance and (ii) the power loss due to multiple reflections in the environment from the transmitter to the receiver.

The average power large-scale path loss for an arbitrary Transmitter-Receiver separation distance d, measured in dB's, namely, $\overline{PL}_d(d)[dB]$, is usually modeled by [1, 2]

$$\overline{PL}_d(d)[dB] = \overline{PL}(d_0) + 10n\log\left(\frac{d}{d_0}\right), \tag{1}$$

where n is the path loss exponent describing the rate of power loss due to distance ($n = 2$ for free space, unobstructed path), $\overline{PL}(d_0)$ is a reference power loss at distance d_0, say, $d_0 = 1$ m for indoors and $d_0 = 100$ m to 1 km for

outdoors transmission.
The power loss due to multiple reflections is usually modeled by a multiplicative process [1]. This is because, at each reflection of the transmitted signal from a surface, a fraction of its power is absorbed by the surface while the remaining carries beyond reflection. Thus, the instantaneous power loss at the receiver due to M reflections, $\mathbb{PL}_r$, is usually modeled by

$$\mathbb{PL}_r \triangleq \prod_{m=1}^{M} \mathbb{PL}_r^m, \tag{2}$$

where $\mathbb{PL}_r^m$ is the reflection coefficient of the mth reflection along a given path, representing the power absorbed by the reflectant surface which, unless one pursues ray tracing techniques, $\{\mathbb{PL}_r\}_{m=1}^{M}$ and M are modeled as random variables.

Thus the overall long-term path loss (in dB) for a particular path is thus modeled by [1]

$$\mathbb{PL}(d)[dB] = \overline{\mathbb{PL}}_d(d)[dB] + 10 \log \mathbb{PL}_r, \tag{3}$$

where d represents the total length of the path. For the direct line of sight (LOS) path, d represents the direct distance between the transmitter and the receiver and $M = 0$ for there are no reflections and thus the second term in (3) is zero.

For a large number of reflections, in view of the Central Limit Theorem (CLT), the decibel power loss due to multiplications $X \triangleq 10 \log \prod_{m=1}^{M} \mathbb{PL}_r^m = 10 \sum_{m=1}^{M} \log \mathbb{PL}_r^m$, gives rise to a normally distributed random variable X. That is $\mathbb{PL}_r = e^{X/c}$, is log-normally distributed, where c is a constant of logarithm conversion. In other words, in view of the CLT the multiplication of random variables gives rise to a log-normal distribution for the same reason that addition of random variables gives rise to normally distributed random variables. That is, over large areas [6], the power loss is modeled by the log-normal density

$$f(r) = \frac{1}{\sqrt{2\pi}\sigma_X r} \exp\left(- \frac{(\ln r - \bar{X})^2}{2\sigma_X^2} \right), \quad E[r] = \exp\left(\bar{X} + \frac{\sigma_X^2}{2} \right), \tag{4}$$

i.e., $r = e^X, r \geq 0$, which implies that the overall power loss in decibels is normally distributed $X \sim N(\bar{X}; \sigma_X^2)$. The statistics $(\bar{X}, \sigma_X^2)$ of this distribution are often obtained by matching them to the observed data. In particular, the random variable X is usually taken as having zero mean, which by (3) implies that the overall long-term path loss varies around the long distance power loss, $\overline{\mathbb{PL}}_d(d)[dB]$. However, realistically, the mean and variance for a given transmitter-receiver location are unknown random processes, due to dynamic changes of the propagation structure and the relative motion between the transmitter and the receiver.

Since, for a particular receiver location the number of reflections is random and the power loss is logarithmically distributed in dB's, around the average large scale path loss model, (3) is general enough to take into account the shadowing or long-term effects of signals fading for locations that have different propagation path structures.

Consequently, in a multipath environment, the long-term instantaneous power variations of each path from the transmitter to any neighborhood of a receiver are, to a large extend, due to multiple reflections along each path, giving rise to the multiplicative behavior, in addition to average power loss due to the distance traveled of each path. This means that the power loss for each path, n, is again given by (3),

$$\mathbb{PL}(d_n)[dB] = \overline{\mathbb{PL}}_d(d_n)[dB] + 10 \log \mathbb{PL}_{r_n}, \tag{5}$$

where the subscript n is used to distinguish between the different paths.

In view of this, the time-spread of the channel can be represented by a multiplicative impulse response model, which in its time-invariant form is given by [6]

$$\begin{aligned} h(t) &= \sum_{n=1}^{N} r_n e^{j\phi_n} \delta(t - t_n) \\ &= \sum_{n=1}^{N} \left(\Pi_{m=0}^{M_n} r_{n,m} \right) e^{j \sum_{m=0}^{M_n} \phi_{n,m}} \delta\left(t - \sum_{m=0}^{M_n} t_{n,m}\right). \end{aligned} \tag{6}$$

Here the set $\{r_n, \phi_n, t_n\}_{n=1}^{N}$ describes the attenuation, phase change and time delay of different paths.

The relation between the random variable $r_{n,m}$ in (6) and (5) is

$$r_n = e^{(1/c)\overline{\mathbb{PL}}_d(d_n)[dB]} \Pi_{m=1}^{M_n} (\mathbb{PL}_{r_n}^{m}) = r_{n,0} \Pi_{m=1}^{Mn} r_{n,m},$$

where the parameter set $\{r_{n,m}, \phi_{n,m}, t_{n,m}\}_{m=1}^{M_n}$ describes the attenuation, phase and time delay due to the LOS propagation path (eg. $r_{n,0}$) between the $(m-1)$th and the mth of the M_n reflections in the nth path from the transmitter to the neighborhood of the receiver. Thus, the amplitude of the received signal other than the LOS path is subject to reflection power loss proportional to $\Pi_{m=1}^{M_n} r_{n,m}$ where $r_{n,m}$ is the attenuation coefficient at the mth reflection of the nth path (e.g., representing power absorbed by the reflectant surface) [1].

Experimental data analyzed in [6] support the point of view that this multiplicative, multipath impulse response model is general enough to describe short-term fading effects in addition to long-term effects.

4.5.3 Log-Normal Multipath Models: Time-Varying

An important assumption in the introduction of the above time-invariant multipath model is spatial homogeneity, although different experiments arrive to distributions which experience time-varying statistics. In order to capture the time-varying properties of the multipath channel, the impulse response model (6) should be replaced by a time-varying analog. The time-varying analog consists of amplitude, phase, delays and number of paths which have explicit time dependence. Thus statistical wide band models for multipath radio propagation treat each point in the 3-D space as linear time-varying filter having an impulse response [12] (see also [1, 5, 8])

$$h(t;\tau) = \sum_{n=1}^{N(0,t)} r_n(t,\tau)e^{j\Phi_n(t,\tau)}\delta\big(\tau - \tau_n(t)\big). \tag{7}$$

Here $h(t;\tau)$ is the response of the channel at time t, due to an impulse applied at time $t-\tau$. $N(0,t)$ is the number of multipath components up to time t, while the set $\{r_n(t,\tau), \Phi_n(t,\tau), \tau_n(t)\}_{n=1}^{N(0,t)}$ describes the random time-varying distortion, phase shift and arrival time of the different paths. In particular, the attenuation component, $r_n(t,\tau)$, corresponds to the path having been subjected to a delay $\tau_n(t)$. $\Phi_n(t,\tau)$ is a function of the phase as well as the Doppler shift. The low-pass equivalent response of the channel to an arbitrary complex-valued input signal $S_\ell(t)$ (assuming an additive white noise channel) is given by

$$\begin{aligned} y_\ell(t) &= \int_{-\infty}^{\infty} h(t;\tau)S_\ell(t-\tau)d\tau + n_\ell(t) \\ &= \sum_{n=1}^{N(0,t)} r_n(t,\tau_n(t))e^{j\Phi_n(t,\tau_n(t))}S_\ell(t-\tau_n(t)) + n_\ell(t), \end{aligned} \tag{8}$$

where $\{n_\ell(t)\}_{t\geq 0}$ is a complex-valued white noise process. If we consider next the transmission through the channel of a narrow band signal with an equivalent low pass signal envelope representation $\{S_l(t)\}_{t\geq 0}$ modulated at the carrier frequency ω_c, namely, $Re\{S_l(t)e^{j\omega_c t}\}$, then the received signal due to the narrow-band transmitted signal is $Re\Big\{y_\ell(t)e^{j\omega_c t}\Big\}$.

4.5.3.1 *Log-Normal Power Loss Model*

The discussion in this section is geared towards the derivation of models which describe the dynamical behavior of the power loss $\{r_n(t,\tau)\}_{n\geq 1}$; the arrival times and the number of paths will be treated in Section 4.5.4. The statistical properties (first and second moments) and probability distributions of these models are computed in each case. To simplify the notation, in the sequel we shall represent $r_n(t, \tau)$ by $r_n(t)$.

In subsequent developments we shall need the definition of the standard Brownian motion [13].

Definition 3.1 *Let* $\left(\Omega, \mathcal{A}, P; \{\mathcal{F}_t\}_{t\geq 0}\right)$ *be a basis probability space with filtration. The random process* $\{W(t)\}_{t\geq 0}$ *is a Brownian motion relative to* $\{\mathcal{F}_t\}_{t\geq 0}$ *if*

1. $W(0) = 0, P - a.s.;$
2. $\{W(t)\}_{t\geq 0}$ *has stationary independent increments, and the increments* $W(t) - W(s)$ *are normally distributed with*

$$E\Big[W(t) - W(s)\Big] = 0, \qquad Var\Big(W(t) - W(s)\Big) = \sigma^2|t - s|;$$

3. *For almost all* $\omega \in \Omega$ *the function* $W(\cdot, \omega) : [0, \infty) \rightarrow \Re$ *is continuous* $P - a.s.$

When $\sigma = 1$ *the process* $\{W(t)\}_{t\geq 0}$ *is called a standard Brownian motion.* □

4.5.3.1.1
The Geometric Brownian Motion Model

Consider a transmitted signal arriving at receiver under the assumption that is goes through many reflections.

Let $0 < t_1 < t_2 < t_3 < ... < t_m$ denote the observation instances at the transmitter of a continuous wave signal, and let $S_n(t_j)$ denote the power loss coefficient associated with the jth observation time, $1 \leq j \leq m$ of the nth path wave. Here $S_n(t_j)$, (which is the analog of $\mathcal{PL}_{r_n}$ indexed by t_j), represents power loss coefficient along the nth path at time t_j, $1 \leq j \leq m$. Suppose that the propagation environment from the transmitter to the receiver is such that a subset of the power loss coefficients associated with $S(t_j)$, correspond to those of $S(t_{j-1})$, $1 \leq j \leq m$. Then to a first-order approximation the percent change (a condition which will be removed when we formulate more general models)

$$\frac{S_n(t_2)}{S_n(t_1)}, \ldots, \frac{S_n(t_m)}{S_n(t_{m-1})}, \tag{9}$$

are independent random variables. Moreover, these reflection coefficients are non-negative and usually exhibit oscillatory behavior consisting of exponential growth intermittent with exponential decay over the long run. These properties support the geometric Brownian motion as a first-order model for the reflection model. Therefore, we can let $S_n(t) = e^{X_n(t)}$, where $\{X_n(t)\}_{t\geq 0}$ is a Brownian motion with non-zero drift, described by the Itô differential equation:

$$dX_n(t) = \gamma_n dt + \delta_n dW_n(t), \qquad X_n(0) \in \Re, \tag{10}$$

where $\{W_n(t)\}_{t\geq 0}$ is a standard Brownian motion which is assumed be independent of $X_n(0)$. It is also assumed that $\vartheta_n \triangleq (\gamma_n, \delta_n)$ are random parameters (possibly depending on τ) which are independent of the Brownian motion and $X_n(0)$. Thus, for a given ϑ_n and $X_n(0) = X_{n,0}$, the distribution of $X_n(t)$ is Gaussian with mean $\gamma_n t + X_{n,0}$ and variance $\delta_n^2 t$. When the parameters (γ_n, δ_n) are zero then $S_n(t) = S_n(s) = S_n(0) = e^{X_n(0)}$, $\forall t, s \in [0, \infty)$ and $S_n(t)$ will be a log-normal random variable, provided $X_n(0)$ is a normally distributed random variable. That is, we arrive at the traditional model for long-term fading.

If (γ_n, δ_n) are not random, then the geometric Brownian motion does not capture the properties of the reflection power loss and that of long distance power loss, and thus it is insufficient. Specifically, it violates the property that long-term power loss variations vary slowly around a mean value which represents the standard distance power loss (although a generalization of (γ_n, δ_n) to time-varying functions will). However, with properly chosen distributions of the parameters (γ_n, δ_n) this model can match the desired property of a slowly varying mean. Nonetheless, its simplicity validates its subsequent discussion and forms the core for the more general model which is introduced shortly.

An application of the stochastic Itô differential rule to $S_n(t) = e^{X_n(t)}$ yields the Itô stochastic differential equation (SDE)

$$dS_n(t) = S_n(t)\Big[\Big(\gamma_n + \frac{\delta_n^2}{2}\Big)dt + \delta_n dW_n(t)\Big], \quad S_n(0) = e^{X_n(0)}. \tag{11}$$

Clearly the random parameters (γ_n, δ_n) can be used to model the propagation environment of each path.

From (11) we also write

$$\frac{dS_n(t)}{S_n(t)} = (\gamma_n + \frac{\delta_n^2}{2})dt + \delta_n dW_n(t). \tag{12}$$

The parameter $(\gamma_n + \frac{\delta_n^2}{2})$ is the instantaneous conditional expected percentage change of the power loss per unit time, δ_n^2 is the instantaneous conditional variance per unit time, and $S_n(0)$ is the power loss at time zero. Thus, (12) can be solved to yield

$$S_n(t) = S_n(0)\exp\Big(\gamma_n t + \delta_n W_n(t)\Big), \quad S_n(0) = e^{X_n(0)}. \tag{13}$$

Taking logarithms of both sides we have

$$\log S_n(t) = \log S_n(0) + \gamma_n t + \delta_n W_n(t). \tag{14}$$

Therefore, if we let

$$X_n(t) \triangleq \log S_n(t), \quad S_n(t) = e^{X_n(t)}, \tag{15}$$

then $\log S_n(t)$ evolves like a Geometric Brownian motion, that is, a Brownian motion with non-zero drift coefficient γ_n [14]. From (14), (15) we deduce that for a given ϑ_n and $X_n(0) = X_{n,0}$, $\{X_n(t)\}_{t\geq 0}$ is a special case of an Ornstein-Uhlenbeck process and thus $\{S_n(t)\}_{t\geq 0}$ is a log-normal process. Moreover, the model (11) which describes the long-term effects of power loss in each path has the following empirical properties for fixed ϑ_n and $X_n(0) = X_{n,0}$.

1. $\{S_n(t)\}_{t\geq 0}$ has continuous trajectories and $\{\log S_n(t)\}_{t\geq 0}$ has stationary independent increments, that is, $\Big(S_n(t) - S_n(s)\Big)/S_n(s), t > s$ is independent of the past history of the signal $\{S_n(\tau); 0 \leq \tau \leq s\}$, and $\Big(S_n(t) - S_n(s)\Big)/S_n(s)$ is identically distributed to $\Big(S_n(t-s) - S_n(0)\Big)/S_n(0)$. This is justified provided that the reflectant surfaces are made of the same material.
2. $\{S_n(t)\}_{t\geq 0}$ is log-normally distributed random process. In particular, $X_n(t) = \log S_n(t)$ has increments $X_n(t+s) - X_n(s)$ which are normally distributed with mean $\gamma_n(t-s)$ and variance $\delta_n^2(t-s)$, which are independent over disjoint intervals, $X_n(t_4) - X_n(t_3)$, $X_n(t_2) - X_n(t_1)$, $t_4 > t_3 \geq t_2 > t_1 \geq 0$. Thus, the increments of $\{X_n(t)\}_{t\geq 0}$ are independent, stationary and identically distributed. Furthermore, $X_n(t)$, is Markovian.
3. The local variance of the signal $\{S_n(t)\}_{t\geq 0}$ increases when its realization increases and/or the volatility parameter δ_n increases.

From the property of normally distributed increments we have for $t > s$:

$$\begin{aligned} Prob\Big\{X_n(t) &\leq X_{n,t} | X_n(s) = X_{n,s}, \vartheta_n\Big\} \\ &= \int_{-\infty}^{\left(X_{n,t} - X_{n,s}\right)} \frac{1}{\sqrt{2\pi\Big(t-s\Big)\delta_n^2}} . \exp\Big(-\frac{\big(y - \gamma_n(t-s)\big)^2}{2(t-s)\delta_n^2}\Big) dy \\ &= \int_{-\infty}^{\left\{X_{n,t} - X_{n,s} - \gamma_n(t-s)\right\}/\delta_n} f(y, t-s) dy, \qquad t > s, \end{aligned} \tag{16}$$

where

$$f(x,t) = \frac{1}{\sqrt{2\pi t}} e^{-\frac{x^2}{2t}}. \tag{17}$$

Theorem 3.2 *The probability density of S_n at time t conditional on ϑ_n is given by*

$$f_{\vartheta_n}(S_{n,t}, t) = \int_0^{\infty} f_{\vartheta_n}(S_{n,t}, t; S_{n,s}, s) f_{\vartheta_n}(S_{n,s}, s) dS_{n,s}, \qquad t > s, \tag{18}$$

where $f_{\vartheta_n}(S_{n,t}, t; S_{n,s}, s)$ *is log-normal given by*

$$f_{\vartheta_n}(S_{n,t}, t; S_{n,s}, s) = \frac{1}{\sqrt{2\pi(t-s)\delta_n^2} S_{n,t}} \exp\left(-\frac{\left(\log(\frac{S_{n,t}}{S_{n,s}}) - \gamma_n(t-s)\right)^2}{2(t-s)\delta_n^2}\right),$$
$$S_{n,t}, S_{n,s} > 0.$$

Remark 3.3 *Let* $X_n(0)$ *be Gaussian. The density of* S_n *at time* t *conditional on* ϑ_n *is also log-normal given by*

$$f_{\vartheta_n}(S_{n,t}, t) = \frac{1}{\sqrt{2\pi p_n(t)} S_{n,t}} \exp\left(-\frac{\left(\log S_{n,t} - m_n(t)\right)^2}{2p_n(t)}\right), \quad S_{n,t} > 0,$$

where $\{m_n(t)\}_{t\geq 0}$ *and* $\{p_n(t)\}_{t\geq 0}$ *are the mean and variance of the Gaussian process* $\{X_n(t)\}_{t\geq 0}$ *given* ϑ_n.

The conditional expected value and second moment of $S_n(t)$ *at time* t *conditional on its current value at time* s, $t > s$ *and* $\vartheta_n \triangleq (\gamma_n, \delta_n)$ *are*

$$E\left[S_n(t) | S_n(s) = S_{n,s}, \vartheta_n\right] = S_{n,s} \exp\left(\gamma_n(t-s) + \frac{\delta_n^2}{2}(t-s)\right), \tag{19}$$

$$E\left[S_n^2(t) | S_n(s) = S_{n,s}, \vartheta_n\right] = S_{n,s}^2 \exp\left(2\gamma_n(t-s) + 2\delta_n^2(t-s)\right). \tag{20}$$

Remark 3.4 *Note that it is straightforward to generalize the Geometric Brownian motion model to one which has time-varying parameters* $\{\gamma_n(t), \delta_n(t)\}_{t\geq 0}$. *In this case,*

$$dS_n(t) = S_n(t)\left[\left(\gamma_n(t) + \frac{\delta_n^2(t)}{2}\right)dt + \delta = ta_n(t) dW_n(t)\right]. \tag{21}$$

4.5.3.1.2
The Mean-Reverting Log-Normal Process

Consider next an alternative more precise model for long-term fading, which assumes the same type of power loss as for the geometric Brownian motion, namely, $S_n(t) = e^{X_n(t)}$. However, instead of $\{X_n(t)\}_{t\geq 0}$ being a geometric Brownian motion, it is assumed that it is generated by a mean-reverting version of an Ornstein-Uhlenbeck process given by

$$dX_n(t) = \beta_n\Big(\gamma_n - X_n(t)\Big)dt + \delta_n dW_n(t), \qquad (22)$$
$$\beta_n > 0, \quad \delta_n > 0, \quad \gamma_n = \overline{PL}_n(d_n)[dB].$$

Here $\vartheta_n \triangleq (\beta_n, \gamma_n, \delta_n)$ (which may depend on τ) are random parameters and $\{W_n(t)\}_{t\geq 0}$ is a standard Brownian motion process which is independent of $X_n(0)$. We will show next that the first term in (22) implies a long-term adjustment towards the value γ_n, where β_n is the speed of the adjustment. The long-term mean, γ_n, models the average large-scale path loss for an arbitrary Transmitter-Receiver separation distance d_n incorporated into (22).

To determine the behavior of $\{S_n(t)\}_{t\geq 0}$ we first examine the behavior implied by (22).

For a given ϑ_n, the solution of (22) starting at $X_n(s) = X_{n,s}$ is

$$X_n(t) = e^{-\beta_n(t-s)}\Big(X_{n,s} + \gamma_n\big(e^{\beta_n(t-s)} - 1\big) + \delta_n e^{-\beta_n s}\int_s^t e^{\beta_n \tau} dW_n(\tau)\Big).$$

Consequently, $\{X_n(t)\}_{t\geq 0}$ is a Gaussian random process, and the conditional expected value and variance of X_n at time t conditional on its current value at s, $t > s$ and ϑ_n are

$$\begin{aligned} m_n(t) &\triangleq E\Big[X_n(t)|X_n(s) = X_{n,s}, \vartheta_n\Big] \\ &= e^{-\beta_n(t-s)}\Big(X_{n,s} + \gamma_n\big(e^{\beta_n(t-s)} - 1\big)\Big), \end{aligned} \qquad (23)$$

$$p_n(t) \triangleq Var\Big(X_n(t)|X_n(s) = X_{n,s}, \vartheta_n\Big) = \delta_n^2\Big(\frac{1 - e^{-2\beta_n(t-s)}}{2\beta_n}\Big). \qquad (24)$$

As t becomes large, the random process $\{X_n(t)\}_{t\geq 0}$ converges in law to a Gaussian random variable with mean $\gamma_n = \overline{PL}_n(d_n)[dB]$ and variance $\frac{\delta_n^2}{2\beta_n}$, that is,

$$f_{\vartheta_n}(X_{n,\infty}, \infty; X_{n,s}, s) = \frac{1}{\sqrt{2\pi\frac{\delta_n^2}{2\beta_n}}}\exp\Big(-\frac{(X_{n,\infty} - \gamma_n)^2}{2\big(\frac{\delta_n^2}{2\beta_n}\big)}\Big). \qquad (25)$$

As expected the steady-state conditional mean is the long-term mean of $\{X_n(t)\}_{t\geq 0}$. This implies that for large transmitter receiver separation distance d_n, the power loss is dominated by the average large scale path loss, since the contribution due to reflections will be negligible.

From (22), the dynamics of S_n are obtained by the Itô differential rule [13] to yield

$$dS_n(t) = S_n(t)\Big(\beta_n\Big[\gamma_n - X_n(t)\Big] + \frac{1}{2}\delta_n^2\Big)dt + S_n(t)\delta_n dW_n(t). \tag{26}$$

Clearly, for a given ϑ_n the joint process $\{S_n(t), X_n(t)\}_{t\geq 0}$ is a Markov process although $\{S_n(t)\}_{t\geq 0}$ is not a Markov process. Since $X_n(t) = \log S_n(t)$, the distribution of $S_n(t)$ is log-normal and it is straightforward to derive the density and the moments for $S_n(t)$ (by following the exposition of Section 4.5.3.1.1).

Remark 3.5 *Note that an alternative interpretation of (22) is to write $X_n(t) = \gamma_n + \tilde{X}_n(t)$, where $\{\tilde{X}_n(t)\}_{t\geq 0}$ is a Gaussian random process of Ornstein-Uhlenbeck type which may have a zero mean and thus does not possess a mean-reverting structure. In this case, γ_n will be the analog of $\overline{\mathbb{PL}}_d(d_n)[dB]$ and $\tilde{X}_n(t)$ will be the analog of $10\log \Pi_{m=1}^{M_n} \mathbb{PL}_{r_n}^m$ of Section 4.5.2.1.*

Corollary 3.6

1. *The density of S_n at time t conditional on its current value at time s, $t > s$ and ϑ_n is given by*

$$f_{\vartheta_n}(S_{n,t}, t; S_{n,s}, s) = \frac{1}{\sqrt{2\pi p_n(t)}S_{n,t}} \exp\Big(- \Big(\log\frac{S_{n,t}}{S_{n,s}} - m_n(t)\Big)^2 / 2p_n(t)\Big), \quad S_{n,t}, S_{n,s} > 0.$$

where $\{m_n(t), p_n(t)\}_{t\geq 0}$ are given by (23), (24), respectively.

2. *The conditional expected value and variance S_n at time t conditional on its current value at time s, $t > s$ and ϑ_n are given by*

$$E\Big[S_n(t)|S_n(s) = S_{n,s}, \vartheta_n\Big] = S_{n,s}^{e^{-\beta_n(t-s)}} \exp\Big(\gamma_n\big(1 - e^{-\beta_n(t-s)}\big) + \frac{\delta_n^2}{4\beta_n}\big(1 - e^{-2\beta_n(t-s)}\big)\Big),$$

$$E\Big[S_n^2(t)|S_n(s) = S_{n,s}, \vartheta_n\Big] = S_{n,s}^{2e^{-\beta_n(t-s)}} \exp\Big(2\gamma_n\big(1 - e^{-\beta_n(t-s)}\big) + \frac{\delta_n^2}{2\beta_n}\big(1 - e^{-2\beta_n(t-s)}\big)\Big).$$

3. *As t becomes large the density of S_n at time t conditional on its current value at time s, $t > s$ and ϑ_n has a steady-state density given by*

$$f_{\vartheta_n}(S_{n,\infty}, \infty; S_{n,s}, s) = \frac{1}{\sqrt{2\pi \frac{\delta_n^2}{2\beta_n}} S_{n,\infty}} \exp\Big(- \Big(\log\frac{S_{n,\infty}}{S_{n,s}} - \gamma_n\Big)^2 / 2\Big(\frac{\delta_n^2}{2\beta_n}\Big)\Big).$$

Proof. Follows from the earlier discussion. □

Hence, the log-normal model described by (26) approaches the log-normal random variable as time increases.

Remark 3.7 *Similarly as before, we point out that a generalization to time-varying parameters* $\{\gamma_n(t), \delta_n(t)\}_t \geq 0$ *is possible.*

In this case (22) is given by

$$dX_n(t) = \beta_n(t)\Big(\gamma_n(t) - X_n(t)\Big)dt + \delta_n(t)dW_n(t), \tag{27}$$

where $\vartheta_n \triangleq \Big(\beta_n(t), \gamma_n(t), \delta_n(t)\Big)_{n\geq 1}$ *are measurable with respect to the* σ*-algebra* $\mathcal{F}_t$ *(possibly generated by* $\sigma\{W_n(s);\quad 0 \leq s \leq t\}$*, or a bigger* σ*-algebra). Again, for fixed* $X_n(s) = X_{n,s}$ *the solution of (27) is*

$$X_n(t) = e^{-\beta_n(t,s)}\Big(X_{n,s} + \int_s^t e^{\beta_n(\tau,s)}\beta_n(\tau)\gamma_n(\tau)d\tau + \int_s^t e^{\beta_n(\tau,s)}\delta_n(\tau)dW_n(\tau)\Big),$$

$$\beta_n(t,s) \triangleq \int_s^t \beta_n(\tau)d\tau.$$

Moreover, letting $S_n(t) = e^{X_n(t)}$ *we obtain*

$$dS_n(t) = S_n(t)\Big(\beta_n(t)[\gamma_n(t) - X_n(t)] + \frac{1}{2}\delta_n^2 = (t)\Big)dt + S_n(t)\delta_n(t)dW_n(t).$$

Assuming $\vartheta_n(t)$ *is a deterministic function of t, then*

$$E[X_n(t)|X_n(s) = X_{n,s}] = e^{-\beta_n(t,s)}[X_{n,s} + \int_s^t e^{\beta_n=(\tau,s)}\beta_n(\tau)\gamma_n(\tau)d\tau],$$

$$Var[X_n(t)|X_n(s) = X_{n,s}] = e^{-2\beta_n(t,s)}\int_s^t e^{2\beta_n(\tau,s)}\delta_n^2(\tau)d\tau,$$

from which the first and second moments of $S_n(t)$ *can be determined.*

4.5.4 Statistical Analysis of the Received Signal

In this section, we introduce mathematical models for the arrival times of the individual paths at the receiver, the number of reflections encountered in each path from the transmitter to the receiver and the number of paths.

Consider the transmission of a low-pass signal, $S_\ell(t)$, through the time-varying multipath fading channel (with time-invariant scatters). Each path of the channel has the effect of delaying the original signal, attenuating it and affecting its phase which is represented by its low-pass equivalent time-varying impulse response given by (7) and repeated below by

$$h(t;\tau) = \sum_{i=1}^{N(0,t)} r_i(t,\tau)e^{j\Phi_i(t,\tau)}\delta\big(\tau - \tau_i\big). \tag{28}$$

Throughout this section we assume for simplicity that

$$\Phi_i(t,\tau) \triangleq \phi_i(t) - (\omega_c + \omega_{d,i}(t))\tau + \omega_{d,i}(t)t. \tag{29}$$

Here $\phi_i(t)$ and $\omega_{d,i}(t)$ represent the phase-shift and Doppler-shift respectively corresponding to the path i delayed by τ_i. $N(0,t)$ denotes the number of paths received in the time interval $[0,t)$. The output of the channel when the input is the low-pass signal $S_\ell(t)$ is represented by

$$\begin{aligned} y_\ell(t) = \int_{-\infty}^{\infty} h(t;\tau)S_\ell(t-\tau)d\tau &= \sum_{i=1}^{N(0,t)} r_i(t,\tau_i)e^{j\Phi_i(t,\tau_i)}S_\ell\big(t-\tau_i\big) \\ &\triangleq \sum_{i=1}^{N(0,t)} h_\ell(t,\tau_i;\mathbf{m}_i(t,\tau_i)). \end{aligned} \tag{30}$$

Here we make use of a marked point process, which is defined as a point process with an auxiliary random variable, called the mark, associated with each point, where for each (t,τ_i) the mark $\mathbf{m}_i(t,\tau_i)$ takes values in the space of marks, say, $\mathbf{M}$. The marks $\{\mathbf{m}_i(t,\tau_i)\}_{t\geq 0} \triangleq \{r_i(t,\tau_i), \phi_i(t), \omega_{d,i}(t)\}_{t\geq 0}$ are arbitrary random processes associated with the *ith* occurrence time τ_i.

4.5.4.1
Path Arrival Times: Non-Homogeneous Poisson Process

We examine the arrival times from their macroscopic point of view, as we try to capture the overall characteristics of the signal. We make this assumption even if past experimental data disagree with this model [9]. We attribute these discrepancies to the fact that the arrival times were examined from the microscopic point of view in isolation of the remaining characteristics of the received signal. This is because the independent increment and orderliness properties of the Poisson process which makes it inadequate in matching experimental data of the arrival times, do not imply that the received signal $y_\ell(t)$ will have these properties. More general models such as doubly stochastic Poisson processes which are neither orderliness neither independent increment processes are studied in [11], and they hinge upon our understanding of the current model.

In fact our analysis shows that many of the statistical properties of the received signal can be explained using a non-homogeneous Poisson counting process. These include the second-order statistics, the power delay profile and Doppler spread of the channel and the Gaussianality of the received signal.

Notation 4.1

1. $|c|^2 \triangleq c^* c$, *where* $c \in C$ *is complex and "*" denotes complex conjugation;*
2. *For* $\rho \in C^n$, *where* $\rho_{R_i} \triangleq Re(\rho_i)$ *and* $\rho_{I_i} \triangleq Im(\rho_i), 1 \leq i \leq n$ *denote the real and imaginary components of* ρ, *respectively. The complex derivatives with respect to* ρ *and* ρ^* *are defined in terms of real derivatives as follows:*

$$\frac{\partial}{\partial \rho_i} \triangleq \left(\frac{\partial}{\partial \rho_{R_i}} - j\frac{\partial}{\partial \rho_{I_i}}\right)/2, \quad \frac{\partial}{\partial \rho_i^*} \triangleq \left(\frac{\partial}{\partial \rho_{R_i}} + j\frac{\partial}{\partial \rho_{I_i}}\right)/2, \quad 1 \leq i \leq n.$$

Let $\left(\Omega, \mathcal{A}, P\right)$ be a complete probability space equipped with filtration $\{\mathcal{F}_t\}_{t\geq 0}$ and finite-time $[0, T_s], T_s < \infty$, on which the following random processes are defined.

$$r_i : [0, T_s] \times \Omega^r \times \Omega^\tau \to \Re, \quad \phi_i : [0, T_s] \times \Omega^\phi \to \Re, \quad \omega_{d,i} : [0, T_s] \times \Omega^{\omega_d} \to \Re,$$
$$\tau_i : \Omega^\tau \to \Re, \quad N : (0, T_s) \times \Omega^N \to \{1, 2, 3, \ldots\}, \quad \mathbf{m}_i \triangleq (r_i, \phi_i, \omega_{d,i})$$

in which $N(0, T_s)$ is a non-homogeneous Poisson counting process with law

$$\text{Prob}\left\{N(0, T_s) = k\right\} = \frac{\left(\int_0^{T_s} \lambda(t)dt\right)^k}{k!} \exp\left(-\int_0^{T_s} \lambda(t)dt\right), \ \lambda(t) \geq 0, \forall t \in [0, T_s], \tag{31}$$

$$E[N(0, T_s)] = \int_0^{T_s} \lambda(t)dt, \quad E[N^2(0, T_s)] = (\int_0^{T_s} \lambda(t)dt)^2 + \int_0^{T_s} \lambda(t)dt.$$

We note that if the Poisson parameter λ is constant the arrival times $\{\tau_i\}_{i\geq i}$ are homogeneous Poisson points and hence $\Delta\tau_i \triangleq \tau_{i+1} - \tau_i$ are exponentially distributed with law $\text{Prob}\{\Delta\tau_i \leq x\} = 1 - e^{-\lambda x}, x \geq 0$.

This section investigates the statistical properties of the causal time-varying response of (30), which means that $= h_\ell(t, \tau_i; \mathbf{m}_i(t, \tau_i)) = 0$, for $\tau_i > t$, for every i.

Theorem 4.2 *Let* $\{N(0, t)\}_{t\geq 0}$ *be a non-homogeneous Poisson counting process and* $\{\tau_i\}_{i=1}^{N(0,t)}$ *its points.*
For any points τ_i *satisfying* $0 \leq t_1 \leq t_2 \leq \ldots, t_n \leq t$, *the density of* $\{\tau_j\}_{j=1}^n$ *conditional on* $N(0, t) = n$ *is*

$$f_\tau(t_1, t_2, \ldots, t_n \mid N(0, t) = n) = n! \prod_{i=1}^{n} \frac{\lambda(t_i)}{\int_0^t \lambda(\sigma)d\sigma},$$

which is the density of the order-statistics of the sample of n observations taken from the uniform distribution on $[0, t)$.
Under the condition that $N(0, t) = n$, *the points* $\{\tau_i\}_{i=1}^{N(0,t)}$ *of the Poisson process are iid with law*

$$f_{\tau_i}(\tau) = \frac{\lambda(\tau)}{\int_0^t \lambda(\sigma)d\sigma}, \quad 0 \leq \tau \leq t \leq T_s, \quad 1 \leq i \leq n. \tag{32}$$

Thus given $N(0, t) = n$, *this is the density of the ith occurrence time,* τ_i, *of a non-homogeneous Poisson process on the interval* $[0, t)$.

4.5.4.1.1
Distribution and Moment Generating Functions

In this section we derive expressions for the moment generating functions of the received low-pass equivalent signal.
Distribution and Moment Generating Functions. Let $v_\ell : [0, T_s] \to C$ and define

$$y_{\ell,v}(T_s) \triangleq \int_0^{T_s} y_\ell(\tau) dv_\ell(\tau). \tag{33}$$

The probability density function and characteristic functions of the complex random process $y_{\ell,v}(T_s)$ are, respectively, defined by

$$f_{y_{\ell,v}}(x_\ell, T_s)dx_\ell \triangleq E\left[I_{\{y_{\ell,v}(T_s)\in dx_\ell\}}\right], \quad \Phi_{y_{\ell,v}}(v_\ell, T_s) \triangleq E\left[e^{jRe\left(y_{\ell,v}(T_s)\right)}\right],$$

where $I_{\{.\}}$ is the indicator function.

Theorem 4.3 *Consider model (30)-(31). Assume that for each* $t \in [0, T_s]$ *the marks* $\{\mathbf{m}_i(s, \tau_i), 0 \leq s \leq t\}_{i \geq 1}$ *satisfy the independence and iid property*

$$\begin{aligned} & f\Big(\{\mathbf{m}_i(\sigma, \tau_i); 0 \leq \sigma \leq t\}_{i=1}^{N(0,t)} | \tau_1, \tau_2, \ldots, \tau_{N(0,t)}, N(0, t)\Big) \\ &= \Pi_{i=1}^{N(0,t)} f\Big(\{\mathbf{m}_i(\sigma, \tau_i); 0 \leq \sigma \leq t\} | \tau_1, \tau_2, \ldots, \tau_{N(0,t)}, N(0, t)\Big) \\ &= \Pi_{i=1}^{N(0,t)} f\Big(\{\mathbf{m}_i(\sigma, \tau_i); 0 \leq \sigma \leq t\} | \tau_i\Big), \end{aligned} \tag{34}$$

where $f\Big(\{\mathbf{m}_i(\sigma, \tau_i); 0 \leq \sigma \leq t\} | \tau_i\Big)$ *are identical.*
Let $v_\ell : [0, T_s] \to C$ *be such that*

$$\int_0^{T_s} \int_0^{T_s} f_\ell(\alpha, \beta) dv^*(\alpha) dv(\beta) < \infty, \tag{35}$$

where f_ℓ is defined by

$$f_\ell(\alpha,\beta) \triangleq \int_0^{min(\alpha,\beta)} \lambda(\tau)E\Big[h_\ell^\star(\alpha,\tau;\mathbf{m}(\alpha,\tau))h_\ell(\beta,\tau;\mathbf{m}(\beta,\tau))\Big]d\tau$$
$$+\int_0^{\alpha} \lambda(\tau)E\Big[h_\ell^\star(\alpha,\tau;\mathbf{m}(\alpha,\tau))\Big]d\tau \;\times\; \int_0^{\beta} \lambda(\tau)E\Big[h_\ell(\beta,\tau;\mathbf{m}(\beta,\tau))\Big]d\tau. \tag{36}$$

The characteristic function of $y_{\ell,v}(T_s)$ is

$$\Phi_{y_{\ell,v}}(jv,T_s) \triangleq E\Big[\exp\Big\{jRe\Big(\int_0^{T_s} y_\ell(\sigma)dv_\ell(\sigma)\Big)\Big\}\Big]$$
$$= \exp\Big\{\int_0^{T_s}\lambda(\tau)E\Big[\exp\Big\{jRe\Big(\int_0^{T_s} h_\ell(\sigma,\tau;\mathbf{m}(\sigma,\tau))dv_\ell(\sigma)\Big)\Big\}-1\Big]d\tau\Big\}. \tag{37}$$

Moreover,

$$\Psi_{y_\ell}(\rho,t) \triangleq \ln E\Big[e^{jRe\left(\rho^\star y_\ell(t)\right)}\Big] = \sum_{k=1}^{\infty} j^k \frac{\gamma_{\ell,k}(t)}{k!}, \qquad \rho\in C, \tag{38}$$

where

$$\gamma_{\ell,k}(t) = \int_0^t \lambda(\tau)E\Big[Re\Big(\rho^\star h_\ell(t,\tau;\mathbf{m}(t,\tau))\Big)\Big]^k d\tau, \tag{39}$$

is the $\gamma_{\ell,k}(t)$ cummulant of $y_\ell(t)$ and

$$E\Big[y_\ell(t)\Big] = (-2j)j\tfrac{\partial}{\partial\rho^\star}\gamma_{\ell,1}(t) = \textstyle\int_0^t \lambda(\tau)E\Big[h_\ell(t,\tau;\mathbf{m}(t,\tau))\Big]d\tau,$$
$$E\Big[y_\ell^\star(t)\Big] = (-2j)j\tfrac{\partial}{\partial\rho}\gamma_{\ell,1}(t) = \textstyle\int_0^t \lambda(\tau)E\Big[h_\ell^\star(t,\tau;\mathbf{m}(t,\tau))\Big]d\tau, \tag{40}$$
$$Var\Big(y_\ell(t)\Big) = (-2j)^2j^2\tfrac{\partial}{\partial\rho}\tfrac{\partial}{\partial\rho^\star}\tfrac{\gamma_{\ell,2}(t)}{2!} = \textstyle\int_0^t \lambda(\tau)E\Big[|h_\ell(t,\tau;\mathbf{m}(t,\tau))|^2\Big]d\tau.$$

Here expectations involving the marks are for fixed τ.

Corollary 4.4 *Consider model (30)-(31) and (34) holds.*
The joint characteristic function of $y_\ell(t_1),\dots,y_\ell(t_n)$ is

$$\Phi_{\mathbf{y}_\ell}(\rho_1,t_1;\dots;\rho_n,t_n) \triangleq E\Big[\exp\Big\{jRe\Big(\rho^\star y_\ell(t)\Big)\Big\}\Big],$$

$$= \exp\left\{ \int_0^{T_s} \lambda(\tau) E\Big[\exp\Big(j \sum_{i=1}^{n} \Big[\rho_{R_i} Re\Big(h_\ell(t_i, \tau; \mathbf{m}(\sigma,\tau)) \Big) + \rho_{I_i} Im\Big(h_\ell(t_i, \tau; \mathbf{m}(\sigma,\tau)) \Big) \Big] \Big) - 1 \Big] d\tau \right\}$$

$$= \exp\left\{ \int_0^{T_s} \lambda(\tau) E\Big[\exp\Big(jRe\Big(\rho^\star \mathbf{h}_\ell(t, \tau; \mathbf{m}(\sigma,\tau)) \Big) \Big) - 1 \Big] d\tau \right\}, \tag{41}$$

where $\mathbf{y}_\ell = \Big(y_\ell(t_1), \ldots, y_\ell(t_n) \Big)' \in C^n$, $\mathbf{h}_\ell(t, \tau; \mathbf{m}(\sigma,\tau)) = \Big(h_\ell(t_1, \tau; \mathbf{m}(\sigma,\tau)), \ldots, h_\ell(t_n, \tau; \mathbf{m}(\sigma,\tau)) \Big)'$. *The joint-moment generating function of the complex random variables* $y_\ell(t_1), \ldots, y_\ell(t_n)$ *is*

$$E\Big[\prod_{i=1}^{n} \Big(y_\ell^\star(t_i) \Big)^{k_i} \prod_{i=1}^{n} \Big(y_\ell(t_i) \Big)^{m_i} \Big]$$

$$= (-2j)^{\sum_{i=1}^{n} (k_i + m_i)} \prod_{i=1}^{n} \Big(\frac{\partial}{\partial \rho_i} \Big)^{k_i} \prod_{i=1}^{n} \Big(\frac{\partial}{\partial \rho_i^\star} \Big)^{m_i} \Phi_{\mathbf{y}_\ell}(\rho_1, t_1, \ldots; \rho_n, t_n)|_{\rho=0}.$$

Proof. The derivations are similar to those of Theorem 3.1. □

Correlation and Covariance. Clearly, the joint statistics of the random variables $y_\ell(t_1)$ and $y_\ell(t_2)$ are obtained from Theorem 3.1, by letting $d\nu_\ell(\sigma) = \Big(\rho_1^\star \delta(\sigma - t_1) + \rho_2^\star \delta(\sigma - t_2) \Big) d\sigma$, $\rho_1 \in C$, $\rho_2 \in C$, $0 \le t_1, t_2 \le T_s$,

$$\Phi_{y_{\ell,\nu}}(j\nu_\ell, T_s) = E\Big[e^{jRe\Big(\rho_1^\star y_\ell(t_1) + \rho_2^\star y_\ell(t_2) \Big)} \Big] \stackrel{\triangle}{=} \Phi_{y_\ell}(\rho_1, t_1; \rho_2, t_2); \tag{42}$$

which is by definition the joint characteristic function of $y_\ell(t_1)$ and $y_\ell(t_2)$. Substituting into (37) yields

$$\Phi_{y_{\ell,1}, y_{\ell,2}}(\rho_1, t_1; \rho_2, t_2) = \exp\left\{ \int_0^{T_s} \lambda(\tau) E\Big[\exp\Big\{ jRe \sum_{i=1}^{2} \rho_i^\star h_\ell\Big(t_i, \tau; \mathbf{m}(t_i, \tau) \Big) \Big\} - 1 \Big] d\tau \right\}. \tag{43}$$

Consequently, using (43) and the causality assumption we deduce the covariance and correlation functions of the random variables $y_\ell(t_1)$ and $y_\ell(t_2)$:

$$C_{y_\ell}(t_1, t_2) \stackrel{\triangle}{=} R_{y_\ell}(t_1, t_2) - E\Big[y_\ell^\star(t_1) \Big] E\Big[y_\ell(t_2) \Big] = \int_0^{\min\Big(t_1, t_2 \Big)} \lambda(\tau) E\Big[h_\ell^\star(t_1, \tau; \mathbf{m}(t_1, \tau)) h_\ell(t_2, \tau; \mathbf{m}(t_2, \tau)) \Big] d\tau, \tag{44}$$

$$\begin{aligned} R_{y_\ell}(t_1,t_2) &\triangleq E\Big[y_\ell^\star(t_1)y_\ell(t_2)\Big] \\ &= \int_0^{\min\left(t_1,t_2\right)} \lambda(\tau)E\Big[h_\ell^\star(t_1,\tau;\mathbf{m}(t_1,\tau))h_\ell(t_2,\tau;\mathbf{m}(t_2,\tau))\Big]d\tau \\ &+ \int_0^{t_1} \lambda(\tau)E\Big[h_\ell^\star(t_1,\tau;\mathbf{m}(t_1,\tau))\Big]d\tau \\ &\times \int_0^{t_2} \lambda(\tau)E\Big[h_\ell(t_2,\tau;\mathbf{m}(t_2,\tau))\Big]d\tau. \end{aligned} \tag{45}$$

The function $f_\ell(.,.)$ in (36) is identified to be the correlation function and the condition (35) ensures existence of $\int_0^T jRe\Big(y_\ell(\sigma)dv(\sigma)\Big)$ in mean-square sense and that its second moment is finite.

Central Limit Theorem. The next theorem states that as λ increases, the received signal is approximately Gaussian.

Theorem 4.5 *Consider model (30)-(31) and (34) holds.*
Consider the joint-characteristic function of $y_\ell(t_1),\ldots,y_\ell(t_n)$

$$\begin{aligned} &\Phi_{y_\ell}(\rho_1,t_1;\ldots;\rho_n,t_n) \\ &= \exp\Big\{\int_0^{T_s}\lambda(\tau)E\Big[\exp\Big\{\sum_{i=1}^n jRe\Big(\rho_i^\star h_\ell(t_i,\tau;\mathbf{m}(t_i,\tau))\Big)\Big\}-1\Big]d\tau\Big\}. \end{aligned} \tag{46}$$

Assume $E\big[y_\ell(t)\big]<\infty, \mathrm{Var}\big(y_\ell(t)\big)<\infty, 0\leq t\leq T_s$ *and that*

$$\frac{\Gamma_\ell(t_1,t_2,t_3)}{\sqrt{\sigma_{y_\ell}(t_1)\sigma_{y_\ell}=(t_2)\sigma_{y_\ell}(t_3)}}, \tag{47}$$

tends to zero, uniformly on $[0,T_s]\times[0,T_s]\times[0,T_s]$, *as the rate* $\lambda(t)$ *tends to infinity.*
Here

$$\begin{aligned} \Gamma_\ell(t_1,t_2,t_3) &= \int_0^{\min\left(t_1,t_2,t_3\right)}\lambda(\tau)E\Big[\mid Re\Big(\rho^\star h_\ell(t_1,\tau;\mathbf{m}(t_1,\tau))\Big) \\ &Re\Big(\rho^\star h_\ell(t_2,\tau;\mathbf{m}(t_2,\tau))\Big)Re\Big(\rho^\star h_\ell(t_3,\tau;\mathbf{m}(t_3,\tau))\Big)\mid\Big]d\tau, \\ &\sigma_{y_\ell}(t_i)=\sqrt{\mathrm{Var}\Big(y_\ell(t_i)\Big)}. \end{aligned}$$

Then the joint characteristic function of the centered random variables

$$y_{\ell,c}(t_i)\triangleq\frac{y_\ell(t_i)-Ey_\ell(t_i)}{\sigma_{y_\ell}(t_i)},\quad 1\leq i\leq n, \tag{48}$$

is in the limit, as the rate tends to infinity, Gaussian with joint-characteristic function

$$\lim_{\lambda\to\infty} \Phi_{y_{\ell,c}}(j\rho_1, t_1; \ldots; j\rho_n, t_n) \triangleq \lim_{\lambda\to\infty} E\Big[\exp\Big\{\sum_{i=1}^{n} jRe\Big(\rho_i^\star y_{\ell,c}(t_i)\Big)\Big\}\Big]$$

$$= \exp\Big\{-\frac{1}{2}\int_0^t \lambda(\tau)E\Big[\sum_{i=1}^{n}\Big[\frac{\rho_{R_i}}{\sigma_{y_\ell}(t_i)}Re\Big(h_\ell(t_i,\tau;\mathbf{m}(t_i,\tau))\Big)$$

$$+\frac{\rho_{I_i}}{\sigma_{y_\ell}(t_i)}Im\Big(h_\ell(t_i,\tau;\mathbf{m}(t_i,\tau))\Big)\Big]^2 d\tau\Big\}. \tag{49}$$

Proof. (46) is obtained by (41). (49) is obtained by expanding the joint characteristic function of the centered random variables in power series and letting $\lambda \to \infty$, see [11]. For real signals with marks having no explicit (t, τ) dependence this is derived in [16]. □

4.5.4.2
Channel Autocorrelation Functions and Power Spectra

In this section we employ the previous results to compute the frequency characteristics and the standard autocorrelation and power spectra of time-varying channels. First, we compute the Fourier transform of the impulse response to obtain the time-varying transfer function of the channel. Second, the Fourier transform on the channel autocorrelation function with respect to the path delay τ gives rise to delay power spectrum. A second Fourier transform on the resulting function with respect to time variations gives rise to a quantity which characterizes the Doppler spread of the channel. These ideas are developed in [17] (see also [5]).

We consider the time-varying impulse response, given by (28) and (29) with

$$\omega_{d,i}(t) = 2\pi f_{d,i}(t), \qquad f_{d,i}(t) = f_m \cos\vartheta_i(t), \qquad f_m = v/\lambda_c. \tag{50}$$

That is, the incoming plane wave is vertically polarized traveling in the $x-y$ plane, making an angle $\vartheta_n(t)$ with the x-axis; a generalization to 3-D (see [3]) is straight-forward. The mobile travels in the x-axis direction with velocity v. λ_c is the wavelength of the incident wave.

The autocorrelation function of $h(t;\tau)$ is defined by

$$E[h(t_1;\tau_1)h^\star(t_2;\tau_2)]$$

$$= E\Big[\sum_{i=1}^{N(0,t_1)} r_i(t_1,\tau_1)e^{j\Phi_i(t_1,\tau_1)}\delta(\tau_1-\tau_i)$$

$$\times \sum_{i=1}^{N(0,t_2)} r_i(t_2,\tau_2)e^{-j\Phi_i(t_2,\tau_2)}\delta(\tau_2-\tau_i)\Big]$$

$$= E\Big[\sum_{i=1}^{N\left(0,min(t_1,t_2)\right)} r_i(t_1,\tau_1)r_i(t_2,\tau_1)e^{j\left(\Phi_i(t_1,\tau_1)-\Phi_i(t_2,\tau_1)\right)}\delta(\tau_1-\tau_i)\Big]$$

$$\delta(\tau_2-\tau_1) + E\Big[\sum_{\substack{i=1\\ i\neq j}}^{N(0,t_1)}\sum_{j=1}^{N(0,t_2)} r_i(t_1,\tau_1)e^{j\Phi_i(t_1,\tau_1)}r_j(t_2,\tau_2)e^{-j\Phi_j(t_2,\tau_2)}$$

$$\delta(\tau_1-\tau_i)\delta(\tau_2-\tau_j)\Big]. \tag{51}$$

In most radio transmission media, the attenuation and phase-shift of the channel associated with path delay τ_1 is uncorrelated with the attenuation and phase-shift of path delay τ_2 making the second term of (51) negligible. This is called an uncorrelated scattering (US) channel. In the sequel we assume an US channel and (34) holds. However the removal of the US assumption does not affect the calculation because we obtain (44) by subtracting the mean from (45).

4.5.4.2.1
Uncorrelated Scattering

1. *Autocorrelation in Time Domain*

$$E[h(t_1;\tau_1)h^\star(t_2;\tau_2)] = \Phi_c(t_1,t_2;\tau_1)\delta(\tau_2=-\tau_1),$$

where

$$\Phi_c(t_1,t_2;\tau)$$
$$= E\Big[\sum_{i=1}^{N\left(0,min(t_1,t_2)\right)} r_i(t_1,\tau)r_i(t_2,\tau)e^{j\left(\Phi_i(t_1,\tau)-\Phi_i(t_2,\tau)\right)}\delta(\tau-\tau_i)\Big].$$

This represents the correlation of two impulses applied at times $(t_1-\tau)$ and $(t_2-\tau)$. In computing the above autocorrelation function we employ (44) and the uncorrelated scattering condition.

$$\Phi_c(t_1,t_2;\tau)$$
$$= \int_0^{min(t_1,t_2)} \lambda(\tau)E\Big[r(t_1,\tau)r(t_2,\tau)e^{j\left(\Phi(t_1,\tau)-\Phi(t_2,\tau)\right)}\Big]\delta(\tau-\beta)\,d\beta$$
$$= \lambda(\tau)E\Big[r(t_1,\tau)r(t_2,\tau)e^{j\left(\Phi(t_1,\tau)-\Phi(t_2,\tau)\right)}\Big].$$

2. *Power Delay Profile or Delay Power Spectrum*
   The power delay profile of the channel is defined by

$$\Phi_c(t;\tau) \triangleq \Phi_c(t_1,t_2;\tau)|_{t_1=t_2=t} = \lambda(\tau)E\left[r^2(t,\tau)\right]. \tag{52}$$

This expression represents the average power output of the channel as a function of the time delay, τ and the observation time t. In the frequency domain it is represented by its delay power spectrum

$$\Phi_c(t;\Delta f) \triangleq \mathcal{F}_\tau\{\Phi_c(t;\tau)\} = \int_{-\infty}^{\infty} \lambda(\tau)E\left[r^2(t,\tau)\right]e^{-j2\pi\Delta f\tau}d\tau. \tag{53}$$

Notice that the rate of the Poisson process, $\lambda(\tau)$, is an important parameter which shapes the power delay profile through multiplication, while it shapes the delay power spectrum through convolution.

3. *Frequency Variations of the Channel*
   By taking the Fourier transform of the time-varying impulse response we obtain the time-varying transfer function of the channel

$$\begin{aligned} h(t;f) &\triangleq \mathcal{F}_\tau\{h(t;\tau)\} = \int_{-\infty}^{\infty} h(t;\tau)e^{-j2\pi f\tau}d\tau \\ &= \int_{-\infty}^{\infty}\sum_{i=1}^{N(0,t)} r_i(t,\tau)e^{j\Phi_i(t,\tau)}\delta(\tau-\tau_i)e^{-j2\pi f\tau}d\tau \\ &= \sum_{i=1}^{N(0,t)} r_i(t,\tau_i)e^{j\Phi_i(t,\tau_i)}e^{-j2\pi f\tau_i}. \end{aligned} \tag{54}$$

Clearly,

$$E[h(t;f)] = \int_0^t \lambda(\tau)E[r(t,\tau)e^{j\Phi(t,\tau)}]e^{-j2\pi f\tau}d\tau. \tag{55}$$

4. *Autocorrelation in Frequency Domain (Spaced-Frequency, Spaced-Time)*
   The autocorrelation of the time-varying impulse response of the channel, $h(t;f)$, is defined by

$$\begin{aligned} &\Phi_c(t_1,t_2;f_1,f_2) \triangleq E[h(t_1;f_1)h^*(t_2;f_2)] \\ &= E\Big[\sum_{i=1}^{N(0,t_1)} r_i(t_1,\tau_i)e^{j\Phi_i(t_1,\tau_i)}e^{-j2\pi f_1\tau_i} \\ &\qquad \times \sum_{i=1}^{N(0,t_2)} r_i(t_2,\tau_i)e^{-j\Phi_i(t_2,\tau_i)}e^{j2\pi f_2\tau_i}\Big] \end{aligned}$$

$$= \int_0^{min(t_1,t_2)} \lambda(\beta) E\left[r(t_1,\beta) r(t_2,\beta) e^{j\left(\Phi(t_1,\beta)-\Phi(t_2,\beta)\right)}\right]$$
$$e^{-j2\pi(f_1-f_2)\beta} d\beta$$
$$= \Phi_c(t_1,t_2,\Delta f); \qquad \Delta f \triangleq f_1 - f_2, \tag{56}$$

where we evaluated it using (45) and the uncorrelated scattering condition. (56) shows that the US channels are WSS in the frequency domain which confirms what was shown in [17].
Note that we may rewrite the above expression as

$$\Phi_c(t_1,t_2;\Delta f) \triangleq \mathcal{F}_\tau\{\Phi_c(t_1,t_2;\tau)\}$$

which may be viewed as the Fourier transform of the autocorrelation, $\Phi_c(t_1,t_2;\tau)$. Setting $t_1 = t_2 = t$ we have an expression for the Fourier transform of the power delay profile or delay power spectrum given by

$$\Phi_c(t;\Delta f) \triangleq \mathcal{F}_\tau\{\Phi_c(t_1,t_2;\tau)\}|_{t_1=t_2=t} = \int_0^t \lambda(\beta) E\left[r^2(t,\beta)\right] e^{-j2\pi\Delta f\beta} d\beta$$

5. *Time Variations of the Channel*
(Scattering Function, Doppler Power Spectrum)
The time variations of the channel give rise to the Doppler shift which is evidenced by the broadening of the channel spectrum. Here we start by assuming that the channel is WSS with respect to time variations, and thus the autocorrelation in the frequency domain may be written as $\Phi_c(\Delta t;\Delta f) \triangleq \Phi_c(t+\Delta t, t; f+\Delta f, f)$
The scattering function is defined as follows

$$S_c(\lambda;\Delta f) \triangleq \mathcal{F}_{\Delta t}\{\Phi_c(\Delta t;\Delta f)\} = \int_{-\infty}^{\infty} \Phi_c(\Delta t;\Delta f) e^{-j2\pi\lambda\Delta t} d\Delta t$$
$$= \int_{-\infty}^{\infty}\int_0^t \lambda(\beta) E\left[r(t+\Delta t,\beta) r(t,\beta) e^{j\left(\Phi(t+\Delta t,\beta)-\Phi(t,\beta)\right)}\right]$$
$$e^{-j2\pi\Delta f\beta} d\beta e^{-j2\pi\lambda\Delta t} d\Delta t. \tag{57}$$

Setting $\Delta f = 0$ in (57) we obtain the Doppler power spectrum of the channel

$$S_c(\lambda) \triangleq S_c(\lambda;\Delta f)|_{\Delta f=0} = S_c(\lambda;0) = \int_{\infty}^{\infty} \Phi_c(\Delta t;0) e^{-j2\pi\lambda\Delta t} d\Delta t.$$

Further simplifications of the above computations require specific assumptions on the distribution governing the various random processes.

Example 4.6
This example is introduced for illustrative purposes. Consider an WSSUS channel with $r_i(t,\tau) = r_i(\tau)$, $\phi_i(t,\tau) = \phi_i$ and $\vartheta_i(t) = \vartheta_i$ which satisfy (34) and are mutually independent

1. *Signal envelop: Ricean distribution which includes the existence of a strong direct path.*

$$\begin{aligned} p_r(r(\tau)) &= \frac{r(\tau)}{\sigma^2} e^{-(r^2(\tau)+s^2(\tau))/2\sigma^2} I_0\left(\frac{r(\tau)s(\tau)}{\sigma^2}\right), \quad r \geq 0, \\ E[r^2(\tau)] &= 2\sigma^2 + s^2(\tau), \qquad Var[r^2(\tau)] = 4\sigma^4 + 4\sigma^2 s^2(\tau), \end{aligned}$$

where $I_0(.)$ is the modified Bessel function of the first kind of zero-order, $I_0(x) = (\frac{1}{2\pi}) \int_0^{2\pi} e^{x\cos u} du$ (see Fig. 4.5.1).

2. *Signal phase: Uniform distribution. $p_\phi(\phi) = \frac{1}{2\pi}$, $\phi \in [0, 2\pi]$.*
3. *Angle of incoming signal: Uniform distribution. $p_\vartheta(\vartheta) = \frac{1}{2\pi}$, $\vartheta \in [0, 2\pi]$.*

1. *Autocorrelation in Time Domain (see Fig. 4.5.2)*

$$\begin{aligned} \Phi_c(\Delta t; \tau) &= \lambda(\tau) E\left[r(t,\tau) r(t+\Delta t, \tau) e^{j\left(\Phi(t,\tau) - \Phi(t+\Delta t, \tau)\right)}\right] \\ &= \lambda(\tau) E[r^2(\tau)] E\left[e^{j2\pi f_m \Delta t \cos(\vartheta)}\right] \\ &= \lambda(\tau) E[r^2(\tau)] \frac{1}{2\pi} \int_{-\pi}^{\pi} \cos(2\pi f_m \Delta t \cos(\vartheta)) d\vartheta \\ &= \lambda(\tau) E[r^2(\tau)] J_0(2\pi f_m \Delta t) \end{aligned}$$

2. *Power Delay Profile or Delay Power Spectrum (see Fig. 4.5.1)*

$$\begin{aligned} \Phi_c(0; \tau) &\triangleq \Phi_c(\Delta t; \tau)|_{\Delta t = 0} = \lambda(\tau) E[r^2(\tau)] \\ \Phi_c(0; \Delta f) &\triangleq \mathcal{F}_\tau\{\Phi_c(0;\tau)\} = \int_{-\infty}^{\infty} \lambda(\tau) E[r^2(\tau)] e^{-j2\pi\Delta f \tau} d\tau. \end{aligned}$$

3. *Autocorrelation in Frequency Domain (Spaced-Frequency, Spaced-Time)*

$$\begin{aligned} \Phi_c(\Delta t; \Delta f) &\triangleq \mathcal{F}_\tau\{\Phi_c(\Delta t; \tau)\} \\ &= J_0(2\pi f_m \Delta t) \int_{-\infty}^{\infty} \lambda(\beta) E[r^2(\beta)] e^{-j2\pi(\Delta f)\beta} d\beta \end{aligned}$$

4. *Time Variations of the Channel (Scattering Function, Doppler Power Spectrum) (see Fig. 4.5.3)*

$$S_c(\lambda; \Delta f) \triangleq \mathcal{F}_{\Delta t}\{\Phi_c(\Delta t; \Delta f)\} = \int_{-\infty}^{\infty} \Phi_c(\Delta t; \Delta f) e^{-j2\pi\lambda\Delta t} d\Delta t$$

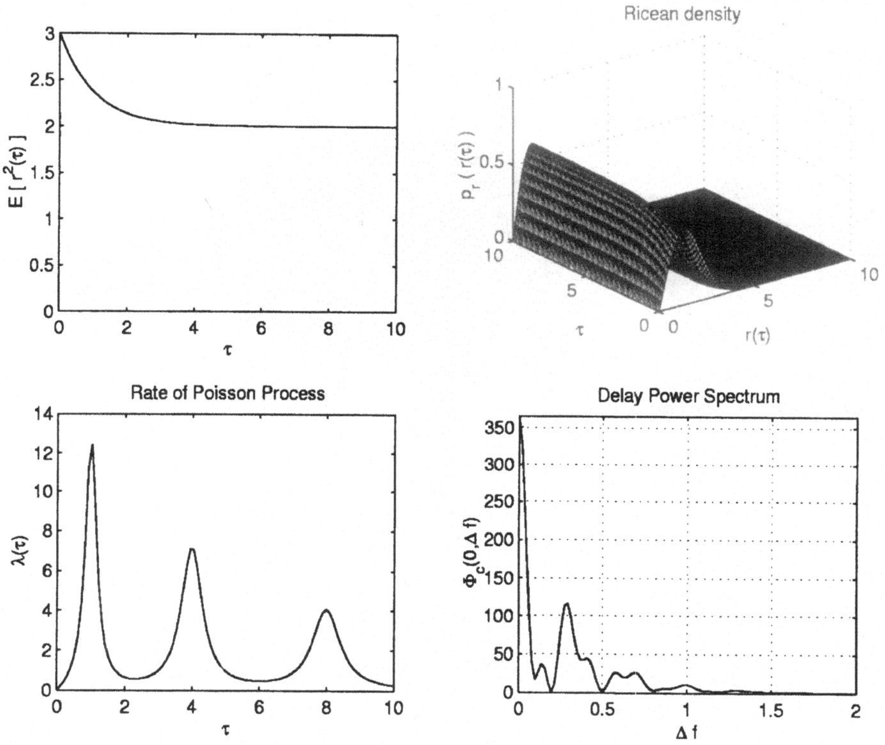

Fig. 4.5.1. Mean and densitiy of Ricean distribution; Time varying Poisson rate and Delay power spectrum

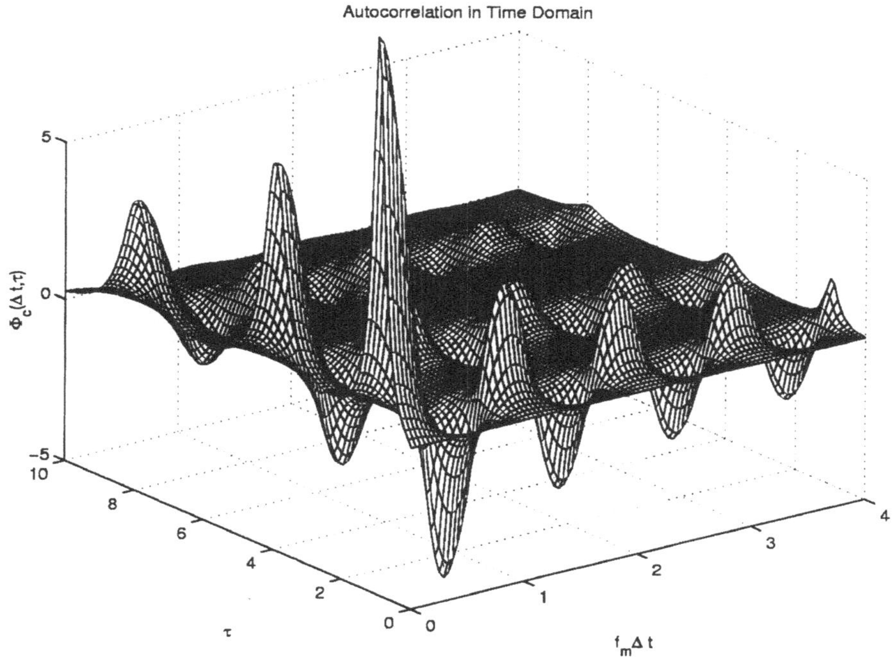

Fig. 4.5.2. Autorcorrelation in Time Domain

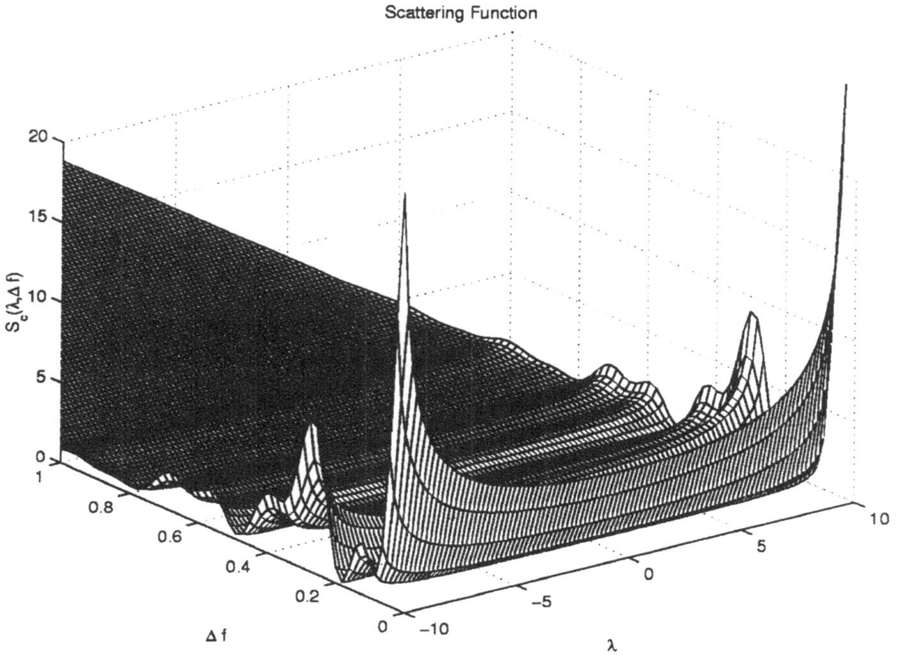

Fig.4.5.3. Scattering Function

4.5.5 Conclusion

This paper employs stochastic differential equations emerging from Ornstein-Uhlenbeck processes to model the long-term fading effects of power loss in multipath wireless channels. The arrival times of the individual paths are modeled using point processes and more general counting processes. For the case of Poisson processes, a general moment generating function is derived from which various statements concerning the properties of the received signal are derived.

References

[1] K. Pahlavan and A. H. Levesque. *Wireless Information Networks*. New York, Wiley-Interscience, 1995.

[2] T. S. Rappaport. *Wireless Communications*. New Jersey, Prentice-Hall, 1996.

[3] T. Aulin. A modified model for the fading signal at a mobile radio = channel. *IEEE Transactions on Vehicular Technogoly*, = 43:2935–2971, 1979.

[4] R.H. Clarke. A statistical theory of mobile radio reception. *Bell Systems Technical Journal*, 47:957–1000, 1968.

[5] J.G. Proakis. *Digital Communications*. McGraw-Hill, New York, 1995.

[6] A. J. Coulson, G. Williamson, and R. G. Vaughan. A statistical basis for lognormal shadowing effects in = multipath fading channels. *IEEE Transactions on Communications*, 46(4):494-502, = 1998.

[7] D. C. Cox, R. Murray, and A. Norris. 800MHz attenuation measured in and around suburb houses. *AT&T Bell Laboratory Technical Journal*, 673(6), 1984.

[8] A. D. M. Saleh and R. A. Valenzuela. A statistical model for indoor multipath propagation. *IEEE Journal on Selected Areas in Communications*, 5(2):128–137, 1987.

[9] H. Suzuki. A statistical model for urban radio propagation. *IEEE Transactions on Communications*, 25:673–680, 1977.

[10] P.L. Rice. Mathematical analysis of random noise. *Bell Systems Technical Journal*, 24:46–156, 1944.

[11] C. D. Charalambous. Statistical analysis of multipath fading channels: Counting processes, second-order statistics and central limit theorems. *Preprint*, 1999.

[12] G. L. Turin. Communication through noisy, random-multipath channels. *IRE Convention Record*, pages 154–166, 1956.

[13] R.J. Elliott. *Stochastic Calculus and Applications*. Springer-Verlag, New York, 1982.

[14] S. Karlin and H. M. Taylor. *A first Course in Stochastic Processes*. Academic Press, 1975.

[15] C.D. Charalambous and N. Menemenlis. Stochastic models for long-term multipath fading channels and = their statistical properties. IEEE Transactions on Communications Systems (submitted), 2000.

[16] E. Parzen. *Stochastic Processes*. SIAM, Classics in Applied Mathematics, 1995.

[17] P. Bello. Characterization of random time-variant linear channels. *IEEE Transactions on Communications*, (11):360–393, = 1963.

5 Channel Characterization & Modelling

5.1 Introduction

The transition from second generation mobile systems to third generation naturally involved the necessity towards smaller cell radii. Therefore the demand and standards on measurement systems and wave propagation models have changed in dependance on the properties of the propagation environment models. Two different fundamental methods of modeling can be distinguished. Those based on statistical methods and are dependent on channel measurements and those based on deterministic methods and are dependent on ray-tracking techniques.

In section 5.2 indoor radio channels are defined using both of the above methods. In section 5.3 a mobile channel characterization and availability is analyzed using statistical as well as diversity techniques. In section 5.4, LEO mobile satellite channels are defined taking, however, info consideration time variation due to satellite motion.

5.2 Wideband Modeling of Time-Variant Indoor Radio Channels

Frank Layer, Torsten Englert, Ralf Kattenbach and Henning Früchting

5.2.1 Introduction

From the early days of using radio waves to transmit messages, the investigation of radio channel properties has been an important and interesting topic of research. In contrast to the early high-power and low-frequency broadcasting systems with high ranges nowadays the focus moves to low-power and high-frequency communication systems with low ranges. This tendency towards smaller cell radii has been moving the scientific area of interest from rural macro-cells and urban micro-cells to in-building pico-cells. Therefore also the demands and standards on measurement systems and wave propagation models have changed in dependence on the properties of the propagation environments and systems. The background for this trend and the increasing importance of the knowledge of channel properties, especially of indoor environments, is on the one hand the fact that a very high percentage of mobile calls are made from or to persons inside buildings, on the other hand the rapidly increasing popularity of wireless data transmission systems, for example in supermarkets, assembly shops or offices. In Europe, besides the frequency ranges at 900 MHz with the application GSM 900, which has been topic of numerous investigations in the past, currently the frequency bands around 1.8 GHz (DECT, GSM 1800) and 5.2 GHz for wireless indoor data transmission using wireless local area network applications, like for example HIPERLAN, are of increasing interest. For the determination of channel properties there is only one alternative to time-consuming and expensive channel measurements - the computer-based modeling of the radio channel, which is already intensively used in practice. Since indoor areas are very complex environments with regard to wave propagation, the demands on the models are immense. The objective of these models is a good reproduction of the radio wave propagation considering scattering, diffraction, reflection and shadowing. These effects are dependent on the geometrical and the electromagnetic characteristics of the environment and show a substantial variation [1] in real environments. At the end, the quality of a model depends on the precision of the observed characteristics, the method of analysis (considered details, comprehensiveness) and the used approximations and simplifications. This especially applies if time-variant indoor radio channels are considered, as the channel properties may be slowly changing during the observation time, for example due to movements. Depending on the application, two different fundamental methods of modeling can be distinguished: on the one hand the statistical modeling, which is based on channel measurements and mainly used to get an insight into the channel properties of a typical class of indoor environments, on the other hand the deterministic modeling (which is mainly applied using ray-tracing techniques) to predict the wideband channel characteristics of specific indoor environments based on the building geometry. In the following, a representative for each of these modeling methods will be introduced and applied in two subsections. In order to build up a database for the statistical modeling approach and to be able to verify the deterministic predictions, a reliable measurement method is necessary and therefore will be presented first.

5.2.2 Swept-Frequency Measurement Set-Up

For the present investigations, a frequency domain measurement method - the swept-frequency method [36] with a moving receiver [15] - has been chosen, since the environment is kept static during all measurements and therefore a real-time recording of the response of the channel is not necessary. On the other hand, the system offers the consideration of arbitrary bandwidths at a very high accuracy and high dynamic range. The measurement set-up can be seen in Fig. 5.2.1. The main item is a network analyzer with an S-parameter test set to which the transmitter and receiver antennas are connected. The antennas are of discone type, which has a radiation pattern similar to that of a $\lambda/2$-dipole. In contrary to the static transmitter antenna Tx, the receiver antenna Rx is located on a positioning system, which is driven from a PC by a stepper motor. After a calibration, the network analyzer measures the parameter $S_{21}(f)$ (which is identical to the transfer function $T(f)$ of the channel) within the bandwidth of - in this case - 600 MHz around the desired center frequency in 401 discrete steps. Taking into consideration the impacts of the windowing function for the Fourier transform [15], this measurement bandwidth results in a resolution of $\Delta\tau = 3.3\,\text{ns}$ on the delay time axis. After each sweep the transfer function is stored on a hard disk of the PC and the receiver is moved over the distance of $\lambda/8$ to the next position. With this method, on the length of 1.84 m of the positioning system 89 (at a center frequency of f_c = 1.8 GHz) and 256 (at 5.2 GHz) consecutive transfer functions are measured, from which the corresponding impulse responses can be calculated offline using discrete Fourier transforms. The resulting space-variance of the transfer function is considered to be a "time-variance in static environment". This consideration is reasonable in view of the theory of signals [15] (signals are also time-variant) and is valid if there are no moving scatterers in the propagation environment. Space-variance and time-variance in static environment then can be linked by determining a constant velocity of the receiver, which in the following is assumed to be v = 1 m/s.

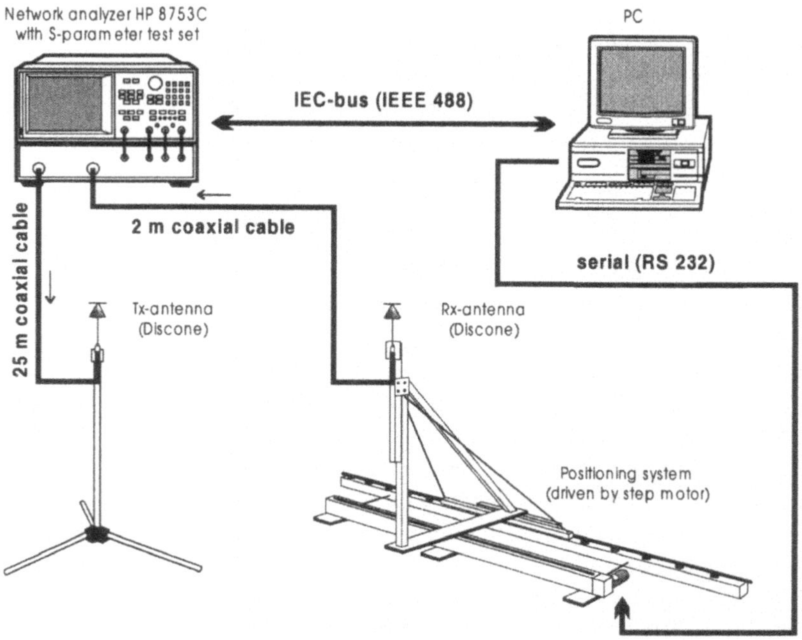

Fig. 5.2.1 Measurement set-up for swept-frequency measurements

5.2.3
Statistical Modeling Of Wideband Indoor Radio Channels Considering Deterministic Effects

5.2.3.1
Fundamentals Of Statistical Modeling

The description of outdoor mobile radio channels by means of their statistical properties [1, 31, 32] is generally accepted. Adopting these methods for indoor mobile radio channels appears to be questionable at first glance, due to the great amount of deterministic influences in indoor environments. However, with an increasing number of propagation paths the system functions show a rather random behavior, which speaks for a statistical description of the indoor radio channel [15].

A commonly used mathematical formulation of the Time-Variant Impulse Response $h(t,\tau)$ by using weighted delta functions is the following:

$$h(t,\tau) = \sum_i a_i(t) \cdot e^{j\varphi_i(t)} \cdot \delta(\tau - \tau_i(t))$$

with (1)

$a_i(t)$ ... time - variant amplitude

$\varphi_i(t)$ ... time - variant phase

$\tau_i(t)$ ... time - variant delay time

i ... pathnumber

Using this equation, there are three quantities that have to be described by statistical methods: path amplitudes $a_i(t)$, path phases $\varphi_i(t)$ and delay times $\tau_i(t)$. A series of publications deal with the topic of statistical modeling of indoor radio channels (e. g. [10, 28]); however the investigations of different teams of researchers show different and in some cases even contrary results. Essential contrary results can be found for the description of path amplitudes by theoretical distributions [11]. Further discrepancy can be found for the description of arrival times, due to the condition of clustering of arriving paths [11]. Path phases are often assumed to be uniform distributed [31, 32, 10, 11]; however there has been a lack of justifying this assumption by evaluations of measurements [11].
One major deterministic effect is the occurrence of time-variant changes in delay time of paths due to changes in propagation length by moving transmitters, receivers or scatterers. This effect is also present in outdoor environment, but especially in indoor environment it becomes more distinct and cannot be neglected. For this reason, a detection of the receiving paths that has to be as exact as possible forms the basis for an accurate evaluation of the statistical characteristics of Time-Variant Impulse Responses.

Path Detection in the Time-Variant Impulse Response

Detection of paths in the Time-Variant Impulse Response can be completely done by computer algorithms. Starting point for the detection of paths is the Delay Power Density Spectrum $P_h(\tau)$, which can be calculated by:

$$P_h(\tau) = \frac{1}{t_2 - t_1} \int_{t_1}^{t_2} |h(t,\tau)|^2 \, dt \qquad (2)$$

From its calculation (2) the Delay Power Density Spectrum can be interpreted as a mean squared impulse response. Due to the averaging, noise influences are reduced in this function and the received paths can be detected as local maxima. Fig. 5.2.2a shows a calculated Delay Power Density Spectrum, in which the detected paths appear distinctly in the logarithmic graph of this function. Due to movements of either transmitter, receiver or scatterers, the indoor radio channel has to be considered to be a time-variant system and thus the arrival times also become time-variant, i. e. $\tau_i \rightarrow \tau_i(t)$. This fact has been considered by a so-called "Path-Tracking"-algorithm when detecting the paths in the Time-Variant Impulse

Response $h(t,\tau)$. Starting with the detected paths in the Delay Power Density Spectrum as initial values, a more accurate detection of maxima is realized in the impulse response measured at half of the covered distance under consideration of the maximum possible shift in delay. In the next step, a detection of maxima is performed successively in adjacent impulse response samples considering the maximum shift in delay for movements within $\lambda/8$ between two measurement points. In Fig. 5.2.2**b** an exemplary result of the algorithm is depicted in a contour plot of the Time-Variant Impulse Response. For weak paths however the algorithm reaches its limits and thus slight deviations from the actual course occur, as can be seen for example in Fig. 5.2.2b at a delay time $\tau \approx 70$ ns.

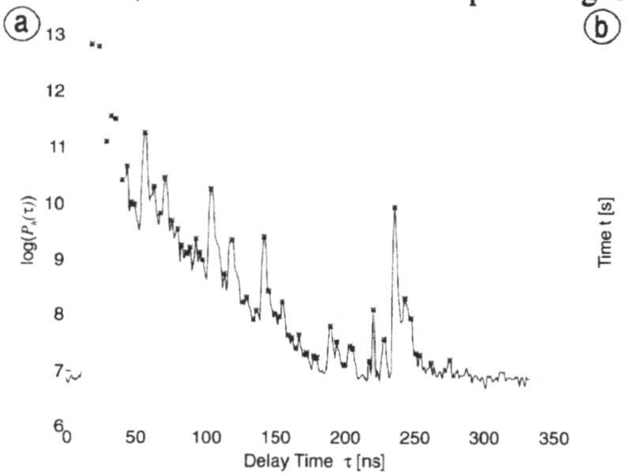

Fig. 5.2.2 a) Logarithmic Delay Power Density Spectrum $P_h(\tau)$ with detected paths (*),
b) Contour plot of a Time-Variant Impulse Response $h(t,\tau)$ with detected paths (*)

In the literature, usually two general approaches for a detection of paths can be found:

- assumption of constant delay times for all paths [1]
- independent detection of paths in single impulse response samples [31, 32, 10]

Due to the occurring simplification of the physical propagation process, both approaches can only be used for the description of special cases of the indoor mobile radio channel. The first approach e. g. completely neglects the variations in delay described before. In the second approach weak paths (e. g. paths in a fading gap) are neglected, and thus the number of paths can vary between two impulse response samples. In comparison with these approaches the path-tracking algorithm therefore yields results that are much nearer to what results from the physical propagation process.

Statistical Evaluation of Measurements

A mathematical description of the Time-Variant Impulse Response $h(t,\tau)$ is achieved by a sum of weighted delta-impulses (1). The delta-function in (1) causes the occurrence of single paths at distinct delay times τ_i. However, it has to be noted that strictly speaking this is only valid for an infinite bandwidth. In this case, each multipath component could be resolved and would then reveal a constant amplitude and merely a fluctuation of the phase. If the bandwidth is limited, the delta-function turns into a *sinc*-function and the superposition of unresolved multipath components results in time-variant fluctuations of the amplitude. From those considerations it can be seen that the statistical evaluation and modeling of the Time-Variant Impulse Response strictly speaking is only valid for channels with limited bandwidth, and thus the description by (1) in context with statistical modeling implies some kind of simplification.

Using the detected paths from the path-tracking algorithm, the amplitudes, phases and delay times of single paths can be extracted and investigated with respect to their statistical distribution functions. For the fitting of theoretical distribution functions for amplitudes, phases and delay times to the measured data (each consisting of 88 or 89 impulse response samples for $f_c = 1.8$ GHz) a modified simplex algorithm [25] is used. Fig. 5.2.3 shows the optimum fit of two distribution functions to amplitudes of a path. In this case the fit of a Rayleigh distribution yields insufficient results (Fig. 5.2.3a), whereas the fitted Nakagami distribution almost matches exactly the measured data (Fig. 5.2.3b). The optimization criterion is the least mean square error. The remaining mean square error

$$e^2 = \frac{1}{N}\sum_{i=1}^{N}(x_{i\,\text{actual}} - x_{i\,\text{desired}})^2 \qquad (3)$$

can be used as a quality criterion and is given in Fig.5.2.3 for each of the

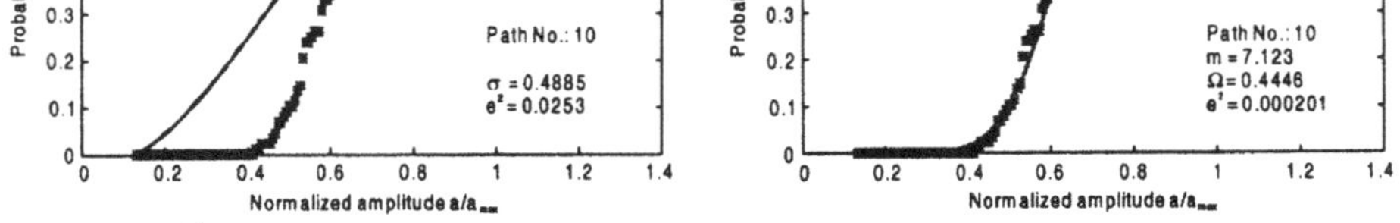

evaluations.

Fig. 5.2.3: Fitting a) of a Rayleigh distribution and b) of a Nakagami distribution to the measured data
(* = measured data, — = fitting curve)

Investigations of several fitting curves have shown that a remaining error of $e^2 < 0.5 \cdot 10^{-3}$ can be classified as an extremely good fit. In order to get a comprehensive representation of the errors, the frequency distributions of the

remaining mean square error as well as the frequency distributions of the difference of the mean square error to the minimum error of all investigated distributions are determined. By using these distributions, an efficient classification of the achieved goodness-of-fit can be done.
A total amount of 63 measurements at $f_c = 1.8$ GHz with same height (1.30 m) of transmitting and receiving antenna in different office-, corridor- and residential environments under line of sight and non line of sight conditions were investigated [15, 16, 17]. In these 63 measurements, 3268 received paths in total were detected with the path tracking algorithm described in the previous section.

a) Distribution Functions for the Path Amplitudes

For the description of the time selective amplitude fading, five different distribution functions have been investigated with respect to their fit to the measured data ($f_c = 1.8$ GHz).

a.1 Rayleigh-Fading

The Rayleigh distribution

$$f_a(a) = \frac{a}{\sigma^2} \cdot e^{-\frac{a^2}{2\sigma^2}}, \quad a \geq 0 \tag{4}$$

describes the amplitude fading for the case of a superposition of scattered components with approximately identical power. Fig. 5.2.4 shows the frequency distribution of the mean square error when fitting a Rayleigh distribution to measured data. More than 15% of the fitted curves[1] show an error greater than $2.5 \cdot 10^{-3}$. Although there is a good fit for some paths, it has to be noted that these are mainly weaker paths, which are rather insignificant in modeling. The Rayleigh distribution thus seems to be not very appropriate for the modeling of path amplitudes in indoor environment.

[1] The number of errors that are greater than $2.5 \cdot 10^{-3}$ is depicted in the upper right corner of Figures 4a, 5a, 6a, 7a, 8a, 9 and 10b.

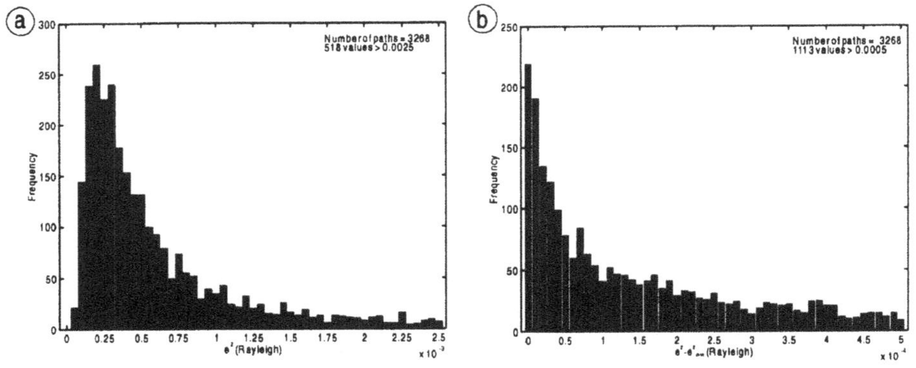

Fig. 5.2.4 Frequency distribution a) of errors e^2 when fitting a Rayleigh distribution to path amplitudes,
b) of the difference of errors to the minimum error from all investigated distributions when fitting a Rayleigh distribution to path amplitudes

a.2 Lognormal-Fading

The Lognormal distribution

$$f_a(a) = \frac{1}{\sqrt{2\pi} \cdot \sigma \cdot a} \cdot e^{-\frac{(\ln a - \mu)^2}{2\sigma^2}}, \quad a > 0 \tag{5}$$

has often been used to describe large scale variations of the signal amplitudes [11]. Both the frequency distribution of the mean square error in Fig. 5.2.5a and the frequency distribution of the difference of the mean square error to the minimum error of all distributions (Fig.5.2.5b) show even worse results than for the Rayleigh distribution. However, there is a smaller amount of extremely bad fits (84 values ≙ 2.6%). A detailed comparison for the Rayleigh- and Lognormal distribution for selected paths shows a better fit for the Lognormal distribution to paths with higher amplitudes (which are essential for modeling). From this point of view, the Lognormal distribution could be stated to be more suitable for the description of amplitude fading than the Rayleigh distribution. However, as will be seen in the following, other distribution functions yield even better results in many cases.

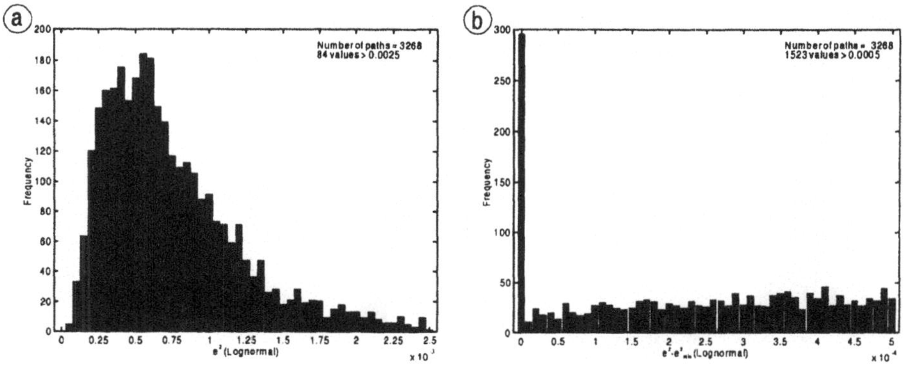

Fig. 5.2.5 Frequency distribution a) of errors e^2 when fitting a Lognormal distribution to path amplitudes,
b) of the difference of errors to the minimum error from all investigated distributions when fitting a Lognormal distribution to path amplitudes

a.3 Rice-Fading

The Rice distribution

$$f_a(a) = \frac{a}{\sigma^2} \cdot I_0\left(\frac{a \cdot v}{\sigma^2}\right) \cdot e^{-\frac{a^2+v^2}{2\sigma^2}}, \quad a \geq 0 \tag{6}$$

represents a Rayleigh scattering process with a superimposed dominant component. In general, both the frequency distributions Fig. 5.2.6a and 5.2.6b show very good fittings, which are distinctly better than the results for the Rayleigh- and the Lognormal distribution. The number of bad fittings (see Fig. 5.2.6a) is small (55 values ≙ 1.7%) and also the deviation to the minimum error of all investigated distributions (Fig.5.2.6b) turn out to be small. Accordingly, the Rice distribution describes the statistical properties of amplitude fluctuations very well.

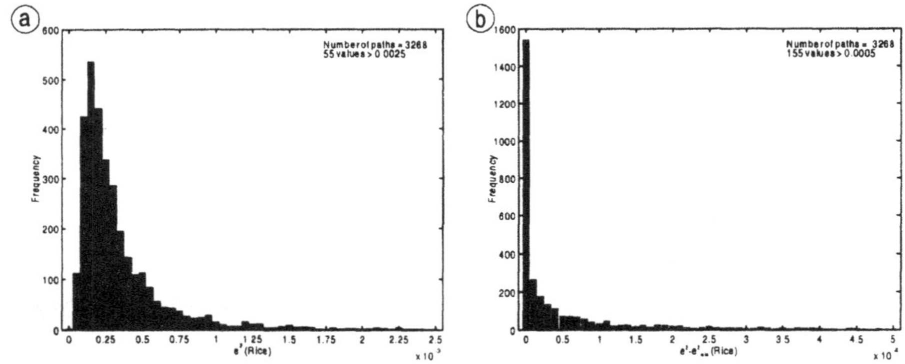

Fig. 5.2.6: Frequency distribution a) of errors e^2 when fitting a Rice distribution to path amplitudes,
b) of the difference of errors to the minimum error from all investigated distributions when fitting a Rice distribution to path amplitudes

a.4 Nakagami-Fading

The Nakagami distribution has the ability to approximate the distribution functions described in subsections *a.1 Rayleigh-Fading*, *a.2 Lognormal-Fading* and *a.3 Rice-Fading*. Furthermore, there is a plausible physical explanation [1, 24] for the use of the Nakagami distribution:

$$f_a(a) = \frac{2 \cdot m^m \cdot a^{2m-1}}{\Gamma(m) \cdot \Omega^m} \cdot e^{-\frac{m \cdot a^2}{\Omega}}, \; a \geq 0 \tag{7}$$

This function represents an approximate solution for the resulting amplitude distribution when superimposing a great number of vectors with random amplitudes and phases ("random-phasor problem") [1, 24, 5]. This describes the effect of the superposition of multipath components, as occurring for a radio channel, in a general form.

Both frequency distributions in Fig. 5.2.7 reveal again very good fittings, which are comparable to those for the Rice distribution. Again there is only a small number of poor fittings (see Fig. 5.2.7a: 40 values ≙ 1.2%). Important to note is the similarity of the frequency curves for Rice (Fig.5.2.6a) and Nakagami (Fig.5.2.7a): both curves are nearly congruent, so in this case the Nakagami distribution appears to be only used for approximating the Rice distribution, although Nakagami distribution is capable of approximating many other forms of fading. Also the comparison of the frequency distributions of errors to the minimum error for all investigated distribution functions (Fig. 5.2.6b and 5.2.7b) shows a distinct similarity of both distributions, with the Nakagami distribution producing slightly better results. The better results can be explained with the more

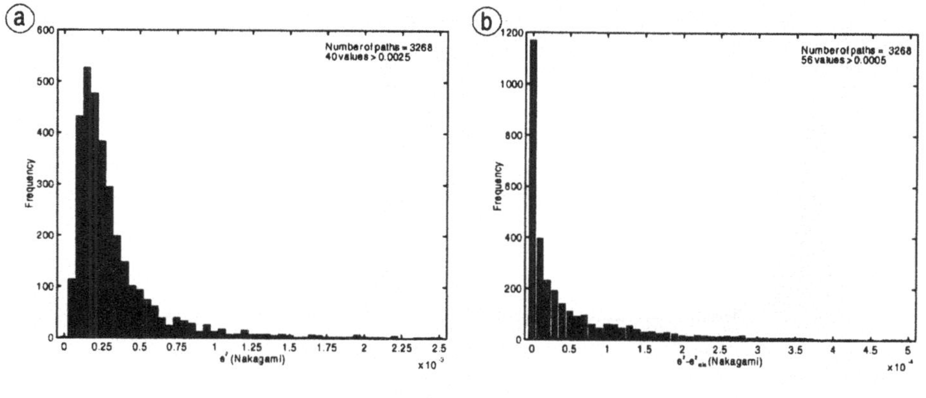

Fig. 5.2.7 Frequency distribution a) of errors e^2 when fitting a Nakagami distribution to path amplitudes,
b) of the difference of errors to the minimum error from all investigated distributions when fitting a Nakagami distribution to path amplitudes

universal applicability of the Nakagami distribution, due to its ability to approximate other fading forms.

a.5 Weibull-Fading

As Hashemi reports in his paper [11], some authors successfully used the Weibull distribution

$$f_a(a) = \frac{\gamma}{\beta} \cdot \left(\frac{a}{\beta}\right)^{\gamma-1} \cdot e^{-\left(\frac{a}{\beta}\right)^{\gamma}}, \quad a \geq 0 \tag{8}$$

for the description of amplitude fading. In contrast to the distributions considered up to now, yet there is no plausible explanation for using this function. The results in Fig. 5.2.8 confirm the use of this function for describing amplitude fading. In general the fitting for Weibull yields even better results than for Nakagami. The curves of the frequency distributions of errors (Fig. 5.2.8) show again a great similarity to the curves for Rice (Fig. 5.2.6) and Nakagami (Fig. 5.2.7). From this fact it can be assumed, that the Weibull distribution can also approximate the Rice distribution. Furthermore, the good fitting of the Weibull distribution may lead to the conclusion that the Weibull distribution possibly occurs also as a solution for the random-phasor problem.
As it has been shown in the previous subsections, the Nakagami distribution is suitable quite well for describing the amplitude fading for the frequency band around 1.8 GHz. Investigations at 5.2 GHz show that the Nakagami distribution is also suitable for this frequency band [8]. Due to the missing physical explanation

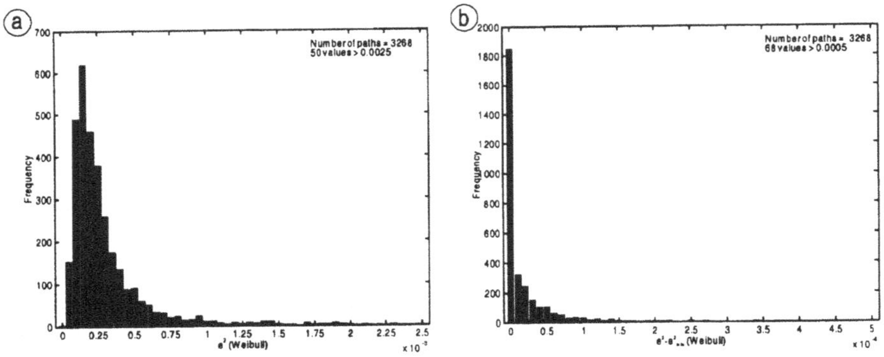

Fig.5.2.8: Frequency distribution a) of errors e^2 when fitting a Weibull distribution to path amplitudes,
b) of the difference of errors to the minimum error from all investigated distributions when fitting a Weibull distribution to path amplitudes

for using the Weibull distribution, the Nakagami distribution should be preferred for the description of amplitude fading, because of its physical plausibility.

b) Distribution Function for the Phase Values

In conformity with other authors [31, 32, 28, 11], path phases are assumed to be uniform distributed values. From a physical point of view, an explanation of this assumption can be given when taking into account the relation between the phase and the covered distance: changes of the covered distance in the order of λ will result in phase changes in the order of 2π, which leads to the assumption of a uniform distribution for longer covered distances.
The uniform distribution approach shows very good results (see Fig. 5.2.9) for the 3268 investigated paths: for nearly all paths the uniform distribution matches the measured data very well. The results thus give the confirmation for the use of a uniform distribution for describing the phase distribution of measurements, which (according to [11]) was missing up to now.
It has to be noted that the amplitude and phase values are not independently distributed. It rather turns out from additional investigations in the Doppler domain that adjacent samples are highly correlated [15], which then has to be taken into account in modeling [9].

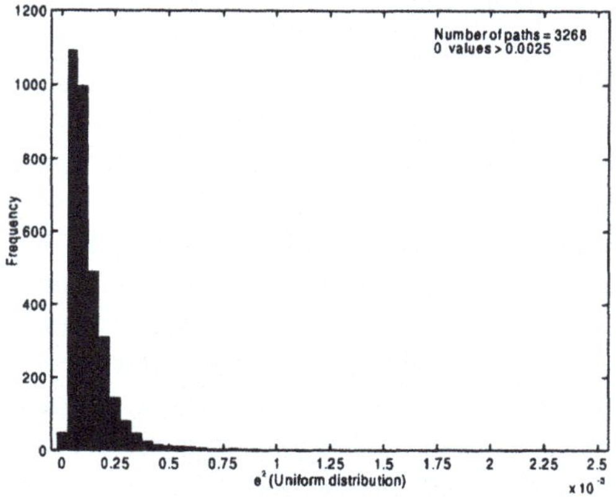

Fig. 5.2.9 Frequency distribution of errors e^2 when fitting a uniform distribution to path phases

c) Distribution Functions for the Delay Time

Instead of an investigation of the arrival times, the so called "interarrivaltime" has been investigated. The interarrivaltime can be calculated by the difference of the delay between two paths:

$$\zeta_i = \tau_i - \tau_{i-1};\ i \geq 2 \tag{9}$$

with

i ... pathnumber

For describing the arrival of paths, standard- or modified Poisson processes are frequently used [11]. Assuming a standard Poisson process with constant arrival rate λ leads to an exponential distribution for the interarrivaltimes:

$$f_\zeta(\zeta) = \lambda \cdot e^{-\lambda\zeta}, \zeta \geq 0 \tag{10}$$

The assumption of a non-constant arrival rate in the form of

$$\lambda(\zeta) = \frac{1}{\beta^\gamma} \cdot \zeta^{\gamma-1} \tag{11}$$

will lead to a modified Poisson process with Weibull distributed interarrivaltimes:

$$f_\zeta(\zeta) = \frac{\gamma}{\beta} \cdot \left(\frac{\zeta}{\beta}\right)^{\gamma-1} \cdot e^{-\left(\frac{\zeta}{\beta}\right)^\gamma}, \zeta \geq 0 \tag{12}$$

The detailed derivation of the relation between a non-constant arrival rate λ according to (11) and the resulting Weibull distribution for the interarrivaltimes as well as physically plausible explanations for the suitability of this distribution are given in [15] and [16].

Using the exponential distribution for a description of the statistical distribution of interarrivaltimes reveals poor matching, because nearly all errors are above the value of $0.5 \cdot 10^{-3}$, which has been declared as a limit for a good fit. The related frequency distribution of errors is depicted in Fig.5.2.10a. Using a Weibull distribution, the fitting results are much better, as can be seen from Fig. 5.2.10b. In a first and somewhat more intuitive reasoning these essentially better fitting for the Weibull distribution can be explained by the greater flexibility of the distribution (two parameters in contrast to only one parameter for the exponential distribution). But going more into detail, there is also a physically plausible explanation when taking into account the underlying propagation process [15, 16].

Fig. 5.2.11 shows the two-dimensional frequency distribution of the investigated 5606 impulse response samples. A distinct condensation occurs for the pair of values $\gamma = 2$ and $\beta = 0.8 \cdot 10^{-8}$, which from equation (11) leads to a linear relationship between the arrival rate of paths and the interarrivaltime.

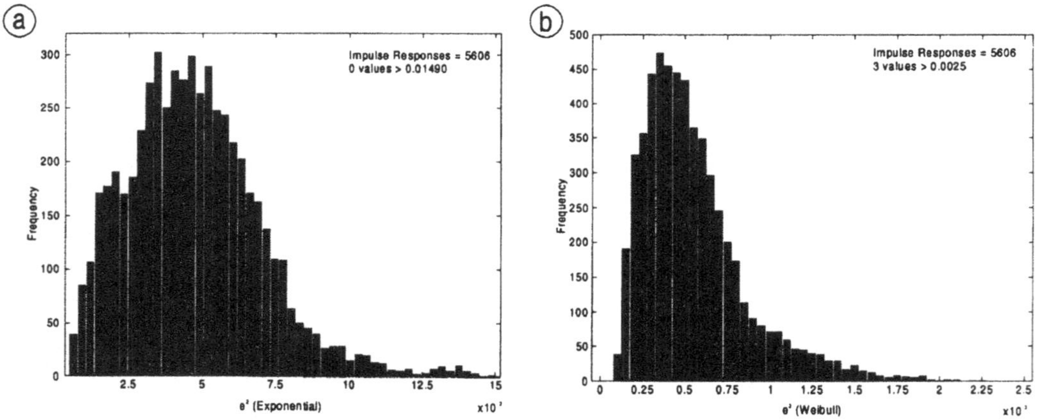

Fig. 5.2.10 Frequency distribution of errors e^2 when fitting a) an exponential distribution to path interarrivaltimes,

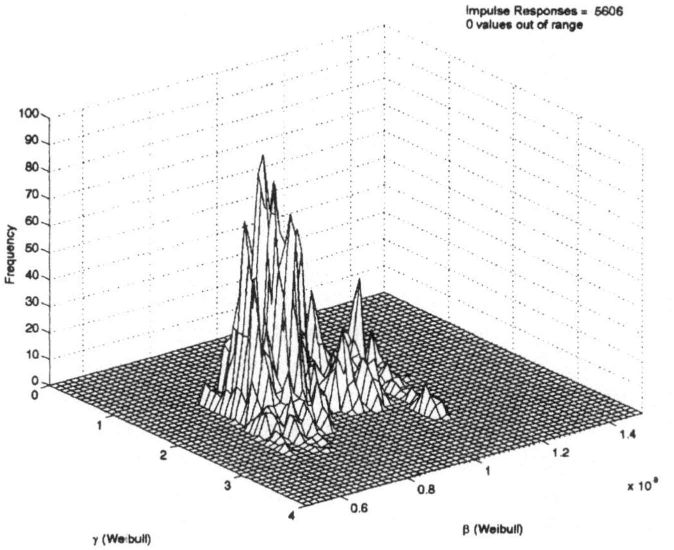

b) a Weibull distribution to path interarrivaltimes

Fig. 5.2.11 Parameter distribution when fitting a Weibull distribution to path interarrivaltimes

5.2.3.2
Modeling Approach

As already mentioned before, the correlation between adjacent impulse responses has to be taken into account in modeling. For this reason the described modeling approach, which was proposed in [15, 17], is splitted in two parts: the first part is the statistical modeling of arrival times, paths amplitudes and paths phases; in the second part the deterministic influences are considered by filtering the statistical values with appropriate Doppler spectra.

In the statistical part, the arrival times τ of single paths are generated by using a Weibull distribution for the interarrivaltimes. These paths of the Time-Variant Impulse Response are weighted by statistically independent amplitude values (Nakagami distributed) and statistically independent phase values (uniform distributed). Due to generating statistically independent values, they are also uncorrelated and thus white processes. A Fourier transform of the statistically generated Time-Variant Impulse Response $h(t,\tau)$ with respect to the time t yields the Doppler-Variant Impulse Response $s(\nu,\tau)$ [15]:

$$s(\nu,\tau) = \int_{-\infty}^{+\infty} h(t,\tau) \cdot e^{-j2\pi\nu t} dt \tag{13}$$

Theoretically, each path in this function has a constant spectrum versus the Doppler axis ν, due to the generation of statistically independent values for the amplitudes and phases of the Time-Variant Impulse Response. However, this is only valid for continuous random variables and thus the spectrum of the created discrete function is not perfectly white. In order to improve the results, therefore a whitening filter is applied additionally.

The second step of the modeling comprises a filtering of the Doppler-Variant Impulse Response with realistic Doppler spectra in order to consider the correlation between two adjacent impulse response samples. These realistic Doppler spectra are taken from a wideband measurement of the Time-Variant Transfer Function *T(t,f)* in a representative indoor environment. A two-dimensional Fourier transform and a calculation of the absolute squared value leads to the Scattering Function $P_s(\nu,\tau)$, which contains the Doppler spectra for different arrival times τ_i. For an extraction of the single Doppler spectra from this Scattering Function, the detection of paths is performed in the Delay Power Density Spectrum $P_h(\tau)$, which can be derived from the measurements by using equation (2). This function is predestinated for a detection of paths, since it can be interpreted as a mean squared impulse response and, due to the averaging, the paths appear as distinct local maxima in this function (cf. Fig. 5.2.2a). With this information, each path (real- and imaginary part) in the statistically generated Doppler-Variant Impulse Response is multiplied with the corresponding spectrum from the "deterministic" Scattering Function. For obtaining values that correspond as good as possible to reality, the amplitude of each path of the statistically generated function is normalized to its maximum value, so the multiplication with the deterministic (in this case measured) function leads to values that are closest to the real ones. In the last step of the simulation, the filtered Doppler-Variant Impulse Response is inverse Fourier transformed, to get the Time-Variant Impulse Response. Fig 5.2.12 shows a block diagram of the described modeling approach.

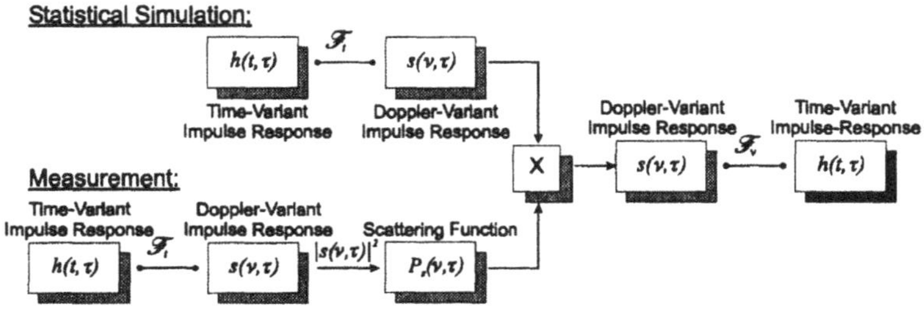

Fig. 5.2.12 Block diagram of the proposed modeling approach

5.2.3.3
Verification Of The Approach

In order to verify the proposed approach, a comparison between a simulation and an equivalent measurement is given in this section.
First of all, some general remarks about the environment, the measurement, propagation effects in indoor environments and the potentialities for the interpretation of system- and correlation functions [15] will be discussed. After that, a detailed comparison of three selected functions, each for measurement and simulation, will be given.

General Remarks

The presented simulation and the related measurement describe the properties of an indoor environment as depicted in **Fig. 5.2.**. A meeting room (cf. also 5.2.20) with dimensions 4.42 m x 7.64 m and a height of 3.5 m was used for the positioning of the transmitting antenna. The receiver Rx was placed in the middle of an office room with the same height and dimensions 5 m x 2.20 m.
In multipath propagation, different delay times refer to different path lengths, which can be determined in the Time-Variant Impulse Response $h(t,\tau)$, the Doppler-Variant Impulse Response $s(\nu,\tau)$, the Scattering Function $P_s(\nu,\tau)$ and the Delay Power Density Spectrum $P_h(\tau)$ by peaks versus the delay axis τ. For the presented investigations (measurement and simulation) a bandwidth of 600 MHz was used, which yields a theoretical resolution of 1.67 ns. The actual resolution in the measurement and simulation results (Fig. 5.2.14-19) however is 3.3 ns, due to additionally using a window function before transferring to time domain. In the graphs of the Scattering Function (Fig. 5.2.16, 5.2.17), the Doppler Frequency ν, when normalized to its maximum value $\nu_{\max}$, i. e. $\nu / \nu_{\max} = \cos(\alpha)$ reveals an information about the azimuth angle of incidence α of the respective path relative to the Rx movement direction.

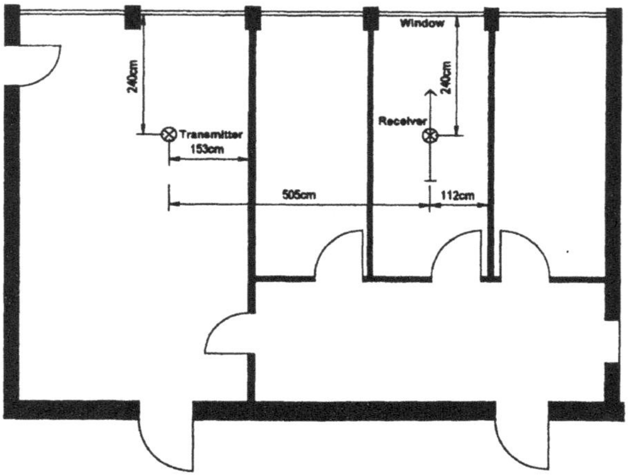

Fig. 5.2.13 Ground plan of the measurement site with positions of transmitter Tx and receiver Rx
(the arrow indicates the direction of movement)

Parameters for the Simulation

The path-tracking algorithm (see section "Path Detection in the Time-Variant Impulse Response") detected 49 paths in the measurement, thus the simulation consists of the same number of paths. Interarrivaltimes of these paths are modeled by a Weibull distribution, path amplitudes are modeled by the Nakagami distribution and path phases are uniform distributed (the parameters for all distributions correspond to those extracted from the measurement). For the filtering of the statistical function with a realistic spectrum, the Scattering Function from the measurement is used.

Comparison of the Time-Variant Impulse Responses

Fig. 5.2.14 shows a plot of the absolute value of the Time-Variant Impulse Response for the measurement. The first path arrives under non line of sight conditions (cf. Fig. 5.2.13) at a delay time of about 20 ns, which coincides well with the distance between transmitter and receiver. All paths show a distinct fluctuation versus the time axis t. This fluctuation is caused by a superposition of signal components with approximately the same path length, which cannot be resolved as single paths due to the limited bandwidth of 600 MHz. There is also a distinct decay versus the delay axis τ; the last noticeable path arrives at $\tau \approx 110\,\text{ns}$.

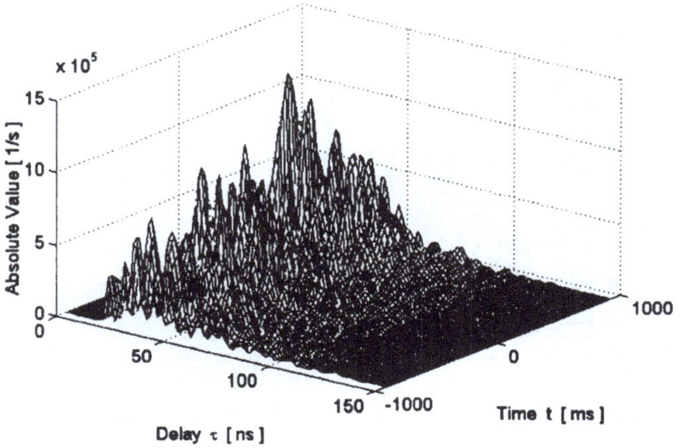

Fig. 5.2.14 Time-Variant Impulse Response of the measurement

In Fig. 5.2.15 the Time-Variant Impulse Response of the simulation is depicted. In general it has a characteristic comparable to the measurement: all paths reveal a distinct time variance and there is also a distinct decay versus the delay axis τ . The simulation clearly shows the single simulated paths that have been generated at discrete delay times. Worth of note is the influence of the correlation between two consecutive impulse response samples over the time t. In contrast to most of the other statistical models, which show a spiky nature of the Time-Variant Impulse Response (for example [26]), the simulation in Fig. 5.2.15 rather shows a smooth curve over the time axis t, which results from the consideration of the preceded value. Additionally the bandwidth of the simulation is limited to 600 MHz, which can be seen in the widening of the single impulses over the delay axis τ . A comparison of the amplitude values for both Time-Variant Impulse Responses shows that they are about in the same range.

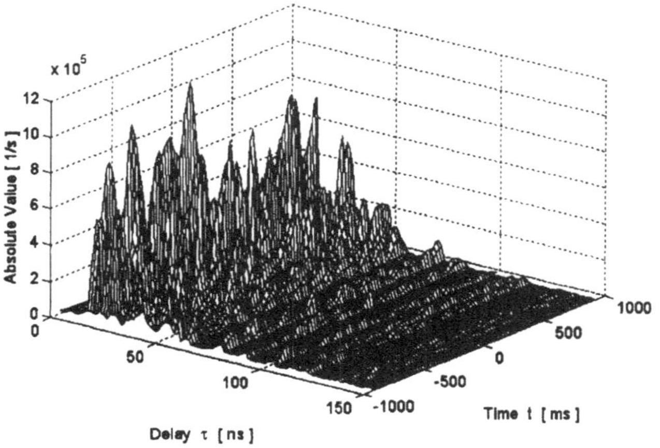

Fig. 5.2.15: Time-Variant Impulse Response of the simulation

The main difference between measurement and simulation can be seen in the behavior of the amplitudes versus the time axis *t*. Due to modeling the amplitudes by only using their statistical distribution function, the position of maxima and minima is random and thus naturally cannot coincide with the measurement. Besides, this is not intended at all, since statistical models usually find application for a simulation of a specific *kind* of environment in a general constellation of a mobile radio system.

Comparison of the Scattering Functions

In Fig. 5.2.16 the Scattering Function of the measurement is depicted. This plot shows a more or less spiky nature with signal components distributed over the whole range of normalized Doppler shifts. However, three main directions of incidence can be noticed: most of the power is received under an angle of incidence of $\alpha = 90° \, (\nu / \nu_{max} = 0)$, which corresponds to the direction of the direct path. Further dominant signal components arrive under angles of incidence of $\alpha \approx \pm 180° \, (\nu / \nu_{max} \approx \pm 1)$, which are paths through the opened door and their reflections from the window at the end of the office (see also Fig. 5.2.13). In addition, the Scattering Function shows that most of the transmitted power arrives at the receiver with delay times smaller than 50 ns.

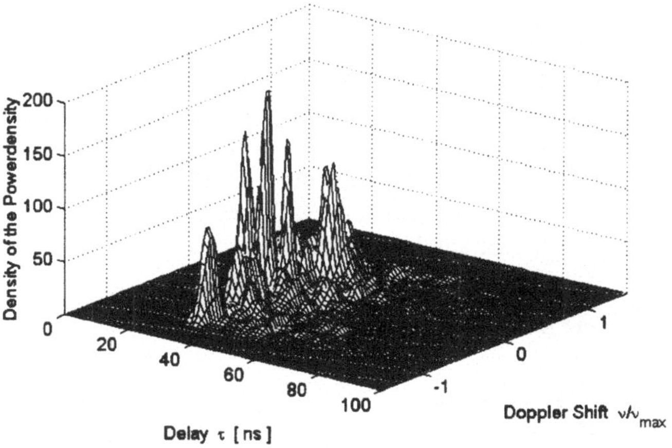

Fig. 5.2.16 Scattering Function of the measurement

Fig. 5.2.17 shows the graph of the simulated Scattering Function. At first glance it can be found that the simulated function reflects the general characteristics of a realistic environment. Comparing the simulation with the measurement, there is a certain correspondence, except for some details.

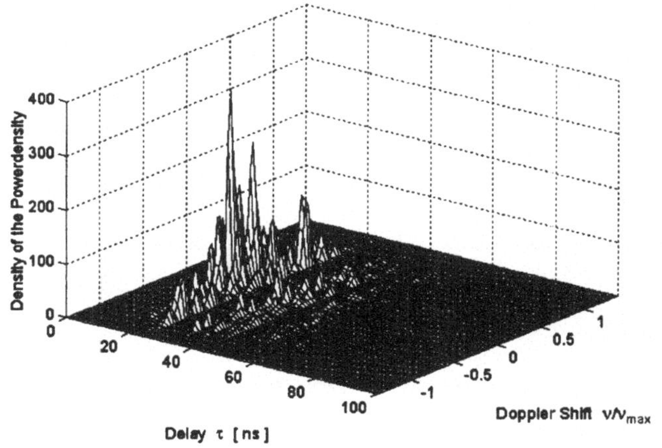

Fig. 5.2.17 Scattering Function of the simulation

As already for the impulse response, the simulation of discrete paths can clearly be seen in the Scattering Function. Regarding the angles of incidence, the simulated direct path ($\alpha = 90°$) coincides very well with the measurement, whereas the signals from the front and back ($\alpha \approx \pm 180°$) are stronger attenuated compared with the direct path of the simulation. However, the magnitudes of the simulated function generally are greater (approximately factor 2) than the amplitudes of the measurement. The first effect can be explained by the fact that the discrete delay times are generated statistically and thus do not exactly correspond to the delay times of the measurement. When applying the filtering, some of the statistically generated paths are thus multiplied with the slope rather than the maximum of the corresponding Doppler spectrum and therefore the magnitude is reduced. The second effect results from the demand that the simulated function should have the same power as the equivalent measurement. Due to the concentration of signal components to discrete delay time intervals (see Fig. 5.2.17), this condition can only be satisfied by higher amplitudes.

Comparison of the Delay Power Density Spectra

Fig. 5.2.18 and Fig. 5.2.19 show the Delay Power Density Spectra $P_h(\tau)$ of the measurement and the simulation, respectively. From its calculation (see equation (2)) this function can be interpreted as a mean squared impulse response. Another interpretation is that of being the power density distribution over the delay time τ [15]. On account of statistically generated arrival times, the value of the Mean Delay m_τ of the simulation shows a distinct deviation when compared with the value from the measurement. On the other hand, the standard deviation of this functions, more commonly known as r.m.s. Delay Spread σ_τ [15], shows very good correspondence between measurement and simulation.

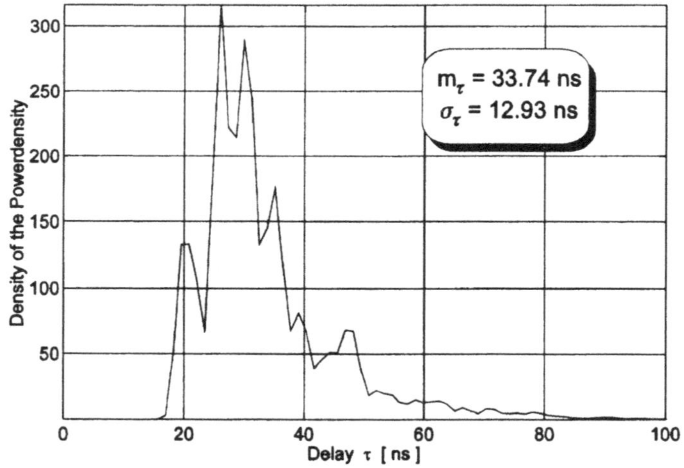

Fig. 5.2.18 Delay Power Density Spectrum of measurement with calculated Mean Delay m_τ and Delay Spread σ_τ

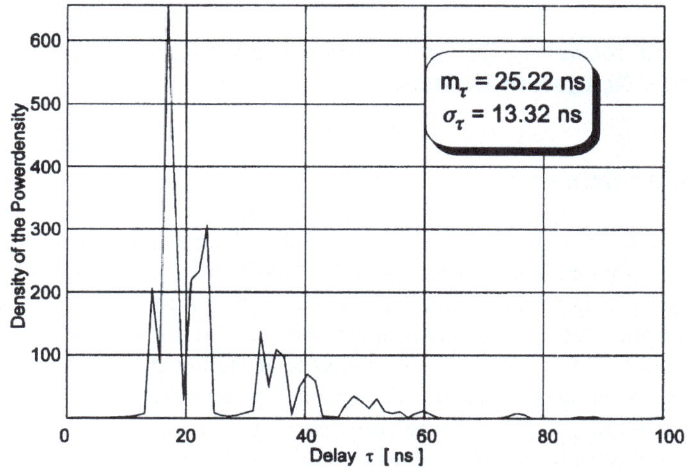

Fig. 5.2.19: Delay Power Density Spectrum of simulation with calculated Mean Delay m_τ and Delay Spread σ_τ

5.2.3.4 Conclusions For Statistical Modeling

A modeling approach for indoor mobile radio channels has been presented in the previous sections. Specific feature of this method is the consideration of the correlation between two adjacent path values, which, in this form, is not given in other statistical models. The approach uses the well-known modeling of arrival times, amplitudes and phases by statistical distributions, but additionally the deterministic properties are considered by filtering with appropriate Doppler spectra, which are taken from wideband measurements. As an alternative to wideband measurements the realistic Doppler spectra can also be calculated from a deterministically generated Scattering Function, for example by using ray-tracing algorithms [23]. Thus the presented modeling approach is a fusion of statistical and deterministic modeling. This hybrid considers the distinct deterministic effects in indoor environments, which cannot be described by a purely statistical model. On the other hand, a purely deterministic modeling does not consider statistical influences of indoor environments, which of course are also present. A comparison of the simulated Time-Variant Impulse Response, the Scattering Function and the Delay Power Density Spectrum with the functions calculated from the measurement shows that the simulation can be used very well for a description of the characteristics of the indoor mobile radio channel.

5.2.4 Ray Optical Modeling Of Time-Variant Indoor Radio Channels And Impacts Of Scattering Objects

5.2.4.1 *Introducing Remarks*

Computer-aided ray optical predictions of the channel properties of indoor and outdoor environments have become an essential and favorable help for the design of all kinds of digital mobile speech and data transmission systems and for the choice of appropriate components and networks. For the modeling of the indoor radio channel at 5.2 GHz and the analysis of dominant propagation effects in specific environments deterministic wideband modeling approaches are used, since the derived complex channel impulse response yields predictions about the complete information of the physical channel properties of a specific environment. This information includes the delay times, magnitudes and directions of incidence of the multipath components at the receiver, for instance, but also the angles of departure of the involved paths at the transmitter can be predicted, which is even an advantage compared to measurements. Numerous efforts have been made to develop such models for the prediction of characteristic values or functions that describe the multipath properties of space-variant indoor radio channels in the GHz- and the 10 GHz-range, for example [30, 4, 6, 2, 34, 23]. These methods usually perform the better the less complex the propagation environment is. However, real indoor environments may have an arbitrarily complex interior structure due to any kind of indoor obstacles, like for example furniture. Questions arise whether ray optical approaches are still applicable for real indoor environments that contain significant furnishing and whether the results from an empty environment are still valid.

Former work deals with the topic of reflection and transmission properties of building structures very detailed but does not investigate the impacts of typical furnishing explicitly. First reason is that in most cases the main intention is to verify modeling approaches that are sufficient to predict narrowband field strength tendencies, since a wideband characterization is not required, second reason being that for lower frequencies the wave lengths are too large to allow a consideration of smaller obstacles, which is not necessarily the case for higher frequencies. However, not only building materials but also furnishing may be a reason for scattering, diffraction or penetration and therefore for interference. In the investigations in [18] the focus is rather set on building material parameters at higher frequencies, in [29] furthermore windows, window shades and movable partitions are included. In [6] the reflection properties of objects with finite dimensions are investigated, but this publication does not deal with furnishing explicitly, whereas in [14] there is a suggestion that details inside of buildings should be taken into consideration in models. In the ray optical model in [4] only the rough geometry including doors and windows is considered. Finally, in [2] some pieces of metallic furniture are considered in a ray-launching model and the impacts are found to be significant, however, these investigations are not based on

measurements and furniture with properties of metal is not typical for many propagation environments.
In the following, the first step to handle the arisen questions regarding the influences of furniture on the radio channel will be to investigate the impacts of typical indoor obstacles by evaluating wideband measurements in indoor environment with and without furniture [19]. The investigations will be carried out at 5.2 GHz - a frequency band which is planned for wireless local area network applications, for instance. The results of a measurement campaign that has been carried out to investigate influences of some typical indoor obstacles on the mobile radio channel and therefore the receiving conditions at a space-variant receiver will be evaluated. In a second step the achieved findings will be considered in a ray-tracing model and exemplary results will be compared to wideband measurements in real indoor environment with some furnishing [22].

5.2.4.2
Impacts Of Furniture On The 5 Ghz Indoor Radio Channel

Wideband measurements of the indoor radio channel at 5.2 GHz are evaluated to get a basic insight into the impacts of furnishing on the receiving conditions.

Measurement Scenarios

A meeting room has been chosen for the measurements as an example for a typical office environment. The constellations used are given in Fig. 5.2.20. The empty room of 7.64 m x 4.42 m only consists of floor and ceiling, walls, windows, windowsill, doors and heaters (dark gray in Fig. 5.2.20). Furthermore, a small washbasin is located at one corner of the room. This corner is tiled from the floor up to a height of 1.40 m. Moreover, two rows of neon lights (which were switched off during all measurements) are fixed at the ceiling. For the investigations with furniture, all components and objects that are usually part of the interior decoration were placed within the room. The essential components are sketched in light gray in Fig. 5.2.20. These are for example several tables and chairs, shelves, a TV and a window shade outside the window. Furthermore the room is divided by a movable partition wall. Behind this wall there is an area that temporarily can be used as a workshop. Among other things there are a workbench, several electric devices and mechanic tools. For both the scenario with and without furniture the two Tx-Rx-constellations given in Fig. 5.2.20 have been used to get four different wideband measurements to evaluate. Both the transmitting and receiving antennas were discone, which have been placed at the same height of 1.60 m with vertical polarization. This is not a generalizing but a very typical Tx-Rx-constellation. The dynamic range of the measurement system in this case was about 85 dB, which is sufficient for short ranges.

Evaluation and Comparison of Wideband Measurements with and without Furniture

The absolute values of the Time-Variant Impulse Responses of the Tx_1-Rx_1-scenario are shown in top-views in Fig. 5.2.21a (without furniture) and Fig. 5.2.21b (with furniture), respectively. The magnitudes are colored according to a linear scale to be able to distinguish between different dominant signal contributions and to consider interactions of different components. In both figures the time-dependent changes of the delay times of different multipath components can be seen. The axis are the propagation delay of the signal components and the equivalent time for the movement of the receiver. Generally speaking both functions look very similar. It can be noticed that the line of sight signal shows interference with multipath components (reflections at the upper wall in Fig. 5.2.20 and at the floor), which results in a periodic structure in the magnitudes along the time axis due to the coherent superposition of these components. This pattern can also be noticed when the furniture is in the room, even though some pieces of furniture are obstructing both reflected signal paths. For a further comparison, the Delay Power Density Spectra are taken into consideration, Fig. 5.2.22a and Fig.5.2.22b, respectively. These functions show the averaged power density of the signal components over the delay time [15]. In the range up to 10 dB below the power of the direct signal the functions are nearly identical. Impacts of the furniture can be seen beyond this mark. The Mean Delay m_τ and the Delay Spread σ_τ slightly decrease when using this scenario.

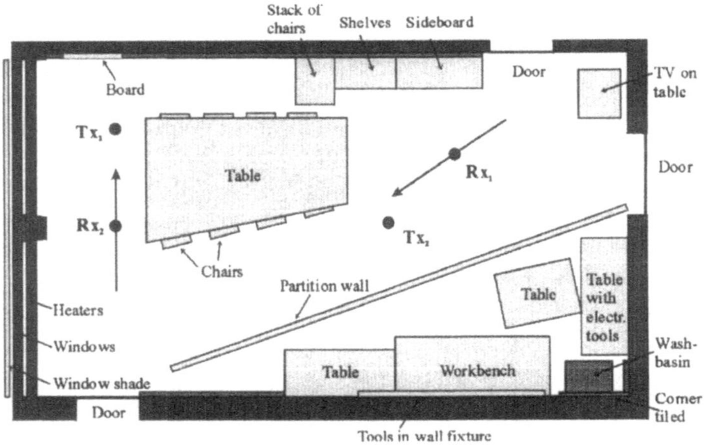

Fig. 5.2.20 Indoor scenario for the two measurement constellations with positions of the transmitter Tx and directions of movement of the receiver Rx, additional furniture is colored in light gray

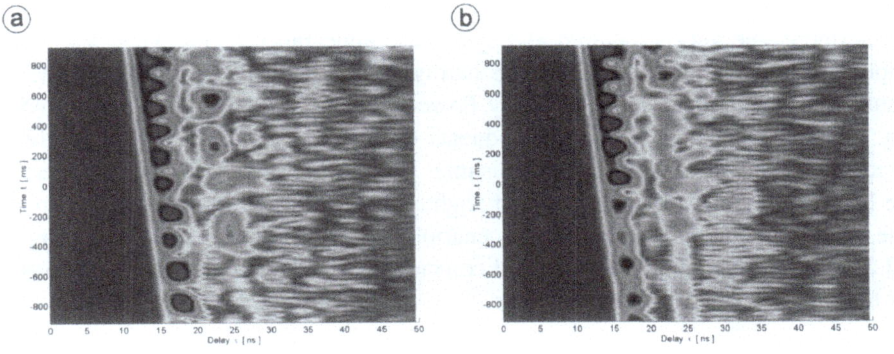

Fig. 5.2.21 Time-Variant Impulse Response in top-view, linear scaling, a) *without* furniture in the room
and b) *with* furniture in the room, scenario Tx_1-Rx_1

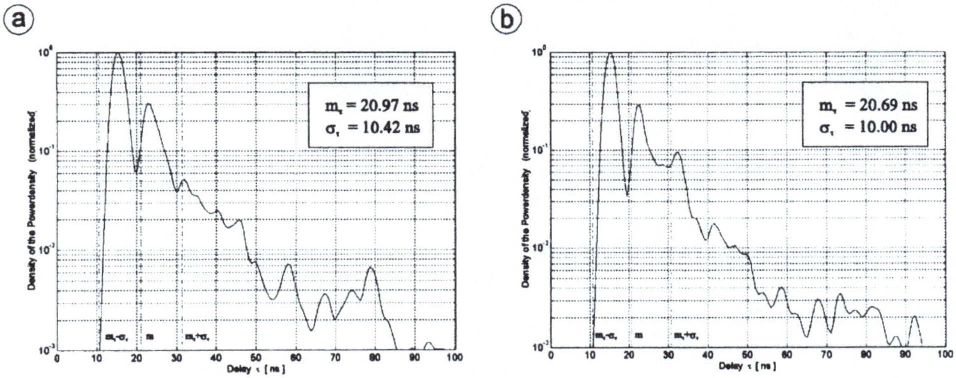

Fig. 5.2.22: Normalized Delay Power Density Spectrum, logarithmic scaling, with Mean Delay m_τ and
Delay Spread σ_τ a) *without* furniture in the room and b) *with* furniture in the room, scenario Tx_1-Rx_1

The same investigations have been carried out using the second scenario Tx_2-Rx_2. In the Time-Variant Impulse Responses, Fig. 5.2.23a and Fig. 5.2.23b, quite different characteristics can be found for the two scenarios with and without furniture.The direct signal at $\tau = 14$ ns in Fig. 5.2.23a does not show any fluctuations, whereas the direct signal in Fig. 5.2.23b is interfering with a multipath component resulting from a single reflection at the partition wall.

Furthermore, the very structured and symmetric appearance of the Time-Variant Impulse Response in Fig 5.2.23a can hardly be found in Fig. 5.2.23b. Further results can be derived from the Delay Power Density Spectra, Fig. 5.2.24a and Fig. 5.2.24b, respectively. Only components that are more than 10 dB below the direct signal show changes due to furnishing. In contrary to the first scenario, here the Mean Delay and the Delay Spread slightly increase when including furniture. The Scattering Functions for the first scenario are depicted in Fig. 5.2.25a and Fig. 5.2.25b, respectively, and show a good coincidence when comparing the functions for the cases with and without furniture.

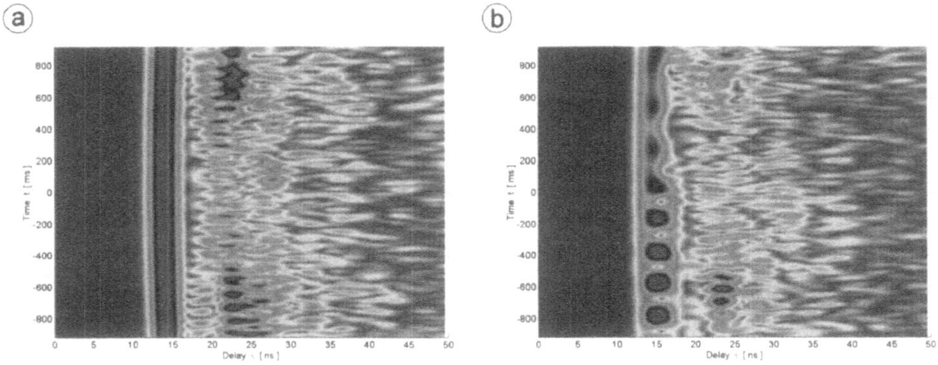

Fig. 5.2.23 Time-Variant Impulse Response in top-view, linear scaling, a) *without* furniture in the room

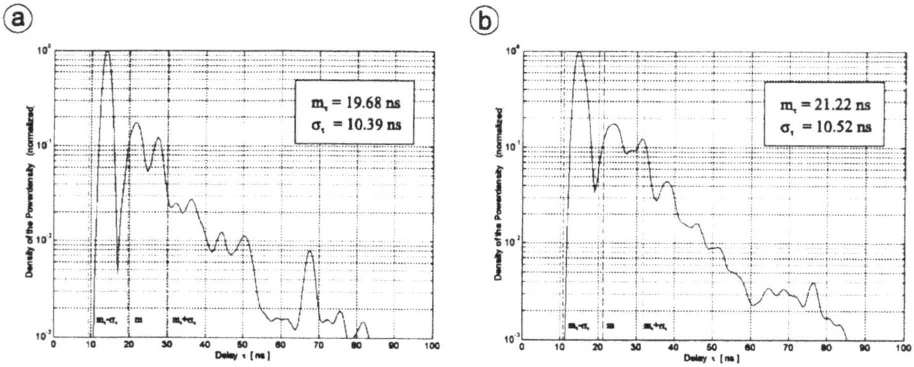

and b) *with* furniture in the room, scenario Tx_2-Rx_2

Fig. 5.2.24 Normalized Delay Power Density Spectrum, logarithmic scaling, with Mean Delay m_τ and Delay Spread σ_τ , a) *without* furniture in the room and b) *with* furniture in the room, scenario Tx_2-Rx_2

As mentioned before, the normalized Doppler shift ν/ν_{max} is identical to the cosine of the azimuth angle of incidence of the signal components at the receiver [15]. Generally speaking, with furniture there are more contributions at the receiver. This can be seen from the larger dark areas in the Scattering Function, Fig 5.2.25b. Some distinct peaks additionally appear, some peaks get wider when taking into consideration the furniture. These facts indicate that there are more scattering processes if there is furniture in the room, which is plausible since there are more obstacles at which scattering can occur. Evaluating the Scattering Function using a ray-tracing algorithm [22], the dominant signal components (up to 20 dB below the direct signal) are assigned to their propagation paths. In Fig. 5.2.25 these components are marked by circles in the Scattering Functions and are drawn as geometric propagation paths in the ground plan. Some of these contributions overlap and interfere coherently due to the limited bandwidth of the measurement set-up and therefore are identified as one component at the receiver.

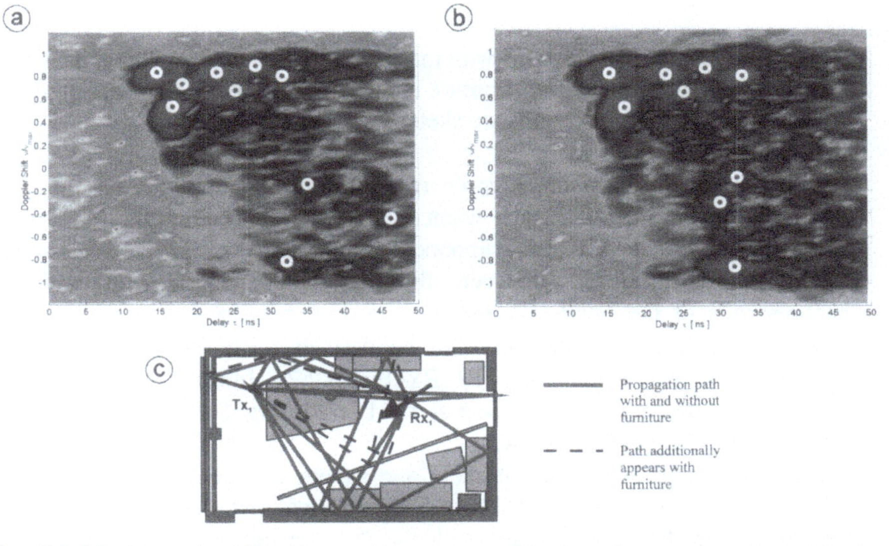

Fig. 5.2.25: Scattering Functions with main contributions in top-view, logarithmic scaling, a) *without* and b) *with* furniture in the room, scenario Tx_1-Rx_1, significant propagation paths are sketched in the ground plan (c)

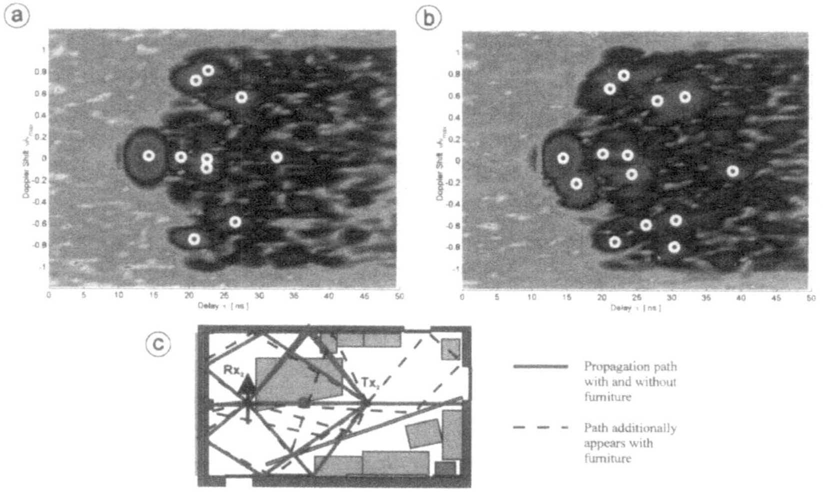

Fig. 5.2.26: Scattering Functions with main contributions in top-view, logarithmic scaling, a) *without* and b) *with* furniture in the room, scenario Tx_2-Rx_2, significant propagation paths are sketched in the ground plan (c)

Many significant paths (for example resulting from scattering at walls in the empty room) remain significant when taking into consideration the furniture. Some strong and several weak components additionally appear in the Scattering Function with furniture. However, the main propagation paths and basic characteristics of the Scattering Function of Fig. 5.2.25a remain significant in Figure 25b. The same statements can be derived from the evaluation of the Scattering Functions for scenario Tx_2-Rx_2,Fig. 5.2.26. In this case additionally a distinct local maximum can be found at $\tau = 16\,\text{ns}$ in Fig. 5.2.26b resulting from a single reflection at the partition wall.

Further interactions with this obstacle and others can be found by evaluating further components (see Fig. 5.2.26). Even the window shade can be detected separately from the windows, although the distance between both is only a few centimeters and the resolution of the measurement set-up is much too small to resolve this in the first place. However, since the corresponding propagation paths include two scattering processes (at the windows/shade and at the walls), they have quite different propagation delays and the multipath components can be resolved separately.

Summary and Conclusions from the Measurement Campaign

Wideband measurements that have been carried out in a meeting room have been used to evaluate the impacts of furnishing on wave propagation at 5.2 GHz for a typical office scenario. It has been found that with furniture more different scattering processes are detected which contribute to the received signal. The channel parameters Mean Delay and Delay Spread for both scenarios slightly change, but with different signs depending on the propagation environment and constellation of Tx and Rx. One important thing to be noticed is that nearly all significant components from the cases without furniture remain significant in the cases with furniture. Additionally some dominant peaks and many weak components appear in the Scattering Function with furniture. From this it can be seen that there is furnishing that should be included in deterministic modeling approaches to improve the prediction quality. In the present case (vertically polarized antennas with identical height) especially the scattered paths resulting from the large obstacles in the height of the antennas contribute significantly to the received signal. Therefore, these components should be taken into consideration in ray optical approaches, if possible, to improve the prediction quality. However, further measurements with different environments, antennas and polarizations will have to be carried out and evaluated to verify and substantiate the obtained results.

5.2.4.3 Ray Optical Modeling With Significant Interior Structures

Using computer-based ray-tracing algorithms, the modeling of the propagation mechanisms and the propagation scenario has to be carried out very conscientiously. A compromise has to be found between expenditure and nearness to reality with regard to the efficiency of the model in practical use. In the following, a three-dimensional ray-tracing algorithm [23] using the image source method (for example [6]) is used to apply the results obtained by the preceding investigations on the impacts of furnishing.

Description of the Modeling Approach

The ray optical description is valid as long as the wavelength is small with respect to the dimensions of regarded obstacles. In view of the indoor objects to be found typically this means that at higher frequencies the modeling of the environment can (or even must) be carried out more detailed than at lower frequencies. The indoor environment can either be modeled by a two-dimensional ground-plan to which a constant height is assigned to, with even floor and even ceiling, or by combining three-dimensional objects. Depending on the application the pros and cons of both kinds of modeling have to be considered carefully: on the one hand this can mean a very easy and efficient handling of the building data by importing a blueprint in pixel- or vector-based data format, which implies the restriction to a modeling of obstacles that have at least almost the height of the rooms and nearly constant extents (for example doors, shelves, huge windows, etc.); on the other hand a very realistic and detailed three-dimensional modeling of the environment may suffer from disadvantages regarding the efficiency. Also for reasons of efficiency usually only the modeling of essential wave propagation mechanisms like transmission through homogeneous obstacles and multiple specular reflections is taken into consideration. In practice, in the GHz- and 10 GHz-range scattering from discontinuities (diffraction) only has to be considered if there is a transition from line of sight to non line of sight or vice versa, since in most other cases the diffracted contributions may be neglected in comparison to other multipath components. For the consideration of rough, smaller-sized scattering surfaces alternative approaches may be used to include also scattered contributions in other directions than the specular [20]; furthermore, for very inhomogeneous the media wall-types may be classified from measurements according to their reflection- and transmission properties and used in the ray optical models by look-up tables [21]. However, with respect to the efficient realization of the modeling on computers, a confinement to only dominant signal components is reasonable, as on the one hand these contributions have the major impacts on the reception quality at the receiver and on the other hand ray optical approaches have the property that the calculation time and memory requirements rise drastically with the number of considered obstacles in the propagation environment and with the number of interactions of the electromagnetic waves at these objects [13]. Strictly speaking, the used Fresnel theory of reflection and transmission is only valid for flat, smooth surfaces with infinite extensions. However, in most applications the influence of finite dimensions of even scatterers (for example walls) may be neglected if the considered obstacle covers the first few Fresnel zones of the incident electromagnetic waves [6] and if the roughness of its surface is small with respect to the wavelength [3]. Adopting these assumptions, the modeling of multiple transmissions and reflections in the following is done by an algorithm using the "characteristic matrix" [35]. In this approach, which is used in the field of liquid crystal cells, the Fresnel plane wave transmission and reflection coefficients of multilayer-structured obstacles are calculated considering angle of incidence, material parameters (thickness of each layer of an obstacle, complex permeability and permittivity) and polarization of the incident wave. Scattering in non-specular directions is neglected.

In conformity with the performance of the swept-frequency measurement system described before, the receiver is moved along a straight line using discrete steps in the model, while the environment is kept static. The space-variant channel properties are determined in 255 intervals of $\lambda/8$ on a total length of 1.84 m to obtain a sufficient oversampling with respect to the maximum occurring Doppler frequency. The achieved space-variance also in this case will be interpreted as a "time-variance in static environment" assuming a constant velocity of 1 m/s of the receiver. The wideband characteristics of the impulse response can either be achieved by time-domain modeling of the signal strengths and delays of the multipath components [23] or with a direct modeling of the Time-Variant Transfer Function in frequency domain [7] by simulating the functioning of the swept-frequency measurement system in analogy. Using the time-domain modeling approach, which is usually found in the literature, a direct comparison of the measured and simulated space-variant impulse response is only possible by taking into consideration the finite bandwidth of the measurement set-up. Also with regard to wideband applications it is reasonable to consider the fact that every radio broadcasting system has a certain finite bandwidth. That means, every application only uses a more or less wide interval of the infinitely wide frequency spectrum of the radio channel. For that reason it is not sufficient only to calculate the discrete arrival times of single multipath components at the receiver, as in reality they cannot be resolved with this infinite resolution anyway. The "smeared" appearance of a measured impulse response along the delay time axis is realized in the model if the finite measurement bandwidth is taken into consideration [15]. Basic reflections on the Fourier transform, which are given in the following Table 5.2.1, illustrate this fact.

Table 5.2.1: Impacts of bandlimitation on the signal in time-domain based on the Fourier transform

Frequency domain	Time domain
$X(f) = a = const\|_f$	$x(t) = a \cdot \delta(t)$
$X(f) = \begin{cases} a & for \quad \|f\| < B/2 \\ 0 & for \quad \|f\| > B/2 \end{cases}$	$x(t) = a \cdot B \cdot \frac{\sin(\pi B t)}{\pi B t}$

The delay times of different signal components that are obtained from a ray-tracing algorithm [23] can be interpreted as Dirac pulses in time domain with an infinite bandwidth each. However, real channels have a finite bandwidth due to the bandlimitations in transmitter and receiver which has the effect of a limitation in the magnitudes and a deformity of the pulses to *sinc*-functions in time domain. The resulting broadening of the pulses of each multipath component leads to a coherent superposition (and thus interference) of the different signal components. An example for the impacts of the bandwidth on the prediction results is given in Fig. 5.2.27. In Fig. 5.2.27a the absolute value of a simulated Time-Variant Impulse Response is shown with a signal bandwidth of B = 600 MHz. The same signal is shown in Fig. 5.2.27b, but using a much higher bandwidth (and therefore higher resolution) of B = 2.2 GHz instead. Due to the implicit bandlimitation of the measurement results in the measurement set-up the simulated results thus must also be weighted by the bandwidth of the measurement system according to:

$$h_b(t,\tau) = F^{-1}\{F\{h(t,\tau)\} \cdot T_{id}(f)\}$$

In this formula $h_b(t,\tau)$ is the bandlimited simulated channel impulse response, $T_{id}(f)$ is a rectangular transfer function with the effective bandwidth of the system, $h(t,\tau)$ is the simulated channel impulse response with infinite bandwidth, t the time during the movement of the receiver and τ the delay time of the multipath components. F and F^{1} describe the Fourier transform and the inverse Fourier transform, respectively. Therefore, in these considerations the measurement set-up can be seen as a special case of an application with a high bandwidth. For a more reliable prediction of the absolute values of the magnitudes of each signal component further impacts have to be taken into consideration, like the space

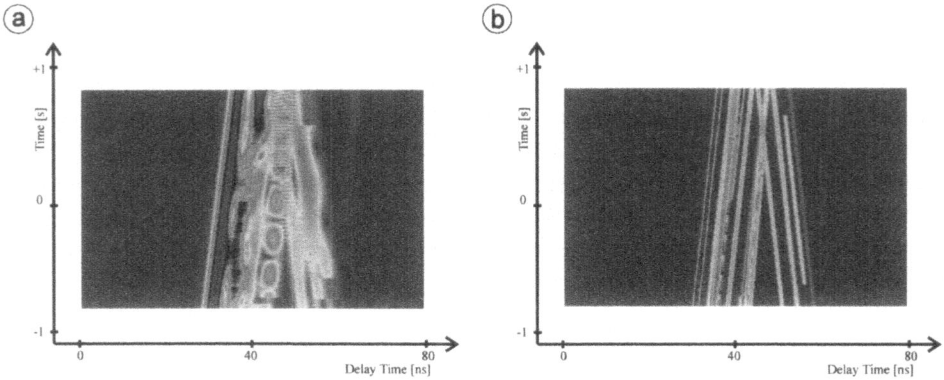

Fig. 5.2.27: Absolute values of an exemplary simulated Time-Variant Impulse Response with bandwidth of a) 600 MHz and b) 2.2 GHz; smaller bandwidth results in wider pulses and therefore distinct coherent superposition and interference

limitation of the Rx-movement and the properties of digital signal processing used for the conversion of the magnitudes [15] as well as the antenna gains and radiation patterns. The impacts of the involved antennas are considered by weighting the predicted rays with the known stationary antenna pattern. The antennas may be tilted and turned in any direction in the model, for example to investigate directional or polarimetric aspects.

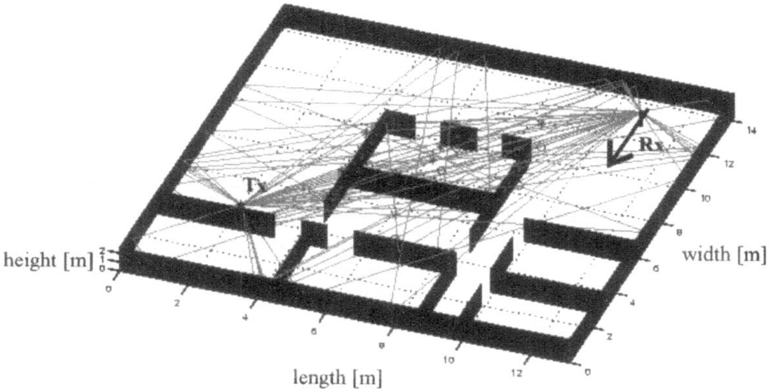

Fig. 5.2.28: Example for deterministic indoor scenario [23] with transmitter Tx and moving receiver Rx and some of the calculated propagation paths (shown up to the second order of reflection)

Finally, prediction results are compared to wideband measurements to validate the entire model. The ray optical modeling approach for the prediction of time-variant indoor radio channels has already been verified in "deterministic indoor environment" (an air-raid shelter without furnishing, windows or doors, see Fig. 5.2.28) at 1.8 GHz in the first place [23]. However, the impacts of real indoor objects have not been taken into consideration in [23].

Description of the Verification Scenario

For the verification of the simulation results in the 5 GHz range, measurements at 5.2 GHz have been carried out in a classroom. The measurement- and simulation-scenario is given as a ground plan in Fig. 5.2.29. The room has a length of 13.13 m, a width of 11.15 m and a height of 3.50 m. The front and the back side of the classroom are walls made from stone and concrete, additionally a huge closet with glass doors can be found at one wall and a board at the opposite one. The left and right side of the room contain huge windows and four posts each. Furthermore many tables, chairs, a washbasin and small heaters (below the windows) are placed in the room. These smaller objects are neglected in the model according to the findings from the measurements on the impacts of furniture. Propagation effects from and to the outside of the room are neglected as well, since only the wall containing the wooden door is not an outer wall and propagation effects through this wall are supposed to be neglectable when considering only dominant signals within this room. The investigations in deterministic indoor environments at 1.8 GHz in [23] have shown that especially in line of sight cases the main contributions are provided by the direct signal and by paths containing one specular reflection. Additionally, propagation paths with two reflections are considered in the present investigations. Reflections of order three or higher usually do not contribute significantly to the dominant arriving signal magnitudes in line of sight cases.

The ray optical method is applicable, since the wavelength is 5.8 cm and the extensions of the indoor obstacles (except for the posts with a width of about 30 cm) are much bigger, which is a necessary condition for the validity of ray-tracing. Therefore unacceptable prediction errors due to this fact are not expected.

The electromagnetic material properties of the walls and indoor obstacles are dependent on the frequency. Since for the frequency band at 5.2 GHz unambiguous values are not available, the material parameters for the simulations are estimated taking into consideration the literature [3, 12] and are given in Table 5.2.2. All materials are assumed to be homogeneous, ideally even, layered (if necessary) and dielectric with a neglectable conductivity. The values of the parameters are assumed to be constant within the bandwidth of interest. Further parameters for both measurement and simulation are given in Table 5.2.3.

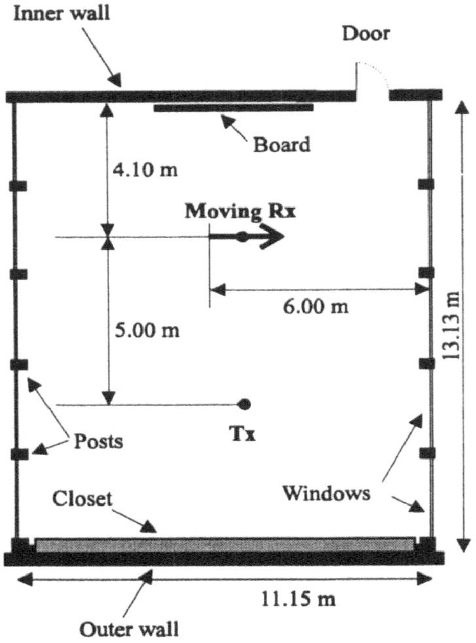

Fig. 5.2.29: Indoor scenario with transmitter Tx and direction of movement of the receiver Rx

Table 5.2.2: Material parameters used for the simulation

Material	Permittivity	Thickness
Floor, ceiling, outer wall (concrete)	5.5-j0.3	0.4 m
Inner wall, posts (concrete)	5.5-j0.3	0.2 m
Windows (glass, air, glass)	6-j0.04 1-j0 6-j0.04	0.003 m 0.02 m 0.003 m
Board (slate)	7-j0.3	0.05 m
Door (wood)	2.5-j0.03	0.05 m
Closet (wood)	2.5-j0.03	0.02 m
Glass door of closet (glass)	6-j0.04	0.003 m

Table 5.2.3: Parameters used for measurement and simulation

Parameter	Value
Power Tx	15 dBm
Tx-/Rx-antenna gain	0.65 dBi
Height of the antennas	1.3 m
Polarization Tx/Rx	vertical/vertical
Bandwidth	600 MHz

Analysis of Propagation Effects

As mentioned before, for the comparison of measurements and simulations it is necessary to consider the impacts of the measurement set-up on the results. In this case it is easier to implement the properties of the measurement set-up in the simulations than excluding these properties from the measurements. The main influences of the measurement system result from its limited bandwidth, therefore the simulation results are bandlimited with the bandwidth of the measurement system, which results in broader and coherently superimposing *sinc*-like pulses. Furthermore the magnitudes are calculated taking into consideration propagation loss, polarization, loss due to reflection and transmission, digital signal processing and space-limitations [15], so the deterministic character of the model is preserved. When comparing the top-views of the Time-Variant Impulse Responses qualitatively for simulation (Fig. 5.2.30a) and measurement (Fig. 5.2.30b), the influence of time-variance on the multipath components can be seen.

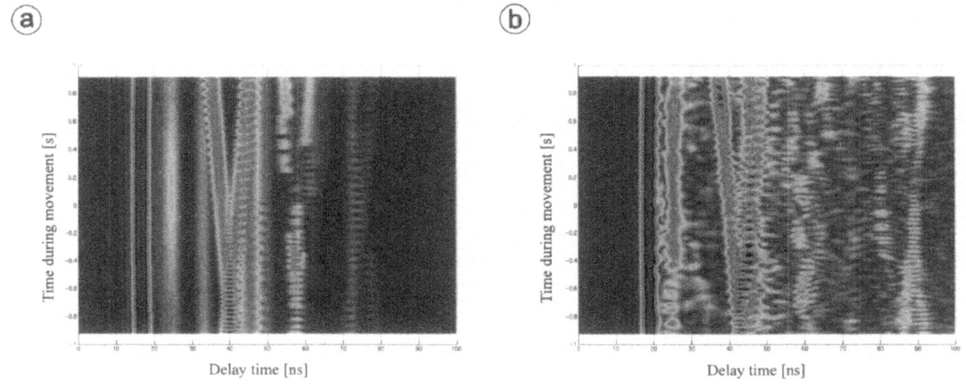

Fig. 5.2.30 Absolute values of the Time-Variant Impulse Response derived from a) simulation and b) measurement, top-view

Again the magnitudes in Fig. 5.2.30a and Fig. 5.2.30b are colored according to a linear scale for a more critical comparison. Due to the additional bandlimitation of the simulated impulse response (Fig. 5.2.30a) several contributions which have nearly the same delay time interfere coherently. This can be seen from the small-scale fluctuations along the ordinate for example at $40\text{ ns} \le \tau \le 45\text{ ns}$ both for measurement and simulation. The dominant contributions and their characteristics are predicted properly. Despite the agreement that can be found, the magnitudes still suffer from uncertainties concerning the exact gain and antenna pattern of the antennas and the material properties of the obstacles that are penetrated by the waves or reason for scattering. Furthermore, the magnitudes of some multipath components are predicted too small to be resolved with linear scaling or are even missing due to the neglect of higher order reflections or due to neglects concerning the room structure. The direct signal at about 18 ns in each figure is in both cases the strongest path and has nearly a constant magnitude for both cases. However, the delay times are not exactly the same which can be explained with the fact that the calibration of the measurement set-up has to be carried out without the antennas. Consequently, the antennas and their lead-ins strictly speaking have to be assigned to the radio channel in the measurement, which results in a slightly higher propagation delay in the impulse response for the measurement [15]. Both facts can also be noticed from the Delay Power Density Spectra in linear scaling (Fig. 5.2.31a and Fig. 5.2.31b) which are obtained by averaging the power densities of all 256 consecutive impulse responses along the ordinate in Fig. 5.2.30a and Fig. 5.2.30b, respectively.

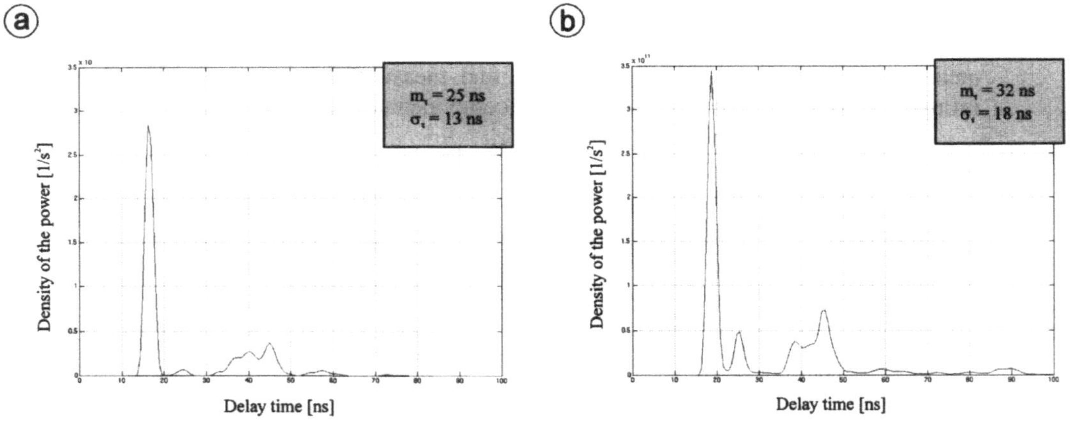

Fig. 5.2.31: Delay Power Density Spectrum obtained from a) simulation and b) measurement
with Mean Delay m_τ and Delay Spread σ_τ

The calculated Mean Delay m_τ and the r.m.s. Delay Spread σ_τ both are smaller for the simulation, mainly due to missing components of higher order reflections and the aforementioned deviation due to the calibration method. In Fig. 5.2.32a and Fig. 5.2.32b the Scattering Functions gained from simulation and measurement, respectively, are shown with linear scaling for the magnitudes.

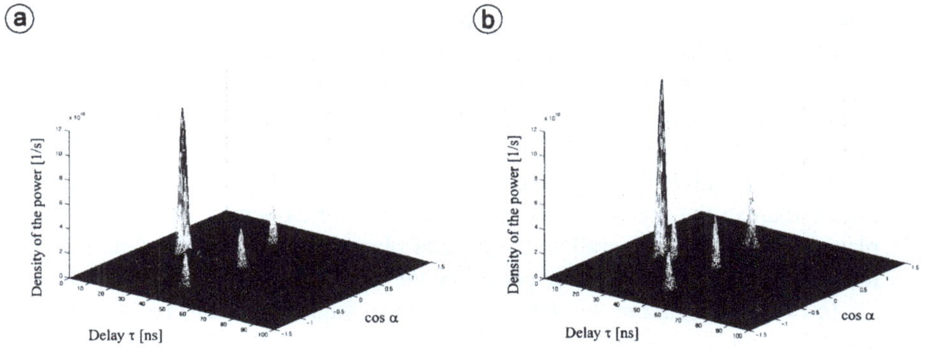

Fig. 5.2.32: Scattering Function derived from a) simulation and b) measurement

From this function the power densities and azimuth angles of arrival of different multipath components can be extracted. Again, the delay axis describes the time delay of the multipath components; the values on the normalized Doppler shift axis are identical to the cosine of the azimuth incidence angle at the receiver. Apart from a distinct basic similarity it can be noticed that nearly all components - even the direct signal, which is the strongest - are predicted with a magnitude that is too small. This leads to the supposition that possibly the gain of the antennas, which has been calculated according to a theoretical formula [15], has been chosen too small in the simulation. Furthermore, the magnitudes of the components that contribute from reflected and obstructed signal paths suffer from the above-mentioned uncertainties concerning the material parameters of the obstacles, which is a basic problem of deterministic modeling. Comparing the coordinates of the single peaks still a quite good agreement can be found. Nearly all important contributions are modeled properly. Using the ray-tracing algorithm, the evaluation of the delay times (which leads to the length of each propagation path) and of the normalized Doppler shift $\cos(\alpha)$ (which leads to the azimuth angle of incidence α at the receiver relatively to its movement direction) of each peak offers the possibility to assign them to the physical propagation paths and to identify the involved scatterers or obstacles ("Path Identification"). Apart from the direct line of sight signal, which can be noted as the strongest peak, there are three significant peaks at delay times of $38\text{ ns} \le \tau \le 44\text{ ns}$ both in measurement and simulation. The contribution at $\tau = 42\text{ ns}$ and $\cos(\alpha) = -0.85$ results from to a single-reflected path from the third window on the left side in Fig. 5.2.29. The single-reflected ray from the corresponding right window can be found at $\tau = 38\text{ ns}$, $\cos(\alpha) = 0.9$, which slightly deviates since the geometry is not exactly symmetric. At the delay time of $\tau = 46\text{ ns}$, $\cos(\alpha) = 0$, the reflection contribution from the board can be seen. An evaluation of this multipath contribution using a much higher bandwidth of 2 GHz (i. e. a higher resolution on the delay time axis) shows, that this peak consists of two independent contributions, one from the board and one from the wall behind (which is attenuated when penetrating the board). This is not resolved by the used measurement set-up, since the achievable resolution using a bandwidth of 600 MHz is about 3.3 ns, which corresponds to an additional path length of at least 1 m. Furthermore it can be noticed that the time delay of the corresponding peak in the simulation (Fig. 5.2.32a) is somewhat higher than in the measurement, Fig. 5.2.32b. Consequently, the board has been placed too close to the wall in the ground plan of the simulation. One noticeable thing is the deviation of the peaks in the area around $24\text{ ns} \leq \tau \leq 33\text{ ns}$ and $\cos(\alpha) = 0$. Calculating the delay time for the single-reflected ray from the floor, a delay to the direct path of about 2 ns can be found. Since this is less than the resolution of the measurement system and the magnitudes of the reflected and the direct path are very different, the former cannot be resolved. However the delay of the first order reflection at the ceiling is about 5.5 ns later and can be detected at the time delay $\tau = 24\text{ ns}$, $\cos(\alpha) = 0$. In the measurement, a distinct peak can be found, whereas in the simulation the magnitude is much smaller. This shows that the ceiling in reality is reflecting much better than in the model and that the material parameters for the ceiling have not been chosen appropriately. The small peak at $\tau = 33\text{ ns}$ and $\cos(\alpha) = 0$ in the simulation (Fig. 5.2.32a) can hardly be found in the measurement (Fig. 5.2.32b). Obviously this peak results from two double-reflected rays, each with one reflection at the floor and one at the ceiling, which can also be concluded from the angle of arrival and the delay time. Since obviously the reflection properties of the ceiling have been chosen too pessimistic, those of the floor must have been much too optimistic in the model to even compensate for this. This leads to the conclusion, that the thin floor covering (which was neglected in the model) results in an immense reflection attenuation. At a time delay of $\tau = 60\text{ ns}$ and $\cos(\alpha) = 0$ finally the reflection from the closet and the wall behind it can be found. Since the differences in distance of the paths containing one reflection (at the wall, the wooden back and the glass door of the closet) are below 1 m, they cannot be resolved as single peaks and result in one broad contribution. Using a higher bandwidth in the measurement, these components could be identified as two (or even three) peaks. Further paths can be detected in the Scattering Functions of both measurement and simulation. These components have much smaller power densities and therefore are less important.

Summary and Conclusions from the Deterministic Modeling Approach

Using a deterministic approach, the dominant propagation paths have been modeled and evaluated in a real line of sight indoor environment. The impacts of furnishing have been small (some additional scattering, which in this case leads to a broadening of some pulses) or even negligible (hardly any significant contributions missing due to neglect of tables, chairs and heaters) in the chosen classroom environment. Qualitative information has been obtained, whether the material parameters of involved obstacles have been chosen well or badly, thus an optimization or determination of these parameters is possible and therefore also topic of further investigations. Moreover it has been found that a very important condition for good results in the simulation is the exact determination of the location of transmitter and receiver and the exact true to scale drawing of the ground plan and the involved objects. Even slight deviations may lead to very different results. This is understandable, since also small-scale fading effects are modeled, which depend upon an accuracy of the coordinates of much less than one wavelength. Therefore a reliable prediction of these small-scale fading characteristics seems to be the more problematic the more complex the environment is. For very inhomogeneous, unstructured indoor objects there is no efficient analytical method to describe the interactions with the radio waves. For these cases measurement-based effective attenuation coefficients [21] are proposed to be used in look-up-tables to still be able to use ray optical approaches. Since these reflection- and transmission coefficients are only absolute values without any phase information and since the coherent Fresnel theory is only valid for large-sized objects, the ability to predict small-scale fading effects decreases with an increasing number of small-sized, irregular details in the propagation environment. The reason is, that the range of validity for coherent reflection or scattering theories [27, 33] cannot be met for such objects. Incoherent methods do not predict the phase information, but as the impacts of small-sized irregular surfaces have to be considered by *incoherent* methods, the phase may be varied statistically or randomly in the model with respect to the observation time. This on the other hand results in a loss of information about the Doppler shifts in the Scattering Function [20]. Therefore incoherent methods can neither be used to predict the small-scale fading effects resulting from "deterministic" wave superposition nor for the extraction of the directivity of the components from the Scattering Function. However, the impacts of incoherent contributions due to furnishing or complex obstacles in the impulse responses of the investigated environment seem to be significant if not only dominant signal paths are evaluated and thus their consideration in future semi-deterministic modeling approaches is justified.

REFERENCES

1. Braun W. R., Dersch U., "A Physical Mobile Radio Channel Model", *IEEE Transactions on Vehicular Technology*, Vol. 40, pp. 472-482, May 1991.

2. Chen S.-H., Jeng S.-K., "An SBR/Image Approach for Radio Wave Propagation in Indoor Environments with Metallic Furniture", *IEEE Transactions on Antennas and Propagation*, Vol. 45, No. 1, January 1997.

3. Cichon D. J., *Ray Optical Modeling of Wave Propagation in Urban Micro- and Pico-Cells* (in German), doctoral thesis at the University of Karlsruhe (T.H.), 1994.

4. Cichon D. J., Wiesbeck W., "Indoor and Outdoor Propagation Modeling in Pico Cells", in Proceedings of *5th International Symposium on Personal, Indoor and Mobile Radio Communications*, IEEE PIMRC'94, Den Haag, Netherlands, pp. 491-495, 1994.

5. Dersch U., Rüegg R. J., "Simulations of the Time and Frequency Selective Outdoor Mobile Radio Channel", *IEEE Transactions on Vehicular Technology*, Vol. VT-42, pp. 338-344, August 1993.

6. Dersch U., Zollinger E., "Propagation Mechanisms in Micro-Cell and Indoor Environments", *IEEE Transactions on Vehicular Technology*, Vol. 43, No. 4, pp. 1058-1066, 1994.

7. Englert T., Layer F., Kattenbach R., Früchting H., "Deterministic Wideband Modeling of the Time-Variant Transfer Function", Proceedings of *10th International Symposium on Personal, Indoor and Mobile Radio Communications,* IEEE PIMRC'99, Osaka, Japan, 12.-15. September 1999.

8. Englert T., Kattenbach R., "Statistical Analysis of Wideband Measurements in Indoor Environment at 5.2 GHz" (in German), Proceedings of *ITG-Fachtagung "Wellenausbreitung bei Funksystemen und Mikrowellensystemen"*, Weßling, Germany, pp. 245-256, May 1998.

9. Englert T., Kattenbach R., Früchting H., "Statistical Modeling of Wideband Indoor Radio Channels considering Deterministic Effects", Proceedings of *1999 Int. Workshop on Mobile Communications focused on UMTS & IMT-2000*, Chania, Crete, June 1999.

10. Hashemi H., "Impulse Response Modeling of Indoor Radio Propagation Channels", *IEEE Journal on Selected Areas in Communications*, Vol. 11, pp. 967-978, September 1993.

11. Hashemi H., "The Indoor Radio Propagation Channel", *Proceedings of the IEEE*, Vol. 81, pp. 943-968, July 1993.

12. Hippel A. v., *Dielectric Materials and Applications*, Library of Congress Cataloging in Publication Data, Technology Press of MIT, 1954/1995.

13. Huschka T., "Ray Tracing Models for Indoor Environment and their Computational Complexity", in Proceedings of *5th International Symposium on Personal, Indoor and Mobile Radio Communications*, IEEE PIMRC'94, Den Haag, Netherlands, pp. 486-490, 1994.

14. Ju K. M., Chiang C. C., Liaw H. S., Her S. L., "Radio Propagation in Office Buildings at 1.8 GHz", in Proceedings of *7th International Symposium on Personal, Indoor and Mobile Radio Communications*, IEEE PIMRC'96, Taipei, Taiwan, pp. 766-770, 1996.

15. Kattenbach R., *Characterization of Time-Variant Indoor Radio Channels by Means of their System- and Correlation Functions* (in German), doctoral thesis at the University of Kassel, Aachen: Shaker Verlag, 1997.

16. Kattenbach R., "Statistical Distribution of Path Interarrivaltimes in Indoor Environment", in Proceedings *48th Annual Vehicular Technology Conference*, IEEE VTC'98, pp. 548-551, 1998.

17. Kattenbach R., Englert T., "Investigation of Short Term Statistical Distributions for Path Amplitudes and Phases in Indoor Environment", in Proceedings *48th Annual Vehicular Technology Conference*, IEEE VTC'98, pp. 2114-2118, 1998.

18. Langen B., Lober G., Herzig W., "Reflection and Transmission Behaviour of Building Materials at 60 GHz", in Proceedings of *5th International Symposium on Personal, Indoor and Mobile Radio Communications*, IEEE PIMRC'94, Den Haag, Netherlands, pp. 505-509, 1994.

19. Layer F., Früchting H., "Investigation on the Impacts of Furniture on the 5 GHz Mobile Indoor Radio Channel", Proceedings of *1999 Int. Workshop on Mobile Communications focused on UMTS & IMT-2000*, Chania, Crete, June 1999.

20. Layer F., Früchting H., "Modeling Time-Variant Wideband Indoor Radio Channels with Rough Scattering Surfaces: A Comparison", Proceedings of *European Wireless'99 Conference*, Munich, Germany, October 1999.

21. Layer F., Kattenbach R., Früchting H., "Determination of Reflection and Transmission Properties of Wall-like Objects by Evaluating the Scattering Function", Proceedings of *Vehicular Technology Conference'99*, IEEE VTC'99 Fall, Amsterdam, Netherlands, September 1999.

22. Layer F., Kattenbach R., Früchting H., "Modeling and Analysis of Dominant Propagation Effects in Real Indoor Environment at 5.2 GHz", in Proceedings of *9th International Symposium on Personal, Indoor and Mobile Radio Communications*, IEEE PIRMC'98, Boston, USA, pp. 630-634, 1998.

23. Layer F., Kattenbach R., Früchting H., "Ray Optical Approach for a Wideband Indoor Wave Propagation Model with Moving Receiver", in Proceedings *8th International Symposium on Personal, Indoor and Mobile Radio Communications*, IEEE PIMRC'97, pp. 135-139, 1997.

24. Nakagami M., *The m-Distribution - A General Formula of Intensity Distribution of Rapid Fading*, in W.C. Hoffmann (ed.): Statistical Methods in Radio Wave Propagation, London: Pergamon Press, 1960.

25. Nelder J. A., Mead R., "A simplex method for function minimization", *Computer Journal*, Vol. 7, pp. 308-313, 1964/65.

26. Rappaport T. S., Seidel S. Y., Takamizawa K., "Statistical Channel Impulse Response Models for Factory and Open Plan Building Radio Communication System Design", *IEEE Transactions on Communications*, Vol. 39, pp. 794-807, May 1991.

27. Ruck G. T., Barrick D. E., Stuart W. D., Krichbaum C. K., *Radar Cross Section Handbook,* Volume II, New York: Plenum Press 1970.

28. Saleh A. A. M., Valenzuela R. A., "A Statistical Model for Indoor Multipath Propagation", *IEEE Journal on Selected Areas in Communications*, Vol. SAC-5, pp. 128-137, February 1987.

29. Sato K., Manabe T., Ihara T., Saito H., Ito S., Tanaka T., Sugai K., Ohmi N., Murakami Y., Shibayama M., Konishi Y., Kimura T., "Measurement of Reflection and Transmission Characteristics of Interior Structures of Office Building in the 60 GHz Band", in Proceedings of *7th International Symposium on Personal, Indoor and Mobile Radio Communications*, IEEE PIMRC'96, Taipei, Taiwan, pp. 14-18, 1996.

30. Seidel S. Y., Rappaport T. S., "Site-Specific Propagation Prediction for Wireless In-Building Personal Communication System Design", *IEEE*

Transactions on Vehicular Technology, Vol. 43, No.4, pp. 879-891, November 1994.

31. Suzuki H., "A Statistical Model for Urban Radio Propagation", *IEEE Transactions on Communications*, Vol. COM-25, pp. 673-680, July 1977.

32. Turin G. L., Clapp F. D., Johnston T. L., Fine S. B., Lavry D., "A Statistical Model of Urban Multipath Propagation", *IEEE Transactions on Vehicular Technology*, Vol. VT-21, pp. 1-9, February 1972.

33. Ulaby F. T., Moore R. K., Fung A. K., *Microwave Remote Sensing - Active and Passive,* Volume II, Reading, USA: Addison-Wesley Publishing Company Inc., 1982.

34. Valenzuela R. A., "Ray Tracing Prediction of Indoor Radio Propagation", in Proceedings of *5th International Symposium on Personal, Indoor and Mobile Radio Communications*, IEEE PIMRC'94, Den Haag, Netherlands, pp. 140-144, 1994.

35. Wöhler H., *Analytical and Numerical Methods for Calculation of Electro-Optical Properties of Liquid Crystal Cells* (in German), doctoral thesis at the University of Karlsruhe (T.H.), pp. 51-54, 1990.

36. Zaghloul H., Morrison G., Fattouche M., "Frequency Response and Path Loss Measurements of Indoor Channel", Electronic Letters, Vol. 27, pp. 1021-1022, June 1991.

5.3 Satellite Diversity: a Technique to Improve Link Performance and Availability for Multicoverage Constellations

Pierpaolo Loreti and Michele Luglio

5.3.1 Introduction

The role and the importance that the satellite has in mobile communication scenario are well defined. In fact the third generation systems (UMTS/IMT2000 for example), to provide real global coverage and services, are conceived as two integrated or interworking segments: the terrestrial and the satellite one.

To play an essential role, other than providing global coverage, the satellite component must offer very reliable links. In fact, due to both propagation and co-channel interference impairments, which make the link weak, global and continuous coverage is only theoretically provided.

Many techniques are under study or being implemented, trying to mitigate the main impairments due to the mobile channel. To reach this goal the concept of diversity can be adopted and implemented. The useful signal can follow more than one path through different satellites covering the same area, increasing link availability [1] or improving signal to noise ratio, opportunely combining the signals in the rake-like receiver (see section 5.1.4).

The diversity order can be theoretically arbitrary (depending on constellation architecture) even if in practical systems dual diversity is being to be implemented for the first time [2].

Multiple satellite communication can be one of the most effective means to improve performances, even though system complexity increases too.

This chapter describes the satellite diversity concept and the relative system architecture. Moreover it aims at characterising the theoretical availability improvement in case of an arbitrary number of satellites assuming all the links statistically independent. It also provides time domain simulation results on system performance improvement in terms of link availability and carrier to interference ratio (C/I). In fact diversity represents a practical means to counteract shadowing and to co-channel interference.

5.3.2 Generalities

Advanced mobile communication systems must be characterised by large set of offered services and great capillarity in coverage. The former can be achieved offering advanced multimedia capability. The latter can be ensured implementing a system with a satellite component perfectly integrated with the terrestrial one.

To ensure real global coverage, as well as great link reliability and quality, is mandatory. To reach this goal, two critical aspects are worthy to be particularly cared. The first concerns mobile channel impairments, such as shadowing (caused by the environment). The second regards co-channel interference (due to signals transmitted by all the users in visibility of the serving satellite, in the up link, and of all the satellites in visibility of the wanted user, in the down link [3]), very critical when using multibeam coverage. Satellite diversity can be a practical means to mitigate the impairments due to both of them, using two or more links to establish communication.

UMTS/IMT2000 is the first system conceived to provide multimedia services utilising both components in an integrated scenario aiming at enhancing coverage and capacity. Also other systems are being to be designed allowing the presence of two interworking segments. The deployment of multisatellite constellations in non-geostationary circular orbits is a possible architecture [4] if aiming at small, low power consuming and low cost terminals and at optimising intersegment handover delays. Also geostationary solutions are pursued [5-7], not all belonging to UMTS/IMT2000 scenario. In any case the radio-mobile transmission channel strongly limits link performance.

In fact link availability is usually not so good due to the shadowing, depending on the low elevation angles of visibility. The angles are time variant because of the user mobility, in case of geostationary constellations, or of the relative motion between user and satellites, in case of non-geo orbits constellations.

In this scenario the implementation of satellite diversity can be a means to meet the requirements, especially if looking at the meaningful achievable results in terms of both link availability and C/I improvement.

Link availability can be increased as compared to the single link case, being the probability that two (or more) links would be simultaneously obstructed the product of individual probabilities, if statistically independent links are assumed. Actually the advantage achieved by using diversity only theoretically matches the last assumption because it depends on the constellation structure and on the azimuthal correlation among links.

Also interference is a main problem strongly impacting overall performance of such systems. The increase of the useful received power due to the combination of two (or more) signals from different satellites makes the carrier-to-interference ratio (C/I) greater, thus improving link availability too. Thus, the presence of overlapping beam areas, which in general makes sharply increase interference level, turns out to improve performance.

Several analyses, models and simulations based on theoretical approaches are present in the literature comparing performances in terms of link quality and capacity with single link reception case [8-12].

Many studies have been carried out to characterise the radio-mobile channel and to evaluate the performance of the new satellite systems. Usually in these models a finite state diagram describes the channel evolution. Each state corresponds to a different physical situation (shadowed, unshadowed, etc) and a statistical representation describes the transitions among different states.

Diversity technique has been proposed (and in some case adopted [2]) for systems utilising Code Division Multiple Access (CDMA). At present only dual diversity has been considered both at theoretical and at implementation level but, considering the constellation architecture of some LEO [13] or GEO [6, 7] systems, the use of satellite constellations allowing the implementation of multiple diversity can be considered technically feasible.

In this chapter a generalised model addressing the availability statistics in a multiple diversity scenario is presented. The model represents the generalisation of the two Markoff states satellite channel model and, to simplify the analysis, assumes that all the channels are statistically independent.

To quantify advantages on real systems, the results of simulations performed on a 48-satellites system in low earth circular orbits will be shown.

5.3.3 Satellite Diversity Architetture

In land mobile communication systems path diversity consists in taking advantage from the multipath propagation due to the great number of reflections caused by the surrounding environment. The different signals are opportunely combined in the Rake receiver [14] to get the best signal to noise ratio improvement. In satellite systems, where typically the multipath component is negligible, diversity is exploited utilising more than one satellite-user link to perform communication, thanks to the intrinsic capabilities offered by CDMA access scheme [8].

This technique, improving signal to noise ratio, allows to get lower BER, lower transmitted power (EIRP), lower required E_b/N_0, or higher link margins (see Fig. 5.3.1). Otherwise, if propagation conditions, depending on the time varying link geometry, make one or more links drop, for example due to obstacles, the other link(s) may hold the communication (see Fig. 5.3.2). Anyway the adoption of diversity makes C/I level improve, assuming a constant number of channels per satellite with respect to the single satellite scenario, without penalising overall system capacity [9].

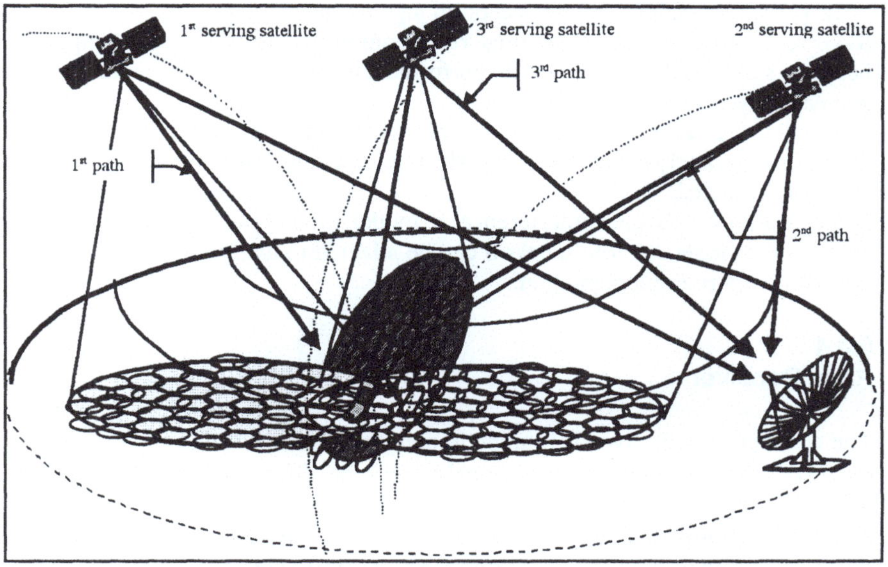

Fig. 5.3.1. Performance improvement in satellite diversity scenario

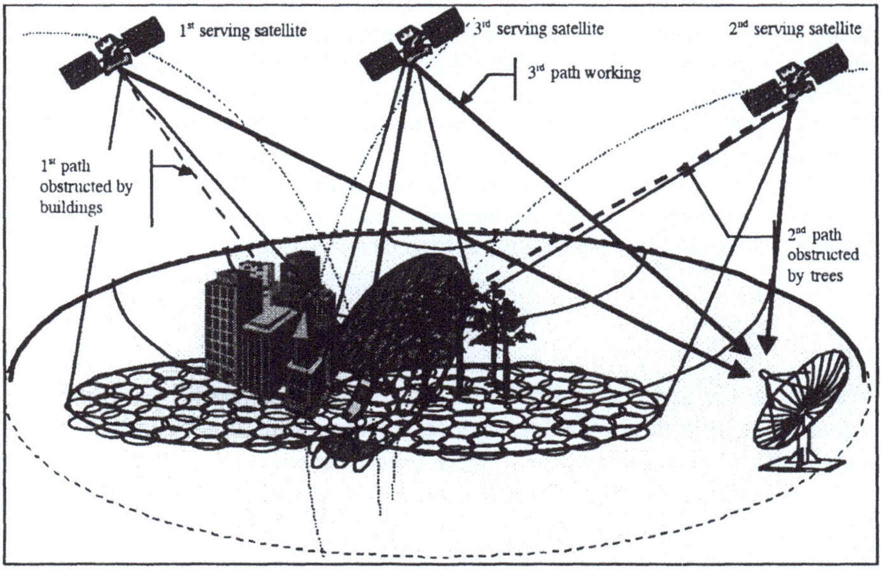

Fig.5.3.2. Availability improvement in satellite diversity scenario

Capacity improvement evaluation for a LEO system implementing diversity is presented in [10]. In [15] results on the link margin improvement using multiple satellite diversity are presented.

In [8] static analysis of the reception signals in the down link, assuming ideal Rake receiver performing maximal ratio combining, is addressed. Comparisons between single and double satellite performance are also reported in terms of BER and capacity.

In [11], [12] and [16] multiple signals reception, with particular regard to the down link, is analysed and some performance comparisons with single satellite case are presented.

Regarding co-channel interference, a model [3, 17] suitable for time domain analysis for single satellite scenario can be adapted to analyse the new scenario in case of diversity implementation.

5.3.4
Rake Receiver

The concept of path diversity can be implemented thanks to the capability offered by a particular receiver to opportunely combine different signals coming from different paths.The Rake receiver, proposed in [18], has been conceived to counteract multipath fading for terrestrial systems, usually strongly affected by this kind of impairment. It is based on the concept that the reflected signals can be distinguished and so can be suitably combined, taking in this way advantage from multiple propagation.

In satellite scenario the multipath component is usually negligible compared with the direct path component. In this case a Rake-like receiver can be used to combine more than one direct signal coming from as many satellites as if the different signals are seen as fictious multipath components.

The signals can be combined according to three main algorithms: selection combining, equal gain combining, maximal ratio combining. In any case the output of the combiner is a linear combination of the signals on the single paths.

In case of selection combining the best signal among the *N* available is selected at each instant. In this case the output can be carried out by:

$$U_o = \sum_{k=1}^{S} c_k u_k \tag{1.1}$$

where S is the number of satellites used to exploit diversity, U_o is the output signal, u_k is the k-th contributor, the coefficients c_k are all equal to zero but that corresponding to the selected path which is equal to one.

In case of equal gain combining the different signal are combined in the receiver each with equal weight. In this case the combination can be expressed according to the following expression:

$$U_0 = \frac{1}{S}\sum_{k=1}^{S} u_k \tag{1.2}$$

resulting in an average of different signals coming from the different paths.

In case of maximal ratio combining the different signals are linearly combined and the different weights are related to the channel impulse response for the relative path. In this case the combination can be expressed as in (1.1), where the coefficients c_k are equal to the complex conjugate of the impulse response coefficients relative to the kth path. Usually an estimate of these parameters is utilised. The signal to noise ratio can be expressed as

$$SNR = \frac{\left|\sum_{k=1}^{S} c_k u_k\right|^2}{\sum_{k=1}^{S} |c_k|^2 E\{n_k^2\}} \tag{1.3}$$

where n_k is the noise plus interference signal.

In case of satellite systems, demodulator architecture based on one rake receiver per path is preferred rather than a single receiver.

A critical issue related to signals combining is that each followed path has different length due to the different relative position user-satellite-gateway. To balance the propagation delay difference, which is time variant especially in case of non geostationary constellations, appropriate buffers must be foreseen. The dimension of the buffers will depend on the worst case delay difference and on the maximum utilised data rate.

For two non geostationary constellations, Globalstar [2] and ICO [19], taking into account the minimum elevation angle (10° for Globalstar and 28.5° for ICO), the variability of the worst case delay difference ranges from about 14 ms to about 16 ms.

5.3.5 Satellite Mobile Channel Characterisation

5.3.5.1 Uncorrelated Channel Case

Electromagnetic propagation is in general characterised by a direct component and by a diffuse component. In case of satellite propagation the former is called Line Of Sight (LOS) component and the latter multipath component each showing different statistical behaviour. Many models have been developed to analytically characterise the mobile satellite propagation channel [20-25]. The main and most accepted of them consider the direct component to follow a Rician statistic and the multipath component to follow a Raileigh or Raileigh/log-normal statistics.

The shadowing process can be characterised by the transition from a good state (presence of LOS) to a bad state (shadowing). Among the proposed models, those based on a Markoff process are considered in this context.

For each of three different types of identified environment (urban, suburban, rural) and for each set of elevation angle, outage probability thresholds are provided (Table 5.3.1), defined on the basis of propagation measurements [21].

The Rice factor R, dependent on environment and elevation angle, can be achieved linearly interpolating measured data according to (1.4):

$$R = r_{k0} + r_{k1} * \alpha \tag{1.4}$$

where α is the elevation angle, r_{k0} and r_{k1} are environment dependent coefficients.

When satellite diversity is introduced the two-state model is modified in an $M = 2^S$ states model with S the number of involved satellites.

In [26] and [27] a combined model for two satellites diversity (4 states model) is presented taking also into account statistical dependence of the two channels.

In [28] a three-state propagation model developed for Land Mobile Satellite Sistems (LMSS) has been presented. The state A is assumed in case of LOS, the state B is assumed in case of slight blockage by trees or by small obstacles and the state C in case of full shadowing. The probability density function adopted for each state are the Nagakami-Rice fading model, the Loo model (lognormal for the direct component and Rayleigh for the reflected one) and Raileigh respectively, adapted to the case of diversity implementation. The advantages in terms of link availability for theoretical system architecture and simulation results, taking into account user mobility, are presented too.

5.3.5.2
Correlated Channels Case

The analysis in uncorrelated channels conditions represents at the same time the easiest and the most general way to deal with satellite diversity. It represents also the most theoretical approach being the performance upper bound, in terms of improvements, for the system.

Table 5.3.1. Outage probability thresholds

Angles (degrees)	Urban	Suburban	Rural
<10	0.85	0.6	0.4
10-20	0.79	0.41	0.22
20-30	0.71	0.3	0.14
30-40	0.6	0.18	0.07
40-50	0.48	0.11	0.03
50-60	0.3	0.07	0.01
60-70	0.08	0.04	0.001
70-75	0.03	0.02	0.0001
75-80	0.001	0.005	0.0001
80-90	0.0001	0.0001	0.0001

In fact it is not easy to find out a general model taking into account the correlation among the links because it strongly depends on the particular environment (real or simulated) chosen for the analysis. In other words also the correlation behaviour is a not stationary stochastic process showing time and spatial variability.

In [26] and [27] correlation dependent on azimuth separation among the different paths is considered, reducing the diversity advantages with respect to the theoretical independent channel model.

In [29] the problem of evaluating the correlation coefficient for two paths as a function of the azimuth separation utilising measured data is addressed. An empirical mathematical model, validated by measurements campaign performed in different environments, is derived to evaluate the correlation coefficient for an earth station looking at two satellites separated in azimuth and with different elevation angles.

In [30], after a critical overview of the main satellite diversity studies, the computation of reliable correlation coefficient for different environments (starting from urban) has been approached by means of simulation. To reach this goal two different approaches have been followed, both starting from the same definition of a stochastic process. The former [31] aims at an explicit formulation for the correlation between shadowing states associated with pairs of satellites. The latter [32] performs the analysis in terms of the azimuth correlation sector, a parameter indicating the average angular separation for two satellites to become uncorrelated in a certain environment.

The simulation results in [30] offer a set of values of the correlation coefficient and of the azimuth correlation sector carried out in a canyon like environment with fixed street width and building height.

In [33] some results of an extension of the Rice-lognormal model proposed in [23] are shown in the case of satellite diversity taking into account the correlation among channels.

5.3.6
The Finite State Markoff Satellite Channel Model

Double diversity can be characterised by the four states Markoff model presented in [26] which represents an extension of the two-state model (see Fig. 5.3.3), presented in [21], to the case of two satellites. The basic assumption is that each link can assume only two possible states, good (G) or bad (B), depending on the presence or not of the line of sight for each. In Fig. 5.3.3 p_{XY} represents the transition probability from the state X to the state Y.

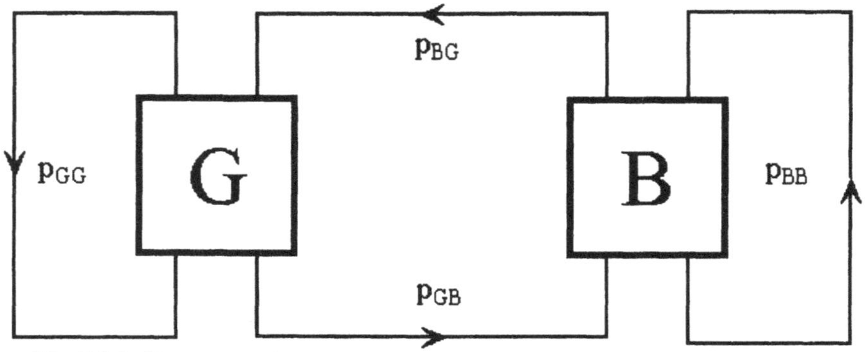

Fig.5.3.3. Two-state channel model

The state and the transition probabilities depend on the relative position of satellite and user and in particular on the elevation angle between them. For each satellite a transition probability matrix $P_s[p_{xy}]$ can be defined (s is the satellite index) as in (1.5):

$$P_s = \begin{bmatrix} p_{BB}^{(s)} & p_{BG}^{(s)} \\ p_{GB}^{(s)} & p_{GG}^{(s)} \end{bmatrix} \tag{1.5}$$

5.3.7 The Multi-State Markoff Satellite Channel Model

The model presented in the previous section can be extended and generalised to the case in which the number of satellites is arbitrary [34]. The importance of the resulting model and the achieved results is that they are of general validity not depending on the particular constellation.

5.3.7.1 Transition Probabilitiy

If the user can work simultaneously with S satellites, assuming the two state model for each link, the system can assume $M = 2^S$ states. The states can be opportunely grouped in $N = S+1$ states, where the jth state ($j \in J = \{0, \ldots, N-1\}$) represents the state in which j satellites are available.

The jth state includes a number of elementary states given by

$$\binom{S}{j}.$$

The transition probability matrix describing the Markoff process is reported in (1.6).

$$P = \begin{bmatrix} p_{00} & \cdots & p_{0k} & \cdots & p_{0N} \\ \cdots & \cdots & \cdots & \cdots & \cdots \\ p_{h0} & \cdots & p_{hk} & \cdots & p_{hN} \\ \cdots & \cdots & \cdots & \cdots & \cdots \\ p_{N0} & \cdots & p_{Nk} & \cdots & p_{NN} \end{bmatrix} \tag{1.6}$$

To figure out each term p_{hk} ($h, k \in J$), the model assumes that all the S channels are statistically independent and that all of them are characterised by a two state Markoff model.

The N-state Markoff satellite channel model is completely described if all the terms of (1.6) are calculated and the initial state probabilities are given.

The probability to pass from the state h to the state k is given by:

$$p_{hk} = \sum_{h=1}^{\binom{S}{h}} \sum_{k=1}^{\binom{S}{k}} \left(\prod_{n=1}^{a} p_{BB}^{(n)} \prod_{n=a+1}^{b} p_{BG}^{(n)} \prod_{n=b+1}^{c} p_{GB}^{(n)} \prod_{n=c+1}^{S} p_{GG}^{(n)} \right) \tag{1.7}$$

where a is the number of satellites remaining in the state B, (b-a) is the number of satellites changing the state from B to G, (c-b) is the number of satellites changing the state from G to B, (S-c) is the number of satellites remaining in the state G.

5.3.7.2
State Probability

Assuming statistical independence among different signals and that all S satellites are geometrically visible (the probability to have only a subset of the S satellites in visibility is not predictable because it depends on the constellation geometry), the probability to stay in each of the N states is given by:

$$p_j = \sum_{k=1}^{\binom{S}{j}} \prod_{i=1}^{j} p_{gik}(\alpha_i) \prod_{i=j+1}^{S} p_{bi\bar{k}}(\alpha_i) \tag{1.8}$$

where j is the number of satellites offering good visibility, p_{gi} is the probability that the i-th satellite is in good visibility, p_{bi} is the probability that the i-th satellite is not visible, k is the disposition of the visible satellites, $\bar{k}$ is the disposition of the non visible satellites in the whole set and α_i is the visibility angle of the i-th satellite.

Each of the N states corresponds to the different possible dispositions of j elements in a S-dimension set.

5.3.7.3
The 3-State Model

For sake of simplicity the whole set of N states can be further grouped identifying three main states. In this approach three Markoff states are assumed:

A. no satellite is available;
B. at least one satellite is available but not all;
C. all satellites are available.

Thus:

$$P_A = p_0 \tag{1.9}$$

$$P_B = \sum_{j=1}^{S-1} p_j \tag{1.10}$$

$$P_C = p_S. \tag{1.11}$$

The theoretical probability to having service is $P_B + P_C$, corresponding to the probability to have at least one satellite in visibility. Actually, if the number of satellites in visibility is lower than the implemented diversity order, the link margin may not be sufficient to close the link.

Fig. 5.3.4 shows a three-states diagram of the system evolution.

The transition probability matrix [**P**(x, y)], (x, y ∈ {A, B, C}), is represented by (1.12).

$$P_3 = \begin{bmatrix} p_{AA} & p_{AB} & p_{AC} \\ p_{BA} & p_{BB} & p_{BC} \\ p_{CA} & p_{CB} & p_{BB} \end{bmatrix} = \begin{bmatrix} \prod_{s=1}^{S} p_{BB}^{(s)} & \sum_{k=1}^{N-1} p_{0k} & \prod_{s=1}^{S} p_{BG}^{(s)} \\ \sum_{k=1}^{N-1} p_{k0} & \sum_{k=1}^{N-1}\sum_{h=1}^{N-1} p_{kh} & \sum_{k=1}^{N-1} p_{kN} \\ \prod_{s=1}^{S} p_{GB}^{(s)} & \sum_{k=1}^{N-1} p_{Nk} & \prod_{s=1}^{S} p_{GG}^{(s)} \end{bmatrix} \tag{1.12}$$

5.3.7.4
Numerical Results

Implementing (1.8) it is possible to figure out the theoretical probability to exploit diversity or to have at least one satellite in visibility in different multiple satellites scenarios.

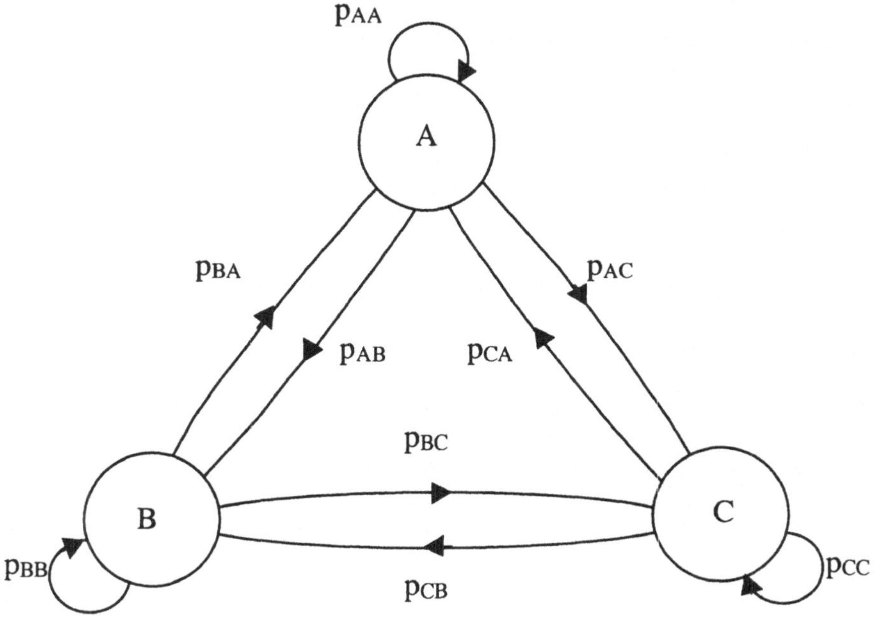

Fig.5.3.4. Three-State diagram.

The considered scenarios regard system architectures optimised to provide diversity with 2 up to 6 satellites. The elementary probability to have each satellite in visibility is extracted from Table 5.3.1 according to the matrix of Table 5.3.2 for each of the three environments. The followed criterion is that less satellites are available to exploit diversity, lower is the average elevation angle. For the S = 2 case the chosen set of probability corresponds to the results of simulations carried out on Globalstar constellation.

Table 5.3.2. Satellite average angle assumptions in different diversity scenarios

Angles	6	5	4	3	2
<10	x	x	x		
10-20	x	x		x	
20-30	x		x		x
30-40		x		x	
40-50	x		x		x
50-60		x		x	
60-70	x		x		
70-75		x			
75-80	x				
80-90					

In Fig. 5.3.5 the probability to exploit diversity with k satellites (p_K) in urban environment for all the five different scenarios is shown. In Fig. 5.3.6 the probability to have at least k satellite in visibility equal to:

$$\sum_{j=k}^{S} p_j ,$$

where p_j is calculated as in (1.8), still in urban environment, is shown.

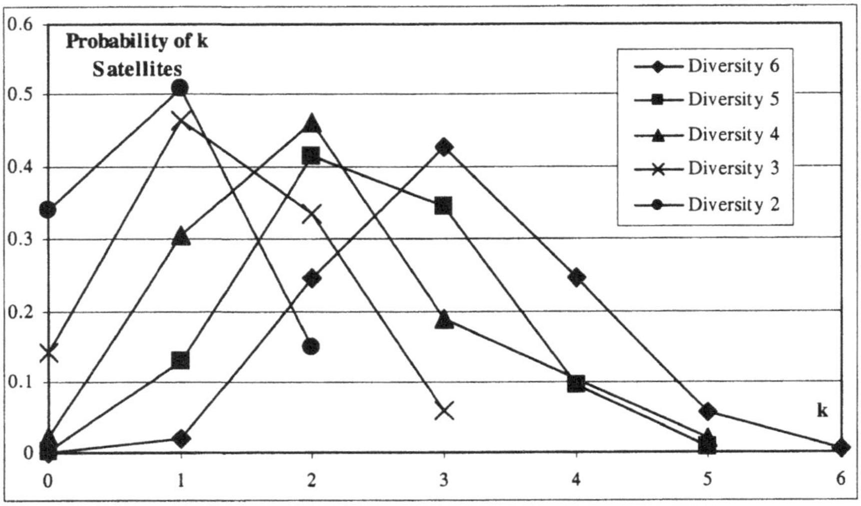

Fig.5.3.5. Probability to use k satellites in urban environment

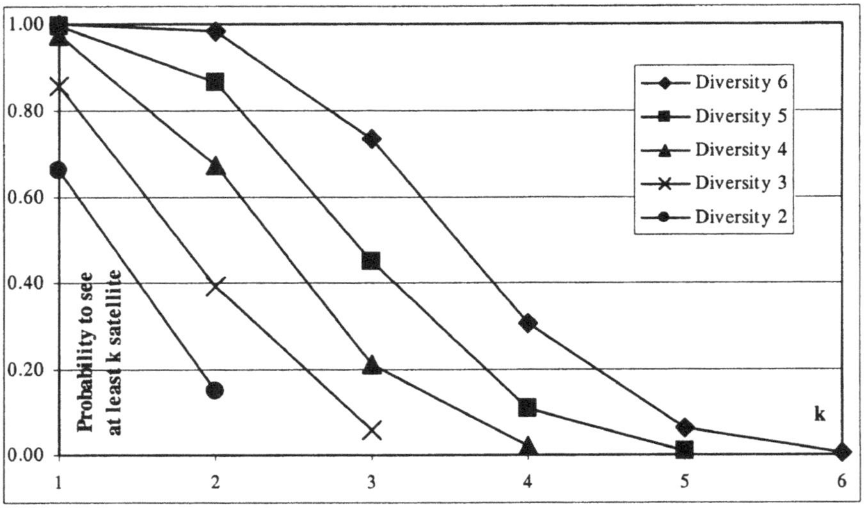

Fig. 5.3.6. Avaliability in case of k satellites in urban environment

5.3.8
Impact of Diversity Implementation on System Performance

The benefit of introducing satellite diversity to increase link availability and to evaluate the probability to exploit diversity can be quantified through a statistical analisysis depending on selected constellation architecture.

As an example in [1] a constellation at 1414 km height, 8 orbital planes, 6 satellites per orbit, orbit inclination 52° with respect to the equatorial plane, at sampling period of 11.4 s, over 400 orbital periods and environment parameters given in [21] is considered. The gateway is located at 42°N latitude and 13°E longitude. The users are simulated as a grid of points uniformly distributed on a sphere. The results are referred to the average over the set of users (~ 20) located in a ring around the gateway, at a distance between 700 and 800 km.

5.3.8.1
Availability Improvement

Table 5.3.3 provides the probability to have no satellites in visibility, from both user and gateway, when more than one satellite is in visibility with 10° minimum elevation angle for the user and 5° for the gateway.

Fig. 5.3.7 shows the unavailability improvment as function of the minimum required coverage angle, for urban environment. Of course the real improvement is expected to be lower because if not all the satellites are in visibility the link may not be closed. Note that the improvement between 2 and 3 satellites is not so meaningful because the selected constellation (Globalstar) is optimised to provide double diversity.

Table 5.3.3. Unavailability with 10° minimum elevation angle

Environment	One satellite	Two satellites	Three satellites
Rural	4.23 %	0.47 %	0.24 %
Suburban	13.1 %	3.21 %	1.98 %
Urban	62.13 %	46.33 %	38.06 %

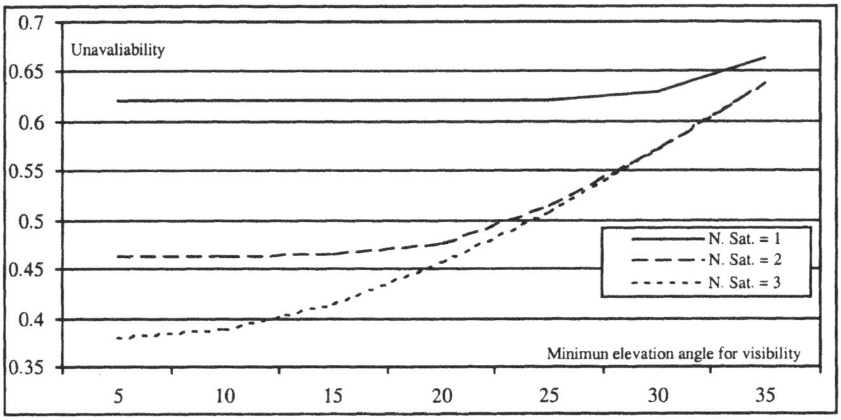

Fig.5.3.7. Diversity advantage in urban environment

5.3.8.2 Diversity Exploitation Probability

To exploit diversity is of course necessary to have more than one satellite in visibility contemporarily.

Fig. 5.3.8 shows the probability to exploit diversity in urban environment. The abscissa is still the minimum elevation angle to consider a satellite in visibility.

In urban environment the probability to have in visibility more than one satellite is lower with respect to the other two environments due to worse channel conditions.

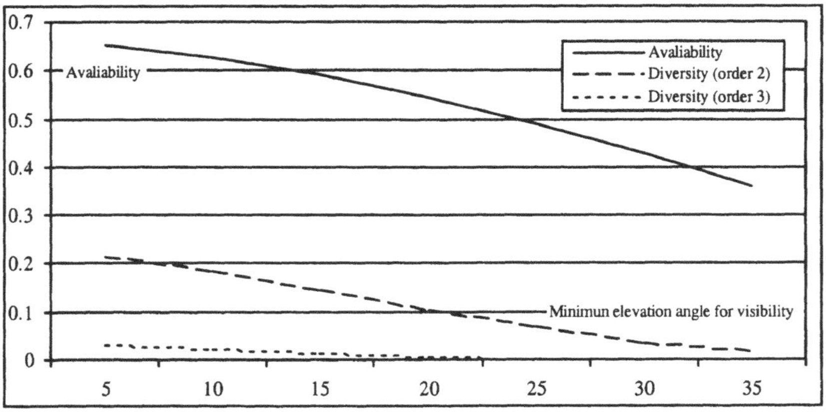

Fig.5.3.8. Diversity probability in urban environment

5.3.8.3
Co-channel Interference for Multi-Satellite Multi-Spot Coverage

Co-channel interference can be one of the main limiting factors, both in terms of link availability and capacity [35].

Moreover, when constellations in circular orbits are used, overlapping coverage of different satellites represents one of the major causes of interference independently on the access scheme, unless diversity is implemented.

In the up-link, the useful and interfering signals are generated by user terminals and received at the satellite antenna serving the wanted user. In this case, each signal goes through different paths, thus experiencing different propagation conditions. When more than one satellite is present in the interference scenario, also terminals served by other satellites contribute to interference.

In the downlink, interference is experienced at the mobile terminal antenna. As regards the serving satellite, both the useful and the interfering signals are radiated from the satellite antenna through the same path (the fluctuations experienced by the signals are correlated). The main lobe radiates useful and interfering signals (if any, in CDMA case) of the serving spot, while interfering signals from other spots are radiated by the side lobes of the relative beam. When more than one satellite contribute to interference, the signals from not serving satellites are only interference, coming from main or side lobes, unless diversity strategy is adopted.

In case of single satellite scenario, interference can be carried out utilising a general model [3, 17] suitable for time domain analysis. In case of diversity implementation, the interference scenario is modified and the model can be adapted as a consequence [1]. The analysis is performed on only one user (wanted) and the others are all potential interferers. One or more (with diversity) satellites serve the wanted user, both in up and down link, while the others are potential interferers.

5.3.8.4
C/I Improvement

The improvement of C/I ratio due to diversity implementation makes link availability increase. In this case one orbital period, sampling period of about 132 s, the same number of channels with and without diversity, and a processing gain equal to 32 dB is assumed in [1]. Multibeam satellite antennas with 49 beams have been adopted. To combine the different signals a simplified maximal ratio combining algorithm has been utilised in such a traffic load condition that total interference $I \gg N = kTB$.

Table 5.3.4 shows the improvement of the average and minimum values of C/I using diversity. Fig. 5.3.9 and 5.3.10 show the C/I as a function of time for up and down link respectively.

Table 5.3.4. Improvement of average and minimum values of C/I using diversity

	C/I up link (dB)			*C/I down link (dB)*		
	No div.	Div. 2	Div. 3	No div.	Div. 2	Div. 3
Minimum	-0.3	5.57	8.04	2.03	3.36	4.35
Average	1.98	6.97	9.55	3.54	4.73	5.26

5.3.9 Impact on System Capacity

The impact that satellite diversity has on system capacity depends on how we define capacity. If capacity is defined as the number of simultaneously available channels, it is obvious that it decreases proportionally with the number of satellites used to exploit diversity. Otherwise, if capacity is intended as the amount of traffic effectively carried, then it can even increase due to the intrinsic capability of diversity to improve link quality and to get lower probabilities of call dropping [9].

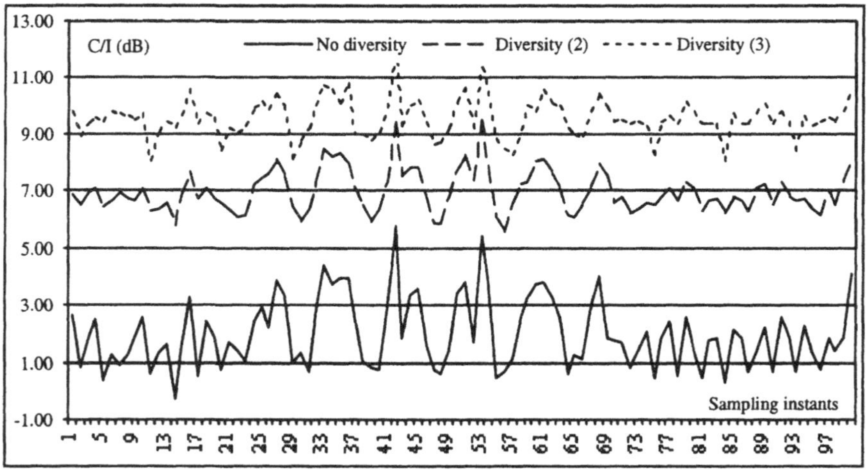

Fig.5.3.9. Up link C/I (dB) time variation with one, two and three satellites

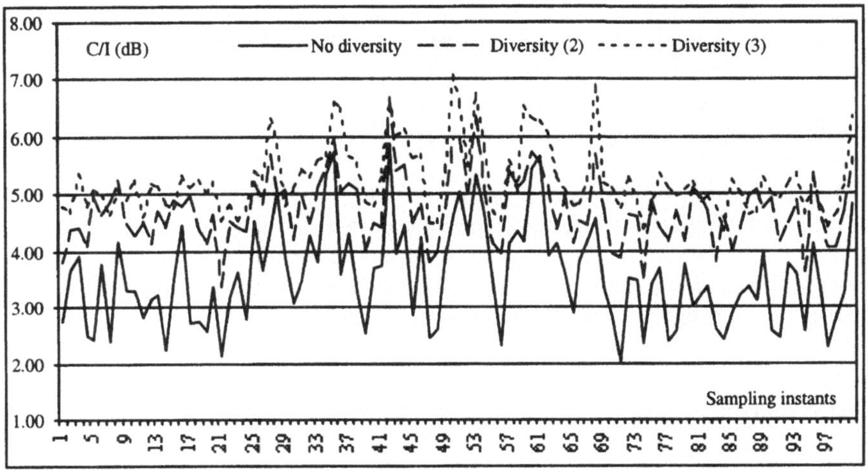

Fig.5.3.10. Down link C/I (dB) time variation with one, two and three satellites

Of course the impact on capacity is also strongly dependent on the environment. In fact where the channel behaviour is quite good (as in rural areas), the impact is not so relevant, while in urban areas where the channel is quite critical, the impact of diversity is more relevant, corresponding to the better improvement on system performance.

The evaluation of capacity is even more complex if referred to multiservice systems due to the fact that each user can simultaneously require different data rates or different services with different QoS.

Simple analyses have been performed in [9] considering single service, constant bit rate and CDMA access. Another way to perform analyses is to consider spectral efficiency as the ratio between total data rate carried by the system and the total available bandwidth.

5.3.10 Network Aspects

The implementation of diversity has an impact also on network functionality especially in terms of signalling exchange and in terms of handover management.

First of all, it is obvious that in this case the frequency of handover occurrence increases if compared with the single satellite case [9].

The use of diversity, along with the adoption of CDMA access scheme, allows the implementation of soft handover. In this way when one of the links drops the communication is held by the other S-1 links in the time interval to replace the dropped satellite link.

The critical aspects concern the criteria to choose the new satellite, if any in visibility when one drops, and the signalling exchange to manage the procedure.

One possible criterion can be geometrical (the best elevation angle satellite) even if not practical to implement. In fact nobody is able to instantaneously measure the elevation angle between user and each satellite in visibility.

Another possibility can be to select the satellite with the best channel conditions (which only theoretically coincides with the previous) or lowest attenuation.

Finally, perhaps the best approach would be to select the satellite offering the best SNR, taking also into account the interference and thus the traffic load.

The traffic has to be re-routed through the new path by the serving gateway, impacting the delay difference among the different paths, having to handle the traffic flow during the handover execution phase.

5.3.11 Conclusions

In order to provide even more advanced services and to gain an important role in the global communication scenario, satellite systems need to enhance performances. In a scenario of integration with terrestrial systems they need to show comparable performance ensuring real global coverage and services.

Satellite diversity can be an advanced technique to reach this goal, at the cost of system complexity.

The improvement in availability and in signal to noise ratio makes this technique very attractive to reach some very ambitious targets.

The possibilities offered by the implementation of such technique have been analysed, along with some of the main drawbacks.

References

1 P. Loreti, M. Luglio, "Analyses and Simulations on Performance Improvement due to Satellite Diversity Implementation", *1999 International Workshop on Mobile Communications*, Crete, Greece, pp. 83-91, June 1999.

2 R. A. Wiedeman, A. J. Viterbi, "The Globalstar Mobile Satellite System for worldwide Personal Communications", *Proc. of 3rd Int. Mobile Satellite Conference (IMSC '93)*, pp. 285-290, June 1993.

3 P. Loreti, M. Luglio, "Time Domain Interference/Capacity Analysis for non-GEO Satellite Constellations", *3rd European workshop on Mobile Personal Satcoms (EMPS '98)*, Venice, Italy, pp. 63-72, November 1998.

4 G. Caire, R. De Gaudenzi, G. Gallinaro, M. Luglio, R. Lyons, M. Ruggieri, A. Vernucci, H. Widmer, "Development and Validation of Wideband CDMA IMT-2000 Physical Layer for Satellite Applications", *International Mobile Satellite Conference (IMSC 99)*, Ottawa, Ontario, Canada, pp. 213-219, June 1999.

5 A. Franchi, E. Trachtman, L. Christodoulides, J. Sengupta, "Multimedia via Inmarsat", *IEEE Multimedia Magazine*, pp. 15-19, vol. 6, n. 4, Oct.-Dec. 1999.

6 F. Carducci, G. Losquadro, V. Podda, "The Euroskyway Worldwide System Providing Broadband Service to Fixed and Mobile End-Users", *4th Ka Band Utilization Conference*, Venice, Italy, pp. 33-40, , November 1988.

7 Elizondo, F. Gargione, R. Gobbi, K. Shockey: "The Astrolink satellite system overview", *2nd Ka Band Utilization Conference*, Florence, Italy, pp., September 1996.

8 R. De Gaudenzi, F. Giannetti, "DS-CDMA Satellite Diversity Reception for Personal Satellite Communication: Satellite-to-Mobile Link Performance Analysis", *IEEE Transactions on Vehicular Technology*, pp. 658-672, vol. 47, No. 2, May 1998.

9 G. E. Corazza, C. Caini, "Satellite Diversity Exploitation in Mobile Satellite CDMA Systems", *IEEE Wireless Communications and Networking Conference*, New Orleans, LA, Sept. 1999.

10 H. M. Aziz, R. Tafazolli, B. G. Evans, "Gain in Capacity by Using Satellite Diversity in a CDMA Based S-PCNs", *AIAA '96*.

11 B. R. Vojcic, L. B. Milstein, R. L. Pickholtz, "Downlink DS CDMA Performance Over a Mobile Satellite Channel", *IEEE transactions on Vehicular Technology*, pp. 551-560, vol. 45, No. 3, August 1996.

12 S. G. Glisic, J. J. Talvitie, T. Kumpumächi, "Design Study for a CDMA-Based LEO Satellite Network: Downlink System Level Parameters", *IEEE Journal on Selected areas in Communications*, pp. 1796-1807, vol. 14, No. 9, December 1996.

13 M. A. Sturza, F. Ghazvinian, "Teledesic Satellite System Overview", *Proc. of Ka Band Utilization Conference*, Rome, Italy, pp. 231-238, October 1995.

14 J. G. Proakis, *Digital Communications*, Mc Graw Hill International, 3rd edition, 1995, pp. 797-806.

15 N. Hart, T. Goerke, A. Jahn, "A Discussion on Mobile Satellite System and the Myths of CDMA and Diversity Revealed", *International Mobile Satellite Conference (IMSC '95)*, Ottawa, pp. 469-475.

16 T. Sakai, K. Kobayashi, S. Kubota, M. Morikura, S. Kato, "Soft-Decision Viterbi Decoding with Diversity Combining for Multi-Beam Mobile Satellite Communication Systems", *IEEE Journal on Selected areas in Communications*, pp. 285-290, vol. 13, No. 2, February 1995.

17 Vatalaro, G. E. Corazza, C. Caini, and C. Ferrarelli, "Analysis of LEO, MEO, and GEO Global Mobile Satellite Systems in the Presence of Interference and Fading", *IEEE Journal on Sel. Areas in Communications*, pp. 291-300, vol. 13, no. 2, February 1995.
18 R. Price and P. E. Green, "A communication technique for multipath channels", *Proceedings of IRE*, pp. 555-570, vol. 46, Mar. 1958.
19 P. Poskett, "Satellite Systems Architectures", *Proc. of 2nd European Workshop on Mobile/Personal Satcoms (EMPS '96)*, Rome, Italy, pp. 485-500, October 1996.
20 D. Parsons, *The mobile Radio Propagation Channel*, Pentech Press London.
21 Lutz, D. Cygan, M. Dippold, F. Dolainsky, W. Papke, "The Land Mobile Satellite Channel - Recording, Statistics and Channel Model", *IEEE Trans. on Vehic. Techn.*, Vol. VT - 40, May 1991.
22 C. Loo, "A Statistical Model for a Land Mobile Satellite Link", *IEEE Trans. on Vehic. Techn.*, pp. 122-127, Vol. VT - 34, Aug. 1985.
23 E. Corazza, F. Vatalaro, "A Statistical Model for Land Mobile Satellite Channels and its Application to Nongeostationary Orbit Systems", *IEEE Trans. on Vehic. Techn.* , pp. 738-741, Vol. VT - 43, Aug. 1994.
24 F. Vatalaro, "Generalised Rice-Lognormal Channel model for Wireless Communications", *Electronic Letters*, pp.1899-1900, Vol. 31, No.22, Oct.1995.
25 E. Corazza, A. Jahn, E. Lutz, F. Vatalaro, "Channel Characterization for Mobile Satellite Communications", *Proc. of the First European Workshop on Mobile /Personal Satcoms (EMPS 1994)*, Rome, Italy, pp.225-262, 1994.
26 E. Lutz, "A Markoff Model for Correlated Land Mobile Satellite Channels", *International Journal of Satellite Communications*, pp. 333-339, vol. 14, February 1996.
27 Bischl, M. Werner, E. Lutz, "Elevation-dependent Channel Model and Satellite Diversity for NGSO S-PCNs", *VTC '96*.
28 Y. Karasawa, K. Kimura, K. Minamisono, "Analysis of Availability Improvement in LMSS by means of Satellite Diversity Based on Three-State Propagation Channel Model", *IEEE transactions on Vehicular Technology*, pp. 1047-1056, vol. 46, No. 4, Nov. 1997.
29 P. P. Robet, B. G. Evans, A. Ekman, "Land Mobile Satellite Communication Channel Model for Simultaneous Transmission from a Land Mobile Terminal via two Separate Satellites", *International Journal of Satellite Communications*, pp. 139-154, vol. 10, 1992.
30 M. Vázquez-Castro, C. Tzaras, S. Saunders, F. Pérez-Fontán, "Shadowing Correlation for Mobile Satellite Diversity", *Joint COST 252-255 Meeting*, Tolouse, May 8-10, 1999.
31 C. Tzaras, S. R. Saunders, B. G. Evans, "A Phisical-Statistical Propagation Model for Diversity in Mobile Satellite PCN", *VTC 1998*.
32 F. P. Fontán, M. A. Vázquez, S. Buonomo, E. Kubista, A. Paraboni, B. Arbesser, B. Baptista, "A Methodology for the Characterisation of Environmental Effects on Global navigation Satellite System (GNSS) Propagation", *International Journal of Satellite Communications*, pp. 1-22, vol 16, 1998.
33 F. Mazzenga, F. Vatalaro, "Channel Modelling and Performance Evaluation of EO Systems", *AP2000 Davos, Switzerland*, April 9-14, 2000.

34 P. Loreti, M. Luglio, "A Generalized N-State Model to characterize satellite diversity for arbitrary number of satellites in case of uncorrelated channels", *submitted to IEEE Tranactions. On Communications.*
35 P. Loreti, M. Luglio, "Interference Evaluations and Simulations for Multisatellite Multispot Systems", *submitted to IEEE Journal on Selected Areas in Communications.*

5.4 LEO satellite mobile channels: modeling by time sequences of Finite State Markov Channels

F. Babich, G. Lombardi, E. Valentinuzzi And O. E. Kelly

5.4.1 Introduction

The use of Low Earth Orbit satellites (LEOs) for the development of a wireless communication network is favorable when coverage is required for large geographical areas that have low population density and/or complex topography. For those requirements, standard cellular systems are impractical to deploy. IRIDIUM and GLOBALSTAR mobile systems have recently started working and have awakened interest in LEO based wireless systems. As in land mobile systems, improved knowledge of the mobile channel properties brings opportunity for increased system capacity and other system optimizations [1].

This chapter focuses on the digital narrow-band mobile satellite channel [2]. Signal propagation mechanisms from the satellite to the subscriber are very similar to those in land-mobile cellular systems, so that a simple Ricean model with Clarke spectrum may be considered [2, 3, 4]. However, satellite motion introduces time variation of the Rice factor, K, and of the fading rate, $f_D T$, where f_D is the maximum Doppler spread and T is the bit duration. Detailed investigations of the time varying satellite channel are found in [5, 6, 7, 8].

The following analysis considers the effect of time variations in the fading rate under the simplifying assumption that the Rice factor K remains constant. That assumption holds for example when obstacles near the mobile subscriber do not cause variations in the direct path (highways, open places) [5]. In this work, the resulting channel is approximated by a time sequence of digital channels with constant Doppler spread, each of which is modeled as a Finite State Markov Channel (FSMC) by applying the Context Tree Pruning (CTP) algorithm [9, 10, 11]. It is straightforward to introduce a time variation of K in the model. CTP is suitable for the construction of FSMC models for digital stationary channels with memory, as the digital terrestrial mobile channel. In Section 5.4.2, the variation of the channel Doppler spread is examined, while in Section 5.4.3 remarks are made about Context Trees and the CTP algorithm. The fitting of CTP based models to the channel is the object of Section 5.4.4. Some results about PED evaluation are discussed in Section 5.4.5, in order to give an insight about the effectiveness of

CTP in satellite channel modeling, with respect to usual time-continuous models. Some final observations are made in Sections 5.4.6 and 5.4.7.

5.4.2 Doppler Spread Variation

Satellite motion induces a Doppler spread on the subscriber received signal whose instantaneous value $f_D(t)$ depends on the orbital geometry and dynamics. The high speed of LEO satellites (in the range 1-10 km/s) allows the motion of subscribers with respect to the Earth to be neglected in evaluations of $f_D(t)$.

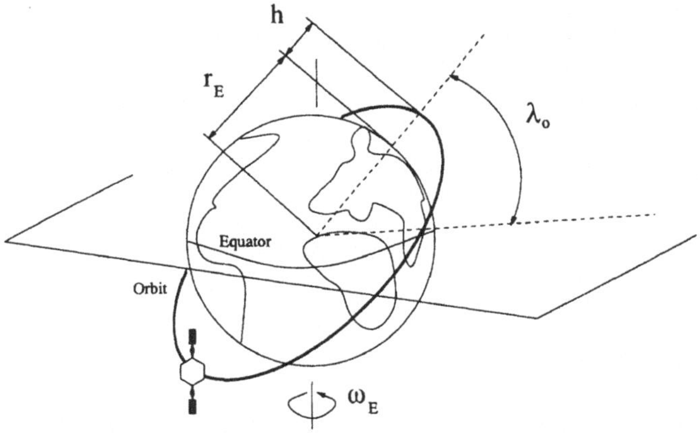

Fig.5.4.1. Geometry and Notation of Satellite Orbit

If the satellite shown in Fig.5.4.1 has a nearly circular orbit at height h (m) above Earth, orbit inclination λ_o (rad) from the Equatorial plane and maximum elevation θ_{max} (rad) at the observation point on Earth (at the time t=0), one has the following efficient approximation for $f_D(t)/f_c$ [12] (f_c is carrier frequency):

$$\frac{f_D(t)}{f_c} \cong -\frac{\omega r_E}{c} \frac{\eta\left(\frac{r_E}{h}+1\right)\sin\omega t}{\sqrt{1+\frac{2r_E}{h}\left(\frac{r_E}{h}+1\right)(1-\eta\cos\omega t)}}, \tag{1.1}$$

with:

$$\omega = \omega_s - \omega_E \cos\lambda_o \text{ (rad/s)}, \tag{1.2}$$

$$\omega_s = \frac{1.995\cdot 10^7}{(r_E+h)^{1.5}} \tag{1.3}$$

$$\eta = \cos\left[\cos^{-1}\left(\frac{r_E}{r_E + h}\cos\vartheta_{max}\right) - \vartheta_{max}\right], \tag{1.4}$$

where $c = 3 \cdot 10^8$ (m/s) is light speed in vacuum, $r_E = 6.37 \cdot 10^6$ (m) and $\omega_E = 7.27 \cdot 10^{-5}$ (rad/s) are Earth radius and angular speed, respectively.

Equation (1.1) is valid under the hypothesis that the satellite, during its visibility time, has a nearly circular uniform motion around Earth center in the Earth-Centered Fixed (ECF) frame, that is comoving with Earth. It is sufficient that the visibility time of satellite is much shorter than satellite rotation period and that the satellite angular speed, ω_s, in the Earth-Centered Inertial (ECI) frame (integral with fixed stars), is much larger than ω_E [1]. In fact, if φ is the latitude of the subsatellite point, the satellite angular speed in ECF is evaluated in [12], as follows:

$$\omega_{ECF}(\varphi) = \omega_s \sqrt{1 - 2\frac{\omega_E}{\omega_s}\cos\lambda_0 + \left(\frac{\omega_E}{\omega_s}\right)^2 \cos^2\varphi},$$

and the following approximation can be made:

$$\omega_{ECF}(\varphi) \cong \omega_s \sqrt{1 - 2\frac{\omega_E}{\omega_s}\cos\lambda_0} \cong \omega,$$

where $\sqrt{1+x} \cong 1 + x/2$, if $|x| << 1$, and ω and ω_s are given by (1.2) and (1.3), respectively.

Let us consider the satellite Iridium 11 of IRIDIUM constellation [14]. It has a polar orbit with h=780 (km) and λ_0=86.5 (degrees), so that ω_s=14.35ω_E>> ω_E. Its linear speed in ECF frame is about $(r_E+h)\omega$=7.1 (km/s). With reference to its visibility from Venice, Italy (lat. 45.45° N, long. 12.35° E), it is obtained $\theta_{max}=5\pi/36$ (=25 degrees), for the orbit 7849 (Dec. 15th, 1998). In Figs. 5.4.2a and 5.4.2b the evolution of $f_D(t)$ and $\theta(t)$, respectively, are given (θ (t=0)= θ max and f_c=1623.925 MHz): notice that the visibility window for this case is ≈ 800 s, that is very small with respect to the revolution period ≈ 1 h 40 m).

[1] In [13] ECI and ECF frames are exactly defined.

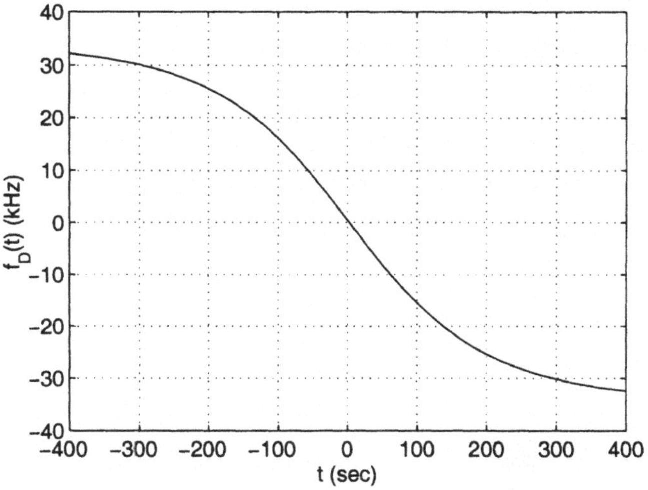

(*a*)

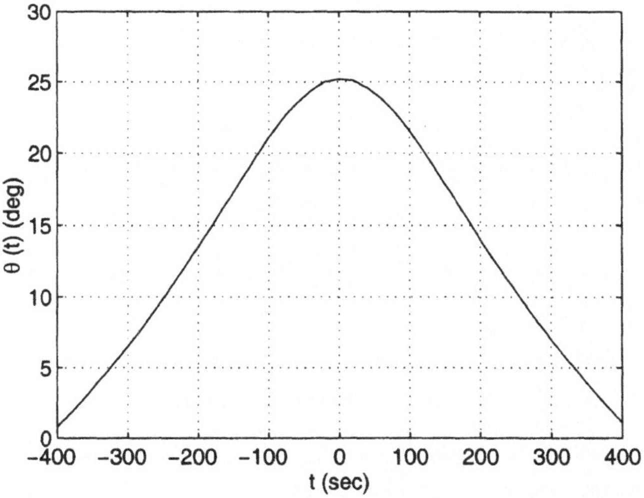

(*b*)

Fig.5.4.2. Time evolution of Doppler shift (*a*) and of elevation (*b*) for the satellite Iridium 11 (Venice, Italy, Dec. 15, 1998, orbit 7849).

5.4.3 Context tree models

The next paragraphs review context tree (CT) models and the context tree pruning algorithm introduced in [15]. Let S_F be a finite alphabet and consider a discrete stationary source producing sequences of symbols in S_F, $x^n = \{x_i\}_{i=1}^n \in S_F^n$, $x_i \in S_F$ (n is a positive integer).

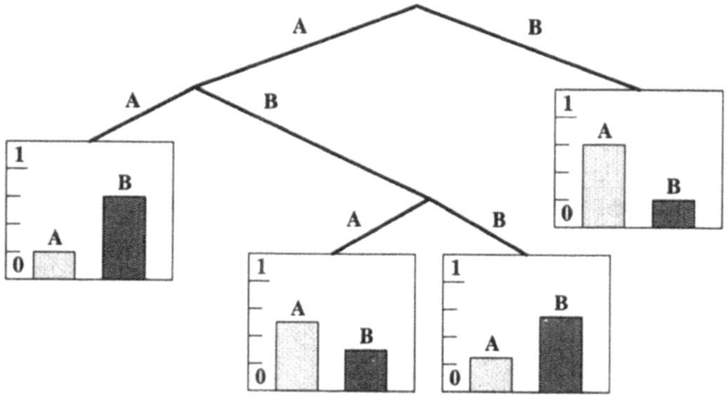

Fig.5.4.3 Example of Context Tree: alphabet S_F={A, B}, context set C={AA, ABA, ABB, B}. The histograms at each leaf represent the conditional probabilities $P(A|c)$, $P(B|c)$=1-$P(A|c)$, $c \in C$. Depth M=2

A *CT source model* is defined by a parameter vector β and a *context set* C. The conditional probability vector $\beta = \{P(x|c)\}_{x \in S_F, c \in C}$ specifies the probability mass function of the next symbol x, given that the current context of the process is c. The *context* of the process is its recent history, i.e. a short sequence of symbols in S_F. Denote the l-th context by $c_l = \{c_l(i)\}_{i=1}^{m_l}$, $c_l(i) \in S_F$, $m_l \geq 1$, $l = 1,\dots,|C|$ [2]. The process definition must in all cases be sufficient to assign the conditional probability of the occurrence of a symbol, and this imposes two constraints on the context set C: For sufficiently large n, the context c must be *defined* for every x^n and it must be *unique*. The uniqueness requirement disallows any context in C from being the prefix of another context in C, i.e. for any two contexts $c_p, c_q \in C$, it never happens that $c_p(i)=c_q(i)$ for all i=1,2,...,min(m_p, m_q). The definiteness requirement assures that, for any sequence x^n, there exists in C a context of length m that matches the last m characters of x^n. Denote that context

[2] The notation $|W|$, where W is a set or a sequence, denotes the size of the set or the length of the sequence, respectively.

$c(x^n)=\{x_{n-k+1}\}_{k=1}^{m}$. From this construction it follows that $P(x_{n+1}|x^n)=P(x_{n+1}|c(x^n))$. $M=\max\{m_l\}_{i=1}^{|C|}$ is the *CT depth* or *order*.

The name "Context Tree" arises because the context set C may be expressed as a tree, as in Fig. 5.4..3, such that definiteness is assured if the tree is complete and uniqueness is assured if contexts correspond to tree leaves (and not to interior nodes). Each context is a path from root (most recent) to leaf (least recent). To illustrate β, the same diagram shows the probability mass function $P(x|c)$ as a bar chart at leaf c.

5.4.3.1 Context Trees and Markov Models

A CT model considers finite history and therefore may be embedded in a Markov chain. For example, if all of leaves have depth M (formally $m_l=M$, $\forall l=1,...,|C|$) then the model is a M-th order Markov model on S_F. In general, a CT model admits an equivalent representation as a first order Markov chain with state set S_M, made by finite symbol sequences in S_F, that are *context extensions*. It may happen that $|c(x^{n+1})|>|c(x^n)|+1$, that is, the new context requires more letters than are available in the old context plus the most recent letter. In the CT formulation, that additional historical information remains available in x^{n+1}, but a Markov chain requires that all necessary historical information be available in the *states*, hence the need to form some states by extending the corresponding contexts.
By construction and the prefix property, for any state sequence $s_k=\{s_k(i)\}_{i=1}^{|s_k|}\in S_M$ there exists a unique context $c\in C$ that is a *prefix* of state s_k, i.e. $c=\{s_k(i)\}_{i=1}^{|c|}\in S_M$, $|c|\le|s_k|$ [16]. Define a function pref to generate all prefixes of a sequence: pref(x_n)=$\{\phi, x_1, x_1x_2, \ldots, x_1\ldots x_n\}$. The states set must be closed under the operation that determines the new state s' from the current state s and most recent letter x; that is $s'=\text{pref}(\{xs\}\cap S_M)$ must yield a unique element of S_M for all s, x (clearly $s'(1)=x$).

More than one valid Markov chain state set can be derived from a given context set, but the following algorithm finds the smallest such state set. Begin with $S_{M,0}=C$ and let $S_{M,k}$ be the state set at the k-th iteration. Given a candidate state $s\in S_{M,k}$, one of the following two cases holds:

- If $\forall x\in S_F$, $\text{pref}(\{xs\})\cap S_{M,k}$ is non-empty, then $s\in S_{M,k+1}$;
- If $\exists\bar{x}\in S_F$ such that $\text{pref}(\{\bar{x}s\})\cap S_{M,k}=\phi$ then $\forall s'\in S_{M,k}$ satisfying $\{\bar{x}s\}\in\text{pref}(s')$[3], define $s''=\{s'(i)\}_{i=2}^{|s'(i)|}$ and let $s''\in S_{M,k+1}$, and let $s\notin S_{M,k+1}$. Repeat for all the possible $\bar{x}\in S_F\in S_{M,k}$.

[3] There exists such an s' because the tree associated to $S_{M,k}$ is complete.

Iteration terminates when $S_{M,k} = S_{M,k+1}$. The resulting state set is $S_M = S_{M,k}$. The transition matrix $\mathbf{T}_M$ is given by: $P_M(s'|s)=P[s'(1)|c(s)]$ where $c(s) = \text{pref}(s) \cap C$ if $s' \in \text{pref}(\{s'(1)s\})$, otherwise $P_M(s'|s)=0$. $P_M(s'|s)$ is evaluated for all s, s' in S_M.

It results that $|S_F| \le |C| \le |S_M|$ and $N_\beta = |C|(|S_F|-1) \le N_M(|S_M|-1)$, where N_β and N_M are the number of parameters required for a complete description of the conditional probability vector $\boldsymbol{\beta}$ over the context set C and of Markov transition matrix $\mathbf{T}_M$ over the state space S_M, respectively. For the tree in Fig.5.4..3 we have $S_M = \{s_i\}_{i=1}^{5} = \{AA, ABA, ABB, BA, BB\}$, and:

$$T_M = \begin{bmatrix} P(A|c_1) & 0 & 0 & P(B|c_1) & 0 \\ P(A|c_2) & 0 & 0 & P(B|c_2) & 0 \\ P(A|c_3) & 0 & 0 & P(B|c_3) & 0 \\ 0 & P(A|c_4) & 0 & 0 & P(B|c_4) \\ 0 & 0 & P(A|c_4) & 0 & P(B|c_4) \end{bmatrix}$$

It is easy to see that a CT model admits also an equivalent representation as a M-th order Markov model on S_F. The main advantage of a CT model is that it generally requires fewer parameters for its description than an equivalent Markov model (in the best case, the smallest equivalent Markov model and the context tree model may be coincident).

5.4.3.2
Context Tree Pruning algorithm

Given a sequence $\mathbf{x}=x^n$ (*training sequence*) originated from a source with unknown properties, the Context Tree Pruning algorithm is optimal in estimating the best fitting CT model (C,β), when a maximum order M is assigned *apriori*. The basic structure of the algorithm is the following (more details are in [15]):

1. from the sequence $\mathbf{x}$, a temporary context set C' and the conditional probabilities $P_x(x|c)$, $c \in C'$ are estimated by measuring the relative frequency of transitions:
$$P_{\mathbf{x}}(x|c) = \frac{n_{\mathbf{x}}(x|c)}{\sum_{y \in S_F} n_{\mathbf{x}}(y|c)}, c \in C';$$
2. the child-parent test statistic (based on Kullback-Leibler distance) is generated at each child node of the tree:
$$\Delta(cy) = \sum_{x \in S_F} n_{\mathbf{x}}(x|cy) \log \frac{\Pr(x|cy)}{\Pr(x|c)};$$
3. the tree is *pruned*, with the following rules:
$$\Delta(cy) \ge T_P = 2(|S_F|+1)\log_2(n+1),\ |c| \le \min\left[\frac{\log_2 n}{\log_2 |S_F|}, M\right]$$

4. the tree is *completed*: If a node is retained (i.e. not pruned), then every node on the path from that node to the root is retained. If one child of a parent is retained, then all the children of that parent are retained. The construction of C is finished.
5. parameter vector $\boldsymbol{\beta}$ is estimated from frequencies observed in the training sequence of transitions defined by C.

If the source is not a CT source, when M and n are increased toward infinity, the algorithm estimates a sequence of trees approximating the source with increasing detail (theoretically, a CT with infinite depth would be required).

5.4.4 Channel Modeling

A simple Ricean model with Clarke's spectrum can be used to represent the flat mobile satellite channel. Such model is normally used to describe the terrestrial flat fading channel [17, 18]: the effect of satellite motion is partially considered by modeling the time variation of the Doppler spread, $f_D=f_D(t)$, as consequence of satellite motion (the speed of the mobile subscriber is negligible compared to satellite speed).

Fig.5.4.4 shows a plot of *fading rate* $|f_D(t)|T$ for the satellite Iridium 11 ($t>0$, Section 5.4.2), where $T=20$ µs is the bit duration (corresponding to IRIDIUM bit-rate equal to $1/T=50$ kbit/s).

Examining Fig..5.4.4 one sees that during the visibility time the fading rate decreases from its maximum value (≈ 0.7, $\theta=0$) to a null value ($\theta=\theta_{max}$) and then increases back to the maximum. Such variation, as shown in Fig. 5.4.4 can be well approximated by a saturation law. The fading goes through different regimes: for $t<0.1$ (s), it is $|f_D(t)|T<10^{-3}$ ("*slow fading*"); while for $t>100$ (s), it is $|f_D(t)|T>10^{-1}$ ("*fast fading*"). In the former case, a model of the digital narrow-band fading channel as first-order Finite State Markov Channel (FSMC) is proposed by Wang [19], while in the latter case, a memoryless model of the channel is acceptable [20]. Higher order models can be introduced to represent intermediate regimes [10].

5.4.4.1 Markov modeling

Stationary Case. If a constant power transmission is assumed and f_D is constant, the Signal-to-Noise Ratio (SNR(t)) may be written as:

$$\mathrm{SNR}(t)=|\tilde{z}(t)|\overline{\mathrm{SNR}}, \tag{1.5}$$

where $\overline{\mathrm{SNR}}=\mathrm{E}\{\mathrm{SNR}(t)\}$ ($\mathrm{E}\{\cdot\}$ is the expectation operator), and $\tilde{z}$ is a complex Gaussian process, with independent in-phase and quadrature components (each having an autocorrelation coefficient equal to $\rho(\tau)=J_0(2\pi f_D\tau)$), and $E\{|\tilde{z}|\}=1$. $|\tilde{z}|$ follows Rice distribution (let K be Rice factor) [17, 18].

A Markov model for SNR(t) at the portable end on the flat mobile channel can be obtained as follows. Let SNR(t) be sampled with a given rate T_s and quantized with respect to a set of thresholds $\{A_i\}_{i=1}^{L-1}$ [10, 19], giving a symbol sequence $\{\text{snr}_l\}, \{\text{snr}_l\} \in S_F = \{k\}_{k=1}^{L}$ where $\text{snr}_l = k$, if $A_{k-1} \leq \text{SNR}(lT_s) < A_k$ ($A_0 = 0$, $A_L = \infty$). The sequence of {*fading states*} $\{\text{snr}_l\}$ has infinite memory, so that a CT model with depth $M \rightarrow \infty$ would be required for an exact fit [10, 21]. Nonetheless the level of fidelity provided by small values of M proves to be sufficient to obtain useful results in the application described in Section 5.4.5.

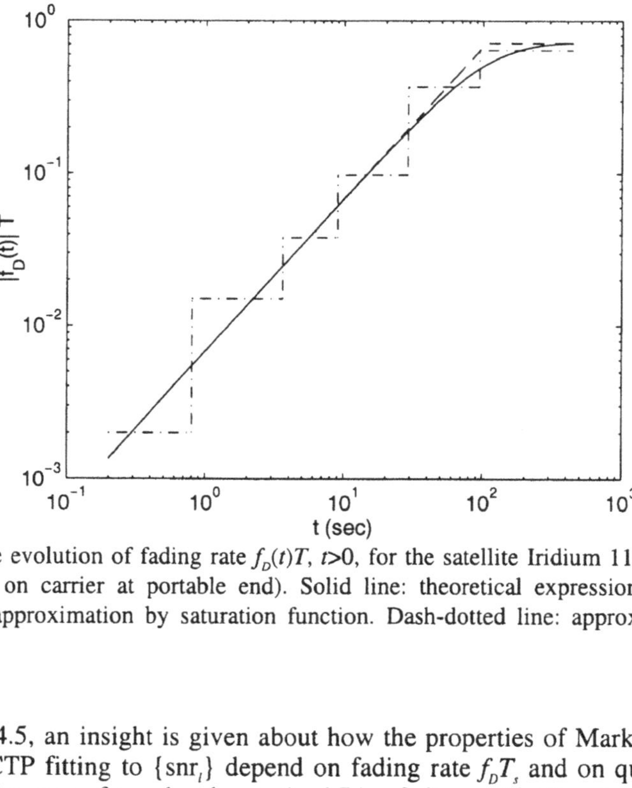

Fig.5.4.4 Time evolution of fading rate $f_D(t)T$, t>0, for the satellite Iridium 11 (T=20μs is the bit-period on carrier at portable end). Solid line: theoretical expression (Eq. 1.1). Dashed line: approximation by saturation function. Dash-dotted line: approximation by quantization.

In Fig. .5.4.5, an insight is given about how the properties of Markov models selected by CTP fitting to $\{\text{snr}_l\}$ depend on fading rate $f_D T_s$ and on quantization thresholds. The case of two-level quantized Rice fading, with Rice factor K=5, is considered (the quantization threshold, in dB, is $\overline{\text{SNR}_{\text{dB}}} - F$, where F *fade margin*). For CT fitting, the training sequence has a length $N = 5 \cdot 10^5$ and the maximum depth is M=7. In Fig.5.4.5a a contour plot of the selected tree depth is drawn, while in Fig. 5.4.5b the number of contexts in each model is displayed. When quantization thresholds are not too far from average SNR and fading rate is neither too low nor too high ($10^{-2} < f_D T_s < 0.4$), models with order higher than one are selected.

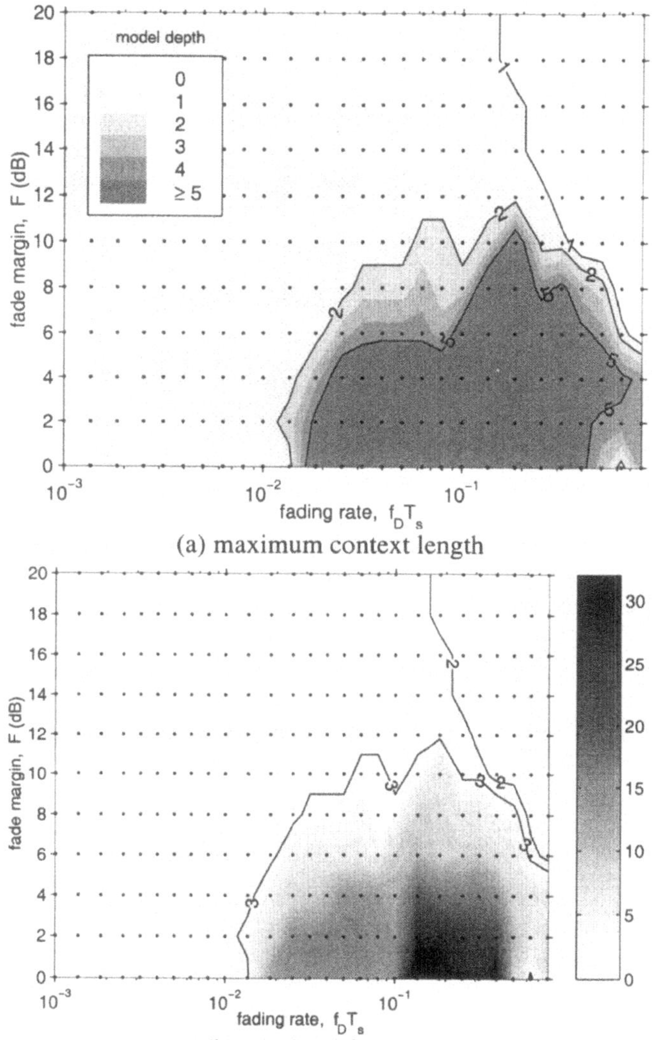

(a) maximum context length

(b) number of contexts

Fig. 5.4.5 CT model properties for two-level quantized Rice fading (Rice factor K=5, alphabet size L=2) as function of fading rate $f_D T_s$ and of fade margin F. Each data point represents a CT model (with maximum depth M=7), fitted to $5 \cdot 10^5$ simulated samples.

Observe the abrupt passage from higher order models to first-order models, as $f_D T_s$ decreases. The memory of the fitted model appears to be diminishing as the memory of the random process increases. In fact, the amount of training data required to fit a random process to a fixed fidelity grows exponentially with process memory. Observed here is the mechanism whereby CTP correctly enforces the limitations imposed by the amount of available training data. Overfitting is thereby avoided. First-order model is generally very close to the one introduced by Wang [19] for Rayleigh fading and extended to Ricean fading in

[22, 23] (see Appendix). The Markov model of slow fading is suitable in considering phenomena that are faster than the fading. (Hidden Markov Models are useful to evaluate slow phenomena in slow fading [21].)

After building a CTP based Markov model for $\{snr_l\}$, an FSMC model for the digital channel is obtained by associating *crossover probabilities* to each state (extended context) of the model, assuming modulation format does not depend on channel state [16]. Let modulation be binary (the extension to multi-level modulations is straightforward).

When the fading state is $snr=\varphi\in S_F$, assuming fading constant within the bit, the average crossover probability associated to snr is [19]:

$$e(\varphi)=\text{prob}(\text{bit error}\mid\varphi)=\frac{1}{\pi_\varphi}\int_{A_\varphi}^{A_{\varphi+1}} e(\sigma)f(\sigma)d\sigma \tag{1.6}$$

where $\pi_\varphi=\text{prob}(\text{snr}=\varphi)=\int_{A_\varphi}^{A_{\varphi+1}} f(\sigma)d\sigma$, $f(\sigma)$is the probability density function of SNR at the receiver (non-central χ^2), and $e(\sigma)$ is the Bit Error Rate (BER) on an AWGN channel with SNR equal to σ.

The current interval of the SNR is known from the fading state s:snr=snr(s)=$s(1)\in S_F$. Crossover probability models the probability of error on a BSC associated to each interval of SNR values. In the construction of the model, two states that result in the same snr value have the same crossover probability. The additional historical information in the states is not applied to the crossover calculation; the CT based model is *only* a description of the evolution of {snr}.

Non-Stationary Case. If the transmitted power is constant, but Doppler spread is a function of time $f_D=f_D(t)$, a model for the time-continuous process SNR(t) can be obtained, quantizing $|f_D(t)|$ by a set of thresholds $\{f_{D,i}\}_{i=1}^N$, and approximating $f_D(t)$ by step functions. In time interval $\tau=\tau_i$, where $f_D=f_{D,i}$, a stationary model having $f_D=f_{D,i}$ is used to represent SNR on the channel. A time sequence of time-continuous models results. To describe the digital channel in each time interval τ_i, a Finite State Markov Channel (FSMC$_i$) with constant parameters is obtained from the corresponding time-continuous model according to the methods of the previous section: the corresponding fading rate is $f_{D,i}T$. Each FSMC$_i$ may be modeled either by first-order (Wang), or by higher-order (CTP), or by memoryless models, depending on the value of $f_{D,i}T$.

In the case considered here, the thresholds are $\{f_{D,i}T\}_{i=1}^6=\{2\cdot10^{-3}, 1.5\cdot10^{-2}, 3.8\cdot10^{-2}, 9.8\cdot10^{-2}, 3.7\cdot10^{-1}, 6.4\cdot10^{-1}\}$ (dash-dotted line in Fig.5.4.4). For each model of the sequence, the value of the Rice factor and the average SNR are $K=10$ and $\rho=10$ (dB), respectively: in [5] such values have been measured on highways. Variations of both parameters due to satellite motion are neglected. Moreover, the sampling time of time-continuous processes is $T_s=T=20$ μs and quantization thresholds are four and logarithmically equispaced ($\{A_i\}_{i=1}^4=\{\rho\text{-}12, \rho\text{-}9, \rho\text{-}6, \rho\text{-}3\}$ dB). The modulation format on the digital channel is assumed to be BPSK and crossover probabilities are evaluated in each state from (1.6). The maximum CTP model depth is fixed to M=8 and the length of training sequence is $N=5\cdot 10^7$.

5.4.5 PED evaluation

As a sample application, the Packet Error Distribution (PED) has been evaluated for each model FSMC$_i$, defined in the time interval τ_i. The PED, $P(m,n)$, is defined as the probability of having m errors in a block of length n. The choice n=64 (bits) has been made so that it is not necessary to consider effects of excessively long memory. This value of n is near to the packet length at the output of many standard coders. For a single FSMC model, the PED calculation is specified by the state set S_M, the transition matrix **T** over S_M, and the crossover probability vector **e** over S_M (as specified in Section 5.4.4.1: $\mathbf{e}=\{e(s_k(1))\}_{k=1}^{|S_M|}$, where $S_M=\{s_k\}_{k=1}^{|S_M|}$. Calculation follows the recursive algorithm proposed in [24, 25]:

$$\begin{gathered}\underline{G}(m,n)=\underline{G}(m,n-1)\mathbf{Q}+\underline{G}(m-1,n-1)\mathbf{P},\\ \mathbf{Q}=\mathrm{diag}(\mathbf{e})\mathbf{T},\qquad \mathbf{P}=[\mathbf{I}-\mathrm{diag}(\mathbf{e})]\mathbf{T},\\ P(m,n)=\underline{G}(m,n)\underline{1}^{\mathrm{T}}\end{gathered}\tag{1.8}$$

with initial conditions:

$$\underline{G}(1,1)=\underline{p}\mathbf{P},\quad \underline{G}(0,1)=\underline{p}\mathbf{Q},\quad \underline{G}(m,n)=0,\ m<0,\tag{1.9}$$

where $\underline{G}(m,n)=\{G_k(m,n)\}_{k=1}^{|S_M|}$ is a row vector[4], diag(**v**) is the diagonal matrix having **v** as diagonal vector, **I** is the identity matrix, $\underline{1}^{\mathrm{T}}$ is a column vector made by ones, $\underline{p}=\underline{p}\mathbf{T}$ is the stationary distribution associated to **T**[5], and $\underline{0}$ is a null row vector. In Figs. 5.4.6 and 5.4.7 results obtained from Wang model, CTP model, and simulation (based on filtering method [26]) are compared. Figs. 5.4.6a, 5.4.6b and 5.4.7 refer to "slow", to "intermediate" and "fast" fading cases, respectively. Notice that for "fast" fading (last two values of f_DT), Wang model cannot be applied and the following binomial expression, based on a memoryless model, is used for $P(m,n)$:

$$P_{\mathrm{fast}}(m,n)=\binom{n}{m}p_b^m(1-p_b)^{n-m},$$

where p_b is the average bit error probability on the fading channel.

---

[4] Let $\underline{b}$ be a sequence of n bits, starting at a generic step w (bit decisions are supposed to be independent each from other), and $s(w+n+1)$ be the state of the FSMC at the step $w+n+1$: $G_k(m,n)$=prob(m errors in $\underline{b}$} and $s(w+n+1)=s_k$).

[5] Note that $\underline{p}\neq\underline{\pi}$, where $\underline{\pi}$={prob (φ=k)} $k=1\ldots L$, as defined in Section 5.4.4.1

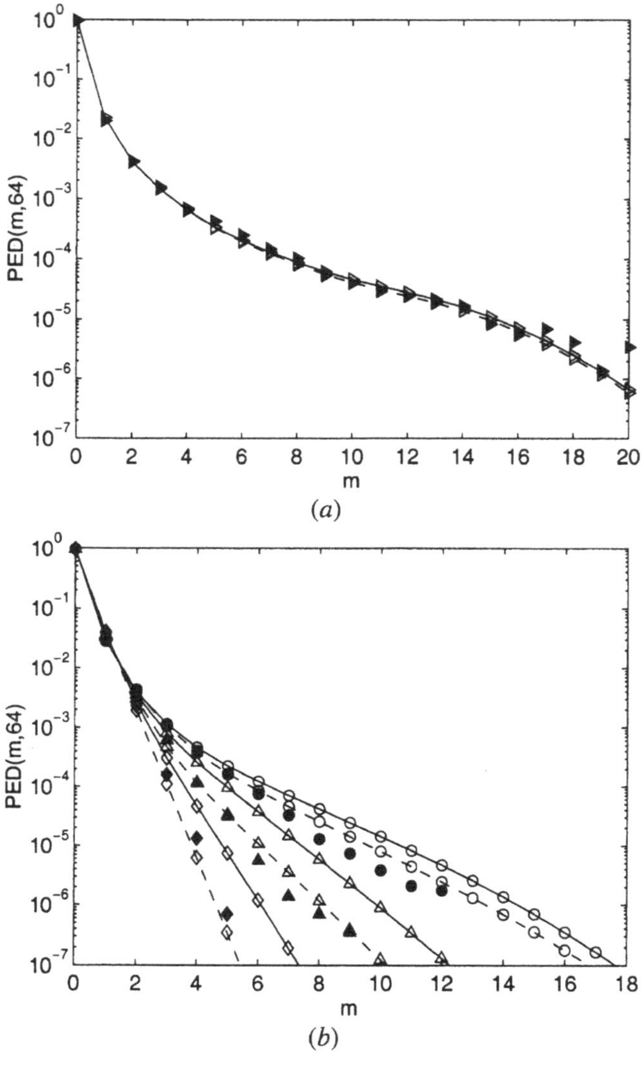

Fig. 5.4.6 Packet Error Distribution with 64 bit packet length, PED(m,64), as function of error number, m. (a): slow fading; 'Δ' $f_DT=2\cdot10^{-3}$. (b): intermediate fading; 'o': $f_DT=1.5\cdot10^{-2}$; 'Δ' $f_DT=3.8\cdot10^{-2}$.; '◊' $f_DT=9.8\cdot10^{-2}$. Empty markers refer to theory (solid line: Wang; dashed line: CTP) and filled markers to simulation.

Wang model gives correct results only for $f_DT < 10^{-3}$ (slow fading), when also CTP is not capable to extract higher order models, while for all other values only general CTP models allow correct prediction. In the case of fast fading, results provided by CTP are close to the ones supplied by a memoryless model, that may be considered acceptable.

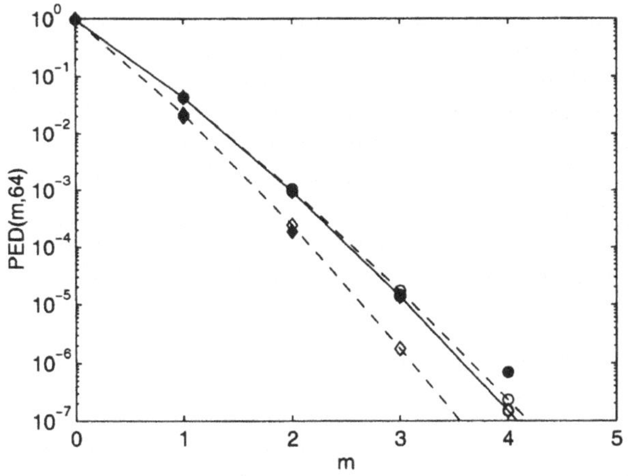

Fig.5.4.7 Packet Error Distribution with 64 bit packet length, PED(m,64), as function of error number, m. Fast fading. 'o' $f_D T=3.71\cdot10^{-1}$, '◊' $f_D T=6.45\cdot10^{-1}$. Empty markers refer to theory (solid line: Wang; dashed line: CTP) and filled markers to simulation.

5.4.6 Discussion

The digital narrow-band satellite fading channel exhibits noticeable time variation of its fading rate $f_D T=f_D T(t)$. If other causes of channel non-stationarity are negligible, an approximate model for the channel is obtained by quantizing $f_D T(t)$ inside satellite visibility time window and applying a stationary model, like Finite State Markov Channel (FSMC), within each time interval where $f_D T(t)$ is assumed constant.

Only for a part of visibility window can fading be approximated as either "slow" or "fast"[6] thereby allowing channel properties to be evaluated with sufficient accuracy by means of a first-order FSMC model, like the one proposed by Wang [19], or with a memoryless model, respectively. In the remaining regime, application of the Context Tree Pruning (CTP) algorithm leads to FSMC models, capable to include the effect of channel memory beyond first-order (embedded Markov models [27, 28]). Such models provide more accurate results than first-order and memoryless models that are special cases.

A general and tractable digital channel analysis is allowed by FSMC models given that their Markov nature permits employing all related theory. They lead to easy theoretical evaluation of several useful parameters (as shown with the example of PED) or to readily setting-up system optimization procedures (e.g. when dealing with multiple access techniques [27, 28]). On the contrary, direct use of theoretical Clarke's model would require simulation or cumbersome evaluation.

[6] Fast fading approximation is also invoked when using interleaving [17, 20].

CTP facilitates a reasonably faithful evaluation of system parameters in situations where only the effect of "short-term" channel memory is considered significant. This happens with Packet Error Distribution (PED) when packet length is small enough, and with protocol optimization, which can be performed also with limited channel knowledge (sometimes even a first-order Markov model, in place of a memoryless model, leads to results that are satisfactory) [27, 28]. When the effect of distant events must be included (as in error burstiness analysis), FSMC models may become too complex or impractical to generate, in which case other tools such as Hidden Markov Models [21] may be more suitable.

5.4.7 Summary

In the framework of mobile satellite communications, suitable models for the digital channel are required in system project and optimization. In this chapter, a time sequence of Finite State Markov Channel (FSMC) models, each based on the so-called Context Tree Pruning (CTP) algorithm, is proposed to provide a theoretical representation of the digital channel for narrow-band Low Earth Orbit (LEO) mobile satellite communications. The time variation of Doppler shift on carrier due to satellite motion is considered, being a significant cause of channel non-stationarity, and a model is accordingly defined to represent the channel during satellite visibility period. As a model application, Packet Error Distribution (PED) is evaluated from model and compared with simulations.

Appendix: First-Order (Wang) Markov Models

A review is made about the first-order FSMC model, used to describe the digital channel with stationary Ricean fading and introduced by Wang in the special case of Rayleigh fading [19] (in [11, 23] the extension to the Ricean case is provided). Such model requires the hypothesis of low fading rate ("slow" fading: $f_D T_s \ll 1$): using notations from Section 5.4.4.1, fading is nearly constant on a time interval equal to T_s, the state set is $S_M = S_F = \{k\}_{k=1}^{L}$ and transitions between non-adjacent states are not possible, that is $t_{i,j}$=prob($\mathrm{snr}_{l+1}=j|\mathrm{snr}_l=i$)=0, if $|i-j|\geq 2$.

The evaluation of $t_{i,i\pm 1}$, given by:

$$t_{i,i\pm 1} = \frac{\mathrm{prob}(\mathrm{snr}_{l+1} = i \pm 1, \mathrm{snr}_l = i)}{\pi_i},$$

where $\pi_i = P(\mathrm{snr}_l = i)$, is made as follows [11]. Let consider the time interval $\Delta = [iT_s, (i+M)T_s]$, with M integer. On average, $MT_s N_{i+1}$ transitions across A_{i+1} with positive slope occur in Δ, where N_i is the average transition rate of SNR(t) across the level A_i [29], with positive *or* negative slope. Given that at most *one* transition with positive slope is allowed in the time slot T_s, M transitions at most may occur in Δ. Then, from frequency definition of probability [30], one has:

$$\text{prob}(\text{snr}_{l+1} = i+1, \text{snr}_l = i) \approx \frac{MT_s N_{i+1}}{M} = N_{i+1} T_s \,.$$

Considering transitions with negative slope across *Ai*, one has similarly prob(snr_{l+1}=i-1,snr_l=i)≈$N_i T_s$. Thus, the elements of the transition matrix are:

$$\begin{aligned}
t_{i,j} &\approx 0, && |i\text{-}j| \geq 2, \\
t_{i,i+1} &\approx \frac{N_{i+1}\, T_s}{p_i}, && i = 0, \ldots, L\text{-}2, \\
t_{i,i-1} &\approx \frac{N_i\, T_s}{p_i}, && i = 1, \ldots, L\text{-}1, \\
t_{i,i} &= 1 - \sum_{\substack{j=0,\\ j\neq i}}^{L-1} t_{i,j}\,, && i = 0, \ldots, L\text{-}1,
\end{aligned} \tag{A.1}$$

Previous arguments do not depend on the statistical distribution of SNR(t).Let us assume the channel is subject to Ricean fading, so that a=SNR(t) follows a non-central χ^2 distribution with two degrees of freedom (2-NC-χ^2), with average P_t and Rice factor K. Then, the expression of N_i and π_i are [29]:

$$N_i = f_D \sqrt{2\pi\sigma^2 A_i}\; f(A_i) \tag{A.2}$$

$$\pi_i = \int_{A_i}^{A_{i+1}} f(z)dz$$

with:

$$f(z) = \frac{1+K}{P_t}\, I_0\!\left(2\sqrt{\frac{K(1+K)z}{P_t}}\right) \exp\!\left(-K - \frac{1+K}{P_t}z\right) U(z), \tag{A.3}$$

where $U(z)$ is Heaviside step function and σ^2=P_t/(K+1). From Equations (A.1) and (A.2), it results that the elements of matrix **T** are proportional to the fading rate $f_D T_s$.

Observe that, for 2-NC-χ^2 distribution, the diffuse mean power, σ^2, is related to the average values E{a} and E{a^2} by:

$$\sigma^2 = E\{a\} - \sqrt{2E^2\{a\} - E\{a^2\}} \tag{A.4}$$

Two models approximating the given one can be introduced, observing that a 2-NC-χ^2 random variable can be approximated by a Gamma or Weibull one, having the same mean and variance [16, 31]. The probability density function of such distributions are given by:

$$f_\Gamma(z) = \frac{z^{m-1}}{\beta^m \Gamma(m)} \exp\!\left(-\frac{z}{\beta}\right) U(z), \tag{A.5}$$

$$f_W(z) = \frac{m z^{m-1}}{\beta^m} \exp\left[-\left(\frac{z}{\beta}\right)^m\right] U(z), \tag{A.6}$$

respectively, where $\Gamma(\cdot)$ is the Gamma function, m the shape factor and β the scale factor. The relationships between the parameters of the 2-NC-χ^2 distribution and of the two approximating ones are summarized in Table A.1.

Table A.1: Relationships between the parameters of some distributions

r.v.	2-NC- $\chi^2 \longrightarrow$ r.v.	r.v $\longrightarrow$ 2-NC- χ^2.
Γ	$\beta = \lambda P,\ m = \frac{1}{\lambda},\ \lambda = \frac{1+2K}{(K+1)^2}$	$P = m\beta$ $K = m - 1 + \sqrt{m(m-1)}$
W	$\beta = \frac{P}{\Gamma\left(1+\frac{1}{m}\right)}$ $\frac{\Gamma\left(1+\frac{2}{m}\right)}{\Gamma^2\left(1+\frac{1}{m}\right)} - 1 = \frac{1+2K}{(1+K)^2}$	$P = \beta\Gamma\left(1+\frac{1}{m}\right)$ $K = \mu - 1 + \sqrt{\mu(\mu-1)}$ $\frac{1}{\mu} = \frac{\Gamma\left(1+\frac{2}{m}\right)}{\Gamma^2\left(1+\frac{1}{m}\right)} - 1$

The FSMC based on Gamma approximation is obtained letting $f(z) \approx f_\Gamma(z)$ into Eq. (A.2) and evaluating σ^2 from Eq. (A.4) with $E\{a\}=m\beta$ and $E\{a^2\}=m(m+1)\beta^2$ while the FSMC based on Weibull approximation is evaluated using $f(z) \approx f_W(z)$ and $E\{a\}=\beta\Gamma(1+1/m)$, $E\{a^2\}=\beta^2\Gamma(1+2/m)$. Note that the Weibull based model has a simple closed form.

Glossary

AWGN Additive White Gaussian Noise channel.
BER Bit Error Rate.
BSC Binary Symmetric Channel.
ECF Earth-Centered Fixed frame.
ECI Earth-Centered Inertial frame.
CT Context Tree
CTP Context Tree Pruning algorithm.
FSMC Finite State Markov Channel.
LEO Low Earth Orbit satellite.
PED Packet Error Distribution.
SNR Signal-to-Noise Ratio.

References

1. A. Goldsmith and P. Varaiya, "Capacity, Mutual Information and Coding for Finite-State Markov Channels" , *IEEE Trans. Inform. Theory*, vol. 42, no. 3, pp. 868-886, May 1996.
2. R. M. Gagliardi, *Satellite Communications*, Van Nostrand Reinhold, New York, NY, 2nd edn. 1991.
3. C. Loo, "A Statistical Model for Land Mobile Satellite Link" , *IEEE Trans. Veh. Technol.* , vol. VT-34, no. 3, pp. 122-127 , Aug. 1985.
4. G. E. Corazza and F. Vatalaro, "A Statistical Model for Land Mobile Satellite Channels and Its Application to Nongeostationary Orbit Systems" , *IEEE Trans. Veh. Technol.* , vol. 43, no. 3, pp. 738-742 , Aug. 1994.
5. E. Lutz, D. Cygan, M. Dippold, F. Dolainsky, and W. Papke, "The Land Mobile Satellite Communication Channel-Recording, Statistics and Channel Modeling", *IEEE Trans. Veh. Technol.* , vol. 40, no. 2, pp. 375-386 , May 1991.
6. F. P. Fontán, J. P. Gonzâlez, M. J. S. Ferreiro, M. A. V. Castro, S. Buonomo, and J. P. Baptista, "Complex Envelope Three State Markov Model Based Simulator for the Narrow-band LMS Channel", *Int. J. of Wireless Sat.* Commun. , vol. 15, no. 1, pp. 1-15 , 1997.
7. B. Vucetic, "An Adaptive Coding Scheme for Time Varying Channels", *IEEE Trans. Comm.*, 39, No.5, 653-663, May 1991.
8. B. Vucetic, J. Du, "Channel Modeling and Simulation in Satellite Mobile Communication Systems", *IEEE J. Sel. Areas Commun.*, 10, No. 8, 1209-1218, Oct 1992.
9. F. Babich, O. E. Kelly, and G. Lombardi, "A Variable-Order Discrete Model for the Mobile Propagation Channel", in *Broadband Wireless Communications*, M. Luise and S. Pupolin, Eds., London, UK, 1998, pp. 259-266, Springer -Verlag.
10. F. Babich, O. E. Kelly, G. Lombardi, "A Context-Tree Based Model for Quantized Fading", *IEEE Communications Letters*, Vol. 3, No. 2, pp. 46-48, February 1999.
11. F. Babich, G. Lombardi, E. Valentinuzzi, "Context Tree Prunning modelling for LEO Satellite Channels", *International Workshop On Mobile Communications*, June 24-26 1999, Creta.
12. I. Ali, N. Al-Dhahir, and J. E. Hershey, "Doppler Characterization for LEO Satellites", *IEEE Trans. Commun.* , vol. 46, no. 3, pp. 309 - 313, Mar. 1998.

13. W. L. Pritchard, H. G. Suyderhoud, R. A. Nelson, *Satellite Communication Systems Engineering*, Prentice-Hall, Eiglenwood Cliffs, New Jersey, 2nd edn., 1993.
14. J. V. Evans, "Satellite Systems for Personal Communications", *Proc. IEEE*, 86, No. 7, 1325-1341, July 1998.
15. M. J. Weinberger, J. J. Rissanen, and M. J. Feder, "A Universal Finite Memory Source", *IEEE Trans. Inform. Theory*, vol. 41, no. 3, pp. 643-652, May 1995.
16. G. Lombardi, *Measurement, Simulation and Analysis of Indoor Mobile Channel*, PhD thesis, Dip. di Elettrotecnica, Elettronica e Informatica, Univ. of Trieste, Trieste, Italy, 1998 (in Italian).
17. R. Steele (Ed.), *Mobile Radio Communications*, Pentech Press, London, 1992.
18. J. D. Parson, *The Mobile Radio Propagation Channel*, Halsted Press, New York, Toronto, 1992.
19. H. S. Wang and N. Moayeri, "Finite-State Markov Channel: A Useful Model for Radio Communication Channel" , *IEEE Trans. Veh. Technol.*, vol. 44, no. 1, pp. 163-171, Feb. 1995.
20. J. G. Proakis, *Digital Communications*, McGraw--Hill, New York, 1989.
21. W. Turin and R. Van Nobelen, "Hidden Markov Models for Fading Channels", *IEEE J. on Sel. Areas in Commun.*, vol. 16, n. 9 pp. 1809-1817, Dec. 1998.
22. F. Babich, G. Lombardi, "Generalized first-order Markov models for the Rician fading channel", *International Workshop On Mobile Communications*, June 24-26 1999, Creta.
23. F. Babich, G. Lombardi, "A Markov Model for the Mobile Propagation Channel", *IEEE Trans. on Vehicular Technology*, vol. 49, no. 1, January 2000.
24. L. N. Kanal A.R.K. Sastry, "Models for channels with memory and their application to error control," *IEEE Proc.*, vol. 66, no 7, pp. 724-744, July 1978.
25. H. Bischl and E. Lutz, "Packet Error Rate in the Non-Interleaved Rayleigh Channel" , *IEEE Trans. Commun.*, vol. 43, no. 2/3/4, pp. 1375-1382, Feb./Mar./Apr. 1995.
26. J. I. Smith, "A Computer Generated Multipath Fading Simulation for Mobile Radio", IEEE Trans. Veh. Technol., vol. VT-24, no. 3, pp. 39-0, Aug 1975.
27. O. E. Kelly, F. Babich, "Protocol optimization using discrete channel statistic", *IEEE International Conference on Universal Personal Communications (ICUPC)* 1998, Florence, Italy, 5-9 October 1998.
28. F. Babich, O. E. Kelly, "Design of adaptive transmission protocols for the fading channel", *International Workshop On Mobile Communications*, June 24-26 1999, Creta.
29. W. C. Jakes, *Microwave Mobile Communications*, Wiley-Interscience, New York, 1974.
30. A. Papoulis, *Probability, Random variables, and Stochastic Processes*, McGraw-Hill, New York, third edn., 1991.
31. G. L. Stüber, *Principles of Mobile Communication*, Kluwer Acad. Publ., Dordrecht, NL, 1996.

6 System Architecture

6.1 Introduction

Converging economic and technology drivers influence which architectures and components are used in the design of radio products for wireless base stations. In some markets, 3G mobile systems will be the only cellular service, by passing current infrastructure architectures entirely. This development forces infrastructure manufacturers to develop common hardware and software platforms for the Universal Wireless Market and this necessitates a common hardware architecture reconfigured by software. Network architectures such as the multi-layer cellular system play a vital role as far as improved quality of service and higher capacity over the traditional single-tier cellular network under the same resource constraints.

In section 6.2 efficient analog to digital converters, digital to analog converters, digital receive and transmit signal processors are presented as a vehicle to software radio tranceivers development. The software radio concept enables necessary flexibility for second and third generation of wireless systems. The general trend of wireless wide-band transceivers points to continuing reduction of radio frequency processing at the expense of increased digital signal processing. At the end of 20th century the state-of-the-art cost efficient data converters perform data conversion up to 250 MHz Intermediate Frequency. Intermediate Frequency data converters and digital signal processors (Receiver / Transmitter) for software defined radio base station architectures are presented covering major digital wireless standards. It focuses on the widely accepted necessity of hardware compatibility between second and third generation wireless basestations.

Section 6.3 presents an overview of current research and published work on the performance of multi-layer cellular architectures and the handover performance in such networks. In a multi-layer cellular network, macrocells with a large coverage area, overlay microcells or picocells with smaller coverage areas. This network architecture, if well engineered, can provide improved QoS and higher capacity over the traditional single-tier cellular network under the same resource constraints. The small coverage of microcells/picocells, and the consequent multipath and shadow fading characteristics, mean that handover events in future systems will occur at a much higher rate. Therefore efficient handover algorithms are a cost-effective way of enhancing the capacity and QoS of cellular systems.

6.2 Transceiver Components and Architectures for Software Defined Radio 3G Base-stations

Dimitrios Efstathiou

6.2.1 Introduction

Third generation wireless technologies in co-existence of current second generation communication systems will be able to provide high data rate wireless services. Particularly, multiple data rates, multimedia communications, variable rate speech coding, adaptive modulation, adaptivity to link quality, service mix and variable division of spectrum, all require flexibility within the same air-interface standard and / or programmability among different air-interface standards [1] [2]. Third generation wireless systems pose new requirements in terms of dynamic range and sampling frequencies compared to previous generations [3]. As a response to the technical challenges surrounding the realisation of third generation systems and their compatibility to existing second generation systems, a number of enabling technologies have emerged, including wide-band radio frequency (RF) components, very fast signal conversion technologies, high performance digital signal processing devices, and software concepts (e.g. real-time software download, application program interface etc.). Resulting wireless transceiver architectures are critically dependent on the performance of mixed signal components, with analogue-to-digital converters (ADC) and digital to analogue converters (DAC) being limiting performance factors. General trend of wide-band transceivers points to continuing reduction of radio frequency processing at the expense of increased digital signal processing.

Multi-mode and multi-carrier receivers and transmitters can be realised with low cost, enhancing the attractiveness of system roaming, flexibility to support new services and even autonomous selection of air interface standard according to the environment and user needs [4]. Lower-cost base stations can improve system flexibility and enhance the availability of new services. Software defined radio can provide an efficient bridge from the second generation to the third generation wireless systems and become one of the key elements to the realisation of the potential of the third generation market.

The development of cost efficient analogue to digital converters, digital to analogue converters, Digital Receive and Transmit Signal Processors is called upon to play a crucial role in the concept of software radio transceivers. Software radio concept enables necessary flexibility and innovation for existing and next generation of wireless systems.

Converging economic and technology drivers influence which architectures and components are used in the design of radio products for wireless basestations. Operators in growing markets such as Latin America and Asia, especially in China, are depended on cost reductions in the existing digital cellular infrastructure to make services economically viable, while capacity limitations, especially in Japan, and wideband data services fuel the drive toward third generation (3G) deployment. In some markets, 3G will be the only cellular service, bypassing current infrastructure architectures entirely. This combination of demand for lower cost 2G systems and the initial deployment of 3G systems forces infrastructure manufacturers to develop common hardware and software platforms for the universal wireless market based on software radio concept.

This chapter discusses common hardware platforms - reconfigured by software - for second and third generation basestations. It addresses the critical functional blocks - Digital Receive and Transmit Signal Processors, Analogue to Digital and Digital to Analogue converters. - It addresses as well the technology challenges for receiver and transmitter realisation of multi-standard re-configurable basestations. In section 6.2.3 a desirable and a more practical approach of software defined radio is presented. In section 6.2.4, analogue to digital converters (ADCs), digital to analogue converters (DACs) requirements and specifications are reviewed showing the implications on system design. In the same section major characteristics of IF (Intermediate Frequency) Digital Receive/Transmit Signal Processors are reviewed as well. In section 6.2.5 major functional and performance parameters of ADCs, DACs and Digital Receive/Transmit Signal Processors are illustrated using specific products. In section 6.2.6 various receiver and transmitter architectures are revisited from the technology aspect. Wide-band architectures for single-carrier and multi-carrier wireless systems - based on advanced components presented in section 6.2.4 and 6.2.5 - are addressed. Important trade-offs in implementation will be emphasised having in mind specifications of the wireless systems. Applications are presented by revisiting existing transceiver architectures in the framework of next generation wireless standards.

6.2.2 What is Software Defined Radio?

What is Software Radio? The most literal translation of software radio would be a radio where signals on the antenna, or perhaps at an intermediate frequency are digitised with a high performance ADC and sent to a receiver (base-station, computer, mobile phone etc.). Once digitised software code would be used to select a RF channel and demodulate the signal (Fig. 6.2.1a.) While this is a worthy goal, it is only now becoming practical for some second-generation specific applications. Third generation systems set the demand for very fast DSPs. Architecturally, DSPs have to improve their performance by means of higher levels of parallelism, incorporating multiple data paths and execution units in order to cope with the computational burden of third generation systems. Some functions of a W-CDMA receiver such as matched filtering and de-spreading, are computationally intense, and are likely to be implemented as dedicated ASICs or at least a hardware accelerators within a DSP core. The peak computational demand of a software radio for a 3.84 MHz W-CDMA mode is about 1600 MIPS. An FPGA or DSP-core based ASIC could deliver the required computational capacity

Historically, the relatively low number of prevalent standards, as well as the state of the art and the high cost of key components, has limited the benefits and use of software radios. The second generation of wireless systems has offered a variety of different modulation formats and multiple-access technologies to be covered by a single radio. In addition, dual mode operation and compatibility requirements with analogue systems make this task even more challenging. Often the main goal of current developments in dual-mode (eventually triple-mode) transceivers covering drastically different data rates and modulations formats is reduction of cost, power and size.

During the past five years, the software radio has emerged from demonstrations in military research to become a cornerstone of seamless evolution among different air-interface standards and different generations of communication systems [5].

A practical definition of software defined radio includes radios with a set of pre-defined hardware modules (ASIC's, FPGA's) that are selectable through software for common hardware for several different systems. These modules provide multi-rate signal processing functions (decimation, interpolation), digital down / up conversion capabilities and fixed or programmable coefficient filters.

A more reasonable name for this desired technology would be a digital re-configurable radio (in this chapter software radio and re-configurable radio are considered exchangeable terms). Note that digital receivers can be designed to receive digitally modulated signals as well as analogue FM signals. As with a software radio, an ADC is used to digitise the signal at the antenna or at an intermediate frequency [28]. However, instead of processing the digitised data solely in software, a variety of flexible re-configurable Application Specific Integrated Circuits (ASICs) or Field Programmable Gate Arrays (FPGAs) and general purposed Digital Signal Processors (DSPs) are used to reduce system power dissipation, size and cost (Fig.6.2.1b). In effect, the filters and demodulation/modulation that would have run on a transceiver have been generalised and committed to silicon with programmable characteristics. Thus a practical software radio is one where selected functions have been committed to silicon with enough flexibility to reconfigure for a variety of different air-interface standards.

Therefore, pre-defined hardware modules enable the efficient realisation of transceiver functions in terms of power consumption, minimal component count and compactness. Economic and technology drivers influence existing software defined radio architectures and restricts re-configurable components in baseband and IF sections. However, Integrated Circuits manufactures are convinced that emerging advances in semiconductor industry will make available re-configurable components for RF section in the near future.

A fundamental concept of software radio is to host all the software (DSP and otherwise) on programmable processors (Digital Receive/Transmit Signal Processors). All aspects of the air interface including the channel waveforms are thus defined in software and implemented in real time through software and/or hardware. Emerging cellular base stations favour larger granularity hardware modules. Digital IF signal processing provides repeatable manufacturing, higher accuracy, and improved flexibility than comparable high dynamic range analogue designs. The software radio approach reduces the hardware complexity of a cell site from several racks of discrete single-channel radios to one or two selves of open architecture.

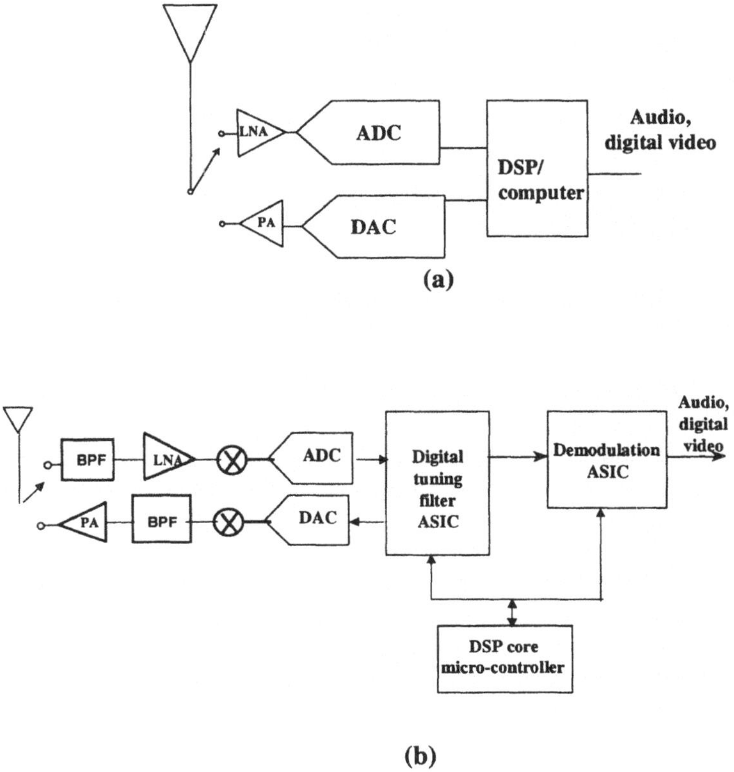

Fig.6.2.1 Software Defined Radio

This work focuses to the IF section of a re-configurable radio for base-stations. The IF processing section translates the information bearing signal(s) between base-band and IF [6]. The IF stage of a software radio typically consists of a data conversion stage (ADC, DAC), a digital rate change stage, cascaded filtering stages, a numerically controlled oscillator (NCO) and modulator stage (DRSP, DTSP). In some cases, the rate change and modulation stages can be combined into a single functional block.

The development of cost efficient ADCs, DACs, digital down / up frequency converters and programmable digital filters, is crucial for the software radio transceivers. Following the functional architecture model for 3G systems outlined in [7], re-configurable modules represent critical functional entities based on hardware with certain flexibility (programmable filters, different communication ports etc.). The components presented in this chapter are intended to meet the needs of base-station manufacturers for hardware flexibility and reuse, increased channel capacity, lower cost per channel and quicker time to market product availability.

6.2.3 Requirements and Specifications for SDR Mixed Signal Components.

In general, the amount of radio frequency (RF) processing continues to reduce at the expense of increased digital signal processing. Wireless transceiver architectures are critically dependent on the performance of mixed signal components, with ADC as a limiting performance factor [8] [9]. With the maturity of high speed analogue to digital conversion technology moving towards the goal of 14-16 bit resolution, greater than 85 dB dynamic range and a sampling rate in excess of 80 MHz, the development of wide-band software radio systems becomes more feasible.

Before addressing the performance of integrated receiver and transmitter components, it is important to emphasise critical mixed signal (Analogue - to - Digital and Digital - to - Analogue conversion) component functionalities and specifications. Data converters' impact on the performance of the wireless systems is presented and the requirements of the wireless system which influence component selection are addressed.

6.2.3.1 Analogue - To - Digital Converter Specifications

Characteristics of mixed signal components, relevant to the development of software radios, largely depend on the architectural choices. In particular, wireless systems may employ base-band sampling, low intermediate frequency (IF) sampling (sampling of the full band while the desired signal is positioned at low IF) and bandpass sampling at higher IF.

Band-pass sampling (also known as sub-sampling, or harmonic sampling) is an attractive technique that is applied in radio receivers [10] [11]. It intentionally aliases the band-pass signal to perform down conversion to the base-band or to some low carrier frequency, most often selected to be $f_s/4$ for the most efficient implementation of the In-Phase and Quadrature (I & Q) digital down conversion to base-band. This procedure is analogous to the mixing process in a radio front-end. Digital filtering is performed to eliminate aliases outside the band of interest. Major benefits of IF sampling include the reduction of the number of components at radio IF stage (I&Q down converters, base-band filters), elimination of DC offset problem, elimination of I&Q imbalance problem and increased flexibility for filtering in digital domain.

The maximum ratio of signal power to the total quantisation noise power or signal to noise ratio (SNR) is usually referred as the *dynamic range* of the converter. For a sinusoidal input signal dynamic range of the ideal converter is

$$SNR_Q[dB] = 1.76 + 6.02b$$

and maximum signal to noise power density in dB is given by

$$SN_0R_Q = -1.249 + 10\log(f_s) + 6.02b$$

If the maximum signal frequency is smaller than half the sampling frequency, further filtering can be applied and maximum theoretical SNR due to quantisation noise only is given by

$$SNR_Q[dB] = 1.76 + 6.02b + 10\log\left(\frac{f_s}{2f_{\max}}\right)$$

assuming a sinusoidal input with an amplitude equal to full scale range FSR. This shows that over-sampling can increase the maximum possible SNR, which is often referred as the processing gain.

The accurate specification of ADC is given by *Signal-to-Noise Plus Distortion Ratio* (SNRD), where distortion is included in the noise part. SNRD is a function of the number of bits, analogue input frequency and signal level. *Effective Number of Bits* (ENOB) is defined as the number of bits required in an ideal ADC so that the mean squared noise power in the ideal ADC is equal to the SNRD in the real ADC.

In typical wireless scenarios signals that are outside of the pass-band may generate products that will fold into the passband. The mechanism of inter-modulation (IM) product generation is similar to the one in mixers, except that in this case interfering signal interacts with sampling frequency. When this occurs all of the quantisation noise may be concentrated in few IM products. The magnitude of this effect depends on non-linearity of ADC. Typical value of IM level is about 10 dB below the total quatisation noise, even if there is over-sampling [10]. For a sinusoidal input to an ADC, *Spurious Free Dynamic Range* (SFDR) is defined as the ratio of the sinusoidal signal root mean-squared (rms) power to the peak power of the largest spurious signal in the ADC output spectrum, measured over the entire first Nyquist zone. SFDR allows a designer to assess how well an ADC can simultaneously detect a weak signal in the presence of a strong signal. This is important specification for ADC used in radio receivers. For an ideal ADC, maximum SFDR occurs at full-scale input level. In practical ADCs maximum SFDR occurs at input level 3 to 6 dB below the full-scale input level. Prediction of SFDR for practical ADCs is difficult and measurements are required for its characterisation. It is important to note that SFDR considers all sources of distortion, regardless of their origin [12].

There is a large difference between the SFDR and the SNR of an ADC. The power of the residual error includes quantisation noise, random noise and non-linear distortion within the entire Nyquist band, so it can be much higher than the peak power of the largest spurious product. Hence, SFDR can be much larger than the SNR. SFDR specification is very useful for applications where signal bandwidth is much smaller than Nyquist bandwidth so over-sampling can be used to reduce the noise power. However, the spurious component may still fall into the desired band and does not necessarily improve with digital filtering. The SFDR performance of an ADC is greatly improved by adding dither to the ADC. The addition of out-of-band noise (dither) can improve SFDR well into the noise floor. Although the amount of dither is converter-specific, the technique applies to all ADCs in which static deviation from 1-LSB step between adjacent codes, limits performance, rather than ADC specific issues, such as slew-rate limitation [13].

In wide-band radio receiver applications IM distortions can result from multiple input signals. Different methods of ADC specification may be required. The usual method in this case is to generate multiple tone input signals (up to 48 tones) and to specify a SFDR as the ratio of the combined power of all of the multiple tones to the peak power of the largest spurious signal in the ADC output spectrum. The *noise power ratio* (NPR) is useful in applications such as mobile cellular radio where the spectrum consists of many narrow-band channels and where adjacent channel interference can degrade system performance. NPR defines how well an ADC can limit cross-talk between channels.

Harmonically related errors can further reduce the dynamic range of an ADC. For a signal of frequency f_0 aliased harmonics appear at frequencies $|\pm Kf_s \pm nf_0|$, $K = 1, 2, 3, \ldots$, where n is the order of the harmonic. The second and the third harmonics are the ones most usually specified, although some other harmonic may be specified as the worst harmonic. Harmonic distortion is specified with the input signal near full-scale (0.5 - 1 dB below the full scale to prevent clipping) [12]. Converter linearity can be described in terms *of total harmonic distortion* (THD). At low frequencies harmonic distortion generally results from unequal step sizes in ADC transfer functions. At higher frequencies it results from slew-rate, signal feed-through, non-linear resistors and capacitors. THD increases with either amplitude or frequency increase of the input signal [14]. *Total harmonic distortion plus noise* (THD+N) is the ratio of the rms value of fundamental signal to the mean value of total power of all harmonics plus all noise components (excluding DC) [12]. It is important to specify the bandwidth over which the noise is measured. For noise measurement bandwidth [0, $f_s/2$] THD+N is equal to SNRD.

In band-pass sampling applications where the maximum input frequency is higher than the sampling frequency, important specification is *analogue input bandwidth*. It can be characterised for full-power and small-signal conditions. While there is no universal definition, the most common defines it as the frequency where the amplitude of the ADC output falls to 3 dB below the maximum output level. Wide full-power bandwidth ensures a flat output when ADC is used in sub-sampling applications. As input frequencies approach full-power bandwidth other converter characteristics (THD, SNR) may start to vary considerably as the signal approaches the 3 dB frequency [14].

Distortions related to sampling time play a critical role in the wireless applications, sometimes being the limiting factor, more important than the dynamic range of the converter due to quantisation noise. *Acquisition time* is the time required for the circuit to charge if switched to a constant voltage. This error is usually specified to some high degree of accuracy. It is important for sample and hold (S&H) circuit to fully charge to a signal value and transition time must be allowed before switching to the hold mode. *Settling time* after the hold command is given as the time required for the circuit to disconnect and for any switching transients to die out. This time must be allowed before the ADC conversion command is given [10].

Aperture jitter is the variation in time of the exact sampling instant. It presents the most critical parameter for the band-pass sampling application. Aperture jitter is caused by external sources like sampling clock, or by internal sources when the S&H sampling switch fails to open at the precise time intervals. It effectively results in a noise level that can be characterised by the ratio of the sine wave power to the noise power caused by jitter (in $f_s/2$ bandwidth) given by

$$SNR_J[dB] = 20\log\left(\frac{1}{2\pi f_0 \sigma_a}\right)$$

where f_0 is the frequency of the sine wave being sampled and σ_α is the rms aperture jitter in seconds, assumed to be flat within $f_s/2$ bandwidth. Recent results on the aperture jitter requirements for the wide-band signals (e.g. CDMA signals) indicate that the loss due to the aperture jitter is much smaller for wide-band signals compared to that predicted for classical narrow-band signal analysis [15].

A number of factors may influence the required dynamic range of the converter. The most important is the quality of the link (usually expressed in Bit error Rate, BER) which effectively translates into overall SNR requirements, from which converter SNRD can be derived. One way to specify the required dynamic range is to determine the headroom of the converter, defined as the ratio of the input signal power required to saturate converter and the minimum signal level required for the given BER performance. Few specific factors, inherent to wireless systems, have to be considered in this process. The *peak-to-average ratio* of the input signal depends both on modulation format as well as filtering. This factor is critical for higher-level modulation schemes like M-QAM and for multi-carrier signals. *Interference level* at the input of the converter is the most critical factor determining the dynamic range. Two issues have to be considered: the level of the interference compared to the desired signal level and the frequency of the interference, which may be restrictive for the aperture jitter requirements. The level of interference is usually much higher than the signal and should not produce saturation of the converter. One of the design practices is to provide enough headroom for ADC to exceed the out-of-band rejection required within the bandwidth of the analogue filter [10]. At the same time a *fading margin* may be required. While automatic gain control (AGC) can compensate for the slow fading (shadowing), small-scale fluctuations due to multi-path are usually combated by combination of diversity, equalisation and coding. If these functions are to be performed digitally, the dynamic range of the converter must accommodate effects like burst-to-burst variation in TDMA systems.

6.2.3.2 Digital-To-Analogue Converter Specifications

As much as ADCs are critical for the receive path of the wide-band receiver, Digital-to-Analogue Converters (DACs) are essential for the transmitter path in software radio transceivers. New applications in wireless communications pose new requirements not only on component performance but also to the way the components are specified [9]. Standard time-domain specifications for linearity and resolution need to be supplemented with additional parameters to describe the performance in the frequency-domain. The performance of the DAC under the specified system operating conditions will not merely affect the performance of the target system, but will actually determine whether a particular modulation scheme or system architecture can be implemented at all.

Communication systems are most often analysed and characterised in the frequency domain. As a result, a high-speed DAC is primarily evaluated using specifications and characterisation data pertaining to its frequency-domain (ac linearity) performance with secondary consideration to its static (dc) and dynamic (time-domain) specifications. This is because both static and dynamic non-linearities manifest themselves in the frequency domain and contribute to a DAC's large-and small-signal distortion performance beyond what would be predicted by the process of sampling and quantisation [16].

Although the actual effects of dynamic non-linearities on a DAC's frequency domain performance are nearly impossible to predict, certain observations can be made. The *glitch impulse*, historically considered a key specification in DAC signal-reconstruction applications, is a measure of the size of the earliest part of the transient generated when a DAC switches between two output levels. Glitch impulse is commonly associated with various time skew errors within the DAC - non-simultaneous action of the current switches - and is usually measured as the area under the transient. Its code dependency, as opposed to its mere existence, is just one of several sources of frequency-domain distortion.

Output *slewing*, if it has code-dependent rise and fall times, is also a major contributor to distortion. In fact, the code dependency of a DAC's settling time has far more impact upon its frequency domain performance than the settling time itself. Output slewing often limits the performance of DACs in multi-tone applications with high peak-to-average signal ratio of a reconstructed waveform [16]. In applications like CDMA, dynamic errors ultimately define the noise floor of the spectral mask beyond what is predictable from the effects of quantisation. After settling, the steady-state static non-linearities are dominant at lower frequencies ultimately setting the limit on DAC performance.

Manufacturers typically describe DAC imperfections with four parameters: offset, gain error, differential non-linearity (DNL), and integral non-linearity (INL). Output *offset* is usually defined as a constant dc offset in the transfer curve of analogue output versus digital input, and therefore has no effect on the frequency-domain characteristics of the output. *Gain* defines the full-scale output of the DAC in relation to its reference circuit and represents a power scaling issue in the transmit path of most communication systems. Digital communication systems based on quadrature modulation schemes - which use two DACs to create separate base-band I (in-phase) and Q (quadrature) signals - should evaluate both gain and offset errors because of the need for tight matching of the I and Q channels.

Differential non-linearity (DNL) and *integral non-linearity* (INL) are static non-linearity specifications, which provide some useful information about a DAC's ultimate limit in frequency-domain performance. DNL is the worst-case deviation from an ideal 1-LSB step between adjacent codes; it is typically measured in least-significant bits (LSBs) and may be positive or negative. DACs having a DNL specification of greater than 1 LSB are non-monotonic. Both INL and DNL specifications are very important for multi-carrier or spread-spectrum applications. The reconstructed output in these applications spans a wide dynamic range and typically exhibits a high peak-to-average ratio in the time domain. As a consequence, these applications often exercise a DAC's small-scale linearity throughout the DAC static transfer function.

In the next few paragraphs some of the various common DAC specifications are reviewed. *Spurious-free dynamic range* (SFDR), a very often quoted DAC specification, defines the difference, in decibels, between the rms power of the fundamental and the largest spurious signal within a specified frequency band. The spur does not have to be harmonically related to the fundamental (although it often is) nor does the frequency band necessarily have to include the DAC's baseband Nyquist zone - the frequency range from dc to half the update rate.

Signal-to-noise-and-distortion (SNRD) ratio is the most encompassing frequency-domain specification since it includes all of the noise and distortion that falls within the Nyquist zone. SNRD is defined as the ratio of the power of the fundamental to the power of the rms combination of the noise and distortion, expressed in decibels. SNRD provides valuable information about a DAC's dynamic range. DAC noise sources such as DNL errors, thermal noise, digital feed-through, and clock jitter can all degrade a DAC's noise performance, thereby reducing its dynamic range. A qualitative measure of a DAC dynamic range is the DAC effective number of bits (ENOB) - its useful resolution under the conditions of the SNRD measurement - given by

$$ENOB = \frac{SNRD - 1.76}{6.02}$$

Total harmonic distortion (THD) is a more complete specification of a DAC's distortion, or ac linearity performance, since it includes most of the harmonically related spurs as well as aliased - folded back - harmonics within the DAC's

Nyquist bandwidth. The THD is the ratio of the rms sum of the first five or six harmonics to the rms value of the output signal, which is usually expressed in decibels. In practice, the measured THD will often be 1-3 dB below the SFDR (to Nyquist) measurement obtained from the same spectral plot because the dominant spur typically contains more than half of the total rms distortion power. The dominant spurious signal and harmonic at higher output frequencies typically is a second or third harmonic, which could also be aliased back into the Nyquist bandwidth.

6.2.3.3 Digital Intermediate Frequency Signal Processors specifications

To assure that the level of performance for a specific air-interface standard is achieved within a Digital IF Signal Processor (DIFSP) the NCO - which is located at the front end of a DIFSP and performs the tuning and quadrature split - or separation of In-phase and Quadrature signal components - must have adequate noise floor and overall spurious performance and sufficient tuning resolution to assure maintenance of signal integrity [17]. This is because spurs generated by the local oscillator are convoluted with the input signals through the mixing process.

For example, if there were two input signals at 90 dB separation, spurs at -100 dB generated by the NCO would convolute with the larger signal to produce images only 10 dB below the smaller signal. Whether or not these images could be attenuated by the filtering process depends in large part on the location of the image itself, which is largely dependent on the overall frequency plan of the system.

The overall tuning *resolution of the NCO* - defined by the number of bits within the circuit portion of the circuit - is also important because it will determine the centre point of the bandwidth being processed [17]. Errors in tuning accuracy can result in truncation of the signal being processed, which is obviously undesirable. Once the tuning and mixing functions have been completed, the complex data signal (In-phase and Quadrature components) has been properly conditioned for the initiation of the data rate change and filtering process.

Another important function provided by a DIFSP is the multi-rate signal processing [17]. For base-station applications in which multiple transmission protocols are being processed by a piece of hardware, the presence of a *resampler* (or data rate change unit) provides the base-station designer with a good tool to simplify overall system design. For example, the IS-95 CDMA system requires precise timing related to the 1.2288 Mcps (synchronised to GPS). In order to receive an IS-136 TDMA (24.3 ksps) signal with the same ADC without a data rate change unit, a 398.131 Msps common multiple of both system data rates would need to be used - assuming the DIFSP output data rate is four times the chip rate to assist in the correlation process. The situation only gets worse in a receiver design accommodating both third generation family standards (3.84 Mcps and 3.6864 Mcps) and the mature GSM (270.8333 ksps) with a common clock frequency well above one gigahertz. With the emphasis being placed on software defined radio (multi-standard, multi-mode) base-stations, the presence of a data rate change unit (multi-rate processing) enables basestation designers to design their systems for the highest data rate possible for the protocols being considered and use the DRSP to adjust data rates down as needed to meet air-interface specific needs.

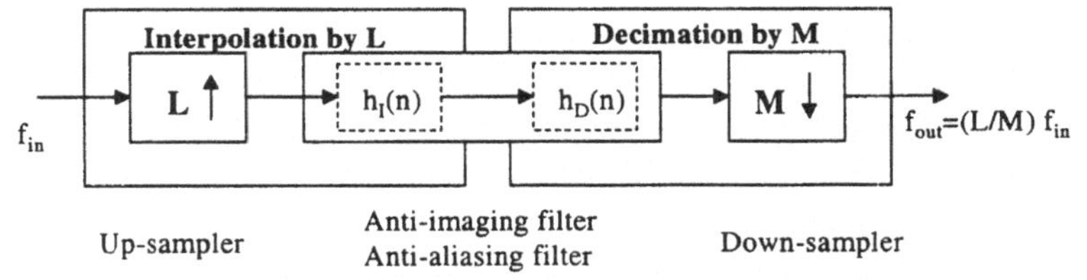

Fig.6.2.2 Block diagram of a multi-rate filter.

Multi-rate signal processing is applied whenever there is a significant ratio between the sample rate of a sampled data signal and the bandwidth of that same signal [18]. A resampler allows the output sample rate to have an integer or non-integer relationship to the input sample clock. Furthermore, it allows for the use of a common sample clock to be used when receiving multiple signal types. A resampler can be thought as a multi-rate filter. Multi-rate filters can be used to increase, to decrease or to preserve the sample rate of a signal being processed. A filter can have two or more sample rates, an input rate and an output rate. It is possible that the filter operates internally at rates that are neither the input nor the output rate. The unique feature of a multi-rate scheme is that the resampling operation occurs within the filter rather than after the filter for a down-sampler or prior to the filter for an up-sampler [19]. A multi-rate filter is composed by an interpolator followed by a low-pass filter and a decimator (Fig.6.2.2). The interpolation process creates images of the original signal at fixed multiples of the original data rate. The low-pass filter can be recursive (IIR) or non-recursive (FIR). Most filter designs are non-recursive for ease of access to linear phase performance. The decimation process allows for non-integer decimation.

Typically, there are three stages to the design of digital filter responses for passband filters. First, the ideal filter response is specified. Next, a floating-point response is designed. Finally the floating point coefficients are quantized to yield a fixed point response. The *coefficients resolution* - the number of bits that a floating point coefficient is quantized - plays crucial role to the DIFSP filtering performance. Converting floating-point coefficients into fixed point coefficients requires quantizing the coefficients and calculating the frequency response of the filter to verify that the filter meets the required specification. If it does not, the number of coefficients bit has to be increased.

6.2.4 Transceiver Components for Software Defined Radio

Air-interfaces specify architecture and component requirements, especially for data converters. Specific air-interface standards such as GSM and IS-95 determine the performance requirements for any given transceiver architecture. Adjacent channel blocking specifications, minimum sensitivity and operator spectrum allocations, are among the concerns that a radio architect must consider when specifying the radio design and data converters requirements.

Recent advances in high-speed data converters and digital IF components enable the realization of re-configurable radio systems. Advanced mixed signal and digital IF components provide improved flexibility and increased performance with respect to dynamic range and selectivity. Traditionally, data converters emphasis has been put on ADCs. Recently, DAC specifications have been revisited in the light of wide-band multi-carrier transmission requirements. In addition, components that provide multi-rate signal processing functions (decimators, interpolators), digital down/up conversion capabilities and a certain level of programmability FIR filters enable the efficient realization of transceiver functions in terms of power consumption, minimal component count and compactness.

DSPs with features optimized for communication applications complete the digital wide-band transmit and receive path. These DSPs are optimized for computational numeric processing applications, with large on-chip memory, a rich instruction set, and support for multitasking and parallel processing capability. Details of the advanced DSPs are not covered in this work. Technology requirements and software engineering considerations for the software radio can be found in [20]. In the following sections IF mixed signal and digital components are described and illustrative performance figures are shown for base-station transceivers design.

6.2.4.1 Receiver Components

ADC functionality can be covered either by flash or sub-range or sigma-delta converters. While sigma-delta converters may be preferable for mobile terminal applications, in terms of power consumption and available oversampling, they are not suitable for multi-channel base station applications [21]. Major approaches usually belong to one of two categories: a) perform sub-sampling operation at IF after channelisation in the analogue domain or b) digitize a wide portion of the band at IF and perform channelisation digitally [11]. Both approaches will be described using a dual channel with gain ranging ADC (AD6600) and a wide-band IF sampling ADC (AD6640) as examples.

A dual-channel mixed signal receive chip (AD6600) [22] is designed to directly sample signals at frequencies up to 250 MHz. The device includes an 11 bit 20 Msps ADC, input amplifiers, automatic gain ranging circuitry, a 450 MHz

Track and Hold (T/H), digital Received Signal Strength Indicator (RSSI) outputs, references and control circuitry. Each of the two input channels has its own attenuation stage and attenuation levels are set by the on-chip RSSI circuitry. It can be configured for several different applications. The main application of the chip is the dual diversity TDMA or CDMA receiver, where each analogue channel originates from a different antenna. However, applications can be extended to two independent IF signals. Finally, the ADC can sample one input signal at the full clock rate of the chip to achieve maximum processing gain from over-sampling.

The block diagram of a dual-channel with gain ranging ADC is depicted in Fig.6.2.3. It consists of three stages. The first stage includes a pair of phase compensated attenuators, followed by the output selection multiplexer. The attenuation is controlled with on-chip switches between 0 and -30 dB in 6 dB steps. The matching between the gain settings is better than 0.5% and phase mismatch is less than 0.2 degrees. The input impedance does not change with attenuator settings, so there is no amplitude modulation (AM) to phase modulation (PM) distortion. The second stage is a wide bandwidth 11 bit ADC. The third stage includes a high speed synchronous peak detector and RSSI control interface. The peak detector has five reference points, each 6 dB lower than the previous one. 6 dB of digital hysteresis eliminates level uncertainty due to noise and amplitude variations at the threshold points. The RSSI tracks the IF envelope of the input signal, one clock cycle before the conversion is made. It always allows 6 dB of ADC headroom to prevent clipping, until the last attenuator is selected.

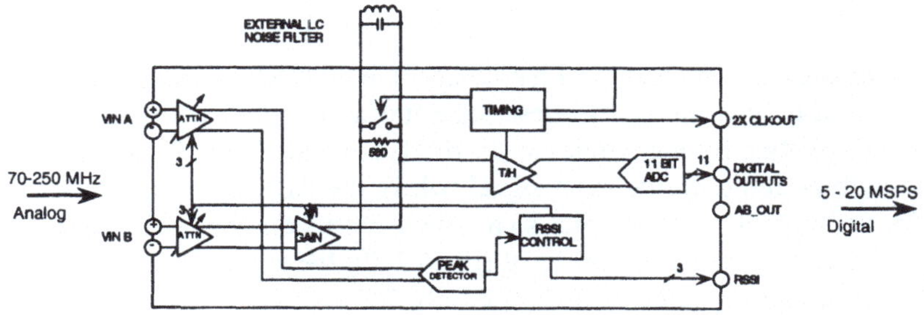

Fig6.2.3 Block diagram of dual channel gain ranging ADC with RSSI.

Aperture jitter performance is of particular importance for the over-sampling converter. The theoretical SNR at 250 MHz and 0.7 psec jitter is limited to 61 dB. However, lower aperture jitter of 0.3 psec can be achieved. Noise sources in the device include not only quantisation noise, but also gain ranging and ADC thermal noise. SNR due to all noise sources is 61.3 dB, which is extended by 30 dB due to gain ranging. Final dynamic range is application dependent due to the amount of over-sampling available. Analogue input bandwidth is 1GHz, worst spur -70 dBc for a 100 MHz input, and -53 dBc for a 250 MHz input at -1 dBFS.

State-of-the art high speed, high performance, monolithic 14-bit analogue-to-digital converters are technology of choice for base stations. As an example AD6644 [23] is a 65 Msps converter designed to address the requirements of wide-band multi-channel receivers. It maintains 95 dB spurious - free dynamic range (SFDR) over a bandwidth of 25 MHz. All necessary functions, including track-and-hold (T/H) and reference are included on-chip to provide a complete conversion solution. This ADC provides a signal-to-noise ratio of 75 dB. The block diagram of the converter is depicted in Fig.6.2.4. It employs a three-stage sub-range architecture. All necessary functions, including track-and-hold (T/H) and reference are included on-chip. The AD6644 provides CMOS compatible digital outputs at 65 Msps.

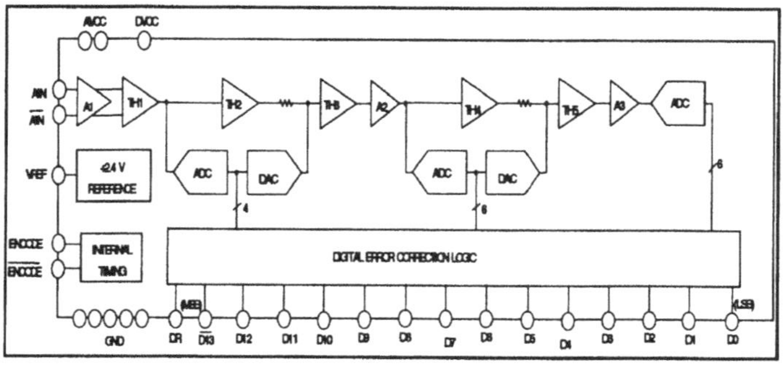

Fig.6.2.4 Block diagram of wide-band sampling converter.

Worst harmonics (second and third harmonics) are 80 dBc, and fourth and higher order are 85 dBc. Important to wireless communications applications is the multi-tone SFDR specified for eight tones at -20 dBFS to be 90 dBFS. Fig.6.2.5a shows the case with no dither to the input signal, where the ADC performance shows a noise floor of around -100 dBc, plus some spurs throughout the spectrum at -80 to -90 dBc. By adding dither to the signal (Fig.6.2.5b) the spurs - that rise from the noise floor - are greatly reduced [24]. The fully differential analogue input stage is specified for frequencies up to 70 MHz thus enabling IF sampling of wide bandwidths 5-25 MHz.

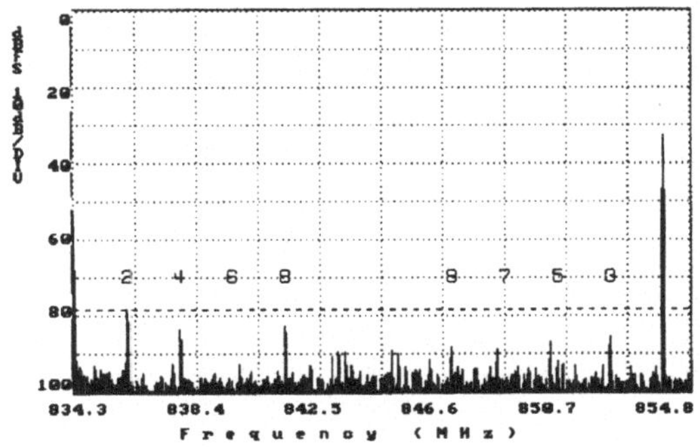

(a)

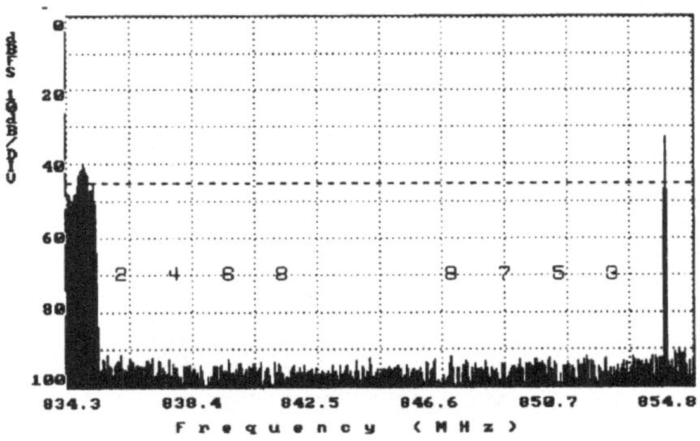

(b)

Fig.6.2.5 SFDR performance of wide-band sampling converter.

Regardless of the IF sampling strategy - narrow-band versus wide-band and single-channel versus multi-carrier - digital receive signal processors are utilized to complete the receive chain and provide the final IF stage of a receiver in a digital domain before the DSP processing [11]. DRSPs are designed to bridge the gap between high speed ADCs and DSPs. Current implementations of digital communication system receivers use analogue filters to select the desired channel followed by a Nyquist rate analogue-to-digital converter. To achieve greater programmability in a multi-standard receiver, it would be attractive to select the desired channel in the digital domain. To remove the analogue channel-selection filtering completely would require a wide dynamic range ADC. The over-sampled ADC would digitise both the desired and undesired channels and remove the undesired channels using a digital decimation filter of a DRSP. State of the art DRSPs operate at an input signal rate of 62.5 - 80 Msps.

A typical digital receive signal processors has one or four channels and each channel consists of three cascaded signal-processing elements: a frequency translator from intermediate frequency to base-band section, cascaded fixed coefficient decimating filters, and a programmable coefficient decimating filter (single channel DRSP Fig. 6.2.6a and multi-channel DRSP Fig..5.2.6b). Most of DRSPs support three input data modes, single channel real, single channel complex and dual channel real. Technical characteristics of the state of the art programmable digital DRSPs, (Analog Devices, Graychip, Harris Semiconductor) are present in [25] [26] [27].

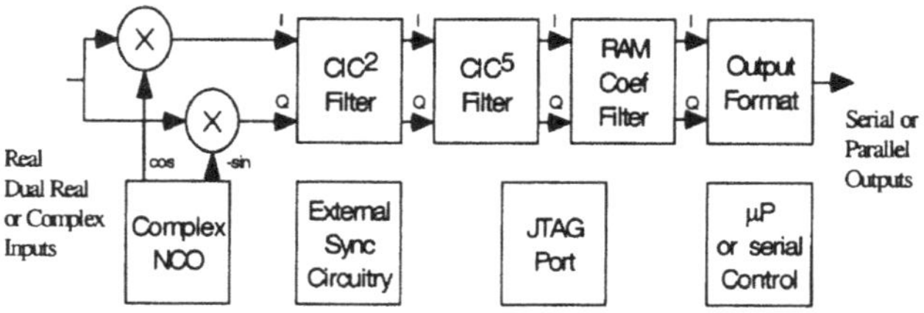

(a)

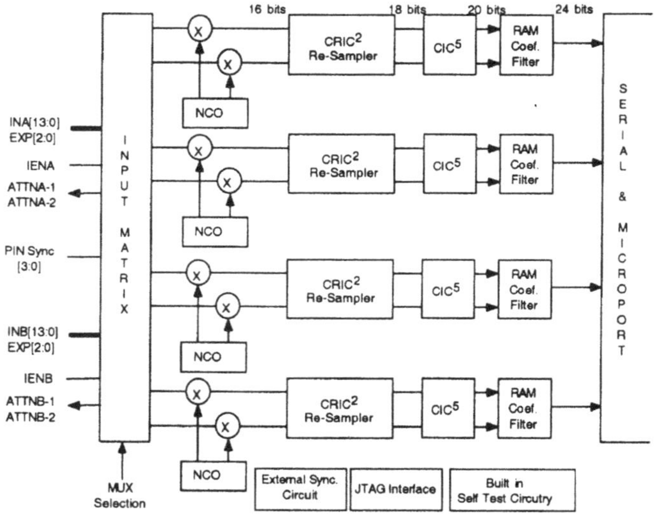

(b)

Fig. 6.2.6 Block diagram of single-channel and quad-channel DRSPs.

The high resolution Numerically Controlled Oscillator (NCO) allows a single carrier to be selected from a high speed data stream. Frequency translation in a DRSP is accomplished with a 32-bit complex NCO. Real data entering this stage is separated into in-phase and quadrature components. This stage translates the input signal from a digital intermediate frequency (IF) to base-band. Phase and amplitude dither may be enabled on-chip to improve spurious performance of the NCO. A phase offset word is available to create a known phase relationship between different DRSP channels (multi-channel devices). Today's technology provides quadrature matching of the sine and cosine waveforms to a great resolution of 0.012 Hz. Since the sampling is done at IF, there is no gain imbalance in the base-band I and Q prior signal digitisation. Using identical digital logic on both paths ensures identical gain and phase throughout the digital processing down to the LSB. Thus the signal of interest located within the wide-band input to the demodulated without the performance degradation of the phase mismatch or I/Q gain imbalance. State of the art DRSPs have NCO with worst spur better than –105 dBc.

High dynamic range decimation filters with wide decimation rates allow both narrow-band and wide-band carriers to be extracted. The overall filter response for a DRSP is the composite of all stages of decimating filters, which are implemented differently by various Integrated Circuits manufacturers. Filtering is usually divided into multiple filter stages - usually three stages - to provide anti-alias, matched filtering and decimation for multi-standard applications. Among many filter designs there are three general filter structures that are used most commonly, a Cascaded Integrator Comb (CIC), a Halfband filter and a Finite Impulse Response (FIR) filter [17]. Each has its own strengths and weaknesses.

The CIC represents an efficient filter implementation that is characterised by a gradual transition band roll off - defined by the number of stages used in the implementation - very good alias attenuation in the band of interest and a relatively flat pass-band over a limited range. The downside is that its output must be over-sampled - typically by at least eight times - in an attempt to assure that the bandwidth of interest stays in the flattest portion of the passband possible, as well as to maximise alias attenuation characteristics of the filter. It is usually implemented as a fixed coefficient filter with programmability being provided only in terms of number of stages and the amount of decimation desired. For these reasons, the CIC is normally the first filter used in the chain because it provides an excellent method not only of decimating/interpolating the bandwidth of interest, but also in attenuating out-of-band signals that fold back in because of Nyquist sampling.

The Halfband filter is also typically implemented as a fixed coefficient filter; however, it also represents a less efficient implementation than the CIC. Its frequency response is characterised by a very flat pass-pand. The output must also be over-sampled - typically by a minimum factor of 4 - and the normal implementation calls for a decimation ratio of 2.

The poly-phase FIR is the most flexible filter implementation. It is fully programmable, has a low over-sample ratio and typically supports integer decimation factors. Although less efficient than the CIC because it needs more coefficients, however, given enough available coefficients can provide the user with steep transition bands and good out-of-band attenuation. An important specification for the FIR filter is the number of bits in the coefficient available as it defines its out-of-band attenuation performance. Programmable FIR filter provides matched filtering in a single stage and re-configuration for software radio. Besides matched filtering, A DRSP provides the capability of a shared execution of some intense computational processes for frequency offset and symbol timing estimation.

Filtering characteristics of the state-of-the-art DRSPs are presented in this paragraph. In *AD6624* [25] there are two fixed coefficient, high speed decimating filters. The first filter (a second order resampling CIC) reduces the sample rate by a programmable rational ratio of L/M, where $L \leq 512$, $M \leq 4096$ and $L < M$. Following CIC2 is the second fixed-coefficient (fifth order CIC) decimating filter. This filter further reduces the sample rate by a programmable ratio from 1 to 32. Following the CIC filters is a 160-coefficient FIR filter with decimation rates programmable from 1 to 32. *HSP50214* [26] uses a decimating (4-32) fifth order CIC filter to the data at the output of the NCO. The CIC filtered data is processed by up to 5 decimate by 2 half-band filters. The half-band filters are followed by a 255 coefficient programmable FIR filter. *GC4114* [27] mixer outputs are filtered and decimated by 8-16,384 using a fourth order CIC filter. The CIC outputs are filtered by two stages of filtering. The first stage is a 21 coefficient decimate by 2 compensating filter with two sets of fixed coefficients. The second stage decimation by 2 or 4 programmable FIR filter uses 63 coefficients.

6.2.4.2 Transmitter Components

In the past wide-band transmitters were usually devoted less attention than the receiver counterpart. In light of emerging third generation wireless systems, multi-channel Digital Transmit Signal Processors (DTSP) are of importance for multi-standard base station allowing multiple modulation schemes, programmable shaping filter and programmable interpolating filters. Furthermore, some of the functions necessary for transmit beam-forming can be integrated and efficiently realised depending on the chosen architecture. A DTSP allows for configuration of predefined operation modes, with capability of user defined transmission schemes and transmission modes [9].

Digital Transmit Signal Processors (alternatively called programmable digital up-converter) allow easy base-station re-configuration for software radio. A quad-DTSP (Fig.6.2.7) is designed to bridge the gap between DSPs and high speed DACs. It has three signal processing elements: a programmable Coefficient Interpolating Filter, Programmable Cascaded Integrator Comb (CIC) Interpolating filter, and a Numerically Controlled Oscillator/tuner (NCO). Technical characteristics of the state of the art programmable DTSPs, (Analog Devices, Graychip, Harris Semiconductor) are presented in [28] [29] [30]. Wide-band DTSPs comprise N (N ≤ 4 in current state of the art devices) identical digital up-converters operating up to 65 Msps complete with synchronisation circuitry and cascadable wide-band channel summation. Complex coefficient multiplication at the output provides transmit phased array capability.

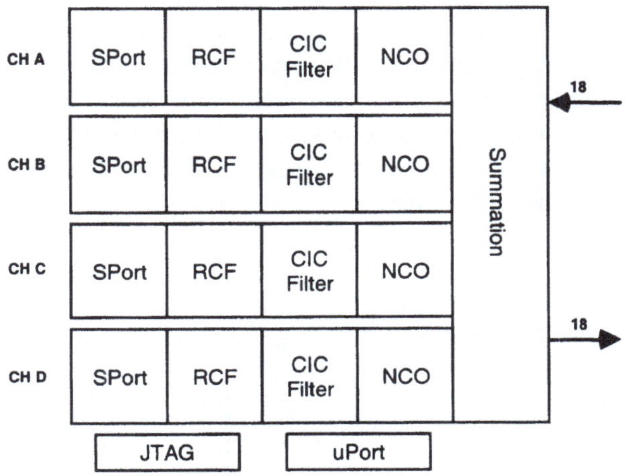

Fig.6.2.7 Block diagram of a quad DTSP.

The wide range of interpolation factors in each filter stage makes a DTSP useful for creating both narrow-band and wide-band carriers in a high-speed sample stream. The interpolating filters remove unwanted images of signals sampled at a fraction of the wide-band rate. For large interpolation factors, thi DTSP processing gain allows a 12 - 14 -bit DAC to express the sum of multiple signals sampled at a lower rate without significantly increasing the noise floor about each carrier. In addition, the programmable coefficient initial filtering stage allows anti-imaging and pulse shaping functions to be combined in a single, cost effective filter. A range of integer factor related to different manufacturer implementation could interpolate the output of a programmable coefficient filter. The interpolating programmable filter pre-compensates for the droop associated with the CIC filters that follows it. Some DTSPs support pulse shaping and interpolation using an on-board Look Up Table for a variety of digital modulation schemes as GMSK and QPSK. The Look-Up-Table is downloaded from a base-station micro-controller to the re-configurable programmable coefficient filter. A typical DTSP provides further interpolation with multi-stage high speed cascaded integrator comb (CIC) filter. The CICs have a programmable response designed to achieve the flattest CIC stop-band response at the lowest possible power consumption. The output rate of the multi-stage CIC filter stage is equal to the master clock rate.

A high resolution NCO (32 bits) allows flexibility in frequency planning and supports both digital and FM air interface standards. The high speed NCO is used to tune a quadrature sampled signal to an IF channel, or the NCO can be directly frequency modulated at an IF channel. Multi-carrier phase synchronisation pins and phase offset registers allow intelligent management of the relative phase of the independent RF channels. This capability supports the requirements for phased array antenna architectures and management of the wide-band peak / power ratio to minimise clipping at the DAC. In state-of-the-art DTSPs spurs are lower than the spurs of reconstructing DAC that will follow in the transmitter chain. The wide-band input and output ports allow multiple DTSPs to be cascaded into a single DAC (Fig.6.2.8). The master clock for the entire system is based on the DAC clock rate (up to 65 Msps).

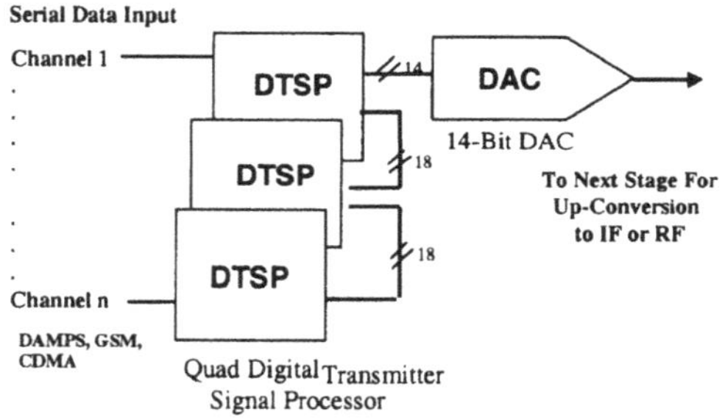

Fig.6.2.8 DTSP Daisy chaining for additional channels.

Filtering characteristics of current state-of-the-art DTSPs are presented in this paragraph. *AD6622* [28] uses one programmable FIR filter of 128 coefficients providing an interpolation factor of 4, 5, 6 and 8. AD6622 provides further interpolation by a factor of 1 – 1,024 with two stages of high speed cascaded integrator comb filter. The CICs have a programmable response designed to achieve the flattest CIC stop-band response at the lowest possible power consumption. The output rate of the CIC filter stage is equal to the Master Clock rate. *GC4114* [29] uses two programmable FIR filters. In the first FIR 63 coefficient filter input data is interpolated by a factor of 2. The second interpolate by 2 filter is a 31 coefficient compensating filter which pre-compensates for the droop associated with the CIC filter that follows it. The third cascaded integrated comb (CIC) filter interpolates by another factor of 8 – 16,384. *HSP50215* [30] has a programmable 256-coefficient FIR filter. The filter serves to both pulse shape and interpolate the sample rate to 4, 8 and 16 times the input sample rate. The output of the programmable filter is routed into a interpolation filter. The interpolation filter interpolates by a factor set in the re-sampling NCO. The output rate of the CIC filter stage is equal to the Master Clock rate.

To complete the wide-band digital transmission path a high-performance DAC (optimised for SFDR and THD) is required. Example of such a DAC is the AD9754 with 14-bit resolution and 125 Msps (Fig..6.2.9) [31]. The wide dynamic range of a high-performance DAC is of great benefit in multi-tone applications where many channels of information are being transmitted over several MHz of bandwidth and the peak-to-average output signal ratio is high, with average output signal level at a fraction of the full scale range. Typical DC specifications include 2.5 LSB INL and 1.5 LSB DNL. Offset error is 0.025% of the full scale range (FSR) and gain error is less than 2% of FSR. Dynamic characteristics include output settling time of 35 nsec and glitch impulse of 5 pV-sec.

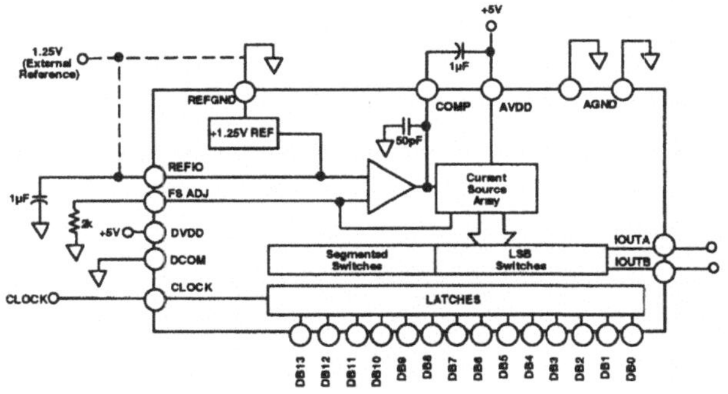

Fig.6.2.9 Functional DAC block diagram.

SFDR of the AD9754 is characterized as a function of the output frequency for different input signal levels in Fig.6.2.10. Many wide-band signals are not used at the full scale. In fact amplitudes that are close to full scale have very low probability and larger SFDR can be used for the transmission signals.

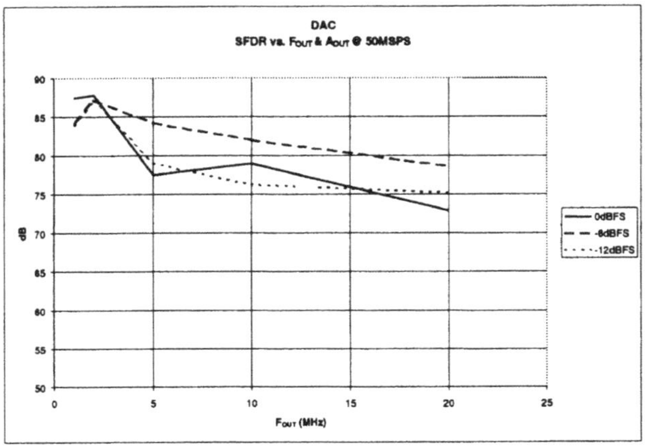

Fig.6.2.10 SFDR of a high performance DAC for 50 Msps.

Multi-carrier software radio architectures take the benefit of interpolating Digital-to-Analogue Converters. State-of-the-art DACs (AD9772) can reconstruct full-scale waveforms with bandwidths as high as 63.3 MHz (Fig.6.2.11) while operating at an input data rate of 150 Msps [32]. This is 14-bit oversampling device optimized for base-band or IF waveform reconstruction applications requiring exceptional dynamic range. It is a low distortion DAC with a digital interpolation filter (interpolation by 2) and clock multiplier. For base-band applications, the digital interpolation filter provides a low-pass response hence providing up to a three-fold reduction in the complexity of the analogue reconstruction-filter (Fig.6.2.12a). It does so by multiplying the input data rate by a factor of two while simultaneously suppressing the original upper in-band image by more than 73 dB. For direct IF applications, the 2x digital interpolation filter response can be reconfigured to select the upper in-band image (i.e. high pass response) while suppressing the original base-band image (Fig. 6.2.12b). To increase the signal level of the higher IF images and their pass-band flatness in direct IF applications, the AD9772 also features a "zero stuffing" option in which the data following the 2x interpolation filter is up-sampled by a factor of two by inserting mid-scale data samples.

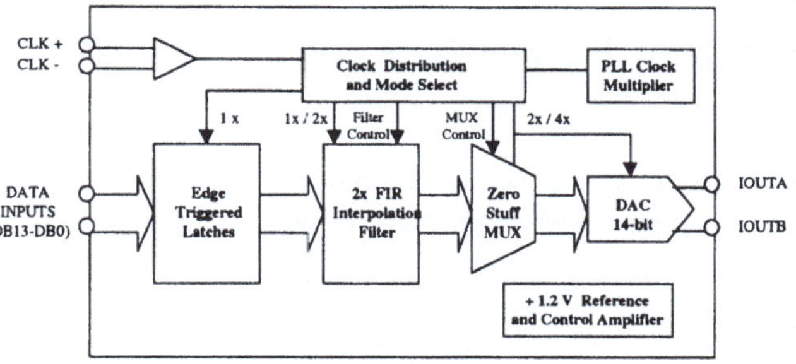

Fig.6.2.11 Block diagram of an interpolating DAC.

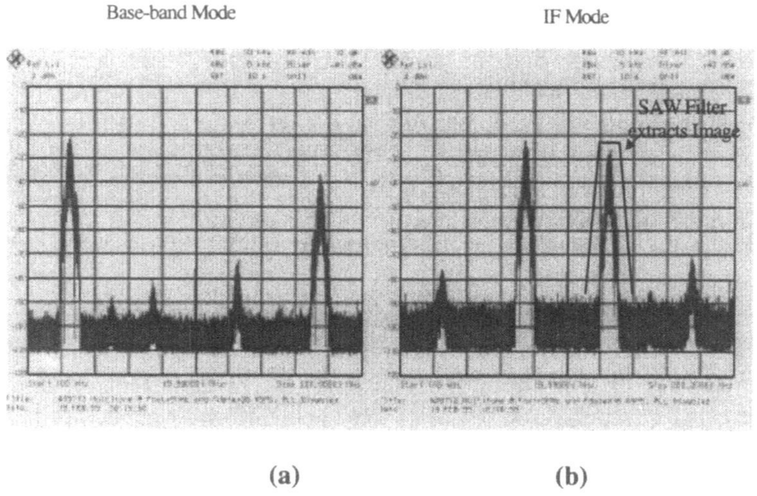

Fig.6.2.12 Base-band versus IF mode of an interpolation DAC.

6.2.5 Software Radio Transceiver Architectures

Demands on today's cellular infrastructure equipment are different than those that were placed on systems deployed more than ten years ago. The ability to handle emerging 3G standards and maintain compatibility with existing analogue and digital standards is critical. The rapid expansion of existing and new networks by cellular operators has resulted in further cell-site constraints that require the use of smaller and less costly basestation systems. In the following paragraphs the evolution of basestation transceiver architectures from the early single-channel baseband data conversion to the recent software radio IF data conversion architectures are presented. Transceiver designs with key components enabling software radio are discussed.

6.2.5.1 Receiver Architectures

Traditional receiver architectures use analogue circuits for signal down-conversion to a baseband signal. This architecture (superheterodyne receiver) uses double IF analogue down-converters with baseband sampling ADC's i.e up to 1 MHz bandwidth with sample rates up to 10 Msps (Fig.6.2.13a). This architecture then migrated to double IF, I and Q demodulation using baseband or low IF sampling ADC technology borrowed from handset integration development.

The next jump in receiver architecture evolution, (Fig.6.2.13b) has been to more direct, low/high IF sampling I and Q demodulation using high speed, wide-band, high resolution ADCs. These ADCs would have been typically 10 to 12 bit resolution, sample rates up to 20 Msps, with input bandwidths up to the Nyquist frequency. These architectures are reducing the size of radios by eliminating analogue IF mixer and filter stages with these functions being carried out using digital signal processing.

The next stage of this evolution is now rapidly moving towards even more wide-band architectures, (Fig .6.2.11c, 6.2.11d) using the state of the art ADC technology for high IF sampling (250 MHz) imposed by base-station manufactures for multi-mode receivers [7] [33] [34]. To minimise complexity of the receiver the analogue front-end of the receiver is fixed (non-parametrisable) and system clock for the ADC and digital signal processing hardware is the same (fixed) [21]. Bandwidth of the analogue front-end is determined by the widest bandwidth of the air-interface standard supported. Out-of-band signals should be attenuated enough so that aliases in the band of interest are well below desired signal level. An important implication is that ADC has to provide adequate analogue input bandwidth to accommodate IF signals [9]. These are sub-sampling techniques with even more digital signal processing carried out to perform channel selection, digital down conversion and channel filtering before interfacing to high speed DSP's for demodulation.

Base-band Sampling (Super-heterodyne)

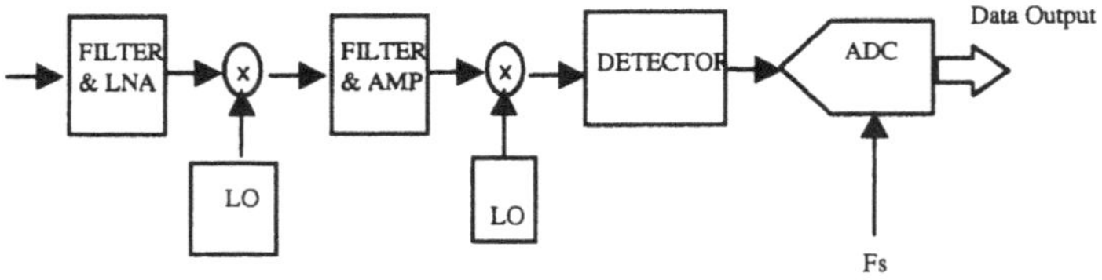

(a)

Base-band, Analogue Down-conversion

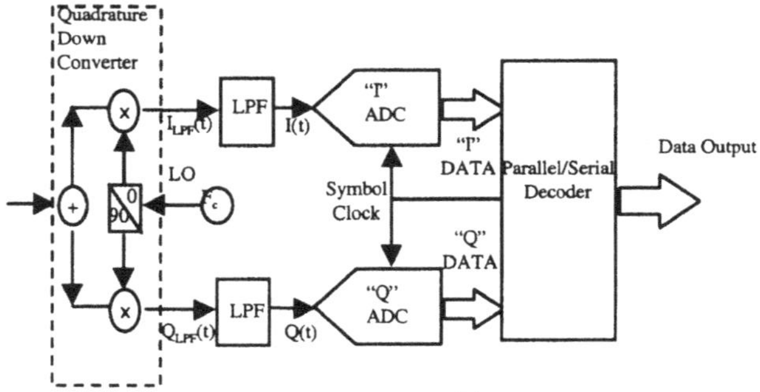

(b)

IF Under-Sampling (Direct IF)

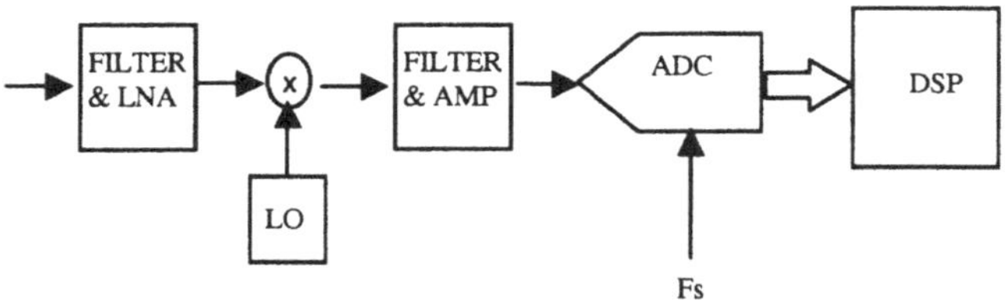

(c)

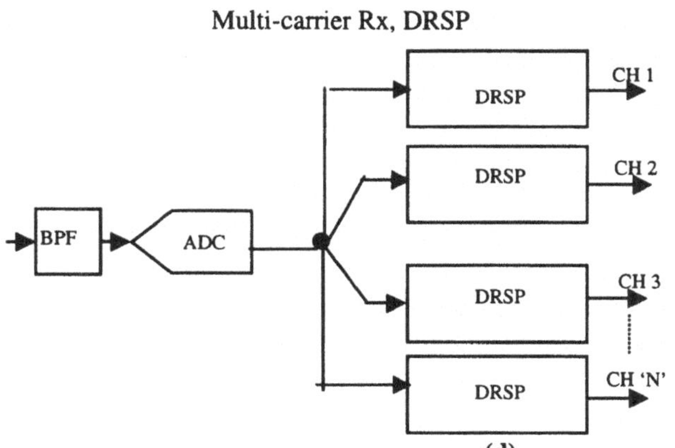

Fig.6.2.13 Receiver architectures.

Existing second and third generation base-station architectures based on IF sampling, use one radio per radio-frequency (RF) carrier to process cellular and personal communication signals. In narrow-band single channel applications a high over-sampling ratio is desired. This spreads the quantisation noise of the ADC over a wider spectrum and allows the digital filtering of the DRSP to remove much of this noise. This increases the effective dynamic range of a receiver if the sampling rate is chosen such that the signal harmonics and/or inter-modular distortion (IMD) products fall out of the band of interest. In this case these spurs could be filtered by the DRSP and the quantitisation noise would be the dominant dynamic range limitation of the proposed receiver solution. A DSP is then used to perform the demodulation of the digital channel. This has the advantage of allowing for in system configuration options and can even allow for improved modulation techniques to be applied in the future. If many carrier frequencies are used this leads to a high degree of hardware replication.

Recent advancements in RF and digital signal processing (DSP) technologies lead to new base-station architecture trends that process entire bands of cellular signals using a single tuner and extensive DSP techniques. Wireless systems designers refer to this as wideband software defined radio. Selection of the bandwidth to be digitised depends on the maximum sampling rate and dynamic range of the ADC. Selection of the digitisation bandwidth versus the sampling rate of the converter ultimately defines the oversampling processing gain. Analogue front-end has direct impact on the ADC dynamic range [7]. In the case of single-channel IF sampling analogue front-end provides sufficient attenuation of multiple signals, and blocking attenuation becomes critical for the dynamic range. In the case of wide-band IF sampling front-end linearity is critical for inter-modulation generation and selectivity is critical for blocking attenuation. A high sampling rate in excess of 65 Msps would allow receivers to capture a wide swath of the RF or IF spectrum in a multi-carrier set-up. Meanwhile, a 14-bit converter offers a four-fold improvement in dynamic range over 12-bit converter, making it possible to capture very weak RF signals - from distant handsets in the cell - in the presence of stronger ones. The wide-band approach allows the software radio to meet the spectral requirements of many different standards using a single analogue RF front end. The wide-band approach also offers a multi-channel advantage over the single-channel approach moving the transceiver design towards the realisation of bandwidth-on-demand. In the following paragraphs, two software radio receiver architectures are presented which enable co-existence of older analogue wireless systems and matured digital second-generation wireless systems, enabling the seamless evolution to third generation systems.

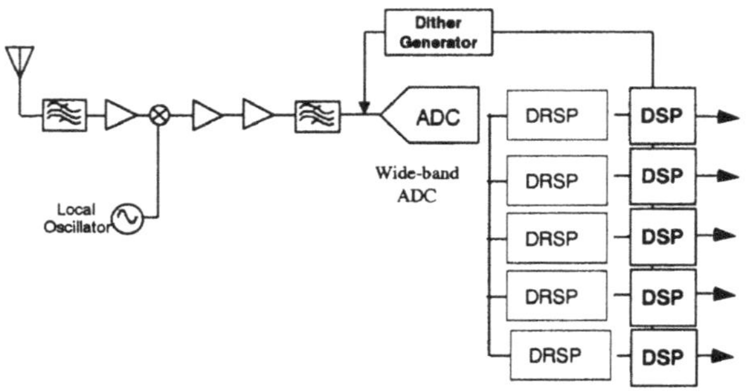

Fig.6.2.14 Receiver architecture for software radio.

Unlike a traditional single-channel receiver, the majority of channel selectivity in a wide-band multi-carrier receiver is performed by a digital receive signal processor that typically follows the ADC. Fig.6.2.14 shows a GSM multi-carrier base station transceiver where a wide-band 12-bit 65 Msps ADC is used instead of a narrow-band ADC [15]. The ADC samples a bandwidth of 25 MHz at IF of 70 MHz. Several DRSPs are pipelined off of a single ADC and used to tune in and isolate separate channels from the digitised wide-band data. The digitisation of this wide bandwidth allows many channels to be digitised using the same ADC and IF circuitry. Each DRSP channel processes only one GSM carrier and will process the data at the ADC sample rate. Each AD6624 output rate is 270.8333 ksps (1 sample per symbol). The total decimation rate achieved within a DRSP channel is 240. It is split among the first filter stage (second order CIC filter with a decimation ratio $M_{CIC2}= 2$), the second filter stage (fifth order CIC filter with a decimation ratio $M_{CIC5}=20$) and the he programmable RCF filter (92 coefficient FIR filter, providing a decimation ratio $M_{RCF}= 6$).

The DRSP provides a great deal of selectivity by mixing down a channel of interest and filtering the out of band noise and adjacent channels. Unlike the single-carrier solutions it is much more difficult to place the spurious content out of the band of interest because more of this bandwidth is used due to the larger number of carrier channels. The aliased spurs of one channel are likely to fold back on itself or another channel. This places a greater requirement on the Spurious Free Dynamic Range (SFDR) of the ADC than in the single-carrier case. The SFDR is then usually the limiting factor of the wide-band multi-carrier system. Provided that the ADC has sufficient SFDR (greater than 75 dB) for micro-cell GSM basestation then the DRSP can use process gain by filtering the out of band noise and adjacent channel power. This increases the SNR of the digital data stream. Digital signal processing techniques are then used to demodulate the digital data. The decimation of the DRSP is a more critical factor for multi-carrier than it is for single-carrier, because the digital data rate must be lowered such that it can be processed by a DSP.

In order to improve ADC's SFDR dither is used in presented architecture (Fig. 6.2.14). The addition of out-of-band noise can improve SFDR well into the noise floor. Dither takes the coherent spurious signals generated within the ADC and randomises them so they appear as additional noise in the floor of the converter. However, the overall SNR of the converter degrades by an amount less than 1 dB.

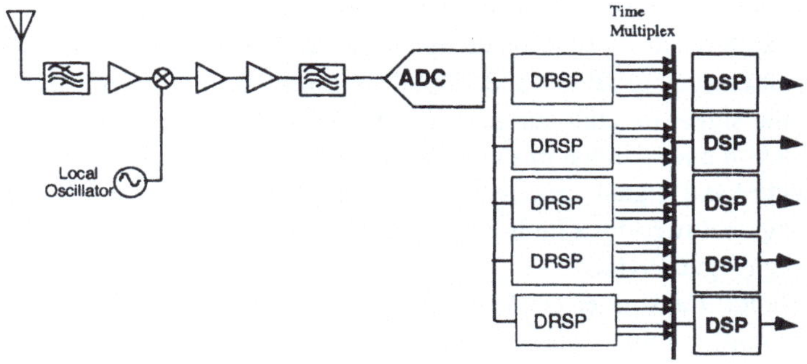

Fig.6.2.15 Multi-carrier CDMA receiver structure.

A receiver based on a wide-band ADC (AD6644) which is shown in Fig.6.2.15, is designed to pass several IS-95 CDMA channels. Analogue front-end contains SAW filters. The wide-band 14-bit 65 Msps ADC samples a bandwidth of 25 MHz at IF of 70 MHz. AD6644 is followed by pipelined DRSPs. Each DRSP is configured as one quad channel receiver and would be tuned to an IS-95 carrier. The input rate at the DRSPs is 58.9824 Msps (48 × 1.2288 Mcps). One NCO in each DRSP translates the desired signal to base-band. Rejection of strong narrow band channels outside our multi-channel band is required for preventing them from adversely affecting our channels of interest when they alias in their band after the ADC. The digital matched filtering is implemented at the programmable coefficient filter of the DRSPs. AD6624 total decimation is 48, with the first filter stage providing a decimation ratio M_{CIC2}= 2, the second filter stage providing a decimation ratio M_{CIC5}=12 and the programmable RCF filter, a 48 coefficient FIR filter, providing a decimation ratio $M_{RCF}= 2$. At the output of a DRSP the signal is de-spread back to the original message bandwidth at the DSP, giving an increase in the SNR which is called the 'processing gain' and is a measure of the chip rate to the information bit rate. Digital techniques such as RAKE receiver and FEC can be implemented at the DSP bank.

6.2.5.2 Transmitter Architectures

Traditional transmitter architectures use analogue circuits for signal up-conversion to an IF frequency [11] [35]. In this architecture digital base-band signal generation is followed by quadrature DACs (I-channel, Q-channel) and analogue up-converters, where signal pulse shaping and anti-alias filtering is done in analogue domain (Fig. 6.2.16a). To improve transmitter system performance, signal pulse shaping can be done digitally followed by interpolating filters (Fig.6.2.16b). Digital pulse shaping eliminates analogue filter miss matching for I and Q channels.

In a different design approach signal up-conversion to an IF is done in digital domain. In this architecture a base-band signal generation is followed by digital pulse shaping, interpolating filters and digital up-conversion (Fig. 6.2.16c). A DAC converts the signal to analogue domain. Analogue filters are still critical in signal reconstruction and removal of out-of-band signals. In alternative transmitter architecture, wide-band signal processing is accomplished by Digital Transmit Signal Processing, followed by a high-performance wide-band DAC and single IF analogue filtering (Fig. 6.2.16d).

Three base-station transmitter configurations for wide-band multi-channel and single-carrier re-configurable radio are presented in the following paragraphs. These transmitter configurations use efficiently the advanced features of DTSPs and DACs presented in section 5.2 enabling common hardware platform for second and third generation digital wireless systems.

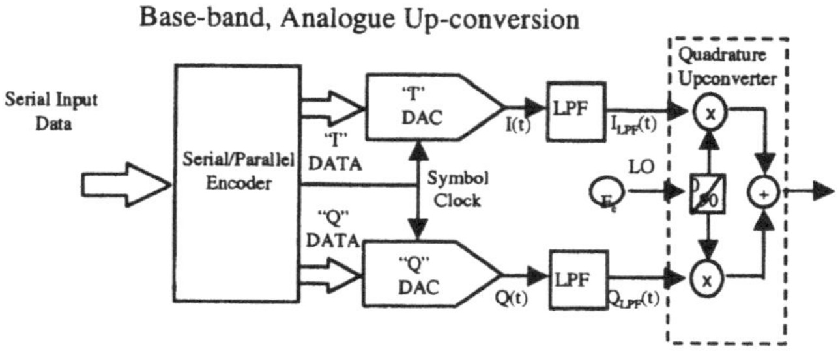

(a)

Base-band with Digital Interpolation

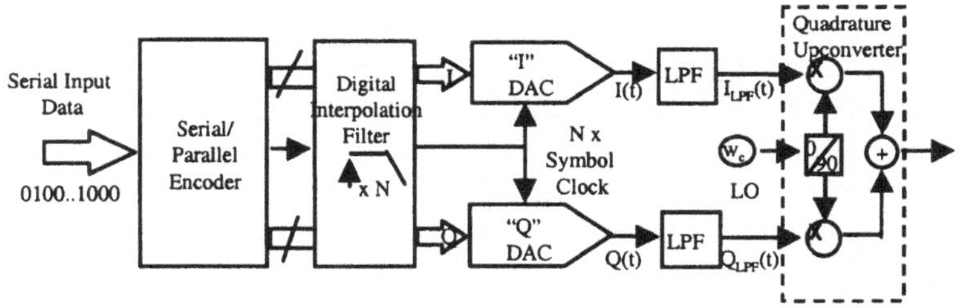

(b)

DTSP with Direct IF Output

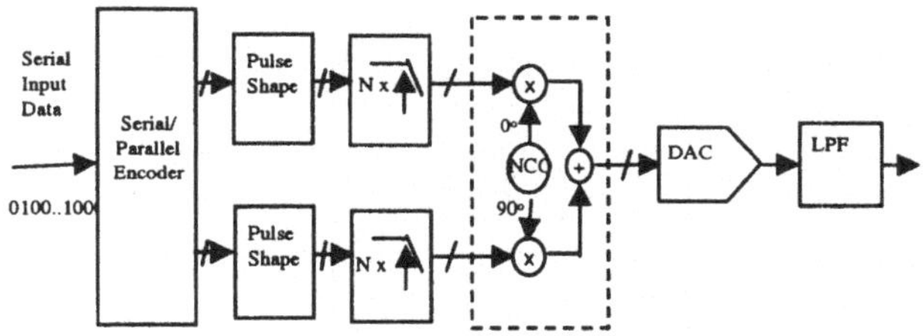

(c)

Multi-carrier DTSP

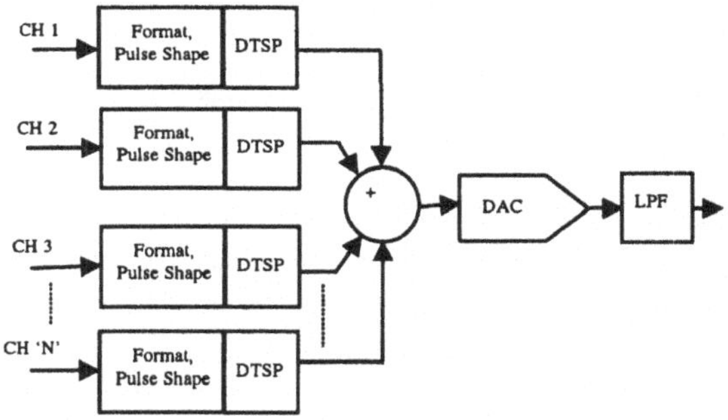

(d)

Fig.6.2.16 Transmitter architectures

A GSM transmitter is presented using a multi-channel system architecture where IF up-conversion is done in a digital manner (Fig. 6.2.17). The GMSK complex signal rate at the NCO input is R_s = 65 Msps (R_s = 240 × 270.8333 ksps). The programmable coefficient filter interpolates the input signal by a factor of 6 and pre-compensates for the CIC filter roll-off in the pass-band. Its impulse response is 8 symbols and the filter is a FIR implementation with 48 coefficients. The output of the FIR filter is routed to a gain adjust multiplier in order to perform the ramp up and ramp down power mask over a time slot, which is specified by the GSM radio standards [36]. The two stage CIC filter provides a total interpolation rate of $L_{TOT} = 40$. The interpolated GMSK signal is up-converted to an IF = 22 MHz by a sine/cosine sequence generated by the NCO. The signal is converted to an analogue signal by a 14-bit DAC (AD9772) in order to complete the wide-band transmit signal processing. The analogue signal is further filtered, amplified, and up-converted to the desired radio frequency. Spurs of the DTSP are much lower than the spurs of the reconstructing DAC.

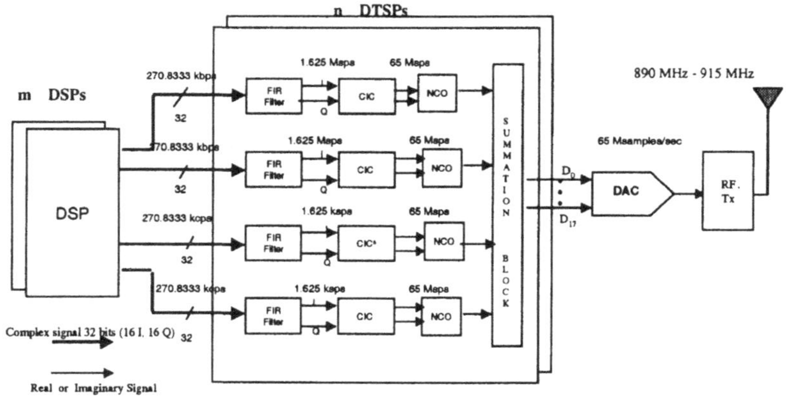

Fig.6.2.17 Multi-channel GSM base-station generic block diagram.

In a second application a quad DTSP processes two IS-95 (CDMA) carriers. A quad DTSP is configured as two pairs of channels. Each channel provides independent serial data inputs that are connected to the serial port of DSP chips through a FPGA that formats input data accordingly. The IS-95 data rate (1.2288 Mcps) of a transmitted carrier is distributed to two DTSP channels at a rate of 0.6144 Mcps from a FPGA. A root raised cosine pulse-shaping filter is used for QPSK modulated inputs. The filter impulse response is 12 symbols long with 48 coefficients. Here QPSK samples are interpolated by a factor 8 per DTSP channel resulting to a sample rate of 4.9152 Msps. The output of the pulse shaping filter is routed through a gain adjust multiplier to high speed Cascaded-Integrator-Comb filters, where data is interpolated by a factor 12, to a data rate of 58.9824 Msps. The interpolated CDMA signal is up-converted to an IF = 28 MHz by a sine/cosine sequence generated by each NCO. The outputs of the four up-converters are summed and scaled on-chip to output two IS-95 carriers to a high performance DAC (AD9772).

A quad-DTSP can be configured to generate a W-CDMA signal by distributing the input signal over multiple channels of the DTSP. For example, a 3.84 Mcps input signal could be distributed to all four channels of a DTSP (Fig.6.2.18). By staggering the timing of each channel, a complete wide-band channel can be generated. If a linear input of 32 bits (16 bits I, and 16 bits Q) is distributed in all four channels, then the input rate per channel is 30.72 Mbits/sec (0.96 x 32). The programmable coefficient filter is an approximation to a square root raised cosine with a roll-off factor α = 0.22. The programmable coefficient filter also pre-compensates the CIC roll-off in the pass-band. The programmable coefficient FIR filter impulse response is 16 symbols with 32 taps. Each channel of the DTSP is loaded with all 32 taps. Each channel operates on different input data at different timing instances. The four channels are sequenced in time by loading different values in each start hold off counters. Each channel is started with synchronised offsets.

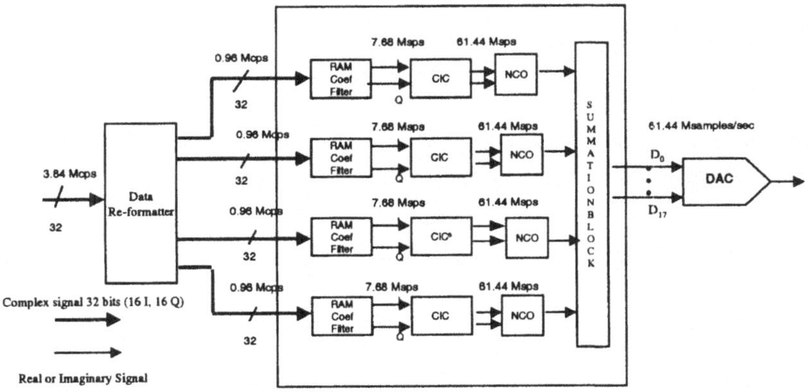

Fig.6.2.18 DTSP Configuration for UMTS (W-CDMA).

In this implementation each of four channels calculates a quarter of the composite pulse shaped signal with an actual interpolation rate of L_{FIR} = 8 per channel and a composite interpolation rate of L_{CFIR} = 2. The two Cascaded-Integrator-Comb (CIC) filters provide a total interpolation rate of L_{CIC} =8. The total interpolation achieved for a W-CDMA channel is L_{TOT} = 16. Each NCO is programmed with the same frequency in order to up-convert the base-band signal. The Summation Block sums the four channels to build a composite pulse shaped signal (single channel). Fig. 6.2.19 represents the composite transfer function (RCF and CIC). This filter passes a – 50dBc image at 6 MHz offset from the channel centre. This image could be further rejected by a SAW filter placed in the analog transmit path. Filter rejection could be improved by relaxing the ripple specifications.

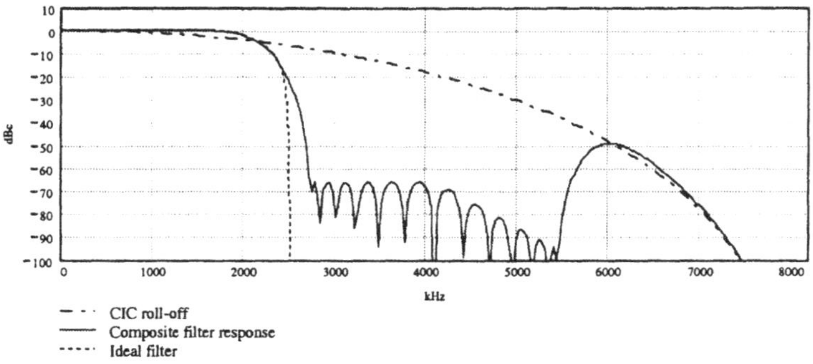

Fig.6.2.19 DTSP Composite Filter Response for UMTS (W-CDMA).

6.2.6 Acknowledgments

The author would like to thank Dr. Z. Zvonar, Mr. J. Camp, Mr. B. Brannon, and Mr. P. Hendricks for their valuable comments and suggestions.

6.2.7 References

1. Software Radio Workshop, organised by European Commission, Brussels, May 1997.
2. Universal Mobile Telecommunications Systems Conference, UMTS' 99, Software Defined Radio, a Pre Congress Seminar C, Monaco, June 1999.
3. R.Karimi and B.Friedrichs: Wideband digital receivers for multi-standard software radios. *Advanced Communications Technologies & Services (ACTS) Mobile Communications Summit 1997*, pp 392-398, Aalborg, Denmark.
4. D. Efstathiou, R. Clark, T. Gratzek: Enabling Technologies for Software Radios. Pre-Congress Seminar C, Software Defined Radio, *Universal Mobile Telecommunications Systems Conference UMTS'99*, Monaco, France, June 7-10, 1999.
5. J. Mitola: Technical Challenges in the Globalisation of Software Radio. *IEEE Communications Magazine*, February 1999, Vol. 37, No. 2, pp. 84-89.
6. D. Chester: Digital IF filter technology for 3G systems: An introduction. *IEEE Communications Magazine*, February 1999, pp. 102-107.
7. H.Erben and P.Crichton: Software radio architecture for UMTS," *Advanced Communications Technologies & Services (ACTS) Mobile Communications Summit 1997*, pp 374-379, Aalborg, Denmark.
8. R.Arnott et al.: Advanced base station technology," *IEEE Communications Magazine*, Vol. 36, No. 2, pp 96-102, February 1998.
9. D.Efstathiou and Z.Zvonar: Enabling Technologies and Components for Multi-Standard Software Radio Base Stations," to appear in *Wireless Personal Communications*, An International Journal, Special Issue on Intelligent Wireless Multimedia Systems, Terminals and Components, 1999.
10. M.Frerking, Digital Signal Processing in Communication Systems, *Van Nostrand Reinhold*, New York, 1994.
11. F. Harris: Digital signal processing in radio receivers and transmitters," *International Journal of Wireless Information Networks,* Vol 5, No 2, pp 133-146, April 1998.
12. R.Groshong and S.Ruscak: Undersampling techniques simplify digital radio. *Electronic Design Magazine,* May 1991.
13. B. Brannon: Digital-radio-receiver design requires re-evaluation of parameters. *Electronic Design Network*, November 5, 1998, pp. 163-170.
14. *Analog Devices*: High Speed Design Techniques. 1996., ISBN-0-916550-17-6.
15. K.Stewart: Effect of sample clock jitter on IF-sampling IS-95 receivers," Proceedings of *1997 International Symposium on Personal, Indoor and Mobile Radio Communications (PIMRC'97),* September 1997, Helsinki, pp 366-370.
16. P.Hendriks, Specifying Communication DACs. *IEEE Spectrum Magazine*, pp 58-69, July 1997.
17. J. Carcia: Software defines wireless radio. *Electronic Engineering Times*, May 17, 1999, pp. 78.
18. F. Harris: Oversampled data simplifies design. *Electronic Engineering Times*, May 17, 1999, pp. 74.
19. T. Hentschel, M. Henker, G. Fettweis: Sample Rate Conversion in Software Radio Terminals. Proceedings of 4th Advanced Communications Technologies and Services ACTS' 99 Mobile Summit, Sorrento, Italy, June 1999, pp. 733-738.

20. A.Drewer: Software engineering considerations for software radios," ACTS Mobile Communications Summit 1997, pp 453-458, Aalborg, Denmark.
21. T.Hentschel and G.Fettweis: Software Radio Receivers," in CDMA Techniques for Third Generation Mobile Systems, *Kluwer Academic Publishers*, pp 257-283, 1998.
22. Analog Devices: AD6600, Dual Channel, Gain Ranging ADC with RSSI.
23. Analog Devices: AD6644, 14-bit, 65 Msps IF sampling ADC.
24 B. Brannon: Wide-dynamic-range ADC converters pave the way for wide-band digital-radio receivers. *Electronic Design Network*, pp. 187-205, November 7, 1996.
25. Analog Devices: AD6624, 80 Msps Quad Digital Receive Signal Processor.
26. Harris Semiconductor: HSP50214 Programmable DownConverter.
27. Graychip: GC4014, Quad Receiver Chip.
28. Analog Devices: AD6622, Four Channel, 65 Msps Digital Upconverter.
29. Graychip: GC4114, Quad Transmit Chip.
30. Harris Semiconductor: HSP50215 Digital UpConverter.
31. Analog Devices: AD9754, 14-Bit, 125 MSPS High Performance D/A Converter.
32. Analog Devices: AD9772, 14-Bit, 150 MSPS TxDAC+ with 23 Interpolation Filter.
33. D. Efstathiou, B. Brannon, T. Gratzek: Software Radio, Fact or Fiction. *Electronic Design Magazine*, December 1, 1998, pp. 117-122.
34. Y. Shinagawa: Software Radio Technologies-from the view point of terminal manufacturer. *First International Software Radio Workshop*, June 1998, Rhodes, Greece, pp. 25-35.
35. D. Efstathiou: Wideband Transceiver Architectures for Multi-Mode Base Stations. Proceedings of 1999 *International Workshop on Mobile Communications*, June 24-26 1999, Chania, Crete, Greece.
36. Digital cellular telecommunications system (Phase 2+); Radio transmission and reception (GSM 05.05 version 7.0.0 Release 1998).

6.3 Mobility Management and Resource Allocation in Multi-Layer Cellular System

Mahboubeh Lohi, Dilshan Weerakoon, A.H. Aghvami

6.3.1 Introduction

The rapid increase in the size of the wireless mobile community and its demand for high speed multimedia communications stands in clear contrast to the rather limited spectrum resources that have been allocated in international agreements [1]. In future, network providers and operators have to increase the capacity of their systems to support more users and also to increase the capacity per user, as multimedia services will become common in future. This is mainly due to the convergence of mobile and wireless systems and multimedia systems. In order to increase the capacity to meet the level of demand, several solutions have been proposed and some of them are already in use. A common scheme used in existing cellular systems is Cell Splitting: where a cell subdivided into typically three areas. Here larger cells are split into smaller cells, which can be split into even smaller cells by using decreased transmission power. Very small cells can be employed in high-density areas. However, it should be emphasised that even though cell splitting is expensive, it gives a service providing more flexibility. Other schemes include cell sectoring, frequency hopping combined with discontinuous transmission and power control and reuse partitioning. Although these schemes increase capacity, they do not meet demands in future systems for such matters as transmitters with low transmission power, small number of handovers for fast moving mobile users and easy to install base stations compared to the conventional approach of base station antennas on the top of high buildings. As a logical extension to the conventional cellular system, multi-layer cellular systems have been proposed [1]-[3].

6.3.2 Multilayer cellular system

Fig. 6.3.1. illustrates the concept of a hierarchical multi-layer cellular system. In such systems, there are many different types of cell whose cell radii and shape are determined by the geographical coverage area and the density of mobile users. The upper layer uses a conventional cell with relatively large radius, about 1~35km, with antennas radiating relatively large powers (0.6~10W) from the tops of buildings etc.

The addition of smaller cells to give higher capacity for smaller coverage area, i.e., microcells (less than 500m radius) and picocells (mostly used in indoor environments and public structures such as airports and busy railway stations) can provide the capacity and coverage for the network. In such an integrated cellular system (multi-level cellular system), the main goal is to provide a balance between maximising the number of users per unit area (favours small cells) and minimising the network control and handover rate (which favours larger cells).

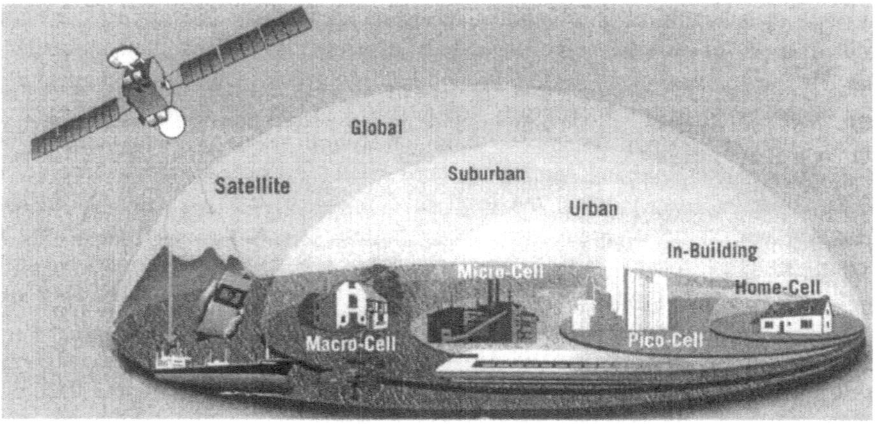

Fig. 6.3.1. Hierarchical cellular system[1]

By having more than one cellular coverage area, the number of handovers can be reduced whilst at the same time increasing system capacity. The downside of a multi-level cellular system is system complexity and extremely high signalling load. Based on the advantages and disadvantages above, a two-layer cellular system is analysed in this paper. Two-layer architecture has to be carefully optimised so that the upper layer and the lower layer can act complementarily. In two-layer cellular systems, the upper layer of macrocells would give overall coverage area, and take control of fast moving mobile users. The lower layer of microcells would focus on slow moving users moving between high-rise buildings, for example; while the picocells focus on the foyer of a theatre, exhibition centre, video conferencing, board meeting rooms etc.

Cellular layer selection is shown below.

[1] This picture is from UMTS forum [20]

Macrocell	⟷	Fast moving users
Microcell	⟷	Slow moving users / users with high bandwidth requirements
Picocell	⟷	Stationary users / users with high bandwidth requirements

In multi-layer cellular environments, the size of different cells has a great impact on propagation path conditions and would restrict many system parameters, especially the maximum user bit rate. When a cell size is large as in the case of a macro-cellular system, path loss and delay spread tend to be large. Hence, the transmission power of the terminal is limited and its symbol rate is limited to a smaller value, thereby the user bit rate is also limited. On the other hand, in picocells/microcells, due to reduced path loss and lower delay spread, higher bit rates are likely to be possible [2]. Propagation conditions in microcells are highly dependant on the geographical nature of the coverage area. In a business district, width of the street, moving obstacles, height of buildings and so on affect propagation.

6.3.3 Call Admission Control

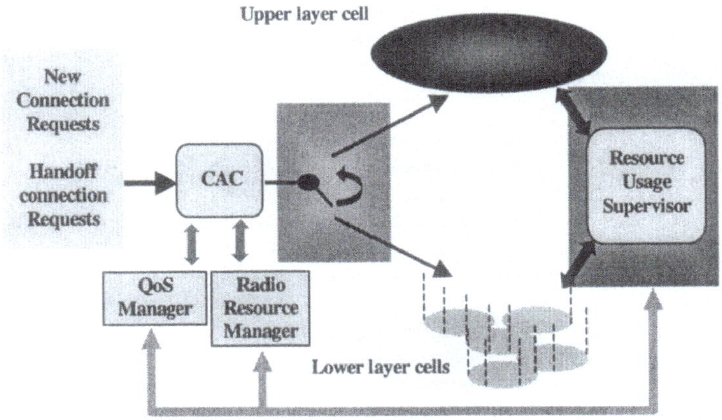

Fig. 6.3.2. Call Admission Control Process

The main objective of the call admission control (CAC) process is to compare the required QoS and capability to meet requirements. Several call admission control algorithms have been proposed for multi-layer cellular systems. Some of them are simple to implement while others are more complex. In a multi-layer cellular network, maximum bandwidth utilisation can be achieved if the CAC admits new mobile users with high bit rates and low mobility to the lower layer cell available while users with low bit rates and high mobility are given admission to the upper layer. But in reality, at the time of call origination it is only possible to know the transmission rate requirements (from the traffic descriptor) and not the mobility pattern. Hence, alternative CACs are required. The traffic descriptor is an important part of the CAC process. The traffic descriptor contains the traffic characteristics of the call, such as minimum and maximum bit rate, whether transmission type is constant bit rate, variable bit rate or available bit rate. This information is used to identify high bit rate users (video, teleconference, web browsing) and low bit rate users (voice, email, paging etc.).

During the decision making process, CAC communicates with both QoS manager and Radio Resource Manager. QoS manager deals with issues related to the QoS requirements of the mobile user while Radio Resource Manager deals with radio resource issues such as available bandwidth, buffers, quality of radio links etc.

Implementation complexity — Since call admission control is a real-time traffic control procedure, it is necessary to reduce processing volume. Admission control should be made simple enough to perform in a real-time fashion. Also, complicated hardware and/or software mechanisms in implementing admission control should be avoided.

6.3.3.1

Overflow call admission control

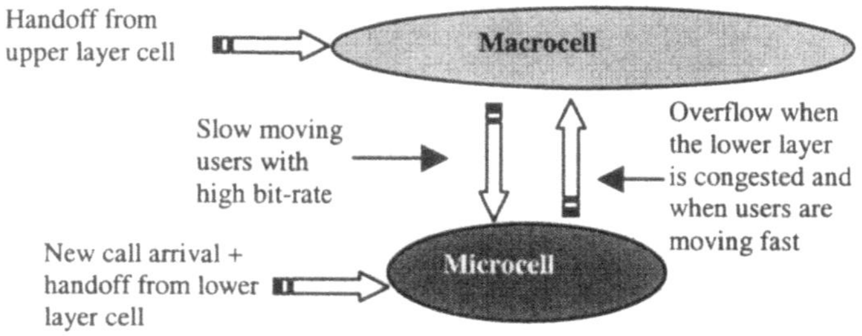

Fig. 6.3.3. Overflow CAC process

In the overflow CAC process, for a new mobile user opening up in a multi-layer environment, priority of cell use is given to the lowest layer (picocell/microcell) available. This can handle most users. Once the call is accepted and moving around the cell, a mobile user is re-assigned between the layers where necessary, to reduce handoffs and also to utilise bandwidth more efficiently in each layer. The system can use either unidirectional or bi-directional scheme to do this task.

6.3.3.2
Class based call admission control

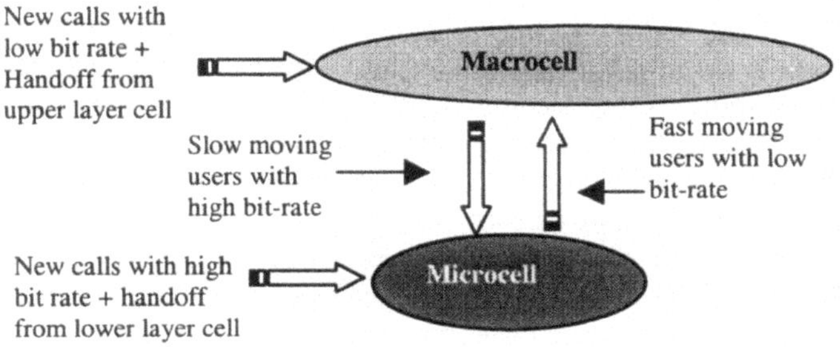

Fig. 6.3.4. Class based CAC process

This Call Admission Control scheme is based on the service class type. Here, at the time of call origination, mobile users requiring more capacity (video, teleconferencing etc.) are given admission to the lower layer cell which can handle more users while users with low capacity requirements (voice, paging etc.) are given admission to the upper layer cell. When a call is in progress, slow and fast moving users are moved between the layers to reduce the number of handovers.

6.3.4
Unidirectional and Bi-Directional Hierarchical Systems

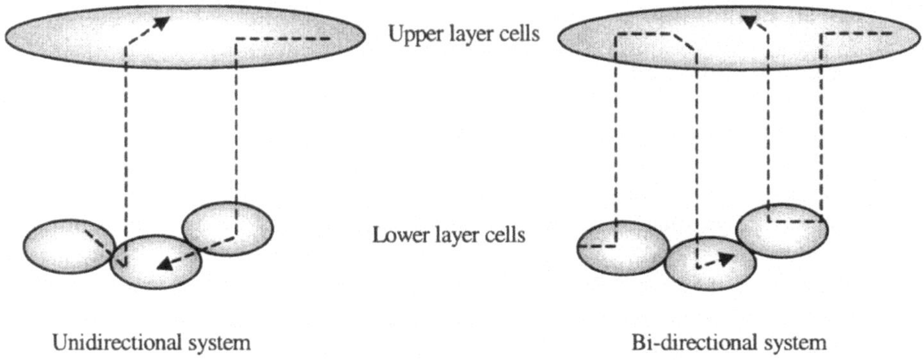

Fig. 6.3.5. Unidirectional and bi-directional hierarchical system

In order to move mobile users between the layers, two main schemes are considered in the hierarchical system: unidirectional and bi-directional. First, it is important to understand why it is necessary to move mobile users between the layers. As mentioned above, optimum performance of hierarchical systems can be achieved if fast moving mobile users and user's with low bit rate requirements are joined to the upper layer while slow moving users and user's with high bit rate requirements are joined to the lower layer. In real mobile networks, a particular vehicular user may change his/her speed randomly depending on the location. For example, a user may move slowly on a road with high traffic density and tend to move fast if he/she joins a motorway. Similarly, a user may require greater bandwidth when downloading web pages to his/her wireless laptop and may require less or no bandwidth while reading a web page. In addition, future systems will also support some high bit rate applications with QoS re-negotiation capability. Here, during a handover to a cell if there are not enough radio resources in the new cell, the user may be supported with reduced QoS where less bandwidth is allocated in the new cell. Finally, if a mobile user in the lower layer cell moves to an area where it is not covered by lower layer cells then regardless of speed and user bit rate, the mobile users need to move to the upper layer.

In the unidirectional hierarchical systems, users are only allowed to move between layers once. The unidirectional scheme is commonly used together with an overflow call admission control process. This scheme reduces system complexity and signalling load.

In bi-directional hierarchical systems, a mobile user can move between the upper and the lower layers at any time. This gives a mobile user the advantage of being in the correct layer all the time. This scheme requires continuous monitoring of the all active mobile users. The function of "Radio Usage Supervisor" (see Fig. 6.3.2) is to monitor all the active mobile users and place them in the correct layer. The "Radio Usage Supervisor" communicates with both Radio Resource Manager and QoS manager before moving users between the layers in order to reduce unnecessary handovers and call dropping probability.

6.3.5 Speed Estimation

In mobile networks, fast moving mobiles may have a high probability of dropped calls due to excessive delay. One of the solutions to this problem is to use overlaying macrocells for fast moving mobiles. This method necessitates a method of estimating a mobile user's velocity. Velocity estimation can be achieved either by mobile station assisted base station estimation or mobile station estimation. There are various conventional mobile user speed estimation methods such as Rayleigh fading, level crossing rate (LCR), zero crossing rate (ZCR), covariance method [4], cell dwell time, Doppler effect etc.

Rayleigh fading has a relatively good performance in a non-noise environment. In a environment with high rising building or mountains, Rayleigh fading estimation does not work accurately. Therefore use of this method is limited to non-noise environments. Level crossing rate, zero crossing rate and covariance methods perform well under practical urban co-channel interference conditions [4]. Speed estimation using cell dwell time can be implemented easily compared to other schemes. This scheme take into consideration the time spent by a mobile user in a cell before handoff to an adjacent cell together with a specific negative offset to the received power level for a particular time period.

6.3.6 Channel Allocation Scheme in Multi-layer Cellular System

The main aim of the multi-layer cellular system is to improve the bandwidth utilisation compared to a single layer cellular system. In order to achieve this, efficient frequency management and channel assignment is required. The function of frequency management is to distribute the total available bandwidth in dynamic or fixed fashion to each cell. Allocation of specific channels to base-station(s) or to mobile stations is the function of channel assignment.

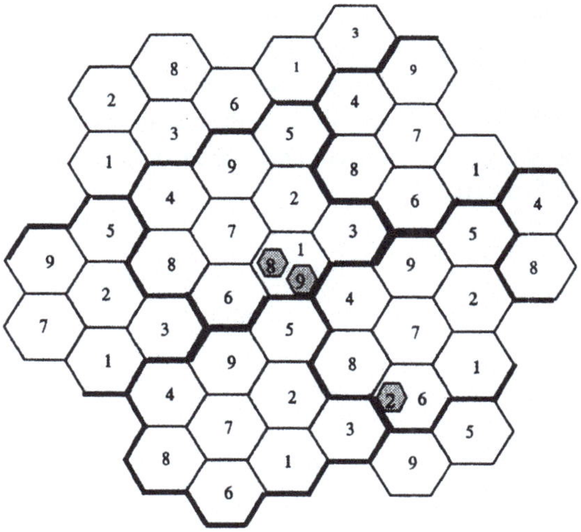

Fig. 6.3.6. Hexagon Shape Multi-Layer Cellular Coverage

Fig. 6.3.6 shows a cellular coverage area with hexagonal shaped cells, which are numbered with the frequencies they are using. Hexagons can be packed side by side without any overlapping and examination of a set of hexagons shows that they can be packed in clusters such that no two adjacent cells use the same set of frequencies. The cluster size η is determined by the condition:

$$\eta = i^2 + ij + j^2 \quad \{i \geq j \quad i,j \geq 0 \quad i,j \in N\}$$

Assuming that the co-channel interference is only from the adjacent cells, we can reuse the same set of frequencies in adjoining clusters as shown in Fig. 6.3.6. Now let us expand this concept to investigate how we can reuse the same frequencies within the same cluster. In the case of 9-cell cluster (Fig. 6.3.6) there is no co-channel interference within cell 1 from set of frequencies used in cell 8 and cell 9 (assuming co-channel interference in only from the first tier). Therefore within macrocell 1 we could have a picocell or a microcell underlaying with the same set of frequencies used in macrocell 8 and/or macrocell 9.
This technique is only valid for systems with clusters of $\eta \geq 9$. The amount of spectrum increment from the above technique above can be estimated as below:

Total number of channels available in the system= n
Number of macrocells per Cluster = η
Number of channels per macrocell = n / η
Number of co-channel interference free macrocells per cluster
$= \eta - 7 \quad (\eta \geq 7)$
Number of co-channel interference free channels per macrocell
$= ((\eta - 7)n / \eta) + (n / \eta)$
Percentage increment per cluster = $(\eta - 7) * 100$

This is the theoretical maximum capacity increment per cluster provided both upper and lower layers use the same TDMA-FDMA access technique. We can see that each cell will have additional $((\eta - 7)n / \eta)$ channels per macrocell, which means we could have pico/micro cells underlaying each macrocell with $((\eta - 7)n / \eta)$ channels provided that total coverage area of pico/microcells is less than the coverage area of macrocell.

6.3.7 Adaptive Bandwidth Allocation by Dynamic Partitioning

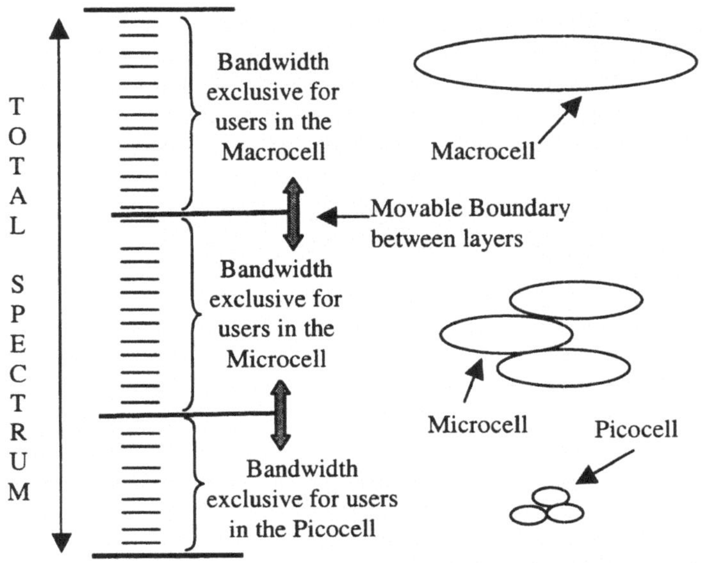

Fig. 6.3.7. Adaptive Bandwidth Allocation

Fig. 6.3.7 shows the concept of adaptive bandwidth allocation with dynamic partitioning between the upper and the lower layers of a multi-layer cellular system. As shown above, a partition, of which one band is allocated to the lower and the other to the upper layer, divides the whole frequency spectrum available in the coverage area, which gives each layer its own set of frequencies. The main objective of this scheme is to improve the bandwidth utilisation while guaranteeing the QoS of calls in progress. The performance of this system depends on the position of the partition limit between the layers at a given time. The partition position at any given time depends mainly on five factors: call arrival pattern for each service class in each layer; mean channel occupancy time by each service class; estimation of number of handoff calls; number of active calls (each service class) and QoS requirements such as minimum acceptable blocking/dropping probability of high priority services. In the case of spitting the bandwidth may lead to loss of trunking efficiency, but this can be improved by having movable boundaries between bandwidth separations.

Since the position of the boundary between layers is a function of time and depends on the parameters (see above), which are also a function of time, the system needs to monitor all cells continuously and keep a record of their performance. This will increase the signalling load of the system.

6.3.8
Intelligent Bandwidth Sharing Scheme

Under intelligent bandwidth sharing scheme where bandwidth allocation employs fixed channel allocation for the upper layer and dynamic channel allocation for the lower layer. If a mobile user in the upper layer requests a channel, the channel controller allocates a free channel from a pool of channels. If the request is from the lower layer, a channel with the minimum Channel to Interference Ratio (CIR) margin within the base station's coverage area is assigned.

The All-Channel Concentric Allocation (ACCA) scheme proposed in [10] for single layer cellular system is a suitable candidate for an intelligent bandwidth sharing scheme.

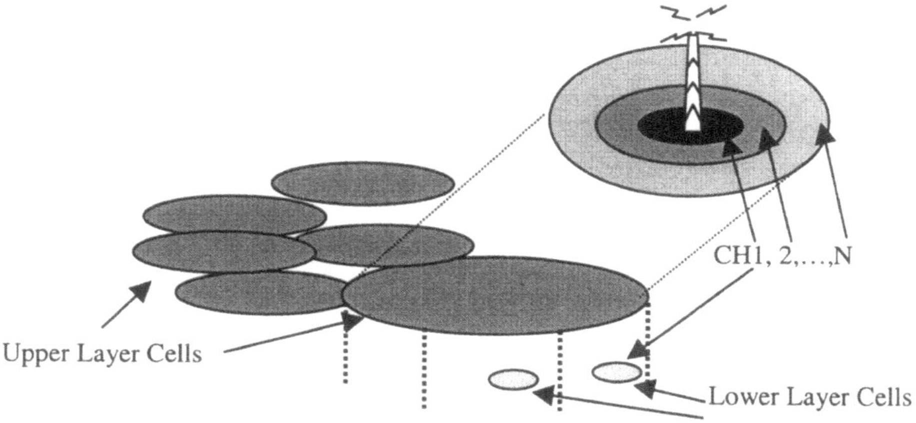

Fig. 6.3.8. Principle of the all-channel concentric allocation

Under the ACCA scheme, the upper layer cell is divided into N concentric regions; each region has its own channel allocation method. Here, each mobile is assigned a channel belonging to the concentric region in which that channel is allocated, and has a specific desired signal level corresponding to the channel location. Therefore, each channel has its own reuse distance determine from the desired signal level. When allocating channels to cells in the lower layer, channels having CIR below a certain threshold within that area will be allocated depending on their position relative to the macrocell base station and the coverage area of the cell. The overall spectrum efficiency of this scheme can be improved by using advanced power control techniques. This scheme requires continuous monitoring of all the mobile users to avoid any co-channel interference if a mobile moves to an area where the channel is already being used.

6.3.9
Cellular Systems and handover

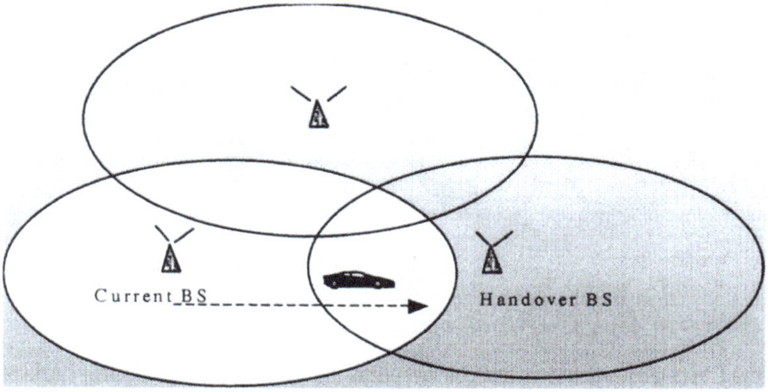

Fig.6.3.9. Handover process

Handover is the mechanism that transfers an ongoing call from one base station (BS) to another as a user moves through the coverage area of a cellular system as shown in Fig. 6.3.9. Therefore it must be fast and efficient to prevent the quality of service (QoS) from degenerating to an unacceptable level. This is probably the most sensitive aspect of mobility provision, and is an essential element of cellular communications, since the process chosen for handover management will affect other mobility issues. A core requirement of future wireless networks, which rely on Quality of Service (QoS) criteria, is to limit handover dropping or cell overload probabilities to predefined levels which are almost independent of load condition. Due to radio spectrum limitations, future wireless systems will have to consist of microcells/picocells so that handover events in future systems will occur at a much higher rate than in today's macro-cellular systems. Smaller cell systems allow greater spectral reuse, larger capacity, and the use of low-power lightweight hand-held user devices. However, small cell systems induce an increase in the number of cell boundary crossings by mobile users. The required exchange of supervisory messages can be burdensome. Systems that employ microcells with overlaying macrocells have been suggested as one way to address these issues [section 2].

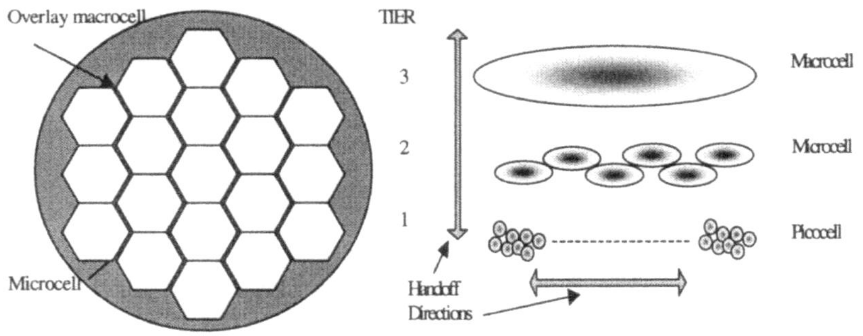

Fig. 6.3.10. Hierarchical cellular systems

To reduce the forced termination of calls in progress, handover calls are given priority access to channels in both the microcell and macrocell levels. The number of handovers in an overlay system is much less than in a single layer microcell system because fast-moving vehicles can be connected in the overlay macrocell. Also in an overlay system the infrastructure costs are reduced because only high traffic load areas require additional base stations. Efforts to quantify the performance of particular approaches have become common in publications and attempt to answer questions such as:

- What is the duration of service interruption during a handover?
- How many handovers are made, and are they necessary to meet the QoS?
- Does an algorithm need to know whether a user is in the macrocell/microcell or picocell plane to interpret measurements of the current and candidate base stations?

In practise, however, handover algorithms with a specific set of parameters cannot perform uniformly well in different communications systems such as macrocells, microcells, and overlay systems as well as special architectures such as multi-channel bandwidth systems and evolutionary architectures. In the following subsections some of the essential issues concerning the processing of handovers are discussed.

6.3.10 Handover Management

Handover management is the process of maintaining the connections to and from a mobile node, as it moves from the region covered by one mobile access point to another. There are a variety of schemes that have been proposed for handover management like the path extension, anchor rerouting, hop-by- hop backtracking, distributed hunt, loose select etc. The performance metrics that would determine the efficiency of the handover management algorithm are: time, cell-loss, signalling and scalability.

Time - As the mobile moves from the region of one mobile access point to another, the time taken to effect the handover successfully is a very good measure of the efficiency of the algorithm. This handover time determines the suitability of the algorithm for implementation in networks, carrying real time traffic like video and audio.

Cell-Loss - Certain handover algorithms result in cell-loss. The number of cells lost in the process of effecting a handover is another good performance metric. Typically, certain algorithms designed to support real time traffic compromise on the cell loss. Cell-loss can be avoided by buffering the cells. Buffering cells is acceptable for data traffic, but for real time traffic, it must be kept to a minimum.

Signalling – A number of messages are exchanged and must be kept to a minimum. This is a good performance metric.

Scalability - The algorithm should scale well with increasing number of nodes in the system. The cell loss, if any, and the handover time should be controlled.

The performance metrics [13] used to evaluate handover algorithms are as follows:

Table 6.3.1. Performance metrics

Call blocking probability	The probability that a new call attempt is blocked
Call dropping probability	The probability that the call drops due to a handover failure
Rate of handover	The number of handovers per unit time which combines with the average call duration
Unnecessary handover probability	The probability of having more than one handover by a particular handover algorithm when the existing radio link is still adequate.
Delay	The distance the mobile moves from the point at which the handover should occur to the point at which it does.
Hanover blocking probability	The probability that a handover attempt is blocked

Handover Detection Strategies

Handover decision is made and initiated based on measurement. Different systems use different approaches to execute handover processes and these are characterised by handover protocols. The terminal continuously measures the level of signal in current channels and compares each with others in different channels. Based on the measurement of Received Signal Strength (RSS), a handover decision is made and initiated by the base station or the mobile terminal. Therefore mobile terminals continuously measure the level of the RSS in the current channel and compare it with others in the different channels.

There are three basic types of handover protocols:

Mobile-Assisted HandOver (MAHO), in this method the handover is more decentralised. Both the mobile station and the base station supervise the quality of the link (i.e. RSSI, WEI, and sometimes quality). The mobile station does a Received Signal Strength Indication (RSSI) measurement of neighbouring base stations. This handover strategy is employed by the high-tier GSM cellular standard, and the mobile station transmits the measurement results to the base station twice a second. The decision as to when and where to execute the handover is still made in the network (i.e., the base station and MSC).

Network-Controlled HandOver (NCHO), in this method, the base station monitors the signal strength and quality from the mobile station, and when these deteriorate below some threshold, the network arranges for a handover to another base station. The network examines all the surrounding base station to monitor the signal from the mobile station and report the measurement results back to the network. The network then chooses a new base station for the handover and informs both the mobile station through the old base station and the new base station. NCHO is employed by the low-tier CT-2 Plus and the high-tier Advanced Mobile Phone Service (AMPS) cellular system.

Mobile-Controlled HandOver (MCHO), in this method the mobile station continuously monitors the signal strength and quality from the accessed base station and several handover candidate base stations. MCHO is the most popular technique for low-tier radio systems and is employed by both DECT and PACS.

The decision making process of handover may be centralised or decentralised. The handover decision may be made at the mobile station (MS), base station (BS), or mobile switching centre (MSC). For MCHO, NCHO, and MAHO, handover failure can occur for a number of reasons. Some of them are listed below:

- The network takes too long to set up the handover after the handover has been initiated.
- There is no available channel on the target base stations.
- The target link fails in some way during the execution of handover.
- Handover is denied by the network, either for lack of resources, or because the portable has exceeded some limit on the number of handovers which may be attempted in some period of time.
- In some other systems handovers can fail due to resource blocking (e.g. DECT).

6.3.11 Radio Link Measurement

The radio link measurement process determines the need for handover and the target or new channel for transfer. Three measurements are used to determine the quality of the channel:

1. Word Error Indicator (WEI), is an indication of whether the current burst was demodulated properly in the portable.
2. Quality Indicator (QI), is the estimate of the "eye opening" of a radio signal, which relates to the signal-to-interference (S/I) plus noise ratio, including the effects of dispersion.
3. Received Signal Strength Indication (RSSI), is a measure of co-channel interference power and ratio. Distance-dependent fading, lognormal or shadow fading, and Rayleigh or multipath fading affects RSSI measurements.

A number of different handovers can be identified as:

- Soft hand-over: In a soft hand-over, the new path is activated before the old path is destroyed, therefore a mobile can simultaneously communicate with more than one base stations during the handover, resulting in two active information streams to and from the mobile terminal [40]. Soft handover is more complicated and requires some Code-Division Multiple Access

(CDMA) systems because of the near/far problem and the need for precise power control.

- Hard hand-over: On the other hand, hard handover is characterised by a mobile having a radio link with only one base station at any time.
- Seamless hand-over: It lies somewhat between soft and hard handovers. In this case, both new and old paths during handover carry information, but the old path is used for a brief period only. The mobile terminal only switches over to the new path and drops the old one when the new path is stable.
- Backward hand-over: In a backward hand-over, the handover is predicted ahead of time and handover execution can be initiated via the old base station.
- Forward hand-over: In a forward handover, a mobile user moves so abruptly to a new base station that the handover can only be initiated after the mobile has associated itself with the new base station.

6.3.12 Handover Processing in Cellular Environment

Published performance evaluations of handover studies use several initiation criteria and assume the signal is averaged over time to remove rapid fluctuations due to the multipath nature of the radio environment. The handover initiation criteria analysed in the literature are based essentially on four variables: The length and shape of the averaging window, the threshold level, and the hysteresis margin. The following discussion explains various approaches:

1. Relative signal strength: Selects the strongest receiving base stations at all times [13]. This method stimulates too many unnecessary handovers when the current base station signal is still adequate.
2. Relative signal strength with threshold: allows a user to hand over only if the current signal is less than a threshold (sufficiently weak) and the other is the stronger of the two.

In practice this method is not used alone and the effectiveness of its threshold depends on prior knowledge of the crossover signal strength between the current and candidate base stations. This is shown in Fig. 6.3.11. As you will see the effect of a threshold depends on its value compared to the signal strengths of the two base stations at the point at which they are equal. Some specific situations are:

- T1: Threshold is higher than compared value; this scheme performs exactly like the relative signal strength scheme and therefore handover occurs at position 1.
- T2: Threshold is lower than crossover value; the mobile will delay handover until the current signal level crosses the threshold at position 2.

T3: The delay may be so long that the mobile drifts far into the new cell. This reduces the quality of the communication link and may result in the call being dropped and may also causes additional interference to co-channel users.

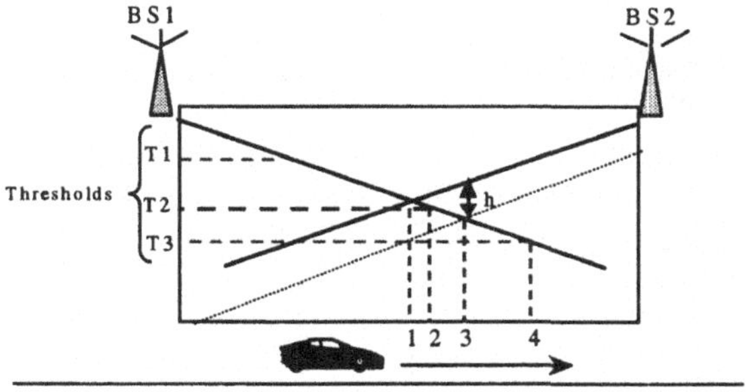

Fig. 6.3.11. Relative signal strength with thresholds

3. Relative signal strength with hysteresis: allows a user to hand over only if the new base station is sufficiently stronger (by a hysteresis margin, h, in Fig. 6.3.11) than the current one. In this case handover will occur at position 3. The results in [21],[23] are useful for determining the averaging interval and hysteresis level to achieve and optimum balance between the number of unnecessary handovers and delay in handover for a simplified scenario in which an mobile station travels along a straight line from one base station to another at a constant velocity.
4. Relative signal strength with hysteresis and threshold: handover happens if the current signal level drops below the threshold and the target base station is stronger than the current one by a given hysteresis margin [15].
5. Prediction handover: the decision is made based on the expected future value of the received signal strength. This technique shows better results in simulation than all the above methods, in terms of a reduction in the number of unnecessary handovers [14].

In [8], the Received Signal Strength (RSS) of a mobile terminal is given by the equation:

$$RSS_{dB} = -10 * \xi * \log(R)$$

Where R is the distance from the serving base station to the mobile terminal and ξ is the propagation path loss coefficient. In the above equation, only the propagation path loss is considered for determining the signal strength. This is not a bad approximation as Rayleigh fading is assumed to be averaged out and lognormal fading is nearly constant over any short period of time.

6.3.13 Velocity Based Handover Algorithm in Multi-Layer System

In multi-layer systems, handover between cellular layers is based either on average velocity and bandwidth usage of a mobile user or when there is no coverage area by the current layer. For all velocity based handover algorithms, velocity estimation is needed. This estimation can be either mobile-assisted base estimated or mobile estimated. The decision on whether a mobile user is classified as a slow or fast moving user will depend on the cell radius, R and the speed threshold Vo for users in a cell, which in turn is related to the cell dwell time threshold through the following equation [9].

$$\tau_o = \frac{\pi R}{2V_o}$$

In urban and sub-urban scenarios, a mobile terminal during its roaming can speed up, when the street is empty, or slow down (or even stop) whenever there is traffic on the road or at traffic lights. Therefore in most cases, the magnitude of the velocity takes on a wide range of values. Hence, it is not always possible to use velocity estimation to determine the most suitable layer. Another problem encountered by velocity based handover is the ping-pong effect. An alternative algorithm is therefore needed for highly variable traffic environments.

First, it is necessary to study the behaviour of a moving vehicle in a highly variable traffic environment. At the beginning and at the end of a journey, the speed of a vehicle is zero, thus the average velocity is zero. Another problem is that in high-density traffic, vehicles are compelled to slow down more frequently and the waiting time at traffic lights is high. As a result a mobile terminal's true average speed cannot be obtained. Therefore, with only the average velocity of a vehicle, the correct cellular layer cannot be determined. Now consider the rate of change of velocity or acceleration. Since a moving vehicle is very unlikely to make a rapid acceleration or deceleration, the acceleration over a certain time period gives a better approximation to the mobile user's behaviour.

The acceleration of a mobile user can be determined by:

$$\overline{a(t)} = \frac{1}{(t+\Delta t)-t}\int_{t}^{t+\Delta t} a(t)\,dt$$

Where a(t) denotes the magnitude of the acceleration at time t and $\overline{a(t)}$ is the time average of a(t) starting from t to t+•t. Since the integral of a(t) is speed

$$\overline{a(t)} = \frac{1}{(t+\Delta t)-t} * v(t)\Big|_t^{t+\Delta t} = \frac{v(t+\Delta t)-v(t)}{(t+\Delta t)-t}$$

Where v(t) is the speed of the mobile terminal at time t and Δt is the time interval. In order to obtain the average acceleration, over every time interval Δt, the acceleration is averaged and stored for τ seconds more (τ >> Δt). If a large enough number of samples of vehicle speeds is taken, it is possible to use the central limit theorem to approximate the acceleration of any mobile terminal.

If the time interval Δt and time duration τ is properly chosen, the variation of $\overline{a(t)}$ against time gives the mobile terminal's speed variation. The acceleration variation alone does not give the mobile terminal's actual behaviour. In order to improve the algorithm, a second condition i.e., magnitude of the velocity itself, should be added. Using this information, we propose a new handover algorithm between layers.

If (*Average_velocity_for_last_τ_sec* > $v^i_{\min}$)
  and (*Average_acceleration_for_last_τ_sec>0*)
    Then Handoff to a higher layer *(if available)*
Else If (*Average_velocity_for_last_τ_sec* < $v^i_{\min}$)
  and (*Average_acceleration_for_last_τ_sec<0*)
    Then Handoff to a lower layer *(if available)*

Where, $v^i_{\min}$ is the minimum velocity in the ith layer. The overall performance of this scheme can be improved by selecting a suitable value for " τ ". This algorithm offers a solution to the ping-pong problem pointed out previously.

6.3.14 The Impact of Handover on QoS

There are two key issues for how handover affects the QoS of an existing connection. First, when a network accepts a connection to a fixed endpoint, the requested QoS is made available to the connection during its lifetime with a high probability of delivery in the absence of network failure. The second issue is minimising the effects of QoS disruption during the process of handover. One way to minimise QoS disruption during handover is to ensure a "lossless handover". In this method, all cells in transit during the handover process are buffered at either a switch or a radio port within the network to maintain in-sequence cell delivery to the mobile terminal. For more information refer to [16]. For some applications, the cell transfer delay and delay variation may be more important than avoiding cell loss. In this case, depending on the allowable delay variation, the current radio port may transfer the buffered cells to the new radio port without waiting for the marker cell to arrive.

6.3.15 Handover Prioritisation

In a cellular system forced termination of an on going call is less desirable than blocking of new calls. Therefore in order to reduce the failure of handover, one way is to prioritise handover. In a heavy traffic situation a significant

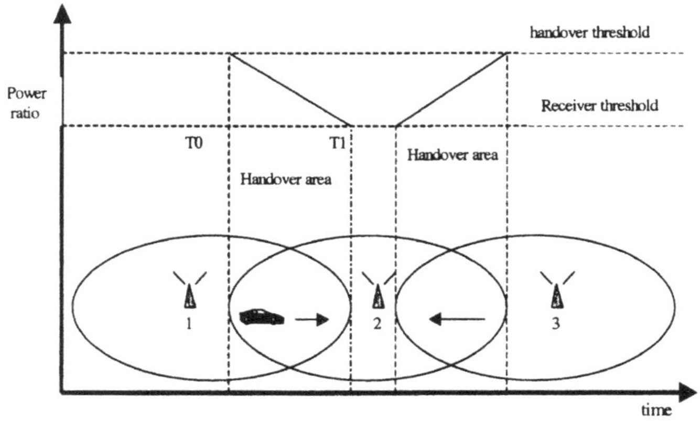

Fig. 6.3.12. Queuing of handover in handover area

improvement handover performance can be obtained this way. Channel assignment strategies with handover prioritisation have been proposed to reduce the probability of forced termination. Two prioritising schemes are guard channels and queuing.

Guard channels: Improves the performance by reserving a fixed or dynamically adjustable number of channels exclusively for handover. The remaining channels can be shared equally between handovers and new calls. However, this method has the risk of utilising spectrum inefficiently. Efficient usage of guard channels requires the determination of an optimum number of guard channels, information on the traffic pattern of the area, and estimation of the channel occupancy time distributions. Another disadvantage is a reduction in the total carried traffic due to the fact that fewer new calls are granted channels.

Queuing of handover: In this scheme no new call is granted a channel before the handover requests in the queue are served. When the power level received by the base station in the current cell reaches a certain threshold (the point where the power level received by the base station in a neighbouring cell has started to exceed the power received by the current base station), the call is queued for service from a neighbouring cell. The call remains queued until either an available channel in the new cell is found or the power from the base station in the current cell drops below a second threshold (the point at which the received power from the current base station is at the minimum acceptable level). If the call reaches the second threshold and a new channel has not been found, then the call is terminated. Queuing handover requests is made possible by the existence of the time interval that the mobile station spends between these two thresholds. This interval defines the maximum allowable waiting time in the queue. Based on the traffic pattern and the expected number of handover requests, the maximum size of the handover queue could be determined. The handover area as shown Fig. 6.3. 12, is the area where the ratio of received power levels from the current and the target base stations is between the handover and the receiver thresholds [39].

In [25], prioritised handover schemes have been analysed. It was assumed that the probability density function (pdf) of the speed of cell-crossing terminals is the same as the pdf of the terminal speed speeds inside the cells.

In [39], the performance of microcellular mobile communication systems such as handover priority system, overlaying macrocell system, and handover priority with overlaying macrocell system are evaluated and compared to that of standard microcellular system by using spectrum efficiency, grade of service, and total efficiency as evaluation criteria. The computational results suggest that the combined system (handover priority with overlaying macrocell system) have higher efficiency over the wide range of the offered traffic.

6.3.16 Analysis of Handover Performance

Analytical, simulation and emulation approaches are the three basic mechanisms that have been used to evaluate the performance of handover algorithms. For real-world situations, the analytical approach is complex and mathematically intractable.

However the analytical approach can quickly give a preliminary idea about the performance of some handover algorithms for simplified handover scenarios. However the actual handover procedures are quite complicated, this makes the analytical approach less realistic and for real-world situations, this approach is complex and mathematically intractable. In reference [41] analytical model is proposed to study the traffic performance of a microcell/macrocell overlay for personal communication systems architecture.

The emulation approaches use a software simulator, consisting of a handover algorithm, to process measured variables (e.g., received signal strength (RSS) and bit-error rate (BER)). Actual propagation measurement based simulation has the advantage of giving better insight into the behaviour of the radio channels and more accurate data. The main disadvantages are that this approach requires periodic measurement efforts and is not suitable for a comparison of different handover algorithms on the same platform. Reference [30] uses measured data in handover simulation and was found that the optimal handover threshold level was different for different sites.

The simulation approach is the most commonly used handover evaluation mechanism. In the literature several simulation models suitable for evaluation of different types of handover algorithms under different deployment scenarios have been proposed and used [21]-[23]. This approach provides a common testbed for comparison of different handover algorithms and also provides insight into the behaviour of the system. The combination of the analytical and simulation approaches can be very powerful as software simulation provides fast, easy, and cost effective evaluation and analytical approach gives insight into handover behaviour quickly.

6.3.17 Conclusion

In this paper some of the key features in multi-layer cellular systems and their impact on radio resource management and handover schemes have been discussed. Under radio resource management, spectrum sharing between layers is an important issue in multi-layer systems. Adaptive bandwidth allocation by a dynamic partitioning scheme, that adapts to traffic variations and QoS requirements, seems to present a promising possibility for multi-layer systems. Bandwidth sharing, using fixed and dynamic channel allocation schemes, also improves bandwidth utilisation significantly due to bandwidth reuse within the upper layer. Handover events in future multi-layer cellular systems will occur at a much higher rate compared to today's macro-cellular systems. Therefore efficient handover algorithms can enhance system capacity and service quality cost effectively. Handover algorithms with a specific set of parameters cannot perform uniformly well in different communication system deployment scenarios since these scenarios impose different restrictions and unique environments on the handover process. The analytical, simulation and emulation approach have been used to evaluate the performance of handover algorithms. Handover prioritisation can improve handover related system performance. Adding microcells/picocells, thus creating a multi-layered network, has been shown to allow continued capacity expansion and to reduce handover failures on the platform already provided by the macrocellular network.

Reference

1. D.D. Weerakoon., B. Jafarian, A.H. Aghvami , "Radio Resource Management in Multi-Layer Cellular Environment with Multimedia Services", WPMC'98, Yokosuka Japan, Nov 1998.

2. F. Kojima, S. Sampei, N. Morinaga, "An Intelligent Radio Resource Management Scheme for Multi-Layered Cellular Systems with Different Assigned Bandwidths", IEICE Trans Commu., Vol E81-B, No 12., December 1998.

3. X. Lagrange, "Multitier Cell Design", IEEE Communications Magazine. August 1997.

4. R. Ramjee, R. Nagarajan, D. Towsley, "On optimal Call Admission Control in Cellular Networks", Baltzer Journals, March 1995.

5. D. Brockmeyer, H. Halstroem, "The life and work of A.K. Erlang", Acta Polytechnic Scandinavica 1960, No 287.

6. O.T.W. Yu, V.C.M. Leung,"Adaptive Resource Allocation for Prioritised Call Admission over an ATM-Based Wireless PCN", IEEE JSAC, Vol.15, No.7, Sep 1997.

7. T.S. Yum, W.S. Wang, "Hot spot traffic relief in cellular concept", IEEE JSAC Vol 11, 1993, pp. 934-40.

8. L. Ortigazo-Guerrero, A.H. Aghvami, "On the Optimum Partitioning in a Microcell/Macrocell Cellular Layout with Overflow", GLOBECOM 1997, pp 991-994.

9. W.C.Y. Lee, Y.S. Yeh, "On the estimation of the second-order statistics of log normal fading in mobile radio environment", IEEE Trans. Comm., Vol 22, No 6, pp 809-873, 1974.

10. T. Takenaka, T. Nakamura, Y. Tajima, "All channel concentric allocation in cellular system", IEEE International Conference for Communications (ICC) 1993, pp 920-924.

11. K. Ogawa , T. Hattori, and H. Yoshida, "Optimum Multi-layered cell Architecture for Personal Communications Systems with High Degree of Mobility", Proc. VTC, 1994, Stockholm, Sweden, June 8-10, 1994, pp. 644-48.

12. M. D. Austin and G. L. Stuber, "Velocity Adaptive handover Algorithms for Microcellular Systems", Proc. IEEE ICUPC, 1993, Ottawa, Canada, Oct., pp.793-97.

13. M. Gudmundson, "Analysis of Handover Algorithms, "Proc. Vehicular tech. Conf., pp. 537-42, 1991.

14. V. Kapoor, G. Edwards, and R. Sankar, "Handoff criteria for Personal Communication Networks", Proc. ICC, pp. 1297-1301, 1994.

15. R. Vijayan, J.M. Holtzman, "The Dynamic Behaviour of Handoff Algorithms", Proc. ICUPC'92, paper 2.03, 1992.

16. N.D. Tripathi, J. H. Reed and H. F. Vanlandingham, "Handoff in Cellular Systems", IEEE Personal Communications, December 1998, pp. 26-37.

17. G. P. Pollini, "Trends in Handover Design", IEEE Communications Magazine, March 1996, pp. 82-90.

18. A.J.M. Ransom, "Handoff Considerations in Microcellular Systems Planning", PIMRC 1995, pp. 804-808.

19. N. Zhang, J. Holtzman, "Analysis of handoff algorithms using both absolute and relative measurements", Proc. 44th IEEE VTC, vol. 1, 1994, pp.82-86.

20. "Monilennium", UMTS Forum news letter, No.5 Nov 98.

21. R. Vijayan and J.M. Holtzman, "Sensitivity of handoff algorithms to variations in the propagation environment", Proc. 2nd IEEE ICUPC Ottawa, Canada, Oct. 1993.

22. R. Vijayan and J.M. Holtzman, "A Model for Analysing Handoff Algorithms", IEEE Trans. VTC, Vol. 2, Aug. 1993, pp.351-56.

23. R. Vijayan and J.M. Holtzman, "Analysis of handover algorithm using nonstationary signal strength measurements", Proc. IEEE GLOBECOME, Orlando, FL, Dec. 1992

24. N. Zhang and J.M. Holtzman, " Modelling and analysis of handoff algorithms in multi-cellular systems", Proc. 44th IEEE VTC, Vol. 1, 1994, pp. 82-86.

25. D. Hong and S. S. Rappaport, " Traffic model and performance analysis for cellular mobile radio telephone systems with prioritised and non-prioritised handoff procedures"!, IEEE Trans. VTC. Vol. VT-35, No. 3, Aug. 1986, pp. 77-92.

26. R. Rezaiifar, A. M. Makowski and S.Kumar, "Optimal control of handoffs in wireless networks", Proc. 45th IEEE VTC, 1995, pp. 887-91.

27. T. Kanai, M. Taketsugu, and S. Kondo, "Experimental Digital Cellular System for Microcellular Handoff", Proc. 38th IEEE VTC, June 15-17, 1998, pp. 170-75.

28. M. D. Austin and G. L. Stuber, "Directed biased handoff algorithm for urban microcells", Proc. 44th IEEE VTC, 1994, pp. 101-5.

29. H. Xie and S. Kuek, "Priority handoff analysis", Proc. 43rd IEEE VTC, 1993, pp. 855-58.

30. S. Chia and R.J. Warburton, "Handover criteria for a city microcellular radio systems", Proc. 40th IEEE VTC, 1990, pp. 276-81.

31. N. W. Whinnett, "Handoff between dissimilar systems: general approaches and air interface issues for TDMA systems", Proc. 45th IEEE VTC, 1995, pp.953-57.

32. M. E. Anagnostou and G. C. Manos, "Handover related performance of mobile communication networks", Proc. 44th IEEE VTC, 1994, pp. 111-14.

33. G. Liodakis and P. Stavroulakis, "A novel approach in handover initiation for microcellulaar systems", Proc. 44th IEEE VTC, 1994, pp. 1820-23.

34. G. Senarath and D. Everitt, "Comparison of Alternative Handoff Strategies for Micro-cellular Mobile Communication Systems", Proc. 44th IEEE VTC, 1994, pp. 1465-69.

35. K. Ivanov and G. Spring, "Mobile speed sensitive handover in a mixed cell environment", Proc. 45th IEEE VTC, 1995, pp. 892-96.

36. L-R. Hu and S.S. Rappaport, "Personal Communications Systems Using Multiple Hierarchical Cellular Overlays", Proc. IEEE ICUPC'94, San Diego, CA, 1994, pp.397-401.

37. O. E. Kelly, and V. V. Veeravalli, "A Locally Optimal Handoff Algorithm", PIMRC '95, Toronto, Canada, September 1995, pp. 809-13.

38. S. Tekinay, B. Jabbari, "A Measurement-Based Prioritisation Scheme for Handovers in Mobile Cellular Networks", IEEE JSAC, Vol. 10, No. 8, October 1992.

39. M. Inoue, H. Morikawa, M. Mizumachi, "Performance Analysis of Microcellular Mobile Communication Systems", VTC'94, pp. 135- 139

40. B. Jabbari et al., "Network Issues for Wireless Communicaitons", IEEE Communication Magazine, Jan. 1955.

41. Y-B. Lin, L-F. Chang, and A. Noerpel, "Modelling hierarchical microcell/macrocell PCS architecture", Proc. ICC, 1995, pp. 405-9.

42. I Katzela, M. Naghshineh, "Channel assignment schemes for Cellular Mobile Telecommunication Systems: A Comprehensive Survey", IEEE Personnel Communications, June 1996, pp. 10-31.

43. S.W. Halpern, "Reuse Partitioning in Cellular Systems", IEEE Tran. on Vehicular Tech., 1993, pp 322-27.

44. R.C.V. Macario, "CELLULAR RADIO Principles and Design", Second addition, Antony Rowe Ltd.

45. William C.Y. Lee, "Mobile Cellular Telecommunications – Analog and digital systems", Second addition, McGraw-Hill.

46. M. Frullone, "Analysis of optimum resource management strategies in layered cellular structures", Proc. of ICPUC 94, pp 371-375.

47. D. Weerakoon, N. Georganopoulos, B. Jafarian, A.H. Aghvami, R. Tafazoli," Optimum Radio Resource Partitioning in a Multi-Layer Cellular System", Fourth ACTS Mobile Communications Summit, Conference Proceedings, June 1999, pp. 495-500.

48. K. Ivanov and G. Spring, "Mobile Speed Sensitive Handover in a Mixed Cell Environment," Proc. of the IEEE Vehicular Technology Conf., Chicago, 1995, pp. 892–896.

49. X. Lagrange and P. Godlewski, "Performance of a Hierarchical Cellular Network with Mobility-Dependent Hand-Over Strategies," Proc. of the IEEE Vehicular Technology Conference, Atlanta, 1996, pp. 1868–1872.

50. M. Lohi, D. Weerakoon, A.H. Aghvami, "Trends in multi-layer cellular system and handover design", IWMC'99, Crete Greece 1999.

51. D. Zeghlache, F. Khan, and D. Calin, "On the Performance of High to Low Level Handoff in a Hierachical Personal Communication System," Proc. of the ACTS Mobile Telecommun. Summit, 1996, pp. 218--223.

52. K. L. Yeung and S. Nanda, "Channel Management in Microcell/Macrocell Cellular Radio Systems," IEEE Trans. on Vehicular Technology, vol. 45, no. 4, 1996, pp. 601–612.

7 Traffic Characterization & Management

7.1 Introduction

The continuous improvement is the performance of personal computers, workstations and distributed multimedia applications requires more and more sophisticated telecommunications network infrastructures. The Asynchronous Transfer mode (ATM) has been adopted as the transport technology to support multimedia services characterized by high speed transmission requirements. The next step in the evolution process of these networks is the total integration of ATM, Multimedia and wireless technologies. This process also involves the development of distributed traffic monitoring for the dynamic allocation of the capacities of ATM wireless channels of heterogeneous traffics.

In section 7.2, a Distributed Traffic Monitoring Algorithm (DTMA) is introduced which utilizes raw nodal data as well as decisions conveyed by neighboring nodes, to detect changes in traffic characteristics and rates and to adapt allocations of channel capacities accordingly. The pertinent traffic characterizations are represented by conditional probabilities, learned by a binary stochastic neural network. The DTMA system induces algorithmic couplings which accelerate the decision process, as compared to the TMA: the rate by which correct decisions are made increases, at the expense of increased rate in erroneous decisions as well. The DTMA outperforms the TMA significantly, as the Kullback-Leibler distances between the various traffic processes increase. Key operational parameters of the DTMA are learned on-line to obtain pre-specified performance levels.

In section 7.3, a mechanism, which is threshold-based to privilege handover requests over new call requests, to decide the admission of multimedia calls using the Available Bit Rate (ABR) mode already standardized by the ATM Forum for transmission on wired ATM networks is proposed. The calls considered are multimedia because they are constituted by a number of monomedia streams that have to be active together. Each user declares, as provided for in the ABR, a Minimum Cell Rate (MCR) and a Peak Cell Rate (PCR) for each media, in terms of the number of radio channels required. A multimedia call can be accepted only if the network can guarantee the minimum requirements, that is, the MCR, to all the component monomedia streams. If, on the other hand, there are free channels, they are assigned, according to a proportional strategy, to the media that before were transmitting with a number of channels less than the PCR. An analytical model is presented to evaluate the performance of this strategy. Finally, the model is used to analyze the influence of the system parameters in a case study: a wireless access network to the ATM backbone in the office area of a factory.

7.2 Real-Time Learning For Traffic Characterization And Distributed Traffic Monitoring In Wireless ATM Networks

Anthony T. Burrell and Titsa P. Papantoni-Kazakos

7.2.1 Introduction

We focus on Asynchronous Transfer Mode (ATM) networks, where traffic Quality of Service (QOS) and efficient network utilization need to be simultaneously satisfied, and where the characteristics of the traffic are time-varying.

We consider digital wireless LANs in the presence of the heterogeneous traffic environments as induced for instance by multimedia traffics. We assume that the LANs contain base-stations and can thus employ centralized scheduling at the transmission stage. We also consider environments where traffic rates are time-varying and we focus on the effective accommodation of the heterogeneous traffic at the transmission stage, with the simultaneous efficient utilization of the transmission channels. This objective is mainly accomplished by a statistically sound Traffic Monitoring Algorithm (TMA) which is implemented by a higher level protocol and allocates channel capacity dynamically to the various heterogeneous traffics in accordance to changes of their rates that it detects. The TMA may be superimposed on either a Framed CDMA (FCDMA) or a Framed Time Domain-Based (FTDB) multiplexing technique, for the transmission of the multimedia traffic. In reference [3], the TMA-FCDMA and a TMA-FTDB combination schemes were evaluated and compared for some traffic scenarios, in terms of the channel utilization factors, traffic rejection rates, and traffic delays performance metrics. It was found that the FCDMA outperforms the FTDB as transmission speeds and lengths of transmitted messages increase.

The Traffic Monitoring Algorithm (TMA) considered in this chapter was first introduced in reference [1] and was further expanded, analyzed and utilized in references [2], [3], [4], as a centralized monitoring mechanism operating on local external traffic arrival data. In conjunction with a specific transmission policy per traffic class, the TMA induces a traffic rejection rate versus wasted capacity rate tradeoff, whose quantitative balance may be controlled via the selection of the TMA operational parameters: a set of decision thresholds. In refernce [6], a learning algorithm was proposed and analyzed for the on-line adaptation of these thresholds, when the TMA operates on only local traffic arrival data. In the present chapter, the TMA is extended to operate in a distributed environment. The learning algorithm for the on-line adaptation of the thresholds is included. In addition, a binary stochastic neural network is proposed for the learning of the traffic characteristics utilized by the distributed TMA.

7.2.2

The Distributed TMA

As in references [2] and [3], let us consider frame structures for the allocation of channel resources to different traffic classes (see the figure below for Time Domain-Based Capacity Allocation). The objective of a traffic monitoring algorithm, such as the TMA, is to adjust the boundaries of the allocations per frame according to the demands of the detected corresponding traffic

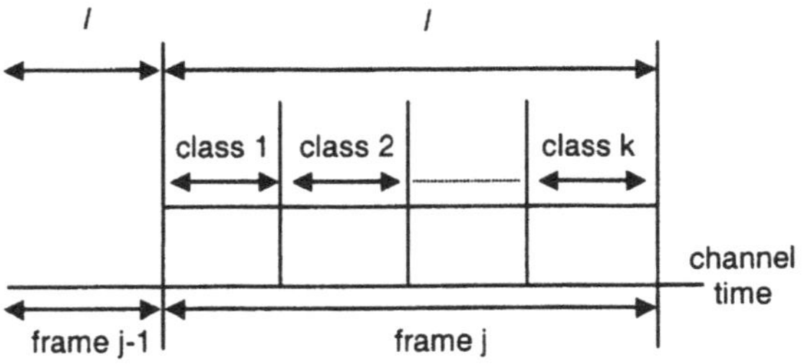

Fig. 7.2.1. Time Domain-Based Capacity Allocation

Let us consider the TMA as applied to a single traffic class, and let us then assume the simple case where the statistics of the traffic are described by one of two possible parametrically known processes, μ_0 and μ_1. The objective of the TMA is then to monitor the active process, where the latter may be alternating between the above two processes. Let $g_j\ (n_1, \dots, n_q)$; j=0, 1 denote the joint distribution of packet arrivals in q consecutive frames, as generated by the process μ_j, where n_i is the number of arrivals in the ith frame. As described in references [1], [2], [4], the centralized TMA, which operates solely on the local traffic data, consists in this case of two single alternating monitoring algorithms $\{g_{1-j} \rightarrow g_j, \eta_{1-j}\}_{j=0,1}$ where η_{1-j} denotes the threshold associated with the monitoring of the $g_{1-j} \rightarrow g_j$ shift. Let the sequence $\{S_i\}_{i \geq 0}$ be the points (corresponding to beginnings of frames) where shift decisions are made by the algorithm, and let at some S_i the distribution $g_{1-j}(\bullet); j=0, 1$ be decided by the centralized TMA as just starting to act. Then, immediately, the single algorithm $\{g_{1-j} \rightarrow g_j, \eta_{1-j}\}$ starts operating. Let $V_{1-j,j,}$ (r) denote the operating value of this algorithm at its r-th algorithmic adaptation. Then, the operating values of the centralized TMA are sequentially updated as follows.

$$V_{1-j,j}(0) \triangleq 0$$

$$V_{1-j,j}(r+1) = \tag{1}$$

$$= \max\left(0, V_{1-j,j}(r) + \log \frac{g_j\left(n_{r+1} | n_1, \ldots, n_r\right)}{g_{1-j}\left(n_{r+1} | n_1, \ldots, n_r\right)}\right)$$

At S_{i+1}, when the threshold η_{1-j} is first crossed by the algorithm, it is decided that the distribution $g_j(\bullet)$ has started acting, and the $\{g_j \rightarrow g_{1-j}, \eta_j\}$ algorithm that monitors a $g_j \rightarrow g_{1-j}$ shift is simultaneously activated.

We note that during time periods when the $\{g_{1-j} \rightarrow g_j, \eta_{1-j}\}$ algorithm operates, the TMA basically assumes that the active process is μ_{1-j}. The log ratio in (1) is the updating step of the centralized TMA.

Let us now consider the case where, in addition to the local data, the TMA also utilizes data from other network nodes, in an indirect fashion. Specifically, let the TMA also utilize decisions induced by the TMAs employed at neighboring nodes. Let the number of the neighboring nodes be N, indexed from 1 to N, and let $u_{i,1-j}$; j = 0, 1, i =1, ..., N, be defined as follows:

$$u_{i,1-j} = \begin{cases} 1; & \text{if the ith node decides that the process } \mu_{1-j} \text{ is active} \\ 0; & \text{if the ith node decides that the process } \mu_j \text{ is active} \end{cases} \tag{2}$$

We then propose a Distributed TMA (DTMA) whose only difference from the centralized TMA lies in the updating step. Indeed, we propose that the log ratio in (1) be substituted by the updating step,

$$\sum_{i=1}^{N} w_i u_{i,j} + \log \frac{g_j(n_{r+1} | n_1, \ldots, n_r)}{g_{1-j}(n_{r+1} | n_1, \ldots, n_r)} \quad (3)$$

where $\{w_i\}$ are constants whose objective is to weigh the contribution of the various TMAs according to their respective performance characteristics. Let α_i and β_i denote respectively the traffic rejection rate and the wasted capacity rate induced by the TMA employed by the ith neighboring node. Then, drawing from some parallelisms with the models considered in reference [5], we propose the following form of the set $\{w_i\}$:

$$w_i = \log \frac{(1-\beta_i)(1-\alpha_i)}{\beta_i \alpha_i}; \; i = 1, \ldots, N \quad (4)$$

We note that the weights $\{w_i\}$ in (4) are positive iff the sum of the traffic rejection and the wasted capacity rates is less than one. We also note from (3) that the decisions of a neighboring TMA contribute then positively to the acceleration of a $g_{1-j} \rightarrow g_j$ shift decision (by increasing the size of the updating step), when these decisions point to the μ_j process as being active.

Remarks: The capacity allocation policy/multiplexing technique/TMA combination proposed and studied in references [1] and [2], as well as the distributed extension introduced in this paper, induce a tradeoff between traffic rejection rates and wasted capacity rates. The balance of this tradeoff is controlled by the value of the decision thresholds $\{\eta_m\}$ used by the TMAs. In references [1], [2], [3], and [4] these thresholds have been selected statically, with only indirect connection to the induced traffic rejection and capacity waste rates.

In reference [6], the centralized TMA in (1) is considered and the thresholds are "learned" on-line to attain a given quantitative balance of the tradeoff between traffic rejection rates and wasted capacity rates.

The learning algorithm in reference [6] can be applied for the on-line adaptations of the decision thresholds deployed by the DTMA as well.

7.2.3 Performance Characteristics Of The DTMA

Let $N_{1-j,j}$; j=0,1 denote the extended stopping variable induced by the DTMA algorithm, that is $N_{1-j,j} \triangleq inf\{r = V_{1-j,j}(r) \geq \eta_{1-j}\}$, with updating step number as in (3). Let also $I_{1-j,j}$ denote the Kullback-Leibler number of the process μ_j with respect to the process μ_{1-j}. Then, as can be generalized from reference [4], under general mixing conditions satisfied by the processes μ_0 and μ_1, the performance of the DTMA is as follows:

$$E\{N_{1-j,j}|\mu_j\} \approx \frac{\log \eta_{1-j}}{\sum_{i=1}^{N} w_i + I_{1-j,j}} \quad , \text{ as } \eta_{1-j} \to \infty$$

$$E\{N_{1-j,j}|\mu_{1-j}\} : \begin{cases} \approx \left[\sum_{i=1}^{N} w_i - I_{j,1-j}\right]^{-1} \log \eta_{1-j}; \text{ if } \sum_{i=1}^{N} w_i > I_{j,1-j} \\ \geq 2^{-1}\eta_{1-j} \quad ; \text{ if } \sum_{i=1}^{N} w_i < I_{j,1-j} \end{cases} \quad , \text{ as } \eta_{1-j} \to \infty$$

(5) and (6) respectively

At the same time, the performance of the TMA in (1) is as shown below.

$$E\{N_{1-j,j}|\mu_j\} \approx \frac{\log \eta_{1-j}}{I_{1-j,j}}, \text{ as } \eta_{1-j} \to \infty \qquad (7)$$

$$E\{N_{1-j,j}|\mu_{1-j}\} \geq 2^{-1}\eta_{1-j}, \text{ as } \eta_{1-j} \to \infty \qquad (8)$$

Comparing expressions (5) and (6) with expressions (7) and (8) we observe that the DTMA is clearly superior to the TMA when the constants $\{w_i\}$ that reflect the contributions of neighboring nodes are such that their sum $\sum_{i=1}^{N} w_i$ is smaller than the minimum between the two Kullback-Leibler numbers I_{01} and I_{10}. Then, as compared to the TMA, the DTMA decreases the asymptotic expected stopping times for correct decisions by factors of $\left[\sum_{i=1}^{N} w_i + I_{01}\right]^{-1} \sum_{i=1}^{N} w_i$

and $\left[\sum_{i=1}^{N} w_i + I_{10}\right]^{-1} \sum_{i=1}^{N} w_i$, respectively, while it maintains the exponentially longer asymptotic expected times for erroneous decisions.

7.2.4
On-Line Threshold Learning

For completeness, we include in this section the learning algorithm in reference [6], for the algorithmic thresholds $\{\eta_0,\eta_1\}$. The algorithm may be applied to the neighboring nodes as well and it is proven to minimize the wasted channel capacity rate, subject to a given upper bound on the traffic rejection rate. Let us define:

N_n : Number of message arrivals in the n-th channel frame.
R_n : Number of message rejections in the n-th channel frame.
$\hat{\eta}_{j,n}$: The adaptation of threshold η_j at the end of the n-th channel frame.
δ : Upper bound on the attainable traffic rejection rate
$\hat{p}_n$: Probability of rejection, estimated at the end of the n-th channel frame.
N_n^c : Cumulative number of message arrivals by the end of the n-th channel frame.
I_{01} : The Kullback-Leibler information number of the process μ_1 with respect to the process μ_0.

We propose the simultaneous $\hat{\eta}_{j,n}$; $j = 0,1$ and $\hat{p}_n$ adaptations, described by the algorithm below.

Algorithm:

Initial values:
Select an initial triple $\left(\hat{\eta}_{0,0}, \hat{\eta}_{1,0}, \hat{p}_0\right)$ of values and a positive integer m.

Computational Steps:
(a) Given computed values $\left(\hat{\eta}_{0,n}, \hat{\eta}_{1,n}, \hat{p}_n\right)$, compute $\hat{\eta}_{0,n+1}$, and $\hat{\eta}_{1,n+1}$ as follows:

$$\hat{\eta}_{0,n+1} = \hat{\eta}_{0,n} -$$

$$-\frac{\hat{\eta}_{0,n}}{\hat{\eta}_{1,n}} \frac{\left[2\log\hat{\eta}_{0,n} + \hat{\eta}_{1,n} I_{01}\right]^2}{2I_{01}} \frac{\hat{p}_n(1-\hat{p}_n)(\hat{p}_n-\delta)}{(\hat{p}_n-\delta)^2+\delta(1-\delta)} \tag{9}$$

$$; if \ n = k + \frac{m}{2}k(k-1), \ k = 1,2,...$$

$$\hat{\eta}_{0,n+1} = \hat{\eta}_{0,n} \ ; otherwise$$

$$\hat{\eta}_{1,n+1} = \hat{\eta}_{1,n} +$$

$$+\frac{\left[2\log\hat{\eta}_{0,n} + \hat{\eta}_{1,n} I_{01}\right]^2}{2I_{01}\log\hat{\eta}_{0,n}} \frac{\hat{p}_n(1-\hat{p}_n)(\hat{p}_n-\delta)}{(\hat{p}_n-\delta)^2+\delta(1-\delta)} \tag{10}$$

$$; if \ n = k + \frac{m}{2}k(k-1), \ k = 1,2,...$$

$$\hat{\eta}_{1,n+1} = \hat{\eta}_{1,n} \ ; otherwise$$

Select a positive value ε, such that the $\hat{\eta}_{0,n+1}$, and $\hat{\eta}_{1,n+1}$ values in (9) and (10) are "corrected" to ε whenever they fall below this value.

(b) Use $\hat{\eta}_{0,n+1}$, and $\hat{\eta}_{1,n+1}$ in the TMA, in conjunction with the arrival process of the traffic and the multiplexing/capacity allocation system, to produce the random variables N_{n+1}, and R_{n+1}. Then, update $\hat{p}_n$ as follows:

$$\hat{p}_{n+1} = \hat{p}_n + \frac{R_{n+1} - N_{n+1}\hat{p}_n}{N_n^c + N_{n+1}} -$$

$$-\frac{2k + mk(k-1)}{2(k+1) + m(k+1)k} \frac{\hat{p}_n(1-\hat{p}_n)(\hat{p}_n-\delta)}{(\hat{p}_n-\delta)^2+\delta(1-\delta)} \tag{11}$$

$$; if \ n = k + \frac{m}{2}k(k-1), \ k = 1,2,...$$

$$\hat{p}_{n+1} = \hat{p}_n + \frac{R_{n+1} - N_{n+1}\hat{p}_n}{N_n^c + N_{n+1}} ; otherwise$$

For some small positive value ζ, the $\hat{p}_{n+1}$ value in (11) is "corrected" to ζ, if $\hat{p}_{n+1} < \zeta$, and to 1-ζ if >1- ζ. □

Remarks: The learning algorithm converges almost surely under general conditions (see reference [6]).

7.2.5
Traffic Characterization Via Stochastic Neural Networks

To this point, we assumed that the conditional probabilities in (3), used in the updating step of the DTMA, are parametrically known. However, this assumption may be unreasonable for the diverse wireless environments. Instead, these probabilities may need to be properly learned by an appropriate stochastic mechanism. In this chapter, we propose the deployment of a neural network which is pre-trained to reproduce discrete and finite memory mappings (or approximate) of conditional probabilities/densities induced by the acting network traffics. The knowledge of such conditional probabilities or densities is instrumental in the development of the distributed TMA that was presented in Section II of this chapter.

7.2.5.1
Discrete Finite Memory Robust Mappings

Let $x_1, ..., x_n$ denote a sequence of observations, representing the traffic environment (such as number of arrivals within fixed-length intervals). Then, given $x_1, ..., x_n$, the objective of the discrete mapping is to predict which one of M distinct regions, the observation x_{n+1} is going to be in. Denoting these regions A_j, j=1, ..., M, a high-performance predictive encoding operation requires in fact the computation of the conditional probabilities which are used to map an observed sequence $\{x_1, ..., x_n\}$ onto each of the regions $\{A_j\}$, with corresponding probabilities $\{p_j(x_1, ..., x_n)\}$. Two problems arise immediately:

(i) Exploding computational load, due to the increasing memory represented by the sequences $(x_1, ..., x_n)$.

(ii) Statistical information on the sequences $(x_1, ..., x_n)$ needed for the computation of the probabilities $\{p_j(x_1, ..., x_n)\}$.

The first problem is resolved, if the increasing memory is approximated by finite, say size-l memory. That is, the increasing computational load is, instead, bounded if the process that generates the observations is approximated by an l-order Markov process. Then, the information loss is minimized when the process is Gaussian (see Blahut reference [7]).

Thus, to reduce the exploding computational load due to increasing data memory, we may *initially* model the process that generates the environmental data or observations by an l-order Gaussian Markov process, whose autocorrelation m x m matrix Q, has components identical to the corresponding components of the original process. We name this initial (Gaussian and Markov) process, *nominal* process.

Starting with our nominal process, but incorporating then statistical uncertainties in our model, we are led to a powerful (qualitatively) robust formalization, which results in a stochastic mapping (see Papantoni-Kazakos *et al.* reference [8]), as follows:

Given observations ($x_1, \ldots, x_n$), use the l most recent observations for the prediction of the next datum x_{n+1}, and defining $y_m^T = [x_{n-m+1}, \ldots, x_n]$, decide that $x_{n+1} \varepsilon A_j$ with probability $q_j(y_m)$, defined as follows,

$$q_j(y_m) = \frac{1}{M} r\,(y_m) + [1 - r(y_m)] p_j(y_m), \qquad (12)$$

where $p_j\,(y_m)$ is the conditional probability of $x_{n+1} \varepsilon A_j$, given $y_m^T = [x_{n-m+1}, \ldots, x_n]$, as induced by the Gaussian and Markov nominal process, and where, for some positive finite constant λ,

$$r(y_m) \triangleq 1 - \min\left[1, \frac{\lambda}{\sqrt{y_m^T Q^{-1} y_m}}\right] \qquad (13)$$

The value of the constant λ in (13) represents the level of confidence to the "purity" of the data vector y_m, in terms of it being generated by the nominal Gaussian process: the higher the value of λ, the higher the level of confidence, where as λ decreases, increased weight on purely random mappings (represented by the probability $\frac{1}{M}$ per region) is induced.

In addition, robust estimation of the autocorrelation matrix Q will be required. The components of the autocorrelation matrix Q should emerge from the statistics of the nominal Gaussian process. It is thus necessary to provide a scheme for the robust estimation of the matrix Q, in which observations generated from the outlier process are rejected in the estimation of the components of the matrix. Special attention should be paid to allow for the existence of the inverse autocorrelation matrix Q^{-1} from the estimates of its components.

The robust prediction expression in (12) was based on a Gaussian assumption for the nominal process that generates the data in the environment, where the latter assumption was the result of an information-theoretical approach to the reduction of the computational load caused by increased past memory. The important robust effects induced by the mapping in (12) remain unaltered, however, when instead, the probability $p_j(y_m)$ in (12) arises from an arbitrary non-Gaussian process, and when its conditioning on y_m is substituted by conditioning on quantized values of the scalar quantity $y_m^T Q^{-1} y_m$. When quantized values are involved, the implementation of the mapping in (12) requires the following stages:

(a) Preprocessing. This stage corresponds to long-term memory and involves the robust preestimation, and storage of the matrix Q^{-1}.

(b) Processing. This stage corresponds to short-term memory. It uses the matrix Q^{-1} from the preprocessing step and the observation vector y_m to: (i) first compute the quadratic expression $y_m^T Q^{-1} y_m$, (ii) then, represent $y_m^T Q^{-1} y_m$ in a quantized form comprised of N distinct values, (iii) finally, use the quantized values in (ii) to compute the corresponding value of the function $r(y_m)$ in (13).

(c) Predictive Mapping. This stage involves the estimation of the probabilities $\{p_j(y_m)\}$ and the computation of the probabilities $\{q_j(y_m)\}$ in (12) using inputs from the processing stage, and the subsequent implementation of the prediction mappings.

The three different stages above are performed sequentially by separate but connected neural structures, named *preprocessing layer*, *processing layer*, and *predictive mapping layer*, respectively. Our focus in this paper is on the latter layer: its structure and its operations. Towards that direction, we first note that, due to the quantization operations at the processing layer, the expression in (12) takes the following form:

$$q_{j\rho} = \frac{1}{M} r_\rho + [1 - r_\rho] p_{j\rho} \, ; \quad \text{for } y_m^T Q^{-1} y_m \to R_\rho \, ; \ \rho = 1, \ldots, N \qquad (14)$$

where $q_{j\rho}$, $p_{j\rho}$, and r_ρ denote, respectively, the probabilities $q_j(y_m)$ and $p_j(y_m)$ and the number $r(y_m)$, when the quantized value of $y_m^T Q^{-1} y_m$ equals R_ρ.

7.2.5.2
The Predictive Mapping Layer

Consider the integer M in (14), and let s be a unique positive integer, such that $2^{s-1}<M\leq 2^s$. Then, in modulo-2 arithmetic, each state *j*, *j*=1, ..., M can be represented by an s-length 0-1 binary sequence $x_1 \cdots x_s$. The state R_ρ is provided as an input to the prediction layer by the processing layer, and the former produces a binary sequence $x_1 \cdots x_s$ as a prediction mapping. Given the state R_ρ, the operations of the prediction layer must be such that, a given prediction sequence $x_1 \cdots x_s$ is produced stochastically with probability

$$q\,(x_1 \cdots x_s / R_\rho) = \frac{1}{M} r_\rho + [1 - r_\rho] p(x_1 \cdots x_s / R_\rho) \qquad (15)$$

where expression (15) is the same as expression (14); when the binary representation of the positive integer j in the latter is $x_1 \cdots x_s$, and where $p(x_1 \cdots x_s / R_\rho)$ is the prediction mapping generated by the nominal process that represents the *actual data environment*. Due to the stochastic nature of the rule in (15), such is also the nature of the predictive mapping layer, whose neural representation corresponds then to a stochastic neural network, first developed by Kogiantis *et al.* reference [9], when the response of each neuron is limited to binary. We proceed with the description of the latter representation.

Let us temporarily assume that the probabilities $p(x_1 \cdots x_s / R_\rho)$ have been "learned" and are known. Without lack in generality, let us also assume that $M=2^s$. The original constraint of binary firing per neuron in the prediction layer leads us to the digital representation of the future states $\{x_1, \ldots, x_s\}$. The design can be accommodated easily in a binary tree structure. In detail, given the observed state R_ρ and the resulting r_ρ value, the mapping $x_1 \cdots x_s$ can be obtained via a stochastic binary tree search, on the 2^s-leaves tree, as follows:
(a) With probability r_ρ a fair tree-search is activated, where the tree-node x_1, $x_1=0, 1$ is visited with probability 0.5, and each of the two tree-nodes branching off a visited tree-node $x_1 \cdots x_k$, $1 \leq k \leq s-1$ is also visited with probability 0.5.
(b) With probability $1-r_\rho$ a generally biased tree-search is activated, where the tree-node x_1 is visited with probability $p(x_1/R_\rho)$, while from a visited tree-node $x_1 \cdots x_k$, $1 \leq k \leq s-1$ the tree-node $x_1 \cdots x_k x_{k+1}$ is visited with probability:

$$p(x_{k+1} / x_1 \cdots x_k, R_\rho) \triangleq p(x_1 \cdots x_k x_{k+1} / x_1 \cdots x_k, R_\rho)$$

where

$$p(x_1 \cdots x_s / R_\rho) = p\,(x_1 / R_\rho) \cdots p(x_{k+1} / x_1 \cdots x_k, R_\rho) \cdots p(x_s / x_1 \cdots x_{s-1}, R_\rho) \qquad (16)$$

Thus, the predictive mapping layer may be viewed as been comprised of a fair-search binary tree and a number of biased-search binary trees, each of the latter corresponding to a specific observation state R_ρ. Given R_ρ, the common fair-search binary tree is activated with probability r_ρ, while, with probability $1-r_\rho$, the biased-search binary tree that corresponds to the state R_ρ is activated, instead; we name the latter tree, the R_ρ tree. The nodes of each of the above binary trees are neurons that "fire," if the corresponding tree-nodes are "visited." Given R_ρ, a specific mapping $x_1 \cdots x_s$ is generated either equiprobably from the fair-search binary tree with probability r_ρ, or from the R_ρ-tree via the sequential stochastic representation in (16) with probability $1-r_\rho$. It is thus in the R_ρ-tree that the probabilities which generate the data of the environment must be "learned" and then used to generate prediction mappings.

Given the observation state R_ρ, consider the R_ρ-tree in conjunction with the sequential stochastic representation in (16) of the corresponding prediction mappings, as generated by the process representing the actual data environment. Let $u_{x_1 \cdots x_k}, 1 \le k \le s$ represent the binary random output of the neuron that corresponds to the node $x_1 \cdots x_k$ of the R_ρ-tree. Then, $u_{x_1 \cdots x_k} = 1$ if and only if $u_{x_1 \cdots x_i} = 1, \forall i \le k$. Thus, the output $u_{x_1 \cdots x_k}$ may be viewed as generated by a product, $W_{x_1} W_{x_2/x_1} \cdots W_{x_k/x_1 \cdots x_{k-1}}$, of mutually independent binary random variables $\{W_{x_i/x_1 \cdots x_{i-1}}\}_{1 \le i \le k}$, whose distributions at the operational stage of the R_ρ-tree must be as follows (in view of (16)):

$$\begin{aligned} P(u_{x_1 \cdots x_k} = 1) &= P(W_{x_1} W_{x_2/x_1} \cdots W_{x_k/x_1 \cdots x_{k-1}} = 1) = \\ &= p(x_1 \cdots x_k / R_\rho) = p(x_1 / R_\rho) \cdots p(x_k / x_1 \cdots x_{k-1}, R_\rho) = \\ &= P(W_{x_1} = 1) P(W_{x_2/x_1} = 1) \ldots P(W_{x_k/x_1 \cdots x_{k-1}} = 1); \quad 2 \le k \le s \\ P(u_{x_1} = 1) &= P(W_{x_1} = 1) = p(x_1 / R_\rho), \end{aligned} \tag{17}$$

where

$$P(W_{x_k/x_1 \cdots x_{k-1}} = 1) = p(x_k / x_1 \cdots x_{k-1}, R_\rho); \quad 2 \le k \le s \tag{18}$$

The above logical arguments and expressions lead to the following neural structure of the R_ρ-tree: (a) The neuron corresponding to the tree-node x_1; $x_1=0, 1$ has a binary random variable W_{x_1} built in, where $W_0=1-W_1$. At the operational stage, the neuron must be activated with probability $p(x_1/ R_\rho)$; thus, $P(W_{x_1} = 1) = p(x_1 / R_\rho)$ then, where $P(W_1 = 1) = 1 - P(W_0 = 1)$. (b) For $k \ge 2$, the neuron corresponding to the tree-node $x_1 \cdots x_k$ has a binary random variable $W_{x_k/x_1 \cdots x_{k-1}}$ built in and fires, if and only if the latter variable takes the value 1 and simultaneously the neuron corresponding to the tree-node $x_1 \cdots x_{k-1}$ fires as well. Thus, the binary neural output $u_{x_1 \cdots x_k}$ is formed as a product $u_{x_1 \cdots x_{k-1}} W_{x_k/x_1 \cdots x_{k-1}}$, where

$$P(u_{x_1 \cdots x_k} = 1) = P(u_{x_1 \cdots x_{k-1}} W_{x_k/x_1 \cdots x_{k-1}} = 1) = P(u_{x_1 \cdots x_{k-1}} = 1) P(W_{x_k/x_1 \cdots x_{k-1}} = 1) \tag{19}$$

and where, at the operational stage of the R_ρ -tree, the probability $P(W_{x_k/x_1\cdots x_{k-1}} = 1)$ must be as in (18). We note that

$$W_{1/x_1\cdots x_{k-1}} = 1 - W_{0/x_1\cdots x_{k-1}}, \ \forall k \geq 2, \ \forall x_1\cdots x_{k-1} \qquad (20)$$

and thus

$$P(W_{1/x_1\cdots x_{k-1}} = 1) = 1 - P(W_{0/x_1\cdots x_{k-1}} = 1), \ \forall k \geq 2, \ \forall x_1\cdots x_{k-1}$$

As it is clear from the derivations and arguments in this section, the parameters of interest in the R_ρ -tree neural network consist of the independent binary random variables W_1 and $\{W_{1/x_1\cdots x_{k-1}}; x_i = 0, 1; 1 \leq i \leq k-1\}_{2 \leq k \leq s}$, whose distributions $p(1/ \quad R_\rho)$ and $\{ p(1/ x_1\cdots x_{k-1,} R_\rho);\ x_i = 0, 1;\ 1 \leq i \leq k-1\}_{2 \leq k \leq s}$, must be "learned" in advance, via interaction with the environment which represents the i th model.

Learning at the predictive mapping layer

Given the R_ρ -tree, we observe that, due to (17), any adaptations of the probability $P(u_{x_1\cdots x_s} = 1)$ *backpropagate* to adaptations of each of the probabilities $P(W_{x_1} = 1), \ \ldots, \ P(W_{x_s/x_1\cdots x_{s-1}} = 1)$. It thus suffices to focus on the learning of the probabilities $\{P(u_{x_1\cdots x_s} = 1)\}$ for the various binary sequences $x_1\cdots x_s$, which correspond to the responses of the output or "visible" neurons in the R_ρ -tree network. For easiness in presentation, let us now consider a fixed sequence $x_1\cdots x_s$ (in conjunction with the fixed observed state R_ρ that represents the R_ρ -tree). Let then p denote the value of the probability $p(x_1\cdots x_s / R_\rho)$, as induced by the environment, and let q denote the value of the probability $P(u_{x_1\cdots x_s} = 1)$. Let the natural number n denote discrete observation time from the beginning of the learning stage, and let $\hat{p}_n$ and $\hat{q}_n$ denote estimates at time n of the probability values p and q, respectively. Finally, let the random variable V_n be defined as equal to 1, if the environmental event $\{x_1\cdots x_s / R_\rho\}$ occurs at time n, and as equal to 0, otherwise, and let $Z_n \underline{\underline{\Delta}} V_n(1 - W_\rho)$, $W_\rho = \begin{cases} 0; \ \text{w.p.}\ 1 - r_\rho \\ 1; \ \text{w.p.}\ r_\rho \end{cases}$. In Kogiantis *et al.* reference [10], a Kullback-Leibler matching criterion between p and q was used in conjunction with Newton's iterative numerical method to develop the supervised learning algorithm stated below.

ALGORITHM

Initial Values: Select an initial value $\hat{q}_1 > 0$, while $\hat{p}_1 = v_1$.

Computational Steps: (a) Given computed value $\hat{p}_n$ and z_{n+1} as above, compute $\hat{p}_{n+1}$ as follows:

$$\hat{p}_{n+1} = \hat{p}_n + \frac{z_{n+1}[1-r_{i\rho}]^{-1} - \hat{p}_n}{n+1} \tag{21}$$

For some small positive value δ, the value $\hat{p}_{n+1}$ is corrected to δ if $\hat{p}_{n+1} < \delta$, and is corrected to $1-\delta$ if $\hat{p}_{n+1} > 1-\delta$.

(b) Given computed values $(\hat{q}_n, \hat{p}_n)$, given z_{n+1}, compute $\hat{q}_{n+1}$ as follows:

$$\hat{q}_{n+1} = \hat{q}_n - \frac{(\hat{q}_n - \hat{p}_n)\hat{q}_n(1-\hat{q}_n)}{(\hat{q}_n - \hat{p}_n)^2 + \hat{p}_n(1-\hat{p}_n)} + \\ + \frac{z_{n+1}[1-r_{i\rho}]^{-1} - \hat{p}_n}{n+1}\left[\frac{\hat{q}_n(1-\hat{q}_n)}{(\hat{q}_n - \hat{p}_n)^2 + \hat{p}_n(1-\hat{p}_n)} + \right]^2 \tag{22}$$

where $\hat{p}_{n+1} - \hat{p}_n = \frac{z_{n+1}[1-r_{i\rho}]^{-1} - \hat{p}_n}{n+1}$ from (10).

For some small positive value ζ, the value $\hat{q}_{n+1}$ is corrected to ζ if $\hat{q}_{n+1} < \zeta$, and is corrected to $1-\zeta$ if $\hat{q}_{n+1} > 1-\zeta$.

In Kogiantis *et al.* reference [9], the following theorem was proved.

Theorem

Let the process which generates the observed data in the environment be ergodic. Let then s denote the probability of the event $\{x_1 \cdots x_s / R_\rho\}$, as induced by the latter process. Then, the supervised learning algorithm converges to the probability s, almost surely, with rate inversely proportional to the sample/iteration size n.

We note that the Theorem, if the process that generates the observed data in the environment is ergodic, and if $\{s(x_1 \cdots x_s / R_\rho)\}$ denote the prediction mappings induced by the latter process, then, via the learning algorithm and with almost sure convergence, the prediction mappings produced by the predictive mapping layer are governed by the probabilities

$$\{q'(x_1 \cdots x_s / R_\rho) \triangleq (1-r_{i\rho})s(x_1 \cdots x_s / R_\rho) + M^{-1}r_{i\rho}\}$$

In Kogiantis *et al.* reference [9], it was found that the learning algorithm converges rapidly to predictive probability mappings that are close to those induced by the environment, even under mismatch network conditions. Specifically, when past dependence decays fast with distance, then, even when the network order is less than the order of the Markovian environmental model, convergence to almost the true process is attained in less than fifty iterations in most cases.

References

1. A. T. Burrell, "Traffic Management in Broadband Integrated Services Digital Networks," Ph.D. Dissertation, Univ. of Virginia, August 1994.

2. A. T. Burrell, D. Makrakis, and P. Papantoni-Kazakos, "Traffic Monitoring for Capacity Allocation of Multimedia Traffic in Broadband Networks," *Telecommunications Systems Journal*, 1998, Vol. 9, pp. 173-206.

3. A. T. Burrell, P. Papantoni-Kazakos, et. al., "Dynamic Capacity Allocation and Hybrid Multiplexing Techniques for ATM Wireless," *Wireless Networks Journal*, 1998, Vol. 3, pp. 307-316.

4. A. T. Burrell and P. Papantoni-Kazakos, "Extended Sequential Algorithms for Detecting Changes in Acting Stochastic Processes," *IEEE Trans. on Systems, Man, and Cybernetics*, Vol. 28, No. 5, Sept. 1998, pp. 703-710.

5. D. A. Pados, P. Papantoni-Kazakos, D. Kazakos and A. Kogiantis, "On-Line Threshold Learning for Neyman-Pearson Distributed Detection," *IEEE Transactions on Systems, Man, and Cybernetics*, Vol. 24, No. 10, pp. 1519-1531, Oct. 1994.

6. A. T. Burrell and P. Papantoni-Kazakos, "Real Time Learning with Dynamic Capacity Allocation for Heterogeneous Traffic Management," *6th International Conference on Telecommunication Systems*, Modeling, and Analysis, March 8-10, 1998, Nashville, Tennessee.

7. R. E. Blahut, "Hypothesis Testing and Information Theory," *IEEE Trans. Inform. Theory*, Vol. IT-20, pp. 405-417, 1987.

8. P. Papantoni-Kazakos, D. Kazakos, and K. Birmiwal, "Predictive Analog-to-Digital Conversion for Resistance to Data Outliers," *Information and Computation*, Vol. 98, No. 1, pp. 56-98, May, 1992.

9. A. G. Kogiantis and P. Papantoni-Kazakos, "Operations and Learning in Neural Networks for Robust Prediction," *IEEE Trans. on Systems, Man, and Cybernetics*, Vol. 27, Part B, June, 1997.

7.3 Analysis of an Admission Mechanism for ABR Correlated Traffic in a Multimedia ATM Wireless Network

A. Lombardo, S. Palazzo, G. Schembra

7.3.1 Introduction

The continuous improvement in the performance of personal computers, workstations and distributed multimedia applications requires more and more sophisticated telecommunication network infrastructures. The Broadband Integrated Services Digital Network (B-ISDN) is the communication network of the future. Its target is to provide an efficient and reliable end-to-end transport for various services. The Asynchronous Transfer Mode (ATM) has been adopted as the transport technology of B-ISDN networks on account of its capability to integrate any type of service, whatever its characteristics and requirements.

The aim of ATM network management is not only *congestion control*, dealing with the problem of reducing load during overload, but also *traffic management* [1], dealing with load reduction under overload or load increase during underload and, more important, ensuring the QoS user requirements are met in spite of varying load conditions. Thus, traffic management is required even if the network is underloaded. For this purpose the ATM Forum specified five traffic categories, called ATM transfer capabilities in ITU-T Recommendation I.371 [2]. These are Constant Bit Rate (CBR), real-time (rt) and non real-time (nrt) Variable Bit Rate (VBR), Unspecified Bit Rate (UBR) and Available Bit Rate (ABR). The CBR and both the VBR categories intend to address applications with specified Quality of Service (QoS) requirements. In these cases traffic parameters corresponding to the traffic connections are negotiated at the call setup and then, once the requested connection is accepted, its QoS is guaranteed throughout the session duration. The UBR service category, on the contrary, needs no guaranteed service requirements. In the middle is the ABR service category, whose major goal is the inexpensive support of applications with low QoS requirements. This category adopts a "closed-loop" mechanism to control the source rate according to the available network resources and only guarantees a Minimum Cell Rate (MCR), specified by the traffic source together with its Peak Cell Rate (PCR). ATM switches can use a Virtual Source/Virtual Destination (VS/VD) feature to segment the ABR control loop into smaller loops. This feature allows feedback from nearby switches to reach sources faster, and hop-by-hop control.

At the same time wireless technology has become very popular, first for telephone applications only, and then for high-speed communications. Many standards have been produced for high-speed wireless local area networks (Wireless LANs), such as IEEE 802.11 [3]. Several important aspects of the next-generation wireless networks with their implications for teletraffic are discussed in [4].

Now, the next step in the evolution of networks is the total integration of ATM, multimedia and wireless technologies, to obtain the so-called Wireless ATM [5-8]. Wireless ATM uses a wired ATM network as a backbone and Base Stations (BSs) for wireless access. Multiple mobile terminals connect to the BSs to share their capacity, which can be expressed in terms of a number of channels. The challenge now is to make the access to an ATM network through BSs transparent, so that users can use the same capabilities and functionalities, irrespective of whether they enter the network directly or through a wireless connection. In [9] the concept of ATM Wireless Access (AWA) was introduced to describe the use of ATM transport capabilities to support multimedia services, characterized by high-speed transmission requirements and limited mobility during communications. The same authors illustrated a possible protocol architecture for AWA in [10], and gave some examples of possible realizations of this architecture to obtain connections for CBR and UBR traffic.

Complete integration of ATM and wireless will only be achieved when efficient traffic management strategies are found, which are able to optimize Call Admission Control (CAC) in a wireless ATM environment, as has already been done for wired ATM networks.

The problem of Admission Control in wireless networks is an important issue and has been dealt with in many papers. In [11] class-based wireless call admission control is proposed, with the aim of allocating different bandwidths to different traffic classes, according to their QoS requirements. In [12] the authors propose distributed call admission control in wireless networks with the purpose of limiting the probability of call loss due to handover in overloaded cells. In [13-14] it was observed that the admission mechanism has to privilege handover requests over new call requests because the former are made by running calls and, therefore, cannot tolerate either loss or delay. One way to realize a priority mechanism is channel reservation, whereby a number of channels in the cell are reserved for access of handover calls only. Nevertheless, although this scheme is very simple and effective, it is wasteful if the amount of handover traffic is very low. A more efficient mechanism is a threshold-based channel-reservation policy [14], also called *Guard Channel Policy* in [15]. It uses a threshold to decide whether to accept new call requests or not, whereas handover requests are always accepted if the required channels are available. Moreover, in [15] this mechanism has been demonstrated to be optimal for the call admission control problem when a linear objective function of the new-call blocking probability and the loss probability due to handover has to be minimized.

In this paper we extend what has been discussed in the previous papers, proposing and analyzing a new admission mechanism which has been specifically conceived for wireless ATM networks supporting ABR multimedia traffic. The traffic is generated by heterogeneous multimedia applications, comprising different media correlated with each other. Correlation is an important factor that has to be taken into account in admission control because media belonging to the same multimedia application have to be accepted together, or not accepted at all. Each media has different bandwidth requirements. For each media, the user requests an MCR and declares a PCR. The admission mechanism decides whether to accept a request or not, taking care not to deteriorate the QoS of other calls below their MCR, and also uses a fixed threshold as in [14] so as to privilege handover requests over those for new calls.

The paper is organized as follows. In Section 2 the system we are analyzing is described, together with the related assumptions. In Section 3 the model is outlined and the performance parameters are defined and analytically evaluated. In Section 4 the effectiveness of the proposed mechanism is assessed through a case study; the mechanism is applied in an indoor environment with a wireless network installed in a factory. Finally, in Section 5, the authors' conclusions are presented.

7.3.2 System description

The system we are focusing on in this paper is a wireless cell, belonging to a cell cluster, with C channels in each cell; the input traffic comprises both new calls and calls coming from neighboring cells in the same cluster due to handover. The admission mechanism we use to privilege handover requests over new call requests is a threshold-based channel-reservation policy. Let σ be the threshold used to decide whether to accept new call requests or not. While handover requests are always accepted if the required channels are available, new call requests are only accepted if the number of busy channels in the cell is less than or equal to σ.

The admission controller, one for each cell in the wireless access network, is an explicit rate switch [16], implementing the third-generation ABR strategy, the Explicit Rate (ER), by which it is possible to indicate directly the bandwidth that can be assigned to the connection by means of the ER field in the Resource Management (RM) cells. Since the proposed strategy is applied between the mobile terminal and the call admission controller, it has to use the VS/VD feature.

The input traffic we consider in this paper is generated by N heterogeneous users, using multimedia applications with a number of heterogeneous media correlated with each other. It is reasonable to assume that in a certain time interval, the one in which we will study the mechanism, the number N of users in a cluster that may involve the tagged cell remains constant. This is the case, for example, of an office, according to the classification proposed in [17]. In fact, an office is a cell with a small set of 'regular' occupants working there who often move to neighboring cells, to communicate with their colleagues for example, or to move to a factory database or to the factory cafeteria.

Intermedia correlation has to be taken into account during the admission of a call because if one of the media is accepted, the others also have to be accepted. In other words, a multimedia call can be accepted if and only if a number of channels at least equal to the minimum required for all the constituent media is available. For the sake of simplicity, in the rest of the paper we will consider calls constituted by only two media, for example audio and video[1], because, once this case has been studied, the more general case is straightforward.

Each user requesting admission in a cell requires an MCR, in terms of a minimum number of channels, L_A for audio, and L_V for video. The network has to guarantee these values throughout the connection in that cell. Besides, users declare a PCR, that is, the maximum number of channels they can use for audio, H_A, and for video, H_V. When an admission request has been accepted, either for a new call or a handover, the network will guarantee a number of channels ranging between L_A and H_A for the audio media, and between L_V and H_V for the video media (Allowable Cell Rate - ACR).

Let us assume that multimedia users arrive in the cell from the neighboring cells with a Poisson distribution. Let λ_H be the arrival rate of the requests due to handover (high priority) and λ_{NC} be the arrival rate of the requests due to new calls (low priority). Let us also assume that all calls, both those requesting handover and new calls, once accepted, require the same negative-exponential distributed channel holding time [13]. Let $1/\mu$ be the mean value of the channel holding time.

The admission mechanism works according to the following rules:

- **Condition (1):** an admission request arriving due to handover is accepted if the number of free channels is higher than or equal to $H_A + H_V$, that is, if it is possible to assign it the maximum number of channels required by the two media. In this case, H_A channels will be assigned to the audio media and H_V channels to the video media. If the request is from a new call, it is accepted if, besides the previous conditions, the number of busy channels is less than or equal to the threshold σ;
- **Condition (2):** an admission request arriving due to handover is also accepted if, although the number of free channels is less than $H_A + H_V$, that is, it is not possible to assign the maximum number of requested channels, it is greater than or equal to $L_A + L_V$. In this case, all the available channels will be assigned, distributing them proportionally to the channel range requested by each media, $R_A = H_A - L_A$ for audio, and $R_V = H_V - L_V$ for video. Since the number of channels to be assigned has to be an integer, it may be that all channels except one will be assigned with this rule; in this case, the last channel will be given to one of the two media at random. If the request is from a new call, it is accepted if, besides the previous conditions, the number of busy channels is less than or equal to the threshold σ;
- **Condition (3):** an admission request arriving due to handover is also accepted if, although the number of free channels is less than $L_A + L_V$, it is possible to take the remaining number of channels away from the other calls active in this cell, in such a way that L_A channels can be assigned to the audio media of the requesting call and L_V to the video media; care must be taken, however, to

[1] Although ABR service class has been introduced for data transmission, it can be applied to real-time applications, provided that traffic sources are able to adapt their output bit rate according to a feedback received by the network.

continue to guarantee the minimum required number of channels for the other active calls. The mechanism whereby channels are taken away from other calls also works on a proportional basis: each media involved in an active call will give up a number of channels proportional to the number of channels it has over and above the minimum. If this proportional part is not an integer, each media will give up a number equal to the minimum integer contained in this part. The remaining channels will be randomly subtracted. If the request is from a new call, it is accepted if, besides the previous conditions, the number of busy channels is less than or equal to the threshold σ .

If none of the previous conditions is fulfilled, the request is rejected and a call loss due to handover or a new-call block occurs. Once a request has been accepted, the call will be dealt with regardless of whether it came from another cell (handover) or was activated in the current cell (new call).

When a call ends or when an active user leaves the cell, a number of channels will be freed. Two situations can occur:

- **Condition (a):** all the other sources are using a number of channels equal to the maximum requested. In this case the channels previously occupied by the call which has just ended are made available for future admission requests;
- **Condition (b):** not all the other sources are using a number of channels equal to the maximum requested. In this case all the channels previously occupied by the call which has just ended are distributed among the active calls not using the maximum number of requested channels. The redistribution will be proportional to the number of channels needed to reach the maximum. Here again, channels which cannot be fairly distributed will be randomly assigned.

7.3.3 System model

In order to evaluate the performance of the system described in the previous section, we now define a continuous-time, Markov-based model. In the rest of the paper we will use the symbols introduced in the previous section with a subscript, to indicate the call being referred to. For example, we will use the symbols L_{A_h}, L_{V_h}, H_{A_h} and H_{V_h} to specify the MCR and PCR values of the hth call, and with the symbols λ_{H_h} and λ_{NC_h} to indicate the admission request rate due to handover and new call, respectively. Let us denote the system state at the instant t as $S(t) = \left(S^{(1)}(t), S^{(2)}(t), \ldots, S^{(N)}(t)\right)$, where $S^{(h)}(t) = \left(C_h^{(A)}(t), C_h^{(V)}(t)\right)$ represents the state of the hth multimedia source, $h \in \{1,2,\ldots,N\}$; $C_h^{(A)}(t) \in \left\{0, L_{A_h}, \ldots, H_{A_h}\right\}$ and $C_h^{(V)}(t) \in \left\{0, L_{V_h}, \ldots, H_{V_h}\right\}$ are the number of channels assigned at the instant t to the audio and the video media of the hth multimedia source, respectively. Let $\Im^{(\Sigma)}$ be the system state feasibility region [18-19], that is, the set containing all the possible states of the Markov chain of $S^{(\Sigma)}(t)$. At a first glance the size of $\Im^{(\Sigma)}$ may appear huge and capable of exponentially exploding with an increasing number of sources. Nevertheless, it is strongly limited at least by the following conditions:

- $\sum_{h=1}^{N}\left(C_h^{(A)}(t)+C_h^{(V)}(t)\right)\le C$
- $C_h^{(A)}(t)=0$ if and only if $C_h^{(V)}(t)=0$, because these are the number of channels used by the audio and video media belonging to the same multimedia call. Therefore, given the fact that either both media or neither of the two are accepted in the cell, there is no possibility of only one of them using the channels.

Moreover, as we will see in Section 4, other conditions, depending on the particular case to be analyzed, limit the system's dimensions even further. The generic term of the system transition matrix $Q^{(\Sigma)}$, representing the transition rate from the state $\underline{i}=\left(i_{A_1},i_{V_1},\dots,i_{A_N},i_{V_N}\right)\in\Im^{(\Sigma)}$ to the state $\underline{j}=\left(j_{A_1},j_{V_1},\dots,j_{A_N},j_{V_N}\right)\in\Im^{(\Sigma)}$ is:

$$q_{\underline{i},\underline{j}}=\begin{cases}\lim\limits_{\Delta t\to 0}\text{Prob}\left\{S(t+\Delta t)=\underline{j}/S(t)=\underline{i}\right\} & \forall\,\underline{j}\ne\underline{i}\\ 1-\sum\limits_{\substack{l\in\Im^{(\Sigma)}\\ \underline{l}\ne\underline{i}}}q_{\underline{i},\underline{l}} & \underline{j}=\underline{i}\end{cases}\tag{1}$$

Let us also define:

- the indicator function, indicating whether the number of busy channels is not greater, or greater than the threshold σ, as:

$$I_\sigma(\underline{i})=\begin{cases}1 & \text{if } \sum\limits_{k=1}^{N}i_{A_k}+\sum\limits_{k=1}^{N}i_{V_k}\le\sigma\\ 0 & \text{if } \sum\limits_{k=1}^{N}i_{A_k}+\sum\limits_{k=1}^{N}i_{V_k}>\sigma\end{cases}\tag{2}$$

- the number of available channels when the state is $S(t)=\underline{i}$ as:

$$D(\underline{i})=C-\sum_{h=1}^{N}\left(i_{A_h}+i_{V_h}\right)\tag{3}$$

- the number of channels the active calls can give up when the state is $S(t)=\underline{i}$, as:

$$E(\underline{i})=\sum_{\substack{k=1\\ i_{A_k}\ne 0}}^{N}\left(i_{A_k}-L_{A_k}\right)+\sum_{\substack{k=1\\ i_{V_k}\ne 0}}^{N}\left(i_{V_k}-L_{V_k}\right)\tag{5}$$

- the number of channels the active calls, except the hth, can still receive when the state is $S(t)=\underline{i}$, as:

$$G_h(\underline{i})=\sum_{\substack{k=1\\ k\ne h \text{ and } i_{A_k}\ne 0}}^{N}\left(H_{A_k}-i_{A_k}\right)+\sum_{\substack{k=1\\ k\ne h \text{ and } i_{V_k}\ne 0}}^{N}\left(H_{V_k}-i_{V_k}\right)\tag{6}$$

Keeping in mind what has been said so far, and in particular, the admission conditions stated in Section 2, we can now obtain the elements of the system transition matrix. We have:

- for condition (1) : $q_{\underline{i},\underline{j}} = \lambda_{H_h} + I_\sigma(\underline{i}) \cdot \lambda_{NC_h}$ when $\exists h$ such that:

$$\begin{cases} \left(i_{A_h}, i_{V_h}\right) = (0,0) \\ D(\underline{i}) \geq H_{A_h} + H_{V_h} \\ \left(j_{A_h}, j_{V_h}\right) = \left(H_{A_h}, H_{V_h}\right) \\ \left(j_{A_k}, j_{V_k}\right) = \left(i_{A_k}, i_{V_k}\right) \quad \forall k \neq h \end{cases} \tag{7}$$

- for condition (2): $q_{\underline{i},\underline{j}} = P_1(\underline{i}) \cdot \left(\lambda_{H_h} + I_\sigma(\underline{i}) \cdot \lambda_{NC_h}\right)$ when $\exists h$ such that:

$$\begin{cases} \left(i_{A_h}, i_{V_h}\right) = (0,0) \\ L_{A_h} + L_{V_h} \leq D(\underline{i}) < H_{A_h} + H_{V_h} \\ j_{A_h} = L_{A_h} + \left\lfloor \dfrac{R_{A_h}}{R_{A_h} + R_{V_h}} \cdot \left(D(\underline{i}) - L_{A_h} - L_{V_h}\right) \right\rfloor + z_A \\ j_{V_h} = L_{V_h} + \left\lfloor \dfrac{R_{V_h}}{R_{A_h} + R_{V_h}} \cdot \left(D(\underline{i}) - L_{A_h} - L_{V_h}\right) \right\rfloor + z_V \\ \left(j_{A_k}, j_{V_k}\right) = \left(i_{A_k}, i_{V_k}\right) \quad \forall k \neq h \end{cases} \tag{8}$$

where $R_{A_h} = H_{A_h} - L_{A_h}$ and $R_{V_h} = H_{V_h} - L_{V_h}$ represent the ranges of the channels required by the audio and video media of the hth call, respectively. The terms z_A and z_V are the number of channels assigned to the audio and video media in a random manner, because they were not assigned during the proportional distribution. Since there can be at most only one channel not assigned during the proportional distribution, both these terms will assume the value zero if there remains no other channel to be assigned at random while, if one channel has to be assigned, one of the two terms will be zero, and the other will be one. This will happen with a probability $P_1(\underline{i}) = 1/2$. Therefore we have:

$$\left(P_1(\underline{i}), z_A, z_V\right) = \begin{cases} \left(\dfrac{1}{2}, z, 1-z\right), \ \forall z \in \{0,1\} & \text{if } \dfrac{R_{A_h}}{R_{A_h} + R_{V_h}} \cdot \left(D(\underline{i}) - L_{A_h} - L_{V_h}\right) \\ & \text{is not an integer} \\ (1,0,0) & \text{if } \dfrac{R_{A_h}}{R_{A_h} + R_{V_h}} \cdot \left(D(\underline{i}) - L_{A_h} - L_{V_h}\right) \\ & \text{is an integer} \end{cases}$$

- for condition (3): $q_{\underline{i},\underline{j}} = P_2(\underline{i},\underline{j}) \cdot \left(\lambda_{H_h} + I_\sigma(\underline{i}) \cdot \lambda_{NC_h}\right)$ when $\exists h$ such that:

$$\begin{cases} \left(i_{A_h}, i_{V_h}\right) = (0,0) \\ L_{A_h} + L_{V_h} - E(\underline{i}) \leq D(\underline{i}) < L_{A_h} + L_{V_h} \\ \underline{j} \in \Im_3(h, \underline{i}) \end{cases} \tag{9}$$

where

$$\Im_3(h,\underline{i}) = \left\{ \underline{l} \in \Im^{(\Sigma)} \middle/ \begin{array}{ll} \left(l_{A_k}, l_{V_k}\right) = \left(L_{A_k}, L_{V_k}\right) & k = h \\ \left(l_{A_k}, l_{V_k}\right) = (0,0) & \forall k \neq h, \text{ if } \left(i_{A_k}, i_{V_k}\right) = (0,0) \\ \begin{array}{l} L_{A_k} \le l_{A_k} \le i_{A_k} \\ L_{V_k} \le l_{V_k} \le i_{V_k} \end{array} & \forall k \neq h, \text{ if } \left(i_{A_k}, i_{V_k}\right) \neq (0,0) \end{array} \right\}$$

The term $P_2(\underline{i},\underline{j})$ is the probability that, for all $k = 1,\ldots,N$, the audio media of the kth call, if active, will give up $\left(i_{A_k} - j_{A_k}\right)$ channels out of the $\left(i_{A_k} - L_{A_k}\right)$ possible, and the video media will give up $\left(i_{V_k} - j_{V_k}\right)$ channels out of the $\left(i_{V_k} - L_V\right)$ possible. It can, therefore, be calculated as follows:

$$P_2(\underline{i},\underline{j}) = \frac{F_h\left(\underline{i},\underline{j}\right)}{\sum\limits_{\underline{l}\in\Im_3(h,\underline{i})} F_h\left(\underline{i},\underline{l}\right)} \quad \text{where}$$

$$F_h\left(\underline{i},\underline{j}\right) = \begin{cases} \prod\limits_{\forall k/\left(i_{A_k}, i_{V_k}\right)\neq(0,0)} \begin{pmatrix} i_{A_k} - L_{A_k} \\ i_{A_k} - j_{A_k} \end{pmatrix} \cdot \begin{pmatrix} i_{V_k} - L_{V_k} \\ i_{V_k} - j_{V_k} \end{pmatrix} & \text{if } \underline{j} \in \Im_3(h,\underline{i}) \\ 0 & \text{if } \underline{j} \notin \Im_3(h,\underline{i}) \end{cases}$$

- for condition (a) of the channel freeing rules: $q_{\underline{i},\underline{j}} = \mu$ when $\exists h$ such that:

$$\begin{cases} \left(i_{A_h}, i_{V_h}\right) \neq (0,0) \\ G_h(\underline{i}) \le i_{A_h} + i_{V_h} \\ \left(j_{A_h}, j_{V_h}\right) = (0,0) \\ \left(j_{A_k}, j_{V_k}\right) = \begin{cases} \left(H_{A_k}, H_{V_k}\right) & \forall k \neq h, \text{ if } \left(i_{A_k}, i_{V_k}\right) \neq (0,0) \\ (0,0) & \forall k \neq h, \text{ if } \left(i_{A_k}, i_{V_k}\right) = (0,0) \end{cases} \end{cases} \qquad (10)$$

- for condition (b) of the channel freeing rules, the channels freed by the hth call, $\left(i_{A_h} + i_{V_h}\right)$, have to be redistributed to the media of the other active calls proportionally to the number of channels they need to reach the maximum requested. So, the number of channels to be assigned to the audio and video media of the kth call is:

$$r_A(k,\underline{i}) = \frac{H_{A_k} - i_{A_k}}{G_h(\underline{i})} \cdot \left(i_{A_h} + i_{V_h}\right) \quad \text{and} \quad r_V(k,\underline{i}) = \frac{H_{V_k} - i_{V_k}}{G_h(\underline{i})} \cdot \left(i_{A_h} + i_{V_h}\right) \qquad (11)$$

Since $r_A(k,\underline{i})$ and $r_V(k,\underline{i})$ may not be an integer, a number of $\lfloor r_A(k,\underline{i}) \rfloor$ channels is given to the audio of the generic source k, and a number of $\lfloor r_V(k,\underline{i}) \rfloor$

channels to the video of the same source. Therefore the remaining channels will be:

$$R(h,\underline{i}) = \sum_{\substack{k=1,\\ k\neq h,(i_{A_k},i_{V_k})\neq(0,0)}}^{N} r_A(k,\underline{i}) - \lfloor r_A(k,\underline{i}) \rfloor + r_V(k,\underline{i}) - \lfloor r_V(k,\underline{i}) \rfloor \quad (12)$$

They will be randomly assigned to the media of the other active calls. If we indicate the number of active calls when the state of the system is $S(t) = \underline{i}$ as N_{act}, the number of media of the active sources in the state $\underline{i}$, except for the hth, are $2\cdot(N_{act}-1)$ and the probability of one possible assignment of the remaining channels is therefore:

$$P_3(h,\underline{i}) = \frac{1}{\binom{2\cdot(N_{act}-1)}{R(h,\underline{i})}} \quad (13)$$

Now we can say that $q_{\underline{i},\underline{j}} = \mu \cdot P_3(h,\underline{i})$ when $\exists h$ such that:

$$\begin{cases} (i_{A_h}, i_{V_h}) \neq (0,0) \\ E_h(\underline{i}) > i_{A_h} + i_{V_h} \\ (j_{A_k}, j_{V_k}) = \begin{cases} (0,0) & \forall k \neq h \text{ if } (i_{A_k}, i_{V_k}) = (0,0) \\ \left(\left(i_{A_k} + \lfloor r_A(k,\underline{i}) \rfloor + z_{A_k}\right), \left(i_{V_k} + \lfloor r_V(k,\underline{i}) \rfloor + z_{V_k}\right) \right) & \forall k \neq h \text{ if } (i_{A_k}, i_{V_k}) \neq (0,0) \\ & \forall z_{A_k}, z_{V_k} \in \{0,1\} \text{ such that: } \\ & \sum_{\substack{k=1,\\ k\neq h,(i_{A_k},i_{V_k})\neq(0,0)}}^{N} (z_{A_k} + z_{V_k}) = R(h,\underline{i}) \end{cases} \end{cases} \quad (14)$$

Finally, by using the matrix $Q^{(\Sigma)}$, we can calculate the steady-state probability array, whose generic element is $\underline{\pi}_{[\underline{i}]}^{(\Sigma)} = \lim_{t\to\infty} \text{Prob}\{S^{(\Sigma)}(t) = \underline{i}\}$, by solving the following system:

$$\begin{cases} \underline{\pi}^{(\Sigma)} \cdot Q^{(\Sigma)} = \underline{0}_{2N} \\ \sum_{\underline{i}\in\Im^{(\Sigma)}} \underline{\pi}_{[\underline{i}]}^{(\Sigma)} = 1 \end{cases} \quad (15)$$

where $\underline{0}_{2N}$ is the column array constituted by $2N$ elements all equal to 0.

Now we can use $\underline{\pi}_{[i]}^{(\Sigma)}$ to calculate the system performance, in terms of call loss probability due to handover, new-call blocking probability and probability density function of the number of channels used by the audio and video media when they are active.

The call loss probability due to handover for the hth source is the probability that the cell load is such that the number of free and/or freeable channels is not sufficient to guarantee even the MCR to the hth source when it is not yet active in the cell and could potentially request admission:

$$P\_loss_h^{(\text{Handover})} = \frac{\sum_{\underline{i} \in \Im^{(\Sigma)} / (i_{A_h}, i_{V_h}) = (0,0),\ D(\underline{i}) + E(\underline{i}) < L_{A_h} + L_{V_h}} \underline{\pi}_{[i]}^{(\Sigma)}}{\sum_{\underline{i} \in \Im^{(\Sigma)} / (i_{A_h}, i_{V_h}) = (0,0)} \underline{\pi}_{[i]}^{(\Sigma)}} \tag{16}$$

The new-call blocking probability for the hth source is the probability that the cell has a number of busy, unfreeable, channels greater than the threshold σ when the hth source is not yet active in the cell:

$$P\_block_h^{(\text{New call})} = \frac{\sum_{\underline{i} \in \Im^{(\Sigma)} / (i_{A_h}, i_{V_h}) = (0,0),\ D(\underline{i}) + E(\underline{i}) < L_{A_h} + L_{V_h},\ I_\sigma(\underline{i}) = 0} \underline{\pi}_{[i]}^{(\Sigma)}}{\sum_{\underline{i} \in \Im^{(\Sigma)} / (i_{A_h}, i_{V_h}) = (0,0)} \underline{\pi}_{[i]}^{(\Sigma)}} \tag{17}$$

Finally, the probability density functions of the number of channels used by the audio and video media of the hth call when they are active are:

$$f_{A_h}(c) = \lim_{t \to \infty} \text{Prob}\left\{C_h^{(A)}(t) = c\right\} = \sum_{\underline{i} \in \Im^{(\Sigma)} / i_{A_h} = c} \underline{\pi}_{[i]}^{(\Sigma)} \qquad c \in \left\{0, L_{A_h}, \ldots, H_{A_h}\right\} \tag{18}$$

$$f_{V_h}(c) = \lim_{t \to \infty} \text{Prob}\left\{C_h^{(V)}(t) = c\right\} = \sum_{\underline{i} \in \Im^{(\Sigma)} / i_{V_h} = c} \underline{\pi}_{[i]}^{(\Sigma)} \qquad c \in \left\{0, L_{V_h}, \ldots, H_{V_h}\right\} \tag{19}$$

Once $f_{A_h}(c)$ and $f_{V_h}(c)$ have been obtained, it is possible to calculate the average number of channels each media of the hth call uses during its activity periods, $\wp_{A_h}$ for the audio and $\wp_{V_h}$ for the video media, as:

$$\wp_{A_h} = \sum_{c = L_{A_h}}^{H_{A_h}} c \cdot f_{A_h}(c) \text{ and } \qquad \wp_{V_h} = \sum_{c = L_{V_h}}^{H_{V_h}} c \cdot f_{V_h}(c) \tag{20}$$

7.3.4 Numerical results

In order to demonstrate the efficacy of the analytical model proposed in the previous sections, we will apply it to a significant case study: a wireless network in the office area of a factory. Let us assume that the whole office area is covered by

Table 1: User parameters

MOBILITY	ACTIVITY	
	Audio	Video
$\lambda_H = 833 \cdot 10^{-6}\ s^{-1}$	MCR: $L_A = 2$	MCR: $L_V = 4$
$\lambda_{NC} = 2 \cdot 10^{-3}\ s^{-1}$	PCR: $H_A = 3$	PCR: $H_V = 6$
$\mu^{-1} = 360$ s		

a cluster of the wireless access network and each room in the area by a picocell, each of which has a diameter less than or equal to about ten meters. Let $b = 5.33$ kbit/s be the capacity of each radio link. Due to the reduced dimension of cells and to the fact that each cell corresponds to a private office, the number of people that can simultaneously reside in a cell and require the use of channels is very limited.

Each cell in the cluster is connected to the same ATM switch through a suitably sized link of fixed capacity. We will study system performance in two different cases: 1) when the number of channels in each cell is $C = 25$; 2) when the number of channels in each cell is $C = 28$. We have chosen these values in order to show how a small increase in the number of cell channels determines great differences in system performance.

Let us assume that all users have the same statistics, as regards both mobility and bandwidth requirements. The characteristics of each source are listed in Table 1 and are kept constant. As far as the mobility statistics are concerned, we consider users who, once they have left the cell, will roam in the cluster and come back, active, to the same cell after an average period of about 20 minutes, that is, with a handover rate of $\lambda_H = 833 \cdot 10^{-6}\ s^{-1}$. Moreover, a user present, but inactive, in the tagged cell will turn to its terminal and request admission for a new call after an inactivity period of about 15 minutes, that is, with a new call activation rate of $\lambda_{NC} = 2 \cdot 10^{-3}\ s^{-1}$. Finally, we assume that the generic user active in the tagged cell will hold the channels for an average period of about 6 minutes, that is, with an average channel holding time of $\mu^{-1} = 360$ s. The requested MCR and PCR will be $L_A = 10$ kbit/s $= 2$ channels and $H_A = 16$ kbit/s $= 3$ channels for the audio media and $L_V = 20$ kbit/s $= 4$ channels and $H_V = 32$ kbit/s $= 6$ channels for the video media.

In both cases ($C = 25$ and $C = 28$) we have calculated the loss probability due to handover, $P\_loss^{(\text{Handover})}$, and the new-call blocking probability, $P\_block^{(\text{New call})}$, versus the threshold σ, ranging from zero channels to C. The results are shown in Figs. 7.3.1a and 7.3.1b. In both the figures, the curve of the loss probability due to handover is not present in the case of $N = 4$ because, in this case, it is equal to zero, given that the minimum number of channels required by the sources, even if they are all active in the cell, is $N \cdot (L_A + L_V) = 24$, while

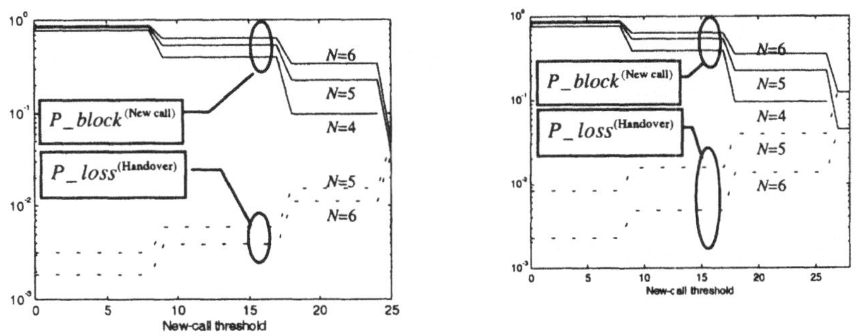

(a) $C = 25$ (b) $C = 28$

Fig. 7.3.1: New-call blocking and handover loss probabilities

the number of channels in the cell is $C = 25$ in Fig.7.3.1a and $C = 28$ in Fig.7.3.1b.

We observe that all the curves in both figures exhibit a staircase shape, due to the fact that small variations of the thresholds do not affect system behavior. As was to be expected, the new-call blocking probability decreases when the threshold increases, but to the detriment of the handover loss probability. The curves coincide when the new-call admission threshold is equal to the number of channels, that is, when no priority is given to active calls coming from neighboring cells due to handover. For this reason, the last segment of the curve is not present in Figs. 7.3.1a and 7.3.1b, because in this case it is null. Finally, we can observe that the larger the number of users, the worse the QoS will be in terms of both blocking and loss probabilities.

Figs.7.3.2, 7.3.3, 7.3.4 and 7.3.5 show the average number of channels used by active users, computed through (20), versus the threshold σ, differentiated for the audio and the video media, in the cases of $C = 25$ (Figs. 7.3.2 and 7.3.3) and $C = 28$ (Figs. 7.3.4 and 7.3.5). While Figs. 7.3.1a and 7.3.1b allow us to analyze the influence of the threshold choice on the admission performance for calls not yet active in the cell, through Figs. 7.3.2, 7.3.3, 7.3.4 and 7.3.5 we can quantify the influence of the threshold choice on already active calls. Of course, the lower the threshold, the more active users are privileged, as they can transmit at the PCR for a longer time. With high threshold value, on the other hand, the probability of having to transmit at the MCR increases.

7.3.5 Conclusions

In this paper we have proposed a threshold-based mechanism in a wireless ATM environment to decide the admission of multimedia calls using the Available Bit Rate (ABR) mode already standardized by the ATM Forum for transmission on wired ATM networks. The mechanism uses a threshold to privilege handover requests over new call requests. The calls considered in the paper are multimedia,

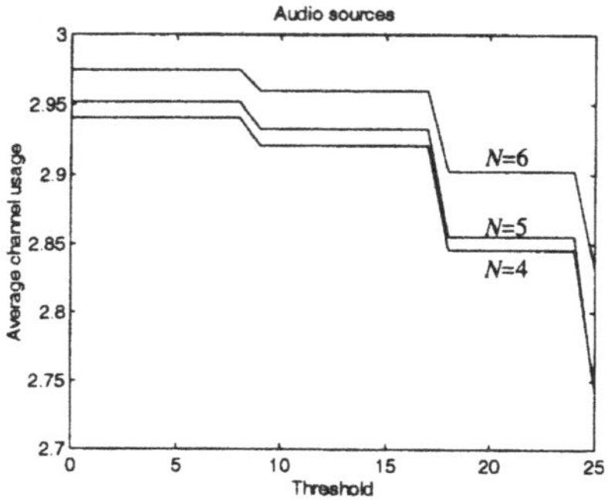

Audio ($C = 25$)
Fig. 7.3.2: Average channel usage ($\wp_A$)

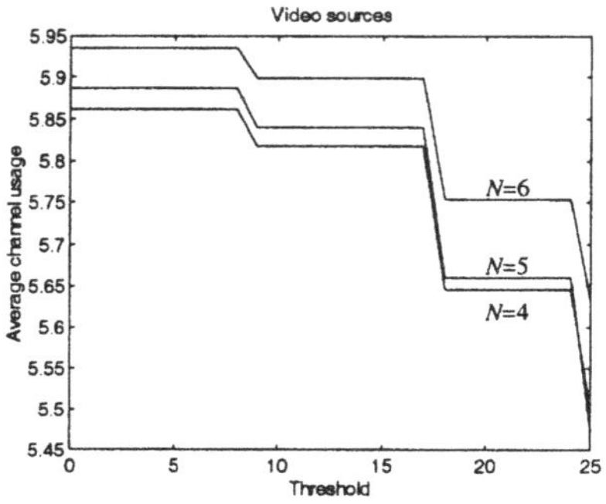

Video ($C = 25$)
Fig. 7.3.3: Average channel usage ($\wp_V$)

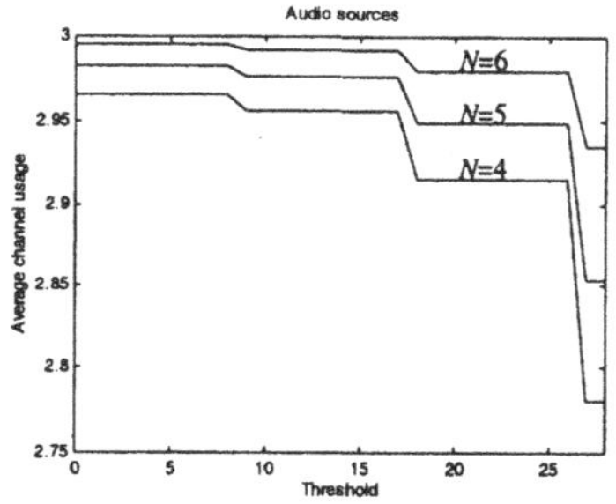

Audio ($C = 28$)
Fig. 7.3.4: Average channel usage ($\wp_A$)

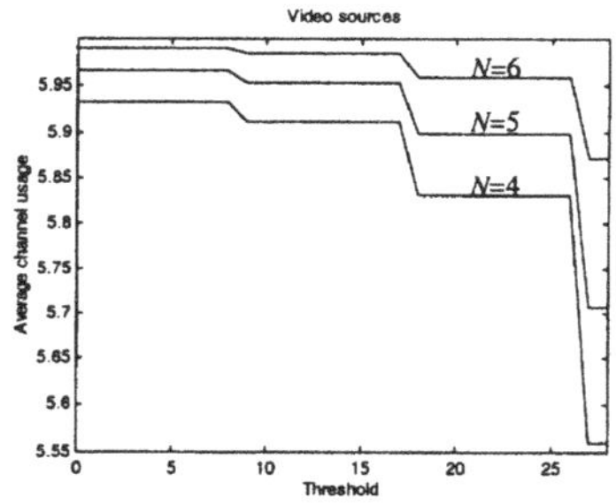

Video ($C = 28$)
Fig. 7.3.5: Average channel usage ($\wp_V$)

that is, they are constituted by a number of monomedia streams that have to be active together. As provided for in the ABR, each user declares a Minimum Cell Rate (MCR) and a Peak Cell Rate (PCR) for each media. An analytical model is presented to evaluate the performance of this strategy. In particular, the model is used to analyze the influence of the system parameters in a case study: a wireless access network to the ATM backbone in the office area of a factory. Through this example we have drawn general conclusions as to the choice, according to the number of users involved in a given cell, of the threshold to be applied to improve the call loss probability due to handover, limiting the number of calls simultaneously active.

References

1. R. Jain, S. Kalyanaraman, S. Famhy and R. Goyal, "Source Behavior for ATM ABR Traffic Management: An Explanation", *IEEE Communications Magazine*, Nov. 1996.
2. ITU-T: "Recommendation I.371 - Traffic Control and Congestion Control in B-ISDN", November 1995.
3. IEEE P802.11D3, *Wireless LAN Medium Access Control (MAC) and Physical Layer (PHY) Specifications*, Piscataway, NJ: IEEE Standards Dept., Jan., 1996.
4. B. Jabbari, G. Mason, "Teletraffic Aspects of Evolving and Next-Generation Wireless Communication Networks", *IEEE Personal Communications*, Dec. 1996.
5. B. A. Akyol, D. Cox, "Rerouting for Handoff in a Wireless ATM Network", *IEEE Personal Communications*, October, 1996.
6. D. Raychaudhuri and N. Wilson, "ATM Based Transport Architecture for Multiservices Wireless Personal Communication Networks", *IEEE JSAC*, October 1994.
7. B. A. Akyol and D. Cox, "Handling Mobility in a Wireless ATM Network", *IEEE Proc. INFOCOM '96*, San Francisco, CA, 1996.
8. R. Yuan *et al.*, "Mobility Support in a Wireless ATM Network", *Proc. 5th Workshop on Third Generation Wireless Info. Networks*, Kluwer, 1995.
9. M. Umehira et al., "An ATM Wireless Access System for Tetherless Multimedia Services", *Proc. ICUPC '95*, Tokyo, Japan, Nov. 1995.
10. M. Umehira, M. Nakura, H. Sato, A. Hashimoto, "ATM Wireless Access for Mobile Multimedia: Concept and Architecture", *IEEE Personal Communications*, Oct. 1996.
11. M. Naghshineh, A. Acampora, "QOS Provisioning in Micro-Cellular Networks Supporting Multimedia Traffic", *IEEE Infocom '95*, Boston, April 1995.
12. N. Naghshineh, M. Schwartz, "Distributed Call Admission Control in Mobile/Wireless Networks", *IEEE Journal of Select. Areas Comm.*, Vol. 14, No. 4, May 1996.
13. D. McMillan, "Traffic Modeling and Analysis for Cellular Mobile Networks", *Teletraffic and Datatraffic, Elsevier Science Publishers (North Holland)*, IAC, 1991.
14. D. McMillan, "Delay Analysis of a Cellular Mobile Priority Queueing System", *IEEE/ACM Trans. on Networking*, Vol. 3, No. 3, June 1995.
15. R. Ramjee, D. Towsley, R. Nagarajan, "On Optimal Call Admission Control in Cellular Networks", *Baltzer, Wireless Networks*, Vol. 3, 1997.
16. H. Saito, K. Kawashima, H. Kitazume, A. Koike, M. Ishizuka, and A. Abe, "Performance Issues in Public ABR Service", *IEEE Communications Magazine*, Nov. 1996.
17. S. Lu, V. Bharghavan, "Adaptive Resource Algorithms for Indoor Mobile Computing Environments", *ACM SIGCOMM '96*, Stanford, CA, August 1996.
18. S. Jordan, P. Varaiya, "Control of Multiple Service, Multiple Resource Communication Networks", *IEEE Transactions on Communications*, Vol. 42, No. 11, Nov. 1994.
19. R. Bolla, F. Davoli, M. Marchese, "Simple Schemes for Traffic Integration at Call Set-up Level in ATM Networks", *Computer Communications*, Vol. 19, 1996.

8 Capacity

8.1 Introduction

The third generation (3G) wireless networks are expected to provide flexible cellular infrastructure covering a variety of service environments. It is expected, therefore, that the need for higher and higher capacity will be a major factor in the provision of new services. It is then natural to expect that the increasing interest in cellular wireless communications makes it important to assess the theoretical and ultimate capacity limitations of cellular mobile radio systems.

In section 8.2 the issue of capacity is studied under the limitation of interference where as in section 8.3 a methodology is presented by which it is possible to analyze capacity sharing between the overlapping cells by dividing frames into time slots proportional in length to the cells allocated capacity.

8.2 Spectral efficiency of cellular mobile radio systems

F. Babich, G. Taricco and F. Vatta

8.2.1 Introduction

The increasing interest in cellular wireless communications makes it important to assess the theoretical and ultimate capacity limitations of cellular mobile radio systems.

It is well known that cellular communication systems can be assumed to be interference-limited. In fact, although frequency reuse increases the spectrum efficiency, it also introduces co-channel interference (CCI), being the same frequency channel used repeatedly in different co-channel cells [1, 2]. The effect of CCI is typically stronger than the effect of noise. Thus, it is often assumed that CCI is the only source of impairment to signal transmission, and noise can be disregarded. As a consequence, CCI determines the maximum data rates achievable.

In this work we evaluate the capacity of the uplink channel (mobile to base station) of a cellular system, with perfect side information (SI) at the receiver, versus the reuse distance and the level of shadowing affecting the interfering signals. Here, we consider the capacity as the average of the maximum mutual information conditioned on the channel gain (i.e., the SI) [3, 4].

The usefulness of some space diversity techniques in order to counteract the fading problem is investigated, too, with reference to the achievable spectral efficiency increasing.

Moreover, the effects of traffic control on system performance are quantified, with perfect SI at the receiver, as a function of the reuse distance and the traffic load. The effects of traffic on spectral efficiency are investigated considering the first two tiers of interferers. This evaluation is compared with a simplified one taking into account the first tier only.

Finally, the application of cell sectorization as a mean to improve a cellular radio system Signal-to-Interference Ratio (SIR), and hence capacity, is investigated. The sectorized system performance is compared with the performance of a conventional system using omni-directional antennas at the base stations (BS).

The remaining parts of this chapter are organized as follows. In the next section the fundamental concepts of a radiomobile system are addressed. Section 8.2.3 deals with the adopted propagation model, whereas in Section 8.2.4 the system model is described. In Section 8.2.5, the evaluation of the spectral efficiency of the communication system under investigation is addressed. Section 8.2.6 presents two different space diversity techniques and their comparison as far as the achievable spectral efficiency gain is concerned. Section 8.2.7 presents numerical results for the spectral efficiency of fully-loaded systems when considering both the first tier and the first two tiers of interferers, highlighting the approximation error deriving from considering the first tier only. The effects of traffic loading on the spectral efficiency are presented, too, considering the first two tiers of interferers. In Section 8.2.8, the application of cell sectorization as a mean to improve a cellular radio system capacity, is investigated. Finally, Section 8.2.9 summarizes the main conclusions.

8.2.2 The cellular coverage

To maximize the efficiency of a radiomobile system, i.e. to increase the number of served users, being the available resources equal, it is necessary that the same physical resource (frequency bandwidth) is used simultaneously by an ever-increasing number of users. This result is obtained by reusing the same carrier frequency (*frequency reuse*) to cover different areas which are separated from one another by sufficient distances so that co-channel interference is not objectionable [2].

Frequency reuse is normally employed in most radio services, but in mobile-telephone services this idea is employed on a more shrunk geographical scale to achieve the maximum spectral efficiency.

8.2.2.1 Cell splitting

The cell splitting technique consists in distributing transmitters of moderate power throughout the coverage area, so that each site primarily covers some nearby subarea, or zone, called *cell*. Each cell is covered by its own transmitter site called *base station*.

If the total allocation of N_c channels is partitioned into N_s sets, then each set will contain nominally $S_c = N_c / N_s$ channels. Each cell uses a different set of channels, which can be reused in other cells separated from one another by sufficient distances. Radio signals are, in fact, subject to path losses proportional to the distance between the transmitter and the receiver, with an exponent (*propagation coefficient*) in general greater than 2. Thus, the interference due to frequency reuse can be limited. In this way, given a frequency bandwidth, it is nominally possible to cover a whatever wide area, or, given an area, to serve a whatever great number of users, by shrinking adequately cells dimensions.

The set of C cells using all the available N_c radio channels, with no frequency reuse, is called *cell cluster*. Cluster dimensioning is obtained through a space efficiency optimization process (frequency reuse on the territory), once given the minimum carrier-to-interference ratio (CIR) tolerated in the system ($[C/I]_{\min}$), the propagation coefficients and the propagation statistics.

To define $[C/I]_{\min}$, used in the optimization process, the technical characteristics of the system (modulation scheme, RF parameters, etc.) and the needed quality level (i.e. the needed fade margin, which is normally equal to the maximum expected fade) must be taken into account. Normally, for FM analogical systems the $[C/I]_{\min}$ value is equal to about 18 dB; for the GSM system this value is about 9 dB [1].

On principle, in the study of cellular coverage techniques, it is not necessary to assume cells dimensions and shape to be uniform and regular [2], because this assumption is not near to the mark. If, as with present-day mobile service, omnidirectional base antennas were used, then each site's coverage area would be roughly circular. Although propagation considerations recommend the circle as cell shape, the circle is impractical for design purposes, because an array of circular cells produces ambiguous areas which are contained either in no cell or in multiple cells. On the other hand, any regular polygon approximate the shape of a circle, and three types of polygons, namely the equilateral triangle, the square and the regular hexagon, can cover a plane with no gaps or overlaps. Thus, a cellular system could be designed with square or equilateral triangular cells, but, for economic reasons, system designers adopted the regular hexagonal shape several years ago.

With reference to the cells dimensions, we can have:

- Ipercells: used in rural areas coverage with a radius greater than 20 km;
- Macrocells: used for motor-way areas coverage with a radius in the range from 1 to 20 km;
- Microcells: used in urban areas coverage with a radius in the range from 0.1 to 1 km;
- Picocells: used for indoor areas coverage with a radius lower than 0.1 km.

To minimize the number of base stations, it is worth using omni-directional base station antennas and hexagonal shape cells, because, given the distance between the fixed transmitter, located at the cell center, and the worst case receiving positions within the cells (vertices of the hexagon), the regular hexagon shape cell is the one with the greatest area.

On the other hand, to minimize the interfering signals effect, and thus to achieve a better channel reuse, it is still possible to use hexagonal shape structures in which the coverage is obtained through directional antennas located at the cell center or at the vertices of the hexagon. In these cases, the covered areas can have a triangular or a rhomboidal shape.

The set of C cells forming a cluster can be expressed by the product $C_1 C_2$, where C_1 is the number of cells needed to cover a hexagon and C_2 is the number of hexagons forming a cluster; for instance, if omni-directional base station antennas are used, being $C_1 = 1$ it is $C = C_2$.

To lay out a cellular system in the sense of determining which channel set should be assigned to each cell, it is worth introducing a co-ordinate system (i, j) allowing to determine, starting from a given cell, all the cells reusing the same radio channels. This co-ordinate system is radially disposed, and the "shift parameters" i and j are non negative integers: in particular, the co-ordinate system origin lies in the reference cell's center, and the general co-channel cell's center is placed in the point (i, j) with $i \geq j$.

In this geometrical structure, the number of hexagons per cluster C_2 turns out to be [2]:

$$C_2 = i^2 + ij + j^2 \tag{1}$$

Since i and j are non negative integers, C_2 can assume only well defined integer values (1, 3, 4, 7, 9, 12, 13, ...).

Being D the distance between the nearest neighboring co-channel cells centers, called *reuse distance*, and R the hexagonal cells radius, it can be shown that the ratio R_u, called *co-channel reuse ratio* is related to the cluster dimension, C_2, as follows:

$$R_u = \frac{D}{R} = \sqrt{3C_2} \tag{2}$$

It follows that, having set the value of the minimum carrier-to-interference ratio tolerated in the system ($[C/I]_{\min}$), the cluster dimension C_2 is determined.

If the system is balanced as far as interference is concerned, i.e. the coverage parameters are such that the interference conditions are the same both in the up-link channel (mobile to base station) and in the down-link channel (base to mobile station), to evaluate the system performance in terms of CIR at the receiver it is enough to analyze only one of the two links, e.g. the up-link channel: this condition is assumed in this work.

Moreover, hexagonal cells are considered with base stations located at their centers: thus, in this case, the worst-case useful signal receiving condition at the base station is when the mobile station moves close to the hexagon vertices (cell border), since in such situation, having set the interference level, the useful signal is minimum.

8.2.2.2
Co-channel interference

In radio mobile systems there are substantially two types of interference:

1. adjacent channel interference, which occurs when energy from the carrier spills over into the adjacent channels;
2. co-channel interference (CCI), which occurs when another transmission on the same carrier frequency impinges upon the receiver.

The dominant form of interference is, however, co-channel interference, which, since it occurs in the same frequency band as the desired signal, cannot be removed by filters. As a consequence, CCI determines the maximum data rates achievable.

Given a symmetrical homogeneous mobile radio system with hexagonal cells (all the cells have equal geometric dimensions and the same propagation characteristics), the number of co-channel interfering cells of the first tier is $K_I = 6$. In fact, the first tier circumference is $2\pi D$ long, and, being the cells on this circumference separated by $\approx D$, we have:

$$K_I = \frac{2\pi D}{D} = 2\pi \cong 6 \tag{3}$$

The six co-channel interfering cells in the second tier cause weaker interference than those in the first tier: this is the reason why in many works the co-channel interfering cells outside the first tier are neglected. For instance, in [5] and [6] a simplified performance evaluation is presented, neglecting the co-channel interferers outside the first dominant tier of interfering cells. In this hypothesis, $N_I = 6$ interferers are supposed to be active in the 6 co-channel interfering cells of the first level at a normalized distance $R_I = R_u = \sqrt{3C}$, where C is the cluster size. In this work this simplified analysis is compared with a more realistic one, considering the first three dominant groups of interfering cells, where the first group forms the so called first tier and the second and third group the so called second tier of interfering cells, as shown in Section 8.2.4.

8.2.2.3
Sectorized antenna systems

Traditionally, radiomobile systems use omni-directional antennas in the horizontal plane at the base station. The co-channel interference in a cellular system may be decreased by replacing a single omni-directional antenna at the base station by several directional antennas, each radiating within a specified sector [7]. Sectorized antenna systems take a traditional cellular area and subdivide it into "sectors" that are covered using directional antennas looking out from the same base station location. A cell is normally partitioned into three 120° sectors or six

60° sectors: as many as six sectors per cell have been used in practical service. When combining more than one of these directional antennas, the base station can cover all directions.

As might be expected, improving antenna directivity at the base station usually results in an increase in the power received at the mobile unit. Moreover, sector antennas reduce the potential interference across the original cell, and they are widely used for this purpose. In fact, by using directional antennas, a given cell will receive interference and transmit with only a fraction of the available co-channel cells. The factor by which the co-channel interference is reduced depends on the amount of sectoring used. For instance, considering a symmetrical homogeneous mobile radio system, with hexagonal cells and directional antennas at the base stations, while maintaining the same load per cell, the number of co-channel interfering cells of the first tier is $K_I = 2$, with a beamwidth of 120°, and $K_I = 1$, with a beamwidth of 60°.

Moreover, if the hypothesis that the useful and the interfering signals have the same p.d.f. both in omni-directional and sectorized systems, the system using directional antennas can work with a lower co-channel reuse ratio, i.e. with a reduced reuse distance. From eq. (2) it follows that a lower co-channel reuse ratio results in a lower cluster dimension, i.e. in a lower number of used channel sets. Since the total allocation of N_c channels is fixed, a lower number of N_s sets implies a higher number of channels per set ($S_c = N_c / N_s$) and per cell site. Each site can thus serve a higher traffic level, in this way reducing the number of cell sites for a given traffic load. On the other hand, given the site number and the traffic load, the interference level at the receiver is lower, since the mobile terminals outside the directional antenna beamwidth would not contribute to the received power.

8.2.3 Propagation model

The adopted propagation model is based on the propagation measurements reported in [8].

The channel gain is the product of three different components:

1. A deterministic path loss depending on the distance between the transmitter and the receiver (see [8]).
2. A slow random attenuation due to surrounding obstacles, modeled by a log-normal distribution with logarithmic standard deviation σ_{sh}.
3. A faster random attenuation, due to the presence of mobile scatterers, modeled by a Rician or a Rayleigh distribution.

8.2.3.1
Deterministic path loss

The deterministic path loss is a large-scale effect, i.e. it causes the received power to vary gradually due to signal attenuation, which is determined by the geometry of the path profile in its entirety.

The large-scale effects determine a power level averaged over an area of tens or hundred of meters and are therefore denoted as the *area-mean* power $\bar{\bar{p}}$. In general it is:

$$\bar{\bar{p}}_{rx} = \bar{\bar{p}}_{tx} G r^{-a} \tag{4}$$

where $\bar{\bar{p}}_{tx}$ is the transmitted area-mean power, $\bar{\bar{p}}_{rx}$ is the received area-mean power, r is the normalized propagation distance between transmitter and receiver (normalized with respect to the cell radius), and G is a constant depending on the transmitting and receiving antennas gains and heights. The exponent a takes values in the range from 2 to 5: these values are supported by the theoretical model for UHF-groundwave propagation.

In this work, both macrocellular and microcellular systems are considered.

For communication in macrocellular networks, eq. (4) holds. For communication in microcellular networks, the path loss for short-range communication is also relevant (if $r\lambda < 4h_r h_t$, where λ is the wavelength, h_r the receiving antenna height and h_t the transmitting antenna height). This requires an extension of the path loss law (4) to include the transition from free-space propagation to groundwave propagation. One solution is a stepwise transition from $a \approx 2$ to $a \approx 4$ at a certain distance g called *turnover* distance or *turning point.*

Therefore, eq. (4) should be modified in:

$$\bar{\bar{p}}_{rx} = \bar{\bar{p}}_{tx} G r^{-a} \left(1 + \frac{r}{g}\right)^{-b} \tag{5}$$

where g is the *turnover* distance normalized with respect to the cell radius. In (5), the exponent a is the basic propagation loss exponent for short distances, while the exponent b accounts for the additional propagation loss exponent for distances beyond the turning point g of the attenuation curve, between 100 and 200 m. The values of the parameters adopted for the macro and micro-cellular case are displayed in Table 8.2.1 (all distances are normalized to cell radius) [8]. For communications in macrocellular networks, (5) still holds with $g = \infty$.

Table 8.2.1: Path loss parameters

Cell Type	a	b	g
Macro	4	-	∞
Micro	2	2	0.67

8.2.3.2
Shadow fading or shadowing

The preceding equations (4) and (5) suggest an exact relationship between path loss and distance. But, in general, buildings are not symmetric and the furnishing are not the same in all directions, and therefore one expects to find somewhat different path losses in different directions. A deterministic model for this variation is not feasible, and therefore statistical models are used.

The cause of this power loss is the obstruction by other objects around the receiver, such as buildings or hills which create radio shadows, and it is usually referred to as *shadow fading* or *shadowing* [9]. The problems of shadowing are most severe in heavily built-up urban centers. Shadowing is a *medium scale* effect: field strength variation occur if the antenna is displaced over distances larger than a few tens or hundreds of meters from the transmitter. Shadowing introduces additional fluctuations of the received signal power, so that this phenomenon can be represented as a slowly fluctuating received power, or *local-mean* power $\bar{p}$, which varies around the area-mean power. The term *local-mean* is used to denote the signal level averaged over a few tens of wavelengths, typically 40λ. This ensures that the rapid fluctuations of the instantaneous received power due to multipath effects are largely removed.

Experiments reported by Egli [10] showed that for paths longer than a few hundred meters, the received (local-mean) power fluctuates with a *log-normal* distribution about the area-mean power, with the area-mean power in the form of (4) or (5), for the macrocellular or the microcellular case, respectively.

The probability density function (p.d.f.) of the local-mean power is thus in the form of:

$$f_{\bar{p}} = \frac{1}{\sqrt{2\pi}\sigma_{\text{sh}}\bar{p}} \exp\left[-\frac{1}{2\sigma_{\text{sh}}^2}\ln^2\left(\frac{\bar{p}}{\bar{\bar{p}}}\right)\right] \tag{6}$$

where σ_{sh} represents the logarithmic standard deviation of the shadowing expressed in the absolute values. The standard deviation s_{sh} expressed in dB is found from $s_{\text{sh}} = 4.34\sigma_{\text{sh}}$.

8.2.3.3 Multipath fading

The radio wave may also be reflected from a hill, a building, a truck, an airplane, or a discontinuity in the atmosphere; in some cases, the reflected signal is significantly attenuated, while in others almost all the radio energy is reflected and very little absorbed [9].

The effect is to produce not one but many different paths between the transmitter and receiver. This is known as *multipath propagation*.

Multipath propagation creates some of the most difficult problems associated with the mobile environment, namely:

1. Random phase shift which creates rapid fluctuations in signal strength known as *multipath fading*.
2. Random frequency modulation due to different Doppler shifts on different paths.

Multipath fading thus has a *small-scale* effect, which leads to rapid fluctuations of the phase and amplitude of the signal if the vehicle moves over a distance on the order of a wavelength or more.

Only narrowband channels are considered: namely, the entire transmission occurs within what is called the *coherence bandwidth* or *correlation bandwidth* [11]. As Jakes defines it, "the maximum frequency difference for which signals are still strongly correlated is called the coherence bandwidth of the mobile radio transmission path" [12]. This implies that the interarrival times of the reflected waves are small compared to the time scale of the variations in the modulating signal. In other words, in narrowband channels, all reflections are assumed to add coherently (phasor addition).

In microcellular mobile radio a dominant path, which may be a line-of-sight (LOS) path, often occurs at the receiver, in addition to many scattered paths. This dominant path may significantly decrease the depth of fading. In this case, the p.d.f. of the received envelope is said to be Rician and the instantaneous received power $p(t)$, expressed in terms of the local-mean power $\bar{p}$ and the Rician K-factor, has a noncentral chi-square p.d.f. [13]:

$$f_p\left(p \middle| \bar{p}, K\right) = \frac{(1+K)e^{-K}}{\bar{p}} \exp\left(-\frac{1+K}{\bar{p}} p\right) I_0\left(\sqrt{4K(1+K)\frac{p}{\bar{p}}}\right) \tag{7}$$

In macrocellular systems, with longer, often obstructed, propagation paths, the power of the direct line-of-sight signal is small compared to the reflected signal power (i.e. $\gamma \to 0$ and thus $K \to 0$). This produces the special case of Rayleigh fading. In this case, the instantaneous received power $p(t)$ is exponentially distributed about the local-mean power:

$$f_p\left(p \middle| \bar{p}, K=0\right) = \frac{1}{\bar{p}} \exp\left(-\frac{p}{\bar{p}}\right) \tag{8}$$

We will assume fading slow enough to be tracked by the receiver, i.e. perfect channel-state information (CSI) to be available at the receiver, and the fading amplitude constant within a signaling interval. Moreover we will assume perfect phase tracking of the phase perturbation process.

8.2.4 System model

The systems considered in our analysis are a symmetrical microcellular mobile radio system and a symmetrical macrocellular mobile radio system with different cluster sizes, being the cell shape hexagonal. Hexagonal cells are approximated by circles of unit radius with the base stations located at their centers. A single user is assumed to be active in each cell. The performance evaluation is related to the uplink channel (mobile to base station).

The desired signal instantaneous received power $p_d(t)$ has a Rician p.d.f. if a microcellular environment is assumed (i.e. a p.d.f. given by (7) and a Rayleigh p.d.f. if a macrocellular environment is assumed (i.e. a p.d.f. given by (8)).

The active user's position, in the reference cell, is assumed to be on the cell border (*worst-case*) or uniformly distributed in the ring $r_0 < R_d < 1$ (*average-case*), where R_d is the user's distance from the cell center and r_0, the closest distance the mobile can be from the BS antenna, is assumed approximately equal to 0.025 for microcellular systems and 0.016 for macrocellular systems. In the *average-case*, the probability density function of R_d is given by:

$$p(R_d) = \frac{2R_d}{1 - r_0^2} \tag{9}$$

In this work, a simplified performance evaluation is presented first, neglecting the co-channel interferers outside the first dominant tier of interfering cells (see, for instance, [5] and [6]). In this hypothesis, $N_I = 6$ interferers are supposed to be active in the 6 co-channel interfering cells of the first level at a normalized distance $R_I = R_u = \sqrt{3C}$, where C is the cluster size[1]. This simplified analysis is then compared with a more realistic one, considering the first three dominant

[1] It should be noticed that, in [6], the authors not only make the hypothesis of neglecting the co-channel interferers outside the first dominant tier of interfering cells, but also assume that the fluctuations in the CIR (Carrier-to-Interference Ratio) are perfectly tracked by the receiver, which is not a hypothesis considered in this work.

groups of interfering cells, where the first group forms the so called first tier and the second and third group the so called second tier of interfering cells. In this second case, $N_I = 18$ interferers are supposed to be active in the 18 co-channel interfering cells of the first two tiers, located at a normalized distance R_I, where

$$R_I = \sqrt{3\left(i^2 + ij + j^2\right)} \tag{10}$$

In (10), i and j are the non negative integer "shift parameters", indicating the position of the general co-channel cell whose center lies at $(u, v) = (i, j)$ in the co-ordinate system the origin of which lies in the reference cell's center. Table 8.2.2 reports the values of (i, j) assumed by the co-channel cells centers, considering the first three dominant groups.

The number of cells per cluster turns out to be:

$$C = i_1^2 + i_1 j_1 + j_1^2 \tag{11}$$

where i_1 and j_1 are the non negative integer "shift parameters", indicating the position of the general co-channel cell of the first tier (i.e. with values (i, j) in the first column of Table 8.2.2).

Table 8.2.2: Co-channel cells positions

I Tier	II Tier	
I Group	II Group	III Group
(1,0)	(1,1)	(2,0)
(1,1)	(3,0)	(2,2)
(2,1)	(1,4)	(4,2)
(3,0)	(3,3)	(6,0)
(3,1)	(2,5)	(6,2)

Assume that the interferers belonging to a single group are located at the same distance R_{int} equal to $R_I - 1$, R_I or $R_I + 1$ (with R_I given by eq. (10)), which will be referred to as *worst-case*, *central*, and *best-case* interference. The *worst-case* interference configuration corresponds to the case where the co-channel interferers are on the near boundary of their respective cells, at a distance $R_I - 1$ from the desired mobile's base station. The *central* interference configuration corresponds to the case where the co-channel interferers are near the center of their respective cells, at a distance R_I from the desired mobile's base station. On the other hand, the *best-case* interference configuration corresponds to the case where the co-channel interferers are on the far boundary of their respective cells, at a distance $R_I + 1$ from the desired mobile's base station.

In this multiuser system, the total interference is given by the incoherent addition of N_I interfering signals, each of which affected by Rayleigh fading and log-normal shadowing. Fading and shadowing affecting each interfering signal are assumed to be uncorrelated (*Suzuki distribution*). In this case, the instantaneous power p.d.f. of each individual interfering signal can be approximated with a log-normal p.d.f. (in the form (6)) [14], with area-mean power $\bar{\bar{p}}_{\text{app}}$ and a standard deviation σ_{app}. The first and second order moments $\bar{\bar{p}}_{\text{app}}$ and σ_{app} of the approximate log-normal p.d.f. and those of the exact Suzuki distribution ($\bar{\bar{p}}_{\text{ex}}$ and σ_{ex}) are related by the following expressions:

$$\begin{cases} \sigma_{\text{app}}^2 = \sigma_{\text{ex}}^2 + \ln 2 \\ \bar{\bar{p}}_{\text{app}} = \dfrac{\sqrt{2}}{2} \bar{\bar{p}}_{\text{ex}} \end{cases} \tag{12}$$

According to the suggestion of Schwartz and Yeh [15], the composite p.d.f. for the total interference sum p_u can be approximated by a pure log-normal distribution with the appropriate moments. Therefore, the p.d.f. for the instantaneous joint interference power $p_u = \sum_{j=1}^{N_I} p_{uj}$ is of the form (6) with the following substitutions:

$$\begin{cases} \bar{p} \rightarrow \bar{p}_u \\ \bar{\bar{p}} \rightarrow 10^{m_n/10} \\ \sigma_s \rightarrow \sigma_n / 4.34 \end{cases} \tag{13}$$

where m_n and σ_n (in dB) are tabulated by Prasad and Kegel in [16] (Table 8.2.2).

8.2.5
Spectral efficiency analysis

As seen in the previous sections, a cellular mobile radio system is essentially interference-limited, because, while the so-called frequency reuse can drastically increase the spectrum efficiency, co-channel interference ultimately determines the data rates and corresponding bit-error rates available to each user. In these systems the effects of interference are typically much larger than the effects of noise. Thus, in this work it is assumed that CCI is the only source of impairment to signal transmission, and noise can be disregarded.

On the basis of these assumptions, the received signal at the base station of the reference cell can be written as the superposition of a useful signal and other interfering signals:

$$Y = \alpha X + \sum_{i=1}^{N_I} \beta_i X_i \tag{14}$$

To summarize the assumptions of the previous sections, the channel gain α is the product of two different components:

1. A deterministic path loss depending on the position of the useful transmitter, namely, $R_d^{-a}\left(1+R_d/g\right)^{-b}$ where R_d is the distance between the useful transmitter and the receiver (see [8]), the exponent a is the basic propagation loss exponent for short distances and the exponent b is the additional propagation loss exponent for distances beyond the turning point g of the attenuation curve.
2. A faster random attenuation, due to the presence of mobile scatterers, modeled by a Rician distribution (with Rician factor K = 4) for microcellular systems, and by a Rayleigh distribution (with $K = 0$) for macrocellular systems.

Given that interfering signals will in general propagate over obstructed paths because of the larger propagation distances, the instantaneous power of each individual interfering signal is assumed to be subject to a combined shadowing and Rayleigh fading. Thus, the channel gains β_i are the product of three different components:

1. A deterministic path loss depending on the position of the interfering transmitter, namely $R_{\text{int}}^{-a}\left(1+R_{\text{int}}/g\right)^{-b}$ where R_{int} is the distance between the interfering transmitter and the receiver.
2. A slow random attenuation due to surrounding obstacles, modeled by a log-normal distribution with logarithmic standard deviation σ_{sh} expressed in dB.
3. A faster random attenuation, due to the presence of mobile scatterers, modeled by a Rayleigh distribution.

Moreover, assume that the transmitted symbols X and X_i have equal power, P.

In the presence of CSI, capacity is the maximum mutual information conditioned on the CSI value, i.e., $C = \max_X I\left(X;Y|\alpha\right)$ [17]. Since it is very difficult to obtain the exact mutual information corresponding to the channel described by equation (14), we will approximate it by a lower bound. Since the channel considered can be regarded as an additive noise channel, its capacity is greater than or equal to the capacity of an additive Gaussian channel with the same "noise" power. Thus, we have the following lower bound:

$$C \geq C' = E_\alpha\left[\log_2\left(1+\frac{|\alpha X|^2}{E\left|\sum_{i=1}^{N_I}\beta_i X_i\right|^2}\right)\right] = E_\alpha\left[\log_2\left(1+\frac{|\alpha|^2}{\sum_{i=1}^{N_I}E|\beta_i|^2}\right)\right] \tag{15}$$

because fading and shadowing affecting each interfering signal are assumed to be uncorrelated and the N_I interfering signals are assumed to add incoherently.

When considering the first tier of interferers only, and the active user's position, in the reference cell, is assumed to be on the cell border (*worst-case*), a closed form expression for the capacity can be obtained for macrocellular mobile radio systems (Rayleigh fading):

$$C' = \int_0^\infty \log_2(1+u\Phi)e^{-u}du = -\exp\left[\frac{1}{\Phi}\right]\mathrm{Ei}\left(-\frac{1}{\Phi}\right)\log_2 e \tag{16}$$

where Ei(x) is the Exponential Integral Function and $1/\Phi = N_I \exp\left(\sigma_{sh}^2/2\right)R_{int}^{-a}$.

With microcellular systems (Rice fading) we obtain a similar result but we have to resort to numerical integration or Monte Carlo simulation. In particular, for microcellular systems (Rice fading) the capacity lower bound can be expressed as:

$$C' = \int_0^\infty \log_2(1+u\Phi)(1+K)e^{-K-(1+K)u}I_0\left(\sqrt{4K(1+K)u}\right)du \tag{17}$$

where

$$\Phi = \frac{\left(1+\frac{1}{g}\right)^{-b}}{N_I \exp\left[\frac{\sigma_{sh}^2}{2}\right]R_{int}^{-a}\left(1+\frac{R_{int}}{g}\right)^{-b}} \tag{18}$$

Fig. 8.2.1 reports the *area spectral efficiency* versus the normalized reuse distance for a microcellular system with fixed logarithmic standard deviation of the shadowing ($\sigma_{sh} = 6$ dB) and no diversity. The *area spectral efficiency* is here defined as the spectral efficiency normalized to the area covered by a single cell, approximately $\pi R_u^2/4$ (see [6]). The area spectral efficiency is shown to be generally increased by decreasing the reuse distance: only for the *worst-case* interference configuration an optimal reuse distance to maximize the spectral efficiency per area unit can be obtained for an intermediate value of R_u (both for the *average-case* (solid curves) and the *worst-case* (dashed curves) active user's position). As far the *worst-case* active user's position is concerned, an optimal reuse distance can be obtained also for the *central* case interference configuration.

Figs. 8.2.2, 8.2.3 and 8.2.4 report the *area spectral efficiency* versus the normalized reuse distance for a macrocellular system with fixed logarithmic standard deviation of the shadowing ($\sigma_{sh} = 0$, 6 and 12 dB, respectively) and no diversity. For the *average-case* active user's position (solid curves) the figures show that the area

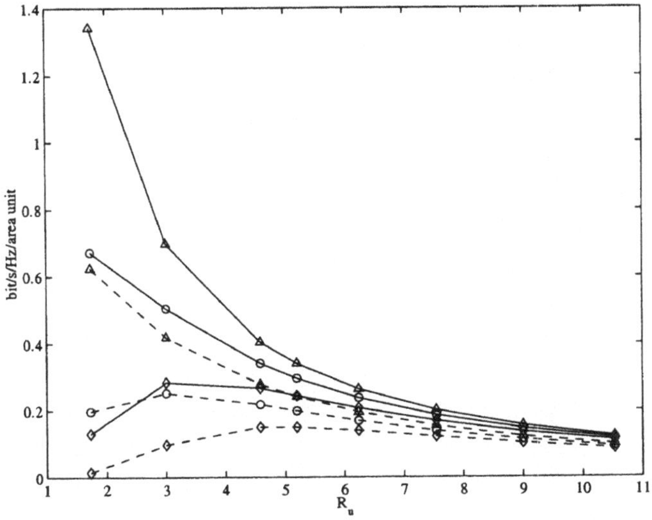

Fig. 8.2.1: Area spectral efficiency versus the normalized reuse distance for microcellular systems with σ_{sh} = 6 dB and without diversity. Curves refer to the *worst* (◊), *central* (o) and *best-case* (Δ) of interference. For each case, the solid curve corresponds to *average-case* active user's position and the dashed curve to the *worst-case* active user's position.

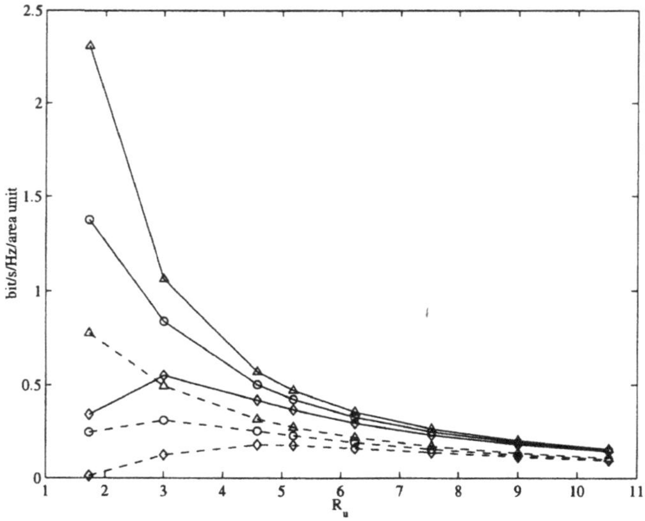

Fig.8.2.2: Area spectral efficiency versus the normalized reuse distance for macrocellular systems with σ_{sh} = 0 dB and without diversity. Curves refer to the *worst* (◊), *central* (o) and *best-case* (Δ) of interference. For each case, the solid curve corresponds to *average-case* active user's position and the dashed curve to the *worst-case* active user's position.

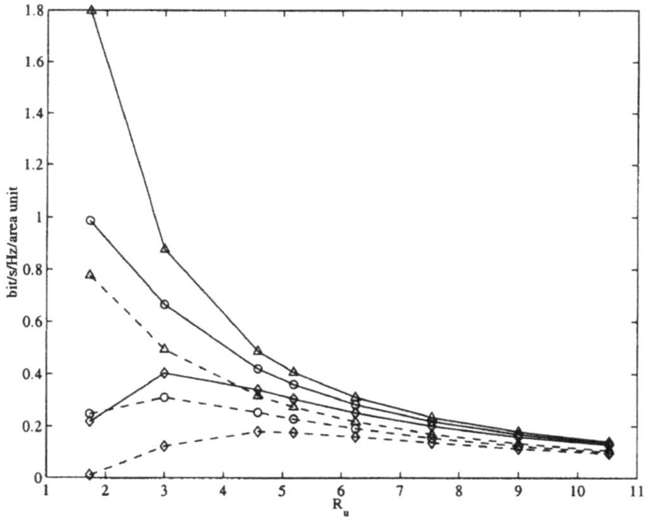

Fig. 8.2.3: Area spectral efficiency versus the normalized reuse distance for macrocellular systems σ_{sh} = 6 dB and without diversity. Curves refer to the *worst* (◊), *central* (o) and *best-case* (Δ) of interference. For each case, the solid curve corresponds to *average-case* active user's position and the dashed curve to the *worst-case* active user's position.

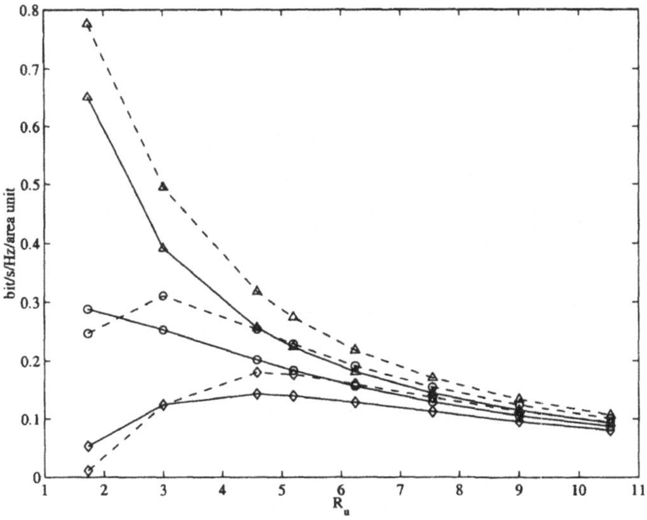

Fig. 8.2.4: Area spectral efficiency versus the normalized reuse distance for macrocellular systems with σ_{sh} = 12 dB and without diversity. Curves refer to the *worst* (◊), *central* (o) and *best-case* (Δ) of interference. For each case, the solid curve corresponds to *average-case* active user's position and the dashed curve to the *worst-case* active user's position.

spectral efficiency attains a maximum in the *worst-case* interference configuration for an intermediate value of R_u. This optimum value of R_u increases with σ_{sh}: for example, for $0 \le \sigma_{sh} \le 8$ dB it corresponds approximately to a cluster size $C = 3$, and for $8 < \sigma_{sh} \le 12$ dB to a cluster size $C = 7$.

8.2.6 Space diversity techniques

In reception, a conventional countermeasure such as the space diversity is supposed to be available in order to avoid deep fades in the received signal.

There are, in reception, *L branches* and thus *L* identically distributed signals all having a Rician or Rayleigh p.d.f. (a *branch* corresponds to one independent signal path in a diversity system). Two linear diversity-combining techniques are considered in the work: Maximal-Ratio and Equal-Gain Combining [18]. If Maximal-Ratio Combining (MRC) is used, we assume *L* independent signals to be present. These signals are weighted proportionally to their signal-to-noise ratios and then summed: before combining them they must be cophased. Therefore, the CIR (Carrier-to-Interference Ratio) of the combined signal is exactly equal to the sum of the CIRs in the various diversity branches. The second technique is the Equal-Gain Combining, in which the signals in the *L* diversity branches are equally weighted. For this combiner to work well, local CIRs should be approximately equal in the various diversity branches.

When considering diversity, we replace equation (14) by the following:

$$Y_j = \alpha_j X + \sum_{i=1}^{N_I} \beta_{ij} X_i \tag{19}$$

for j = 1,...,L where L is the number of diversity branches. Maximal-Ratio Combining and Equal-Gain Combining produce the following equivalent outputs, respectively:

$$Y = \begin{cases} \sum_{j=1}^{L} \alpha_j^* Y_j & \text{(MRC)} \\ \sum_{j=1}^{L} Y_j & \text{(EGC)} \end{cases} \tag{20}$$

Figs. 8.2.5 and 8.2.6 report the spectral efficiency versus the standard deviation of the shadowing for a microcellular and a macrocellular system, respectively, with and without diversity, being the user uniformly distributed in the cell of interest (*average-case*). Figs. 8.2.7 and 8.2.8 report the same results, being the user's position fixed on the reference cell border (*worst-case*).

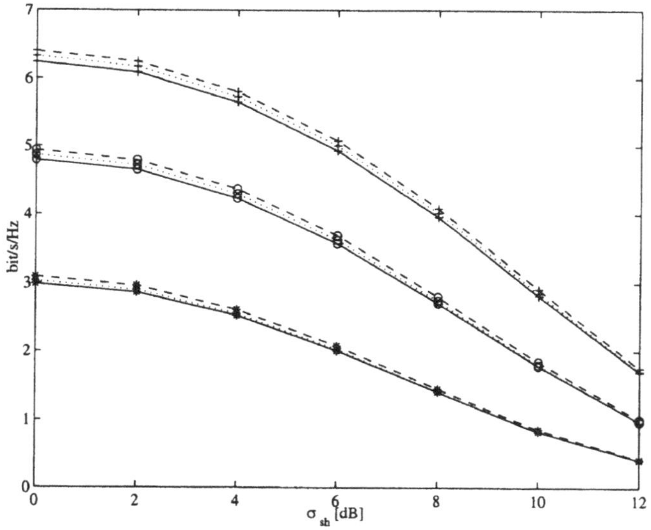

Fig. 8.2.5: Spectral efficiency versus σ_{sh} at fixed cluster size $C = 3$, with and without diversity, for a microcellular system ($a = 2$, $b = 2$, $g = 0.67$, $K = 4$). *Worst* (✻), *central* (o) and *best* (+) case of interference. Solid curves correspond to no diversity, the dashed curves to two-branch diversity with Maximal-Ratio Combining and the dotted curve to two-branch diversity with Equal-Gain Combining. The active user's position, in the reference cell, is assumed to be uniformly distributed in the cell he belongs to (*average-case*).

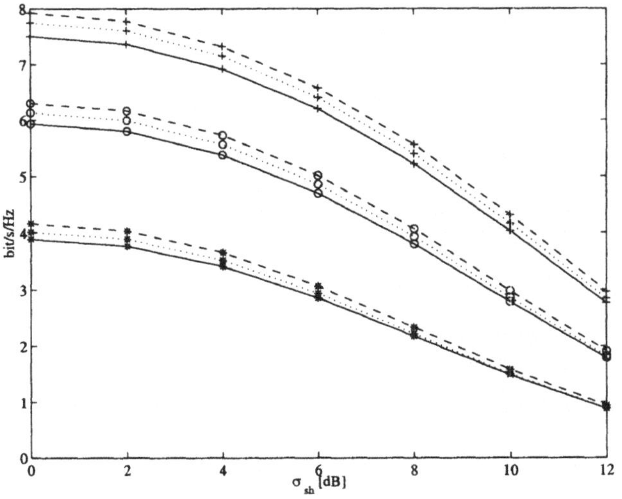

Fig. 8.2.6: Spectral efficiency versus σ_{sh} at fixed cluster size $C = 3$, with and without diversity, for a macrocellular system ($a = 4$, $g = \infty$, $K = 0$). Curve types are described in Fig. 8.2.5.

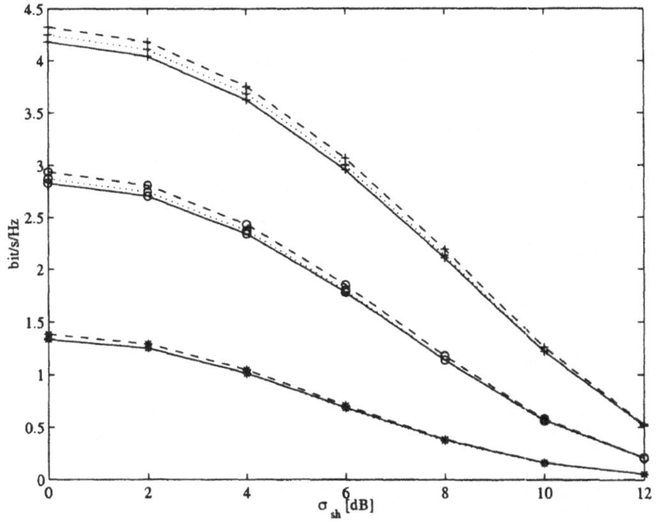

Fig. 8.2.7: Spectral efficiency versus σ_{sh} at fixed cluster size $C = 3$, with and without diversity, for a microcellular system ($a = 2$, $b = 2$, $g = 0.67$, $K = 4$). *Worst* (✳), *central* (o) and *best* (+) case of interference. The solid curves correspond to no diversity, the dashed curves to two-branch diversity with Maximal-Ratio Combining and the dotted curves to two-branch diversity with Equal-Gain Combining. The active user's position, in the reference cell, is assumed to be on the cell's border (*worst-case*).

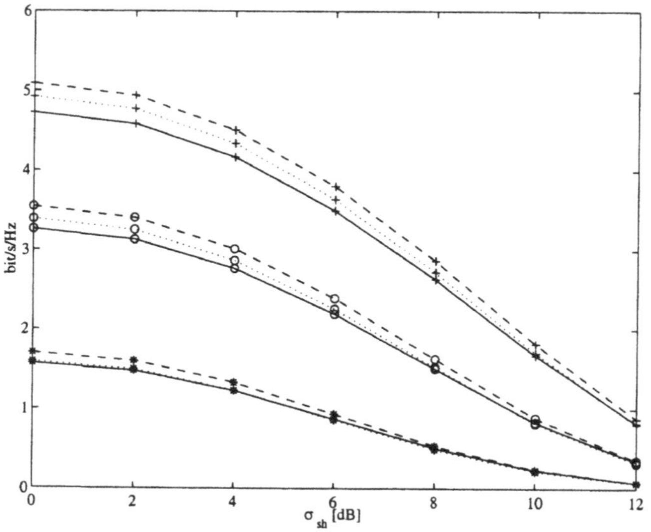

Fig. 8.2.8: Spectral efficiency versus σ_{sh} at fixed cluster size $C = 3$, with and without diversity, a macrocellular system ($a = 4$, $g = \infty$, $K = 0$). Curve types are described in Fig. 8.2.7.

Two-branch diversity is considered. In the figures, the three sets of curves refer to the *worst*, *central* and *best-case* interference with cluster size $C = 3$ ($R_u = 3$). The diagrams show that MRC is the best diversity technique and that two-branch diversity provides limited gain in terms of area spectral efficiency. This can be explained by noting that we are not considering a transmission delay constraint when dealing with Shannon capacity and space diversity can be partly traded off for time diversity (i.e., coding).

8.2.7 Partially-loaded systems

To study the effect of traffic control on the spectral efficiency, a *partially-loaded* system is considered. Traffic control is achieved by means of a multiple access scheme, employing a fixed channel assignment. In this scheme, a free channel, if available, is allocated randomly to a new connection request; otherwise, if no channel is available, the connection is blocked. The serviced channels are assumed to be independent and active with the same probability a_c, that is defined as the normalized traffic per cell. Therefore, the serviced channels and the number of interferers are random variables depending on the traffic loading. The spectral efficiency is estimated by Monte Carlo simulation (with an algorithm that adopts almost the same procedure used in [5]), considering all possible combinations of 18 (or 6 in the single tier case) interferers, each of which is active and contributes to the total interference with probability a_c.

The probability $P_c(i)$ of having a specific number i of active interferers in a group of cells follows a binomial distribution, and is given by:

$$P_c(i) = \binom{6}{i} a_c^i (1 - a_c)^{(6-i)} \tag{21}$$

The probability of having i_1, i_2, i_3 active interferers in the first, second and third group, respectively, is given by:

$$P(i_1, i_2, i_3) = P_c(i_1) P_c(i_2) P_c(i_3) \tag{22}$$

under the assumption of independent interferers.

In particular, the simulation algorithm is composed by the following steps:

1. In each interfering cell of a given group, an interferer is supposed to be active with probability a_c and position $R_I - 1$, R_I or $R_I + 1$, in the three interference configurations considered (i.e. *worst-case*, *central*, and *best-case* interference).

2. One of the possible combinations (i_1, i_2, i_3) of 18 active interferers is considered, i.e. i_1 interferers are supposed to be active in the first group, i_2 interferers are supposed to be active in the second group, and i_3 interferers are supposed to be active in the third group, with i_1, i_2 and i_3 ranging over the integers from 0 to 6, independently.
3. The position of the desired user is picked at random according to (9), if the user is assumed to be uniformly distributed in the ring $r_0 < R_d < 1$ (*average-case*), or fixed on the reference cell border (*worst-case*).
4. The instantaneous desired signal power p_d, with Rayleigh or Rice distribution, is picked at random too, reminding that the Rayleigh distribution is closely related to the central chi-square distribution of order 2, whereas the Rice distribution is closely related to the noncentral chi-square distribution of order 2 [19].
5. The CIR (Carrier-to-Interference Ratio) $\gamma_d(i_1, i_2, i_3)$ of the desired user is calculated using the two-ray path model, considering i_1 active interferers in the first group (at a distance R_{I1}), i_2 active interferers in the second group (at a distance R_{I2}), and i_3 active interferers in the third group (at a distance R_{I3}), where R_{I1}, R_{I2} and R_{I3}, defined by eq. (10), are calculated according to Table 8.2.2 (i.e. with values (i, j) in columns 1, 2 and 3, respectively).
6. The spectral efficiency is calculated as:

$$C'(i_1, i_2, i_3) = \log_2\left(1 + \gamma_d(i_1, i_2, i_3)\right) \tag{23}$$

averaged over $\gamma_d(i_1, i_2, i_3)$.

Repeating the above process (from step 1 to step 6) over 10.000 samples for 100 times, $C'(i_1, i_2, i_3)$ converges with sufficient accuracy, for a given possible combination (i_1, i_2, i_3) of active interferers.

The procedure is repeated for each possible combination of active interferers.

At the end of the process, the spectral efficiency evaluated at step 6 for each combination (i_1, i_2, i_3) of active interferers is averaged as follows:

$$C'_{av} = \sum_{(i_1, i_2, i_3)} P(i_1, i_2, i_3) C'(i_1, i_2, i_3) \quad (24)$$

Here, i_1, i_2 and i_3 range over the integers from 0 to 6, independently. For $i_1 = i_2 = i_3 = 0$ it is assumed that the SNR (Signal-to-Noise Ratio) is equal to 40 dB.

Fig. 8.2.9 shows the spectral efficiency as a function of the reuse factor R_u, as far as the *average-case* active user's position is concerned, for a microcellular system (with parameters $a = 2$, $b = 2$, $g = 0.67$, $K = 4$ and $\sigma_{sh} = 3$ dB), and for a *worst-case*, *central*, and *best-case* interference configuration. A spatially non uniform traffic load is considered, where only the interfering users' transmission is supposed to be non continuous. This allows one to quantify the effects of the interference reduction, taken separately, due to the lowering of traffic load, on system spectral efficiency.

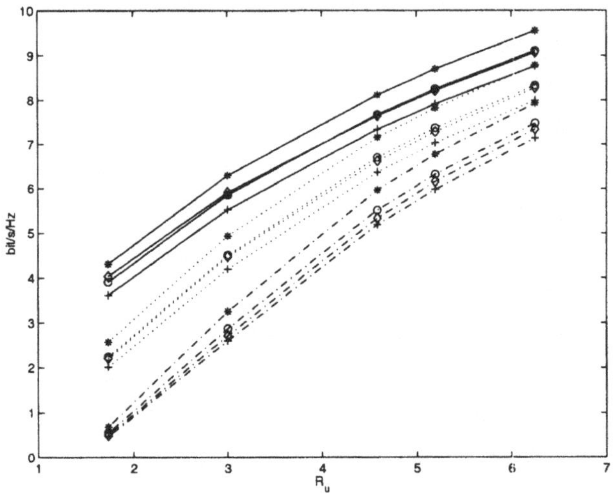

Fig. 8.2.9: Spectral efficiency vs. R_u, for a microcellular system with $\sigma_{sh} = 3$ dB. *Average-case* active user's position. *Best-case* interferer (solid lines), *central* interferer (dotted lines) *worst-case* interferer (dash-dotted lines). Traffic control: $a_c = 1$ ('◊'=6 interferers, '+'=18 interferers), $a_c = 0.8$ ('o'=18 interferers), $a_c = 0.6$ ('✳'=18 interferers).

For the *fully-loaded* system ($a_c = 1$), both the single tier and the two tiers interfering cells models are evaluated and compared. For the *partially-loaded* system ($a_c = 0.8$ and $a_c = 0.6$) only the two tiers model is considered. The figure highlights the approximation error we have by considering the first tier only: indeed, the single tier *fully-loaded* system seems to have about the same efficiency of the two tiers, *partially-loaded* system, with normalized traffic $a_c = 0.8$.

In Figs. 8.2.10 and 8.2.11 is reported the area spectral efficiency, i.e. the spectral efficiency per area unit, versus the reuse distance R_u for a microcellular system and for a *worst-case*, *central*, and *best-case* interference configuration, considering, this time, a spatially uniform traffic load. The area spectral efficiency is given by $\overline{C'} = a_c \frac{4}{\pi R_u^2} C'_{av}$, because the user activity is decreased owing to the partial system loading, and we consider the capacity normalized to the area covered by a single cell, approximately $\pi R_u^2/4$ (see [6]).

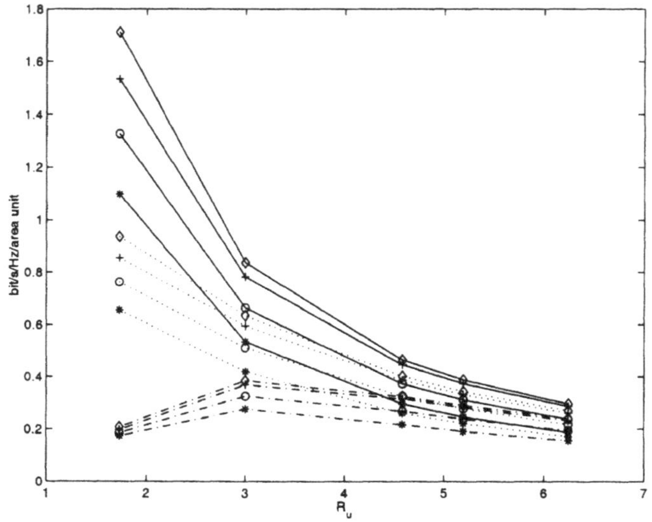

Fig. 8.2.10: Area spectral efficiency vs. R_u, for a microcellular system. σ_{sh} = 3 dB. *Average-case* active user's position. *Best-case* interferer (solid lines), *central* interferer (dotted lines) *worst-case* interferer (dash-dotted lines). Traffic control: $a_c = 1$ ('◊'=6, '+'=18 interferers), $a_c = 0.8$ ('o'=18 interferers), $a_c = 0.6$ ('✳'=18 interferers).

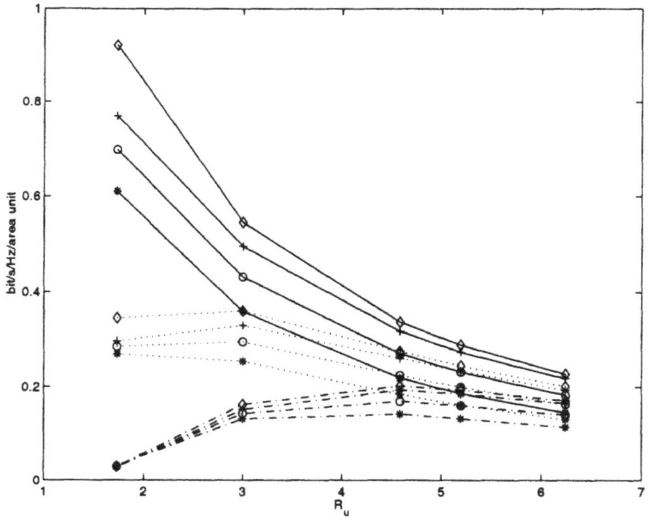

Fig. 8.2.11: Area spectral efficiency vs. R_u, for a microcellular system with σ_{sh} = 3 dB. *Worst-case* active user's position. *Best-case* interferer (solid lines), *central* interferer (dotted lines) *worst-case* interferer (dash-dotted lines). Traffic control: $a_c = 1$ ('◊'=6, '+'=18 interferers), $a_c = 0.8$ ('o'=18 interferers), $a_c = 0.6$ ('✳'=18 interferers).

In Fig. 8.2.10, the active user's position, in the reference cell, is assumed to be uniformly distributed in the ring $r_0 < R_d < 1$ (*average-case*), whereas in Fig. 8.2.11 the active user's position is assumed to be on the cell border (*worst-case*). Also in this case, for the *fully-loaded* system ($a_c = 1$), both the single tier and the two tiers interfering cells models are evaluated and compared, highlighting the approximation error we have by considering the first tier only. For the *partially-loaded* system ($a_c = 0.8$ and $a_c = 0.6$) only the two tiers model is considered.

The curves quantify the fact that the area spectral efficiency of a cellular system increases with the traffic loading. Moreover, the results show that, for the *average-case* active user's position (Fig. 8.2.10), as far as the *central* and *best-case* interference configurations are concerned, the area spectral efficiency of a wireless cellular system is increased by decreasing the reuse distance: only for the *worst-case* interference configuration an optimal reuse distance to maximize the spectral efficiency per area unit can be obtained. For the *worst-case* active user's position (Fig. 8.2.11), an optimal reuse distance can be obtained also for the *central* case interference configuration. This conclusion holds for both *fully* and *partially-loaded* systems.

From results shown in Fig. 8.2.9 it may be derived that traffic control could potentially increase the spectral efficiency by decreasing the normalized traffic itself, since the reduced interference could allow the users to achieve higher rates in *partially-loaded* systems. However, this effect is more than offset by the fact that, if the time wastage due to non continuous transmission is taken into account with a proper definition of the area spectral efficiency (Figs. 8.2.10 and 8.2.11), only a fraction of the allocated bandwidth is used. Therefore, it can be concluded that traffic control has a negative effect on system performance, in agreement with the results in [6].

8.2.8 Sectorized systems

The use of sectorized antennas with $S = 3$ sectors per cell is considered, too. The carriers allocated to each cell are divided into disjoint sets, with each set assigned to a single sector per cell [6]: namely, the bandwidth allocated to a given cell is partitioned into *sector segments*, one for each of the bases antenna sector. The number of co-channel interferers is therefore reduced leading to an increase in system capacity. Namely, for $S = 3$, up to $N_I = 8$ interferers are supposed to be active in the 8 co-channel cell sectors of the first two tiers (see Fig. 8.2.12). To lay emphasis on the effects of sectorization on spectral efficiency, the interferers are assumed, this time, to be uniformly distributed in the cell/sector they belong to.

The effects of sectorization on spectral efficiency are shown in Figs. 8.2.13 and 8.2.14, considering first a spatially non uniform traffic load where only the interfering users' transmission is supposed to be non continuous. This allows one to quantify

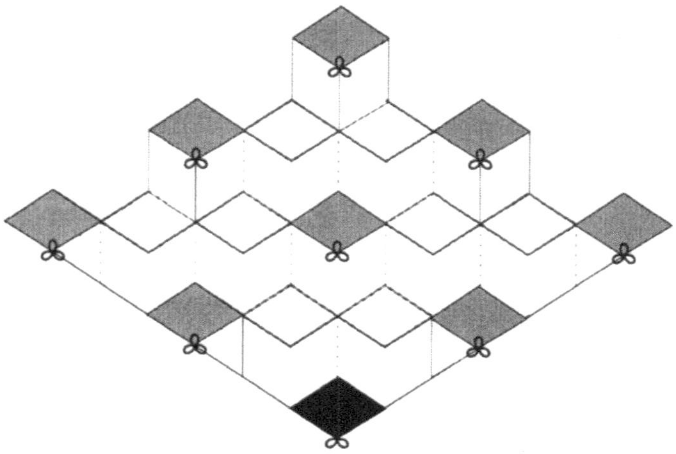

Fig. 8.2.12: Co-channel interfering sectors (cluster size: $C = 3$).

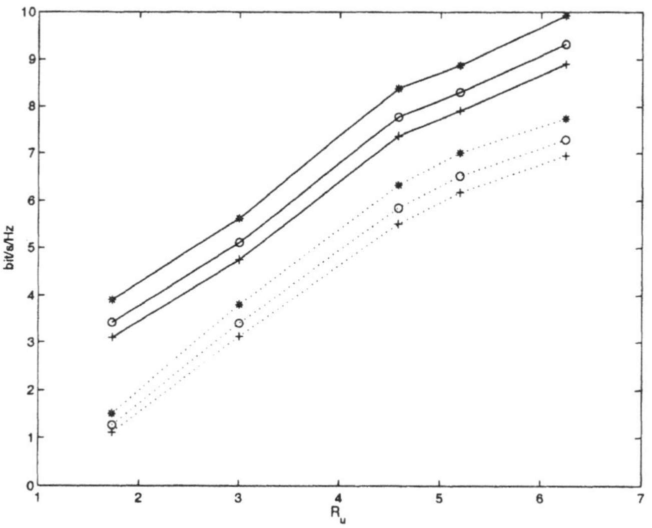

Fig. 8.2.13: Spectral efficiency vs. reuse distance R_u, for a microcellular system ($a = 2$, $b = 2$, $g = 0.67$, $K = 4$ and $\sigma_{sh} = 0$ for the useful signal, $K = 0$ and $\sigma_{sh} = 6$ dB for the interfering signals). The active user's position, in the reference cell/sector, is assumed to be uniformly distributed (*average-case*). Sectorized systems (solid lines), non-sectorized systems (dotted lines). Traffic control: $a_c = 1$ (+), $a_c = 0.8$ (o), $a_c = 0.6$ (*).

the effects of the interference reduction, taken separately, due to the use of sectorization or to the lowering of traffic load, on system spectral efficiency.

In Figs. 8.2.13 and 8.2.14 is reported the spectral efficiency versus the reuse distance R_u for a microcellular system (with parameters $a = 2$, $b = 2$, $g = 0.67$, $K = 4$ and $\sigma_{sh} = 0$ for the useful signal, $K = 0$ and $\sigma_{sh} = 6$ dB for the interfering signals). In Fig. 8.2.13, the active user's position in the reference cell or sector is assumed to be uniformly distributed in the cell/sector he belongs to (*average-case*), whereas in Fig. 8.2.14 the active user's position is assumed to be on the cell/sector border (*worst-case*).

When using sectored antennas, while maintaining the same load per cell, the number of co-channel interferers is approximately reduced by a factor S (with $S = 3$): therefore sectorized systems allow to achieve the highest spectral efficiency (solid lines), with respect to the conventional systems using omni-directional antennas (dotted lines).

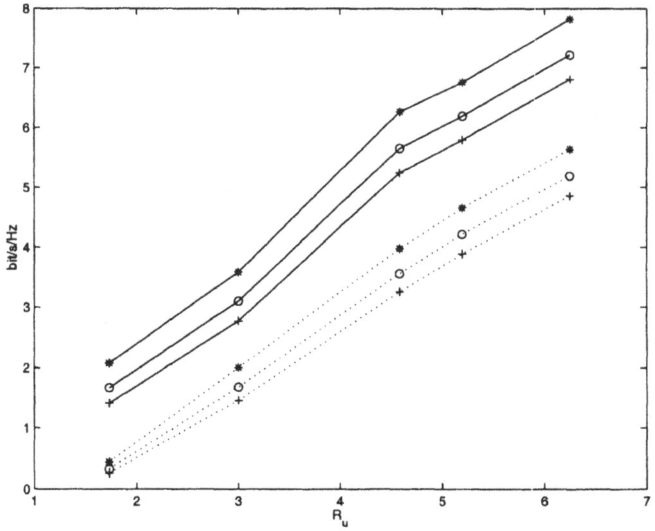

Fig. 8.2.14: Spectral efficiency vs. reuse distance R_u, for a microcellular system ($a = 2$, $b = 2$, $g = 0.67$, $K = 4$ and $\sigma_{sh} = 0$ for the useful signal, $K = 0$ and $\sigma_{sh} = 6$ dB for the interfering signals). The active user's position, in the reference cell or sector, is assumed to be on the cell/sector border (*worst-case*). Sectorized systems (solid lines), non-sectorized systems (dotted lines). Traffic control: $a_c = 1$ (+), $a_c = 0.8$ (o), $a_c = 0.6$ (*).

Moreover, from the figures it may be concluded that, if a spatially non uniform traffic load is assumed (namely, if the time wasted by the desired users, because of non continuous transmission due to the partial system loading, is not taken into account), the reduced interference to the desired users allows these users to achieve higher rates.

In Figs.8.2.15 and 8.2.16 is reported the spectral efficiency per area unit versus the reuse distance R_u for a microcellular system, considering, this time, a spatially uniform traffic load. The area spectral efficiency is given by:

$$\overline{C'} = \frac{a_c}{A} C'_{av} \tag{25}$$

because the user activity is decreased owing to the spatially uniform traffic load hypothesis, and we consider the capacity normalized to the area A covered by a single cell or sector. This area is approximately equal to $\pi R_u^2/4$ (see [6]) for sectorized systems and for the conventional systems using omni-directional antennas.

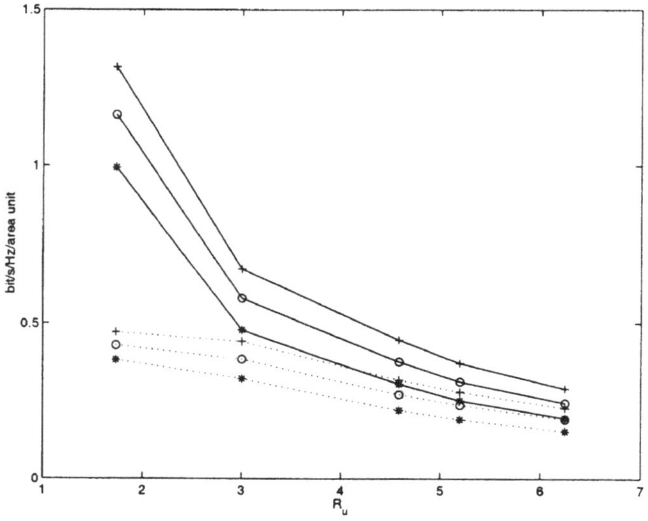

Fig. 8.2.15: Area spectral efficiency vs. reuse distance R_u, for a microcellular system (a = 2, b = 2, g = 0.67, K = 4 and σ_{sh} = 0 for the useful signal, K = 0 and σ_{sh} = 6 dB for the interfering signals). The active user's position, in the reference cell/sector, is assumed to be uniformly distributed (*average-case*). Sectorized systems (solid lines), non-sectorized systems (dotted lines). Traffic control: $a_c = 1$ (+), $a_c = 0.8$ (o), $a_c = 0.6$ ($*$).

In Fig. 8.2.15, the active user's position, in the reference cell or sector, is assumed to be uniformly distributed in the cell/sector he belongs to (*average-case*), whereas in Fig. 8.2.16 the active user's position is assumed to be on the cell/sector border (*worst-case*).

The area spectral efficiency is shown to be generally increased by decreasing the reuse distance. Only for the *worst-case* active user's position (Fig. 8.2.16), an optimal reuse distance can be obtained for omni-directional systems.

For the *average-case* active user's position (Fig. 8.2.15), the best system performance can be obtained when using the reuse distance corresponding to a cluster size C=1. For the *worst-case* active user's position (Fig. 8.2.16), when using a sectorized system the best system performance can be obtained with a cluster size $C = 1$. For omni-directional systems the optimal system performance can be obtained with $C = 3$. These conclusions hold for both *fully* and *partially-loaded* systems.

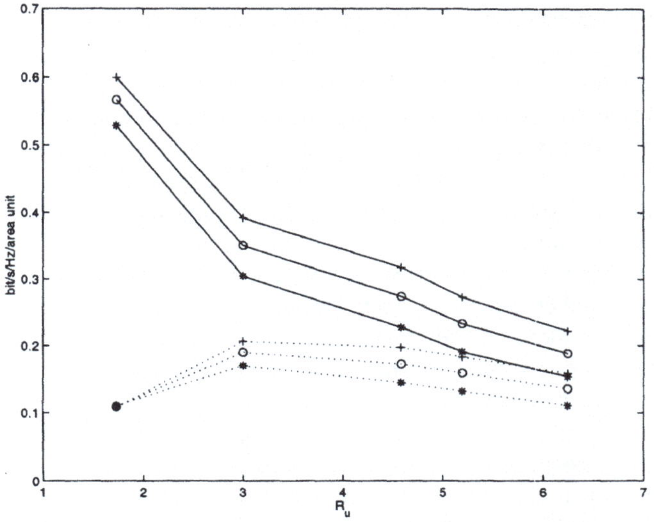

Fig. 8.2.16: Area spectral efficiency vs. reuse distance R_u, for a microcellular system (a = 2, b = 2, g = 0.67, K = 4 and σ_{sh} = 0 for the useful signal, K = 0 and σ_{sh} = 6 dB for the interfering signals). The active user's position, in the reference cell or sector, is assumed to be on the cell/sector border (*worst-case*). Sectorized systems (solid lines), non-sectorized systems (dotted lines). Traffic control: $a_c = 1$ (+), $a_c = 0.8$ (o), $a_c = 0.6$ (∗).

Moreover, the curves of Figs. 8.2.15 and 8.2.16 quantify the fact that the area spectral efficiency of a cellular system generally increases with the traffic loading, i.e. that traffic control has a generally negative effect on system performance. Only for the *worst-case* active user's position and for $C = 1$, traffic control has no effect or a weakly positive effect on area spectral efficiency, if we consider the systems using omni-directional BS antennas (Fig. 8.2.16), since for this cluster size the system area spectral efficiency attains a minimum.

8.2.9 Conclusions

In this chapter an approximation (lower bound) of the capacity of the uplink (mobile to base station) of a cellular system has been obtained, with perfect channel state information (CSI) at the receiver, versus the reuse distance and the level of shadowing affecting the interfering signals. A narrowband wireless communication system has been considered, and its capacity computed as the average maximum mutual information conditioned on the channel gain (i.e., the CSI). The effect of the first two tiers of interferers has been determined and compared to the single tier approximation. The effects of traffic on spectral efficiency have been also investigated by evaluating the *fully-loaded* system performance and the *partially-loaded* system performance. Finally, to quantify the effects of sectorization on system performance, the use of sectorized antennas with S=3 sectors per cell has been considered, maintaining the same load per cell.

A spatially non uniform traffic load has been considered first, where only the interfering users' transmission is supposed to be non continuous. This allows one to quantify the effects of the interference reduction, taken separately, on system spectral efficiency. The results show the system spectral efficiency increase when the normalized traffic is decreased while keeping the desired user's continuous transmission. Moreover, in this case the effects of sectorization on spectral efficiency have been quantified as far as the only decreasing of the co-channel interferers number, due to sectorization, is concerned.

Secondly, a more realistic system has been considered, using a spatially uniform traffic load, and normalizing the spectral efficiency with respect to the area A covered by a single cell or sector. In this case, it has been shown that sectorization increases significantly the system area spectral efficiency (i.e. the spectral efficiency normalized to the area A) with respect to a non sectorized system using omni-directional antennas at the BS, and that this increase is more significant for lower values of the reuse distance.

Finally, the results presented in this work quantify the fact that the area spectral efficiency of a cellular system generally increases with the traffic loading, i.e. that traffic control has a generally negative effect on system performance. In fact, from results shown in the figures it can be derived that traffic control could potentially increase the spectral efficiency by decreasing the normalized traffic itself (since the interference is reduced), but this effect is more than offset by the fact that, if a spatially uniform traffic load is considered, only a fraction of the cell allocated bandwidth is used, because of time wastage due to non continuous transmission of the desired user.

References

1. W. C. Y. Lee, *Mobile cellular telecommunications Systems*, McGraw-Hill, New York, 1989.
2. V. H. Mac Donald, "The cellular concept", *The Bell System Technical Journal*, vol. 58, no. 1, pp. 15-41, 1979.
3. C. E. Shannon, "A mathematical theory of communication", *The Bell System Technical Journal*, vol. 27, pp. 379-423, 1948.
4. C. E. Shannon, "A mathematical theory of communication", *The Bell System Technical Journal*, vol. 27, pp. 623-656, 1948.
5. G. Taricco and F. Vatta, "Capacity of cellular mobile radio systems", *IEE Electronics Letters*, vol. 34, no. 6, pp. 517-518, 1998.
6. M.-S. Alouini and A. Goldsmith, "Area spectral efficiency of cellular mobile radio systems", *Proc. of the 47th IEEE Vehicular Technology Conference - VTC'97*, Phoenix, USA, pp. 652-656, May 1997.
7. G. K. Chan, "Effects of sectorisation on the spectrum efficiency of cellular radio systems", *IEEE Transactions on Vehicular Technology*, vol. 41, no. 3, pp. 217-225, 1992.
8. R. J. C. Bultitude and G. K. Bedal, "Propagation characteristics of microcellular urban mobile radio channels at 910 MHz", *IEEE Journal on Selected Areas in Communications*, vol. 7, no. 1, pp. 31-39, 1989.
9. W. K. Pahlavan and A. H. Levesque, *Wireless information networks*, John Wiley & Sons, Inc., New York, 1995.
10. J. J. Egli, "Radio propagation above 40 MC/s over irregular terrain", *Proceedings of IRE*, pp. 1383-1391, 1957.
11. W. C. Y. Lee, *Mobile communications engineering*, McGraw-Hill, New York, 1982.
12. W. C. Jakes, ed., *Microwave mobile communications*, John Wiley & Sons, Inc., New York, 1974.
13. J.-P. Linnartz, *Narrowband land-mobile radio networks*, Artech House, Inc., Boston, 1993.
14. R. Prasad and J. C. Arnback, "Effects of Rayleigh fading on packet radio channels with fading and shadowing", *Proc. of the IEEE TENCON'89 Conference*, Bombay, India, pp. 546-548, Nov. 1989.
15. S. C. Schwartz and Y. S. Yeh, "On the distribution function and moments of power sums with log-normal components", *Bell System Technical Journal*, vol. 61, pp. 1441-1462, 1982.
16. R. Prasad and A. Kegel, "Improved assessment of interference limits in cellular radio performance", *IEEE Transactions on Vehicular Technology*, vol. 40, no. 2, pp. 412-419, 1991.
17. T. M. Cover and J. A. Thomas, *Elements of Information Theory*, John Wiley & Sons, Inc., New York, 1991.
18. M. Schwartz, W. R. Bennett and S. Stein, *Communications systems and techniques*, McGraw-Hill, New York, 1966.
19. J. G. Proakis, *Digital Communications*, McGraw-Hill, Singapore, 1983.

8.3 Capacity Sharing Between Interfering Cells In CDMA Wireless Networks

Diakoumis Gerakoulis, Wai Chung Chan, and Evaggelos Geraniotis

8.3.1 Introduction

The third generation (3G) wireless networks are expected to provide flexible cellular infrastructure covering a variety of service environments. Currently, while in the process of standardization and development, common air interfaces (CAI) for 3G systems will attempt to accomodate multiple service networks operating in the same frequency band and geographical area. One particular issue related to this problem is the need for underlying cells which may belong to different service providers. Given that there is a public wireless operator offering service in the region, private networks consisting of isolated cells, microcells or clusters of cells may coexist in the same region and frequency bands (underlaying). A cell may be underlaid by private business for use in an office building or factory floor, or by a university in its campus. An underlaying cell may also be permanent, temporary or even mobile (mobile base-station).

The system design for such configuration must then accomodate or eliminate the interference that appears between the overlapping cells. The solution to this problem depends on the access methods used in each cell. Assuming that a cell may either use TDMA/FDMA or CDMA CAIs, we have the following possible cases: (1) Both (over- and under-laying) cells are TDMA/FDMA, (2) One of the cells is TDMA and the other is CDMA and (3) Both are CDMA cells. In the first case where the over- and under-laying cells are TDMA/FDMA the problem may be resolved by frequency reuse or frequency assignments. In the second case where the overlaying cell is CDMA and the underlying cells are TDMA/FDMA (or vice-versa) the interference can be controlled by notch filtering the narrow band TDMA channels as shown in reference [1].

In this paper we only examine the third case where both (the over- and the under-laying) cells are CDMA. The problem in this case, is the interference between the overlapping cells which may be severe since all CDMA cells use the same frequncy band (frequency reuse one). A possible solutions to this problem may be based on interference cancelling if the number of interferors from the other cell is limitted. Another alternative is to provide frequency division between overlapping cells. The method proposed here, presented in the next section, is based on either "capacity division" or "capacity sharing" between overlapping cells. The first approach "eliminates" the interference while the second one"accomodates" the interference between the cells. The use of the first method is appropriate when all cells are outdoors while the second one when the underlaying cells are indoors. Also, the proposed approach has the objective that the existing CAIs (which may be of different CDMA type) will remain undisturbed or with minimum impact on each. An algorithm is also presented which will maximize the "shared capacity" between the overlaying and underlying cells. In section 3 we present the system's capacity and performance evaluation.

8.3.2 System Description

We may consider two possible interference environments: (1) when the underlying cell or cells are outdoors and (2) when the underlying cell or cells are indoors. (The overlaying cell is consider to be outdoors). The method proposed for the first case then will be "Capacity division" while for the second one "Capacity sharing".

The proposed method provides capacity division between overlapping cells having CDMA CAIs.

8.3.2.1 Outdoor Underlaid Cells

In this case since there is strong interference (no physical barrier) between the underlying and overlaying cells. Hence, it is required that signal transmissions between users in each cell are "orthogonally" separated so that interference is eliminated. The approach proposed here is based on *Time-Orthogonal* implementation. This is achieved by dividing each time frame into slots propotional in length to the cell's allocated capacity. Each cell will then transmit on its assigned slot with a transmission symbol rate proportionally higher. (i.e, all bits that would be transmitted in a frame are transmitted now in a time slot). This way the interference between overlapping cells is totaly eliminated since transmissions in each cell takes place only on the assigned time slot. This applies in both the uplink and the downlink. In the uplink all mobiles of the given cell will transmit only on the assigned slot for that cell. Similarly, in the downlink each base station will only transmit on its assigned slot.

The tolal capacity C of the overlaid cell (before introducing any underlying cell) will be divided proportionally between all overlapping cells (overlaid and underlaid) as requested. Then, the capacity of the overlaid cell C_o given K underlying cells, is $C_o = C - \sum_{u=1}^{K} C_u$. Where C_u is the capacity of each underlying cell. The division of the capacity C results from the increase in the

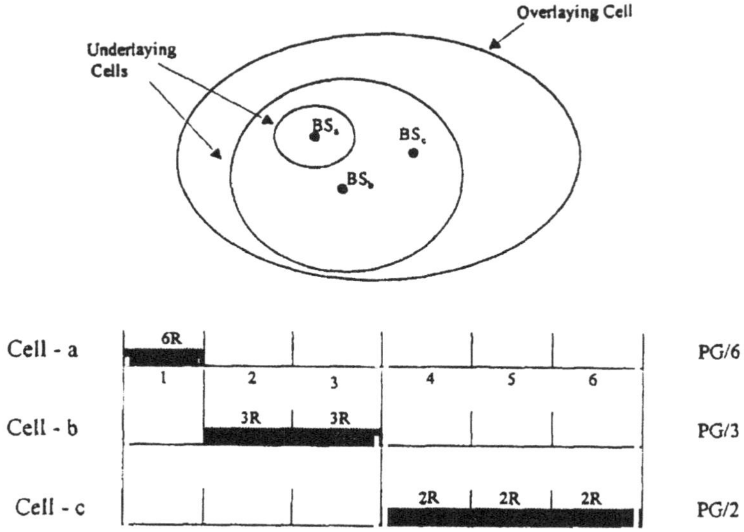

Fig. 8.3.1. Capacity division of overlapping CDMA cells **a**, **b**, and **c** by the "Time-Orthogonal" approach.

bit rate of each overlapping cell which reduces the processing gain and in turn the cell capacity by proportional amount. Figure 8.3.1 shows an example of the proposed concept. We consider 3 overlapping cells. The underlaid cells *a* and *b* are allocated 1/6 and 2/6 of the total capacity C, while the overlaid cell *c* has the remaining 3/6 of C. Cell *a* will then transmit only on the assigned slot 1 at rate $6R$ (6 times higher), cell *b* on slots 2 and 3 at rate $3R$ and cell *c* in slots 4,5, and 6 at rate $2R$. The processing gain (PG) in cell *a* is then reduced to $1/6th$, in cell *b* to $1/3rd$ and in *c* to 1/2. The corresponding capacities of cells *a*, *b* and *c* then are $C/6$, $C/3$ and $C/2$ respectively.

This approach has the advantage that it does not depend of the relative locations between the base stations neither on the relative transmit power between them (or size of the covered area of each cell). Also, the power control loops of each cell operate independently. Given an overlaying cell in operation, an underlying cell may be easily introduced and operate by providing direct communication between the overlaying and underlying base stations (for example by installing a mobile transceiver to the underlaid base station). Then the negotiated capacity allocation will be implemented by speeding up the clock at the assigned rate in each mobile and base station while transmitting in its assigned time slot.

Assuming that each cell fully utilizes its assigned capacity then transmissions will only take place on the assigned slots. If however, the assigned capacity of a cell is not fully utilized, then other cells may use the remaining capacity

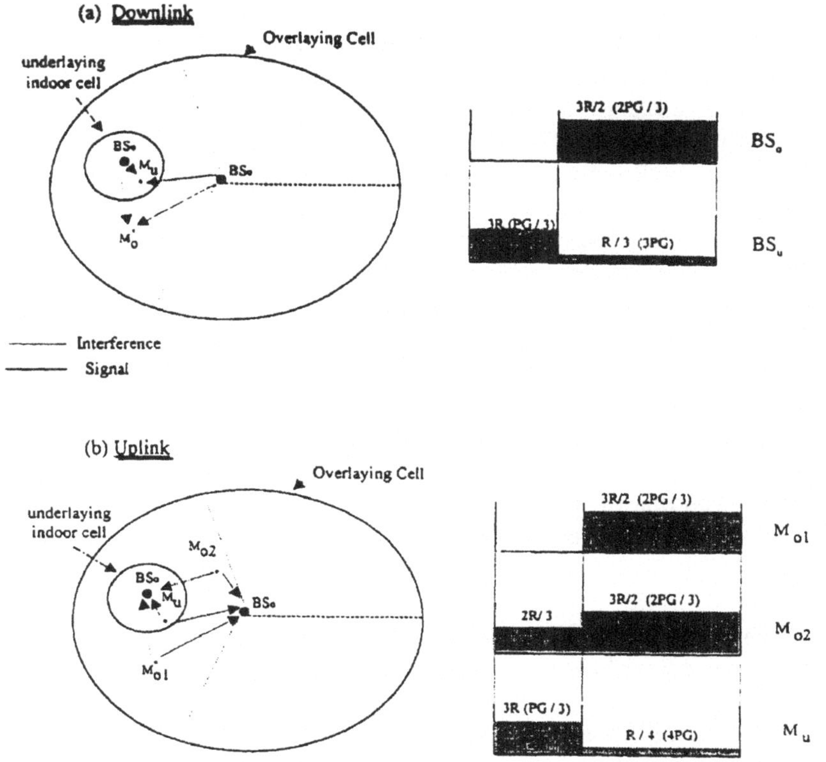

Fig. 8.3.2. Capacity division between Overlaying (outdoors) and Underlaying (indoors) cells. (**a**) Downlink (**b**) Uplink.

by transmitting in the slot assigned to that cell. Hence, a cell may also transmit in a non-assigned slot depending on the interference conditions between the overlapping at that slot. An example is shown in Figure 8.3.2 in which the overlay cell has less users than the available capacity. Then the underlay call may allow certain users to transmit in the non-assigned slot. The rate and the number of users depends on the available capacity. The overlay cell however, has pre-emptive priority on that slot once it receives new call.

8.3.2.2
Indoor Underlaid Cells

In this case we may use either the"Time Orthogonal" approach as in the outdoor case or the "Pseudo-Orthogonal" approach which will be described bellow. The approach that provides the highest capacity is determined (as shown in section 8.3.3) by the degree of isolation (in dBs) between cells . In the Pseudo-

Orthogonal approach we assume that there is always an interference barrier (building walls) which provides sufficient isolation between the indoor (underlaying) cell from the outdoor (overlaying) cell. This approach may then allow to achieve or exceed the total capacity of the Orthogonal approach.

In the Pseudo-Orthogonal approach transmission take place simuntaneously in both cells (which also occupy the same frequency bands). The transmission rates and the number of users in each cell is determined by the required quality of service (QoS). The required QoS is determined by both, a minimum signal to interference ratio (SIR) as well as a minimum transmission rate. The number of users in each cell and the transmission rate of each user will then be determined by the proposed algorithm which attempts satisfy the QoS requirements of each mobile user. The proposed algorithm is based the following actions:

Defensive Action: The transmission rate is reduced so that the SIR is satisfied. That is, by reducing the transmission rate the symbol energy (E_s) is increased so that the processing gain remains unchanged and the SIR (E_s/N_o) meets the QoS requirement. In this case the transmit power also remains unchanged. This action is a defensive responce to an increase of the interference level N_o.

Preventive Action: The transmit power and the transmission rate are reduced so that the required E_s/N_o remains unchanged. This action is needed in order to reduce the interference from this particular transmission. (In order to maintain E_s/N_o the same when the transmit power is reduced, the transmission rate is decreased so that the symbol energy is the same). This action is used in order to prevent the increase of the interference level in other receivers when the mobile is approach these receivers.

The algorithm given below can be applied in either derection, forward (downlink) or reverse (uplink).

The Algorithm

Step 1: A mobile user takes defensive action (DA) if its transmission rate $R > R_{min}$ in order to meet the required signal to interference (SIR) when that falls below a minimum value.

Step 2: A mobile user takes preventive action (PA) as a result of the Joint Power Control (JPC) between the overlay and the underlay cells. The JPC mechanism described and analysed in the next section, attempts to Maximize the Minimum SIR among all mobile users in both cells.

A mobile may take DA when the interference level is increased due to the admission of a new call in either cell. Also mobiles in the underlay cell may take DA when a mobile in the overlay cell is approaching the underlay one. If after a DA a mobile is still not able to achieve the required SIR then the JPC mechanism may order particular mobiles to take PA. A PA may also take place when a mobile in the overlay cell approaching the underlaying cell, is in position of handing-off to the underlaying cell. Assuming that no handoff is allowed, the mobile prevents the handoff conditions to occur by applying PA.

The above algorithm will be successful in meeting the QoS of all users if the number of mobile users in each cell is within a limit. Therefore, the call admission protocol (CAP) is critical to the success of this proccess. The CAP should not allow a new call into the system unless all ongoing calls meet the QoS requirements. Another factor which will impact the QoS or the capacity of each cell is the location or the relative distribution of the mobile users.

Also, the proposed algorithm utilizes a joint power control (JPC) mechanism in which data between the two base station are shared. This is necessary not only to maximize the minimum SIR but also to provide power control stability. (If the power control loops were to operate independently could become unstable).

8.3.3 The Capacity Evaluation

In this section we evaluate the capacity of a underlay and the overlay CDMA cells assuming that both cells are transmitting simultaneously. The capacities for the reverse link (uplink) and the forward link (downlink) are evaluated separately.

The maximum number of simultaneous users are first determined when the underlying cell is turned off. Then, an underlay cell is added to the system which increases the interference level. The number of users in the overlay cells are reduced from the nominal value to a smaller value and the maximum number of users in the underlay cell is then determined.

Let cell $i = 1, 2, \ldots, 7$ denote any overlaying cell in a cluster of 7 cells. Cell 1 is the cell at the center of the cluster and the other six are the cells at the first-tier. Let cell 0 be the (small) underlying cell located inside cell 1,(see Figure 8.3.1).

The signal to noise ratio of a desired signal at a base station of interest is given by

$$SNR(\zeta) = \left[\frac{N_o 10^{\zeta/10}}{2E_b r_c} + \frac{m_\psi}{2N} \sum_k \mathcal{E}\left\{ \frac{P_k}{P_0} \right\} \right]^{-1} \tag{1}$$

where r_c is the FEC encoder rate, E_b/N_o is the bit energy to noise density ratio, $10^{\zeta/10}$ is a log-normal distributed variable accounting for the power control error, N is the processing gain, $m_\psi = 0.5$ is the waveform shaping factor, P_k is the power received at the base station due to an undesired user k, P_0 is the desired power received at the base station, and finally $\mathcal{E}$ represents expectation with respect to the ratio P_k/P_0. The undesired user (interferor) k may be a user from the same cell or from other cells.

The average uncoded bit error probability can then be obtained by taking the expectation with respect to the power control error ζ:

$$P_b = \mathcal{E}\left\{Q\left(\sqrt{SNR(\zeta)}\right)\right\} = \int d\zeta \cdot f(\zeta) \cdot Q\left(\sqrt{SNR(\zeta)}\right) \qquad (2)$$

where $f(\cdot)$ is the log-normal distribution, and $Q(\cdot)$ is defined as

$$Q(x) = \frac{1}{\sqrt{2\pi}} \int_x^{\infty} e^{-u^2/2} du \qquad (3)$$

The term $\mathcal{E}\{P_k/P_0\}$ can be written as

$$\mathcal{E}\left\{\frac{P_k}{P_0}\right\} = \mathcal{E}\left\{\frac{E_k R_k 10^{\zeta_k/10} r_k^{\eta}}{E_0 R_0 10^{\zeta_0/10} r_{0k}^{\eta} L_c}\right\} \qquad (4)$$

The term $E_0 R_0$ is the desired received power, r_k is the distance of the undesired user from its own base station, r_{0k} is the distance of the interfering user from the base station of interest. E_k and E_0 are the energy per bit of the interfering and desired user reprectively and R_k and R_0 are the corresponding data rates. L_c is the *Path loss* signal penetrating the building walls . We assume that signals entering or leaving the bulding with the underlaying cell will experience a loss of L_c dB. η is the path loss exponent.

The Reversen Link Capacity Results

In evaluating the cell capacities the following parameters are used: We assume that the path loss exponent $\eta = 5$ if the signal originates in the underlaying cell and $\eta = 4$ if the signal originates in the overlaying cell. The standard deviation of the log-normal distribution is $\sigma = 2.5$ dB. Making the assumption that the IS-95 CAI is used (in both over- and under-laying cells) the processing gain is $N = 128$, $E_b/N_o = 6$ dB, and the convolutional encoder has the rate $r_c = 1/3$ in the uplink. The length of the overlaying cell radious is $D = 1$ (see Figure 8.3.3). The radius of the underlaying cell is d. The center of the underlay cell is at a distance l from the center of the overlay cell 1. The bit energy and the data rates are assumed to be the same for all the cells. The average bit error probability of the uncoded signal is required to be smaller than or equal to 0.01 while for the coded signal less than 10^{-4}.

The number of users K_0 in the underlaying cell, and K_1 of the overlaying cell are shown in Tables 1 and 2 foe different cell sizes (l, d, *and* D) and different values of path loss at the buliding walls L_c.

The case where $K_0 = 0$ and $K_1 = 30$ users is when the underlaying cell is empty and the overlaying cell is full. In the other cases the total capacity is $K_0 + K_1$

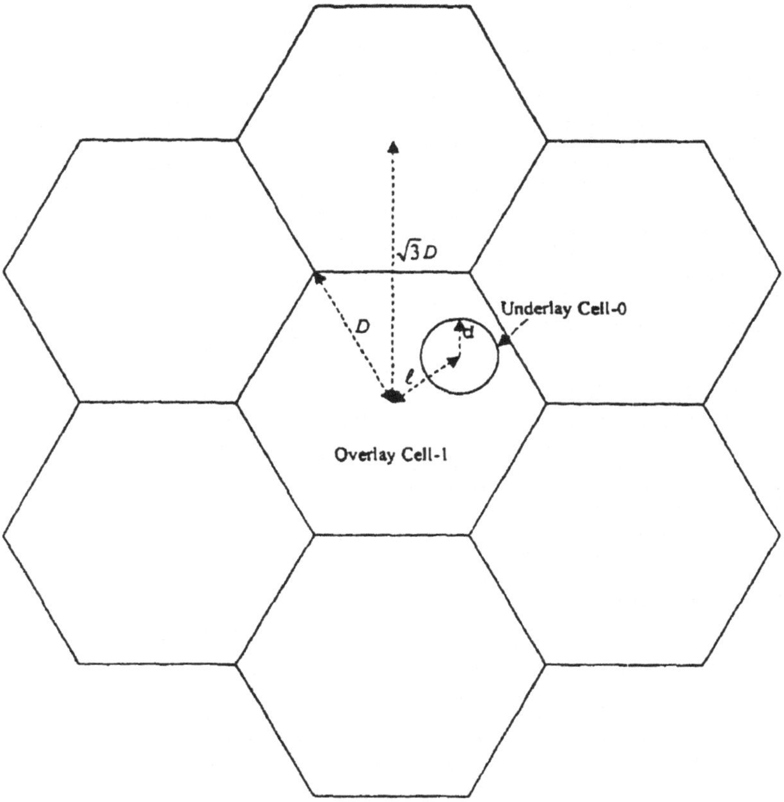

Fig. 8.3.3. The relative dimensions of the overlay and underlay cells.

The capacity is given for the following cases: (1) when $L_c = 1(0db)$ which is the case both under- over-laying cells are outdoors. (2) when $L_c = 2(-2db)$,

Table 1. $l = 0.6$, $d = 0.2$, $D = 1$. Maximum number of instantaneous users in the underlay and overlay cells. L_c is the penetration loss when entering or leaving the underlay cell.

$l = 0.6, d = 0.2, D = 1$										
L_c	1(0db)		2(-3db)			3(-4.5db)				
K_0	0	4	0	4	8	0	1	4	7	10
K_1	30	5	30	10	9	30	16	15	14	13
L_c	4(-6db)									
K_0	0	4	6	8	11	13	15			
K_1	30	20	19	18	17	16	15			
L_c	10(-10db)									
K_0	0	22	23	24	24	25	26			
K_1	30	30	29	28	27	26	25			

Table 2. $l = 0.5$, $d = 0.2$, $D = 1$. Maximum number of instantaneous users in the underlay and overlay cells. L_c is the penetration loss when entering or leaving the underlay cell.

$l = 0.5, d = 0.2, D = 1$																
L_c	1(0db)			2(-3db)				3(-4.5db)								
K_0	0	2	8	0	2	5	8	0	2	4	6	8	11			
K_1	30	7	6	30	14	13	12	30	21	20	19	18	17			

L_c	4(-6db)						
K_0	0	2	4	5	7	8	12
K_1	30	28	27	26	25	24	23
L_c	10(-10db)						
K_0	0	28	29	30	30	31	32
K_1	30	30	29	28	27	26	25

$L_c = 3(-4.7db)$ which is the case of residential homes and (3) when $L_c = 4(-6db)$, $L_c = 10(-10db)$ which is the case of commercial office buldings.

The Forward Link Capacity Results

The forward link capacity has been evaluated in a similar manner and the following parameters are used: $\sigma = 2.5$ dB is the standard deviation of the log-normal distribution, $N = 128$ is the processing gain, $E_b/N_o = 6$ dB is the bit energy to noise density ratio, and $r_c = 1/2$ is the coding rate. The length of the side of the hexagonal cell is $D = 1$. The radius of the underlay cell is d. The center of the underlay cell is at a distance l from the center of the overlay cell 1. The bit energy and data rate are the same for all the cells. The worst case average bit error probability is required to be smaller than or equal to 0.0457 (in order to support 42 users under normal operation).

Table 3. Forward Link. $l = 0.6$, $d = 0.2$, $D = 1$. Maximum number of instantaneous users in the underlay and overlay cells. L_c is the penetration loss when entering or leaving the underlay cell.

$l = 0.6, d = 0.2, D = 1$									
L_c	4 (6 dB)			10 (10 dB)			25 (14 dB)		
K_0	0	1	42+	0	1	42+	0	1	42+
K_1	42	9	9	42	23	23	42	41	41

Table 4. Forward Link. $l = 0.5$, $d = 0.2$, $D = 1$. Maximum number of instantaneous users in the underlay and overlay cells. L_c is the penetration loss when entering or leaving the underlay cell.

$l = 0.5, d = 0.2, D = 1$									
L_c	4 (6 dB)			10 (10 dB)			25 (14 dB)		
K_0	0	1	42+	0	1	42+	0	1	42+
K_1	42	2	2	42	7	7	42	18	18

8.3.4
Discussion of Results and Conclusion

As we observe in the Tables 1 to 4, the maximum capacity of the system of overlay/underlay cells $K^* = K_0 + K_1$, depends on the dB loss L_c, of signal power at the building wall and the number of users in each cell. The capacity K of the overlay cell when the underlay one is empty is $K = 30$ for the uplink and $K = 42$ for the downlink. As we observe in the Tables K^* may be smaller than K or it may exceed K. As shown, for $L_c > 6\ dB\ K^* > K$. Therefore based on the value of L_c or on the comparison of K^* and K we conclude that for outdoor cells ($L_c = 0$dB) or when the indoor cell has $L_c < 6$ dB we apply the *Time-Orthogonal* implementation to the system of overlay/underlay cells. When $L_c > 6$ dB then we apply the *Pseudo-Orthogonal* implementation which offers in this case maximum capacity $K^* > K$. We also observe in Table 2, that for $L_c = 10$ dB the capacity of the underlay cell $K_0 = 32 > K = 30$. This is due to the fact that the interference of adjacent cells (first and second tier) also attenuate at the building walls.

In conlusion, we have presented two basic solutions to the overlay-underlay problem in CDMA, one for indoor and the other for outdoor underlay cell. The *Time-Orthogonal* approach provides flexibility and simplicity in outdoor environment while the *Pseudo-Orthogonal* approach can offer higher capacity in indoor environment. We have also presented an algorithm and a joint power control method which are utilized to maximized the minimum SIR over all users in the system for the Pseudo-Orthogonal case. The overall (overlay/underlay) system capacity has been evaluated for different values of signal path-loss at building walls.

References

[1] L.B. Milstein, D.L. Schilling, R.L. Pickholtz, V. Erceg, M. Kullback, E.G. Kanterakis, D.S. Fishman, W.H. Biederman and D.C. Salerno. "On the Feasibility of a CDMA Overlay for Personal Communications Networks" *IEEE Journal on selected areas in commun.*, Vol 10., No. 4, May 1992 pp 655-667.

APPENDIX A: Reverse Link Analysis

Reverse Link Analysis at the Underlay Cell

Let's consider the performance of the reverse link of the underlay cell 0. The signal to noise ratio received at base station 0 is given by

$$SNR(\zeta) = \left[\frac{N_o 10^{\zeta/10}}{2E_b r_c} + \frac{m_\psi}{2N}\left(\mathcal{E}\{10^{\zeta_k/10}\}(K_0 - 1) + \right.\right.$$
$$\left.\left. + \frac{1}{L_c}\mathcal{E}\{\frac{10^{\zeta_k/10}}{10^{\zeta_d/10}}\}\sum_{k\neq 0} K_k \mathcal{E}_k\{\frac{r_k^4}{r_{dk}^4}\}\right)\right]^{-1} \tag{5}$$

We have considered interferers in different cells separately. The expectation $\mathcal{E}_k$ is with respect to cell k. K_k is the number of users in cell $k = 0, 1, \ldots, 7$.

For the interferers in cell 1, the interference caused to cell 0 is given by

$$\mathcal{E}_k\{\frac{r_k^\eta}{r_{dk}^\eta}\} = \frac{2}{A_1}\Bigg(\int_0^{\alpha^*} d\vartheta \int_0^{r_L(\vartheta)} dr + \int_0^{\alpha^*} d\vartheta \int_{r_U(\vartheta)}^{\frac{\sqrt{3}D}{2\cos\vartheta}} dr + \int_{\alpha^*}^{\frac{\pi}{6}} d\vartheta \int_0^{\frac{\sqrt{3}D}{2\cos\vartheta}} dr + \int_{\frac{\pi}{6}}^{\frac{\pi}{2}} d\vartheta \int_0^{\frac{\sqrt{3}D}{2\cos(\vartheta-\pi/6)}} dr + \int_{\frac{\pi}{2}}^{\frac{5\pi}{6}} d\vartheta \int_0^{\frac{\sqrt{3}D}{2\cos(\vartheta-\pi/2)}} dr + \int_{\frac{5\pi}{6}}^{\pi} d\vartheta \int_0^{\frac{\sqrt{3}D}{2\cos(\vartheta-5\pi/6)}} dr\Bigg) \frac{r \cdot r^4}{[l^2 + r^2 - 2lr\cos\vartheta]^2} \tag{6}$$

where $\sin\alpha^* = d/l$, $r_L(\vartheta) = l\cos\vartheta - \sqrt{d^2 - l^2\sin^2\vartheta}$, $r_U(\vartheta) = l\cos\vartheta + \sqrt{d^2 - l^2\sin^2\vartheta}$, and A_k denotes the area of cell k.

For the interferers in cell $k = 2, 3, \ldots, 5$, the interference caused to cell 0 is given by

$$\mathcal{E}_k\{\frac{r_k^\eta}{r_{dk}^\eta}\} = \frac{1}{A_k}\Bigg(\int_0^{\frac{\pi}{6}} d\vartheta \int_0^{\frac{\sqrt{3}D}{2\cos\vartheta}} dr + \int_{\frac{\pi}{6}}^{\frac{\pi}{2}} d\vartheta \int_0^{\frac{\sqrt{3}D}{2\cos(\vartheta-\pi/6)}} dr + \int_{\frac{\pi}{2}}^{\frac{5\pi}{6}} d\vartheta \int_0^{\frac{\sqrt{3}D}{2\cos(\vartheta-\pi/2)}} dr + \int_{\frac{5\pi}{6}}^{\pi} d\vartheta \int_0^{\frac{\sqrt{3}D}{2\cos(\vartheta-5\pi/6)}} dr\Bigg) \frac{r \cdot r^4}{[l_k^2 + r^2 - 2l_k r\cos((\vartheta - \phi_k)]^2} + \frac{r \cdot r^5}{[l_k^2 + r^2 - 2l_k r\cos((\vartheta + \phi_k)]^4} \tag{7}$$

where $l_k^2 = l^2 + 3D^2 - 2\sqrt{3}Dl\cos\psi_k$ and $\frac{\sin\phi_k}{l} = \frac{\sin\psi_k}{l_k}$, with $\psi_2 = 0$, $\psi_3 = \pi/3$, $\psi_4 = 2\pi/3$, and $\psi_5 = \pi$. The effect of interferers in cell 6 is the same as that in cell 4 and the same is true for cells 3 and 7.

Reverse Link Performance at the Overlay Cell

Let's consider the performance of the reverse link of the overlay cell 1. The signal to noise ratio received at base station 1 is given by

$$SNR(\zeta) = \Bigg\{\frac{N_o 10^{\zeta/10}}{2E_b r_c} + \frac{m_\psi}{2N}\Bigg[\mathcal{E}\{10^{\frac{\zeta_k}{10}}\}(K_1 - 1) +$$

$$+\mathcal{E}\{\frac{10^{\zeta_k/10}}{10^{\zeta_d/10}}\}\left(\frac{K_0}{L_c}\mathcal{E}_0\{\frac{r_k^5}{r_{dk}^5}\} + \sum_{k\neq 0,1} K_k\mathcal{E}_k\{\frac{r_k^4}{r_{dk}^4}\}\right)\Bigg]\Bigg\}^{-1} \qquad (8)$$

For the interferers in cell 0, the interference caused to cell $k = 1, 2, \ldots, 7$ is given by

$$\mathcal{E}_k\{\frac{r_k^\eta}{r_{dk}^\eta}\} = \frac{2}{A_0}\int_0^\pi d\vartheta \int_0^d dr \frac{r \cdot r^5}{(\sqrt{l_k^2 + r^2 - 2l_k r \cos\vartheta})^5} \qquad (9)$$

For interferers in cell $k = 2, 3, \ldots, 7$, the interference caused to cell 1 is given by

$$\mathcal{E}_k\{\frac{r_k^\eta}{r_{dk}^\eta}\} = \frac{2}{A_k}\Bigg(\int_0^{\frac{\pi}{6}} d\vartheta \int_0^{\frac{\sqrt{3}D}{2\cos\vartheta}} dr +$$

$$+\int_{\frac{\pi}{6}}^{\frac{\pi}{2}} d\vartheta \int_0^{\frac{\sqrt{3}D}{2\cos(\vartheta-\pi/6)}} dr + \int_{\frac{\pi}{2}}^{\frac{5\pi}{6}} d\vartheta \int_0^{\frac{\sqrt{3}D}{2\cos(\vartheta-\pi/2)}} dr +$$

$$+\int_{\frac{5\pi}{6}}^{\pi} d\vartheta \int_0^{\frac{\sqrt{3}D}{2\cos(\vartheta-5\pi/6)}} dr\Bigg) \frac{r \cdot r^4}{[3D^2 + r^2 - 2\sqrt{3}Dr\cos\vartheta]^2} \qquad (10)$$

9 Prototypes Of 3G Systems

9.1 Introduction

The evolution from the second generation mobile systems, basically the GSM standard, to the third generation systems has been under way for the last few years. operators and manufacturers as well as the international standards approving organizations seem to be converging to a likely standard for 3G systems and that is CDMA. Wideband Code Division Multiple Access (WCDMA) seems to be the undisputed standard for the wideband air interface. The reasons are many and well founded and we shall mention here a few. CDMA systems are almost immune to jamming, have higher security and privacy, perform better in multipath propagation and are more flexible with a number of users. These systems also provide higher resolution in ranging and thus become a valuable tool in the Global Positioning System (GPS). There are, however, some critical issues that had to be resolved in order that CDMA become the basis for a 3G standard. These issues involve Timing Acquisition and Network Synchronization in general and signal power control.

In section 9.2 various approaches are analyzed with respect to power control problem whereas section 9.3 and 9.4 cover timing acquisition and network synchronization respectively.

9.2 Power control approaches for WCDMA

MSc. Iuliana Virtej, Prof. Heikki Koivo

9.2.1 Introduction

While no one can predict the future, it is certain that in less than five years - at the start of the 21st Century - the way the world communicates, will be vastly different from now. The speed of mobile multimedia communications and the accessibility of mobile multimedia are two such key changes.

9.2.1.1 WCDMA in brief

9.2.1.2 Third Generation Wireless Communication System

By year 2000, multimedia applications will enter in the world of cellular communications. To support the new services, which will set new demands on the infrastructure for radio communications, a new wireless communication system is being to be developed.

When the number of cellular phone users grows, the need to increase the capacity of the cellular systems becomes more important. New systems have been designed and the systems in use today have been enhanced to meet the capacity requirements.

The development of cellular radio networks has been very fast in the last years. First cellular radio systems were based on Frequency Division Multiple Access (FDMA) technology, and they were called first generation cellular systems. They have mainly supported analogue voice communications. The second generation cellular system was introduced during the last few years. These are fully digital systems, and are based on Code Division Multiple Access (CDMA) technology. One of the second-generation cellular system is IS-95 and is based on the CDMA technology.

For the third generation system, feasibility of both TDMA and CDMA technologies has been studied and research is still going on.

The new standard Universal Mobile Telecommunication System (UMTS) will open a new set of spectrum with new access methods that will serve for increasing the capability of the cellular systems. The WCDMA system is intended as a first step towards commercial systems of the third generation digital cellular systems. It is developed to meet the demands of wireless mobile communication in a true multimedia environment, where high-speed packet data and the Internet bearer service play major roles. Development of the "third-generation" mobile system based on WCDMA technology is being conducted with a view to achieve a global standard, based on the GSM footprint.

9.2.1.3
Requirements for the Third Generation Mobile System

The market for fixed networked multimedia services is growing. It is influenced by the falling personal computer costs, increasing acceptance of information technology (IT) in the office and at home together with the growth of Internet with electronic commerce applications and corporate Intranets. The demands for rapid and remote access to information are increasing, driven by higher productivity and flexibility needs and the need of reducing "dead" time.

These factors will also drive the market for mobile multi-media, and they set new requirements on a wireless system, such as:

- High data rate communication services and asymmetric data transmission
- Support for both packet and circuit switched services, such as Internet traffic and videoconference
- New charging mechanism
- Higher network capacity, new technologies that enhance the spectrum efficiency. Each user will require more capacity compared to today's voice services.
- Support for numerous and simultaneous connections, a user should for example be able to browse Internet and receive a fax or a phone call.
- The future of mobile telephony is now being determined by the global standardization work conducted by public authorities and the industry. In Europe, ETSI is working on the Universal Mobile Telecommunications System (UMTS) - which will become an International Telecommunication Union (ITU) IMT-2000 specification. WCDMA technology is one of the major candidates for the UMTS and IMT-2000 standard.

9.2.1.4
Frequency Spectrum

New spectrum has been allocated in the 2 GHz frequency band for the third generation wireless system.

9.2.1.5
Evolution of Cellular Systems

The role of the third generation cellular systems will be to satisfy the needs for immediate access to people and information both in the office environment and globally.

Requirements on the third generation cellular systems will range from basic speech services to high-speed data services, both circuit switched and packet switched. Data speeds will range from 8 kbit/s up to 384 kbit/s initially, and later, up to 2 Mbit/s. GSM will continue to evolve both before and after the third generation systems are introduced. By the time of commercial launch of IMT-

2000 / UMTS, GSM systems will be able to support up to 384 kbit/s with local mobility and at least 115 kbit/s with full mobility.

IMT 2000 and UMTS will open up a new set of spectrum with a new access method that will further serve to increase the capability of cellular systems up to 384 kbit/s with full mobility and 2 Mbit/s locally.

Because of the need for bandwidth flexibility and the support of multimedia services, which will imply changing bandwidth requirements from one user during a session, WCDMA is the preferred access method for IMT-2000. This access method has also other advantages that are described in the following sections.

In summary, selecting GSM as the core network for the provisioning of third generation cellular services will ensure the widest customer base possible from day one. Subscribers will in the same way enjoy roaming possibilities with the several hundred GSM networks that already exist globally. Continuous evolution of GSM systems also on the GSM radio interface will ensure competitiveness.

9.2.1.6 WCDMA Technology

A new WCDMA concept operating with 5 MHz wide radio channels is being developed to fully utilize the potential benefits of CDMA. This new system will have several advantages compared to current second-generation narrowband CDMA systems.

WCDMA technology features:

- High capacity and increased coverage: up to 8 times more traffic per carrier compared to a narrowband CDMA carrier. This is achieved by up to 100% better usage of the frequency spectrum
- Variable and high speed data rates up to 384 kbits/s in wide area and 2 Mbits/s in local area
- Both packet and circuit switched services are included
- Multiple simultaneous services in each mobile terminal
- Support for Hierarchical Cell Structures (HCS) by introducing a new hand-off method between CDMA carriers.

Development of the third generation mobile system based on Wideband Code Division Multiple Access (WCDMA) technology is being conducted with a view to achieve a global standard for the wideband air interface of the GSM core network. Power (control) is one of the critical radio resources of WCDMA. The transmission can be improved by power control because large number of users can be supported and the quality of transmission can be controlled.

One of the problems in power control is that the radio signals in urban areas are subjected to Rayleigh fading. This results in large variations in the signal strength. Due to this mobiles and base stations consume more power. One possible solution for such problems is improved power control.

Different algorithms have been proposed to cancel the effects of the signal fading in the transmission channel. Centralized algorithms are presented in [8] and [9] in which the optimum power vector for achieving the quality of the transmission is calculated. Distributed algorithms are covered in a large number of papers, e.g. in [10],[11],[14],[15]. Fuzzy algorithms are described in [12] and [13], but practical implementations have not yet been reported. Other related issues are admission control and load control, but all are connected with power control algorithms defined for different control loops: open loop closed loop or outer loop. New approaches for power control are proposed in this paper such as power control command coded for different number of bits, non-linear transmitted power and gain scheduling algorithms. Total number of bits for coding the power control command is considered as a practical constraint. Algorithms are developed based on this assumption. The philosophy behind the new idea is to have more users that reach the quality of the transmission. Another practical constraint, maximum transmitted power, is also considered in simulation studies. In section 2 the system model is explained. The proposed power control algorithms are given in section 3. Simulation results are discussed in section 4 and conclusions in the last section.

9.2.2
Cellular system

The base station and mobiles, the radio channels model, and the radio link model can be considered to be the cellular system. First some notations and notions are introduced. The mobile telephone system is covering a geographic area called cell. Every cell has a base station, which receives and sends data from and to the mobile. Sometimes the word terminal is used instead of mobile station indicating than other data than phone calls can be transferred. The transmission between a mobile and base station is usually organized in duplex way, which determines two types of transmissions. Link transmission from a mobile to base station is called uplink transmission. The opposite direction is downlink transmission. The channel quality is evaluated. The mobile is assigned to a new channel if the quality of the transmission is poor. This is called handover. The signal quality is usually defined by signal to interference ratio (*SIR*). During transmission the radio link between transmitter and receiver is subjected to variations, both in time and space. Slow fading, fast fading and distance fading are defined for radio link. Natural objects cause long term fading in the signal power. Mainly unnatural obstacles like cars and buildings can cause several attenuated echoes of the same transmitted signal. The signal arrives to the receiver with different delays in different phases. These cause short term fading. As the mobile is moving, Doppler effect occurs, so the phase shift is different in every multipath component. Multipath propagation determines fast fading, so called Rayleigh fading.

The radio cellular system considered here has a finite cell size. The length of the call, the position of the mobile in the route and the velocity are calculated. The active user is considered a user that has an ongoing call. So-called cubic convolution interpolation is used as a propagation model to calculate the amplitudes and phases of five propagation paths. The variation of the interpolated signal may look like a signal filtered by classical Doppler filter. The ray-tracing model is used to generate the fading in the system. Cubic convolution is selected as interpolation method to estimate the amplitudes and the phases between the ray traced points. The number of active users is denoted by N_{ms}, total number of calls N_{calls}, total number of fails N_{fails}, number of fails due to bad quality N_{q_fails}, number of synchronous failures N_{no_ch}. The transmitted power for *i*th mobile is denoted by Pt_i, power received from each BS (base station) through each path is Pr_{jL} where *j* represents the number of base stations, and *L* the number of paths, mean received power $\mathbf{Pr}_{mean} = [\mathrm{Pr}_{mean1},\ldots,\mathrm{Pr}_{meanMS}]$, SIR_{target} (signal-to-interference ratio target). *SIR* is calculated by maximal ratio combining processing gain of every MS, SIR_{ij} represents *SIR* calculated between the transmitter MS_i and BS_j, *SIR* at the previous time instant SIR_{pre} and SIR_{preij} represents SIR_{pre} calculated between the MS_i and BS_j, signal quality counter $quality_i$, where $i=1,..,MS$, $j=1,..,BS$. The signal-to-interference ratio in the *k*th path of the radio link between the *i*th mobile and *j*th base station, SIR_{ijk} is expressed in [7] as:

$$SIR_{ijk} = \frac{\frac{W}{R} P_{i,j,k}}{N_0 + \alpha \sum_{l=1, l \neq im}^{L} P_{i,j,l} + \sum_{\substack{m=1 \\ m \neq l}}^{MS} \sum_{l=1}^{L} P_{m,j,l}}, \tag{2.1}$$

where W is the chip rate, R is the data bit rate, N_0 is the thermal noise power, *L* is the number of paths, *MS* is the maximum number of mobiles, *BS* is the maximum number of base stations, α is the orthogonality factor, $P_{i,j,k}$ is the received power at the *j*th base station from *i*th mobile through the *k*th path. The auto-correlation and cross-correlation properties of the direct sequence are assumed similar. The maximum ratio combining technique is used for simulation. The receiver combines the signals propagated through different paths by weighting the strengths of each path and chooses the strongest path of the radio link. Therefore $SIR_{ij} = \sum_{l=1}^{MS} SIR_{ijk}$ gives the total SIR in the link between ith and jth base station. The fading in the channel is generated using the ray tracing method.

To keep the presentation clear for overview of the power algorithms, the information of the channel will be collected in a matrix of the channel denoted by *G* at time instant *t*. Power gain from the base station *j* to the mobile *i* is denoted by g_{ij} at time *t*. Here only downlink transmission is considered. Every transmission is supposed to have a connection matrix for uplink and one for downlink due to different frequency bands and different propagation conditions. Transmitted power of base station *j* is $p_j(t)$. The mobile station will experience a carrier signal power $g_{jj}(t)p_j(t)$, thermal noise N_0 and interference which is the sum of the powers from the other base stations. That is called co-channel interference. Instead of *SIR*, carrier-to-interference ratio *CIR* (C/I) is used in many of the proposed algorithms. This is denoted at *i*th mobile by:

$$CIR_i = \frac{p_i(t) g_{ii}(t)}{\sum_{j=1, j \neq i}^{N} p_j(t) g_{ij}(t) - p_i(t) g_{ii}(t) + N_0} \tag{2.2}$$

CIR is used for quality analysis in most of the proposed algorithms.

9.2.3 Power control algorithms

9.2.3.1
Power control overview- Criteria and Approaches

Power control depends on power control criteria, step size, and dynamic range. Power control criteria are path-loss based power control and quality based power control. The open-loop power control and fast loop power control are used for these criteria. Measuring *SIR* the quality of transmission is derived. Since *SIR* corresponds to the same *FER* (frames error rate) and *BER* (bit error rate), the desired *BER/FER* must be mapped into *SIR* target. This is the role of outer loop (quality loop) and is performed by continuously measuring *FER/BER* and *SIR*.

Power control step size defines how much the power should increase or decrease. The power control will vary in a given interval. All the time power is reported relative to the previous power setting. Dynamic power range depends on the distance between user and base station. The dynamic range is bigger in uplink, approximately 80 dB and in downlink 30 dB. Due to this fact the fast fading and slow fading are compensated during power control. Multi-user detection used with power control is another combination, which relaxes power control requirements. Another point for using power control is to increase the capacity of the system.

This section is composed of two main parts. One part is an overview of algorithms and approaches. The second part presents a couple of new power control approaches, especially power control command coded on different number of bits and gain scheduling approach.

Main algorithms for power control of mobile phones are:

- Centralized algorithms
- Distributed algorithms

Other approaches that have been discussed are:

- Prediction in closed-loop power control
- Neural networks and fuzzy algorithms

A block diagram of power control loop is presented in the figure below. This is implemented using the model presented by Ariyavisitakul in [19]. The power of the transmitter is increased or decreased according to the instantaneous *SIR* measurement. The measurement result is compared with a SIR_{target}. If the measurement result is above the target value (error is negative), the transmitting power of the mobile station is decreased. If the measurement result is above the target value and the current transmitted power is below the maximum transmitted power P_{max}, the transmitted power of the mobile is increased. The transmitter power is updated every T_p seconds with a fixed step.

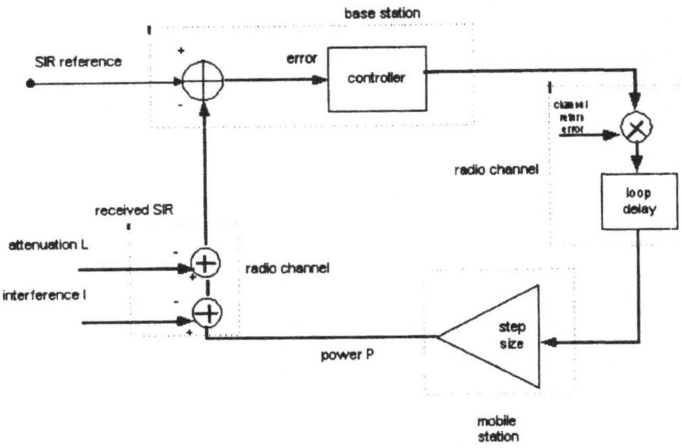

Fig 9.2.1. Closed loop power control

Important characteristics of power control algorithms are centralized or decentralized control structure, quality measure constraints and time delays. In centralized case all the information about the established connections and power gains are known. In decentralized or distributed case only local information is available and therefore power of only one transmitter is controlled.
Quality measures are another aspect of transmission. Speech quality can be measured by the value of *SIR*. This quality measure is intensively used in all works and presented algorithms. Constraints of the transmitted power or of power levels are due to hardware. Transmitted power has an upper and lower limit. Also different standards have various limitations which have to be taken into account. Time delays also influence power control. Measuring and control takes time, which results in loop delays. During transmission both downlink and uplink delays are considered. A typical delay approximation is 1 ms. In practice, it is considered to be constant.

The algorithms proposed up to date are now briefly reviewed. Only downlink connection is considered. Uplink transmission can be derived similarly.

Let us assume that carrier-to-interference ratio *CIR* (C/I) is used as only quality measurement. This is denoted at *i*th mobile by:

$$CIR_i = \frac{p_i(t) g_{ii}(t)}{\sum_{j=1, j \neq i}^{N} p_j(t) g_{ij}(t) - p_i(t) g_{ii}(t) + N_0}, \tag{3.1}$$

The transmission is considered poor for each receiver if *CIR* is under CIR_0 assumed to be the given threshold. The goal of all the algorithms is to provide all receivers sufficient power in such way that the quality of the transmission is fulfilled.

The first proposed algorithm is defined as *centralized* algorithm. The main idea of this type of algorithm is in having *CIR* balanced for controlling the power of different receivers. The satisfactory solution consists of power vector based on the eigenvalue solution. The power vector is called the optimum solution and represents that power, for which all the users have *CIR* above or equal with CIR_o. At instant time relation (3.1) becomes:

$$CIR_i = \frac{R_i}{\sum_{j=1}^{N} R_j - R_i + N_0}$$

$$= \frac{p_i g_{ii}}{\sum_{j=1, j\neq i}^{N} p_j g_{ij} - p_i g_{ii} + N_0} = \frac{p_i}{\sum_{j=1, j\neq i}^{N} p_j \frac{g_{ij}}{gii} - p_i + \frac{N_0}{g_{ii}}}, \text{ for } 1 \leq j \leq N \qquad (3.2)$$

where R_i is the power received from the *i*th transmitter. The transmitted power p_i is used by the base station in cell *i* and g_{ij} is the link gain matrix between the transmitter *i* and receiver *j* at the given moment and the normalized link matrix is $Z_{ij} = \frac{g_{ij}}{g_{ii}}$ [13]. Let $\eta_i = \frac{N_0}{g_{ii}}$ be normalized noise value. The link gain is assumed to depend on path-losses and attenuation factor due to shadowing and is denoted by $g_{ij}=L_{ij}S_{ij}$. Here L_{ij} represents propagation losses which depend of the distance between the transmitter *j* and the receiver *i* at time *t*. The propagation losses are defined as $L_{ij}= r_{ij}^{-4}$. S_{ij} is assumed to be log-normally distributed with a log mean of 0 dB and log-variance of σ^2 dB. It is a standard normal random variable.

Define a power vector $\mathbf{P}=\left[p_1, p_2, \ldots, p_N\right]^T$ such that $CIR_i \geq CIR_0$ for all *i*, $1 \leq i \leq N$. CIR_o is as before [20]. In [20] Zander proves that the largest achievable *CIR*, denoted by CIR^*, is given by $CIR^* = 1/\lambda^* - 1$, where λ^* is the largest real eigenvalue of the positive link gain matrix $\mathbf{Z} = \left[Z_{ij}\right]$. Thus the optimum power vector $\mathbf{P}^*$ is the eigenvector of $\mathbf{Z}$ corresponding to the eigenvalue λ^*. All the mobiles experience the same *CIR** when **P*** is used.

The conclusion is that no other power vector exists to achieve a higher *CIR* for all the receivers. The optimal power control algorithm (PCA) consists of removing as few users as possible and finding the largest submatrix of **Z** for which CIR_o is achievable. The solution is guaranteed by Frobenius theorem of positive definite matrices. The outage probability can be seen as a performance measure. If *N* connections are assumed to be established then the outage probability is defined as:

$$O = \frac{1}{N} \sum_{i=1}^{N} (P(CIR_i \leq CIR_0)) \qquad (3.3)$$

The goal of the algorithm is to minimize the outage probability. This is equivalent to maximizing the number of links in the gain matrix. Two possibilities are treated in the existing papers. One is ignoring noise in this case the power vector is sufficiently large. Another possibility is to take into account noise. In that case the optimal solution, or the optimum power vector, will be the solution of the linear system:

$$CIR_0 = \frac{p_i}{\sum_{j=1, j \neq i}^{N} p_j \frac{g_{ij}}{gii} - p_i + \frac{N_0}{g_{ii}}} \tag{3.4}$$

Distributed algorithms are generalization of the centralized algorithms. The distributed power control algorithms are iterative and *CIR* balancing is faster than in other algorithms. The distributed algorithms experience a dynamic change of the radio link gain matrix and also have discrete power control steps given for every mobile. In [2] a distributed power control algorithm (DPC) is presented for cellular radio systems. The idea is that mobiles adjust their transmitted powers synchronously in discrete time steps. The new transmission power depends on previous power command and the *CIR* estimated one step before. So the new power control command is calculated iteratively and is based only on "local" *CIR* measurements. The algorithm for updating power $p_i(t)$ of the *i*th mobile at time *t* is defined as [2]:

$$p_i(t) = \frac{p_i(t-1)}{CIR_i(t-1)} c \tag{3.5}$$

where *c* is a positive constant.

The algorithm in (3.5) is a simple 'proportional' control scheme. It converges when the optimum power $\mathbf{P}^*$ is achieved. To choose the constant *c* in (3.5) is the hardest part of the design and also a drawback for all the presented algorithms. Also normalization is needed at every time instant, which means delays in practice. The problem of making *c* adaptive remains to be solved.

The solution of the algorithm is shown to be all the time unique and stable. Computation of power is based on local measurements, only on the information from a particular link. The outage probability is demonstrated to be low. This algorithm will be referred as Distributed Power Control (DPC) algorithm. Lin [9] proposed the same type of algorithm where choosing *c* is still remains open. The algorithm is defined as:

$$p_i(t) = cp_i(t-1)\left(\frac{1}{CIR_i(t-1)}\right)^{\xi}, \text{ where } 0 \leq \xi \leq 1 \tag{3.6}$$

$\xi = 1$ gives the same algorithm proposed in [8]. The algorithm presented above is referred as CDPC-II or Constraint Distributed Power Control. The basic idea is to increase or decrease transmitting power (in dB) at the base station based on the difference (in dB) of *CIR* and a constant.

Foschini and Miljanic proved the convergence of the DPC algorithm in [10]. The proof is demonstrated in the noisy case. They showed that the algorithm is totally local in the sense that power and interference measurements of each base-station-mobile-station link are used to evolve power levels on that link. The normalization is not necessary. In this case c is the target that all users tries to reach. When c is found, the optimum power vector exists. The new algorithm is given by:

$$p_i(t) = p_i(t-1)\left(\frac{CIR_0}{CIR_i(t-1)}\right) \tag{3.7}$$

The optimal discrete power vector is obtained also in Distributed Constrained Power Control algorithm proposed by Grandi, Zander and Yates in [2]. The algorithm seems to be an extension of the DPC algorithm. It is defined for all the active users (speech service) which belong to the active user set. The power is updated based on the algorithm presented in (3.8).

$$p_i(t) = \min\left\{\overline{p}, p_i(t-1)\left(\frac{CIR_0}{CIR_i(t-1)}\right)\right\} \tag{3.8}$$

Power $\overline{p}$ represents the upper bound of the transmitted power. If the active users' set at one time instant is defined for all the users, then the algorithm is Synchronous DCPC. Otherwise the algorithm is called Asynchronous DCPC. The solution of the DCPC and ADPC converge to fixed point solution denoted by $\mathbf{p}$:

$$\mathbf{p} = \min\{\overline{\mathbf{p}}, \mathbf{CIR}_0 \otimes (\mathbf{Zp} + \boldsymbol{\eta})\} \tag{3.9}$$

A power vector $\mathbf{p}$ that satisfies the fixed-point solution of (3.9) is referred to as the *continuos stationary power vector*. When the power levels are discrete, the DCPC cannot be explicitly applied. So the discrete version of the previous algorithm is discussed further. The distributed discrete power control (DDPC) algorithm has two objectives:

- To construct a convergence 'envelope', a 'fixed-envelope solution', which corresponds to the 'fixed-point' solution.
- To build an algorithm that converges to the optimal power vector.

A fully distributed power control (FDPC) is proposed in [11]. The main idea in the algorithm is to decide the power level based on CIR calculated for every mobile and a constant Υ. The algorithm is given as

$$p_i(t) = \eta_i(t-1)p_i(t-1), \tag{3.10}$$

where $\eta_i(t) = \dfrac{\min(CIR_i(t), \Upsilon)}{CIR_i(t)}$.

The parameter Υ is determined a priori, so base stations do not need to exchange information to perform the power control algorithm. The same type of algorithm was proposed in dB form and is called FDPC-I. The updating rule is denoted by

$$\mathrm{dB}(p)_i(t) - \mathrm{dB}(p)_i(t-1) = -(\mathrm{dB}\, CIR_i(t) - \mathrm{dB}\, \Upsilon) \tag{3.11}$$

The convergence of the algorithm is discussed in [11].

The main problem with the presented algorithms is to define CIR_0 to which algorithms have to converge. Two possibilities exist. The first case is where the value of CIR_0 is too high. This situation leads to so-called 'party-effect' in the network. The users are increasing their power and in consequence the interference at the other users is increasing. The users that increased their power to the maximum limit may experience other situation. They don't achieve CIR_0, in conclusion they don't have enough power for transmission or they have poor quality of the transmission.
Algorithm, which try to eliminate the drawback, are developed by Almgren, Andersson and Wallsted. The convergence of the algorithm is discussed in [4].
In the other articles the power control algorithm belongs to radio resources allocation algorithms. Power control is one of the radio resources in CDMA. The mentioned type of algorithm is defined when different services are used. Priority is introduced for different services so that all the services will be divided into different classes. Every class will have a different priority. An extension of DCPC is proposed in [13], which leads to a new algorithm called 'Prioritized Gradual Removal' or PGR-DCPC. The target system of the algorithm is the future personal communications system, which will support different classes of wireless services. The idea is to implement a transmitter removal scheme that considers service priority in power control. Transmitter power is a method used for radio resource allocation. If an overload situation occurs on a channel and all transmitters cannot be supported, some of them should be removed (i.e. power off). Power is updated according to DCPC algorithm. The philosophy behind the algorithm is to remove a mobile that belongs to a low priority class. The probability to remove a mobile belonging to a high priority class is smaller than for the mobiles with low priority. The contribution of the algorithm demonstrates the influence of increasing or decreasing *SIR* target for different class of services. A high *SIR* in a high priority class can degrade the quality of the transmission. Simulation results are presented in [15]. As an extension of this work optimization approach is considered in another paper by the same author. The main idea is to formulate the prioritized transmitter removal into a classical optimization problem. Lagrange relaxation technique is applied.

Two articles study the power control using fuzzy algorithms. The limitation of time is not taken into account. In [12] and [13] a new application of FLC (fuzzy logic control) is introduced. This is for power control in cellular systems. The conventional feedback power control algorithms allow base station to send a power command to either decrease or increase each user transmitting signal power level by a fixed power step and then keep the received powers almost equal. The fuzzy PI controller is a discrete controller. It has as inputs the received power error and the difference of error and the output is the transmitted power to the user. Each received signal power will track a nominal step response $r(t)=r_d, t \geq 0$. If $r_d=0$, the problem is called the regulator problem. If $r_d \neq 0$, then we have asymptotic tracking problem. The conventional controllers used are time-optimal bang-bang controller and PI controller. The bang-bang control policy involves switching the control input alternately from one extreme to another, at precalculated time [13]. The control input is used to drive the process output to a desired set point. Viterbi utilised control policy to determine the sign of each fixed transmitting power increment. The precalculated switching times can be very sensitive to model error and noise. The fuzzy controller acts to adjust the switching parameters on line. A variable step cannot be implemented in the mobile hardware. The idea behind the fuzzy algorithms is have a smoother decision signal than in case of conventional controller.

Practical issues are not considered here. Power control algorithm has to take into account that the step size is fixed in sense that fixed power levels are assumed. Also the transmitted power has to have upper limits otherwise the algorithm will have a large overshoot.

Power control, multipath diversity, system bandwidth, and path loss are seen to have an impact on the capacity of the system. The power control schemes are intended to overcome the near-far problem in uplink. The multipath fading at low Doppler frequencies should be mitigated. The algorithms based on *SIR* may improve the performances of the transmission. All these assumptions were for uplink transmission.

In downlink, transmitted power control is used for increasing the capacity of the channel. The transmitted power can be the same in all portables without power control. However, those portables that are near the base station receive a stronger signal from their own base station and less interference from other users and base stations. Reducing the transmitted power to those portables, which are not satisfying the *SIR* requirements, increases the overall capacity. The power control may be designed to follow the slow variations in received *SIR* due to shadow fading and path loss (slow fading) or to adapt to both slow variations as well as fast variations due to the fast multipath fading, in downlink transmission. Diversity is a major contributor to increase the capacity in wireless access systems. At the base station, antenna diversity may be employed to achieve the desired capacity. At the mobile, multipath diversity is more feasible.

Another problem that occurs is imperfect power control. This can influence the capacity of the cell. The aim of the paper is to provide general comments about the implementation of CDMA. The related concepts such as shadowing, call dynamics, voice activation, sectorization are discussed, but these are not taken into account in connection with the implementation issues such as traffic mix, soft handover, DMA Tx/Rx chain. The power control imperfection process occurs due to many variables and is due to the dependence of shadowing.

9.2.3.2
New power control algorithms

Uplink closed-loop power control adjusts the mobile station power control in order to keep the received uplink *SIR* at a given *SIR* target. The same algorithm is applied for downlink transmission where the base station power is adjusted in order to keep the received downlink *SIR* at a given *SIR* target. The transmitted power control command is coded on 1/2/3/4 bits. In [1] and [5] the idea of implementing power control for 1 bit is presented. The number of power control levels will be dependent of the number of bits. Power control steps coded on 2, 3 and 4 bits will determine 4, 8 and 16 transmitted power control commands respectively. The minimum power step is 0.25 dB and maximum is 1.5 dB. The decision is taken based on the variation of the difference ($SIR_{pre\,i} - SIR_{target}$). The difference ($SIR_{pre\,i} - SIR_{target}$) is the error in the closed-loop power control and is denoted by $\Delta\, SIR_i$ for ith mobile. The value of the difference is compared with a fixed value, which represents part of SIR_{target}. The power control is decreased or increased with maximum step when the variation of $\Delta\, SIR_i$ is large compared with the fixed value. The minimum step power control is applied in case that the variation is small. The algorithm is distributed. The power control is calculated for each mobile at every simulation time based on power control error.

Power control of 2 bits. For power control of 2 bits four levels of power are defined, equally distributed as number of steps for increasing and decreasing the level. In the range of 0.25 - 1.5 dB discrete steps of power, $\mathbf{Pc}_{step1}$ and $\mathbf{Pc}_{step2}$ for each mobile are considered. $\mathbf{Pc}_{step1}$ represents the minimum step, 0.25 dB, and $\mathbf{Pc}_{step2}$ maximum step, 1.5 dB. These are collected into vectors and denoted by $\mathbf{Pc}_{step1} = [Pc_{step11}, \ldots, Pc_{step1MS}]$ and $\mathbf{Pc}_{step2} = [Pc_{step21}, \ldots, Pc_{step2MS}]$, where MS represents the total number of mobiles. The definition of power steps presented above shows that each mobile has two power control steps defined and this gives the possibility of increasing or decreasing the power for each one with one value out of two defined. The error probability of the return channel is defined for the ith mobile as $\mathrm{Pc}_{error\_pr\ i}$.

The proposed algorithm is divided into two different phases. In the first phase the sign of the power control correction is decided. In the second part the new transmitted power is calculated.

The sign of the power control is derived based on the following algorithm at the kth moment of simulation:

If ($\Delta\, SIR_i < 0$) then

C = 1 and the power of each mobile is increased with a step chosen in the algorithm.

Else

C = -1 and the power of each mobile is decreased with a step chosen in the algorithm, for $i = 1, \ldots, MS$.

The power control error E is given in the same form as the sign of power control C:

If (q is less than $\mathrm{Pc}_{error\_pr\ i}$) then

E = -1 and the power control step is chosen incorrectly

Else

E = 1 and the power control step is correct,

where q represents a variable for power control error. This is for one bit power control.

Now with the information that is known about increasing or decreasing the power and the information that the chosen power is either well defined or not, the power control command can be formulated. The algorithm takes the form:

1.Initialisation step
2.While(*the number of users is different than zero*) will do{
2.1 Calculate the sign for the power control command C (increasing or decreasing)
2.2 Calculate the sign of the $Pc_{error\_pr\ i}$
2.3 Give the power control command at *k*th time instant:
If $\Delta\ SIR_i \geq 0$ and $\Delta\ SIR_i < (SIR_{target}(k)/2))$
Then $Pt_i(k) = f(Pc_{step1i}, C, E)$
Else if $\Delta\ SIR_i \geq 0$ and $\Delta\ SIR_i \geq (SIR_{target}(k)/2)$
Then $Pt_i(k) = f(Pc_{step2i}, C, E)$
Else if $\Delta\ SIR_i < 0$ and $\Delta\ SIR_i \geq -(SIR_{target}(k)/2)$
Then $Pt_i(k) = f(Pc_{step2i}, C, E)$
Else if $\Delta\ SIR_i < 0$ and $\Delta\ SIR_i < -(SIR_{target}(k)/2)$
Then $Pt_i(k) = f(Pc_{step1i}, C, E)$
Else $Pt_i(k) = Pt_{\max}(k)$
Where $i = 1,\ldots, MS$.
}

The algorithm updates the transmitted power based on the decided power control level and the $Pc_{error\_pr_i}$. The updating function f is non-linear and the constraints taken into account are the total transmitted power and maximum power control level which is between 0.25 dB and 1.5 dB. The non-linear function f is calculated based on the transmitted power Pt multiplied by the decided power control level with E and C. The power is decreased or increased with maximum step when $\Delta\, SIR_i$ is large compared with fixed value and with minimum step when the difference $\Delta\, SIR_i$ is not large. The increasing or decreasing command is given by the variable C which is calculated for every k iteration and for each mobile. The return channel error is calculated in case that the correct power control step is not available.

Power control on 3 and 4 bits. The philosophy of power control algorithms of 3 and 4 bits is similar as above. The power control commands now are codified on 3, respectively 4 bits. The number of power control levels is 8 in case of power control of 3 bits, and 16 levels in the case of 4 bits. For power control of 3 bits 8 levels of power are defined, equally distributed as number of steps for increasing and decreasing. In the range of 0.25 - 1.5 dB discrete steps of power $\mathbf{Pc}_{step1}$ and $\mathbf{Pc}_{step4}$ for each mobile are considered. $\mathbf{Pc}_{step1}$ represents the minimum step, 0.25 dB, and $\mathbf{Pc}_{step4}$ is the maximum step, 1.5 dB, used for power control command. These are collected into vectors and denoted by $\mathbf{Pc}_{step1}=[Pc_{step11},\ldots,Pc_{step1MS}]$ and $\mathbf{Pc}_{step4}=[Pc_{step41},\ldots,Pc_{step4MS}]$, where MS represents the total number of mobiles. The definition of power steps presented above shows that each mobile has three power control steps defined. This gives the possibility of increasing or decreasing the power for each one with one value out of three defined. The algorithm for 3 bits is presented below.

1.Initialisation step
2.While(*the number of users is different than zero*) will do{
2.1Calculate the sign for the power control command C (increasing or decreasing)
2.2Calculate the sign of the $Pc_{error\_pr_i}$

2.3Give the power control command at kth time instant:
If $\Delta\ SIR_i \geq 0$ and $\Delta\ SIR_i < (SIR_{target}(k)/4))$
Then $Pt_i(k) = f(Pc_{step1i}, C, E)$
Else if $\Delta\ SIR_i \geq 0$ and $\Delta\ SIR_i \geq (SIR_{target}(k)/4)$ and $\Delta\ SIR_i < (SIR_{target}(k)/2)$
Then $Pt_i(k) = f(Pc_{step2i}, C, E)$
Else if $\Delta\ SIR_i \geq 0$ and $\Delta\ SIR_i \geq (SIR_{target}(k)/2)$ and $\Delta\ SIR_i < (SIR_{target}(k))$
Then $Pt_i(k) = f(Pc_{step3i}, C, E)$
Else if $\Delta\ SIR_i \geq 0$ and $\Delta\ SIR_i \geq (SIR_{target}(k))$
Then $Pt_i(k) = f(Pc_{step4i}, C, E)$
Else if $\Delta\ SIR_i < 0$ and $\Delta\ SIR_i \geq -(SIR_{target}(k)/4)$
Then $Pt_i(k) = f(Pc_{step4i}, C, E)$
Else if $\Delta\ SIR_i < 0$ and $\Delta\ SIR_i < -(SIR_{target}(k)/4)$ and $\Delta\ SIR_i \geq -(SIR_{target}(k)/2)$
Then $Pt_i(k) = f(Pc_{step3i}, C, E)$
Else if $\Delta\ SIR_i < 0$ and $\Delta\ SIR_i < -(SIR_{target}(k)/2)$ and $\Delta\ SIR_i \geq -(SIR_{target}(k))$
Then $Pt_i(k) = f(Pc_{step2i}, C, E)$
Else if $\Delta\ SIR_i < 0$ and $\Delta\ SIR_i < -(SIR_{target}(k))$
Then $Pt_i(k) = f(Pc_{step1i}, C, E)$
Else $Pt_i(k) = Pt_{max}(k)\}$
Where $i = 1, \ldots, MS$.

Simulation results are presented in section 4.

Gain scheduling concept applied for power control. The second new approach for power control is gain scheduling. Here the power control command is coded on 1 bit and 2/3/4 bits in a split-control range, or a gain-scheduling algorithm is developed.

In traditional control theory gain scheduling is a special kind of open loop adaptation or change of regulator parameters. Split-range controllers that use different sets of parameters for different ranges of the process output can be regarded as a special type of gain scheduling controllers. The main problem in the design of gain scheduling is to find suitable scheduling variables. Gain scheduling has the advantage that the regulator parameters can be changed very quickly in response to process changes. Since no estimation of parameters occurs, the limiting factors depend on how quickly the auxiliary measurements respond to process changes.

Here gain scheduling idea is applied at the system level. The philosophy behind the proposed algorithm is to change the coding of the power control command based on the parameters of the system. The other practical constraint is transmitted power, which is taken into account in the design of the regulator.

The new idea is to implement the power control based on the performances of the system. Here we define the performance of the system to be the number of the total fails. If the number of calls is small then the power control is coded on 1 bit. If the number of calls is increasing and the number of fails is increasing too, the power control command is coded on 2/3/4 bits. Due to this fact the power control steps will be 2/4/8 for increasing and the same number for decreasing the power, equally distributed in the range 0.25 dB - 1.5 dB.

If power control is coded on 1 bit, the corresponding power control step will be one step for increasing and one step for decreasing in the range 0.25-1.5 dB. A fixed step is taken as step size for power control. The power control step is

decided based on the difference between the measured *SIR* and *SIR* target. Relay control is considered for deciding power control steps.
The system should demonstrate that the possibility of choosing different power control achieves a low rate of fails for calls.

The concept of gain scheduling is applied at the system level and the algorithms simulate gain scheduling based on measurements of auxiliary variables. In our case these variables will be number of total initiated calls, and total number of fails. Fig. 9.2.2 displays the gain-scheduling principle.

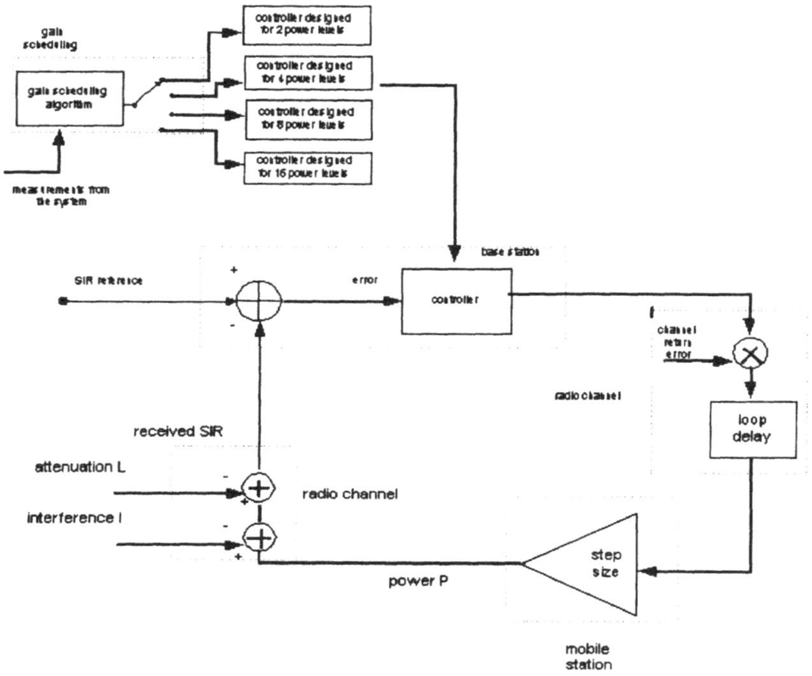

Fig.9.2.2 Gain scheduling applied to power control in mobile communication.

The main algorithm is presented below:

1.Initialisation
2 While (t)
{
3. Generation /*Generation of MS's*/
4.Traffic; /*Generation of traffic for MS's*/
5.Mobility; /* Mobility of MS's*/
6.Propagation;
7.Interference; /* Calculation of interference*/
8.Averaging; /* Averaging filtering*/
9.Synchronisation;/* Synchronisation of MS's*/
10.Admission; /* Admission control*/
11.Quality; /* Quality control*/

```
12.Handover; /* Handover control */
If (number of calls≤10 )
        power2b; /* power control algorithm of 2 bits*/
Else if ((number of calls)>10 and (number of calls≤ 40 ))
        power3b;/*power control algorithm of 3 bits*/
Else if (number of calls>40)
        power4b;/*power control algorithm of 4 bits*/
}
```

The philosophy in the algorithm is as follows. When you have a few mobiles, power control algorithm designed for 2 bits is sufficient, so the power control steps are larger. If the number of calls is increasing the number of bits for representation of power steps should increase, in this case the power is more precisely chosen. The difference between the power control designed for 3 bits and power control designed for 4 bits is not so significant. The simulation results are presented in section 4. The same type of algorithm is used also if the power command is coded on 1 and 2 bits. The algorithm can be sensitive to the chosen variables and also to the their values.

Another algorithm is chosen based on the total number of fails to the system level. Only power control for 1 bit and for 2 bits is used.

```
1.Initialisation
2 While (t)
{
3. Generation /*Generation of MS's*/
4.Traffic; /*Generation of traffic for MS's*/
5.Mobility; /* Mobility of MS's*/
6.Propagation;
7.Interference; /* Calculation of interference*/
8.Averaging; /* Averaging filtering */
9.Synchronisation; /* Synchronisation of MS's */
10.Admission; /* Admission control */
11.Quality; /* Quality control  */
12.Handover; /* Handover control*/
If ((number of fails<20) or (number of mobiles<20))
  power1b        /*1 bit power control */
Else
power2b          /*2 bits power control */;
}
```

The simulation results are presented in section 4.

9.2.4 Simulation results

Simulations results are obtained using NetSim. NetSim is a simulator developed to study Cellular Radio Network Control Algorithms and Planning methods by Telecommunication Laboratory at Helsinki University of Technology [3], and [5]. It provides a simulation environment to study system capacity, coverage and network control algorithms of WCDMA systems. NetSim output files consist of information about call dropping, blocking time and spatial references for these events. Statistics can easily be collected and translated into visual form with the help of MATLAB or other tools.

The main blocks in Fig. 9.2.3 show the links between NetSim and OPNET or COSSAP. It also shows the flow diagram inside the simulator.

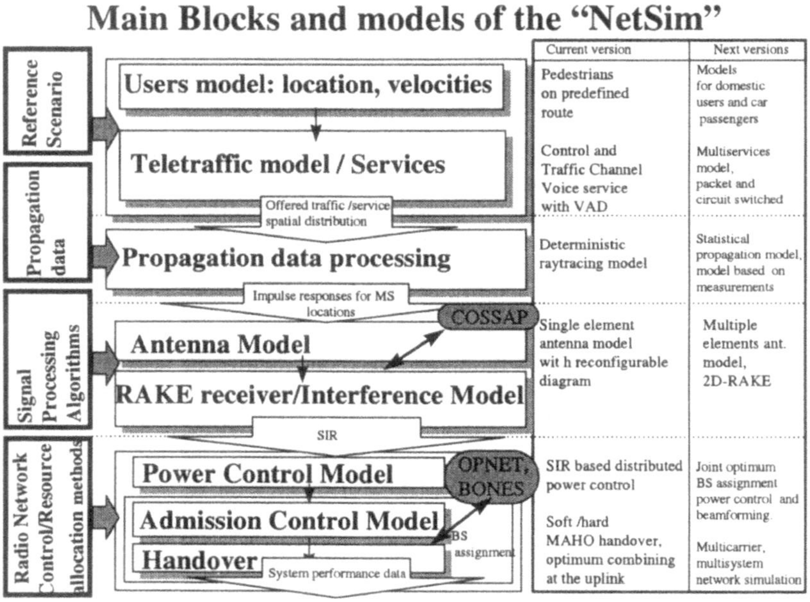

Fig. 9.2.3. Main blocks and models of the NetSim and directions of further developments [1], [3] and [5].

The following assumptions have been made in all simulations: The number of base stations is one, number of transmission paths L = 5, the maximum number of mobiles *MS* is 50 for PC version of simulator. If workstations are used, 3 is the maximum number of base stations and total number of mobiles is 250. Both uplink and downlink transmissions are considered for multi-base stations multi-user environment. Initial value for Pc_{step1i} is 1.0593 and 1.4125 for Pc_{step2i} for each mobile, where *i* varies between 1 and *MS*. Value 1.0593 corresponds to 0.25 dB and 1.4125 to 1.5 dB. If power control command is coded on 3 or 4 bits, then the power control steps have initial values 1.0593 for Pc_{step1i}, 1.1770 for Pc_{step2i}, 1.2947 for $Pc_{step3\,i}$ and 1.4125 for $Pc_{step4\,i}$. The decision in power control algorithm is taken based on the variation of Δ SIR_i. The return channel probability is considered to be 0.01.

Next simulation results are presented, when different decisions of power control commands, of 2, 3 and 4 bits are used. These results were for one base station obtained using NetSim, in uplink transmission (mobiles to base station). The algorithms are compared based on the number of the total fails. A power control algorithm is considered to be better than another, if the total number of fails at the system level is lower. The number of fails has to be as low as possible.

The number of total initiated calls is considered and this is usually bigger than the number of mobiles. A mobile can initiate more than one call until the base station accepts the request. The time of simulation is limited to 50 s for one base station and 50 mobiles, when PC version of the simulator is used. If workstations are used, the simulation time can be between 300 – 600 s. In the first part the PC version is used. The window is called the interval when data is collected. In the ideal case the number of bits for coding the power control is one and the number of fails is 1. The results are shown in Table 9.2.1.

Table 9.2.1. The total number of mobiles, initiated calls, fails, and quality fails for power control of 1 bit.

Total number of mobiles	50
Total number of calls	60
Total number of fails	1
Total number of quality fails	1

Next power control of 2 bits is simulated. The results are presented in Table 9.2.2.

Table 9.2.2. The total number of mobiles, initiated calls, fails, and quality fails for power control of 2 bits.

Total number of mobiles	50
Total number of calls	53
Total number of fails	3
Total number of quality fails	3

For power control of 3 and 4 bits the algorithms have the same structure as for 2 bits. The number of power levels is now 4 and 8 for increasing and the same number for decreasing. For power control of 3 bits, 4 power levels for the range 0.25- 1.5 dB are defined, and 4 for decreasing. The power control command is defined as a non-linear function based on the size of power control step and the calculated error. The number of bits is different because more accurate power control steps are given for each mobile. The simulation results are presented in the next table.

Table 9.2.3. The total number of mobiles, initiated calls, fails, and quality fails for power control of 3 bits over the simulation time

Total number of mobiles	49
Total number of calls	59
Total number of fails	6
Total number of quality fails	6

Comparing the number of fails with the previous algorithm, an increase of them is observed, but this is not significant.

A situation of more power steps for decreasing the power than for increasing the power is considered next. Five steps of power levels for decreasing the power and 3 for increasing are chosen. The results of the simulation given in Table 9.2.4 are better than previous results.

Table 9.2.4. The total number of mobiles, initiated calls, fails, and quality fails for power control of 3 bits

Total number of mobiles	49
Total number of calls	58
Total number of fails	0
Total number of quality fails	0

The same algorithm can be defined for 4 bits power control. The number of the levels of power control steps is 16. First 8 steps for increasing the power and 8 steps for decreasing are considered. The results of the simulation are similar to those in Table 9.2.4. They are summarized in Table 9.2.5.

Table 9.2.5. The total number of mobiles, initiated calls, fails, and quality fails for power control of 4 bits

Total number of mobiles	50
Total number of calls	45
Total number of fails	3
Total number of quality fails	3

The quality of the transmission is measured by average of SIR over one frame for every simulation time. The results using power commands coded for 2, 3 and 4 bits are better than those for one bit power control. In case of one bit power control relay is used. The simulation results are presented in the next figures. The value of the average SIR has to be bigger than 5.012 in order to consider that the quality of the transmission is good.

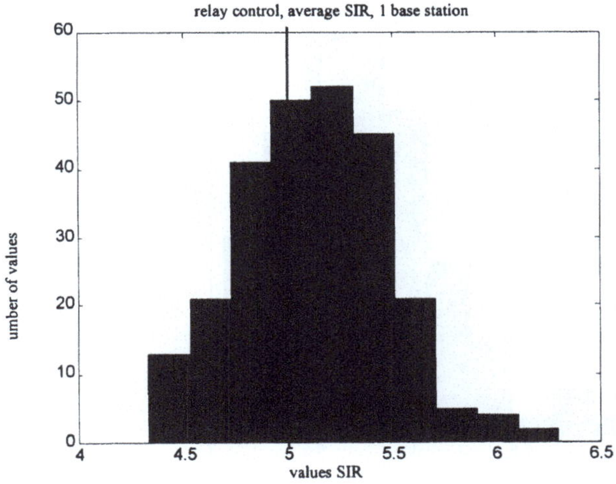

Fig. 9.2.4. Simulation results of 1 bit power control algorithm. Threshold is SIR=5.012.

In Fig. 9.2.5 and Fig. 9.2.6 the quality of the transmission is presented in case 2 and 3 bits. The results are for the case of 1 base station and multi-users.

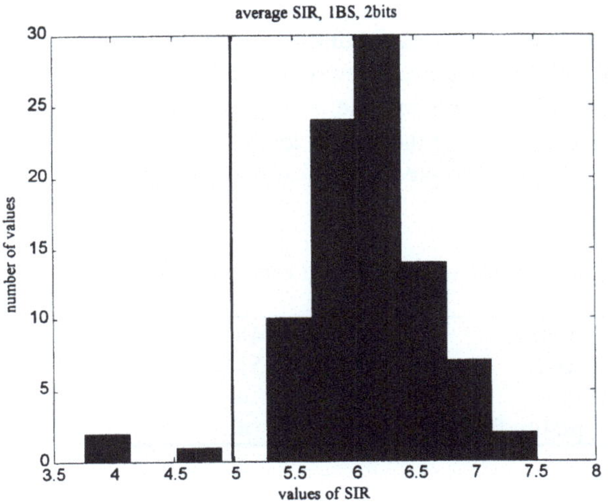

Fig. 9.2.5. Simulation results of the 2-bit power control algorithm. Threshold is SIR=5.012.

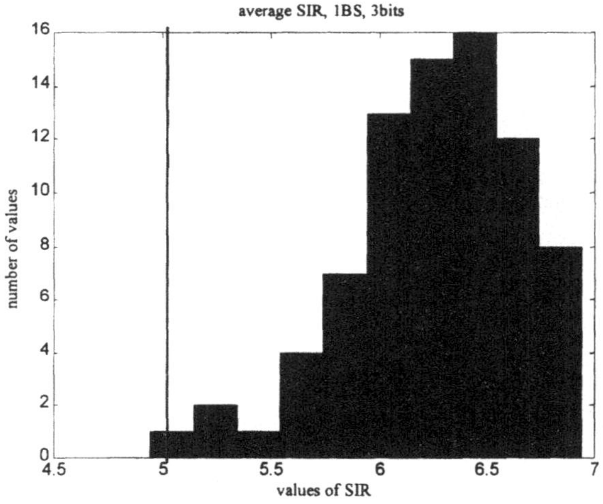

Fig. 9.2.6. Simulation results of the 3-bit power control algorithm. Threshold is SIR=5.012.

The conclusion is that in both cases the number of values (average SIR) below SIR target is less than in case of conventional relay control.

The second part of the simulation results discusses the gain-scheduling algorithm applied first to 1 base station, multi-user environment and then to 3 base stations, multi-user environment. Analysis of the results covers both quality of the transmission and number of fails at the system level.

The simulation results are presented in the following table.

Table 9.2.6. The total number of mobiles, initiated calls, fails, and quality fails for power control of 2/3/4 bits.

Total number of mobiles	48
Total number of calls	51
Total number of fails	4
Total number of quality fails	4

The gain-scheduling algorithm is based on the total number of initiated calls. The results of the simulation are shown in the table above.

Gain-scheduling algorithm is also applied in the case, when power control algorithm is based on the total number of calls and total number of fails. The results are displayed in the next table.

Table 9.2.7. The total number of mobiles, initiated calls, fails, and quality fails for power control of 2/3/4 bits.

Total number of mobiles	47
Total number of calls	62
Total number of fails	7
Total number of quality fails	3

The number of fails is now bigger than in the other cases presented until now. The same type of algorithm can be applied for power control of 1 and 2 bits. The algorithm is choosing the power control using the total number of initiating calls. The results of simulation are presented in Table 9.2.8.

Table 9.2.8. The total number of mobiles, initiated calls, fails, and quality fails for power control of 1 and 2 bits.

Total number of mobiles	48
Total number of calls	52
Total number of fails	3
Total number of quality fails	5

The number of quality fails is now 5. The algorithm is giving almost the same result as the adaptive algorithm of 2/3/4 bits.
The comparison can be done between the number of fails, when the gain-scheduling power control algorithm of 2, 3, and 4 bits and gain-scheduling power control of 1, and 2 bits are applied. The simulation results are presented in Fig. 9.2.7. The numerical results are in the Table 9.2. 9.

Table 9.2.9. The total number fails for power control of 1 and 2 bits and power control of 2/3/4 bits.

Total number of mobiles	48
Total number of fails in case of 2/3/4 bits power control	4
Total number of fails in case of 1/ 2 bits power control	3

In the beginning both algorithms are behaving quite similarly. The number of fails is about the same. After a while gain-scheduling power control for 1 bit and 2 bits is working better than the gain scheduling power control algorithm of 2/3/4 bits. The largest number of fails for 1 and 2 bit gain- scheduling control is 3 over the simulation time. The largest number of fails for the second algorithm is 4. The difference is not significant, but the handover and communication in a complex system are not taken into account.

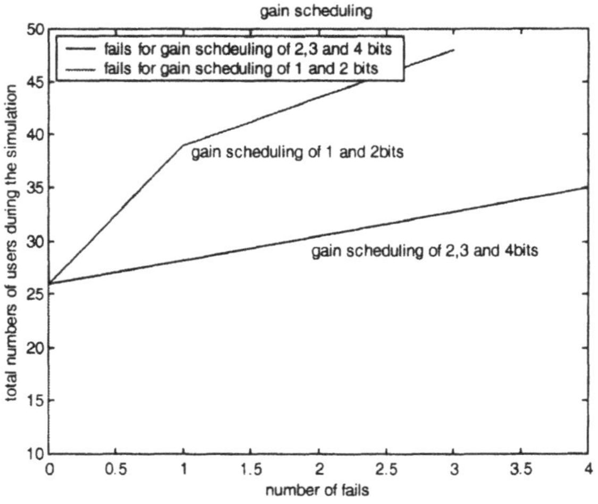

Fig. 9.2.7. Simulation results of gain-scheduling power control.

The results of gain-scheduling algorithm for more than one base station are presented below. The simulations show the influence of the chosen variables in the algorithm. The algorithm can be quite sensitive to the values of the parameters due to on-line measurements performed at the system level. The algorithms are applied in multi-base station multi-user environment. The quality parameter measured is the average SIR. It is computed over the frame for all the mobiles. The transmission is said to be poor, if average SIR is under a certain value. The simulation time in this case is 350s.
In Table 9.2.10 the number of fails is 5. The variable for gain scheduling in this case is the number of fails. If the number of fails is bigger than 5, then power control algorithm is switched from power control of 1 bit to power control of 2 bits.

Table 9.2.10. The total number of mobiles, initiated calls, fails, and quality fails for power control of 1 and 2 bits

Total number of mobiles	100
Total number of calls	150
Total number of fails	55
Total number of quality fails	25

In Table 9.2.11 the number of fails is considered 20. The same variable as in previous case is considered for gain-scheduling algorithm. If the number of fails is bigger than 20, then power control algorithm is switched from power control of 1 to power control of 2 bits.

Table 9.2.11. The total number of mobiles, initiated calls, fails, and quality fails for power control of 1 and 2 bits.

Total number of mobiles	100
Total number of calls	151
Total number of fails	40
Total number of quality fails	15

The quality of the transmission is analyzed as a function of average SIR. The quality of the transmission is good if the value is equal or bigger than 5.012, in linear scale or 7dB in decibels scale. The next results demonstrate that gain scheduling has as good or better performance than conventional relay control.

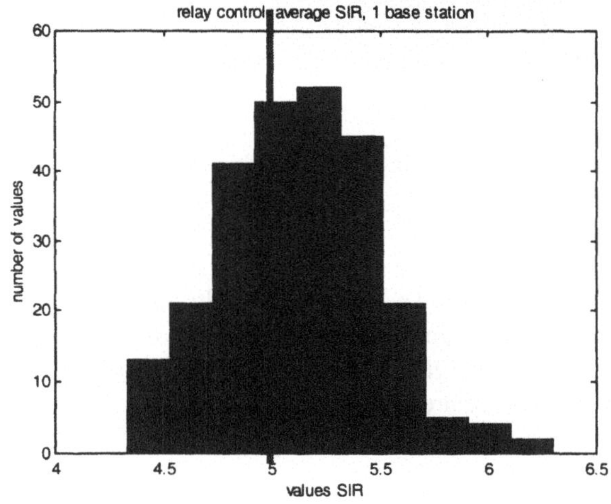

Fig. 9.2.8. Simulation results of power control using relay control. Threshold is SIR=5.012.

The results of the gain-scheduling algorithm in case of 1 BS and multi-user environment are shown in Fig. 9.2.9.

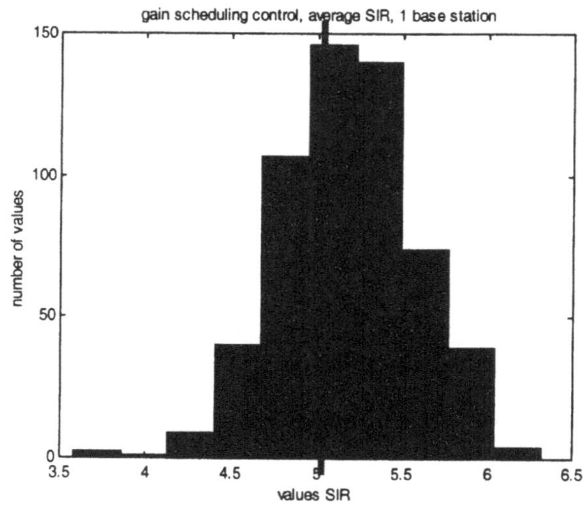

Fig. 9.2.9. Simulation results of gain-scheduling power control. Threshold is SIR=5.012.

The same type of results is shown in case of multi-base station, multi-user environment. Relay control and gain-scheduling algorithm in case of 1 and 2 bits are used. The results in case of relay control are shown in Fig. 9.2.10. In Fig 9.2.11 gain-scheduling results are presented.

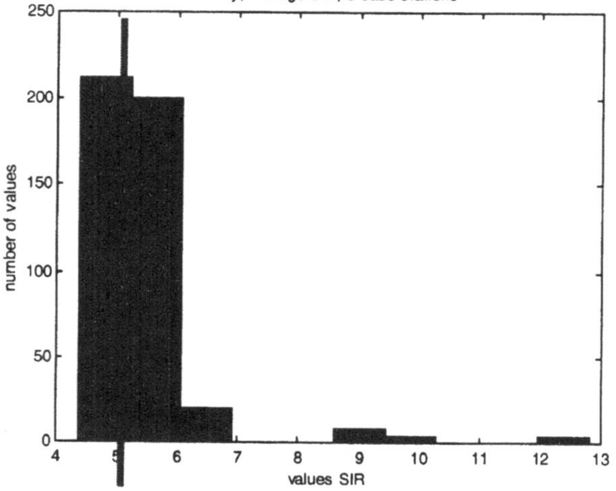

Fig. 9.2.10. Simulation results of power control using relay control. Threshold is SIR=5.012.

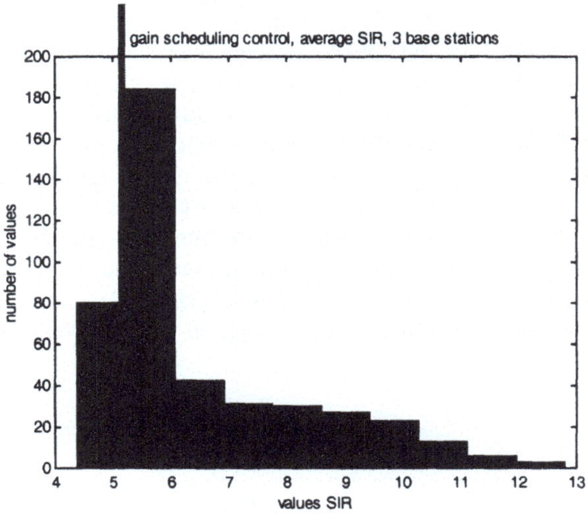

Fig. 9.2.11. Simulation results of gain-scheduling power control. Threshold is SIR=5.012.

From the last results the following conclusion can be drawn. The gain-scheduling algorithm is better than relay control, because the number of fails in the system is about the same as in the traditional power control algorithm, but the quality of transmission based on average SIR is significantly better.

9.2.5 Conclusions

The chapter presents algorithms for power control command in case that the number of bits is varying from 1 to 4. Total number of bits for coding the power control command is considered as a practical constraint. A new power control command is developed based on the idea that power command can be transmitted on 2/3/4 bits and a nonlinear heuristic power control scheme is then applied to power control. A novel gain-scheduling algorithm is also developed at the system level. The coding of power control command gives a possibility to have different power levels. The power control levels are for increasing or for decreasing the power. The dynamic range of power in uplink is 80 dB. The power control algorithms for a command on 2/3/4 bits were tested in the simulator. The PC version of the simulator offers the possibility to have only 1 BS and maximum number of MS's is 50. The uplink transmission is considered here.

The behavior of the algorithms is good in the studied cases. The criterion for comparing the algorithms is the total number of fails, which should be low at the system level. The total number of fails in the system is less than 7, when the total number of calls is varying between 40 and 50. The algorithms are compared with the ideal case, when the power control was coded on 1 bit and the total number of fails was 1. The decisions are taken based on a heuristic control scheme. Another comparison criterion is based on calculating the average SIR over every frame.

The second new approach for power control is the application of gain scheduling for power control to WCDMA system. The decision is based on number of total fails or total number of initiated. This type of algorithm is called gain scheduling. The main problem in the design of gain scheduling is to find suitable scheduling variables. The algorithm can be sensitive to parameter values. The idea is to minimize the number of total fails to the system level. The quality measurements show that the proposed algorithm gives better results than the traditional relay type of control. The proposed algorithms are studied both in a simple environment and also in a more complex one. The transmission is considered poor, if the average SIR is less than SIR target = 5.012. In simulations the proposed algorithms give less values under threshold than conventional relay control.

References

1. ETSI UMTS Terrestrial Radio Access (UTRA) ITU-R RTT Candidate Submission, (1998)
2. Grandhi S.A, Zander J., Yates R.,(1995) "Constrained Power Control", Wireless Personal Communications, vol.1
3. Bergholm P., Honkanen M., Häggman S-G.,(1995) "Simulation of a Microcellular DS-CDMA Radio Network", Proceedings of IEEE VTC, pp. 838 -842.
4. Almgren M., Anderson H., Wallstedt K, (1994)" Power Control in Cellular System", Proc. IEEE Vehicular Technology, Sweden.
5. Pietilä A., (1997) "Development of the Netsim Program", Proceedings of the IRC Workshop'97, pp. 90 -91.
6. Viterbi A.J.,(1995) "Principles of Spread Spectrum Communication", Addison-Wesley, Reading, Mass., USA.
7. Laakso J., Jantti R., Rinne M., Salonaho O., (1998) "Radio Resources Knapsack Packing for WCDMA Air Interference" submitted for publication.
8. Grandhi S.A. , Vijayan R, Goodman D.J, Zander J., (1993) "Centralized Power Control in Cellular Radio Systems", IEEE Transaction on Vehicular Technology, Vol. 42, No 4, November, pp. 466-468.
9. Lin J.C., Lee T.H., Su Y.T., (1994) "Power Control Algorithm for Cellular Radio Systems", Electronic Letter, 3rd February, Vol. 30, No 3, pp. 195-197.
10. Foschini G.J., Miljanic Z., (1993) "A Simple Distributed Autonomous Power Control Algorithm and Its Convergence", IEEE Transaction on Vehicular Technology, Vol. 42, No 4, November, pp. 641-647.
11. Lee T.H, Lin J.C., (1996) "A Fully Distributed Power Control Algorithm for Cellular Mobile Systems", IEEE Journals on Selected Areas in Communications, Vol. 14, No.4, May, pp.692-697
12. Chang P-R. , Wang B.-C.,(1996) "Adaptive Fuzzy Power Control for CDMA Mobile Radio Systems", IEEE Transaction on Vehicular Technology, Vol. 45, No 2, May, pp. 235-236.
13. Chang P-R., Wang B-C., (1996) "Adaptive Fuzzy Proportional Integral Power Control for a Cellular CDMA System with Time Delay", IEEE Journals on Selected Areas in Communications, Vol. 14, No.9, December, pp. 1818- 1827.
14. Andersin M., Rosberg Z., Zander J.,(1994) "Distributed Discrete Power Control in Cellular PCS", Radio Communication Systems, Royal Institute of Technology, Sweden.

15. Kim S.-L., "Optimazation Approach to Prioritized Transmitter Removal in a Multiservice Cellular PCS with Distributed Power Control", Radio Communication Systems, Royal Institute of Technology, Sweden.
16. Åström KJ., Wittenmark B.,(1989) "Adaptive Control", Addison-Wesley Publishing Company.
17. Virtej I., Koivo H., (1999) "Application of Gain Scheduling Concept in Power Control of WCDMA System", Procedeeings of IEEE Vehicular Technology in Communication.
18. Virtej I., Koivo H.,(1999) "New Approach to Power Control for WCDMA System", 1999 International Workshop on Mobile Communication focussed on UMTS & IMT-2000.
19. Ariyavisitakul S.,(1992) "SIR Based Power Control in a CDMA System", IEEE Transaction in Communication, June, pp.868-873.
20. J.Zander, (1992) "Distributed Cochannel Interference Control in Cellular Radio System ", IEEE Transaction Vehicular Technology, Vol. 41, No.1, pp. 57-62, February.

9.3 Timing Acquisition in CDMA Systems

Jyrki Joutsensalo and Tapani Ristaniemi

9.3.1 Introduction

Trends towards multimedia applications and video transmission in mobile environment necessitate high rate data transmission capability over radio interface. In addition, due to the high acceptance of cellular mobile radio systems, capacity limits have already emerged in highly populated areas. Analog mobile telephone systems, such as NMT (Nordic Mobile Telephone) are called first generation systems. They were accompanied by digital second generation systems, for example GSM (Global System for Mobile communications). Those systems are primarily voice oriented; however, especially second generation systems also support low data rate services. For capacity reasons and to be able to offer new services and system features, third generation mobile radio systems are under development. Wideband Code Division Multiple Access (WCDMA or CDMA) technology is a strong candidate for future third generation global wireless mobile communications. WCDMA has selected an air interface solution e.g. in UMTS (Universal Mobile Telecommunication System) standard, which will provide a multitude of services, especially multimedia and high bit rate packed data. CDMA is based on spread spectrum technique. In wireless communications, spread spectrum approach outperforms Frequency Division Multiple Access (FDMA) based NMT and Time Division Multiple Access (TDMA) based GSM. CDMA and spread spectrum have some important benefits:

Jamming is difficult
Security and privacy.
Resistance against the multipath propagation of the signals.
Flexible with the number of users.
High resolution ranging. Global Positioning System (GPS).
Frequency reuse in cellular systems.
Sectorized antennas can be used to increase the diversity.

The difference between FDMA, TDMA, and CDMA depends how to allocate the frequency spectrum to users. All these methods rely on user partitioning in the time-frequency plane. The number of users that can be served is determined by the available frequency slots in FDMA, and by the available time slots in TDMA. In

CDMA, all users are using the same frequency band at all times, as opposed to FDMA and TDMA. In CDMA, the separation of users is carried out by assigning each user with a unique code sequence. Ideally, FDMA, TDMA, and CDMA can offer the same channel capacity. However, in FDMA and TDMA, frequency management is a critical task to carry out. In theory, no frequency management is needed in CDMA, since there is only one common radio channel. In practice, the frequency reuse pattern is smaller in CDMA than in TDMA and FDMA. Moreover, additional users can be served with only a slight degradation of performance, and if multipath propagation is a problem, as happens in mobile environment, the choice of CDMA is especially appropriate [7,12]. Among different realizations of CDMA, the direct sequence (DS) technique has achieved the greatest interest in mobile communications.

The spread spectrum technique has the following essential features:
Radio frequency bandwidth is much larger than information bandwidth.
Data spreading is done by *pseudorandom* signal or *signature waveform* (the *code* in CDMA).
Spreading operation is independent with data.
Spreaded data resembles noise.

The objective of bandwidth expansion is to minimize the impact of noise and disturbances on the transmitted data. The development of spread spectrum techniques started from military needs. Typical properties of spread spectrum system, such as low probability of intercept, inherent antijam capability and message privacy are obviously desirable properties in military communications, but also in mobile communications.

9.3.2 Problems in CDMA implementation

Consider some problems in the CDMA system. First of all, the final objective is demodulation of the desired information signal. The essential requirements are that the hardware/software solutions are computationally simple, and that this solution is statistically quite efficient. The statistical efficiency can be measured e.g. by using probability of acquisition, mean acquisition time, root mean square error, or by the most important way, Bit Error Rate (BER).

One of the most serious problems is Multiple Access Interference (MAI), which arises when several users interfere the others. MAI can be alleviated by increasing the processing gain C, but with fixed chip rate, this decreases the data rate. In addition, *near-far problem* arises when some users are near and the other users are far from the base station. There may be even 90 dB differences between the powers of the different users. If the received powers from different users become too different, a strong user will seriously interfere with the weak ones, even if there is a small correlation between users. The near-far effect is not a problem in the FDMA and TDMA systems, since the orthogonality can be easily achieved in these systems. There are two possible solutions to avoid near-far problem:

Near-far resistant Multi-User Detection (MUD) using knowledge of one or several parameters, such as propagation delay, carrier phase, and received power level [4,5,8,18,19].
Use power control algorithm.

In the first case, the problem is that the MUD needs more complicated solution by joint estimation of the parameters. In the second solution, base station continuously measures the received power from each active user and transmits back a power adjustment command. In this way, the different transmitters can adjust their output power so that the base station receives all signals with equal or nearly equal strength and no user will be completely wiped out by another user. To compensate fast varying propagation channel, power control commands have to be sent very often, which causes overhead to the system. Therefore, it would be desirable to find near-far resistant algorithms to mitigate power control.

Fading is the problem in any wireless systems. It is caused, when the propagation channel in a mobile communication system distorts the transmitted signal. In mobile radio system, transmitted signal propagates through various paths. If extremely short pulse, ideally an impulse, is transmitted over a time-varying multipath channel, the received signal might appear as a train of pulses. When transmitted signal travels over multiple propagation paths, time delay associated to each path is random. Thus, phases of the received multipath components are random. At times phases are adding destructively. When that occurs, the resultant received signal is very small or practically zero. The channel is then said to be in a fade. The rapidity of amplitude variation, and accordingly the distance of two fades in time are affected by the speed of mobile receiver / transmitter. Hence, distinctions of *slow* and *fast* fading are made, where the former is generally understood as a situation, where the channel remains constant during one symbol duration. It is possible to enhance the overall signal-to-noise ratio (SNR) by utilizing time-diversity. RAKE receiver is commonly used to combine the signal energy of multipath components.

Accurate *synchronization* of the receiver carrier's frequency and phase, as well as chip timing, to that of received signal is a prerequisite task before any demodulation can be performed. Chip timing is acquired first. This requires the receiver to have a fairly good estimate for the carrier frequency only. No knowledge of the phase is needed yet [20]. Chip timing acquisition means rough estimation of the chip level timing, with a maximum acceptable error of half chip duration. After this, conventional phase-lock loop (PLL) devices can be used for accurate phase estimation and for carrier frequency refinement. Finally, all the quantities can be tracked by PLL in non-stationary environments. In this chapter, we consider more profoundly chip timing acquisition. Because it refers to the coarse synchronization of the received signal, it basically leads to a hypotheses testing problem. This means testing some possible chip timings whether some of them are correct values or not. The more candidates are tested the more accurate estimation is possible. Fig. 9.3.1 illustrates the problem.

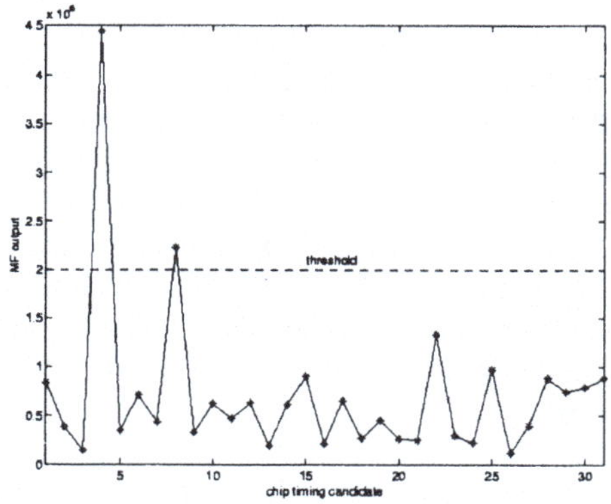

Fig. 9.3.1 Delay spectra for visualization of delay estimation procedure.

In this case, 31 delays are tested (horizontal axis) and they result with certain value (vertical axis), which is related to its suitability for being a correct delay. In the figure, two paths seem to exist, corresponding to 4th and 8th test delay, respectively. If the number of hypotheses were small, one could test all the hypotheses (in this case 31) in parallel manner, and choose the delays corresponding to the peaks of the delay spectra. However, this leads to a problem how many peaks are chosen, since the number of paths is unknown. For example, the 22nd test delay in the figure could correspond to the third path, or it can be just a spurious peak. The problem is more concrete as the number of hypotheses grows, since then delays are generally tested in serial manner. This requires setting a certain threshold value. In serial operation, each incorrect hypothesis (contribution below threshold) is eliminated, before the next value is tested. The threshold setting hence becomes also a problem, especially in highly loaded systems, were mutual interferences of different users may cause more spurious peaks. A good survey of the basic ideas of existing serial and parallel acquision schemes is given e.g. in [15].

There are several ways to synchronize the signal. Maximum Likelihood Method (MLM) is optimal [13], but it is impractical, because it performs joint-synchronization, and thus the complexity of MLM is exponential with respect to the number of users. Joint-synchronization is needed e.g. in optimal MUD [18]. Existing constant pilot signal-based methods [2,22] are still quite complex, or they assume flat fading channels. Therefore, conventional Matched Filter (MF) and pilot-based coherent MF succeed very well in the synchronization with respect to the above-mentioned methods. One sub-optimal approach is subspace-based MUltiple SIgnal Classification (MUSIC) [1,17], which first introduced for direction-of-arrival estimation [16]. The other promising approach is and blind Constrained Minimum Ouptut Energy (CMOE) method [5,9], which has tested in synchronization in [6].

In this chapter, we consider especially differentially coherent (DC) acquisition. DC based acquisition have earlier been studied e.g. in [3,21]. Roughly speaking, DC approach tries to maximize correlation of the MF output process. In our approach, we make the following assumptions. First, the signal to be synchronized has short code, and the other signals have either short or long codes. An example of the short code system is FMA2 [10], while GPS is an example of short / long code system [14]. Using short code acquisition, one initially obtains the reinforced signal having the phase information of the long code to be locked. Second, the desired user receives constant preamble (no data modulation), while the other users have the data modulation on. The cyclic short codes together with constant pilots generate subspace-structured desired signals, which property can be expoited in the detection of the desired signal.

New method using these assumptions is compared with traditional MF-based methods. We show that our method outperforms MF in several situations: (i) in highly loaded environments, (ii) when the number of available symbols is large, and (iii) when channel is fast fading.

9.3.3
Simplified transmission and ideal reception

Consider next the transmission and reception of the spread spectrum signal. In the simplest possible case, we assume that

data is binary.
pseudo-noise is binary.
time is discrete.

no carrier modulation exist.

In the Fig. 9.3.2, it is shown the typical information and pseudo-noise sequences, and in the Fig. 9.3.3, the bandwidth of the information and the spread spectrum signal are shown. The information is spreaded by using *rectangular* chip sequence, also called *pseudo-noise* or *PN sequence*:

$$y[m] = s[m]b[m] \quad , \tag{1}$$

where

$b\lfloor m \rfloor$ is a narrowband *bit* sequence having duration *T*, e.g. in the Fig. 9.3.2: $T = 31$.

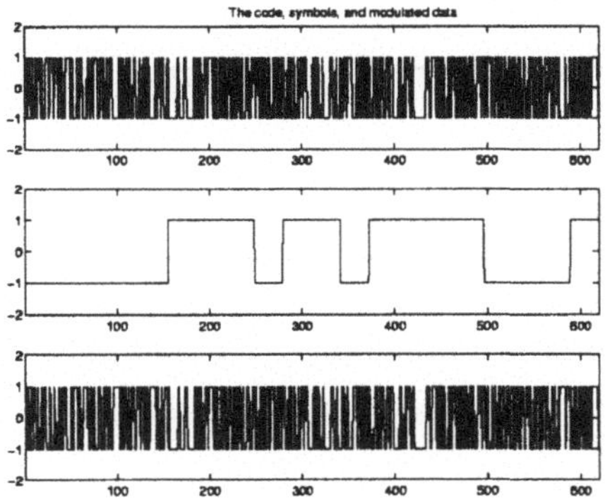

Fig. 9.3.2. An example of the spreading operation: PN sequence (top), information sequence (middle), and spreaded output process (bottom).

is a wideband *chip* sequence having duration T_c, e.g. in the Fig. 9.3..2: $T_c = 1$.

Processing gain $s\lfloor m \rfloor \in \{-1,1\}$ $C = T / T_c$ is a theoretical gain, which measures the increase of
the jamming margin. It has the following properties:

For fixed duration of the signature waveform, its bandwidth is proportional to *C*.
In an orthogonal synchronous DS-CDMA, the number of users that can be supported is less or equal to *C*.
Large *C* contribute to the *privacy* of the system, and
to the interference degration of narrowband jammer.
When *C* is increased, the signals can be transmitted with a lower average power.
Reliability of synchronization increase with *C*.

For example, in UMTS, $T_c \approx 260$ ns , i.e. $W_c = 3.84$ Mchip/s chip rate is used. On the other hand, maximal data rate in indoor office environments is
$W = 2$ Mbit/s, and $T = 500$ ns . In some cases, *C* may be several hundreds. The attractive feature in the spread spectrum systems is that the PN sequence can be generated by very simple *shift registers*. Shift registers can create binary codes of ± 1 's with the properties:

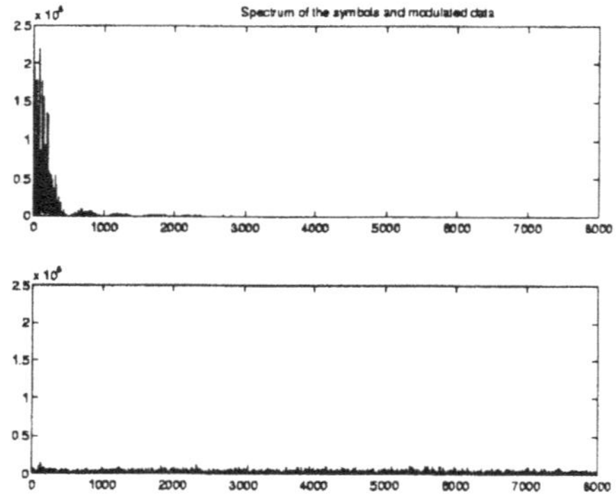

Fig. 9.3.3: Information bandwidth (top), and the bandwidth of the spreaded signal (bottom).

Balance: there is only one of 1's more than −1's.

Run: there are $2(n-1)$ runs of consecutive 1's or −1's, and half of the runs are of length 1, $1/2^2$ of the runs are of length 2, $1/2^3$ of the runs are of length 3, etc.

Correlation: shifted code's correlation with the non-shifted code is -1.

Special PN sequence called *m-sequence* has the period $N = 2^n - 1$, where n may be even 42. It has the following autocorrelation function:

$$E(s[m]s[m+k]) = \frac{1}{2^n - 1}\langle s[m], s[m+k]\rangle = \begin{cases} -1 & \text{, when } k = 0 \\ -1/(2^n - 1), & \text{when } k \neq 0 \end{cases} \quad (2)$$

Therefore, m-sequence is very orthogonal and uncorrelated. In Fig. 9.3.4, the autocorrelation function of m-sequence with large n is shown. Continuous-to-discrete-time conversion is realized by a *chip-matched filter,* which is sampled at the chip rate with a timing offset: $0 \le \tau \le T_c$.

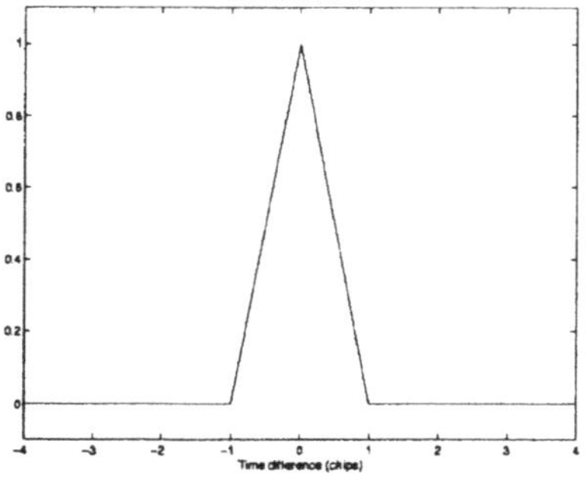

Fig. 9.3.4 Autocorrelation of the m-sequence with large n.

$$r[m] = \int_{mT_c - T_c + \tau}^{mT_c + \tau} p_{T_c}(t - (m-1)T_c - \tau) r(t) dt\,, \tag{3}$$

where $p_{T_c}(.)$ is a *chip waveform*, e.g. in the rectangular case

$$p_{T_c}(t) = \begin{cases} 1 & , 0 \le t \le 1 \\ 0, & \text{otherwise} \end{cases} \tag{4}$$

The received data has the simplified form

$$r[m] = y[m] + w[m] = s[m]b[m] + w[m], \tag{5}$$

where

$s[m]b[m]$ is *desired* signal.
$w[m]$ is *interfering* or *undesired* signal.

For simplicity, assume that the timing is exactly known. Then the desired symbol $b[m]$ is estimated by *despreading* , while interfering signal $w[m]$ is *spreaded* by the same operation:

$$\begin{aligned} o[m] &= s[m]r[m] \\ &= s[m](s[m]b[m] + w[m]) = b[m] + s[m]w[m] \end{aligned} \tag{6}$$

Next one can sum i.e. integrate over symbol duration,

$$\sum_{m=1}^{C} o[m] = Cb[m] + \sum_{m=1}^{C} s[m]w[m] \tag{7}$$

If $s[m]$ is uncorrelated with $w[m]$, then the right hand side term is roughly zero, and the symbol estimate is

$$\hat{b} = sign\left(\sum_{m=1}^{C} o[m]\right) \approx sign(Cb[m]). \tag{8}$$

This is a MF reception approach. The standard receiver structure for DS-CDMA is simply a bank of MFs, each matched to a particular user code. MF is optimum in the MLM sense in the single-user case or if all users are orthogonal. In mobile environment orthogonality cannot in general be guaranteed because transmitted signal propagates through several paths with arbitrary delays. The essential assumptions in the above-described MF approach were:

PN sequence is (roughly) uncorrelated with the interferers.
Exact timing is known.
There is no fading, i.e. channel impulse response is constant with time.

Next we consider more practical CDMA model, where several users exist, fading is fast, and timing is unknown.

9.3.4 CDMA model

The signal model studied in this chapter may in theory be downlink or uplink with fading channel. Consider e.g. downlink, because pilot channels are more suitable in that case. The starting point is that $K-1$ users occupy the CDMA system cell, and a new user, called ``desired user", is making or receiving a call, or arriving from the neighboring cell. The goal is to synchronize the desired user. Two cases can be considered: the desired user has short code, and interfering i.e. existing users have either short or more general long codes. The data models differ, but algorithm to be developed not much. Consider here the short code model. The received data have the asynchronous form

$$r(t)=\sum_{m=1}^{N}\sum_{k=1}^{K}b_{km}\sum_{l=1}^{L}a_{lm}s_k(t-mT-\chi_l)+n(t), \qquad (9)$$

where

a_{lm} is the complex fading factor of the lth, path varying slowly symbol by symbol,

b_{km} is kth user's mth symbol,

$s_k(.)$ is kth user's chip sequence i.e the spreading code. Moreover, $s_k(t)\in\{-1,1\}, t\in[0,T)$, and $s_k(t)=0, t\notin[0,T)$.

χ_l is the delay of the lth path, $\chi_l=d_l+\delta_l$, where $d_l\in\{0,1,...,C-1\}$ and $\delta_l[0,1)$. The delay is assumed to remain roughly constant in the observation interval.

$n(t)$ denotes noise.

C and N are the chip sequence length (i.e. processing gain) and the number of bits, respectively.

The problem is to estimate χ_l corresponding to the strongest paths. First the received data is sampled. Here we operate at the chip-rate, i.e. C equispaced samples per symbol are taken. Since sampling is usually both symbol and chip asyncronous, the collected C-vectors $\mathbf{r}_m$ have the form

$$\mathbf{r}_m=\sum_{k=1}^{K}\sum_{l=1}^{L}a_{l,m-1}b_{k,m-1}\underline{\mathbf{c}}_{kl}+a_{lm}b_{km}\bar{\mathbf{c}}_{kl}+\mathbf{n}_m \qquad (10)$$

where $\mathbf{n}_m$ denotes noise vector, and

$$\underline{\mathbf{c}}_{kl}=(1-\delta_l)\underline{\mathbf{g}}_{kl}(d_l)+\delta_l\underline{\mathbf{g}}_{kl}(d_l+1) \qquad (11)$$

$$\bar{\mathbf{c}}_{kl} = (1-\delta_l)\bar{\mathbf{g}}_{kl}(d_l) + \delta_l \bar{\mathbf{g}}_{kl}(d_l+1) \tag{12}$$

with the ``early" and ``late" parts of the code

$$\underline{\mathbf{g}}_{kl} = \left[s_k[C-d_{l+1}] \ \dots \ s_k[C]\,0 \dots 0\right] \tag{13}$$

$$\bar{\mathbf{g}}_{kl} = \left[0 \dots 0\; s_k[1] \ \dots \ s_k[C-d_l]\right] \tag{14}$$

Define a $C \times 2KL$ dimensional code matrix $\mathbf{G} = [\mathbf{g}_1, \dots, \mathbf{g}_{2KL}]$,

$$\mathbf{G} = \left[\underline{\mathbf{c}}_{11}, \bar{\mathbf{c}}_{11} \dots, \underline{\mathbf{c}}_{KL}, \bar{\mathbf{c}}_{KL}\right] \tag{15}$$

and a $2KL$ source vector

$$\mathbf{a}_m = \left[a_{11,m-1}b_{11,m-1}, a_{11m}b_{11m}, \dots, a_{KL,m-1}b_{KL,m-1}, a_{KLm}b_{KLm}\right]. \tag{16}$$

Then the data (10) can be rewritten as

$$\mathbf{r}_m = \mathbf{G}\mathbf{a}_m + \mathbf{n}_m . \tag{17}$$

In the pure matrix form the representation of the data is

$$\mathbf{X} = \mathbf{G}\mathbf{A} + \mathbf{N}, \tag{18}$$

where

$$\mathbf{X} = [\mathbf{r}_1, \dots \mathbf{r}_N] \tag{19}$$

is the received sampled data matrix,

$$\mathbf{A} = [\mathbf{a}_1, \dots \mathbf{a}_N] \tag{20}$$

is the source matrix containing the symbols as well as fading factors, and

$$\mathbf{N} = [\mathbf{n}_1, \dots \mathbf{n}_N] \tag{21}$$

is the noise matrix. Autocorrelation matrix of the data is

$$\mathbf{R} = E\{\mathbf{r}_m \mathbf{r}_m{}^T\} = \mathbf{G}\mathbf{R}_a\mathbf{G}^T + \sigma^2\mathbf{I} \tag{22}$$

where $\mathbf{R}_a = E\{\mathbf{a}_m \mathbf{a}_m{}^T\}$, and σ^2 is the variance of the noise. Data correlation matrix estimate without scaling is $\mathbf{R} = \mathbf{X}\mathbf{X}^T$.

9.3.5 Acquisition approaches

9.3.5.1 *Some assumptions*

Consider the *differential* correlation matrix

$$\mathbf{C} = E\left\{\mathbf{r}_m \mathbf{r}_{m+2}^T\right\} \tag{23}$$

of the data, where time lag is two symbols. The following assumptions are made:

1. Noise is uncorrelated and zero mean. Then $E\{\mathbf{n}_m \mathbf{n}_{m+p}\} = 0$ for $p \neq 0$.
2. A constant preamble is available for the desired user, user $k = 1$, i.e. $b_{1m} = 1$ for all m.
3. Paths are uncorrelated and zero mean. Then $E\{a_{lm} a_{l'n}\}$ for $l \neq l'$.
4. Channel is slowly varying. Therefore it is assumed that correlation coefficient

$E\{|a_{lm}|\}E\{|a_{ln}|\}$ for sufficiently small $|n-m|$, say $0 < |n-m| \leq 3$.

5. Symbols are uncorrelated and zero mean. Then $E\{b_{km} b_{k',m+p}\} = 0$ for $p \neq 0$.

The source differential correlation matrix with time lag two symbols is equal to

$$\mathbf{C}_a = E\left\{\mathbf{a}_m \mathbf{a}_{m+2}^T\right\} \tag{24}$$

According to Eq. (16), $\mathbf{C}_a$ contains KL block of 2×2 martices, denoted as $E\left\{\mathbf{a}_{klm} \mathbf{a}_{k'l',m+2}^T\right\}$, each of which have the form

$$\begin{bmatrix} E\left\{a_{l,m-1} a_{l',m+1} b_{k,m-1} b_{k',m+1}\right\} & E\left\{a_{l,m-1} a_{l',m+2} b_{k,m-1} b_{k',m+2}\right\} \\ E\left\{a_{lm} a_{l',m+1} b_{km} b_{k',m+1}\right\} & E\left\{a_{lm} a_{l',m+2} b_{km} b_{k',m+2}\right\} \end{bmatrix} \tag{25}$$

This matrix has a different form depending on incides k, k', l, l':
Case 1: $1 = k = k'$. Now all the symbols are equal to 1 and two sub-cases arise: if $l \neq l'$, the channel coefficients do not correlate, and we have $E\left\{\mathbf{a}_{1lm} \mathbf{a}_{1l',m+2}^T\right\} = \mathbf{0}$. If $l = l'$, we have due to slow channel variation

$$E\left\{\mathbf{a}_{1lm} \mathbf{a}_{1l',m+2}^T\right\} = \alpha_l \begin{bmatrix} 1 & 1 \\ 1 & 1 \end{bmatrix} \stackrel{\text{def}}{=} \alpha_l \mathbf{1}, \tag{26}$$

where α_l stands for $E\{a_{lm}a_{\ln}{}^T\}$, with $0<|m-n|\leq 3$.

Case 2: $1=k\neq k'$. Now the desired user's symbols are equal to 1 but others' are random. With $l=l'$ the channel coefficients correlate, but since the user's k symbols are zero mean, we have $E\{\mathbf{a}_{1lm}\mathbf{a}_{k'\ln}{}^T\}=\mathbf{0}$. With $l\neq l'$, the result is the same.

Case 3: $1\neq k,k'$. All the symbols are now random. Since they are uncorrelated by assumption, we directly get $E\{\mathbf{a}_{klm}\mathbf{a}_{k'\ln}{}^T\}=\mathbf{0}$ with any l,l'.

Hence, the correlation matrix $\mathbf{C}_a$ contains only L nonzero blocks of 2×2 in its main diagonal, and all the other elements are zero:

$$\mathbf{C}_a=\begin{bmatrix}\alpha_1\mathbf{1} & \mathbf{0} & \cdots & & & \mathbf{0}\\ \mathbf{0} & \ddots & & & & \vdots\\ \vdots & & \alpha_L\mathbf{1} & & & \\ & & & \mathbf{0} & & \\ & & & & \ddots & \\ \mathbf{0} & \cdots & & & & \mathbf{0}\end{bmatrix}. \tag{27}$$

It is straight forward to calculate that the (sample) correlation matrix of the data with time lag two symbols then equals

$$\mathbf{C}=E\{\mathbf{r}_m\mathbf{r}_{m+2}{}^T\}=\mathbf{G}\mathbf{C}_a\mathbf{G}^T=\sum_{l=1}^{L}\alpha_l\mathbf{c}_{1l}\mathbf{c}_{1l}{}^T, \tag{28}$$

where $\mathbf{c}_{1l}\overset{\text{def}}{=}\underline{\mathbf{c}}_{1l}+\bar{\mathbf{c}}_{1l}$.

Example.

Let us derive the differential correlation matrix using a simple example: assume that $K=2$ and $L=2$. Then the source vectors $\mathbf{a}_m$ and $\mathbf{a}_{m+2}$ are

$$\begin{aligned}\mathbf{a}_m&=[a_{1,m-1}b_{1,m-1},a_{1m}b_{1m},a_{2,m-1}b_{1,m-1},a_{2m}b_{1m},\ldots\\ &\quad a_{1,m-1}b_{2,m-1},a_{1m}b_{2m},a_{2,m-1}b_{2,m-1},a_{2m}b_{2m}]\\ &=[a_{1,m-1},a_{1m},a_{2,m-1},a_{2m},\ldots\\ &\quad a_{1,m-1}b_{2,m-1},a_{1m}b_{2m},a_{2,m-1}b_{2,m-1},a_{2m}b_{2m}]\end{aligned} \tag{29}$$

$$\begin{aligned}\mathbf{a}_{m+2} &= \big[a_{1,m+1}b_{1,m+1}, a_{1,m+2}b_{1,m+2}, a_{2,m+1}b_{1,m+1}, a_{2,m+2}b_{1,m+2}, \ldots \\ &\quad a_{1,m+1}b_{2,m+1}, a_{1,m+2}b_{2,m+2}, a_{2,m+1}b_{2,m+1}, a_{2,m+2}b_{2,m+2}\big] \\ &= \big[a_{1,m+1}, a_{1,m+2}, a_{2,m+1}, a_{2,m+2}, \ldots \\ &\quad a_{1,m+1}b_{2,m+1}, a_{1,m+2}b_{2,m+2}, a_{2,m+1}b_{2,m+1}, a_{2,m+2}b_{2,m+2}\big]\end{aligned} \tag{30}$$

Source differential correlation matrix is

$$\mathbf{C}_a = \begin{bmatrix} \alpha_1 & \alpha_1 & 0 & 0 & 0 & 0 & 0 & 0 \\ \alpha_1 & \alpha_1 & 0 & 0 & 0 & 0 & 0 & 0 \\ 0 & 0 & \alpha_2 & \alpha_2 & 0 & 0 & 0 & 0 \\ 0 & 0 & \alpha_2 & \alpha_2 & 0 & 0 & 0 & 0 \\ 0 & 0 & 0 & 0 & 0 & 0 & 0 & 0 \\ 0 & 0 & 0 & 0 & 0 & 0 & 0 & 0 \\ 0 & 0 & 0 & 0 & 0 & 0 & 0 & 0 \\ 0 & 0 & 0 & 0 & 0 & 0 & 0 & 0 \end{bmatrix} \tag{31}$$

which can be verified using the assumptions 1-5.[1]

In this example, the code matrix $\mathbf{G}$ is

$$\mathbf{G} = \big[\underline{\mathbf{c}}_{11}, \bar{\mathbf{c}}_{11}, \underline{\mathbf{c}}_{21}, \bar{\mathbf{c}}_{21}, \underline{\mathbf{c}}_{12}, \bar{\mathbf{c}}_{12}, \underline{\mathbf{c}}_{22}, \bar{\mathbf{c}}_{22}\big] \tag{32}$$

Data differential correlation matrix is

$$\mathbf{C} = E\{\mathbf{r}_m \mathbf{r}_{m+2}{}^T\} = \mathbf{G}\mathbf{C}_a\mathbf{G}^T = \alpha_1 \mathbf{c}_{11}\mathbf{c}_{11}{}^T + \alpha_2 \mathbf{c}_{12}\mathbf{c}_{12}{}^T. \tag{33}$$

From the structure of the matrix $\mathbf{C}$ we see that it contains only the code and delay information of the first user. The estimate $\hat{\mathbf{C}}$ is obtained in the following way:

$$\hat{\mathbf{C}} = \mathbf{X}_{1:N-2}\mathbf{X}_{3:N}{}^T, \tag{34}$$

where $\mathbf{X}_{a:b}$ contains column vectors of the data matrix $\mathbf{X}$ with indices $a,\ldots,b$.

[1] If the time lag were one symbol, $\mathbf{C}_a$ should contain also the correlations of the undesired users, because in the asynchronous system, adjacent sample vectors contain partly the same symbols.

9.3.5.2 SVD algorithm

Because the signal data have the linear form of (18), the subspace approach can be used for filtering additive noise out. More specifically, our approach whitens the data by taking first a Singular Value Decomposition (SVD)

$$\mathbf{X} = \mathbf{U}\mathbf{\Sigma}\mathbf{V}^T, \tag{35}$$

where $\mathbf{U}$ consists left orthonormal singular vectors, $\mathbf{\Sigma}$ consists singular values as its diagonal, and $\mathbf{V}$ consists right orthonormal singular vectors. Low-rank approximation of the data is $\mathbf{U}_s\mathbf{\Sigma}_s\mathbf{V}_s^T$, where matrices correspond to the largest singular values, and the dimensions of $\mathbf{U}_s$, $\mathbf{\Sigma}_s$, and $\mathbf{V}_s^T$ are $C \times 2KL$, $2KL \times 2KL$, and $2KL \times N$, respectively. By this choice the spanning property $\text{range}(\mathbf{U}_s) = \text{range}(\mathbf{G})$ holds, if no noise exists. The data is *whitened* as follows:

$$\mathbf{Y} \overset{\text{def}}{=} \mathbf{V}_s^T = \mathbf{\Sigma}_s^{-1}\mathbf{U}_s^T\mathbf{X}. \tag{36}$$

Differential correlation matrix

$$\mathbf{D} = E\left\{\mathbf{y}_m\mathbf{y}_{m+2}^T\right\} \tag{37}$$

for whitened data has the form

$$\mathbf{D} = \mathbf{\Sigma}_s^{-1}\mathbf{U}_s^T\mathbf{C}\mathbf{U}_s\mathbf{\Sigma}_s^{-1}. \tag{38}$$

$\mathbf{D}$ is estimated as follows:

$$\hat{\mathbf{D}} = \mathbf{Y}_{1:N-2}\mathbf{Y}_{3:N}^T, \tag{39}$$

where $\mathbf{Y}_{1:N-2}$ is the SVD of $\mathbf{X}_{1:N-2}$, and $\mathbf{Y}_{3:N}$ is the SVD of $\mathbf{X}_{3:N}$. Finally, the delay estimates are obtained from the WDC (Whitened DC) estimator

$$\hat{d} = \arg\max_d \left|\mathbf{g}(d)^T\hat{\mathbf{U}}_s\hat{\mathbf{D}}\hat{\mathbf{U}}_s^T\mathbf{g}(d)\right|, \tag{40}$$

where $\mathbf{g}(d)$ is the circularly shifted code vector

$$\mathbf{g}(d) = \left[s_1(C-d+1),\ldots,s_1(C),s_1(1),\ldots,s_1(C-d)\right]^T \tag{41}$$

If we use the theoretical form of the whitened matrix (38) and apply (28) and (40), we obtain the estimator

$$\mathbf{g}(d)^T \mathbf{U}_s \mathbf{D} \mathbf{U}_s{}^T \mathbf{g}(d) = \sum_{l=1}^{L} \alpha_l \left| \mathbf{g}(d)^T \mathbf{U}_s \boldsymbol{\Sigma}_s{}^{-1} \mathbf{U}_s{}^T \mathbf{g}_{1l} \right|, \tag{42}$$

where codes are tried to fit with the filtered low-rank data instead of the original data.

9.3.5.3 Reference methods

One algorithm to be compared is blind Constrained Minimum Output Energy (CMOE) method [5,9,6]. In [5,9], CMOE was introduced and used for blind interference suppression. Algorithm is a modification of well-known Capon's filter, which minimizes output energy of the disturbing sources while passing the desired source undistorted. The criterion is

$$\hat{\mathbf{c}} = \arg\min_{\mathbf{c}} \mathbf{c}^T \mathbf{R} \mathbf{c}, \quad \mathbf{c}^T \mathbf{g}_1 = 1, \tag{43}$$

where $\mathbf{g}_1$ is the code vector corresponding to the desired user. The criterion leads to the filter

$$\hat{\mathbf{c}}_{\mathrm{CMOE}} = \mathbf{R}^{-1} \mathbf{g}_1 . \tag{44}$$

In [6], it was shown that CMOE has very good synchronization capability. In the synchronization, CMOE method is applied as follows:

$$\hat{d} = \arg\min_{d} \mathbf{g}(d)^T \hat{\mathbf{R}}^{-1} \mathbf{g}(d) . \tag{45}$$

Our simplest methods are Matched filter (MF) and coherent MF. MF simply utilizes the criterion

$$\hat{d} = \arg\max_{d} \mathbf{g}(d)^T \hat{\mathbf{R}} \mathbf{g}(d) . \tag{46}$$

In coherent MF, the criterion is

$$\hat{d} = \arg\max_{d} \left| \mathbf{f}(d)^T \hat{\mathbf{r}} \right| , \tag{47}$$

where $\mathbf{r}$ contains all the data samples in one vector, and $\mathbf{f}(d)$ has the form

$$\mathbf{f}(d) = \left[\mathbf{g}(d)^T, \ldots, \mathbf{g}(d)^T \right]^T . \tag{48}$$

9.3.5.4 Experiments

We compared the methods in the fading downlink environment. System was totally asynchronous. The methods were: WDC (40), CMOE (45), MF (46), and coherent MF (47). Any parameters such as channels, codes of the interfering users, or delays were not assumed to be known. Only the code of the desired user was known. In addition, constant preamble $b_{1_m} = 1$ was used. Gold code with length $C = 31$ was used. The number of paths was $L = 3$ with powers $0, -5$, and $-10\,\text{dB}$. *SNR* was calculated with respect to the path with $0\,\text{dB}$. Therefore, two other paths had quite weak energies (-5 dB and -10 dB) with respect to the noise at low *SNR*s, and thus the simulation environment is quite challenging for estimating the corresponding low energy delays. We noticed that the performance for WDC method (40) became better by selecting the estimated signal subspace dimension $\hat{M} = 2K$ instead of $\hat{M} = 2KL$. This can be justified also by the fact that usually it is impossible to know the number of paths, while number of users can be known by uplink-downlink communication. One hundred simulations were performed with different *SNR*s. The following figures show the results.

In Fig. 9.3.5, probability of acquisition is measured as a function of the number of symbols. Acquisition is occured when the estimation accuracy is smaller than half chip. Here $K = 20$, $L = 3$, $SNR = 10$ dB. In Figure 1.6, mean acquisition time is measured as a function of the number of symbols. Parameters are same as in Figure 9.3.5.

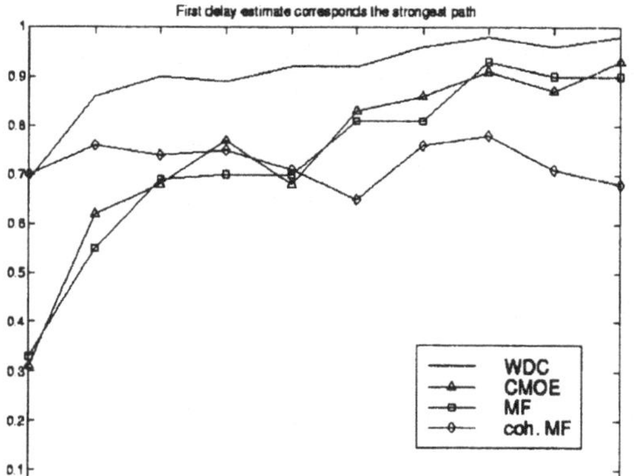

Fig. 9.3.5. Probability of acquisition as a function of the number of the pilot symbols.

In Fig. 9.3.7, the probability of acquisition is measured as a function of the number of users. $N = 500$, $L = 3$, $SNR = 10$ dB. In Fig. 9.3.8, mean acquisition time is measured as a function of the number of users. Parameters are same as in Fig. 9.3.7. In Fig. 9.3.9, root mean-square-error in chips is measured as a function of signal-to-noise ratio. Here it is assumed that the chip index is known. $K = 20$, $N = 500$, $L = 3$. In all experiments, WDC method outperforms other methods. We see that the WDC method is quite robust against the high loading of the systems, because differential correlations with respect to the time are in theory invariant with respect to the interfering users and noise.

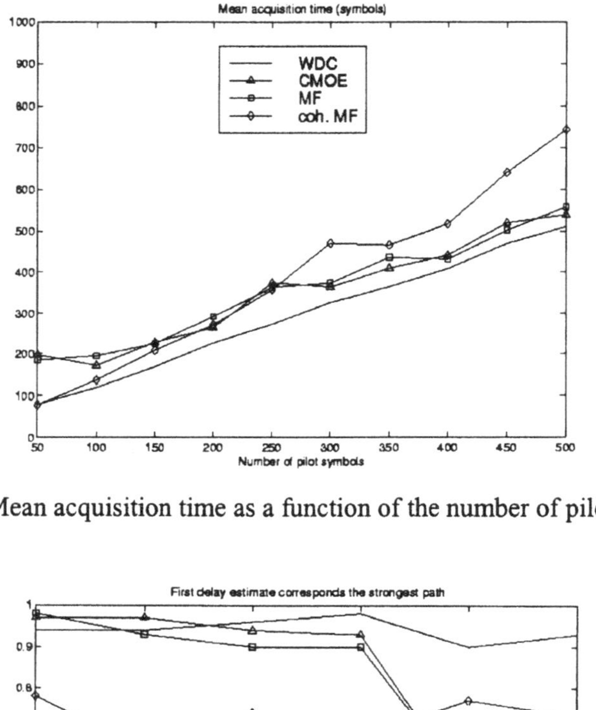

Fig. 9.3.6. Mean acquisition time as a function of the number of pilot symbols.

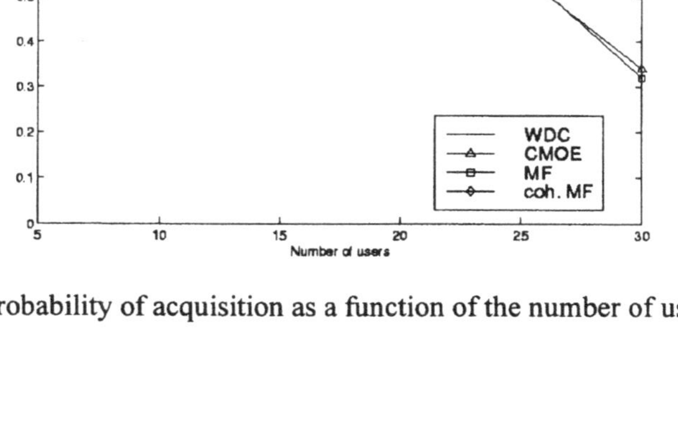

Fig. 9.3.7. Probability of acquisition as a function of the number of users.

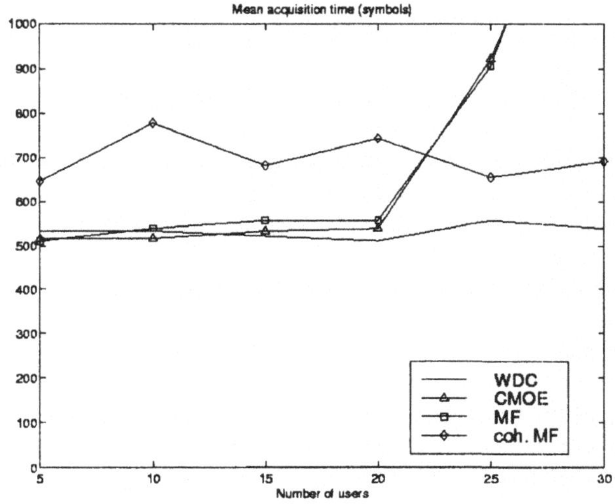

Fig. 9.3.8. Mean acquisition time as a function of the number of users.

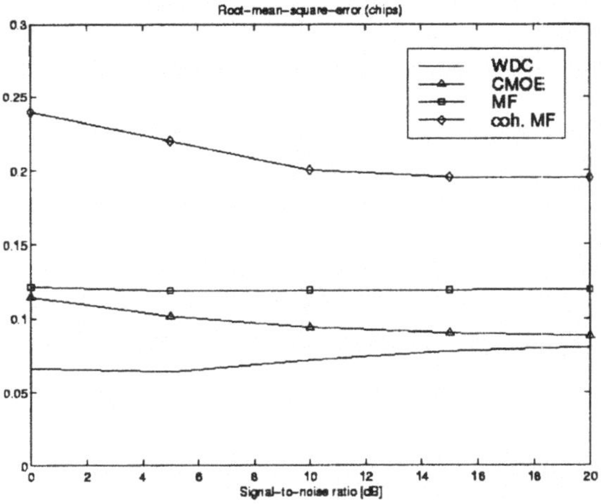

Fig. 9.3.9: Root-mean-square-error as a function of signal-to-noise ratio.

9.3.6 Accurate timing offset estimation

As mentioned before, chip timing acquisition is a first phase in delay estimation process, whose purpose is to get a rough estimate for the delay, say within a chip duration. As is leads to a hypotheses testing problem, a set of code candidates, which correspond to a set of delay candidates, are tested. In practice, the delays are spaced by a half of a chip duration. In acquisition phase, training symbols or pilots can be used, since no actual data transmission is yet done. The roughness of the estimate corresponds to the timing offset of the sampling, i.e. τ in Eq. (3). $\tau = 0$ means that the integration is done over a one whole chip. Otherwise,

$r[m]$ has a contribution of two successive chips in ratio $\frac{\tau}{T_c}$ and $\frac{T_c - \tau}{T_c}$, respectively.

The desired case, $\tau = 0$, is tried to achieve in tracking phase, which immediately follows acquisition. Tracking takes place during actual data transmission. In addition to the refinement of the estimate, tracking is also used to adjustment purposes in non-stationary environments. The most common tracking devise is phase-lock loop (PLL), which, with its delay estimate at hand, continuously correlates the received signal with two kinds of code sequences: the other corresponds to a slightly longer delay and the other equally amount shorter delay. The adjustment of the delay estimate is then obtained by balancing the outputs of those correlators.

In this section we emphasize the possibility of getting a closed form presentation for the timing offset using differentially coherent approach. With finite number of pilot symbols available it is thus possible to get a very accurate delay estimate already in acquistion phase.

9.3.6.1
The algorithm

We adopt the same assumptions as in section 9.3.5. Thus, the time lagged correlation matrix of the received and sampled data has the form of Eq. (28), i.e.

$$\mathbf{C} = E\{\mathbf{r}_m \mathbf{r}_{m+2}{}^T\} = \sum_{l=1}^{L} \alpha_l \mathbf{c}_{1l} \mathbf{c}_{1l}{}^T , \tag{49}$$

where $\mathbf{c}_{1l} \overset{\text{def}}{=} \underline{\mathbf{c}}_{1l} + \bar{\mathbf{c}}_{1l}$. For the sake of clarity, assume first flat fading channel, so that only one path exists with a delay $\chi = d + \delta$. The correlation matrix hence reduces to

$$\mathbf{C} = \alpha_1 \mathbf{c}_{11} \mathbf{c}_{11}{}^T . \tag{50}$$

Suppose standard matched filter is applied to $\mathbf{C}$. Then the shifted replicas of the desired user's code $\mathbf{g}(d)$ (see Eq. (41)) is projected to the data vectors, giving

$$\mathbf{g}(d)^T \mathbf{C} \mathbf{g}(d) = \alpha[(1-\delta)\eta_0 + \delta\eta_1]^2 \tag{51}$$

$$\mathbf{g}(d+1)^T \mathbf{C} \mathbf{g}(d+1) = \alpha[(1-\delta)\eta_1 + \delta\eta_0]^2 \tag{52}$$

where $\eta_i \overset{\text{def}}{=} \mathbf{g}(d)^T \mathbf{g}(d+i)$. From Eqs. (51) and (52) we have then

$$\delta = \frac{\eta_0 \mp P(d)\eta_1}{(\eta_0 - \eta_1)(1 \pm P(d))}, \tag{53}$$

where

$$P(d) = \sqrt{\frac{\mathbf{g}(d)^T \mathbf{C}\mathbf{g}(d)}{\mathbf{g}(d+1)^T \mathbf{C}\mathbf{g}(d+1)}}\,. \tag{54}$$

Eq. (53) is a closed form expression for the timing offset, since the values η_0 and η_1 are a-priori known. For m-sequences, they are C and -1, respectively. In addition, $P(d)$ can be measured from two successive outputs of differential correlator. The result tells also that in the acquisition, it is enough to use delay candidates spaced by one chip duration, not a half of a chip as conventional methods use. With the knowledge got by those delay candidates, the timing offset can be estimated according to Eq. (53).

As the PN code length is increased, it has more and more impulse-like autocorrelation. We can hence quite safely approximate $\eta_1 = 0$, which yields

$$\delta = \frac{1}{1 \pm P(d)}\,. \tag{55}$$

In frequency selective fading channel more paths exists. The correlation matrix $\mathbf{C}$ can be rewritten as

$$\begin{aligned}
\mathbf{C} &= \sum_{l=1}^{L} \alpha_l \mathbf{c}_{1l} \mathbf{c}_{1l}{}^T \\
&= \sum_{l=1}^{L} \alpha_l \left[(1-\delta_l)\mathbf{g}(d_l) + \delta_l \mathbf{g}(d_l+1)\right]\left[(1-\delta_l)\mathbf{g}(d_l) + \delta_l \mathbf{g}(d_l+1)\right]^T \\
&= \sum_{l=1}^{L} \alpha_l \left[(1-\delta_l)^2 \mathbf{g}(d_l)\mathbf{g}(d_l)^T + 2\delta_l(1-\delta_l)\mathbf{g}(d_l)\mathbf{g}(d_l+1)^T + \delta_l{}^2 \mathbf{g}(d_l+1)\mathbf{g}(d_l+1)^T\right]
\end{aligned} \tag{56}$$

Differential correlation then gives

$$\begin{aligned}
\mathbf{g}(d_1)^T \mathbf{C}\mathbf{g}(d_1) &= \alpha_1 \left[(1-\delta_1)\eta_0 + \delta_1 \eta_1\right]^2 + \sum_{l=2}^{L} \alpha_l \left[(1-\delta_l)\eta_{d_1-d_l} + \delta_1 \eta_{d_1-d_l-1}\right]^2 \\
\mathbf{g}(d_1+1)^T \mathbf{C}\mathbf{g}(d_1+1) &= \alpha_1 \left[(1-\delta_1)\eta_1 + \delta_1 \eta_0\right]^2 + \sum_{l=2}^{L} \alpha_l \left[(1-\delta_l)\eta_{d_1-d_l-1} + \delta_1 \eta_{d_1-d_l}\right]^2
\end{aligned} \tag{57}$$

Thus we are left with $2L$ unknowns and only two equations. To overcome that difficulty we could use more dense spacing for the test delays to get more equations. That, however, would complicate the expression. Instead, we again neglect $\eta_i, i > 0$, and obtain Eq. (55). The effect of this approximation is experimentally shown to be minor. The estimation procedure ADC (Accurate DC) is summarized in Table 9.3.1.

Table 9.3.1. ADC-algorithm for accurate delay estimation.

Denote Ω the set of delay candidates.

1. Set $\Omega = \{0,1,\ldots,C-1\}$.
2. For each $d \in \Omega$ do:
2.1. Compute $\hat{\delta}(d)$ according to Eq. (53) or (55).
2.2. Insert $d + \hat{\delta}(d)$ to Ω.
3. Obtain the delay estimate among Ω, which gives maximum to

$$\begin{cases} \left|\mathbf{g}(d)^T \hat{\mathbf{C}}\mathbf{g}(d)\right| \Big/ \left\|\mathbf{g}(d)\right\|^2 \\ \left|\mathbf{c}(d,\hat{\delta}(d))^T \hat{\mathbf{C}}\mathbf{c}(d,\hat{\delta}(d))\right| \Big/ \left\|\mathbf{c}(d,\hat{\delta}(d))\right\|^2 \end{cases}, d = 0,1,\ldots,C-1$$

where $\mathbf{g}(d)$ is of form (41), $\hat{\mathbf{C}}$ is estimated as (34), and $\mathbf{c}(d,\delta)$ is of form

$$\mathbf{c}(d,\delta) = (1-\delta)\mathbf{g}(d) + \delta\mathbf{g}(d+1).$$

9.3.6.2 Experiments

Since the goal is estimation accuracy, we measure the root-mean-square-error (RMSE) of the delay estimates. RMSE measures according to

$$\text{RMSE} = \sqrt{\frac{1}{S} \sum_{i=1}^{S} (\hat{d} - d)}, \tag{58}$$

where S is the number of experiments. In our case, $S = 500$ simulations. The simulation setting was the same as in section 9.3.5.4, including: $N = 500$ pilot symbols, $L = 3$ paths with unequal energies, $C = 31$ chips/symbol, average SNR of 10 dB with respect to single user's strongest path, and $K = 5,\ldots,25$ users. In the experiments we examined the accuracy as the system gets heavily loaded by varying the number of users. The same reference methods were used as earlier. Moreover, they used delay candidates with a step size of $1/100$ chips, in order to see what accuracy could be achieved by those methods. Only one chip spacing was used with ADC. Fig. 9.3.10 tells clearly the achieved accuracy. Due to the approximative nature of (55), ADC does not achieve exactly the level of WCD, but the computational savings are huge. In addition, the other methods are still outperformed.

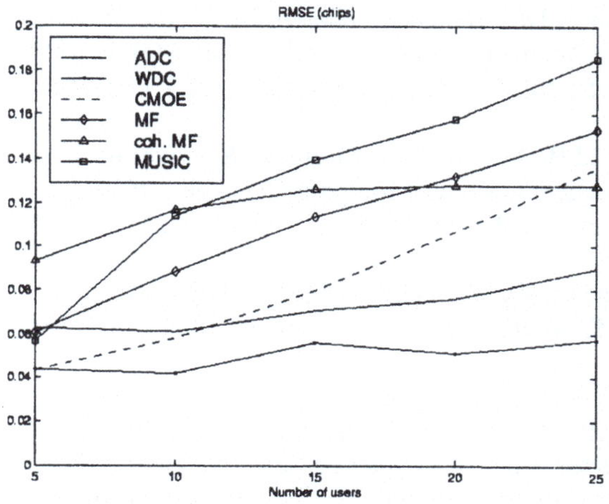

Fig. 9.3.10. RMSE (chips) as a function of number of users.

9.3.7 Conclusions

In this chapter, robust synchronization was studied. It was shown that when constant preamble is sent to the desired user, differential correlation operation efficiently
filters noise and interferers out while preserving desired user's information. Therefore, it is possible to estimate the desired user's delays even when huge number of interferers exists. The method can be applied when uplink contacts downlink at the beginning of the phone call. The price to be paid is that no data information can be sent to the desired user in the acquistion phase.

References

[1] S. Bensley and B. Aazhang, ``Subspace-Based Channel Estimation for Code Division Multiple Access Communication Systems". *IEEE Transactions on Communications*, vol. 44, no. 8, pp. 1009-1020, August 1996.

[2] S.E. Bensley and B. Aazhang, ``Maximum Likelihood Synchronization of a Single User for Code Division Multiple Access Communication Systems," *IEEE Trans. Communications*, vol. 46, no. 3, pp. 392-399, March 1998.

[3] C.-D. Chung, ``Differentially Coherent Detection Technique for Direct-Sequence Code Acquisition in a Rayleigh Fading Mobile Channel," *IEEE Trans. Communications*, vol. 43, no. 2/3/4, pp. 1116-1126, Feb./March/April 1995.

[4] U. Fawer and B. Aazhang, ``A Multiuser Receiver for Code Division Multiple Access Communications over Multipath Channels," *IEEE Transactions on Communications*, vol. 43, no. 2/3/4, pp. 1556-1565, February/March/April 1995.

[5] M.L. Honig, U. Madhow, and S. Verdu, ``Blind Adaptive Multiuser Detection," *IEEE Trans. Information Theory*, vol. 41, no. 4, pp. 944-960, July 1995.

[6] M. Latva-Aho, J. Lilleberg, J. Iinatti, and M. Juntti, ``CDMA Downlink Code Acquisition Performance in Frequency-Selective Fading Channels", *Proc. 1998 IEEE Int. Symp. Personal, Indoor, and Mobile Radio Communications.* (PIMRC'98), Boston, USA, September 8-11, 1998, pp.1476-1480.

[7] W. Lee, ``Overview of Cellular CDMA," *IEEE Transactions on Vehicular Technology}*, vol. 40, no. 2, pp. 291-302, May 1991.

[8] R. Lupas, S. Verdu, ``Near-Far Resistance of Multiuser Detectors in Asynchronous Channels," *IEEE Transations on Communications*, vol. COM-38, no. 4, pp. 496-508, March 1989.

[9] U. Madhow, ``Blind adaptive interference suppression for direct-sequence CDMA , *Proc. IEEE*, pp. 2049-2069, October 1998.

[10] F. Ovesjö at al., ``FRAMES Multiple Access Mode 2 -Wideband CDMA," Proc. 1997 IEEE Int. Symp. Personal, Indoor, and Mobile Radio Communications, (PIMRC'97), September 1-4, 1997, Helsinki, Finland, pp. 42-46.

[11] S. Parkvall, *Near-Far Resistant DS-CDMA Systems: Parameter Estimation and Data Detection*, Ph.D. Thesis, Royal Institute of Technology, October 1996.

[12] R. Pickholtz et al.,``Theory of Spread Spectrum Communications - A Tutorial," *IEEE Transactions on Communications*, vol. 30, no. 5, pp. 855-884, May 1982.

[13] J.G. Proakis, *Digital Communications*, third edition, McGraw-Hill, 1995.

[14] *Proceedings of the IEEE*. Special Issue on GPS, the Global Positioning System. January 1999.

[15] D.V. Sarwate, ``Acquisition of Direct-Sequence Spread-Sprectum Signals", in *Wireless Communications - TDMA versus CDMA*, S.G. Glisic and P.A. Leppänen (eds.), Kluwer Academic Publishers, 1997.

[16] R. Schmidt, ``Multiple Emitter Location and Signal Parameter Estimation", *IEEE Transactions on Antennas and Propagation*, AP-34, pp. 276-290, March 1986.

[17] E. Ström, S. Parkvall, S. Miller, and B. Ottersten, ``Propagation Delay Estimation in Asynchronous Direct-Sequence Code Division Multiple Access Systems", *IEEE Transactions on Communications*, vol. 44, pp. 84-93, January 1996.

[18] S. Verdú, ``Minimum Probability of Error for Asynchronous Gaussian Multiple-Access Channels", *IEEE Transactions on Information Theory*, vol. 32, no. 1, pp. 85-96, January 1986.

[19] S. Verdú, ``Optimum Multiuser Asymptotic Efficieny", *IEEE Transactions on Communications*, vol. COM-34, no. 9, pp. 890-897, September 1986.

[20] A. Viterbi, *CDMA - Principles of spread sprectum communications*, Addison-Wesley, 1995.

[21] M.H. Zarrabizadeh and E.S. Souza, ``A Differentially Coherent PN Code Acquisition Receiver for CDMA Systems," *IEEE Trans. Communications*, vol. 45, no. 11, pp. 1456-1465, November 1997.

[22] D. Zheng, J. Li, S.L. Miller, and E.G. Ström, ``An efficient code-timing estimator for DS-CDMA signals", *IEEE Trans. Sign. Proc.*, vol. 45, no. 1, pp. 82-89, Jan. 1997.

9.4 Network Synchronization in IMT-2000 CDMA Systems

Aleksandar M. Gogic

For my daughter Sofie, may her love of writing endure

9.4.1 Introduction

One of the essential requirements of digital communication systems is the provision of time alignment or synchronization between transmitter and receiver. In digital mobile communication systems, synchronization solutions are exacerbated due to such factors as mobility, multipath propagation, propagation delays, and handoffs. In CDMA systems, there are some unique synchronization issues. Since there is no channel separation by means of time or frequency, receiver must extract timing reference from the compound CDMA signal. Standard correlation techniques are used for this task, whereby receiver searches a known code sequence sent by the transmitter.

Synchronization problem in CDMA is further exacerbated by soft handoff, where the forward link consists of multiple base stations simultaneously transmitting the signal destined for a single mobile station receiver. The same channel data, albeit uniquely modulated, are transmitted on each of the multiple signal paths. For the mobile station to be able to soft combine the forward link signals, they must be time aligned in some fashion. At the very least, these signal components should arrive at the mobile station at approximately the same time, so that the mobile station memory requirements are minimized. This implies that base stations involved in soft handoff must coordinate transmission timing with each other, which can be accomplished in two different ways:

- By synchronizing all base stations to a common timing reference; This aligns base station transmit times to a common clock such as GPS.
- By using round trip propagation delay to provide base station transmission timing advance; This aligns mobile station receive times from multiple base stations, or strictly speaking, the first arriving multi-path components of signals destined to the mobile station.

The two IMT-2000 CDMA proposals, cdma2000 and WCDMA, respectively use each of these two methods.

It is worthy of note that a similar synchronization problem does not exist on the reverse link, since the mobile station signal is received by multiple base stations involved in the soft handoff, and signal reassembled by means of selective combining of signal frames.

In a CDMA system, be it synchronous or asynchronous, each base station must be uniquely identified by means of a code, so that the mobile station can separate signals between base stations. These same codes are also used for time synchronization on the forward link. Fig. 9.4.1 illustrates the cell code concept for cdma2000 and WCDMA. Cell codes are sometimes referred to as "pilot codes".

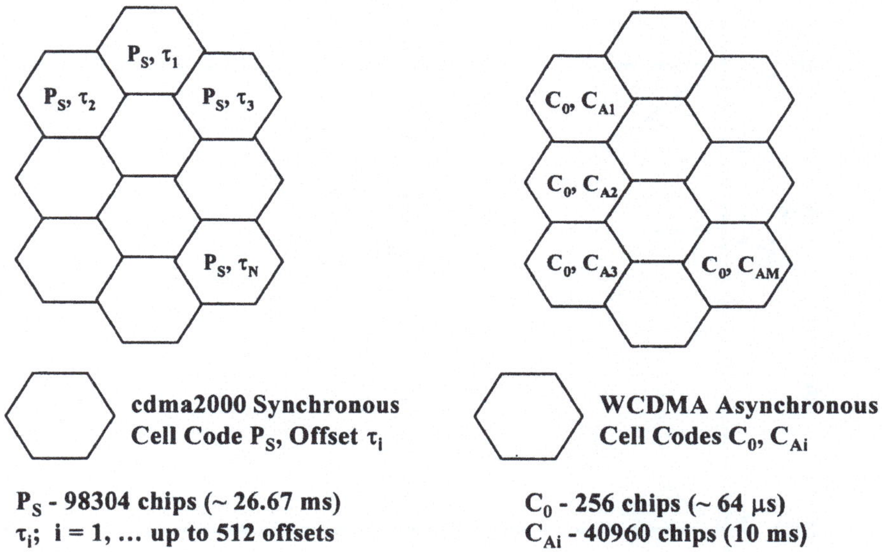

Fig. 9.4.1. cdma2000 and WCDMA Cell Codes

In cdma2000, there is only one unique pilot code labeled P_S in the Figure. Cell separation is accomplished by means of time offsets labeled τ_i. In WCDMA, each cell uses different code labeled C_{Ai}. In addition, a short bursty periodically transmitted code C_0 is introduced to facilitate timing acquisition.

9.4.2 cdma2000 Synchronization Method

Forward link transmission timing in cdma2000 is shown in Fig. 9.4.2. Transmissions from BS1 and BS2 are time synchronized, typically by means of GPS. Due to path length differences, signals arrive to the mobile with time difference ΔT:

$$\Delta T = \frac{d2 - d1}{c} \qquad \text{(Eq. 1)}$$

where c is the speed of light (~$3*10^8$ m/s). To estimate an order of magnitude for ΔT, assume cell radius of 5 km with soft handoff region consuming 30% of linear space between base stations. At extreme ends of the soft handoff region along a straight line connecting the base stations, ΔT amounts to (6500 m - 3500 m)/($3*10^8$ m/s) = 10 μs.

Due to multipath propagation various signal components arrive to the mobile at dispersed times. Fig. 9.4.2 depicts the first arriving components from each of the two base stations. As the mobile moves closer to the halfway point between the two base stations, the time difference gets reduced, as illustrated by $\Delta T'$.

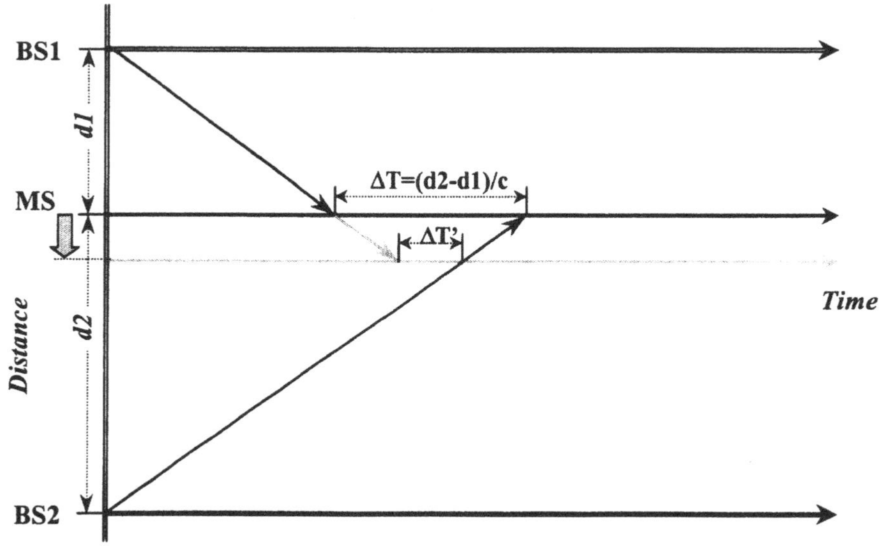

Fig. 9.4.2. cdma2000 Forward Link Transmission Timing

The timing reference in cdma2000 is provided by means of forward link pilot, as mentioned earlier. Pilot code sequence is common for all cells. What distinguishes one cell from the others is the time offset of this sequence, (see Fig. 9.4.3). These so-called PN offsets are labeled τ_1, τ_2, τ_3, … , and each corresponds to a different cell. The pilot code itself, i.e. the psudo-random (PN) sequence of chips, is the same for all base stations and has the length L_p.

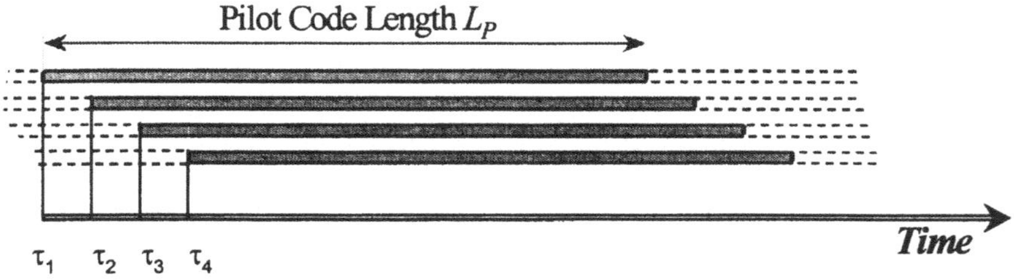

Fig. 9.4.3. cdma2000 Pilot Codes with PN Offsets

9.4.2.1 Initial Search - System Acquisition

To get initially synchronized with the system upon turning power on, the mobile station must acquire two things from the CDMA signal:

- Pilot sequence timing, i.e. the starting point of at least one base station pilot;
- Time offset of the pilot with respect to an absolute timing reference.

Since all cells use the same pilot code, completion of the first task amounts to pilot code search over a segment of length L_p. Code searching techniques are described in more detail later. For now, note that searching for the pilot code does not mean that code match over the entire code length is required. Searching is conducted by means of matching a section of the code. In the worst case, a time segment equal to L_p must be searched. After acquiring a pilot, mobile determines its PN offset τ_X by demodulating cdma2000 Sync Channel, which broadcasts an absolute time reference pegged to the UTC (Universal Coordinated Time).

9.4.2.2 Searching while on Dedicated Channel

When mobile is on call, it must continuously search its serving cell pilot (or multiple pilots if in soft handoff) to collect and coherently combine energy from all significant multipath components arriving to the receiver (see Fig. 9.4.4a). Each of these components is given proportionate weight in the Maximum Ratio Combining (MRC) scheme, as shown in Fig. 9.4.4b. Here, three multipath components of intensities P_1, P_2, and P_3, arriving at times θ_1, θ_2, and θ_3 respectively, are above the threshold P_0, while the fourth component P_4 arriving at time θ_4, is below the threshold and therefore ignored. MRC components are searched over a range of time offsets representing maximum expected dispersion of received signal. To aid in the search process, the base station broadcasts the size of the dispersion window, so that mobile would not needlessly search outside this window. Dispersion windows are measured or estimated in the process of the CDMA system design.

In addition to searching pilot codes of the base station(s) to which it is connected on dedicated channel, mobile must search and measure pilot strengths of neighboring base stations so that it can report to the infrastructure whenever any exceeds handoff threshold. For that purpose each base station broadcasts its neighbor list, informing mobiles which other pilots may be encountered.

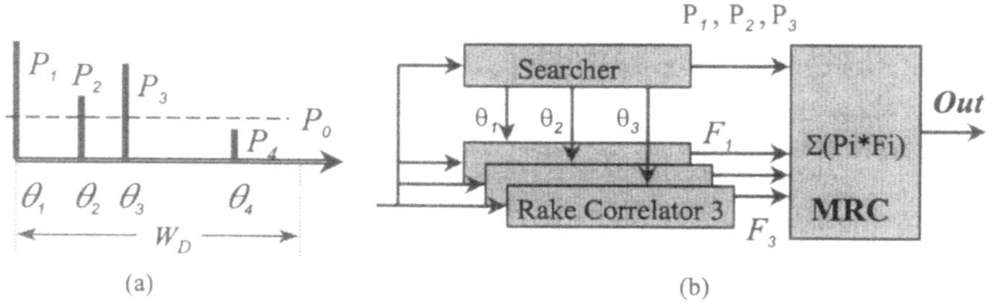

Fig. 9.4.4. (a) Multipath Dispersion; (b) Rake Receiver Structure with Maximum Ratio Combining (MRC)

A neighbor list entry in cdma2000 consists of pilot time offset (τ_i) and neighbor search window ($\mathbf{W}_i$). In contrast to dispersion window, neighbor search window ascribes propagation delay from a neighbor cell, and cell radius governs its upper bound. This can be easily seen from Fig. 9.4.2. Assume that BS1 is the currently serving base station and BS2 is a neighbor. Then Eq. 1 represents propagation delay difference between the serving cell and its neighbor, which should fall within the bounds of the neighbor search window for a good system design. The maximum value of ***d2*** becomes the "cell radius", i.e. the maximum distance from the base station at which it may be placed in the active set of demodulated base stations. When value of ***d2*** approaches the cell radius, and the value of ***d1*** is at its minimum, then the result of Eq. 1 yields the search window. The upper bound for the window is when ***d1*** can be neglected in comparison to ***d2*** , i.e. ***d1*** approaches 0. Eq. 1 reduces to Eq. 2, proving our assertion:

$$\Delta T_{MAX} \cong \frac{d2}{c} \qquad \text{(Eq. 2)}$$

9.4.2.3
Searching in Idle Mode

Yet another form of searching is in idle mode. While monitoring paging channel of the dominant cell, mobile must also measure neighbor pilot strengths to determine if it needs to perform idle handoff and start monitoring paging channel of a different cell. Subsequent to idle handoff, mobile may re-register with the infrastructure, e.g. if it enters a new paging zone. Note that the term "registration" in cdma2000 is equivalent to the term "location update" in GSM.

Idle mode searching can be of significant performance impact, since it affects mobile's stand-by time. The CDMA system normally operates in slotted mode, allowing mobiles to be in reduced power mode most of the time with the receiver chain and much of the rest of the circuitry turned off. The receiver is turned on only during a short time of its assigned time slot. The ratio of sleep and wake times in the mobile's sleep/wake cycle has significant impact on the stand-by time. The

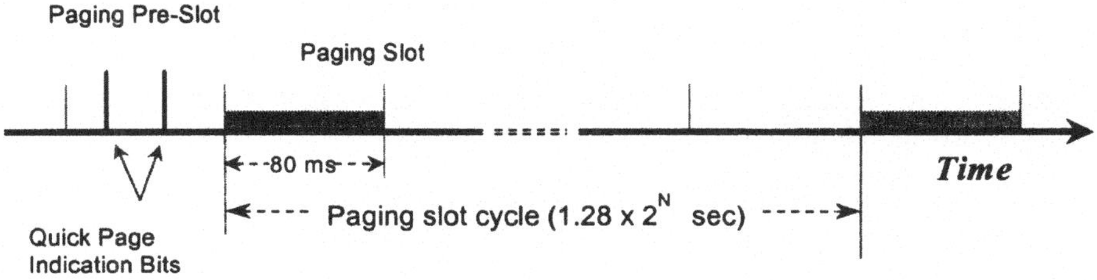

Fig. 9.4.5. cdma2000 Quick Paging Channel (QPCH)

cdma2000 Quick Paging Channel (QPCH) depicted in Fig. 9.4.5 is designed to optimize stand-by time because it minimizes the mobile station wake time.
QPCH allows the wake cycle to be reduced to QPCH single bit time. Quick paging channel consists of two page indicator bits transmitted in the slot preceding the paging time slot. Base station and mobile compute the two bit positions within the QPCH time slot with a hashing function using mobile's IMSI. The reason two bits are used, not one, is to improve diversity in fading and to reduce probability of false page detection by increasing number of possible bit combinations.

When there is a page for the mobile, the bits in the two positions corresponding to its IMSI are set by the base station. The mobile station wakes up and turns its receiver on at the exact time of the first of the two bits. If the bit is not set, the mobile station does not need to check the second bit or the paging slot itself. To improve the degree of confidence with which the mobile station can detect the single bit, the detection threshold can be set high. If the bit is set, or its value is ambiguous, the mobile station goes on to check the second bit in a similar fashion.

With this process the mobile station can reduce its wake time and still maintain low missed page probability. Wake time can be reduced to the first Quick Page Indication Bit in the Paging Pre-Slot whenever the first Quick Page Indication Bit indicates high likelihood of no page for this mobile. This is the most likely case, since the mobile receives pages relatively rarely and the incidence of page indicator bit collision with another IMSI is reasonably small. In those situations idle mode searching becomes the only wake time limitation.

Synchronous mode operation makes it possible for the mobile station receiver to know the states of pilot code phases of neighbor cells, save for the variance of neighbor search windows. It can be easily shown that wake cycle time can be reduced to the QPCH bit time. Since QPCH typically operates at 9600 bps, wake time can be as low as 1/9600 seconds, or 104 μs.

Idle mode search strategy is illustrated in Fig. 9.4.6. Two neighbor pilots are shown for illustrative purposes. Typically neighbor list is much larger. Neighbor pilot PN offsets are labeled τ_1 and τ_2 respectively. They are known to the mobile station, which can therefore compute the state of each of the pilot codes at any instance of time. At the time corresponding to the beginning of the QPCH indicator bit, those states are, at base station reference times, respectively:

$$Pilot_1(T_0) = P(T_0, \tau_1) \qquad \text{(Eq. 3)}$$
$$Pilot_2(T_0) = P(T_0, \tau_2) \qquad \text{(Eq. 4)}$$

However, at the mobile station reference time T_0 those states may vary in the range corresponding to the neighbor search window:

$$[Pilot_1(T_0), Pilot_1(T_0 - W_1)] = [P(T_0, \tau_1), P(T_0 - W_1, \tau_1)] \qquad \text{(Eq. 5)}$$
$$[Pilot_2(T_0), Pilot_2(T_0 - W_2)] = [P(T_0, \tau_2), P(T_0 - W_2, \tau_2)] \qquad \text{(Eq. 6)}$$

The search strategy can therefore be deduced as follows:

1. Always search over the same segment of time $[T_0, T_0 + T_B]$;
2. For each neighbor ***i***, search by correlating to the shifting segment of the pilot code, starting with segment $[P(T_0 - W_i, \tau_i), P(T_0 - W_i + T_B, \tau_i)]$, and

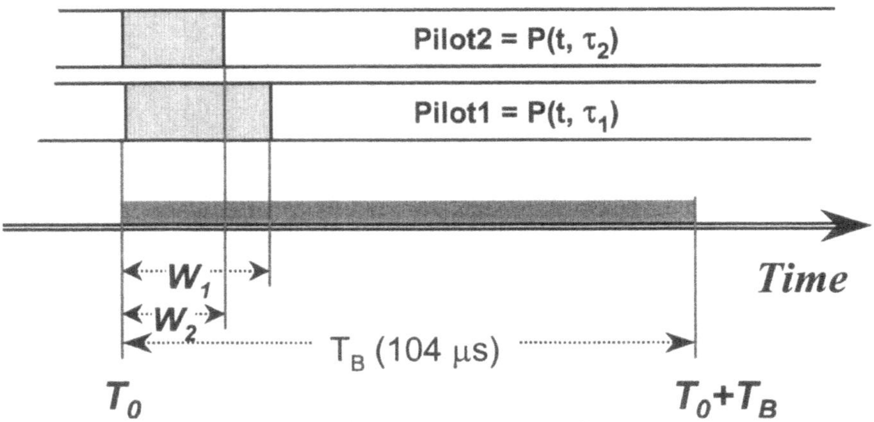

ending with segment $[P(T_0, \tau_i), P(T_0 + T_B, \tau_i)]$

Fig. 9.4.6. cdma2000 Idle Mode Search

The point (a) makes it possible for the mobile station to only turn its receiver on for the duration of the QPCH single bit time, or 104 µs.

It should be noted that the shifting neighbor search segment outlined in point (b) can typically be reduced considerably for some of the neighbors in the neighbor list in the mobile station practical implementation. This can be done by taking advantage of the knowledge of propagation delays acquired in the previous Quick Page time slot search. This is true for neighbors that are strong enough to have been detected in the previous cycle. Since the mobile station position change is rather small within the page cycle period, the search can start at or around the pilot phase computed from its previously measured position, thus maximizing rapidity of detection for those neighbors. For the remaining neighbors however, the entire segment described in (b) must be searched.

9.4.2.4
Summary

As we have seen from the above discussion, cdma2000 pilot structure allows for very simple searching, with the following key characteristics.

- The four kinds of searching: During initial system acquisition, for mutlipath components of its own dedicated channel, for pilots from the neighbor list, and in idle mode, are fundamentally all the same, allowing for a simple modular integrated circuit design with re-allocatable searcher resources;
- Searching in all modes, particularly in the critical handoff mode while on dedicated channel, and in idle mode, is confined to narrow windows associated with propagation delays. This improves search speed, which is crucial for overall system performance.

9.4.3
WCDMA Synchronization Method

Soft handoff forward link dedicated channel transmission timing for WCDMA system is shown in Fig. 9.4.7 BS2 uses timing advance for dedicated channel transmission, so that signals from BS1 and BS2 are time aligned when they arrive at the mobile station. Base stations use round-trip delay measurements to compute timing advance ΔT.

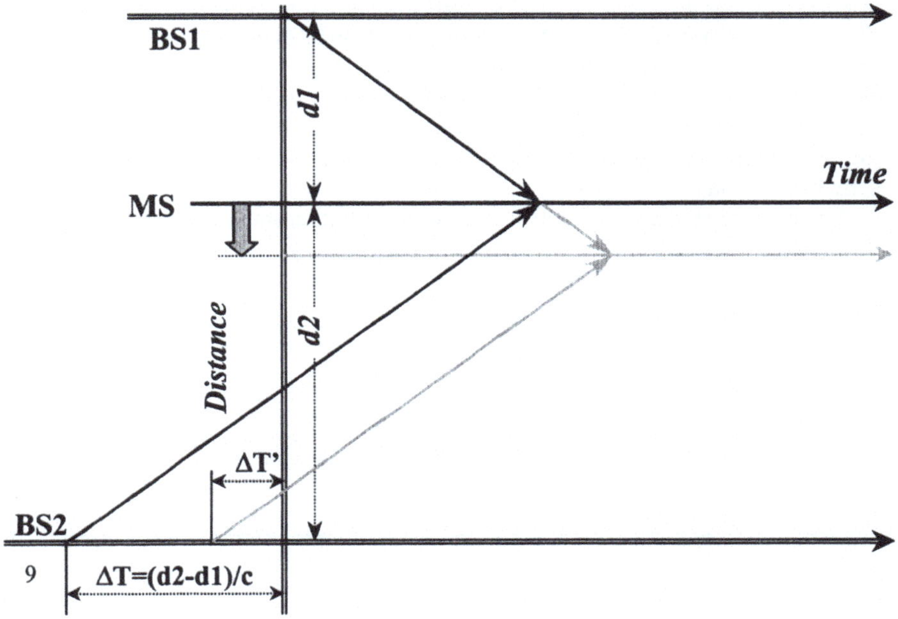

Fig. 9.4.7. WCDMA Forward Link Transmission Timing

In the multipath environment prevalent in mobile systems, only the first arriving signal components can be time aligned. All other components will be dispersed and will arrive later. This is an important point, since time aligning receive signals at the mobile station with the motivation of reducing its memory requirements, can yield only very limited results. Typical memory requirements for turbo or convolutional coders, interleavers, etc., are much higher than the memory hardware savings achievable by time alignment method.

As the mobile station moves and time differences change, the timing advance must also change. As illustrated in Figure 9.4.6, as the mobile station moves closer to the base station BS2, the timing advance gets reduced to ΔT'.

Timing for the common channels in each of the base stations is independent and may drift with respect to each other. This base station relative time drift is another reason why timing advance for the dedicated channels may have to be updated during the call.

9.4.3.1
Initial Search - System Acquisition

Simplified WCDMA forward link common channel structure is shown in Fig. 9.4.8. Upon application of power, mobile station acquires the base station timing from this structure. Primary Search Code SC_1, Secondary Search Code $SC_2^{k,j}$, Broadcast Control Channel (BCCH), and Common Pilot constitute physical entities of interest. Each of them may have different power levels, which is alluded to in the Figure. SC_1 is 256 chips long and comprises a common code for all the base stations. This code is transmitted once every 625 μs, the time slot period T_S which happens to correspond to WCDMA power control rate of 1,600 Hz. Secondary Search Code $SC_2^{k,j}$ is chosen from a set of 16 Hadamard Codes. In addition to identifying the base station scrambling code group (1 of 32 code groups, with 16 codes in each group), this code also provides frame synchronization (j = 1...16).

To accomplish the initial cell search, mobile station must execute the following three steps:

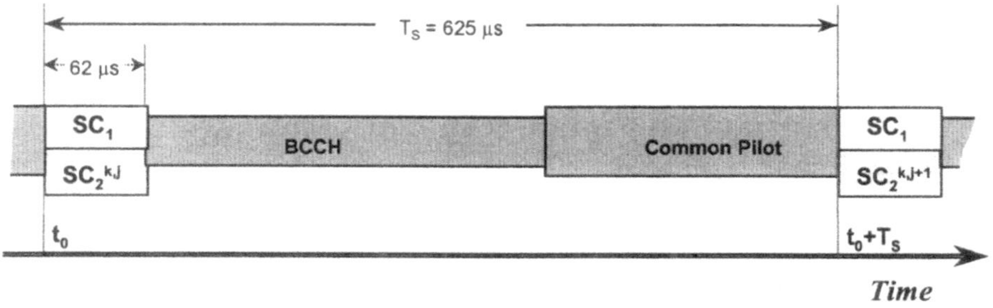

Fig. 9.4.8. WCDMA Forward Link Common Channel Structure

Step 1: Time-Slot and Symbol Timing Detection. Mobile station searches for primary search code SC_1 which is common to all base stations. The result of that search, which is typically conducted with a matched filter, confines further searching to the time-slot reference labeled in Fig. 9.4.8 as t_0, t_0+T_S, etc. It also identifies common channel modulation symbol timing, since the code SC_1 is within defined symbol bounds.

Step 2: Frame Synchronization and Code Group Identification. There are 16 slots within a frame, each identified in the secondary synchronization code $SC_2^{k,j}$. By trial-and-error searching of the secondary code over 32 different possible scrambling code group identifiers, mobile station can identify both the scrambling group and the slot sequence number within the frame (one of 16). Note that this is a different type of search, than in step 1, since mobile station now knows the timing of the code, but must search over a number of possible codes. The term "trial-and-error" search is used to distinguish from the step 1 search style, where mobile knows the code, just doesn't know the starting point (timing). The style of search in Step 1 incidentally is the style employed in cdma2000.

Step 3: Scrambling Code Identification. BCCH and the Common Pilot symbols are scrambled by means of the scrambling code, which now needs to be identified in full. This is accomplished by correlation trial-and-error process over possible codes within the code group (16 codes in each group). The process is similar to step 2, but now searching is conducted over BCCH and Common Pilot. Note that the mobile has in step 1 acquired the bit timing for BCCH and Common Pilot, and can confine search to bit duration, i.e. it can search over these channels even though they carry information not known to the mobile.

It should be noted that the more recent version of WCDMA employs code multiplexed common pilot, not the time multiplexed form shown in Fig. 9.4.8. Code multiplexed common pilot is transmitted simultaneously with BCCH. However this change is of no significant consequence for our discussion.

9.4.3.2 Searching while on Dedicated Channel

As in cdma2000, WCDMA mobile searches pilots of base stations in the active set to achieve coherent detection and perform MRC of signal components. Mobile can search the common pilot (Fig. 9.4.8), or dedicated pilot (Fig. 9.4.9). It may also search both forms of pilot and combine results, if they are both transmitted from the same base station antenna. The best performance is achieved if mobile uses both common and dedicated pilot, because more energy extracted from the pilot yields better estimation of these components and higher detection quality. This is true from the standpoint of detection quality, not overall performance of the system. Higher pilot energy has negative effects on base station power efficiency, and increases interference. Our interest at this time is the search performance, therefore these other issues will be set aside.

Both common and dedicated pilot searching is very similar to that of cdma2000, except that it is confined to time segments, per pilot structure. The search is conducted over a dispersion window, dimensioned to capture all significant multipath components. Where the two CDMA specifications differ significantly is in the area of neighbor searching. Because WCDMA is an asynchronous system, mobile will not know the timing of the neighbor base stations, and must repeat the search steps outlined in the initial system acquisition. The only aid is that it knows potential neighbor scrambling codes and code groups, so it may eliminate some of the trial-and-error attempts of steps 2 and 3. The neighbor list is provided to the mobile station by the infrastructure.

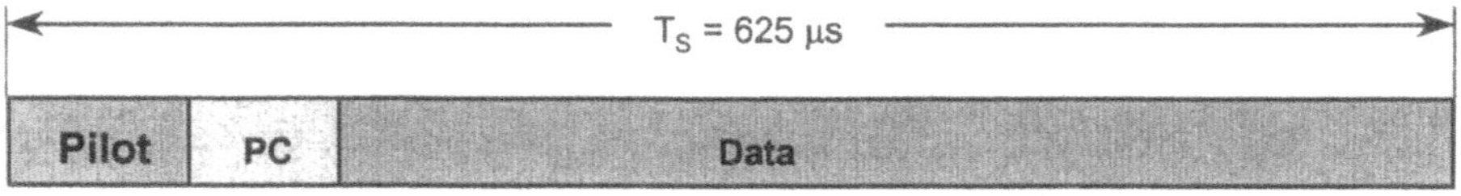

Fig. 9.4.9. WCDMA Forward Dedicated Channel

9.4.3.3 Searching in Idle Mode

WCDMA specification also includes slotted mode paging concept. Mobile station's sleep/wake cycle is governed by the common channel timing of the monitored cell. As explained earlier, mobile measures pilot strength of neighbor base stations. The timing of neighbor cells is not synchronized with the current one. It follows that in the idle mode too mobile must perform the three step search of neighbors to acquire and measure their common pilots. To accomplish this, mobile station receiver must be turned on for at least the duration of one slot cycle (625 μs) to ensure capture of the primary sync code $SC_1$from the neighbor cells. In addition, an intense search effort is required to identify neighbors, decide their timing, and measure their pilot power.

We also can make an observation that the minimum wake time of one slot (625 μs) and search efforts in idle mode are both consequences of common channel structure in WCDMA shown in Fig. 9.4.8. They are not a direct consequence of asynchronous network operation, but an indirect one.

9.4.3.4 Summary

WCDMA structure is a very elegant and it is an effective solution for CDMA network required to operate without an external common time reference. The short primary sync code provides a heartbeat of each base station, on which mobiles can rely to get synchronized to the network. This heartbeat can be used to quickly and effectively acquire system timing, albeit with relatively complex search circuitry for the detection of the primary sync code.

WCDMA searching is largely driven by the asynchronous network operation requirements and consequent forward link structure. Acquisition of the time reference contained in the primary search code SC_1 requires considerable mobile search resources in the mobile station. When this reference is available, such as when searching for multipath components of the dedicated channel, searching effort is comparable to cdma2000. However, when searching for neighbor cells in dedicated mode and idle mode, this reference is not available, and search cycle similar to initial system acquisition is required.

9.4.4
Comparison between Synchronization Methods

9.4.4.1
Searching Performance

As observed in the earlier discussion, the most critical search operation is while mobile station is on a Dedicated Channel, i.e. while transmitting and receiving data. Since mobile station may move, it must continuously search for neighbor pilots, measure their relative strength, and report to the infrastructure if any exceed handoff threshold. This search operation is usually referred to as "handoff search". The handoff search operation is time critical because of compound effects in the following chain reaction:

- Inefficient handoff search results in excessive delays in reporting strong pilot surges;
- This may result in handoff delays;
- Slow handoffs result in increased forward link interference caused by non-demodulated pilots;
- To cope with higher forward link interference, base station(s) on existing link(s) transmit more power on the dedicated channel for the mobile in question;
- In addition to increased interference to all other mobiles, which in turn demand more power on *their* dedicated channels,
- ... this results in lower base station power efficiency and therefore lower system capacity.
- Handoff delays also adversely affects the reverse link by maintaining higher mobile station transmit power due to delayed application of soft handoff gain;
- The reverse link interference at the base station receiver is increased;
- The reverse link system capacity is therefore diminished.

In summary, effects of sluggish handoffs are compounded into potentially a significant loss of system capacity on both the forward and the reverse link. Additionally, sluggish handoffs will likely lead to increased call drop rate.

Initial Search

The three step search strategy for the initial search in WCDMA is very structured, driven by the structure of the forward link in that system, and holds a promise of good performance in terms of initial system acquisition speed. This is necessarily so, since that same strategy, with some simplifications, is applied in the handoff search, which is much more critical and demands higher performance in comparison to the initial cell search. Search performance data from the field in the multi-cell multipath environment with fading are at this time lacking in the public domain. As WCDMA field trials of larger multi-cell systems get on their way, more test data from the field deployments will become available. This is important step in proving out the WCDMA system, since issues associated with searching are extremely complex for a rigorous mathematical analysis.

Taking a closer look at the initial search performance, one can observe that the matched filer type search of WCDMA Step 1 can also be applied on cdma2000 structure. Mobile station would simply apply the matched filter operation on a sub-segment of the pilot code, as shown in Fig. 9.4.10. Moreover, the cdma2000 structure is more flexible, because the sub-segment is not constrained to the length of the SG_1 code, which is fixed in WCDMA. As the searcher capability and IC logic density and speed increase in the future, the segment can increase, resulting in even better performance. This is aptly illustrated in Fig. 9.4.10.

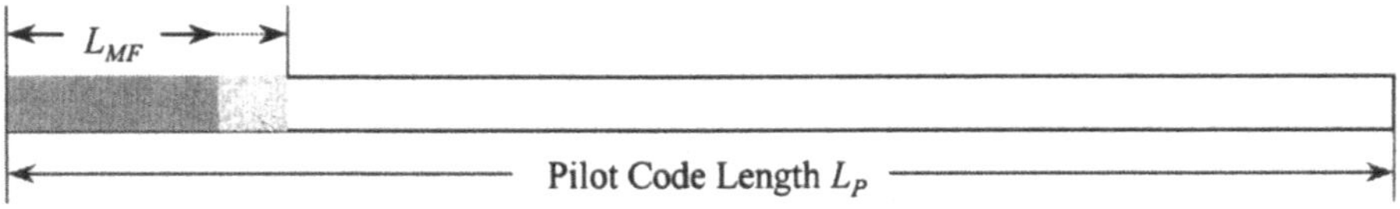

Fig. 9.4.10. Segmented Search Concept in cdma2000

Flexibility of the cdma2000 structure in terms of searching strategy implementation can be exploited even with the current state of the art in logic and DSP processor design. The following paragraphs describe how the search process might be implemented.

Analysis of Initial Cell Search Strategy

To analyze initial search strategy, we introduce the following terminology:

Correlation Segment: Segment of the pilot code used to detect a pilot (corresponds to L_{MF} in Fig. 9.4.10);

Search Segment: Signal segment over which search is conducted, which may designate a section of the signal being used in a search phase.

In cdma2000 cells are grouped in clusters, each consisting of a set of unique pilot code offsets (see Fig. 9.4.1 and Fig. 9.4.3). The largest possible cluster has 512 cells, and corresponds to the value of 1 of cdma2000 parameter PILOT_INC. Systems with smaller clusters can be designed with PILOT_INC set to higher values, with correspondingly smaller cluster sizes. Table 1 lists cell cluster size and minimum pilot shift within a cluster, for various values of PILOT_INC.

Mobile station can exploit this structure when performing initial system acquisition utilizing search method illustrated in Fig. 9.4.10. As stated before, the cdma2000 mobile station has the option of choosing the *correlation segment* length L_{MF} judiciously, in accordance with its searcher speed attributes, rather than being constrained by the common code length (256 chips), as is the case in WCDMA. The mobile station can even employ several different *correlation segment* lengths (L_{MF1} , L_{MF2} , etc.), or deploy varying segment lengths in different phases of the search, for the purpose of optimizing the search speed, as we shall see from the discussion in the forthcoming text.

A prudent initial search strategy would be to search (correlate over) a *search segment* that is long enough to reveal pilot code components from a number of surrounding cells, even if they are very weak. Searcher would make an initial pass over a *search segment* of the received signal with the total length that is smaller than L_P, but with the *correlation segment* that, depending on the mobile's searcher capability and speed, can be much larger than what WCDMA would permit (256 chips). Owing to the increased *correlation segment* length (i.e. larger processing gain), this first pass could reveal pilot components that are quite weak, arriving from more distant cell, not just a dominant cell seen by the mobile station at its given location. Hence the likelihood of detecting these components early in the search process would increase.

Table 9.4.1. Cdma2000 pilot code cluster sizes and time shift increments

PILOT_INC	Cluster Size	Incremental pilot code shift
1	512	52 μs
2	256	104 μs
3	128	208 μs
4	64	417 μs
5	32	833 μs
N	2^{10-N}	$26.67 * 2^{N-10}$ ms

Without delving too deeply into a rigorous statistical analysis, we examine some search cases with the purpose of illustrating the search methodology we are proposing. Referring to Table 1, assume that the system is designed with PILOT_INC value of 4 (cluster size of 64 cells).

Case 1: Strong Dominant Pilot. Assume for a moment that the mobile station is in a location close to a base station and therefore can receive very strong dominant pilot. From Table 1 we can deduce that in a 417 μs long *search segment* of the received signal the probability is 1 in 64 of finding the dominant pilot with a very short *correlation segment* length. Since the signal to interference ratio in this case is very high due to mobile's favorable location, a very short *correlation segment* length may be sufficient for finding the dominant pilot component.

Extending that analytical reasoning, we can conclude that the probability is 1 in 32 of detecting the dominant pilot if the *search segment* is 833 μs, with the remaining conditions unchanged, and so on. We observe that this probability is small for the *search segment* cases given, but the search speed is good, since the *correlation segment* is short, and searcher can cover a lot of ground in a short period of time.

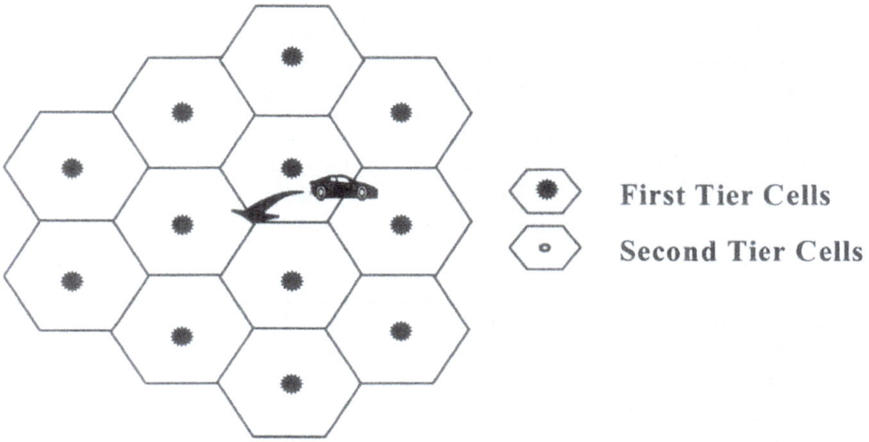

Fig. 9.4.11. First and Second Cell Tiers

Case 2: Absence of Strong Dominant Pilot. In this case the mobile station is located in an area where two or more base station signals are nearly evenly balanced, so that there is not a dominant pilot. This case is illustrated in Fig. 9.4.11. Since the mobile station is far away from any one base station, the received signal levels from the nearby first tier cells are relatively weak. Therefore signal to noise ratios from the second tier cells (more distant ones) are considerably higher than they were in case 1. There may be as many as 9 second tier cells, as illustrated in Fig. 9.4.11 for a simple omni-directional antenna configuration shown.

Let us assume that the mobile is located near a three-way cell junction, so that the three pilots from the first tier of cells are approximately equally strong. Referring again to Table 1, we deduce that the probability is 3 in 64, or 4.7 %, that in a 417 μs long *search segment* a pilot would be detected using a moderately long *correlation segment* size. For the case of an 833 μs long *search segment*, that probability is 3 in 32, or 9.4 %.

Extending this line of reasoning to the second tier cells, the probability is 12 in 64, or 18.8 % that with a long correlation segment a pilot would be detected in a 417 μs *search segment.* Note that both first and second tier cells could be detected using the long *correlation segment*, hence the probability calculation is 3 first tier cells plus 9 second tier cells equal 12 candidate cells out of 64 in a cluster. For 833 μs long *search segment* the probability is 12 in 32, or 37.5 %, and so on. Findings from this simplified probability analysis are summarized in Table 2.

The last examined case, i.e. probability of finding the second tier cell pilots by means of long *correlation segments*, holds the real promise of excellent initial search performance in cdma2000. By a small extension of *the search segment* to just over 3 ms, the probability of pilot detection becomes a certainty, with the qualification that it applies for this particular mobile station location.

Table 2 summarizes the statistics for each of the two examined cases of mobile locations: dominant cell and no dominant cell, broken down to two *correlation segment* lengths in case of no dominant cell. This table already provides a hint of the search strategy in cdma2000, which we will explore in more detail next.

Table 9.4.2. Pilot search probability analysis for cell cluster size 64

Search segment	Search segment	Dominant cell	No dominant cell	No dominant cell
Time	Chips	Short *correlation segment*, e.g. 64 chips	First tier cell, medium *correlation segment*, e.g. 256 chips	Second Tier cell, long *correlation segment*, e.g. 1000 chips
417 μs	1,536	1.6 %	4.7 %	18.8 %
833 μs	3,072	3.1 %	9.4 %	37.5 %
1.67 ms	6,144	6.3 %	18.8 %	75.0 %
3.33 ms	12,288	12.5 %	37.5 %	100 %
6.66 ms	24,576	25 %	75.0 %	100 %

Let us introduce a new term E_S/N_0, (pronounced "ess-no"), or energy of the search filter output relative to noise power spectral density. The other relevant term E_C/N_0 is the conventional one - pilot chip energy to interference power spectral density, which mobile station can readily measure.

From these two quantities one can compute the required search filter integration time. Expressed as M chip periods, this value is obtained as anti-log of E_S/N_0 - E_C/N_0 per Eq. 7, where E_S/N_0 and E_C/N_0 are given in the logarithmic scale.

$$M = 10^{\frac{E_S/N_0 - E_C/N_0}{10}} \qquad \text{(Eq. 7)}$$

Let us take a numerical example to evaluate the order of magnitude of M. Let us assume that the pilot consumes 20 % of the base station power, so that the pilot E_C/N_0 measured by the mobile station is around -7 dB for the dominant pilot case. Let us assume that the required E_S/N_0 for a reliable detection of the pilot is 6 dB. It follows from equation ES, the filter integration period must be at least M = 20 chips. So, with only 20 chips long *correlation segment* the mobile station can detect a strong dominant pilot.

Conversely, one can compute the minimum detectable pilot E_C/N_0 given the *correlation segment* length. If we choose the notation for the *correlation segment* lengths as N_{M1} for the short one, N_{M2} for the medium one, and N_{M3} for the long one, the minimum detectable pilot E_C/N_0 for each of the three cases, assuming E_S/N_0 of 6 dB are given in Table 3. Selecting N_{M3} value of 1000 chips, the detectable second tier or third tier pilots of - 24 dB are achievable.

We can carry our statistical analysis of initial cell search a little further by evaluating relative computational complexity with each of the three values of N_{Mi}. Using single chip integration as a unit, equivalent to a *multiply* and an *add* operation, the number of operations for a search over the segment L using *correlation segment* of M chips is equal to M*L. If the probability of pilot detection within that segment is P1, since searching over additional *search segments* is statistically independent, the total search effort required is N*L/P1. Using Table 2, we compute the search effort for several hypotheses. Several important results of the analysis are summarized in Table 4. First, it indicates that, if there is a dominant pilot, the search effort is considerably lower than for the case of no dominant pilot. This makes intuitive sense, because the favorable signal to noise ratio makes pilot searching faster. Secondly, for the case of no dominant pilot, whether we use shorter (256 chips) or longer (1000 chips) correlation segment, does not significantly affect detection speed. Based on the analysis and Table 4, we summarize and refine the initial search strategy as follows.

Table 9.4.3. Minimum detectable pilot E_C/N_0 given matched filter E_S/N_0 of 6 dB

Correlation Segment Length (chips)	Correlation Segment Notation	Minimum Detectable Pilot E_C/N_0
20	N_{M1}	-7 dB
256	N_{M2}	-18 dB
1000	N_{M3}	-24 dB

Table 9.4.4. Pilot search computational complexity

Receive Signal Strength	Dominant cell	No dominant cell, 1st tier	No dominant cell, 2nd tier
Correlation segment (chips)	20	256	1000
Search Effort (add-multiply)	2.00e6	8.54e6	8.34e6

- Search using a short *correlation segment* over a long *search segment*;
- Simultaneously, search using one or multiple longer *correlation segments* over a shorter *search segment.*

Reduction of *search segment* length comes as a natural consequence of increasing *correlation segment* length, since increased correlation length requires more search effort, and therefore takes longer for the given circuit complexity.

As soon as the mobile station detects one of the pilots, it can expand the search around that code offset, streamlining further search. If a pilot is first detected using the short correlation segment, the mobile has actually detected the dominant pilot, because the short segment would permit detection only of a pilot with good E_C/N_0. Note that searching actually yields the value of E_C/N_0 as a by-product. The longer the *correlation segment*, the more accurate (the better the resolution) of this measurement. If a pilot is first detected using one of the longer correlation lengths, and the E_C/N_0 measurement indicates a strong pilot, chances are the mobile station searcher has stumbled upon the dominant pilot. Further search effort should center around that pilot phase offset, to evaluate all its multipath components prior to demodulating the Sync Channel to determine the PN offset.

Let us take a closer look at some search strategies around the case that a weak pilot is detected using a long *correlation segment*, e.g. using the correlation length of 1000 chips, which can measure pilot traces down to - 24 dB. Clearly, this is not likely to be the dominant pilot, so the mobile station continues to search further. Fig. 9.4.12 depicts schematically as phase diagram the pilot PN offset cluster for the network, as the time offset distance between a given pilot PN offset and all its counterparts. We use the detected weak pilot PN offset as a reference phase and label it PN_0. If PN_0 phase is depicted in the 12-o'clock phase position, then the furthest away PN offset in the cluster is exactly ½ of the phase space around it, or schematically at the 6 o'clock position on this "Pilot Clock". It is readily evident that for no matter what the value of PILOT_INC, this particular phase offset belongs to the cell cluster for the network. Therefore, phase space around it is worth searching, if only to find that this is not the sought value and eliminate it from further search space. Since upon detecting PN_0 the mobile is expecting to detect a stronger pilot, it may search with smaller correlation length, i.e. NM_2.

If the result is less than desirable, i.e. there was no significant pilot energy detected around the 6 o'clock phase space, this process can continue in an analogous fashion at around 3 o'clock and 9 o'clock phase offsets, and so on, until a strong pilot is found. One can see how this process can yield very quickly the strong pilot, and in the process measure many of the neighbor pilot strengths.

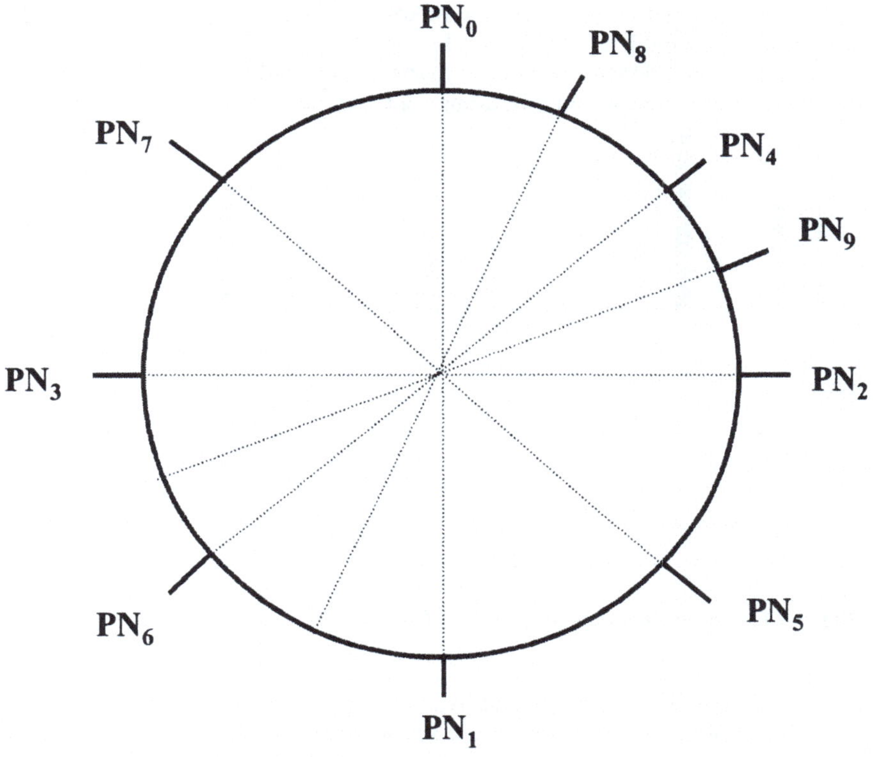

Fig. 9.4.12. Optimum Order of PN Phase Searching upon Initial Detection

In Fig. 9.4.12 PN offsets are enumerated in order in which they should be searched.

To summarize, the cdma2000 pilot segment length L_P is 26.67 ms, which is more than 40 times longer than the WCDMA slot time of 0.625 ms. WCDMA pilot structure is optimized for fast initial acquisition, albeit at some cost due to transmitting synch codes by the base station. However, as seen from the initial search strategy shown, the search space for cdma2000 is much reduced.

Neighbor Searching

Neighbor searching in cdma2000 is essentially the same as when searching its own base station pilot, except that the mobile searches over the neighbor search window at and around PN offsets corresponding to values given in the neighbor list. In WCDMA on the other hand, neighbor searching is akin to initial system acquisition. An extenuating circumstance is that there is a reduced number of search hypotheses in steps 2 and 3, bound by the neighbor list size.

While initial cell search operation of several seconds is permissible for system acquisition, it would not be acceptable in many neighbor searching cases. C/I environment in CDMA is very dynamic, especially in handoff zones. For example, inside multiple story buildings, a movement of several steps can result in

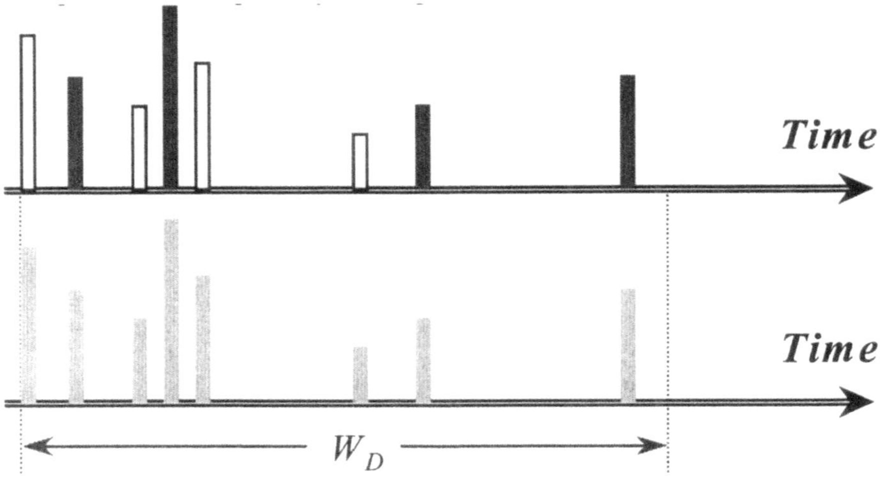

Fig. 9.4.13. Multipath Ambiguity in WCDMA

a drastic change in pilot ratios and handoff requests in quick succession. Handoff delays of several seconds would cause severe system performance degradation.

An additional complication is associated with SC_1 detection. It is plausible that SC_1 codes from a neighbor and own base station, or from two or more neighbors, arrive to the mobile station at almost the same time. Primary code detection in this case can create ambiguity between hypotheses, as shown in Fig. 9.4.13. Upon detection of SC_1 appearances, the mobile station will not know from which base station they stem. This means that the Search Step 2 and 3 hypotheses must be applied for each possible neighbor at each significant multipath component that has been detected in Step 1. This ambiguity exists in neighbor searching on the dedicated channel and in idle mode.

9.4.4.2 Soft Handoff on Common Channels

Possibility to soft combine common channels on the forward link is a distinct advantage of synchronous CDMA systems. In a system such as cdma2000, the Possibility to soft combine common channels on the forward link is a distinct advantage of synchronous CDMA systems. In a system such as cdma2000, the infrastructure can send common channel messages from multiple base stations belonging to a defined cluster. For that purpose, the mobile station maintains Virtual Active Set (VAS) of base stations. VAS is a set of forward link pilots that are received by the mobile station above a certain threshold, intersected with a set of pilots belonging to a page group, as illustrated in Fig. 9.4.14.

The mobile can soft-combine common channels from all the base stations listed in the mobile station's VAS. An opportunity to soft-combine common channels can be thought of as a form of macro diversity on the forward link for a mobile station in idle mode.

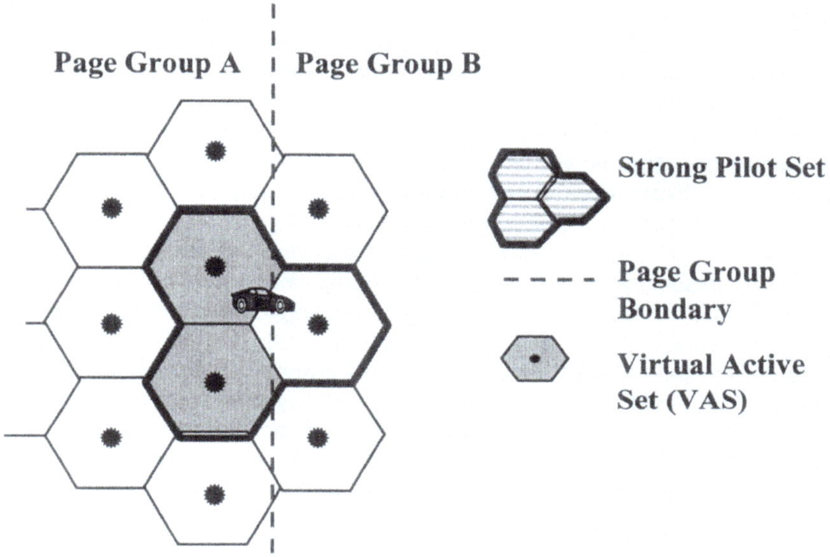

Fig. 9.4.14. Virtual Active Set (VAS) Genesis

To illustrate its importance, consider the ever increasing migration of voice and other forms of communications to their wireless form, otherwise known in the telecom world as "minute migration", i.e. increase in the percentage of minutes of use of the telecommunication networks realized through radio links. As that process is taking place, an increase is occurring in percentage of mobile users inside buildings. This trend will accelerate as data services become more optimized for wireless medium. Yet another important factor of relevance for the point about to be made, is increase in the dynamics of wireless communications. By that we mean the advent of short bursty types of information exchanges, such as SMS, personal calendar assistance updates, broadcast headlines, geographic position location updates etc, as well as control state transition messages invisible to the end user, which are associated with packet switched applications. All this means that there will be a lot more overhead messages such as page messages, MAC control messages, etc. being sent on common control channels.

With this foreword in mind, we conclude that it is of paramount importance to achieve reliable forward link common channel performance, with the particular attention to be given to indoor users. Missed page messages, unreliable MAC control messages, etc. will translate into system performance deterioration and ultimately into revenue loss for the network operator.

With CDMA frequency reuse of one, the number of cells that can penetrate with significant power inside tall buildings can be considerable, leading to a problem colloquially referred to as "pilot pollution". The term usually refers to a situation on the dedicated channel, where interference from non-demodulated cells due to mobile receiver's rake finger limitation, is significant, causing call quality problems and sometimes dropped calls. Although this conventional meaning of pilot pollution is the most visible one to an end user, the analogous case while in the idle mode can be quite devastating for the network performance and user satisfaction. Fig. 9.4.15 shows the deterioration in the common signal E_C/I_0 when one the mobile station receives more "polluting" pilots of considerable power. The diagram assumes interfering base station signals received at exactly the same power as the one being demodulated. Although this exact scenario is unlikely, even worse situations could be quite common due to possible frequent deep fades of the base station signal.

It is evident from the diagram that in the case of no soft handoff on the common channel, as the number of interfering cells increases, the E_C/I_0 for the common channel decreases considerably. For a single interferer of equal power as the useful signal, the drop is 3 dB. A system with the ability to soft-combine common channels would have an improved common fade margin by the same amount. Additionally, if all mobile stations in the system were capable of common channel soft combining, the transmit power of the common channel could be reduced, improving power and spectrum efficiency of the system.

To summarize, key benefits of common channel soft handoff are:

- Base station common channel transmit power can be lowered, taking advantage of soft handoff gain, thus improving power economy and system capacity.
- Immunity to fading and message reliability is increased. Page reliability also results in reduction of paging channel load.

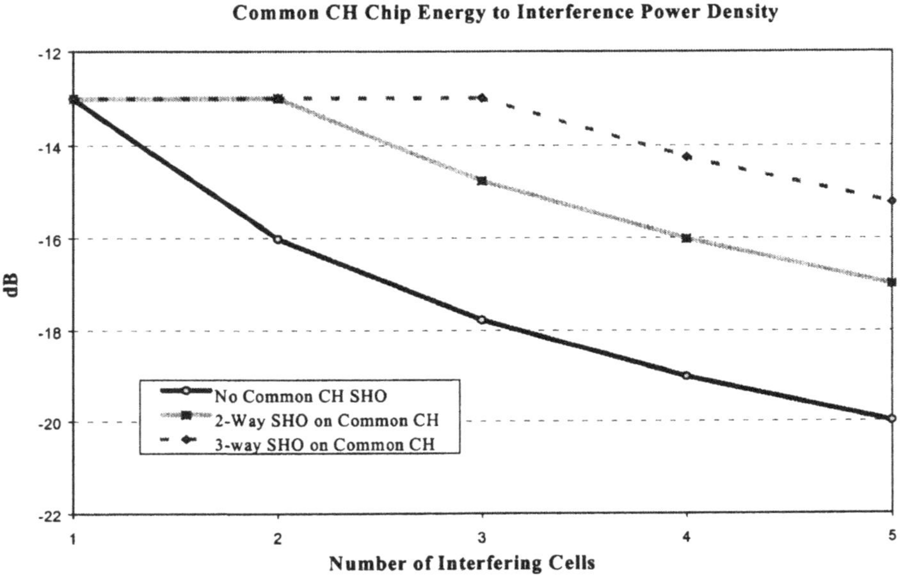

Fig. 9.4.15. „Pilot Pollution“ on the Common Channel without Soft Handoff

9.4.4.3 Battery Life

As already pointed out previously, sleep/wake cycle of idle mode has a significant effect on mobile station standby time. The wake cycle time has a direct bearing on and is the single most important factor of standby time performance. Wake time can be reduced if pages are time aligned, achievable in a synchronous CDMA system. In asynchronous mode, time alignment of pages is impossible, because pages are sent on a common channel.

Synchronous mode allows further improvement in stand-by time by means of Quick Paging Channel illustrated in Fig. 9.4.5, and further elaborated in Fig. 9.4.16 which depict scenarios with synchronous and asynchronous base stations.

As observed from Figure 9.4.16, the quick paging indicators are time aligned in a synchronous system (BS-0 and BS-1), and can be soft combined, improving detection performance in fading. If quick paging were to be implemented in an asynchronous system, the other base station timing can be either late (BS-2) or early (BS-3) relative to the reference base station monitored by the mobile station. The performance would be particularly poor in case of early timing, illustrated by BS-3 in Figure 9.4.11. Let us assume that the base station 0 went into a temporary deep fade, so that the mobile station missed its page. Due to the early timing of BS-3, the page was also missed, which will result in missing the entire paging cycle, creating call setup delay, or missed repeat page possibility.

Coupled with the benefits of soft combining on the common channel, the Quick Paging performance promises to have good reliability and should improve mobile station's standby mode power economy considerably.

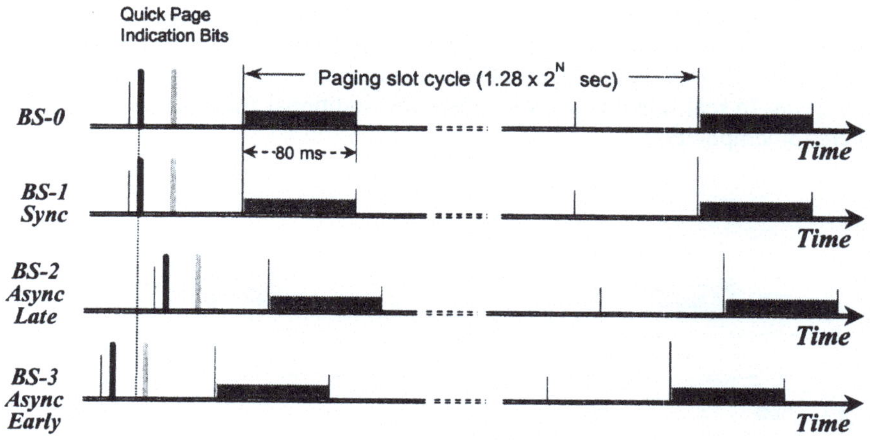

Fig. 9.4.16. Quick Paging in Synchronous and Asynchronous CDMA System

9.4.4.4
MAC Performance

One of the key features of IMT-2000 CDMA wireless systems is a sophisticated Medium Access Control (MAC) for packet radio link management. A well designed MAC makes it possible to schedule data packet transmission among multiple competing mobiles in an efficient manner. Full description of MAC is outside the scope of this chapter. Suffice it to say that poorly designed MAC and packet scheduling mechanism that drives it would result in inefficient utilization of radio interface resources, even if the physical layer is extremely efficient. Radio packet data throughput would suffer as a result.

An illustration of a well-designed MAC is provided in Figure 9.4.12. For the following discussion, we can focus on the forward link, which is probably more critical from the standpoint of performance, particularly for Internet applications. However, observations we are about to make are equally applicable for the reverse link.

Let us assume a base station with 3 wireless packet modems each operating at the rate of 384 kb/s. Each packet is destined for a different mobile station in the short sub-second period of time span depicted. The frame size of 20 ms is assumed. Maximum payload in bytes is indicated in the first frame of every transmitted packet.

In an efficient MAC and packet scheduling design such as the one shown, high rate wireless modems operate continually at network peak load times. Packets are scheduled in such a way that there is no transmission gap between the end of one packet destined for one mobile station and the next one destined for next mobile station.

Fig. 9.4.18 is a simplified timing diagram for the three mobiles served by a base station modem. The three mobiles are shown to be receiving 2, 1, and 3 frames respectively. Before each of the mobile stations involved gets its turn to receive data, it is brought up to a control state labeled "Queued Up State" in the simplified diagram in the Figure. Other states shown are the Active and the Stand-by State. In real implementations, there are additional transitional states, omitted here for simplicity.

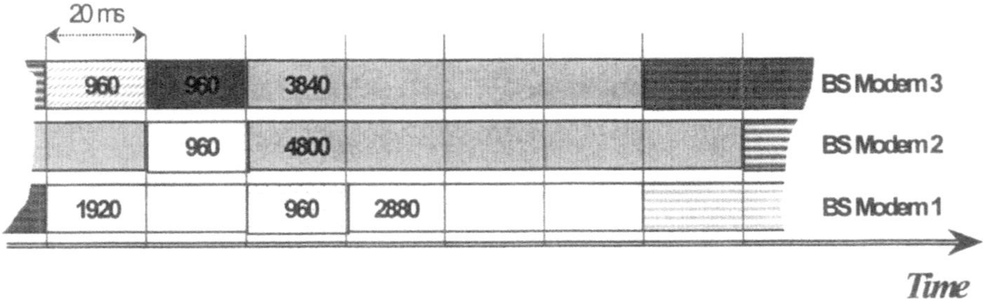

Fig. 9.4.17. MAC Packet Scheduling

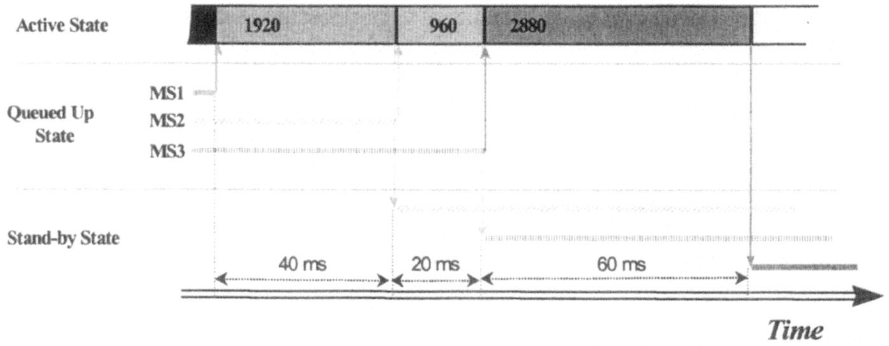

Fig. 9.4.18. Simplified MAC State Transition Timing Diagram

Mobile is in the Queued-Up State typically several seconds before actual data transmission is initiated by the high-speed radio modem. In this state the mobile station exchanges messages with the infrastructure to schedule transmission time, negotiate transmission rate, assign traffic channel codes, etc. If the base station is equipped with a high-gain steerable antennas, it can use this time to compute adaptive antenna parameters.

At the assigned transmission start time, the mobile station transitions to the active state for the duration of transmission. The Fig. 9.4.18 diagram shows that at the end of data transmission, the mobile station transitions to the standby state, but in the actual implementation there may be other intermediate states that the mobile station goes through prior to the standby state. For example, the mobile may dwell for some time in a control hold state, if it detected some transmission errors in the data it received, report them to the base station, and hang on in hopes that a transmission slot be assigned to it again in a quick succession.

We can observe that in the case of a synchronous CDMA system, packet transmissions are governed by the absolute timing of the base station common timing reference. Therefore, when in soft handoff, each of the base stations involved can transmit at the scheduled commonly specified time.

In an asynchronous CDMA system however, packet scheduling for mobiles in soft handoff may be more difficult to accomplish. WCDMA base station can compute transmission timing advance values for each of the mobiles it is serving. If it so happens that a timing advance for mobile station $n+1$ is larger than for the mobile station n (the previously served mobile station) , the base station may not be able to transmit from one to the next without interruption. At the very least, base station power peaking may occur due to overlap in transmit times, in addition to requiring more radio modem resources for the same traffic load.

9.4.4.5 Geographic Location

Advent of geographic location capability in mobile communication networks has been initially motivated by government mandates associated with public safety. Emergency calls in an increasing number of countries are required to supply geographic location information in order to provide better public safety assistance response from such agencies as police, fire protection, disaster assistance, and medical emergency services. Stimulated by these mandates, other geographic location based services are of increasing commercial appeal. They include wireless assisted navigation, location driven advertising and electronic commerce, fleet management, surveillance of minors or delinquents, etc. Wireless network operators are expected to offer an increasing number of ever more sophisticated services based on geographic location. For these services to be useful to consumers, they must be generally more accurate than what is required by the emergency calling mandates. They also must be reliable and ubiquitously available.

GPS Based Techniques

The most widely studied geographic location solutions for mobile communications are based on Global Positioning System (GPS) and involve satellite signal reception at the mobile station. These systems have the advantage of high degree of accuracy, and they are generally independent on the mobile system radio technology, i.e. they are equally applicable to TDMA as they are to CDMA. However, as we shall see, some special features of CDMA systems make these techniques easier to implement and better performing when used in conjunction with synchronous CDMA.

As a response to geographic information request, a GPS capable mobile tunes its receiver to GPS frequency band and performs location determination. For a three-dimensional position determination (latitude, longitude and elevation) it is required that the mobile station receives 4 satellites, which generally can be accomplished in open rural or suburban areas. Three satellites are needed for two-dimensional position determination (latitude and longitude), assuming that a reasonably accurate elevation estimate is available to the mobile. In dense urban areas or indoors, tall buildings may obscure satellites, and signals may be attenuated to the point that GPS reception is poor and unreliable. In those cases it generally takes longer time to integrate the signal to overcome the low signal to noise ratio and obtain the measurement. this can be a lengthy operation, which when the mobile is on the call, can result in call quality degradation that is undesirable or unacceptable, unless a separate receive chain is available in the mobile dedicated to GPS. On the other hand, a separate receive chain in the mobile can be costly, and can result in bulkier hardware. In summary, if solely dependent on its own capability, insufficient number of satellites may be “visible” to the mobile to perform fast and reliable position fix in dense urban environments and indoors.

Satellite Searching

Searching for GPS satellite signals has many parallels to CDMA signal searching, which can be exploited in the mobile station receiver design. In order to accelerate GPS satellite signal search, the infrastructure can assist by providing several different types of information to the mobile station. The following information can reduce GPS search times in the code, frequency, and time domains respectively:

- List of codes transmitted by candidate satellites likely to be visible by the mobile station;

- Doppler shift information, i.e. effects of speed with which each candidate satellite travels relative to the mobile station;
- Code offset and search window times, both related to a priori knowledge of an absolute timing reference used by satellites.

All three types of information require in general a GPS receiver in the base station. GPS receiver can determine these parameters and relay them to the mobile station. The first two may be relayed in a message carried by the cellular radio interface, while the timing information required for satellite code offset and window needs to be more precise and in reference to an absolute time reference. There are several sources of improvement in the system performance afforded by base station assisted position location solution:

- The base station GPS antenna is in a more advantageous location than the mobile station, which may be indoors;
- The base station GPS receiver is not time constrained as the mobile's GPS receiver, and can take more time in performing satellite acquisition functions without much concern for power consumption;
- The base station GPS receiver is not time constrained as the mobile's GPS receiver, and can take more time in performing satellite acquisition functions without much concern for power consumption;
- Base station GPS hardware can be more sophisticated due to lesser cost and size constraints.

These factors point in favor of having GPS receivers available at every base station, which can be used to synchronize the CDMA system.

Let us assume for the moment that despite these advantages the system is not time synchronized, but GPS assistance information is obtained by other means, e.g. by dispatching it from some central point in the network infrastructure. While satellite code list and Doppler data can be furnished, the search time offset and window requires very precise timing information, which can only be obtained by synchronizing the system. Any other means of conveying this information in an asynchronous network would be less precise, and therefore require more effort by the mobile station to extract the GPS data. For example, while timing precision conveyed through a message carried on the radio interface cannot be more precise than several milliseconds, the synchronous CDMA timing is inherently precise within 10 microseconds in the worst case, and typically much less that that.

GPS Search Synchronization Using Round Trip Propagation Delay

Base station conveys time reference parameters (satellite code offset and search window) to the mobile by means of signaling messages.

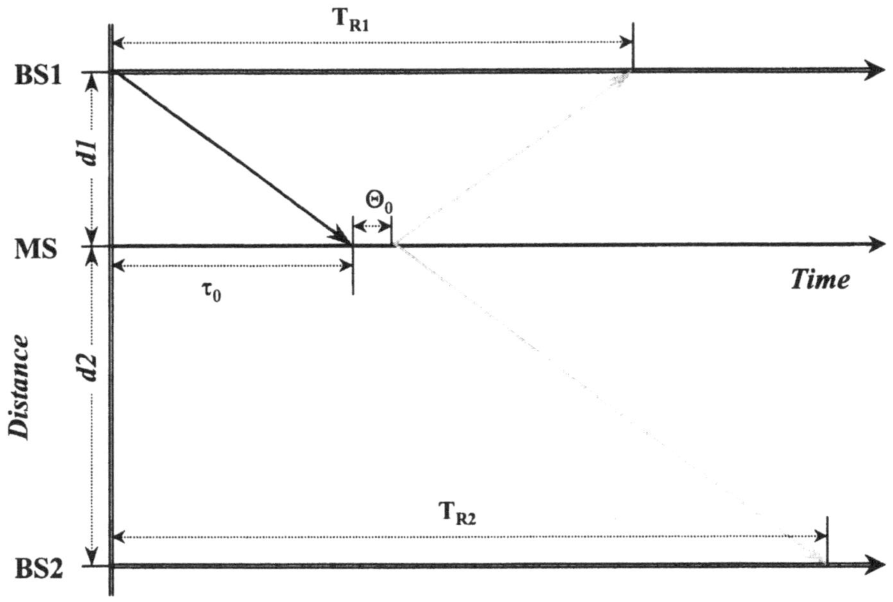

Fig. 9.4.19. Round-trip Delay Computation

Referring to Fig. 9.4.19, mobile's sense of time is shifted compared to the base station due to the propagation delay over the distance between them. In Fig. 9.4.19 the difference is labeled τ and it changes as mobile moves. The infrastructure can estimate τ from the round-trip delay T_{R1} as follows. Mobile station transmission timing is with respect to the earliest arriving demodulated signal component from the base station BS1. Both the identity of the reference base station BS1, as well as the mobile station transmission rules (i.e. delay Θ_0) are known to the infrastructure. Thus BS1 can compute the propagation delay τ_0 as:

$$\tau_0 = (T_{R1} - \Theta_0) / 2 \qquad \text{(Eq. 8)}$$

Infrastructure can account for τ when conveying GPS search timing parameters to the mobile, taking advantage of time synchronization, and thus reduce GPS search window.

Forward Link Triangulation

In addition to using round trip delay for adjusting GPS timing references, this information can be used for base station assisted position location computation by means of Forward Link Triangulation (FLT). Mobile continually searches forward link to detect pilots from all visible base stations. In the process, it can use the earliest arriving components of pilots to measure differences in propagation delay from each of the base stations ΔT_1, ΔT_2, ..., and convey it back to the system. The system can then compute the mobile station distances to all visible base stations:

$$d_0 = \tau_0 * c \qquad \text{(Eq. 9)}$$

$$d_i = (\tau_0 + \Delta T_i) * c; \ldots i = 1, 2, \ldots \qquad \text{(Eq. 10)}$$

Thus, the system will have sufficient data for FLT computations, namely:

- Location (LAT, LON) of the base stations;
- Distance to the nearest base station and at lest two additional base stations.

Triangulation is beyond the scope of this study, but it is illustrated in Fig. 9.4.20. An important point of this analysis is that the triangulation process as described can only be carried out if the CDMA system is synchronized. Then and only then can the mobile station measure the time differences in propagation from various base stations. By combining GPS techniques and FLT, accuracy of position location achievable for the system is estimated to be as summarized in Table 5.

Table 9.4.5. Expected Accuracy of CDMA Geo-Location

Accuracy Component	Unit	Value
Latitude/Longitude	arc-sec	2
Latitude/Longitude	meter	70
Height	m	1
Horizontal Velocity	m/s	0.25
Vertical Velocity	m/s	0.5
Heading	deg	0.35

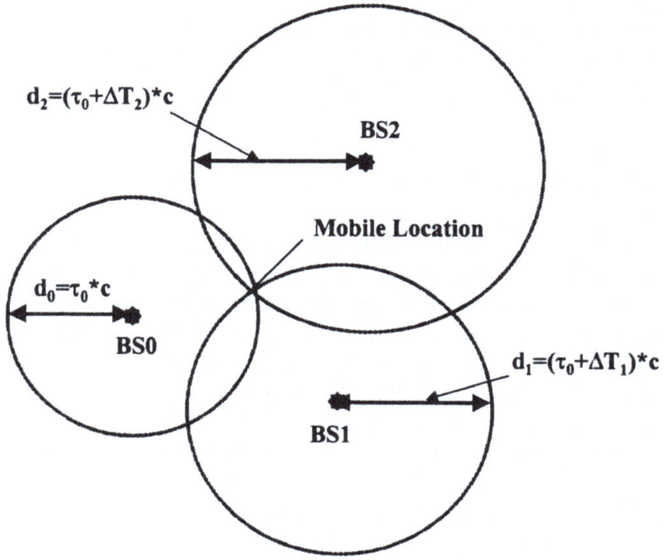

Fig. 9.4.20. Forward Link Triangulation

9.4.4.6
Deployment Flexibility

Synchronization of wireless telecommunication systems is typically done with GPS. Although there are other methods by which common timing reference can be generated and distributed to the base stations, GPS proved to be a cost effective and reliable alternative. Generally, GPS hardware is readily available due to a large and growing number of pervasive applications in the commercial domain. There are a few systems deployed in Russia and the surrounding countries, where the Russian Government-owned equivalent of GPS called GLONAS is used for CDMA system synchronization. However, like GPS, GLONAS relies on satellites, which may not be visible at all base station locations. This is particularly true for indoor installations such as subway stations, underground shopping malls, etc. Existing operators of CDMA systems have successfully solved these problems usually by cabling in GPS signal from a more favorable receiver location. Some alternative synchronization methods have been proposed, but have yet to be proven in a large system installation.

Another motivation for avoiding GPS has been of political nature. Relying on a system that is controlled by one government can cause unease by another. Additionally, although GPS maintenance track record is a good one, for those governments and operators who like to worry about unlikely scenarios, a remote possibility of accidental or intentional unavailability of GPS becomes a source of concern. Due to a combination of these issues, one of the major design requirements for WCDMA was operation in asynchronous mode.

There is one more dimension of operational flexibility afforded through WCDMA structure, although it is not directly linked with synchronization, merely a consequence of an asynchronous structure in WCDMA. Since WCDMA pilots are all different in a cluster of codes, their management in a deployed system is not linked to cell sizes, as it is in cdma2000.

9.4.5
Mixed Synchronization Method

So far we have presented the differences between a synchronous and an asynchronous CDMA system, with cdma2000 and WCDMA as models. We have also closely examined some of the difficulties and challenges that designers of an asynchronous CDMA system must face. This assessment is made mostly from the prospective of a synchronous system, with which we collectively have had much more experience. The final conclusions, on some points more than the others, are tentative. This is particularly true in some of the aspects that were examined, where neither of the two systems has had sufficient field experience, notably on MAC performance. In the final analysis, the quality of operation of either system is in the hands of engineers who implement the details, which sometimes cannot be evaluated solely from the text of the standard. It is therefore feasible that some of the assessments prove incorrect or their significance is diminished by prudent implementations. Additionally, deployment flexibility offered by an asynchronous system adds a different dimension to the debate. It is for these reasons that a proposal is made to "harmonize" 3G CDMA system without removing any of the merits of either of the two synchronization methods described.

The proposed harmonized solution would make most sense if the chip rates of the two systems are compatible with each other, since that would allow soft handoff from one to the other. Since on the current path of the development this is not the case, practical value of some of the aspects of the proposal is diminished. Nevertheless, there are some aspects of it that may still be implementable in future systems containing multiple mobile station receivers, so the idea is presented here with that in mind.

9.4.5.1 Basic Concept

The basic idea is very simple: Each operator makes the decision which cell mode to use, synchronous or asynchronous, with two provisos:

- When running in asynchronous mode, the WCDMA spreading code structure is to be used;
- When running in synchronous mode, the cdma2000 pilot code structure is to be used.

The high level rationale for this approach is as follows. WCDMA solution with the short primary sync (search) code is an effective one if the operator intends to build an asynchronous network. On the other hand, cdma2000 pilot structure with the common code and different base station PN offsets offers an optimum synchronous operation with minimal base station power overhead.

9.4.5.2 Infrastructure Operational Scenarios

Let us further explore operational scenarios with the mixed mode approach. Base stations in a network may support either or both modes. If they support both modes, then each base station operates one mode or the other at any given time. Typically, network operates in the primary mode per operator decision, and may revert to a back-up mode under certain conditions.

For example, an operator may decide to run its network in the synchronous as the primary mode, and use GPS as timing source. If for some reason GPS becomes unavailable, and base stations support asynchronous mode as well, the operation switches to the back-up mode, in this case asynchronous.

A network may also operate partially in synchronous mode (e.g. outdoor base stations), and partially in asynchronous mode (e.g. some or all indoor base stations).

Mobile stations may support either one of the *search modes*, synchronous, asynchronous, or it may have both search modes. Operator would activate for its network mobile stations suitable to the operator's intended mode of operation, for example dual search mode mobiles if it plans to run its network both ways. For heavy roamers, operator may offer activation of dual search mode mobiles even though its own network only runs in one of the two modes.

9.4.5.3
Benefits of Proposed Approach

There are a number of benefits of the proposed solution, including:

- Comprehensive roaming can be achieved with a relatively simple dual search mode mobile station implementation;
- When in synchronous mode, the network runs in the most efficient way possible;
- Dual mode base station is relatively simple since converting from synchronous mode to asynchronous is a matter of turning on primary and secondary sync channels and changing to different pilot codes;
- Single mode base stations are also permissible thus providing economical solutions for operators with less lofty ambitions.

Proponents of both WCDMA and cdma2000 claim advantage of their mode of preference. The proposed approach allows the market to eventually decide which side was closer to forecasting marketplace needs. Or it could be that both were equally responsive to the market demands, which complement each other. On the mobile side too, the incremental cost of dual search mode mobile may be higher than some would like, however, the approach allows the opportunity to use single search mode mobile stations to address lower end of the market not affected by worldwide roaming.

9.4.5.4
Initial Synchronization in Sync/Async Mixed Mode

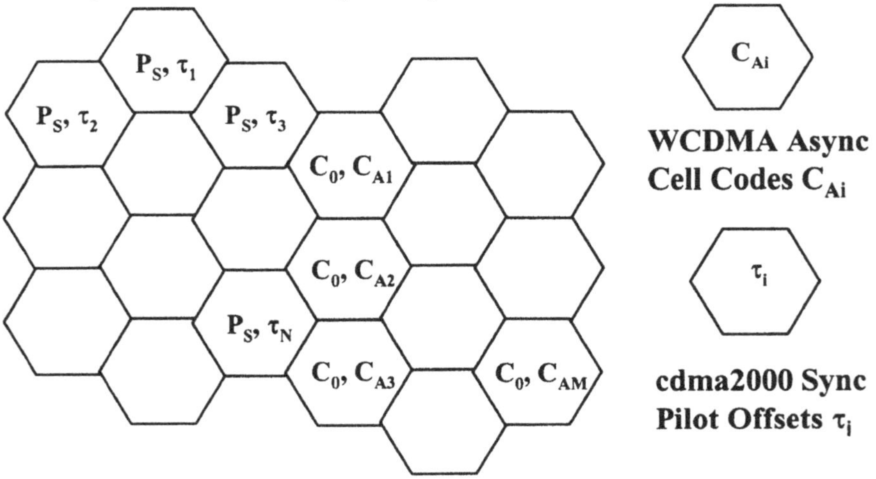

Fig. 9.4.21. Mixed Mode Sync/Async CDMA Network

Synchronous and asynchronous network codes were introduced in Fig. 9.4.1. In the mixed mode network operating partially synchronously and partially asynchronously, base stations of different synchronization styles would operate adjacent to each other, as depicted in Fig. 9.4.21. To summarize, the synchronous portion of the network consists of a number of cells with pilots P_S, each cell in a cluster having a unique PN offset t_i , with *i* ranging from 1 up to 512. The pilot code length is 26.67 ms. The asynchronous portion of the network contains cells characterized by the short common code C_0, which is 256 chips long, and is transmitted 1,600 times per second. In addition, each cell also contains a unique scrambling code labeled C_{Ai}.

We now examine how a mobile station capable of operating in both synchronization modes would work. Dual search mode mobile always starts searching in the mode in which it last operated, since this is the most likely network mode it would encounter. After spending some specified effort and not finding synchronization, mobile switches to the alternate search mode. Acquisition time may be somewhat prolonged for roamers when they first power up in a roaming network of different mode, however this relatively rare occurrence should not be of any overall lasting negative impact on the user. Mobile implementations are also possible in which initial searching is conducted simultaneously for both modes.

9.4.5.5 Initial Synchronization in Sync/Async Mixed Mode

Mixed mode synchronous and asynchronous network handoffs are depicted in Fig. 9.4.22, with Trajectory-1 indicating mobile station movement from the synchronous portion of the network to the asynchronous portion, and Trajectory-2 in the other direction. In the following sections a high level handoff description is given for Trajectory-1 and Trajectory-2, for a mobile station initially in one style of network synchronization moving to the other.

9.4.5.6 Synchronous to Asynchronous Handoff

While served by a sync cell (P_S, τ_1), mobile station is informed of its neighbors, some of which (e.g. C_{A1}) are asynchronous cells. Cell synchronization modes are identified in the neighbor list message broadcast by each base station. In addition to synchronization style of search (offset-window), mobile station also performs asynchronous style of search (matched filter search of common code C_0 then C_{Ai}).

Once the mobile detects an asynchronous neighbor (C_{A1}) with sufficient power, it reports to the infrastructure. Handoff command may ensue, placing the mobile station in mixed mode soft handoff. Mobile station demodulates and soft combines symbols from the two base stations involved, using synchronous (absolute) timing from cell (P_S, τ_1) and a separate timing reference from C_{A1} provided by means of C_0. Mobile station transmit timing is governed by absolute timing of synchronous cell, since synchronous is a special case of asynchronous.

9.4.5.7
Asynchronous to Synchronous Handoff

Asynchronous to synchronous handoff scenario is depicted with Trajectory-2 of Fig. 9.4.22. This style of handoff is somewhat more complex in comparison with Trajectory-1 scenario. This is because the mobile station while in synchronous cell area does not know the timing of its synchronous neighbors.

While in idle mode in an asynchronous cell area C_{A2}, mobile station is informed of its neighbors, some of which, e.g. (P_S, τ_3), are synchronous. In this case mobile station must start synchronous mode searching even before it is on the dedicated channel. Mobile station allocates its searcher resources among the two search styles. Until the mobile station finds synchronization timing, synchronous style search is as for initial system acquisition, i.e. the mobile station must cover the entire code space until it finds timing of the synchronous portion of the system.

There are some techniques for the system to convey the synchronous code timing to the mobile station and therefore accelerate synchronous cell acquisition. However even without deploying those techniques, mobile should be able to find timing of the sync portion of the system. Thereafter, searching is offset-window style, hence much simplified.

If the mobile station gets on the dedicated channel before it acquires synchronization timing, the initial sync search strategy continues, again until timing is acquired. This may result in small manageable call setup delay, however the likelihood of the scenario is small. Once sync pilot (P_S, τ_3) with sufficient power is detected, mobile station reports it to the infrastructure. Handoff command may ensue, placing the mobile station in mixed mode soft handoff. Thereafter the operation is as in the synchronous-to-asynchronous soft handoff case.

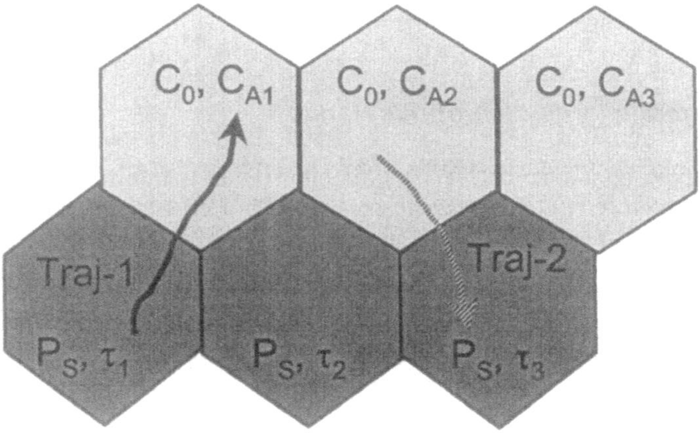

Fig. 9.4.22. Mixed Sync/Async Mode Handoff Scenarios

9.4.6 Conclusion

We have examined two leading CDMA proposals for IMT-2000 wireless systems from a very important prospective of network synchronization. Although it is generally agreed that synchronous mode is less complex and has improved performance, both methods have merits in different domains. It is therefore difficult to make simple judgment which of the two should be chosen by an operator.

Recognizing the situation, it is suggest that a compromise approach be taken between the two choices, which would bring proponents of each choice a step closer without loss of features they aimed to satisfy at the time they set out to develop concepts in the proposed IMT-2000 Radio Transmission Technology. This small step is an important one in that it holds a possibility for full convergence in the future, driven by market conditions, as opposed to prejudging the outcome prior to deploying and testing systems on a grander scale.

It is important to recognize that with the outlined approach both choices would be enhanced with no loss in performance in either case, creating a true win/win situation. There is some added complexity, not to be neglected. However, this added complexity of the compromise solution is hardly existent for single mode systems, which are also feasible, and it is incremental for dual mode infrastructures and mobiles.

The impact of system synchronization on the anticipated burgeoning position location services, including but not limited to government public safety mandates, known in the United States as Enhanced 911 (E-911), has also been examined in some detail, in anticipation that these services will be pivotal in operators decision on how their IMT-2000 systems should run – synchronously or asynchronously. The consensus in the technology circles, that performance benefits will be realized with synchronous systems, has been confirmed with analysis in some detail. The good news for the operators is that both cdma2000 and WCDMA can operate in synchronous mode, and that they can also co-exist in a network. As wireless network architectures start moving toward Internet Protocol models, this will further serve as a unifying force for the mobile systems of next generations.

10 Services

10.1 Introduction

Over the last decade two distinct communication technologies have experienced unprecedented growth and commercial success and become the major vehicles for multimedia services provision. These are the Mobile Systems and the Internet supported by the IP. The development attracted the interest of communications engineers and computer scientists as well as manufacturers and operators. Propelled by the powerful vision of being able to communicate from anywhere at any time with any type of data, a natural meshing of these technologies is taking place. All indicators are that the third generation mobile systems will provide internet access for multimedia services. These possibilities are covered in section 10.2 and 10.3. Section 10.4 covers some reliability issues of wireless networks in general and section 10.5 presents a methodology by which service providers could obtain valuable techno-economic information, as the implementation requirements, both technical and economic, for such systems exceed the capabilities of existing decision support tools. Both last sections are intended to provide quantitative measures for evaluating service quality.

10.2 -Fast Internet and multimedia services: Facing the new challenge

E. Pallis, M. C. B. Smith, A. A. Alexandridis, A. Kourtis, K. Dangakis, P. Kostarakis and G. Kormentzas

10.2.1 Introduction

In the last decade, two distinct communication technologies have experienced unequalled rapid growth and commercial success: mobile communications and multimedia communications. Naturally, the great success in both areas has attracted and inspired many electrical engineers and computer scientists. Propelled by the powerful vision of being able to communicate from anywhere at any time with any type of data, a natural meshing of mobile and multimedia is currently underway. All indicators suggest that, building on advances in network infrastructure, wireless multimedia services will find widespread acceptance over the next decade. Armed with this vision of the future, the objective that shaped current second-generation communication standards have been expanded in third-generation wireless information networks to provide truly ubiquitous and integrated multimedia services. To make the vision of ubiquitous wireless multimedia communications a reality, however, many technical and regulatory problems remain to be solved:

Reliable transmissions over error-prone channels is a key problem for wireless multimedia transmission. For this reason, advanced modulation techniques have to be applied to provide error-free communication. IEEE 802.11 standard provides high immunity to errors, by defining two modulation schemes: Frequency Hopping Spread Spectrum (FHSS) and Direct Sequence Spread Spectrum (DSSS).

Due to the limitations of the available radio spectrum and the enormous amounts of data involved (in case of video transmission), new frequency bands need to be released and new access technologies have to be used. On the one hand, realising the increasing demand of users for new services and faster access to them, the Telecommunications Regulatory Organisations released additional International Scientific and Medical (ISM) frequency bands (like the 2.4GHz, 5.2GHz, etc.) in order to allow manufacturers to develop new products. On the other hand, LMDS (Local Multipoint Distribution System) and MMDS

(Multichannel Multipoint Distribution System) promise broadband access to the provided services.

This chapter demonstrates the feasibility of using the new wireless access technologies (MMDS) with the proposed IEEE standard, and describes a cellular broadband wireless network providing access to fast Internet and multimedia services. The system operates in the ISM frequency band of 2.4GHz, at a data rate of 2Mbps, and uses both FHSS and DSSS modulation techniques. It uses a hierarchical architecture with two core subsystems: the *Service Provider* by the service origin and several *Cell Main Nodes* defining user areas, called *Cells*. For a given number of users (stationary, portable or mobile) in a cell, each user is guaranteed a minimum throughput during multimedia service access. On the one hand, the fact that the system is built around components of standard technology (low cost and easy set-up), makes the system very attractive to both users and service providers. On the other hand, the revenues of such a system are expected to be high. The overall architecture adopted for such a system provides for flexibility and scaleability.

This chapter is organised in four sections:

Section 10.2.2 makes a short review on the existing wired and wireless access technologies, and compares them by way of flexibility, speed, cost, scaleability and time-to-market. It seems that for areas where no cabling infrastructure exists, wireless terrestrial technologies have several advantages over the wired ones. Furthermore, portability, mobility, lower installation and maintenance costs make terrestrial wireless technologies even more attractive.

Section 10.2.3 reviews the IEEE 802.11 standard, as the most reliable way for Point-To-Point and Point-To- Multipoint access in wireless networks. The use of Direct Sequence Spread Spectrum and Frequency Hopping Spread Spectrum techniques, provides high level of security while making the communication channel resistant to noise, interference and jamming.

Section 10.2.4 introduces a prototype system, based on the wireless access technologies, while using all the properties of the IEEE 802.11 standard, which provides fast Internet and multimedia services to a number of users. It describes the architecture of such a system and evaluates its performance under stationary, portable and mobile conditions.

Section 10.2.5 analyses users' feedback, concerning their profile and their willingness to pay for the provided services. Based on this analysis, it comments on the potential for exploiting commercially similar systems in the Greek market. Finally, *Section 10.2.6* concludes by summarising the findings of this chapter.

10.2.2 Wired vs. Wireless: The debate

When the end of the cold war brought to light a quietly potent system called ARPANET, no one could have imagine the implications that it would someday have. So when Internet, as later became known, was thrown open for commercial use, history was made, bringing about a revolutionary shift in information technology. Combine that shift with the rapid expansion of the wireless digital TV and multimedia industry, however, and you get an even more dynamic change: the ability to bring the Information Age to all parts of the world, especially there where the lack of existing infrastructures make access difficult. Though making the transition from video-only services to bundled services that include data transmission is not necessarily an easy one, it is something all wireless system operators and service providers are going to consider as the world gets wired to the Internet.

In this context, the choice of the access technology, handling the connection of the users to the core broadband network, constitutes the main issue. Presently, access systems may be classified as either wired or wireless. For both categories, flexibility, speed, cost and time-to-market are the most important factors that affect their evolution.

10.2.2.1 Wire access systems

There are several wire access systems in the developing market for the widespread use of service access: Public Switched Telephone Network (PSTN), Integrated Services Digital Network (ISDN), Cable and Digital Subscriber Line (xDSL) modems. Cable and Digital Subscriber Line (xDSL) modems are the most promising wired broadband access systems. However, the low flexibility and high installation cost of cable modems and the great dependence of xDSL systems on distance prevent them from achieving significant acceptance and popularity in the domain of broadband services.

PSTN. In the interest of being complete, we mention the PSTN system. In spite of antiquated technology and slow transmission rates, this system is, on a world basis, the most reliable and inexpensive alternative for voice telephony and elemental data transmission. It is this system, after all, through which the current Internet developed via broad access to modems of various speeds (the maximum being somewhat around 56kbps). In addition, it is interesting to note that it is often quicker to carry out a variety of services by simply calling (via the PSTN system) rather than going through the procedure of starting a PC and logging onto a higher speed network. A whole range of commercial services - from ordering pizzas to aeroplane tickets - is done more quickly and has wider public acceptance here.

In spite of a long tradition, however, the future of PSTN is not bright when it comes to services requiring high bandwidth. It is doubtful that any compression technologies will be able to reduce the transmission requirements enough in order to guarantee transmission of even basic services over PSTN in the near future.

ISDN. ISDN, employing the traditional telephone copper infrastructure allowing for both voice and data transmission is widely available in Europe. Demand for this technology is currently quite high. Its ability to provide faster Internet contact as well as a support system for some business and LAN applications are important factors.

However, ISDN must be defined as a medium range technology. Its top speed is only around 128kbps. In addition, ISDN can be difficult to configure and the marketing of the service has been poor. Finally, when seen from the perspective of non-telecom operators, ISDN is rarely a solution they will consider in the development of future systems.

In sum, ISDN represents movement toward higher bandwidth. It provides the user with a moderate increase in bandwidth and through this, it allows them to experience the Internet and other developing services. In the short-run, ISDN will be used increasingly, but in the longer-run the system limitations will likely mean that it be superseded.

Cable Modems. There is quite a lot of industry discussion of the potential of cable modems. Certainly, cable companies have a well-entrenched infrastructure from which to begin deployment. These are devices that function similarly to PSTN modems in that they convert the digital output into analogue signal that is further transmitted through the cable system.

Breadth of services: Since cable networks were originally designed for video transmission, they are ideally suited to broadband transmission of data. Many cable modem trials in U.S. are attempting to deliver 30Mbps to the home (downstream) and 1.5Mbps in the return path (upstream). Data transmission speeds of this order would allow high-quality, two way video to the subscribers of digital cable networks, taking PC-based interactivity to a new level. In Western Europe, mostly in UK, cable telephony has been successfully deployed. Even without a full upgrade to a two-way broadband network, digital cable systems are also well-positioned for digital television signals, or one-way broadband with a telephone return.

Technological attractiveness: Cable modems by themselves pose no technological problems, save perhaps for a lack of standards. If the connection to a PC is an Ethernet card, most tech-savvy users would be able to do their own installations. However, for the service providers, the obstacles are significant.

A two-way Hybrid Fibre Coax (HFC) network must be crated to deliver broadband or telephone services. Though not susceptible foliage and rain, cable systems are very susceptible to noise and interference from electrical disturbances, such as appliances. They also lack the redundancies of the telephone network. The scaleability of cable head-ends and network capacity remain unproved, although equipment vendors claim to have tested the technology to levels where these potential problems can be managed by the operator. Also reliability of cable modems is uncertain, and could be lower than comparable access technologies.

On the whole, migrating an one-way analogue network to a digital duplex system that is positioned to offer broadband and telephony is a very difficult, expensive prospect. The migration is complicated by the lack of experience on the part of cable engineers and managers.

Financial attributes: While cable companies have an advantage in their existing wireline networks, the cost of deploying two-way networks has been the main hindrance to bringing the technology to a mass-market. Firstly, cable has been almost exclusively a residential technology. Without a business-customer base, moving out of the market trial and early-adopter phase will be slow and laborious.

Aside from the actual fibre plant and labour involved in installing an Hybrid Fibre Coax (HFC) network, there are the costs of installing a digital head-end, remote-service terminals and units, testing for noise, new billing configurations for duplex communications, and reliability concerns.

Given that revenues from broadband services cannot be accurately gauged, it is very difficult for cable companies to make solid economic cases for a rapid build-out of two-way networks. The nature of cable modem delivery systems dictates very high start-up costs, with far lower subscriber acquisition costs once reliable duplex systems are in place. Additionally, no subscriber revenues can be generated without the completion of a two-way upgrade. In a global sense, though costs for completely new systems will be somewhat less than their analogue predecessors, it would be an exorbitant undertaking to deploy a digital two-way wireline cable system where there is no existing plant.

Marketing opportunities: One of the major marketing challenges for cable companies to overcome is the perception of poor service from their customers. While the promise of multi-megabit data service will no doubt allay consumers' fears, cable telephony is another matter. The necessity of telephone systems to be relatively noise-free, without latency and redundant may give potential subscribers pause. One solution to this is to offer bundled packages of phone, data, and video services at a low cost.

Probably the main advantage to marketing cable broadband is the low-hassle from consumer standpoint, and the extension of data services such as Web TV to the desktop. One of the steeper challenges will be to connect businesses to cable networks for broadband services.

xDSL The telecommunication companies answer to the problem of their narrow-band copper infrastructure is Digital Subscriber Line (DSL) technology. The ubiquitous presence of twisted-pair copper lines gives xDSL an enormous advantage over other last mile solutions. Also, since DSL is fundamentally based on wires that have been in use for more than 50 years, telecommunication companies will not be stymied by provisioning issues and unfamiliarity of working with newly engineered systems.

Breadth of services: xDSL is essentially a data pipe. Though its early trials were focused on Video-on-Demand (VoD), that use has been left behind in favour of Internet delivery. Depending on the distance of a customer from the service provider or Digital Loop Carrier (DLC), the bandwidth delivered over DSL is scaleable. The rate-adaptive variety of DSL has the ability to change bandwidth speeds depending on line condition and network traffic. The latest modems are able to operate at up to 8Mbps downstream (data to the subscriber) with an upstream transmission rate of 800kbps. These data rates will be possible over a local copper-based loop of up to 3.5km and are more than capable of providing real-time video, at High Definition TV (HDTV) quality. High speed real-time multimedia services have become a reality. The variety of xDSL versions give varying amounts of return loop bit-rates so that someone can have video in the return path. Video-on Demand may become a reality in a few years.

Technological attractiveness: the main technological obstacle of xDSL is its dependence on distance. To receive DSL services, a customer must be no more than 6km away from a service provider central office. Also, these can be no equipment, such as load coils, on the line that is servicing the customer.

XDSL relieves the congested voice network. This is a great advantage for service providers who can monitor the data traffic separately, and free up circuits as well. The combination of DSL and voice traffic over the same line requires the integration of new xDSL management systems.

One other problem plaguing the roll-out of DSL is competing line codes. There are 2 non-interoperable line codes for xDSL equipment: discrete multi-tone (DMT), which has become an ANSI standard, and carrierless amplitude modulation/phase (CAP). Most providers desirous of deploying xDSL are taking a wait-and-see attitude towards the standards issue.

Financial attributes: The beauty of xDSL is that it completely leverages the copper infrastructure. While there are certain provisioning costs such as reliability testing, the only real cost of deploying DSL from a service provider central office is the equipment cost: a pair of modems and a splitter on the customer's premises, and some new equipment in the switch fibre.

xDSL has been used for the last several years as a repeaterless substitute for T1/E1 without much complication. So, xDSL has already been leveraging the business community. Tiered-services over single lines is a natural extension for telecommunication companies, and makes the financial aspect of this technology all the more appealing. Also, xDSL lines relieve traffic from heavily congested voice circuits, which can cut costs.

Marketing opportunities: Splitterless versions of xDSL to the home can be marked as second lines for data. They can also be marketed as a second line for voice and high-bandwidth Internet usage. Business will no doubt be attracted to scaleable, low-tech broadband. The tiered approach to deployment appears best at this stage, slowly evolving 28.8k and 56k infrastructure to higher and higher speeds.

Versions of xDSL: xDSL comes in a seemingly endless number of versions that vary in the capacity carried and reach. In addition, while some versions of the technology allow for symmetrical traffic others are asymmetrical. The four major variants of xDSL are: IDSL, SDSL, ADSL and VDSL.

IDSL (ISDN Digital Subscriber Line) is based on ISDN with which it is fully interoperable.

SDSL (Symmetric Digital Subscriber Line) is a symmetrical technology (the transmission line is a two copper pairs). Its transmission rate is 768kbps in both directions for distances up-to 9km. Since it offers fast, symmetrical performance may present a solution for business subscribers and heavy telecommunication users.

ADSL (Asymmetric Digital Subscriber Line) is the most common type of xDSL and, as the name implies, it provides asymmetrical up and downstream transmission. The obvious target of this technology is those users who received more than they send, a very large percent of users indeed. The actual ratio of incoming to outgoing data is greater than 10:1. At these ratios, ADSL can easily cope with the capacity demands in each direction, particularly if one is speaking of applications such as interactive TV where transmissions from the user into the system are quite small. However, if applications such as the Internet begin to migrate to ADSL one begins to open the need for greater information transport into the system. Current versions of ADSL automatically adapt transmission rates to match the capability of the line. The major advantage of ADSL is that it is the fastest DSL technology with which one can support the maximum 5.5km distance in the loop.

VDSL (Very high-bit-rate Digital Subscriber Line) is a high capacity asymmetric version of DSL that operates at about eight times the rate of ADSL. The major disadvantage is that the range of the technology is quite limited, in the extreme cases this is only about 300 meters. Thus in order to implement this technology fibre optic equipment close to the subscriber is required. This technology is only now arriving.

10.2.2.2
Wireless access systems

The ability to send data to multiple recipients is an inherent characteristic of satellite communications. Such multi-casting over the Internet, however, is accomplished by re-sending the data again and again to each intended recipient. This one-at-a-time process is a highly inefficient network capacity. Yet it is a transmission method that is tolerable for shorter messages such as e-mail. To conserve bandwidth, IP multi-casting handles duplicate transmissions reaching all recipients in a single session. For wide-band services, this multi-casting offers a valuable solution to Net congestion by simultaneously delivering data to an addressed group. Examples are training video and audio conferencing, replicated database information, live video programme material, collaborative work sessions, and other sharing and transfer of files. Along with interactive applications, the clear advantage of multi-casting via wireless transmission technologies promise to play a major role on the Net. End to end system solutions necessarily include hybrid technologies, optimised to overcome traditional shortcomings. Using satellite communications, along with terrestrial facilities, high-speed data is transmitted from self-contained gateways at Post Office Protocols (POP).

Digital satellite. Digital satellite is a system that enjoys a certain popularity. Via a small low price antenna, it affords the user access to high quality broadcast TV signals, the choice of over 100 television channels, and premium/pay-per-view channels at speeds of up to 400kbps. With the further development of return channel capabilities highly interactive services such as Internet can be sent over this system. There are currently digital satellite services available in all major Europe countries, Japan and US. The focus of these services is on the residential market and is intended to compete with cable companies and terrestrial broadcasters. The purchase of the necessary customer premises equipment is generally subsidised.

There are, however, several drawbacks to the system. First, the return channel is difficult to install and is generally only a 14.4kbps PSTN connection. Clashes between the return path function and normal telephony can result. In addition, if the service is intended for those living in rural areas, the phone charges for the return path may be a long-distance call. This means an increased cost to the user.

Digital terrestrial. The terrestrial wireless broadband access systems are classified, according to their type, as MMDS (Multichannel Multipoint Distribution System), or LMDS (Local Multipoint Distribution System). The second type is alternatively called LMCS (Local Multipoint Communication System) or, in Europe, MVDS (Microwave Video Distribution System). Both MMDS and LMDS assume that the user has a fixed terminal with low transportability, due to the use of a rooftop antenna.

- MMDS

MMDS systems operate at frequencies lower than 5GHz with coverage areas (called cells) up to 40km. They are suitable for transmission of video and broadcast services in rural areas where there is no cable infrastructure.

- LMDS

LMDS systems operate at high frequencies (e.g. 24GHz, 42GHz), where large portions of the spectrum are still free. In this case, the coverage area is realised with smaller cells (typically up to a radius of 5km). The extension of this radius requires repeaters to be placed in a LOS (Line Of Sight) configuration. The capability of local coverage, together with the large available bandwidth, make these systems suitable for the distribution of interactive multimedia services.

10.2.2.3 Summary

Above we examined a whole set of alternative access technologies competing for the users' attention and the investors' money. Table 10.2.1 summarises these various wire and wireless technologies, referring to upstream data rate, downstream data rate, and maximum local loop reach.

Summarising, someone can say that, from a consumer perspective, wireless terrestrial access technologies (MMDS/LMDS) are very attractive technologies. MMDS/LMDS technology is scaleable and ideally suited to home and business. When placed in a global perspective, MMDS/LMDS is a perfect technology, enabling poorly wired countries a way of rapidly upgrading their telecommunication infrastructure, while keeping costs comparatively low. The installation and maintenance costs are much less than those of a wired network:

Table 10.2.1 Comparison of various access technologies

Type	Upstream data rate	Downstream data rate	Maximum local loop reach (km)
Analogue modems up to 56kbps	14.4-56kbps	14.4-56kbps	N/A
ISDN	128kbps	128kbps	N/A
IDSL	128kbps	128kbps	5.4
SDSL	768kbps	768kbps	3.6
ADSL	176kbps 224-640kbps	1.54Mbps 6-8Mbps	5.4 3.6
VDSL	640kbps 1.6-2.3Mbps	13Mbps 52Mbps	1.4 0.3
Cable modems	0-768kbps (May require shared media)	30Mbps (May require shared media)	N/A
LMDS	2Mbps	34-40Mbps	5.0

As there is no standardisation of MMDS/LMDS equipment yet, providers must supply proprietary equipment that is installed by their technicians. The requirement of technicians coming to each customer's premises means less work for the consumer. So far, reliable service can be achieved in metropolitan areas, and customers using the technology do not have to learn new skills to take advantage of LMDS.

Once the initial cell is deployed, repeater-hubs will be deployed to create additional cells at a much lower cost. Presently, the costs associated with starting up an MMDS/LMDS service are attractively, low. A higher demand for the service could potentially drive down further the costs of the customer premises equipment. Some telecommunication companies claim that they can achieve greater cell size by reflecting their signals off buildings, and this too may save money. Targeting both a business and residential market is wise in the case of MMDS/LMDS, since business can afford to pay higher costs for the service in its initial deployment stages, giving the service providers room to grow.

In the near-term, wireless terrestrial access technologies have several positives against wired ones. No other technology can deliver such high data rates for such a low cost. In both analogue and digital formats, MMDS/LMDS can provide digital television and interactive multimedia services that matches current wireline offerings. The true calling of MMDS/LMDS, though, appears to be broadband data. MMDS/LMDS could then be used effectively to interconnect LANs, create campus area networks, and serve as a medium for the transmission of ATM and SONET. However, the provision of interactive services in a Point-to-Multi-Point configuration through MMDS/LMDS technologies requires a Multiple Access technique. One widespread standard for wireless multiple Access is the IEEE 802.11 standard, which will be described in the next section.

10.2.3
Wireless LANs: The IEEE 802.11 standard

For some time now, companies and individuals have interconnected computers with Local Area Networks (LANs). This allowed the ability to access and share data, applications and other services not resident on any one computer. The LAN user has at his disposal much more information, data and applications than they could otherwise store by themselves. In the past all local area networks were wired together and in a fixed location as in Fig. 10.2.1.

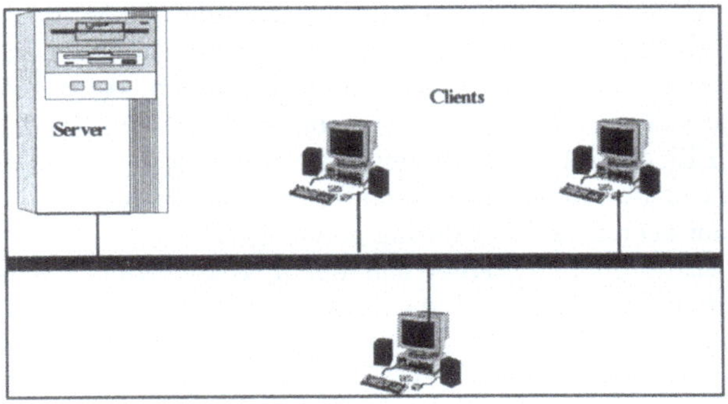

Fig. 10.2.1 Traditional wired LAN.

Why would anyone want a wireless LAN? There are many reasons. An increasing number of LAN users are becoming mobile. These users require that they are connected to the network regardless of where they are because they want simultaneous access to the network. This makes the use of cables, or wired LANs, impractical if not impossible. Wireless LANs are very easy to install. There is no requirement for wiring every station and every room. This ease of installation makes wireless LANs inherently flexible. Another advantage is its portability. If a company moves to a new location, the wireless system is much easier to move than ripping up all of the cables that a wired system would have snaked through the building. Most of these advantages also translate into monetary savings. Fig. 10.2.2 is an example of a wireless LAN.

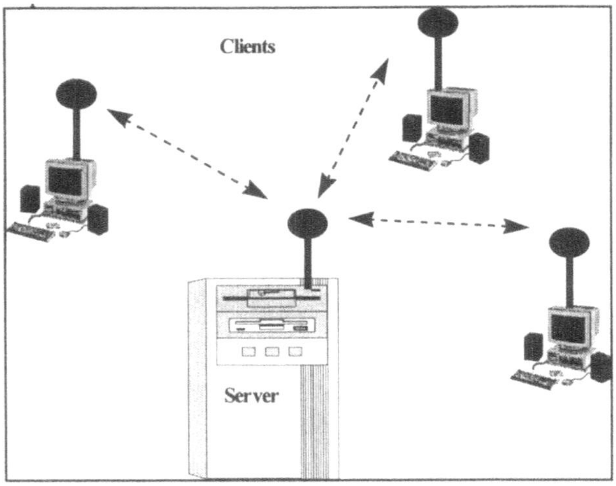

Fig. 10.2.2 Wireless LAN

One widespread standard today underpins much of the commercial wireless LAN market. This is the IEEE 802.11 standard. The IEEE 802.11 specification is a wireless LAN standard developed by the Institute of Electrical and Electronic Engineering (IEEE) committee in order to specify an 'over the air' interface between a wireless client and a base station (server) or Access Point, as well as among wireless clients. Each computer, mobile, portable or fixed, is referred to as a station in IEEE 802.11. The difference between a portable and mobile station is that a portable station moves from point to point but is only used at a fixed point. Mobile stations access the LAN during movement.

Like the IEEE 802.3 Ethernet and 802.5 Token Ring standards, the IEEE 802.11 specification addresses both Physical (PHY) and Media Access Control (MAC) layers.

10.2.3.1 Physical (PHY) layer

There are three media that can be used for transmission over wireless LANs. Infrared, radio frequency and microwave. In 1985 the United States released the Industrial Scientific and Medical (ISM) frequency bands. These bands are 902-928MHz, 2.400-2.4853GHz, and 5.725-5.85GHz and the use of these bands does not require licensing by the Federal Communications Commission (FCC). This prompted most of the wireless LAN products to operate within ISM bands. The FCC did put restrictions on the ISM bands however. The Radio Frequency (RF) systems must implement Spread Spectrum (SS) technology. RF systems must confine the emitted spectrum to a band. Microwave systems are considered very low power and must operate at less than 500mW.

Radio frequency systems must use spread spectrum technology. The main principle of Spread Spectrum communication is that the bandwidth occupancy is much higher than usual. Because of this larger bandwidth the power spectral density is lower, in the channel the signal just looks like noise. The spreading is done by combining the data signal with a code (code division multiple access) which is independent of the transmitted data message. A number of advantages are:

As the signal is spread over a large frequency-band, the Power Spectral Density is very small, so other communication systems are not interfered by this kind of communications. However, the Gaussian Noise level of the communication channel is increased.

Random Access can be dealt with, as a large number of codes can be generated a large number of users can be permitted.

The maximum number of users is interference limited.

Security: without knowing the spreading code, it is (nearly) impossible to recover the transmitted data.

Fading rejection: as a large bandwidth is used the system is less susceptible to distortions.

This technology currently comes into two types: Direct Sequence Spread Spectrum (DSSS) and Frequency Hopping Spread Spectrum (FHSS). A combination of these two is the DS/FH which offers a lot of advantages over the other two, but is not yet commercialised.

Direct sequence spread spectrum (DSSS). With DSSS the transmission signal is spread over an allowed band (for example 25MHz). A random binary stream is used to modulate the transmitted signal. This random stream is called spreading code. The data bits are mapped into a pattern of 'chips' and mapped back into bits at the destination. The number of chips that represent one data bit is the spreading ratio. The higher the spreading ratio, the more the signal is resistant to interference. The lower the spreading ratio, the more bandwidth is available to the user. The FCC dictates that the spreading ratio must be more than ten. Most products have a spreading ratio of less than twenty and the new IEEE 802.11 standard requires a spreading ratio of eleven. The transmitter and the receiver must be synchronised with the same spreading code. If orthogonal codes are used then more than one LAN can share the same band. However, because DSSS systems use wide sub-channels, the number of co-located LANs is limited by the size of those sub-channels.

Frequency hopping spread spectrum (FHSS). This technique splits the band into many small sub-channels (1MHz). The signal then hopes from sub-channel to sub-channel transmitting short bursts of data on each channel for a short period of time, called dwell time. The hopping sequence must be synchronised at the sender and the receiver otherwise information is lost. The FCC requires that the band is split into at least 75 sub-channels and that the dwell time is no longer than 400ms. Frequency hopping is less susceptible to interference because the frequency is constantly shifting. This makes frequency hopping systems extremely difficult to intercept. This feature gives FH systems a high degree of security. In order to jam a frequency hopping system the whole band must be jammed. These features are very attractive to agencies involved with law enforcement or the military. Many FHSS LANs can be co-located if orthogonal hopping sequences are used. Because the sub-channels are narrower than in DSSS, the number of co-located LANs can be greater with FHSS systems.

10.2.3.2
Media Access Control (MAC) layer

The 802.11 MAC layer, supported by an underlying PHY layer, is concerned primarily with rules for accessing the medium. Two network architectures are defined: the Infrastructure Network and the Ad Hoc Network. An Infrastructure Network has a network architecture for providing communication between wireless clients and wired network resources. The transition of data from the wireless to the wired medium is via an Access Point. The coverage area is defined by an Access Point (AP) and its associated wireless clients, and together all devices form a Basic Service Set (BSS).

An Ad Hod network has an architecture that is used to support mutual communication among wireless clients. Typically created spontaneously, an Ad Hoc network does not support access to wire networks, and does not need an AP to be part of the network.

The primary services provided by the MAC layer are as follows:

Data transfer: Wireless clients use Collision Sense Multiple Access with Collision Avoidance (CSMA/CA) algorithm as the media access scheme.

Association: This service enables the establishment of wireless links between wireless clients and APs in Infrastructure Networks.

Re-association: This takes place in addition to association when a wireless client moves from one Basic Service Set (BSS) to another.

Authentication: Authentication is the process of providing a client identity, and in IEEE 802.11 this process takes place prior to a wireless client associating with an AP. By default, IEEE 802.11 devices operate in an Open System, where essentially any wireless client can be associated with an AP without the checking of credentials. True authentication is possible with the use IEEE 802.11 option known as Wired Equivalent Privacy or WEP, where a Shared Key is configured into the AP and its wireless clients. Only those devices with a valid Shared Key will be allowed to be associated to the AP.

Privacy: By default, data is transferred 'in clear'; any 802.11-compliant device can potentially eavesdrop like-PHY 802.11 traffic that is within range. The WEP option encrypts data before it is sent wirelessly, using a 40-bit encryption algorithm known as RC4. The same Shared Key used in authentication is used to encrypt or decrypt the data; thus only wireless clients with the exact Shared Key can correctly decipher the data.

Power management: IEEE 802.11 defines two power modes, an Active Mode, where a wireless client is powered to transmit and receive, and Power Save mode, where a client is not able to transmit or receive, but consumes less power. Actual power consumption is not defined and is dependent upon the implementation.

10.2.4 Wireless provision of fast access to Internet and multimedia services to stationary and portable stations.

Realising the increasing demand of users for new services and fast access to them, the European Union funded a number of projects in the frame of ACTS (Advanced Communications Technologies & Services) research program. The purpose of them is to conduct research and demonstrate trials, which will lead to final recommendations to EU and also to create the infrastructure and develop products that will be available in the market.

One of the ACTS projects is CRABS (Cellular Radio Access for Broadband Services AC 215), which is concerned with cellular radio access to interactive television and broadband services. Its main objective is to develop and demonstrate a cellular radio system in the 42 GHz frequency band, in order to provide digital interactive television and broadband multimedia services through several user and service trials in five areas in Europe. One of the trials, is currently taking place in Athens, Greece and is conducted by National Centre for Scientific Research Demokritos.

Having already investigated the system architecture and application services in the 42 GHz band and taking into consideration the users' requirements for fast access to Internet and multimedia services, the team continues the research and development tasks in portable and mobile environments.

However, current 42GHz technology can not support mobility. For this reason the team developed a broadband cellular wireless network in the frequency band of 2.4GHz, operating at a data rate of 2Mbps. This is an experimental platform, which can offer fast access to Internet and multimedia services to stationary, portable and mobile users. Considering that Internet users are geographically scattered in a large area, it is more convenient to use a cellular system architecture. This is the best architecture for low mobility users, which in this section are called 'portable users'.

10.2.4.1 Overall system architecture

The overall system architecture is shown in Fig. 10.2.3 It consists of a service provider and a number of cells. Each cell consists of a main node and a number of associated users (i.e. those users geographically neighbouring the cell main node). Each cell main node communicates with the service provider through a point-to-point RF link, at 2Mbps, and operates in the frequency band of 2.4GHz using Direct Sequence Spread Spectrum (DSSS) technology. However, the use of half duplex mode equipment and the encapsulation of overhead packets during transmission, result in an actual throughput of 1.2Mbps as measured experimentally (see section 10.2.4.3).

The communication between the service provider and the cell main node follows a Client/Server approach. The service provider answers to requests forwarded by the cell main node asking for Internet provision, or for multimedia services. The Internet provider (ARIADNET) offers access to the Internet, while a multimedia server offers access to multimedia services. The service provider can guarantee to the users a minimal bandwidth for the multimedia services provided through the multimedia server. On the contrary, Internet access follows the usual best effort scheme.

The communication between wired (stationary) users and cell main node is established through the hub of the local LAN via a cable network, at a maximum rate of 10Mbps (Ethernet). The communication between wireless (portable or mobile) users and cell main node is a point to multi-point RF link at 3Mbps, operating in the frequency band of 2.4GHz using Frequency Hopping Spread Spectrum (FHSS) technology. This data rate is equally shared to all wireless users in the cell. The half duplex operations of the equipment and the encapsulation of overhead packets during transmission, result in an actual throughput of 1.7Mbps for the FHSS link as measured experimentally.

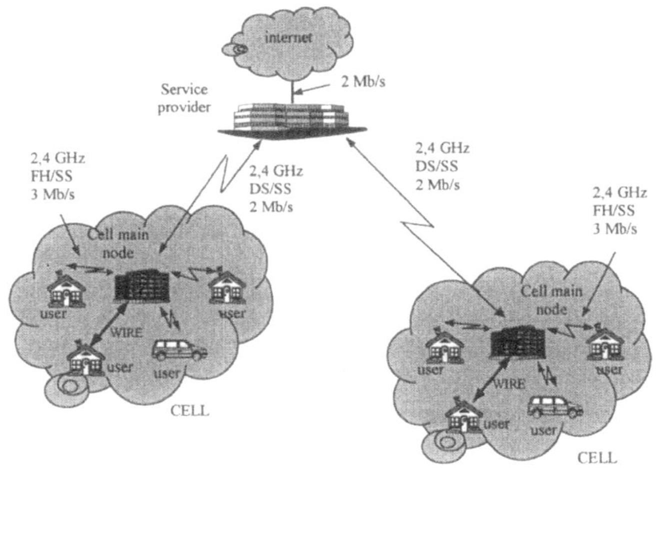

Fig. 10.2.3 Overall system architecture.

Given the two bottlenecks of the network (1.2Mbps and 1.7Mbps), the maximum throughput that may be provided to a user of a cell is 1.2Mbps. This throughput of 1.2Mbps (between service provider and cell main node) is equally shared between all wired users and the group of *all* wireless users. Therefore, the throughput that may be guaranteed during multimedia service access to a wireless user of a cell serving Nw wireless users and Nc cabled users is:

$$Bw = \begin{cases} \dfrac{1.2}{Nw}\ Mbps\ for \quad Nc = 0 \\ \\ \dfrac{1.2}{Nw(Nc+1)}\ Mbps\ for \quad Nc \neq 0 \end{cases} \qquad \textbf{(1)}$$

A wired user may be guarantied throughput at least equal to

$$Bc = \begin{cases} \dfrac{1.2}{Nc}\ Mbps\ for \quad Nw = 0 \\ \\ \dfrac{1.2}{Nc+1}\ Mbps\ for \quad Nw \neq 0 \end{cases} \qquad \textbf{(2)}$$

For example, if a cell main node consisting of one cabled user and two wireless, the system quarantines for multimedia services provision 0.6Mbps to the wired (Bc) and 0.3Mbps to each wireless (Bw).

The throughput during Internet access is not guarantied because there exist other bottlenecks along the information flow, beyond the reach of the system described here.

10.2.4.2 Description of the system

The block diagram of the system that was implemented is shown in Fig. 10.2.4. The network of ARIADNET (which is capable of providing Internet) is connected with the LAN of the service provider via a router, at a rate of 2Mbps. The multimedia services which are offered to the users are hosted in a multimedia server, which is connected to the LAN of the service provider.

From the networking point of view, the users within a cell can be considered to form a Local Area Network (LAN). The LAN of each cell is connected to the LAN of the service provider.

The communication between the LAN of the service provider and the cell is based on a radio Bridge/IP router. The system uses DSSS modulation, which offers the ability of rejecting radio interference. The Bridge/IP router operates in the frequency band of 2.400-2.4835GHz (ISM band) with a data rate of 2 Mbps, and in a half duplex mode. This system collects the data packets it receives from one LAN and retransmits them to the other LAN, in both directions. The Media Access Control (MAC) layer bridge that was used in this trial follows the 802.11 protocol standard of IEEE.

The communication between the wireless users and the cell main node is achieved through a Point-To-Multipoint radio LAN system, based on FHSS technology. The basic configuration of this radio LAN consists of a central unit (the Access Point) and a number of wireless workstations, equipped with an appropriate station adapter. The access point enables the wireless workstations to access the LAN of the cell main node. Workstations are also able to communicate with one another inside the cell coverage area via the access point. Any two wireless stations in two different cells can communicate through their access points.

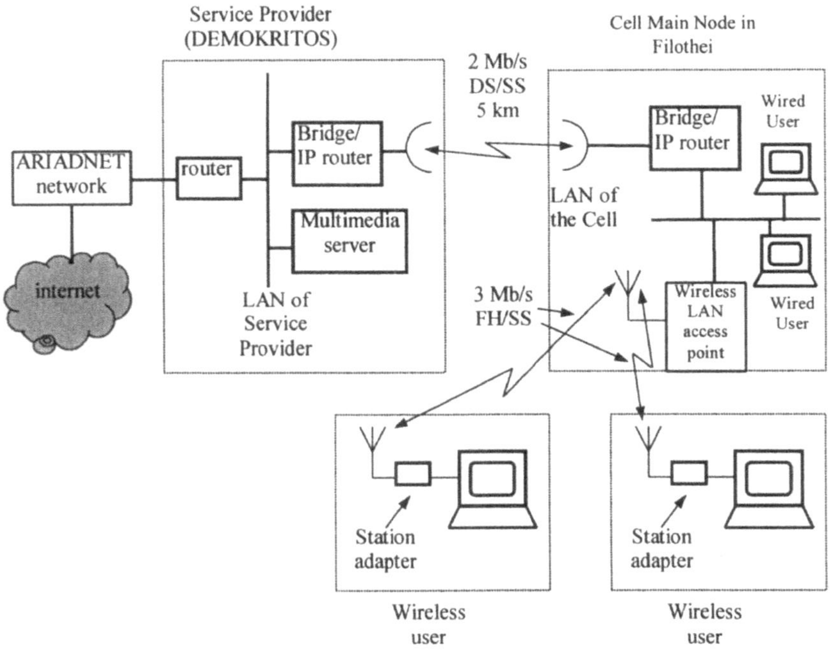

Fig. 10.2.4 Block diagram of the implemented system.

The RF link between the access point and the wireless stations is in the ISM band and the data rate can reach 3Mbps, with automatic fallback to 2 and 1Mbps.

The hardware requirements for a user is a PC, equipped with a common Ethernet card. A wireless user also needs a station adapter, which is externally connected to the PC. This adapter transforms user's request bit-stream, into a RF signal and transmits it to the cell main node. In the reverse direction, the adapter converts the RF reply signal (forwarded by the cell main node) into a bit-stream.

The software requirements include only networking software implementing a TCP/IP stack and a common WWW browser on the top of that. The user can access the services provided by the system using this plain WWW browser.

When a wireless user issues a request for Internet or multimedia services, the corresponding bit stream is transformed by the station adapter to a RF signal, that is collected by the wireless LAN access point in the cell main node. There, the signal is demodulated and forwarded to the bridge/IP router. This device transforms the baseband signal into a RF signal and sends it to the corresponding bridge/IP router of the service provider. There, the signal is again demodulated and forwarded to the LAN of the service provider. If the user's request is for Internet access, then the router forwards the request to the ARIADNET network, which takes the responsibility of processing the user's request and producing the appropriate reply. If the user's request is for multimedia services, the multimedia server produces the suitable reply. In either case the response to the user's request follows a reverse path from the bridge/IP routers to the wireless LAN access point, and from there to the station adapter, where it is converted to the appropriate bit stream.

10.2.4.3
Performance Measurements

An experimental platform of the described system has been implemented. The experimental platform includes:

the service provider located at the premises of NCSR Demokritos (connected to ARIADNET)

one cell main node (located at the Townhall of an Athens suburb, Philothei, at a distance of about 5km from Demokritos under Line Of Sight conditions)

two cabled users (namely C1 and C2), located at Philothei Townhall

two wireless users (namely W1 and W2), located about 10m away from the wireless LAN access point's antenna in Philothei Townhall and

one portable/mobile user (namely P/M) moving around the Townhall.

The cell main node defines a cell with a diameter approximately equal to 1 km for Line Of Sight conditions [4], [5]. The DSSS RF link between Demokritos and Philothei Townhall has an output power of 32mW. Directional antennas of 23dBi gain, in both directions were used. The FHSS RF link between a user and Philothei Townhall has an output power of 100mW and omni-directional antennas of 6dBi gain were used.

A series of experiments were conducted on the platform to verify that the link capacities were fully utilised and that the resources were fairly shared among all users, during stationary and portable/mobile conditions. All measurements during these experiments were made through a performance monitor software tool operating over the MS Windows NT Operating System running on the multimedia server.

Channel sharing. The first experiment was designed to verify that the nominal throughput of 1.2Mbps of the link between the service provider and the cell main node is fully exploited during multimedia service provision. Only a wired user, requesting the download of a video clip 10Mbytes long, was active during this experiment. Based on the design specifications, it was expected that the single user would get all of the 1.2Mbps. Fig. 10.2.5 indeed verifies that the full link speed is utilised.

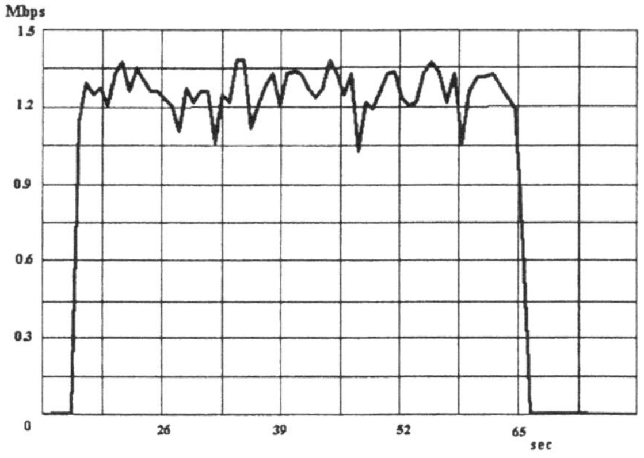

Fig.10.2.5 Service provider to cell main node throughput.

A sequence of repeated similar experiments was conducted to measure the throughput of wired user C1, in the presence of other simultaneous users, all requesting multimedia services. The results appear in Fig. 10.2.6 and are in full accordance with equation (2). It is indeed verified that the throughput of 1.2Mbps between service provider and cell main node is equally shared between all wired users and the group of *all* wireless users. This becomes apparent especially by comparing either the 3rd and 4th, or, similarly, the 5th and 6th, bars in Fig. 10.2.6.

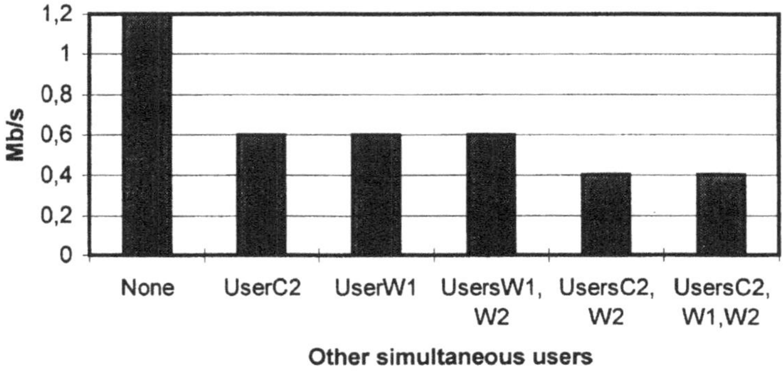

Fig. 10.2.6 Wired User Throughput (User C1)

Another set of measurements traced the throughput engaged by the wireless users. During this set of experiments, only one wireless user was active and requesting the download of a video clip 10Mbytes long, the same as previously. Now, as predicted by equation (1), this user should utilise the full 1.2Mbps capacity of the spread spectrum based wireless LAN. Fig. 10.2.7 shows that the link capacity is indeed utilised.

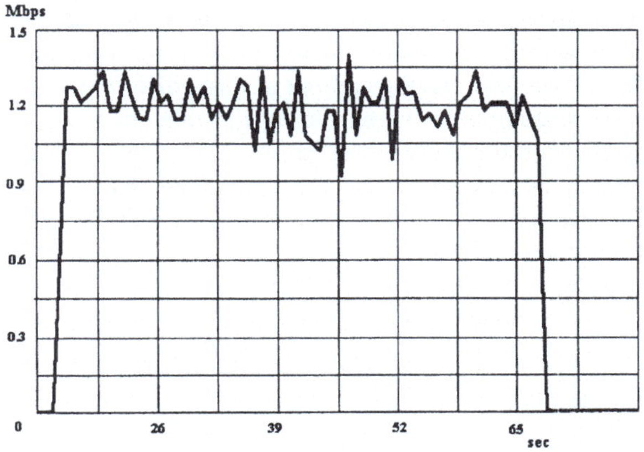

Fig. 10.2.7 Service provider to Wireless User throughput

Figure 10.2.8 depicts the throughput "seen" by a wireless user (i.e., user W1), in the presence of other simultaneous active users. The diagram is in complete accordance with equation (1) and verifies that the 1.2Mbps spread spectrum link is equally shared between all wireless users, while the 1.2Mbps «service provider-to-cell main node» link is equally shared between the wired users and the wireless group of users.

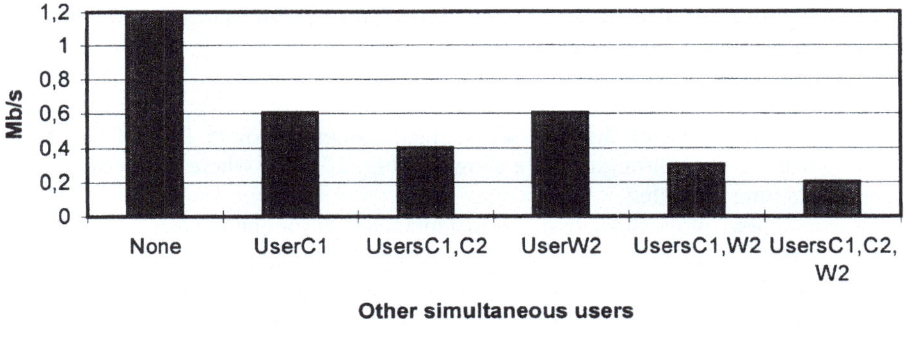

Fig. 10.2.8 Wireless User Throughput (User W1)

Channel utilisation under portable/mobile conditions. The second experiment was designed to verify that the nominal throughput of 1.2Mbps of the link between the service provider and one portable/mobile user (P/M) is fully exploited during multimedia service provision.

The P/M user's PC was placed in a car which could move around the Townhall. The overall throughput of the network was measured during a file transfer procedure, established from the server (at Demokritos) to the user (in the car). Two sets of measurements were carried out. Each one includes three (static) measurements of the network throughput at three different points of the area and a fourth (mobile) measurement taken during the movement of the user following a route passing through the above three points. The test points and the routes of the two measurement sets are shown on the map of the test bed area presented in Fig. 10.2.9. Note that points A and C are in direct LOS conditions with point B (Townhall), while points D and E are not.

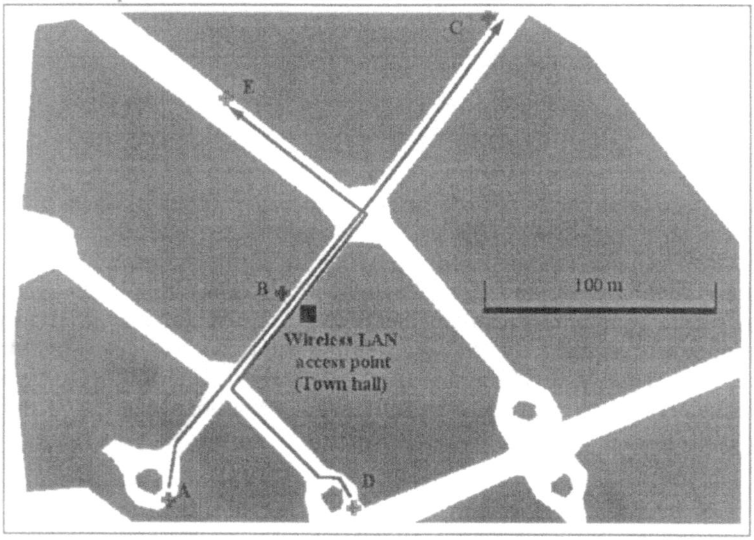

Fig. 10.2.9 Map of the test bed area with the points of static measurements and the routes of mobile conditions measurements.

The results of the first set of measurements (points A, B, C, and movement from A to C through B) are shown in Fig. 10.2.10, where the network throughput Vs time is plotted.

These plots show that the instantaneous throughput varies around an average value. It can be verified that the average throughput at measurement points A, B and C is 871kbps, 972kbps and 935kbps, respectively. The measurement under mobile conditions gives an average throughput of about 828kbps. It can also be verified that during the movement from point A to point C there is a significant degradation of the instantaneous throughput because of the fading of the RF signal.

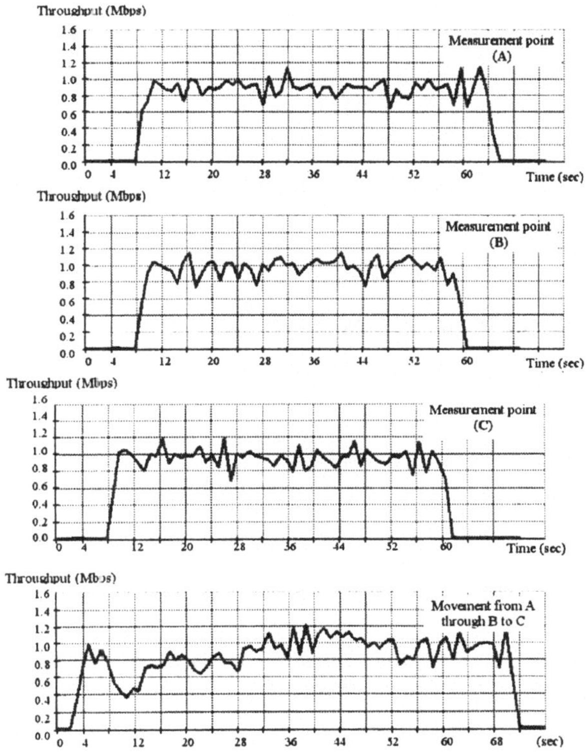

Fig. 10.2.10 Network Throughput Vs Time at measurement points A, B, C and during movement of the portable station from A to C via point B.

The results of the second set of measurements (points D, B, E, and movement from D to E through B) are shown in Fig. 10.2.11. It can be verified that the average throughput at measurement points D, B and E is 635kbps, 972kbps and 564kbps, respectively.

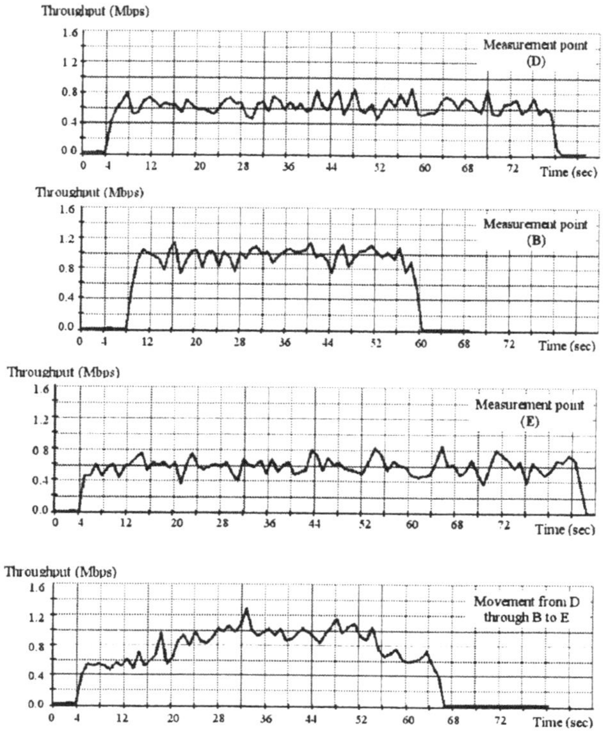

Fig. 10.2.11 Network Throughput Vs Time at measurement points D, B, E and during movement of the portable station from point D to E via point B.

The reduced network throughput at points D and E are due to the no line of sight conditions. The measurement under mobile conditions from point D to point E gives an average throughput of about 776kbps.

The user in the car could also access Internet. However, the network throughput during Internet access could not be measured because there exist other bottlenecks along the information flow, beyond the reach of the system described here.

10.2.5
System's economics

The implemented system, described in subsection 10.2.4.2, served as a demonstration point for collecting data concerning the feedback of users about the provided multimedia services. This subsection presents and analyses these data and elaborates on the potential for commercialisation of the Greek prototype system.

In order to collect reactions from a large sample of users, a number of PCs in the Municipal Centre (MC) connected to the network were offered to visitors. The MC was open for public access three hours per day (18:00 to 21:00 local time), five days per week. It allowed free of charge usage of its equipment and services to a number of real users of different age, educational background, computer skills, and financial status. With respect to this last factor, it should be noted that the area of Philothei is inhabited mainly by high income residents. This bias may need to be taken into account when generalising results presented herein.

From October 1997 to March 1999 the MC was visited 280 times by 85 different persons (some persons visited the MC more than once). The visitors experienced the multimedia services provided by the Greek prototype system and gave considerable feedback concerning the system's appeal and acceptance from their viewpoint. The collected feedback data refer to the pre-mentioned 18 month period. Within this period, the MC was closed for 2 months for summer vacations (August and September 1998), and for another two months due to service and maintenance reasons (January and November 1998).

This section analyses the collected data from a service operator's viewpoint (subsection 10.2.5.1) and comments on the commercialisation potential of the prototype system in the Greek market (subsection 10.2.5.2). Firstly, investigates the subscription fee that a user would be willing to pay (according to the user's profile) for this kind of service. The outcome is used to explore the potential for providing commercial services through a similar system.

10.2.5.1
Feedback of users for the Greek prototype system

Users' profile. This paragraph presents results on the profile of users that visited the MC for trying the offered broadband services (i.e., fast Internet access, and reception of short audio & video clips on demand with a guaranteed minimum throughput). The user profile is a major factor, highly correlated with feedback originating from this user.

In relation to age, the majority of the users was below 30 years old (80% of the total number of visitors). Users above 30 years of age represented 10% of the total; finally, another 10% did not disclose their age. These results show that young persons may constitute a significant portion of the target customer group for any future commercial exploitation of the discussed system.

The results also show that more than two-thirds of the respondents had previous experience in using a PC. Furthermore, it was found that 57% were PC owners, and 60% of them had used Internet before.

The majority of users found fast Internet access and multimedia services (short audio & video clips on demand) very appealing. The attraction of these services is proven by the fact that the number of users that revisited the MC increased over time. Fig. 10.2.12 shows that the number of new users (first bar in each group) decreased as time progressed. In addition, the number of old users (second bar) tended to stabilise. This indicates that, as the system became an established fixture in the MC, those who found the system interesting stabilised and eventually became a core of regular users.

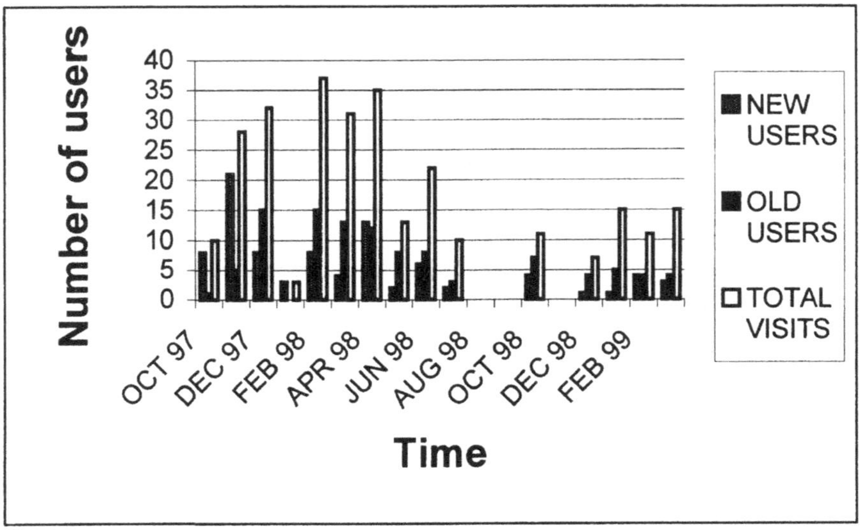

Fig. 10.2.12 New and old users, and total monthly visits to the prototype system Vs time

Furthermore, the recorder of HTTP requests showed that entertainment was the favourite subject of navigation though Internet, as the 33% of the total number of the users' HTTP requests concerned material about entertainment (Table 4.1 displays the percentage of the HTTP requests for the various accessed subjects). It appears that future commercial exploitation of the discussed system has to be based on the highest quality provision of the most appealing service, i.e., entertainment, either through multimedia services, or fast Internet access.

During the demonstration period, as already mentioned, the broadband services of the Greek prototype system were offered to the users of the MC free of charge. When the users were asked if they would pay for these, the overwhelming majority of them (99%) answered that they would pay for fast Internet access and multimedia services. The next subsection discusses the issue in more detail.

Table 10.2.1 Requested information during the demonstration period of the Greek LMDS system

Subject	percentage of HTTP requests
Entertainment	33%
Computing	16%
Education	11%
Games	9%
Art	8%
Sports	6%
Travel	6%
Personal Finance	5%
Health	4%
News	2%

Willingness of users to pay for the provided broadband services. Users of the system had to consider two types of expenditures: the purchase of the required equipment (e.g., a wireless LAN station adapter, an antenna, etc.) and the monthly subscription for accessing the broadband services. Analysis of their feedback concerning these two expenditures is given below.

Fig. 10.2.13 shows the willingness of users to pay for the equipment versus their age. Young users (below 20 years old) were willing to pay any amount of money for the required equipment. Given that persons of this age are rarely self-supported and fully aware of financial issues, this result is of limited significance. Users between 20 and 30 years of age, mainly answered that they would pay an amount of money between 330 to 660USD, without excluding the possibility of spending more money to cover a higher equipment cost. Lastly, older people (above 30 years old) mainly found that a reasonable amount of money should be less than 330USD. Combining these results, it appears that a future commercial system should require home equipment of a cost no greater than 330USD, if full customer acceptance is to be gained.

Fig. 10.2.14 shows the maximum acceptable monthly subscription fee versus age. The figure shows that persons below 20 years old would pay any amount of money; the same comments as for the equipment cost apply. Persons between 20 and 30 years old thought that a monthly subscription of 100USD seemed most reasonable to them. Older users (above 30 years old) thought that the monthly subscription should be less than 66USD. Taking into account that the majority of users of a future system will be young persons (see Subsection 10.2.4.1), the exploration of the potential commercialisation of such a system should be targeted around a monthly subscription of 100USD per user.

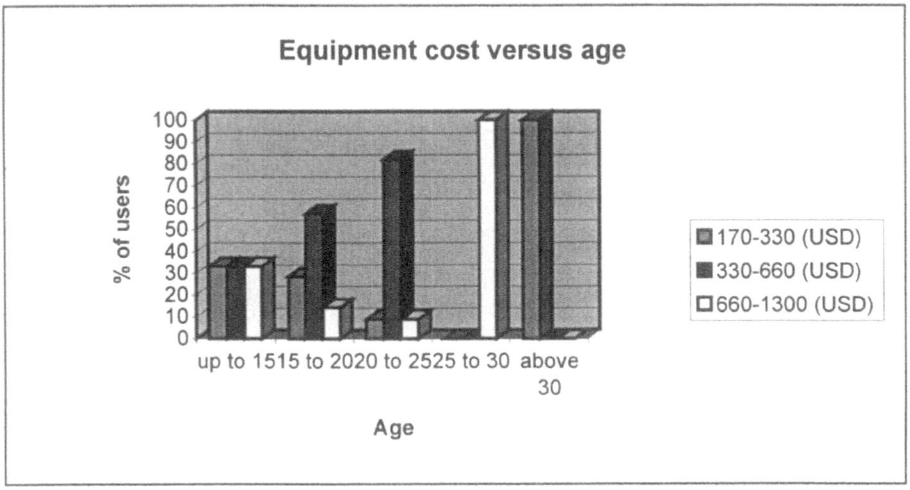

Fig. 10.2.13 User's willingness to pay for equipment Vs age

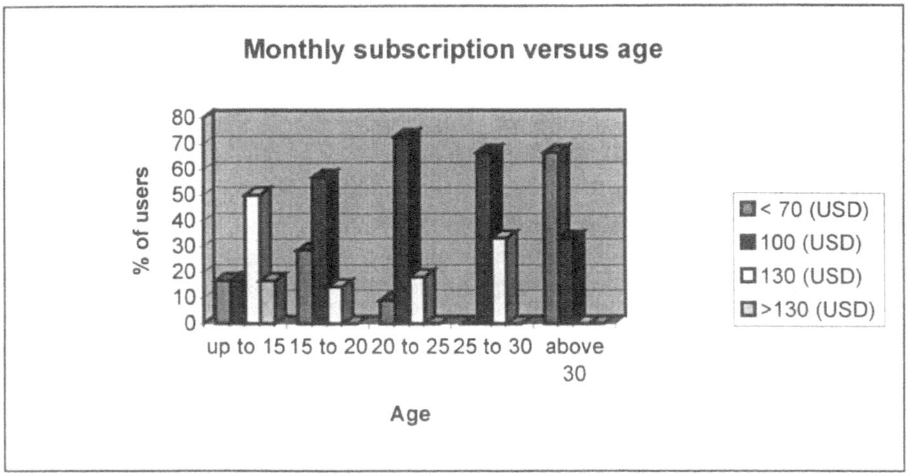

Fig. 10.2.14 Maximum acceptable monthly subscription fee for accessing Internet services Vs age

10.2.5.2 Market potential

This section explores the commercialisation potential of the prototype system (described in Section 10.2.4) in the Greek market, based on the feedback of the Greek users, as presented in previous subsection (10.2.5.1). All the figures of the cost analysis are given in USD.

Note that the discussed cost analysis excludes all expenditures referring to the purchase and installation of home equipment. This is because the users of the Greek prototype system made use of the equipment installed at the MC. However, for future usage of the system, potential users will have to buy the required home equipment (e.g., a wireless LAN station adapter, an antenna, etc.). It is supposed that these users will (be willing to) cover the cost of the equipment required for gaining access to the provided broadband services.

The commercialisation potential of the described system depends on the relation of its capital and monthly operational and maintenance costs (Subsection 10.2.5.1) to its revenues coming by the monthly subscriptions of the interested users (Subsection 10.2.5.2). It is apparent that only if the monthly revenues exceed the monthly expenditures, will the system be profitable and the associated investment feasible.

Capital investment and monthly operational and maintenance costs. The capital investment for the Greek system included expenditures for the purchase of Bridge/IP router (two pieces), router & hub (two pieces), a multimedia Server, and a wireless LAN access point. The total as well as the individual costs of these items are shown in Table 10.2.2.

The monthly operational and maintenance costs of the described system are also depicted in Table 10.2.2. Internet provision requires two charges: one for the PTT that offers the physical connection (usually through leased lines) between its premises and the Internet provider and one for the Internet provider itself. Typical charges per month, for a 2Mb/s link, are 1,500USD for the PTT and 3,500USD for the Internet provider, approximately.

The hardware maintenance of the system requires at least one technician experienced in telecommunication systems (e.g., antennas, transmitters, modulators, receivers etc.) and networking hardware (e.g., servers, routers, PCs etc.). The monthly salary of this person (including social insurance) is about 850USD. The software maintenance of the system requires also one person experienced in systems and networking software (e.g., Windows NT and 95/98, routers programming, etc.). The corresponding monthly salary is approximately the same as the previous one.

The renewal of services (i.e., insertion of new short audio & video clips on the WWW server) demands at least one person with experience in the digitalisation of audio and video clips. The monthly salary of such a person is about the same as previously (850USD). Furthermore, the renewal of services requires two additional charges: one for buying the most recent audio and video clips and one for paying the respective copyright royalties so that broadcasting will be allowed. The cost for buying the top ten audio and video clips in three music categories (i.e., foreign, Greek, classic) is about 1,500USD per month and the monthly cost for copyright royalties is about 300USD. Lastly, fixed expenses (such as secretarial support, rent for the MC premises, power bills etc.) of a sum of approximately 1,500USD should be added to the above costs.

The resulting total capital and monthly operational and maintenance costs of the Greek system are 30,000USD and about 11,000USD, respectively. Therefore, the establishment of a viable and profitable system requires monthly revenues greater than 11,000USD plus a percentage of the capital cost. It should be mentioned that this percentage varies according to the operator's market plans and its financial background. For a common operator, this percentage should approximately result in equipment repayment within 36 months. In this context, the capital investment of 30,000USD requires a monthly revenue of about 940USD (i.e., 833USD instalment + 107USD bank interest).

The following subsection takes into account that, as the feedback suggested, potential users of the Greek system can not afford more than 100USD for a monthly subscription. It explores the system's viability under this assumption.

Table 10.2.2 Estimation of capital investment and monthly operational and maintenance costs

Estimation of capital investment	
ITEM	**COST in USD**
Bridge/IP router (2 pieces)	10,000
router & hub	2,500
a WWW server	15,500
a wireless LAN Access Point	2,000
TOTAL	30,000
Estimation of monthly operational and maintenance costs	
ITEM	**COST in USD**
connection to Internet provider (2 Mb/s)	3,500
leasing of PTT line (2 Mb/s)	1,500
H/W maintenance (1 person)	850
S/W maintenance (1 person)	850
renewal of services (1 person)	850
buying songs and video clips	1,500
copyrights	300
fixed expenses	1,500
TOTAL	10,850 ≅ 11,000

System's viability. It is well known that a good trade-off between quality and cost is the most important factor for the commercial success and the viability of a new system or product. Based on this market tenet, an operator who wants to establish such a system based on the Greek prototype has to consider the quality of the provided services along with the money the user has to pay as a monthly subscription fee.

The implemented Greek prototype is an one-cell system serving a number of users distributed in an area around the cell main node. The number of users in this cell affects the system's efficiency. Efficiency is defined as the factor denoting the number of users in the cell as a percentage of the maximum number of simultaneous users that the cell can serve with the highest QoS (Quality of Service). The highest QoS of such a system is achieved when each user is guaranteed a bandwidth of at least 100kbps (which is the required bandwidth for adequate playback of a short video clip on demand) any time during the day, resulting in an 100% efficient system. Specifically, the proposed prototype may serve up to 20 users, while providing 2Mbps cumulatively to all of them, in order to maintain 100% efficiency.

But efficiency does not always comply with attractiveness (and therefore with viability). Referring to an 100% efficient system (up to 20 users in the cell) based on the Greek prototype, the drawback is that (according to Subsection 10.2.5.2) this system, in order to remain profitable and viable, should have monthly revenues greater or equal to 11,940USD (11,000USD + 940USD). That means that each of the subscribers should pay about 597USD for monthly subscription. However, according to subsection 10.2.5.1, no one would like to become a user/subscriber of such an expensive system. One immediate solution, which makes this system attractive to users and also maintains it profitable, is to increase the number of subscribers. However increasing the number of users to become, say 100, results in a decrease of the system's efficiency (20%) and, therefore, the QoS requirement is not met.

To overcome these problems, a suitable trade-off between efficiency and attractiveness must be made. A multi-cell configuration comes to foreground and seems as a very promising solution. This configuration implies the set up of several cell main nodes (each one defining its own cell area), communicating with the same service provider. In such a configuration the capital cost and the overall number of users/subscribers increase linearly with the number of cells. The monthly maintenance and operational costs remain constant, but only for a moderate number of cells. Should more cells become required, additional staff will be necessary and the increase in maintenance and operational costs will need to be taken into account. This chapter assumes that the system will require only a few cells, without the need for additional staff.

For example, in a 5-cell configuration the capital cost is five times the one-cell configuration, i.e., 150,000USD, while the maximum number of users is raised to 100 (20 per cell) for 100% efficiency. For such a system the monthly revenues should be greater or equal 15700USD [(5*940)+11,000USD], in order to maintain profit and viability, resulting in a 157USD subscription fee. The same system with 40% efficiency (50 users per cell), can serve up to 250 users, resulting in a subscription fee of about 63USD (a preferable fee according to subsection 10.2.4.2).

Figure 10.2.15 shows the trade-off between the number of cells and the monthly subscription fees required to maintain a viable and profitable system. Each thick line corresponds to a different number of users per cell (resulting in different efficiencies), while the dashed one denotes the preferable amount of the subscription fee (100USD), below which the systems becomes also attractive and marketable.

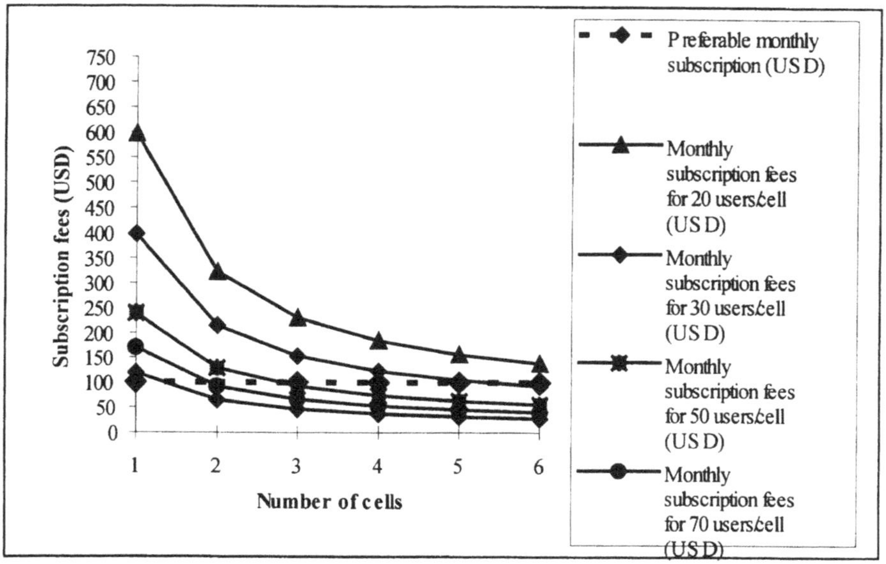

Fig. 10.2.15 Monthly subscription fees versus the number of cells

From Figure 10.2.15, it follows that a configuration using 5 cells suffices to meet the QoS criterion limiting the monthly subscription fee below the maximum acceptable level, while still allowing for profitable commercial exploitation.

More generally, consider a system with no more than C cells so that no additional personnel is needed. Assume an amount of monthly expenses F, plus an amount of monthly expenses per cell B, and suppose that the monthly subscription fee must be no more than M currency units. Then, it may be shown that the number of users (U) in each cell must be at least:

$$U \geq \frac{B + \frac{F}{C}}{M}$$

For a setting of C=6 cells, M=100USD, B=940USD, and F=11,000USD, the above formula results in a requirement for 28 users per cell. Given a link of 2Mbps, each of these users may be guaranteed a bandwidth of 71kbps. The formula may be used to find an optimal trade-off between expenses and QoS requirements (reflected in U), so that the system remains marketable. A similar formula may be devised for larger number of cells, by approximately reflecting B and F, so that costs due to additional personnel may be taken into account.

10.2.6 Conclusions

This chapter presented a system for provision of fast Internet and multimedia services to a number of stationary, portable and mobile users. It reviewed, in brief, the present wired and wireless access technologies, and compared them by way of speed, installation and maintenance cost, time-to-market, flexibility and scaleability.

The use of digital terrestrial access technologies (MMDS/ISM) in conjunction with the IEEE 802.11 standard, provides secure, noise free and reliable transmission, while enabling broadband access to the provided services. A system featuring the above characteristics, can be attractive to users, who want fast access to services from anyplace they are any time during day, while keeping the monthly operating and maintenance costs to a minimum. The installation of such a reliable and cost-effective system requires knowledge of the market's demands; service providers who are willing to invest a lot of money, must aim at long-term revenues.

A prototype system, featuring the above characteristics, was implemented in the context of CRABS European project. It served as a test bed for conducting performance measurements, as well as for collecting users' feedback for the commercialisation potential of the system.

- The performance of the system, which was evaluated under stationary, portable and mobile conditions, show that a user can be guaranteed a network throughput during multimedia service provision, but not during Internet provision. Also, the results show that the system can not guarantee a bandwidth to a mobile user (during Internet or multimedia service provision), due to propagation conditions.
- The potential commercialisation of the system show that the installation of a reliable, high speed and cost-effective system requires a multi-cell configuration, where each cell serves a maximum number of users. This configuration enables for low subscription fees for the user, and high revenues for the service provider.

References

- CRABS AC215, Technical Annex, Aug., 1996.
- CRABS AC215, *Technical Annex,* Aug., 1996. See also http://www.crabs.ariadne-t.gr.
- AC215 CRABS Deliverable D21P1B, AC215/CRABS/WPG2P1/R/K/P/D21P1B, *Specification of detailed systems architecture,* September 1998.
- G. Rosner, "LMDS, cable modems, and xDSL: A Comparative Study", see http://fargo.itp.tsoa.nyu.edu/~rosner/.
- R. Rini, and S. Coran, "The FCC's new deal for MMDS", see: http://wirelessvvd.com/newdeal.html.
- D. Bergen, "LMDS wireless multimedia", see: http://www.wirelessvvd.com/LMDSbergen.html.
- E. Pallis, M.C.B. Smith, A. A. Alexandridis, A. Kourtis, K. Dangakis, P. Kostarakis, «Wireless provision of fast access to Internet and multimedia services to stationary and portable stations, Proceedings of the International Workshop on Mobile Communications 1999, June 24-26, Chania, Greece.
- A. Kourtis, Ch. Mantakas, and V. Pallis, "Broadband Interactive Services and Digital Television at 42GHz", *International Journal of Communication Systems,* January 1999, Vol. 12, pp. 11-22.
- A. Kourtis, and C. Mantakas, "Interactive digital television at 42GHz", *Electronics Letters,* October 1997, Vol. 33, No. 22, pp. 1843-1844.
- Barnhart E., «Recent developments in broadband multipoint terrestrial communication networks», Proceedings of the International Conference on Telecommunications (ICT'98), Vol. II, pp. 361-364, June 21-25 1998, Chalkidiki, Greece.
- G. Kormentzas, V. Pallis, A. Kourtis, K. Kontovasilis, «A Cellular System for Providing Wireless Services», Proceedings of the 3rd ACTS Mobile Communications Summit, Vol. 2, pp. 672-677, June 8-11 1998, Rhodes, Greece.
- A. Kourtis, N. Papadopoulos, and V. Pallis, «A 42 GHz Broadband Wireless Terrestrial System for Digital Interactive Television», Proceedings of the International Conference on Telecommunications (ICT'98), Vol. II, pp. 370-373, June 21-25 1998, Chalkidiki, Greece.
- V. Pallis, A. Kourtis, Ch. Mantakas, N. Papadopoulos, «Provision of Digital Television and Broadband Multimedia Services in the 42 GHz Band», Proceedings of the First International Symposium on Wireless Personal Multimedia Communications (WPMC '98), pp. 99-103, November 4-6 1998, Yokosuka, Japan.

- Jack Glas, «The principles of Spread Spectrum Communication», see: http://www.rfglobalnet.com/library/Papers/files/7/sstech.htm.
- The IEEE P802.1, see: http://grouper.ieee.org/groups/802/11/main.html.
- Andrew S. Tanenbaum, «Computer Networks», International Edition, Third Edition.

10.3 IP Subnet Mobility over General Packet Radio Service

Mika Mizutani and Christian Bonnet

10.3.1 Introduction

The data services through the Internet are getting to be familiarized [1]. A lot of various services become widely used, for example the access to the e-mail server and the web server on the Internet and the Intranet, the access to the private database on the Intranet, the information delivery service using the push technology and the electronic commerce such as banking service and reservation service. As the result, the IP packet traffic is increasing remarkably, and it is forecasted that IP packet traffic is going to be over 160 Gbps in the year of 2001 [2]. So the next generation IP network should become to consist of the network which is advanced from the heterogeneous network such as the telephone network and the mobile network in addition to the growth of the IP network itself. It means the user can access to the data services through not only the IP network but also the above-mentioned heterogeneous network.

On the other hand the mobile network system such as the cellular systems that is called GSM or PDC is paid attention to as the telephone network. Because the mobile subscribers are increasing rapidly, too, and the access with the fixed IP network over the mobile network system is very attractive in order to take advantage of getting the data services anytime and anywhere [3]. So the mobile data network system is required and anticipated strongly.

As the infrastructures of the data service on the current mobile network system that provides the circuit-switched service has the following limitation. First, the transmission speed of Short message Services (SMS), that transmits short data messages reach up to 160 alphanumeric characters long, cannot reach 1 Kbps and the data service over the circuit-switched service cannot reach 10 kbps. Secondly, the circuit-switched service isn't suitable for IP traffic that has a characteristic of the burst traffic. In this service, the radio resource remains being assigned while there is no IP traffic. And the same bandwidth is allocated for the upstream and the downstream even if the subscriber downloads the much information from the Web server. So the circuit-switched service can't use the radio resource effectively when IP packet traffic is transmitted. And thirdly, the cost that is in proportion to the time of occupying the radio resource is a problem to show a tendency to be getting higher.

To solve these problems, the new infrastructures appear. Using the current radio resource efficiently, the High-speed circuit switched data (HSCSD) and GPRS that provides the packet-switched services have been standardized by European Telecommunications Standards Institute (ETSI), and they are expected to be launched in the year of 2000 [4]. The 3rd generation mobile network systems that are called as International Mobile Telecommunications–2000 (IMT-2000) and Universal Mobile Telecommunication System (UMTS) are standardized by International Telecommunication Union (ITU) and ETSI [5][6][7]. And the 3rd generation is able to realizes the wide bandwidth and guarantee the high data transmission speed corresponding to the moving speed of the mobile terminal, the guaranteeing speed is 2 Mbps indoors, 384 kbps in case the moving speed is 10 km/h, and 144 kbps in case of 100 km/h.

In the meantime the IP community recognized the need for the mobility handling between IP subnets as portable computers become more and more popular. This effort resulted in Mobile IP specifications and Ipv6 [8][9]. The mobile data network system such like the GPRS has just considered about the handover between the cells inside the GPRS. And the way to identify the users between the IP community and the mobile network systems is different: the IP address and the mobile subscriber identity. At present the users have to chose between two different solutions in order to combine IP network access and mobility: the GPRS world where the user's terminal address belongs to the address space of a GPRS operator or an Internet, and the IP world where the terminal address belongs to the address space of its home network.

This chapter looks at the integration of IP mobility based on Mobile IP and GPRS to pin in a gap between IP network and the mobile data network system. In this framework the user's terminal retains its home address while roaming under the coverage of a wireless GPRS network. The proposed solution does not impose additional signaling traffic on the radio interface between the terminal and the fixed infrastructure. It has an impact on the GPRS equipment that should handle the functions of the Mobile IP protocol. This leads to architecture the GPRS network not only by defining Routing Areas (RA) but also by grouping the RAs into IP subnets. The expected benefit of this approach is a transparent way to communicate between a roaming terminal and other correspondent terminals.

Section 10.3.2 shows GPRS architecture and the mobility management procedure. Section 10.3.3 describes the way to combine the mobility between IP subnets (IP subnet mobility) with the peculiar mobility management defined in the GPRS. And section 10.3.4 shows the impact of the proposed architecture on the performance of the system.

10.3.2 Outline of GPRS

The purpose of GPRS is to efficiently accommodate burst data sources. Another important goal of the technology is to make it possible for GSM operators to share physical resources between packet data services and other GSM services. It means that the GPRS shares the GSM frequency bands with telephone and circuit-switched service, and makes use of many properties of the physical layer of the original GSM system such as the time-division multiple access (TDMA) frame structure, modulation technique and structure of GSM time slots.

Section 10.3.2.1 shows the GPRS architecture, and section 10.3.2.2 and section 10.3.2.3 describe the Attaching procedure and PDP activation procedure, and the location management procedures as the represent of GPRS mobility management procedures [4][10][11].

10.3.2.1 GPRS Architecture

GPRS provides the packet-switched service to transfer the burst traffic in an efficient manner. GPRS introduces two network nodes in the GSM core network. The Serving GPRS Support Node (SGSN), which is at the same hierarchical level as the MSC, keeps track of the individual MT's location and performs security functions and access control. The SGSN is connected to the base station system (BSS). The Gateway GPRS Support Node (GGSN) provides the interworking with external packet network like the IP network or X.25 network. And the GGSN is connected with the SGSNs via IP-based GPRS backbone network. The HLR is enhanced with GPRS subscription data that includes the packet network address like the IP address. Fig. 10.3.1 shows the GPRS architecture and Fig.10.3.3 shows the protocol structure providing user information transfer.

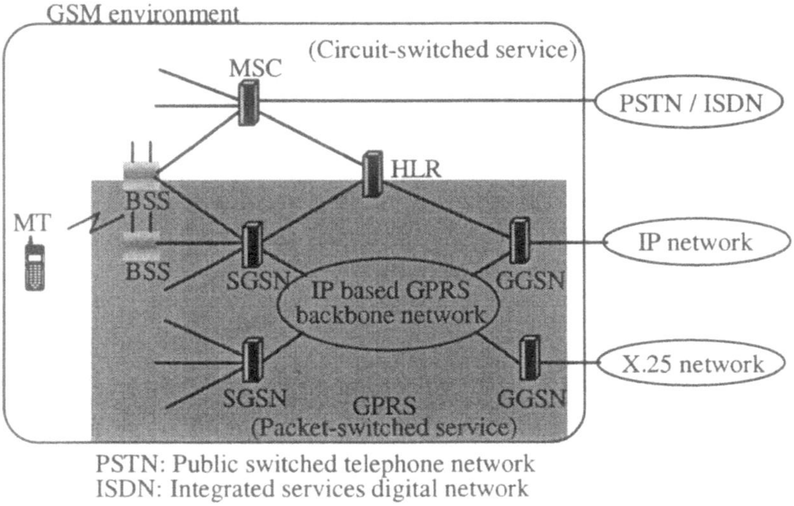

Fig.10.3.1 GPRS architecture

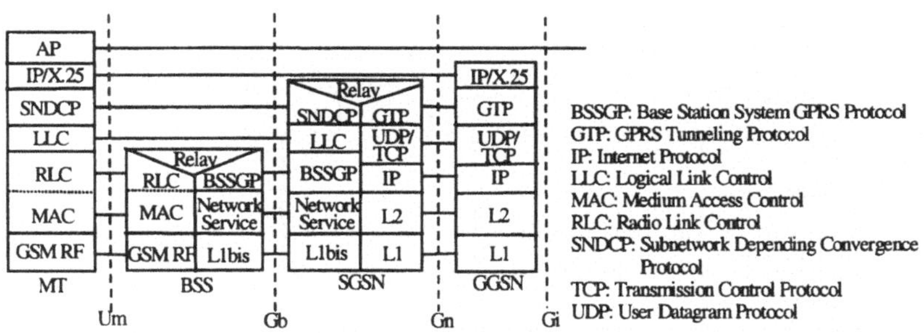

Fig.10.3.2 Protocol structure providing user information transfer

10.3.2.2
GPRS attaching procedure and PDP activation procedure

Before the MT starts transmitting packets, it has to initiate the GPRS attaching procedure and the Packet Data Protocol (PDP) activation procedure. In the GPRS attaching procedure Fig.10.3.5), the MT shall provide its identity. After receiving "Attach Request" with Packet Temporary Mobile Subscriber Identity (P-TMSI) or International Mobile Subscriber Identity (IMSI) from the MT, the SGSN identifies the MT and then the authentication procedure and the equipment checking procedure are initiated between the MT and the HLR via the SGSN. As the result, the SGSN gets the GPRS subscription data that includes the PDP contexts and then creates the mobility management (MM) contexts. The MM Contexts consist of the GPRS subscription data and the mobility information like the identifications of the area and the cell where the MT exists. The GPRS subscription data includes the IMSI and the information of the authentication and the ciphering. The PDP contexts consist of the information of the packet data protocol like the IP address of the MT and the attaching GGSN that are registered in HLR beforehand.

In the PDP activation procedure (Fig.10.3.7), the MT indicates the attaching GGSN and its IP address registered in the HLR. A null IP address means that the MT wishes the dynamic IP address. So the dynamic IP address is assigned through the attaching GGSN in the PDP activation procedure. As the result of the PDP activation procedure, the information to route the IP packet to the MT such like IMSI, IP address of the MT and the attaching GGSN address, is added to the PDP contexts in the SGSN. And the GGSN had the information to route the IP packet as the PDP contexts that includes the SGSN address instead of the GGSN address.

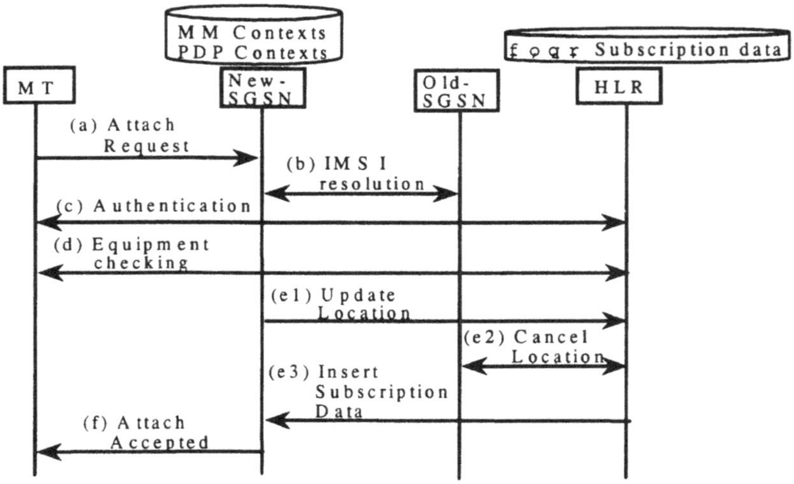

Fig.10.3.3 GPRS attaching procedure

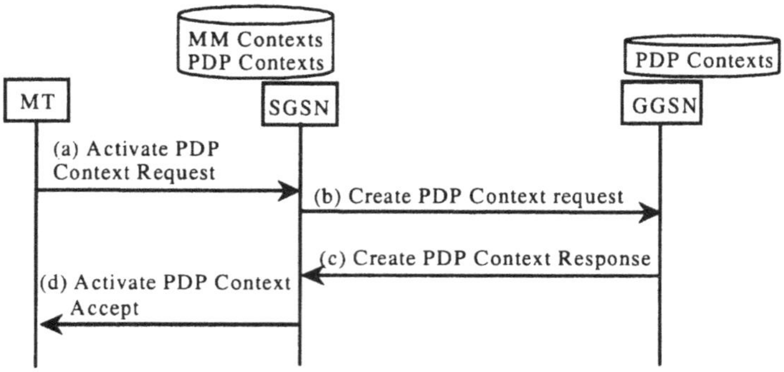

Fig.10.3.4 PDP activation procedure

10.3.2.3
Location management procedure

When the packet is received for a MT, the packet has to be routed to the cell in which the MT is located. In the circuit-switched service, the mobile network system is split up into a number of location areas that consist of a number of cells. In this way the core network can keep a registration of the current location area of each MT, and therefore it will only need to call the mobile within that area. In the packet-switched service of GPRS, the registration area is called as the routing area (RA), and this RA is defined to be smaller than the location area. Because the radio resource is only assigned while packet is transferred, so the frequency of paging per the MT has a tendency to be increasing in comparison with the circuit-switched service.

Whenever the MT receives the packet system information which is broadcasted by the core network, the MT have to check whether it has entered into a new RA by comparing the identification of the RA included in the above-mentioned information. When the MT detects that it has entered a new RA, the MT must initiate the location management procedure. Fig.10.3.9 shows the location management procedure when the MT moves into a new RA that is managed a new SGSN.

As the result of GPRS attaching procedure and the PDP activation procedure, the SGSN that the MT has attached with has the MM Contexts and the PDP contexts. In the location management procedure, the PDP contexts and the MM contexts are created in the new SGSN as follow:

(a) After the MT detects the movement into the new RA, the MT sends "Routing Area Update Request" to the new SGSN.

(b) The new SGSN gets MM contexts and PDP contexts from the old SGSN. And the old SGSN starts the timer and the old SGSN transfer the PDP packets to the new SGSN until this timer expires.

(c) The SGSN updates the PDP contexts in GGSN.

(d) The SGSN registers the new SGSN address in the HLR.

(e) The HLR deletes the GPRS subscription data from the old SGSN. After the timer mentioned in (b) expires, the MM contexts and the PDP contexts are removed from the old SGSN.

(f) The HLR sends the subscription data to the new SGSN.

And after attaching the GGSN and being assigned IP address, even if the MT moves far from the attaching GGSN, the MT uses the same IP address and transfers the IP packets via the same GGSN. In order to support the same environment as the fixed IP network and consider the migration to the next generation IP network, the IP subnet mobility, that provides the IP packet routing mechanism and the terminal movement regardless of the IP network and GPRS, is required. The IP subnet mobility management over GPRS realizes the optimal route selection that means to change the GGSN and the seamless roaming between GPRS and the IP network.

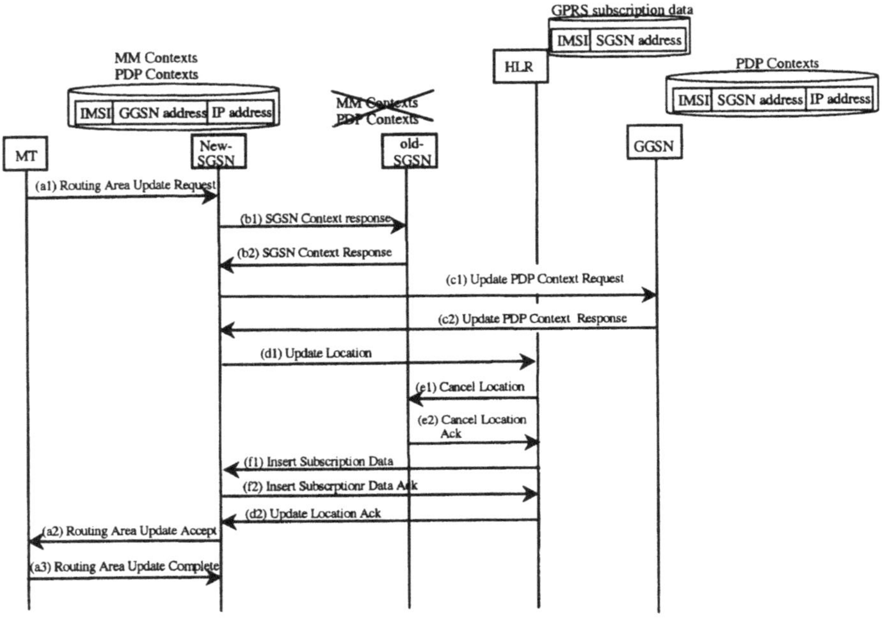

Fig.10.3.5 Location managemetn procedure

10.3.3
IP subnet mobility management over GPRS

In order to realize the IP subnet mobility management over GPRS, the implementation of Mobile IP on GPRS is necessitated. In this case the peculiar mobility management defined in GPRS should be considered to realize the IP subnet mobility management using the Mobile IP efficiently. This section shows how several IP subnets could be allocated in GPRS environment. And the way to combine the GPRS mobility management and the Mobile IP is described.

10.3.3.1
IP subnet concept within GPRS

GPRS should be a part of the IP network, therefore IP subnet concept is required to be introduced into the GPRS environment. To use IP subnet concept within GPRS as same as IP network that consists of several IP subnets, each GGSN manages one IP subnet as shown in Fig.10.3.11 and connects with the IP network directly. And the relationship between the GGSN and the SGSN is defined beforehand. As the result, one IP subnet consists of the several routing areas that are managed by each SGSN, and the all MTs that exists in the same subnet can access with the IP network though the optimal route. So the GPRS core network has to support the following new functions, to select the corresponding GGSN automatically.

(1) When the SGSN receives the request of attaching IP network from the MT, the SGSN make a new connection with the GGSN following the above-mentioned relationship.

(2) When the MT moves into the new RA that is managed by the new SGSN and the new SGSN is allocated in the different IP subnet from the IP subnet where the MT stayed before moving, the new SGSN makes a new connection with the corresponding new GGSN. So the SGSN has to initiate the PDP activation procedure. And the new SGSN deletes the PDP contexts on the old GGSN that manages the old IP subnet.

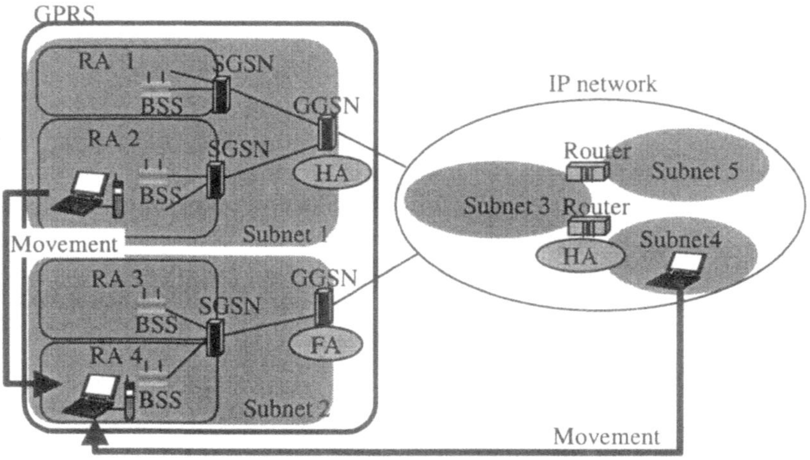

Fig.10.3.6 IP subnet concept within GPRS

10.3.3.2
Combining the IP subnet mobility management and GPRS mobility management

To support the IP subnet mobility inside GPRS and between GPRS and the IP network, the Home Agent (HA) and the Foreign Agent (FA), of which the functions are defined as Mobile IP, have to be assigned on the GGSN in Fig. 10.3.13. The HA is assigned on the GGSN or on the router in the home subnet into which MT belongs. The IP address of the MT is assigned by the home subnet. The FA is assigned on the GGSN in the visiting subnet where the MT moves.

There are following two possible scenarios to identify the movement between the IP subnets:

Scenario 1: the detection of the movement between the IP subnets by the MT, or
Scenario 2: the detection of the movement by the GPRS core network.

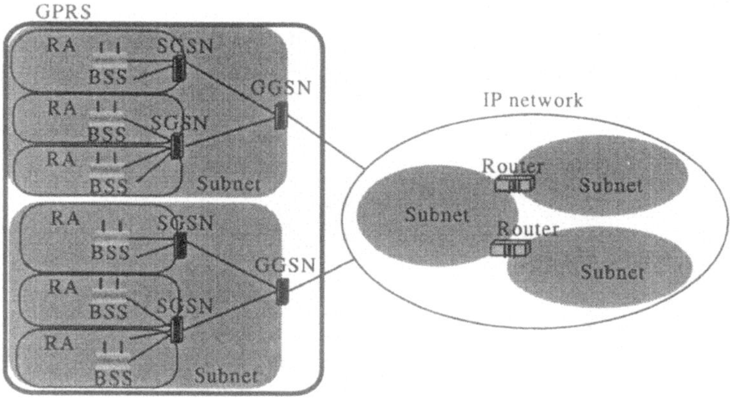

Fig.10.3.7 GPRS structure woth Mobile IP

Scenario 1: The MT detects the movement. In this scenario the MT has to support IP subnet mobility management (MM) function. In Fig. 10.3.15, the IP subnet MM is performed as follow:

(a) When the MT moves into the new RA that belongs into the new IP subnet, the MT only detects the movement into the new RA and initiates the location management procedure.

(b) After receiving "Routing Area Update Request", the new SGSN make a new relationship with the corresponding GGSN according to the IP subnet concept (see section 10.3.3.1).

(c) After the MT receives "Router Advertisement" from the new GGSN, the IP subnet mobility management function on the MT detects the movement into the new IP subnet after the MT compares own IP address and the IP subnet prefix notified by the Router advertisement from the GGSN. And the MT initiates the Mobile IP procedure.

Therefore the control messages related with IP subnet mobility management are transferred on the radio I/F in addition to the control messages of GPRS mobility management. As the result the overhead of the radio I/F is higher than the original GPRS.

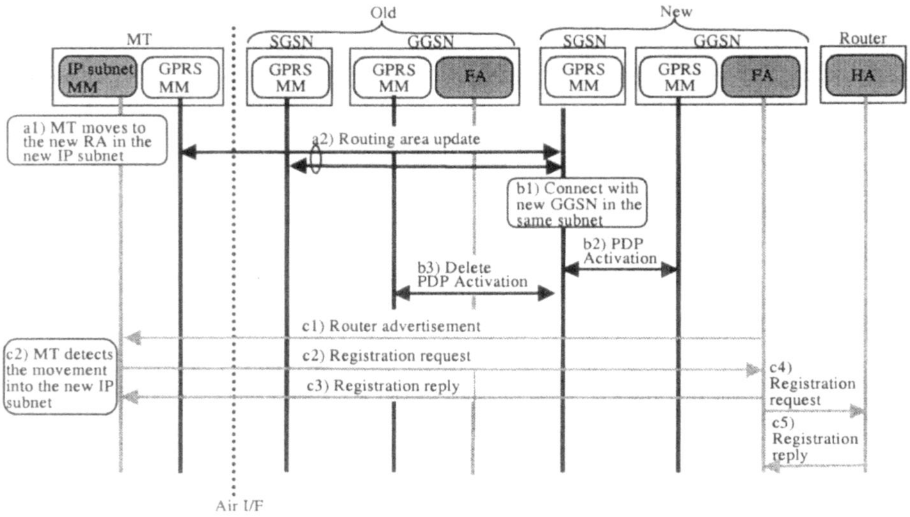

Fig.10.3.8 IP subnet mobility management procedure (Scenario1: the detection by the MT)

Scenario 2: The core network detects the movement. In this scenario the MT only support the GPRS mobility management function that has already prepared by the MT attaching GPRS. But the GGSN has to support the new function, mobility management mapping (MM mapping) function, to detect the movement and initiate the Mobile IP procedure. In Fig.10.3.17, the IP subnet mobility management as follows:

(a) When the MT moves into the new RA that belongs into the new IP subnet, the MT only detects the movement into the new RA and initiates the location management procedure. This procedure is same as scenario 1.

(b) After receiving "Routing Area Update Request", the new SGSN makes a new relationship with the corresponding GGSN according to the IP subnet concept (see section 10.3.3.1). This procedure is same as the scenario 1, too.

(c) The MM mapping function on the new GGSN detects the movement into the new IP subnet as the result of the comparison between the IP address of the MT and the IP subnet prefix address of the corresponding IP subnet. And then this function initiates the Mobile IP procedure.

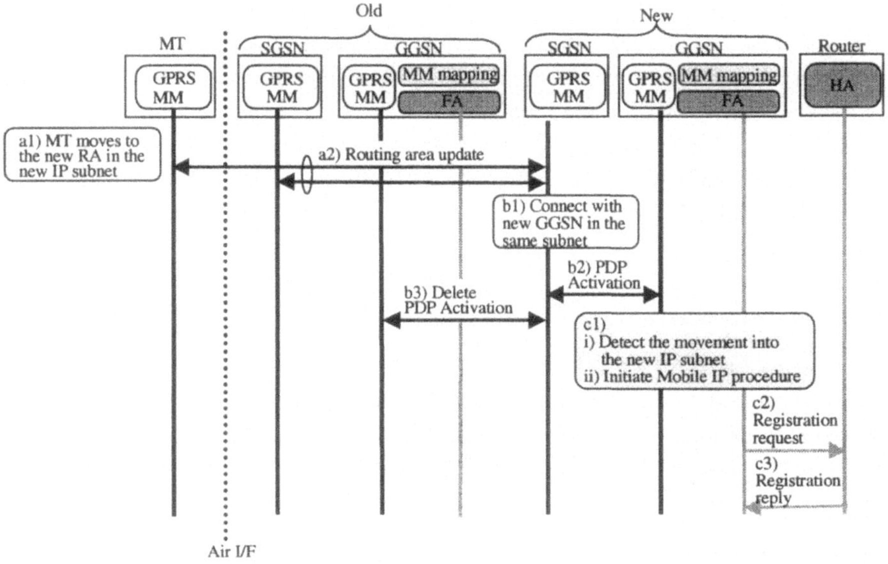

Fig.10.3.9 IP subnet mobility management (Senario2: the detection by the GPRS core network)

The advantages of the detection by the GPRS core network are as follows:

i) the overhead of the radio I/F is as same as the GPRS without IP subnet concept, and
ii) the processing time of IP subnet mobility after the MT moves into the new IP subnet is shorter than the detection by the MT (scenario 1).

After taking into consideration about the MM mapping function on the GPRS core network, the detection by GPRS core network should be selected as the identification of the movement between IP subnets over GPRS environment.

10.3.4 Impact of the IP subnet mobility management on the GPRS

This section presents the impact of the IP subnet concept on the GPRS in terms of access rate to the databases of the GPRS equipment (HLR, SGSN, GGSN) when users are roaming. The numbers of accesses that the HLR, the SGSN and the GGSN must handle are calculated using the simple Fluid flow mobility model [12][13]. The Fluid flow mobility model assumes that the subscribers are moving at an average velocity of v and their direction of movement is uniformly distributed over [0, 2•]. Assuming that the subscribers are uniformly populated with a density of • and the registration area boundary is length of L, the rate of registration area crossing, R is given by

$$R = \frac{\rho v L}{\pi} \qquad (10.3.1)$$

Using the Eq. (10.3.1), the number of the routing area crossing is calculated. The crossing number gives the entering number into the corresponding routing area, and this entering number gives the access rates to one HLR, one SGSN and one GGSN.

10.3.4.1 Assumptions

Table 10.3.1 shows the based assumption of the evaluation. The attached subscribers are distributed equality in the system. In this assumed GPRS system, only one HLR is arranged. And the subscribers who move out from this system aren't considered under the analysis of access rate to the HLR and the GGSN.

Table 10.3.1. Assumption of the mobile data network system

#	Parameters	Values
1	Total number of subscribers	200,000 ~ 20,000,000
2	Percentages of attaching terminals	50%
3	Average moving speed	6km/h
4	Cell radius (hexagonal cell)	1km
5	Number of cells in Routing Area	4, 8, 12
6	Number of cells each SGSN	1200
7	Number of GGSNs	1~12
8	Total number of cells	14,400

The routing area and the IP subnet consists of several cells, N. So the length of boundary is calculated based on [14]. In this paper, Np shows the number of cells on the perimeter, and the number of cell sides exposed is Ns. So if r is the length of a cell side, then the length of the location area perimeter, L is given as

$$L = Ns\, r$$

Np and Ns are given as follows:

$$Np = \begin{cases} 1 & a = 1,\ j = 0 \\ 6j - 3 & a = 0 \\ 6j + 4a - 4 & otherwise \end{cases}$$

$$Ns = \begin{cases} 6 & a = 0 \\ 2Np + 6 & a \geq 1 \end{cases}$$

Use the fact that $N=(j+a)^2+(j+a)j+j^2$. j and a are integer, so N shows the dispersed value.

The number of crossing the routing area, the number of crossing the area managed by the SGSN and the number of crossing the IP subnets are calculated using Eq. (10.3.1) and the above-mentioned assumption. To get the number of accesses to the HLR, the SGSN and the GGSN per second (the access rate), the following movement should be considered (Fig.10.3.19).

1) The movement between the intra RAs that are managed by the same SGSN

Only the SGSN is accessed to update the routing information of the SGSN.

2) The movement between the inter RAs that are managed by the different SGSNs

In Fig. 10.3.9, the old SGSN is accessed to deregister the MT, the new SGSN is accessed to register the MT and the information of GGSN and HLR is updated.

3) The movement between the IP subnets

The movement between the IP subnets means the movement between the inter RAs. In addition, the old GGSN is accessed to deregister the MT and the new GGSN is accessed to register the MT in Fig. 10.3.15 and Fig. 10.3.17.

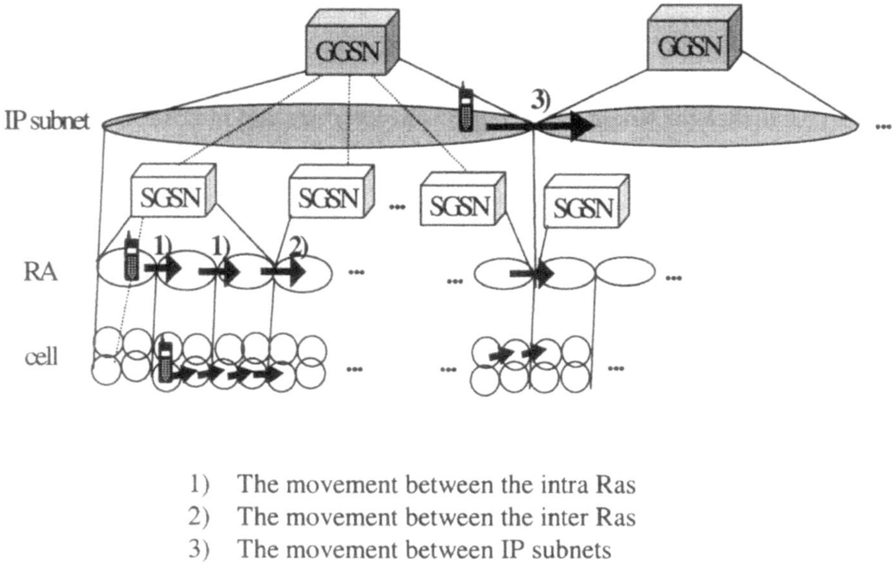

Fig.10.3.10 Definition of movement

10.3.4.2
Results

Firstly, the access rate to the HLR, the SGSN and the GGSN in the original GPRS environment are shown. Then the overhead of the scenario 1 and the scenario 2 supporting IP subnet mobility is compared to the original GPRS operation.

The database access rate in normal GPRS operation 2.1. In this GPRS environment, one HLR and one GGSN are arranged. As the result, the access rate to the HLR has the same value as the access rate to the GGSN. And the number of cells that consist of the RA significantly influences the access rate to the SGSN, and each access rate is increasing remarkably in proportion to increasing the number of attaching subscribers. Therefore the size of RA and the number of attaching subscribers are the important factors to construct the GPRS. In the GPRS environment, Table 10.3.2 shows the example access rates when the attaching subscribers are 1,000,000 people/system. And Fig. 10.3.21 shows the influence of the number of the attaching subscribers on the each database access rate.

IP subnet mobility management overhead. The IP subnet mobility influences only the access rate to the GGSN. This influence comes from the handover between the GGSNs. In case of the original GPRS operation and considering one MT, the access rate to the GGSN equals the number of boundary crossings between the inter RAs not taking into account the access rate that is arisen from the PDP activation procedure: in this case the GGSN is accessed to update the SGSN address after the MT moves into the new RA that is controlled by the new SGSN. If we consider now the case of a GPRS network supporting IP subnet mobility, the total number of SGSN address updates and new SGSN address registration is equal to the number of movements between the inter RAs of the original GPRS operation. The access to delete the information from the old GGSN results in the increment of the access rate to the GGSN. Fig 10.3.23 illustrates the access rates to the GGSN of the scenario 1, the scenario 2 and the normal GPRS operation.

In the normal GPRS operation, the GGSN is fixed while the MT is attached with the GPRS. When the attaching subscribers with each GGSN are assumed to be equal, the access rate to the GGSN is decreasing in proportion to the increasing number of GGSNs.

In the scenario 1 and the scenario 2 that support IP subnet mobility management, the movement between the IP subnets has to be considered. The registration and deregistration caused by the movement between IP subnets influence to the number of accesses to GGSN. The access rate to the GGSN in the scenario 2 gives better result than in the scenario 1. In the scenario 2 GGSN is accessed in the location management procedure, but the scenario 1 should regard the additional access to the GGSN as the Mobile IP procedure. Considering only the access rate to the GGSN, the detection by the GPRS core network (scenario 2) should be selected as the identification of the movement between IP subnets over GPRS environment with IP subnet mobility management.

It is very difficult to consider about the overhead of IP subnet mobility over GPRS and to evaluate whether the IP subnet mobility has an influence on a GPRS operator. The scales of one operator such like the number of GGSNs and SGSNs, the number of subscribers, and the number of RAs give an impact to the IP subnet mobility overhead. And we should think about the connecting time to the Internet through the GPRS and the distribution of subscribers. Increasing the percentage of subscribers, who want the high-speed movement, shall increase the number of handovers between GGSNs. Therefore the high-speed movement has a strong impact on the access rate to the GGSN.

Table 10.3.2. Example database access rate in normal GPRS operation

Attaching subscribers: 1M people/system

Activity	HLR	GGSN	SGSN 4cell/RA	SGSN 8cell/RA	SGSN 12cell/RA
Update/ Registration	40.8	40.8	59.5	46.7	34.0
Deregistration	-	-	3.4	3.4	3.4
Total	40.8	40.8	62.9	50.1	37.4

[times/sec]

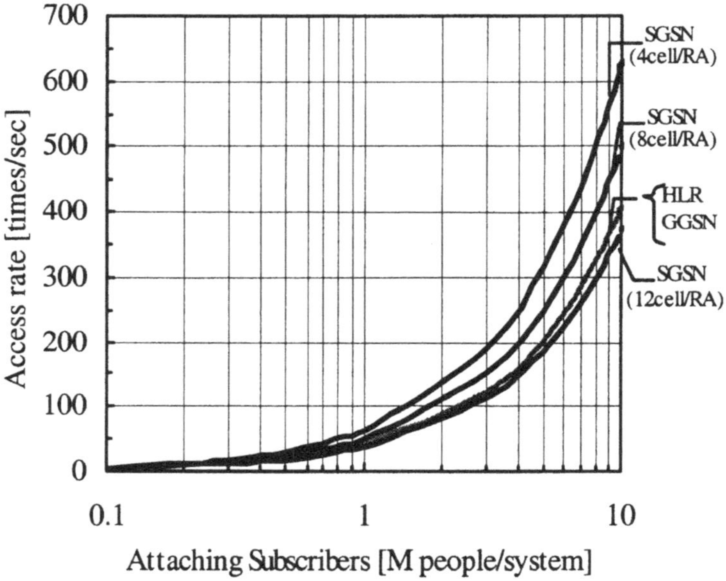

Fig.10.3.11 Influence of attaching subscribers in normal GPRS operation

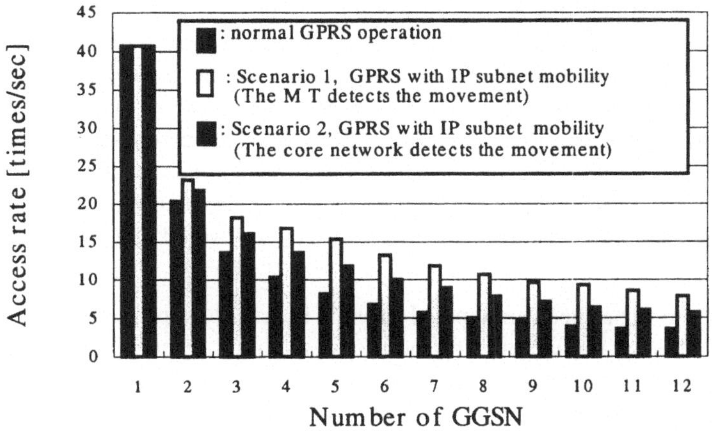

Fig.10.3.12 Access rate to GGSN (Attaching subscribers: 1 M people/system)

10.3.5 Conclusion

In this , the combination of mobility between IP subnets and the peculiar mobility management defined in GPRS is discussed in order to provide the same environment as the fixed IP network on the mobile data network system and to realize the terminal movement regardless of the fixed IP network and the mobile network system. In our proposal, the MT only supports the GPRS mobility management, and the GPRS core network detects the movement between the IP subnets based on the control messages of GPRS mobility management, and initiates the Mobile IP procedure instead of the MT. This combining method isn't influential in any additional control messages transferred on the radio I/F, but requires additional accesses to the GGSN due to the handover between the GGSNs. As the result of supporting IP subnet mobility over GPRS, the roaming between the different GPRS operators and the roaming between the heterogeneous networks are realized using the IP subnet mobility.

We have assumed the fixed IP address that is assigned by the home subnet of the MT. Even if the MT is assigned the dynamic IP address by the IP subnet with which the MT attaches at first, our proposal can provide the IP subnet mobility with same benefit such as no additional control message on the radio I/F.

Our aiming is the flexible access to the IP network, therefore we've proposed to change the access point with the IP network according to the movement of the MT. On the other hand, it has been recognized that the fixed access point have an advantage in providing the high-secure dedicated route to the private network and distributing the attaching subscribers. But it is self-apparent that the users show the tendency to use the Internet to reduce the costs. As the new technologies such like the virtual private network (VPN) that provides the secure route over the Internet has been available, the insufficient ability can be covered along with the progress of IP world.

Mobile IP has been anticipated strongly in order to support the seamless roaming irrespective of the varieties of the access medium: whether the wired or wireless medium and whether the LAN or the mobile networks system such like a cellular network. However the tunneling overhead, the transmission overhead via the HA, the violation of IP subnet concept that is introduced from the existing the roaming terminals in the different IP subnet, and the necessity for the security and the authentication mechanism become subjects to discuss in Mobile IP. The migration from IPv4 to IPv6 can solve some subjects. Because the subnet prefix of which IP address consists is assigned into the roaming terminal, and the route optimization are supported by Mobile IPv6. Taking the using Mobile IP over the mobile data network system in consideration, the following requirements for Mobile IP are cleared: the authentication, the communication with the private network that is protected by the firewall, the dynamic home address assignment, the dynamic home agent assignment and so on [15]. It's high time to evolve Mobile IP into supporting the IP subnet mobility between the heterogeneous networks such like the Internet and the cellular network.

Meanwhile the connecting style to the Internet through the mobile data network system has a tendency to have a variety. Till now, the user only connects to the system while the data and information are gathering. The information delivering service is shown up, so the user need to always keep the attachment with the system as same as the mobile phone requires to be able to catch and initiate a call at any time. And the moving speed, the moving range and the accessing network that the user selects are dependent on each subscriber. Therefore the mobile data network system should have the wide acceptable range in regard to assigning IP address to the mobile terminal, selecting and changing the access point to provide the optimal route. One important aspect is the compatibility with the existing solutions, for instance the compatibility between the peculiar mobility management of GPRS and the IP subnet mobility provided by Mobile IP.

References

1. C. Dhawan, "Remote access network: PSTN, ISDN, ADSL, Internet and wireless, McGraw-Hill, 1998

2. V. G. Cerf, "Keynote speech: Looking Beyond the Millennium," Infocom97, April 1997 (http: //www.infocom97.ckp.or.jp/slides/index.htm)

3. Mobile Internet 98, IBC UK Conferences Limited, April 1998

4. G. Brasche, B. Bernhard, "Concept, Services, and Protocols of the New GSM phase 2+ General packet Radio Service, "IEEE Communication Magazine, August 1997

5. ITU-T Recommendation Q.1701 (Formerly Q.FIN) Version 4.2, "Framework for IMT-2000 networks," May 1998

6. J. S. Dasilva, D. Ikonomou, H. Erben, "European R&D Program on Third-Generation Mobile Communications," IEEE Personal Communication, February 1997

7. ETSI TR 101 111 v.3.0.0.1, "Universal Mobile Telecommunications System (UMTS); Requirements for the UMTS Terrestrial Radio Access System (UTRA)," UMTS 21.01 Version 3.0.1, October 1998

8. C. Perkins, "IP mobility support," RFC 2002, October 1996

9. D. B. Johnson, "Mobility support in IPv6," IETF Internet Draft, Work in progress, March 1998

10. ETSI, "GSM 03.60 General Packet Radio Service: Service Description, Stage 2," v.6.0.0, March 1998

11. J. Cai, D. J. Goodman, "General Packet Radio Service in GSM," IEEE Communication Magazine, October 1997

12. V. K. Garg, J.E. Wilkes, "Wireless and Personal Communication Systems," Prentice Hall PTR, (p.p. 337-355), 1996

13. J. D. Gibson Editor, "The Mobile Communication Handbook," IEEE Press, (p.p. 289-318), 1996

14. E. Alonso, K. S. Meiere-Hellstern, "Influence of Cell Geometry on Handover and Registration Rates in Cellular and Universal Personal Telecommunications

Networks, Proceeding of "8th TTC Specialist Seminar, Universal Personal Telecommunication", (P.P. 261-270), October 1992

15. A. Jonsson, E. Hubbard, J. Malmkvist, A. Roos, "Requirements on Mobile IP form a Cellular Perspective," IETF Internet Draft, Work in progress, February 1999

10.4 Assessing Outage Impact On The Reliability Of Mobile Wireless Networks

Spilios E. Makris and Jay M. Bennett

10.4.1 Introduction

This chapter describes methods that yield an outage index for a network failure in the *mobile wireless* industry segment. The outage index is designed as a measure of the impact of the failure on customer service. The higher the outage index, the greater the impact on the customer. Outage index values are always positive with a small outage having an outage index near zero. The outage index is calculated using data available (or estimated) from service outage reports.

Section 10.4.2 describes the Standards Committee T1 Outage Index for service outages reported by the US wireline industry segment to the Federal Communications Commission (FCC). The section provides the principles for creating the outage index. This is important for understanding the mobile wireless outage index, which is based on the same principles. Section 10.4.3 defines the outage index for the mobile wireless industry segment. Section 10.4.4 lists the data items required to calculate the mobile wireless outage index. Section 10.4.5 presents an example of the use of a mobile wireless outage index worksheet. Section 10.4.6 describes two key principles which guided the construction of the wireline outage index; using examples, it demonstrates how these principles are achieved in the mobile wireless outage index. Section 10.4.7 introduces the use of control charts for monitoring network reliability. It describes how to construct control charts for aggregated outage indexes, sums of outage indexes from outages occurring in a time period. A discussion of normalization to account for network growth is also included.

10.4.2 The Standards Committee T1 Wireline Outage Index

10.4.2.1 History

In the wake of several major Common Channel Signaling (CCS) outages in the United States in 1991, the Federal Communications Commission encouraged the telecommunications industry to examine the reliability of the Public Switched Telecommunications Network and implement a program for its improvement. A part of this initiative was the requirement for major telecommunications carriers to report to the FCC "large" outages of service in their networks. The FCC defined such outages to be ones that potentially affect 30,000 or more customers for 30 minutes or more.[1][2] Since July 1992, these outages have been reported to the FCC and their frequency tracked in quarterly and annual reports by the Network Reliability Steering Committee (NRSC) of the Alliance for Telecommunications Industry Solutions (ATIS). (See [3] and [4] for recent examples.)

For the purposes of these studies, the NRSC wished to have an outage index, a single quantitative measure for the impact of a service outage on customers. The NRSC asked the Standards T1A1.2 Working Group on Network Survivability Performance to develop such an outage index based purely on data reported to the FCC for large outages. Concurrent with this request, Telcordia Technologies had developed its own techniques for measuring customer impact of service outages and had successfully employed them in risk analyses of client wireline networks.[5][6] Based on these techniques, Telcordia Technologies provided contributions to T1A1.2, which proposed an outage index for FCC-reportable outages adapted to data available from reports made to the FCC. These contributions provided the basis for the T1 Wireline Outage Index. [7][8]

The contributions identified several key features required of a useful outage index:

- The outage index should be non-negative.
- The outage index should be monotonic so that more severe outages should have a higher outage index than less severe ones.

- The outage index should be summable so that the impact of network reliability in different time periods can be compared with respect to the sum of the outage indexes of outages within each time period.

The contributions proposed that the outage index should be defined as the sum of outage indexes from each service affected. Each service outage index would be the product of three weights:

1. *Service Weight* which reflects the importance of the service,
2. *Magnitude Weight* which reflects the impact of the number of customers potentially affected by the outage of the service, and
3. *Duration Weight* which reflects the impact of the amount of time that customers were denied service.

While Service Weights were fixed values for different services, the Magnitude and Duration Weights varied according to the data reported to the FCC (i.e., customers potentially affected and outage duration). A notable innovation of the contributions was the use of S-shaped functions to determine these weights from the data.[1] The S-shaped curves allowed outage indexes from a collection of outages to be summed into an aggregated outage index without very large or long outages completely dominating the aggregated sum.

With modifications to the Service Weights and the addition of techniques to determine customers potentially affected from the number of blocked calls reported, T1A1.2 used these Telcordia Technologies contributions to define the outage index for FCC-reportable service outages. After approval by the industry via a Standards Committee T1 Letter Ballot, the T1 Outage Index was published in T1 Technical Report No. 38 in 1994.[10] A year later, T1A1.2 produced T1 Technical Report No. 42 which included further modifications to the T1 Outage Index including the use of a Time Factor to adjust the number of customers potentially affected according to the period of the day in which service was affected.[9] In addition to the Time Factor, Telcordia Technologies also contributed to T1 Technical Report No. 42 statistical techniques for use in analyzing aggregated T1 Outage Indexes. The wireline T1 Outage Index has remained unchanged since then.

---

[1] The original Telcordia contributions provided tables based on the S-curves. A later version of the T1 outage index replacing the tables with the original Telcordia functions was published in Committee T1 Technical Report No. 42. [9]

Since the publication of T1 Technical Report No. 38, the T1 Outage Index has become a standard metric for reliability of wireline telecommunications networks in the United States. Along with the frequency of FCC-reportable outages, the NRSC uses the aggregated T1 Outage Indexes of those outages as a major indicator of network reliability. These aggregated T1 Outage Indexes are analyzed in each quarterly and annual report published by the NRSC. The T1 Outage Index has been recognized outside the United States as well. British telecommunications experts Oodan, Ward, and Mullee found that the T1 Outage Index "is promising and could be adapted and refined for use in a European environment." [11, p. 184].

10.4.2.2 Definition

The T1 Outage Index for an FCC-reportable service outage is :

$$I = \sum_{j=1}^{4} W_S(j) W_D(j) W_M(j) \tag{1}$$

where

- $W_S(j)$ is the Service Weight for Service j,
- $W_D(j)$ is the Duration Weight based on the outage duration for Service j, and
- $W_M(j)$ is the Magnitude Weight based on the customers potentially affected by the outage to Service j.

The Service Weights for each of the four services considered are:

1. IntraLATA[2] Intraoffice Service with Service Weight $W_S = 1$,
2. IntraLATA Interoffice Service with Service Weight $W_S = 2$,
3. InterLATA Interoffice Service with Service Weight $W_S = 2$, and
4. 911 Service[3] with Service Weight $W_S = 3$.

---

[2] LATA: Local Access and Transport Area; one of 196 local geographical areas in the United States within which a local telephone company may offer telecommunications services – local or long distance.

[3] 911 Service: 911 is an emergency reporting system whereby a caller can dial a common number (911) for all emergency services.

The Duration Weight (W_D) measures the impact of outage duration on customers. The Duration Weight is calculated from the following equations based on the duration of the outage of the service:

If the duration is *less than or equal to* 30 minutes:

$$W_D = (Duration\ Minutes / 30)^2 \quad (2a)$$

If the duration is *greater than* 30 minutes:

$$W_D = 2.5 - 1.5\left(\frac{153.54}{123.54 + Duration\ Minutes}\right)^{1.327} \quad (2b)$$

The Duration Weight follows an S-shaped curve, with an inflection point at 30 minutes. The curve rises rapidly for durations up to 30 minutes, and then slowly rises to the asymptote of 2.5 as shown in Fig. 10.4.1. The curve is constructed to level off after outage durations of several hours since by this time customer dissatisfaction has already approached peak levels. It also allows long (and large) outages to have a great weight (when outage indexes are summed) without completely dominating the summation process.

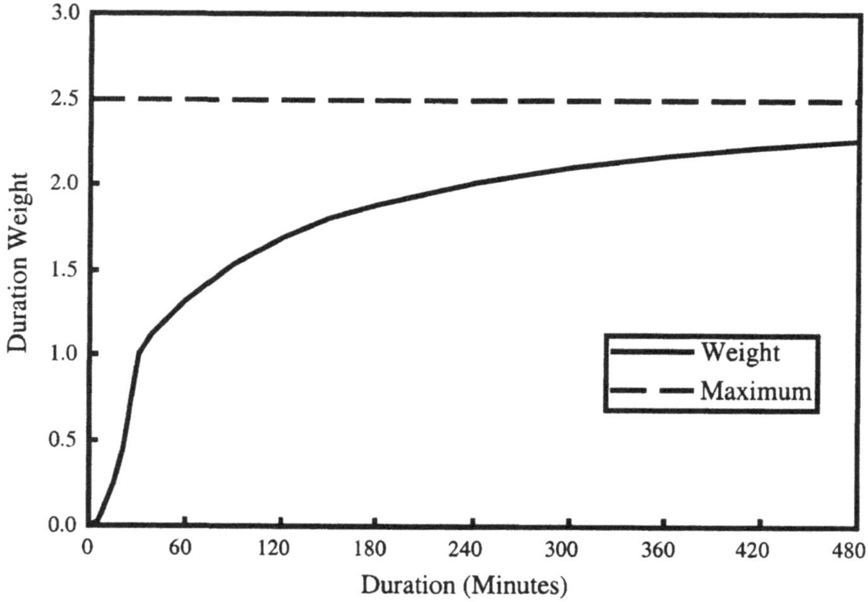

Fig. 10.4.1: S-Shaped Curve for Duration Weight

The Magnitude Weight (W_M) measures the impact of an outage with respect to the customer base affected. The Magnitude Weight follows an S-shaped curve, with an inflection point at 50,000 customers. Thus, the curve rises rapidly up to 50,000 customers, and then slowly rises to the asymptote of 16.67 as shown in Fig. 10.4.2 and 10.4.3.

Using the number of customers affected, W_M is found by using the following equations:

If the number of customers affected is *less than or equal to* 50,000:

$$W_M = \left(\frac{\text{Number of Customers Affected}}{1000} \right)^2 / 3750 \quad . \tag{3a}$$

If the number of customers affected is *greater than* 50,000:

$$W_M = \frac{50}{3} - 16\left(\frac{532.2}{482.2 + \dfrac{Number\ of\ Customers\ Affected}{1000}}\right)^{1.114} \quad . \tag{3b}$$

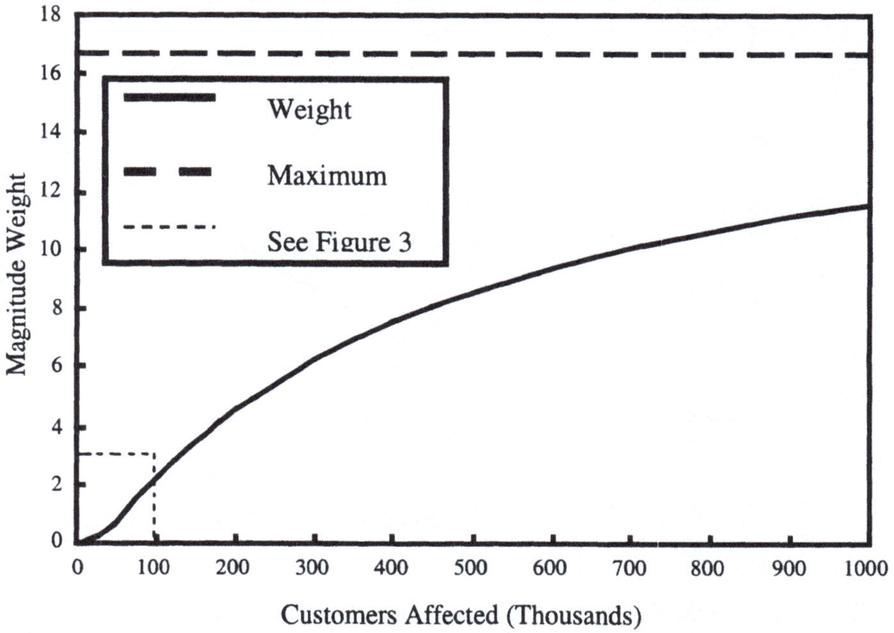

Fig. 10.4.2: S-Shaped Curve for Magnitude Weight

There are two methods for determining the number of customers affected. The *Blocked Calls Method* is used in outages where blocking was less than 100% because the services affected had alternate routing capabilities available. This method uses blocked calls as the basis for customers affected. Typically, if the report to the FCC gives the number of blocked calls as measured at the time of the outage, the number of customers affected is obtained by dividing the number of blocked calls by three; this adjusts for re-attempts of original attempts as a result of high blocking.[4] Because the

[4] A US industry standard assumption is that each blocked first attempt generates on average two additional blocked calls from re-attempts at obtaining a connection. The assumption is based on past studies by Bell Laboratories many years ago. It provided the basis for creating blocked call thresholds of reporting outages to the FCC. [2]

carrier reports to the FCC do not contain a breakdown of blocked calls by service, all blocked calls are treated as a single service with Service Weight 2 (the maximum of IntraLATA Intraoffice, IntraLATA Interoffice, and InterLATA Interoffice Services).

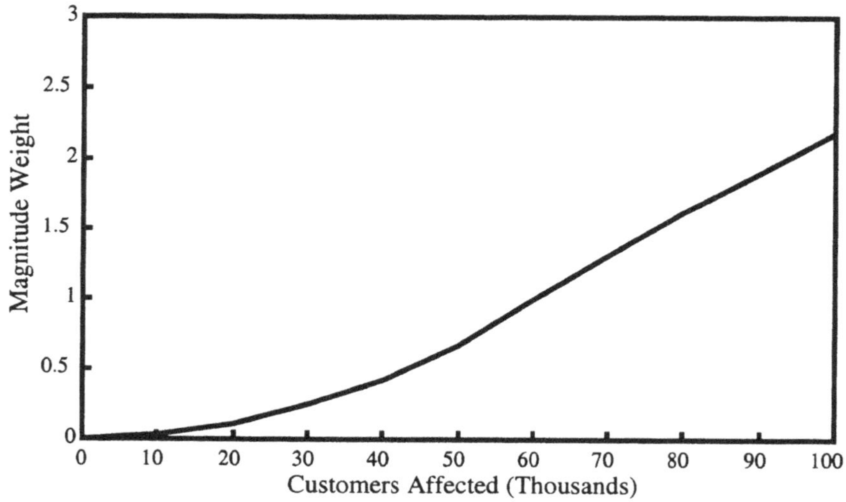

Fig. 10.4.3: Detail of S-Shaped Curve for Magnitude Weight

The *Lines Method* is used in outages where blocking was 100% because the services affected had no alternate routing capabilities available. Typically, such outages occur where customer access to the network originates and where subscriber access lines can be counted and reported to the FCC. Thus, in these situations the number of customers affected is the number of subscriber lines affected. For the purposes of determining the Magnitude Weight using the Lines Method, the number of customers may be reduced by 70%, 90%, or 80% if the outage occurred in evening hours, night hours, or weekend hours (respectively) to adjust for reduced service requests during those hours.

The 911 Service is a special case. Its Magnitude Weight is always determined using the Lines Method and the number of customers affected is never adjusted for the time of day.

10.4.3 Creation of a Mobile Wireless Outage Index

10.4.3.1 General Concepts

The success and rapid acceptance of the T1 Outage Index for FCC-reportable service outages stimulated interest in expanding its concepts to other industry segments. One segment given prime consideration was the mobile wireless industry segment. Recent years have seen tremendous growth in subscription and use of mobile wireless services. The importance of mobile wireless network reliability is critical because of reliance by customers on mobile wireless service for emergency situations.

One of the prime objectives of T1A1.2 in creating an outage index for mobile wireless networks was to preserve consistency with the existing wireline T1 Outage Index for FCC-reportable service outages. While such outages are the largest among all outages, the basic principles on which the T1 Outage Index was created (i.e., the Service Weights and the S-curves for Magnitude and Duration Weights) were felt to be equally applicable to smaller outages.

Thus, T1A1.2 proceeded by adapting the existing wireline T1 Outage Index to a mobile wireless service environment. Upon examination, the major difference observed between wireline and mobile wireless service environments for the purposes of an outage index was the need for mobile wireless customer registration. A mobile wireless customer has to register (indicate readiness to make and receive calls) in order to establish a virtual connection to the mobile wireless service provider. A wireline customer is always registered via the physical connection to the wireline service provider. Once registration has been achieved successfully, wireline and mobile wireless services are almost identical from the customer perspective.

For this reason, the mobile wireless outage index was defined as the sum of two outage indexes, one for registration and one for blocking of registered customer call attempts. Specifically,

$$I = I_{RB} + I_{CB} \quad (4)$$

where:

I_{RB} is the outage index for mobile wireless customers blocked from initial registration, and

I_{CB} is the outage index for mobile wireless registered customers whose call attempts are blocked.

In order to insure consistency with the wireline T1 Outage Index, the calculation of both I_{RB} and I_{CB} used the Magnitude and Duration Weight functions describe earlier. In addition, a mapping of wireline services to mobile wireless services (shown in Table 10.4.1) indicated that the services used in the wireline T1 Outage Index had counterparts in the mobile wireless network architecture. This mapping demonstrated how Service Weights from the wireline T1 Outage Index could be applied in a mobile wireless framework.

Table 10.4.1: Mapping of Wireline Services to Comparable Mobile Wireless Services

Wireline Service	Comparable Mobile Wireless Service	Service Weight (W_S)
IntraLATA Intraoffice	Intra-MSC[5]/PCSC[6] (local mobile-to-mobile)	1
IntraLATA Interoffice	Inter-MSC/PCSC (different MSC/PCSC mobile-to-mobile)	2
InterLATA	Land-to-Mobile and Mobile-to-Land	2
911	911/911	3

[5] Mobile Switching Center or Mobile Telephone Switching Office (MTSO)

[6] Personal Communications Services Center

10.4.3.2 Outage Index for Mobile Wireless Customers Blocked from Initial Registration

A mobile wireless network failure may isolate customers who initially register for mobile access. The outage index for outage of this service is:

$$I_{RB} = 8 \text{ x } W_{DRB} \text{ x } W_{MRB}. \qquad (5)$$

where

- W_{MRB} is the Magnitude Weight based on substituting the number of blocked initial registration attempts for the number of customers affected. The number of blocked initial registration attempts may be estimated from historical data if measured data is unavailable or unreliable.
- W_{DRB} is the Duration Weight based on the duration of time in which initial registration blocking occurred.
- The multiplier 8 is the Service Weight. The inability to register denies the customer access to all of the services considered. Consequently, the appropriate Service Weight is the sum of the weights in Table 10.4.1.

10.4.3.3 Outage Index for Mobile Wireless Registered Customers Whose Call Attempts Are Blocked

In a mobile wireless network failure, mobile customers may be able to register but have their call attempts blocked. The call attempts include both originations by the mobile customer and terminations to the mobile customer. The outage index for outage of this service is:

$$I_{CB} = 3 \text{ x } W_{DCB} \text{ x } W_{MCB} \text{ if access to 911 Service is impacted} \qquad (6a)$$

or

$$I_{CB} = 2 \text{ x } W_{DCB} \text{ x } W_{MCB} \text{ if access to 911 Service is not impacted} \qquad (6b)$$

where

- W_{DCB} is the Duration Weight based on the duration of time in which call blocking occurred.

- W_{MCB} is the Magnitude Weight based on substituting the number of original blocked call attempts for the number of customers affected. The number of original blocked call attempts for calculating the Magnitude Weight can be obtained either by: i)dividing the number of measured blocked call attempts by 3, or ii) estimating the number based on historical counts of call attempts in a typical day during the period of the outage.

The multiplier of 2 in Equation 6b is consistent with the Service Weight used in T1 Technical Report No. 42 [9] when the Blocked Calls Method is used. The factor of 3 used in Equation 6a uses the 911 Service Weight under the worst case assumption that all blocked calls are for access to 911 Service.

10.4.4 Data Items

The calculation of the outage index for an outage requires the following data items. The data items can be grouped by the impact on customers:

(i) Customers Blocked from Initial Registration[7]*:*

S_{RB} = Date and time that customers were first blocked from initial registration

E_{RB} = Date and time that customers were no longer blocked from initial registration

D_{RB} = Duration (in minutes) of time from S_{RB} to E_{RB}

N_{RB} = Number of mobile customers blocked from initial registration during the outage.

(ii) Call Attempts to/from Registered Customers:

S_{CB} = Date and time that calls to/from registered customers were first blocked

[7] When a subscriber is initially registered, the phone informs the network that the subscriber is active and can make/receive calls. The phone continues to periodically register while the subscriber is active. These periodic registrations are not considered in the outage index calculation.

E_{CB} = Date and time that calls to/from registered customers were no longer blocked

D_{CB} = Duration (in minutes) of time from S_{CB} to E_{CB}

N_{CB} = Number of blocked call attempts to/from registered customers during the outage.

In most cases, S_{RB} and S_{CB} will be the same, and E_{RB} and E_{CB} will be the same. Thus, the durations D_{RB} and D_{CB} will be equal. However, the data items and outage index techniques are presented as if these data items are unique for the purposes of generality.

N_{RB} and N_{CB} can be determined from measurements obtained at the time of the outage or from historical data from the past, which is representative of the measurement for the time period in which the outage occurred.

10.4.5 Mobile Wireless Outage Index Example

This section provides an example of calculating the mobile wireless outage index using a worksheet template developed by Telcordia Technologies in a contribution to T1A1.2 [12] and published in T1 Technical Report No. 24A.[13] The data used in the example were generated only as an example and do not represent necessarily that of an actual mobile wireless outage.

In the example, a failure in a mobile wireless network occurred on January 15, 1997. The failure affected service from 5:00 PM to 6:00 PM. Attempts for initial registration were blocked as well as the call attempts of registered customers (which included blocking of calls to 911 Service). During the service outage, 120,000 call attempts from registered customers were blocked as recorded in network measurements. Although the number of blocked initial registration attempts could not be measured, traffic measurements recorded in recent weeks indicated that 10,000 initial registration attempts are typically made in this time period. The first step in the calculation is to transfer the information from this outage report into the Outage Data portion of the worksheet as shown in Table 10.4.2.

Table 10.4.2. Part 1 (Outage Data) of Mobile Wireless Outage Index Worksheet

Mobile Wireless Outage Index Worksheet
Part 1: Outage Data

	Date and Time		Duration (minutes)	Number	
	Start	End		Measured	Historical
Initial Registration Blocking	S_{RB} 1/15/97 5PM	E_{RB} 1/15/97 6PM	D_{RB} 60	N_{RB}	N_{RB} 10,000
Call Blocking	S_{CB} 1/15/97 5PM	E_{CB} 1/15/97 6PM	D_{CB} 60	N_{CB} 120,000	N_{CB}

The second step is to use the Outage Data in the Calculation Summary shown in Table 10.4.3. For Initial Registration Blocking, the number of customers affected is the 10,000 initial registration attempts typically made in this time period. The Magnitude Weight is calculated by substituting this value into Equation 3a:

$$\begin{aligned} W_{MRB} &= \left(\frac{\textit{Number of Customers Affected}}{1000}\right)^2 / 3750 \\ &= \left(\frac{1000}{1000}\right)^2 / 3750 \\ &= 0.0267 \end{aligned} \tag{7}$$

For Call Blocking, the number of customers affected is 40,000 initial call attempts; this value is calculated from the measured blocked call attempts by dividing by 3 in order to account for re-attempts resulting from high blocking percentages during the service outage. The Magnitude Weight is calculated by substituting this value into Equation 3a:

$$W_{MCB} = \left(\frac{Number\ of\ Customers\ Affected}{1000} \right)^2 / 3750$$
$$= \left(\frac{40000}{1000} \right)^2 / 3750$$
$$= 0.4267 \tag{8}$$

Both initial registration and call service were affected for the same duration, one hour. So, their Duration Weights are equal. The Duration Weights are calculated by substituting the duration in minutes into Equation 2b:

$$W_{DRB} = W_{DCB}$$
$$= 2.5 - 1.5 \left(\frac{153.54}{123.54 + Duration} \right)^{1.327}$$
$$= 2.5 - 1.5 \left(\frac{153.54}{123.54 + 60} \right)^{1.327}$$
$$= 1.316 \tag{9}$$

The outage index for initial registration blocking is (see Equation 5):

$$I_{RB} = 8 \text{ x } W_{DRB} \text{ x } W_{MRB} = 8 \text{ x } 1.316 \text{ x } 0.0267 = 0.2811 \tag{10}$$

Since 911 service was affected, the outage index for call blocking is (see Equation 6a):

$$I_{CB} = 3 \text{ x } W_{DCB} \text{ x } W_{MCB} = 3 \text{ x } 1.316 \text{ x } 0.4267 = 1.6846 \tag{11}$$

The outage index for the outage is the sum of these two outage indexes, 1.9657.

Table 10.4.3. Part 2 (Outage Index Calculation Summary) of Mobile Wireless Outage Index Worksheet

Mobile Wireless Outage Index Worksheet
Part 2: Outage Index Calculation Summary

	Customers Affected	Weights		Outage Index
		Duration	Magnitude	
Initial Registration Blocking	N_{RB} 10,000	W_{DRB} 1.316	W_{MRB} 0.0267	I_{RB} 0.2811
Call Blocking	CA_{CB} 40,000	W_{DCB} 1.316	W_{MCB} 0.4267	I_{CB} 1.6846
			OUTAGE INDEX = $I_{RB} + I_{CB}$ =	1.9657

10.4.6 Principles of summation

The T1 Outage Index was constructed to allow the meaningful summation of outage indexes over a set of outages. The S-curves for Magnitude and Duration Weights were constructed under the basic principles that:

- **Principle 1:** The impact of several large outages is *greater* than the impact of a single large outage with an equal duration and number of customers affected, while on the other hand,

- **Principle 2:** The impact of several small outages is *less* than the impact of a single small outage with an equal duration and number of customers affected.

The rationale for these principles is explained by the role of the outage index as a measure of customer attention or perception of the service outages. Small outages, while a nuisance to the customer, are virtually invisible. Th customer is inconvenienced briefly, but the outage quickly passes and the incident is forgotten. As outages grow in size and duration, customer attentiveness increases rapidly, bigger small outages take a greater hold on customer attention. This is the basis for Principle 2. On the other hand, every large outage gains the attention of the customer base. Some large outages are bigger than others, but, in this realm, frequency begins to dominate size and duration. This is the basis for Principle 1.

These principles are carried over from the wireline version to the mobile wireless version. Consider the following examples. For simplicity in exposition, all examples make the following assumptions:

- The outages prevented the measurement of blocked calls. The data for the index was based on historical data. So, customers affected equals the number of blocked calls estimated from historical data.
- Only call attempts to/from registered customers are affected.
- 911 service is not impacted.

Consequently, the outage index for these examples is

$$I = I_{RB} + I_{CB} = 0 + I_{CB} = 2x\, W_{DCB} \times W_{MCB} \tag{12}$$

Table 10.4.4 provides an example of four service outages (Outage A, Outage B, Outage C, and Outage D). All four outages have the same duration of one hour, but affect different number of customers. Outages A, B, and C are large, involving hundreds of thousands of calls blocked. Outage D is even larger than Outages A, B, and C. In fact, the number of calls blocked in Outage D equals the sum of calls blocked in Outages A, B, and C. Standard methods of comparing outage effects would come to the conclusion that the impact of Outage D is the same as that for the set of Outages A, B, and C. After all, Outage D blocked 600,000 calls over one hour and the set of Outages A, B, and C also blocked 600,000 calls over one hour. However, the sum of outage indexes for Outage A, Outage B, and Outage C (35.5) is greater than the outage index for Outage D (24.8). In accordance with Principle 1 above, the outage index is constructed to give greater weight to multiple large outages.

Table 10.4.4. Example of Principle 1 with Respect to Customers Affected

Outage	Customers Affected	Duration (minutes)	Weights		Index
			Magnitude W_{MCB}	Duration W_{DCB}	2 x W_{MCB} x W_{DCB}
A	160,000	60	3.69	1.32	9.71
B	200,000	60	4.53	1.32	11.93
C	240,000	60	5.28	1.32	13.90
Total	600,000	60			35.54
D	600,000	60	9.41	1.32	24.77

The ordering is reversed in accordance with Principle 2 when the outages are small. Table 10.4.5 provides an example of four service outages (Outage A, Outage B, Outage C, and Outage D). As in Table 10.4.4, all four outages have the same duration (10 minutes in this case), but affect different number of customers. Outage D is small involving twelve thousand calls blocked. Outages A, B, and C are even smaller. As in Table 10.4.4, the number of calls blocked in Outage D equals the sum of calls blocked in Outages A, B, and C. Again we find that even though the same number of customers are affected over the outage duration, the sum of outage indexes for the three smaller outages is different than the outage index for the single larger outage. However, now the sum of outage indexes for Outage A, Outage B, and Outage C (0.003) is *less* than the outage index for Outage D (0.009). In accordance with Principle 2 above, the outage index is constructed to give *less* weight to multiple small outages.

Table 10.4.5. Example of Principle 2 with Respect to Customers Affected

Outage	Customers Affected	Duration (minutes)	Weights		Index
			Magnitude W_{MCB}	Duration W_{DCB}	2 x W_{MCB} x W_{DCB}
A	3,000	10	0.00240	0.11111	0.00053
B	4,000	10	0.00427	0.11111	0.00095
C	5,000	10	0.00667	0.11111	0.00148
Total	12,000	10			0.00296
D	12,000	10	0.03840	0.11111	0.00853

10.4.7 STATISTICAL ANALYSIS OF OUTAGE INDEX

10.4.7.1 Background

The major incentive for the creation of an outage index by Committee T1 was the need to identify trends and significant changes in the reliability of telecommunications in the United States. The FCC had required the reporting of major outages from US service providers. Techniques were developed to assess reliability with respect to the frequency of these outages.

A major contribution by the NRSC was the development of control charts to determine whether the reliability of the telecommunications network was in control (stable) with respect to the frequency of FCC-reportable service outages. A *Baseline Year* was selected as the standard for the definition of the stable state of the network. The period from July 1, 1992 through June 30, 1993 (inclusive) was chosen as the Baseline Year. In this period, 165 service outages were reported to the FCC in accordance with thresholds on duration and customers affected. The NRSC decided to report quarterly on the status of reliability in the US telecommunications network, so control charts were created based on quarterly outage frequencies. On average, the Baseline Year had 41.25 outages per quarter. Control limits were constructed based on a Poisson distribution of outages with a mean of 41.25.

Of greatest concern was identifying quarters that had excessive numbers of outages compared to the Baseline Year average. The NRSC established three regions, a "Red" region, a "Yellow" region, and a "Green"region. A quarter with an outage frequency in the Red region indicated that reliability was not in control and immediate investigation by focus teams was needed. A quarter with an outage frequency in the Yellow region is taken as a warning for the future. A quarter with an outage frequency in the Green region indicated that reliability was stable.

The two boundaries of the three regions were based on control limits with different degrees of probability. The boundary between the Red and Yellow regions was defined as the point at which the probability of obtaining such a count (assuming a Poisson distribution with a mean equal to the Baseline Year average) was no more than .005. Similarly, the boundary between the

Green and Yellow regions was defined as the point at which the probability of obtaining such a count (assuming a Poisson distribution with a mean equal to the Baseline Year average) was no more than .025. According to these definitions, the Red/Yellow boundary was set at 59.5 and the Green/Yellow boundary was set at 54.5. Symmetrically, lower limits were also set at 28.5 (probability of being less than this value is less than .025) and 25.5 (probability of being less than this value is less than .005) to monitor significant improvement in network reliability relative to the Baseline Year.

Fig. 10.4.4 is a recent example of the application of the frequency control chart to FCC-reportable service outages.[4] The Red region is the area above the solid horizontal line at 59.5 outages. The Yellow region is the area bounded at the top by the solid horizontal line at 59.5 outages and at the bottom by the dashed horizontal line at 54.5 outages. The Green region is the area below the dashed horizontal line at 54.5 outages. We see that all quarters through the first quarter of 1999 have been in the Green region. One point in the fourth quarter of 1993 was below one of the lower limits indicating the possibility of significant improvement. In general, the control chart indicates stability of network reliability during this period.

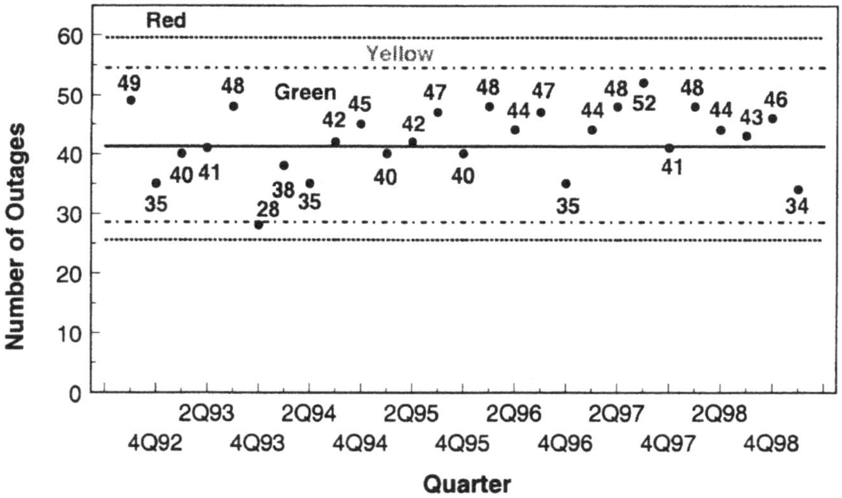

Fig. 10.4.4: Control Chart for Frequency of FCC-Reportable Service Outages

The control chart has additional uses in determining the stability of network reliability. It is possible to determine that network reliability has deteriorated or improved without the use of the control limits. The central horizontal line at 41.25 outages represents the Baseline Year mean. If network reliability is stable, there is a 1/2 chance that a quarter will have an outage frequency above this line and an equal 1/2 chance that a quarter will have an outage frequency below this line. The probability that six quarters in succession *all* have frequencies above the line is $1/2^6 = .016$ which is less than the probability of having a single quarter in the Yellow region. So, six points in row all above the Baseline Year average is a warning that network reliability may have decreased. Using a similar logic, eight quarters in succession all above the Baseline Year average is equivalent to a single quarter in the Red region and is evidence that network reliability is out of control. Note that if the outage frequency in the fourth quarter of 1997 had been one larger, this conclusion would have been reached.

The reason for the creation of the T1 Outage Index was to expand the assessment of network reliability beyond frequency to the impact of outages on network reliability. Defining the T1 Outage Index was the first step. The next step was to develop techniques for its analysis which were similar to those for outage frequency.

10.4.7.2 Outage Index Control Charts

A key part of this effort spearheaded by Telcordia Technologies within the NRSC and at Committee T1 meetings was the development of control charts analogous to those for outage frequency. The basic definitions of the Baseline Year and the Red, Yellow, and Green regions were preserved as was the quarterly time frame of reference.

The difficult part was finding the proper distribution for the sums of outage indexes from a collection of outages. The distribution of outage indexes is highly skewed, a common characteristic of impact metrics. In the Baseline Year, outage index values for FCC-reportable service outages ranged from .0001 to 162. The median outage index value was 2.14 so half the values were between .0001 and 2.14 while the other half were between 2.14 and 162.

Although summing tends to make the distribution less extreme, it was doubtful that the normal distribution, the standard distributional assumption for control charts, would provide the proper level of variability in the sums of outage indexes. The term *aggregated outage index* is used to describe the sum of outage indexes from a collection of outages and will be used from here on.

In order to determine an appropriate distribution for quarterly aggregated outage indexes in the Baseline Year for FCC-reportable outages, a bootstrap simulation was performed. Bootstrap techniques [14] assume that an empirical distribution is a complete description of the distribution modeled. Here, the distribution modeled is that of outage indexes in the Baseline Year. In bootstrap techniques, this distribution is used as if it were the actual distribution of outage indexes.

The other assumption needed is that for the distribution of outage frequency in the Baseline Year. Consistent with the development of the outage frequency control charts, this distribution was assumed to be Poisson with mean 41.25.

To generate an aggregated outage index distribution for a quarter in the Baseline Year, the following steps are followed 1,000 times:

1. Randomly generate a value from a Poisson distribution with mean 41.25. This value N represents the number of outages in the quarter.

2. Randomly select N outage indexes from the 165 outage indexes in the Baseline Year. This selection is done with replacement (implying that the same outage index value can appear twice).

3. Sum the N outage indexes to produce S, an aggregated outage index for this simulated quarter.

This process gives us 1,000 S values which represent the distribution of aggregated outage indexes for a quarter in the Baseline Year. A comparison of the upper quantiles (.995 and .975) of this bootstrap distribution were close to the same quantiles from a Gamma distribution with mean x and standard deviation y. The mean and standard deviation values for the Gamma distribution were calculated from the mean of the Poisson distribution and the mean and standard deviation estimates of the outage index distribution of the Baseline Year.

In general, if μ_N is the mean of a Poisson distribution (in this case, the distribution of outage frequency in a quarter) and μ_Q and σ_Q are the mean and standard deviation of a metric for each occurrence (in this case, the outage index for each outage), then the sum S of the metric (in this case, the aggregated outage index in a quarter) has a distribution with

$$\text{Mean of S} = \mu_S = \mu_N \times \mu_Q$$

and

$$\text{Standard Deviation of S} = \sigma_S = [\mu_N (\mu_Q^2 + \sigma_Q^2)]^{1/2}.$$

(See Reference [15].)

For the Baseline Year,

- the mean of the frequency of outages was $\mu_N = 41.25$,
- the mean outage index was $\mu_Q = 8.737$, and
- the standard deviation of the outage indexes was $\sigma_Q = 18.985$.

Substituting these values into the equations above, we find

$$\text{Mean of S} = \mu_S = 41.25 \times 8.737 = 360.4$$

$$\text{Standard Deviation of S} = \sigma_S = [41.25\,(8.737^2 + 18.985^2)]^{1/2} = 134.2.$$

A Gamma distribution with a mean of 360.4 and a standard deviation of 134.2 has a .995 quantile at 798 and a .975 quantile at 667. These quantiles define the boundaries of the Red and Yellow regions for a quarterly aggregated outage index control chart based on FCC-reportable service outages in the Baseline Year. Similarly, the .005 and .025 quantiles at 107 and 147 respectively define the control limits at the lower end.

Fig. 10.4.5 a recent example of the application of the aggregated outage control chart to FCC-reportable service outages.[4] The control limits calculated above are displayed on the chart. Unlike the frequency chart in Fig.10.4.4 quarter (1Q94) did have a result outside the Green region. Surprisingly, 1Q94 had a relatively low outage frequency which emphasizes the average severity of each outage in the quarter. Recent quarters have been above the mean line more often that not, but not often enough to establish a statistically significant difference from the Baseline Year.

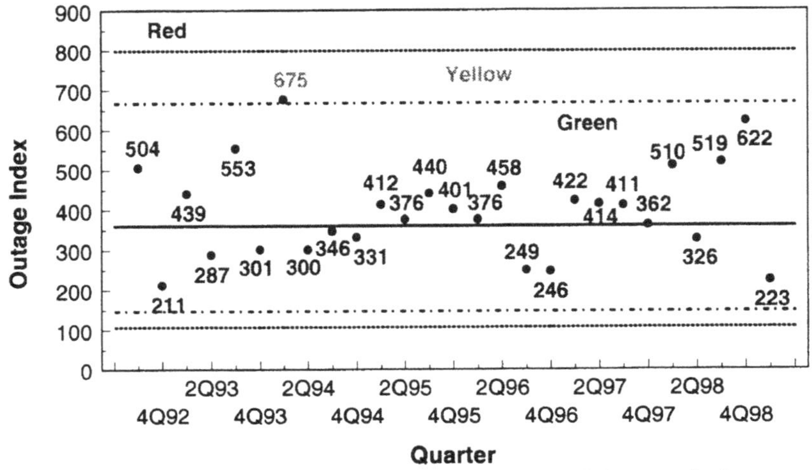

Fig. 10.4. 5: Control Chart for Aggregated Outage Index of FCC-Reportable Service Outages

10.4.7.3 Application To Wireless Networks

Because the wireless version of the outage index has only been developed recently, we do not have examples of its application in control charts. Fig. 10.4.5 and the description of its development in the previous section provides an outline of how a similar control chart for outage indexes can be constructed. The key element of the construction is not to automatically assume a Normal distribution, but to determine the proper distribution possibly using bootstrap simulation as shown above. It is likely that the Gamma distribution will provide the best fit among standard statistical distributions. As a default, this distribution with moments determined from the characteristics of the relevant empirical distribution of wireless outage indexes may be used.

10.4.7.4 Normalization For Network Growth

As networks assume the responsibility of service for increased numbers of customers, they must grow to accommodate those customers without sacrificing performance. Increased growth implies more equipment and possibly greater concentration of customers on equipment. Both of these effects can produce increases in aggregated outage indexes over time. More equipment provides a greater chance for some piece of equipment in the network to break and so increases the frequency of outages. Greater concentration of customers implies that when outages do occur they are more severe because higher numbers of customers are affected.

This growth can have an effect on the proper interpretation of control charts. An increase in aggregated outage index over time may be the result of network growth, not on a decrease in network reliability. One way to adjust for network growth is to normalize outage index values with respect to the time window used as the standard to construct the control chart.

For FCC-reportable service outages, Telcordia Technologies developed normalization techniques adopted by Committee T1.[9] The techniques are based on measures of subscriber lines and carried calls during the Baseline Year and in each subsequent year.

10.4.8 Conclusion

The T1 Outage Index has become an industry standard for measuring the impact of major service outages in the wireline segment of the telecommunications industry. The techniques used in its calculation can be extended to other segments of the industry. This chapter demonstrates how this was accomplished for the mobile wireless segment. These techniques have the advantage of being consistent with the T1 Outage Index in the wireline segment. Users of this outage index for the mobile wireless segment also have the opportunity to apply statistical techniques described in T1 Technical Report No. 42. [9] These techniques include normalization for network size and calculation of confidence intervals. While these techniques were developed for T1 Outage Indexes in the wireline segment, their principles are equally valid for the outage index in the mobile wireless segment as well.

REFERENCES

1. FCC Report and Order 92-58, CC Docket No. 91-273 (7 FCC Record 2010), Federal Communications Commission, Washington, DC, adopted February 13, 1992, released February 27, 1992.

2. *Amendment of Part 63 of the Commission's Rules to Provide for Notification by Common Carriers of Service Disruptions*, Federal Communications Commission, CC Docket No. 91-273, August 1, 1994.

3. Network Reliability Steering Committee, *Annual Report 1998*, Alliance for Telecommunications Industry Solutions, Washington, DC, 1998.

4. Network Reliability Steering Committee, *Macro-Analysis: 1st Quarter 1999*, Alliance for Telecommunications Industry Solutions, Washington, DC, 1999.

5. E. S. Tollar, "Risk Assessment in Telecommunications of CCS Networks," *IEEE Network Operations and Management Symposium*, Volume 1, February 1994.

6. E. S. Tollar and J. M. Bennett, "Network Outage Impact Measures for Telecommunications," *Proceedings of Symposium on Advances in Computers and Communications*, Alexandria, Egypt, IEEE, June 1995, p. 120-126.

7. E. S. Tollar and J. M. Bennett, "Network Survivability," Committee T1 Standards Contribution T1A1.2/93-045, November 8, 1993.

8. E. S. Tollar, J. M. Bennett, and J. D. Healy, "A Proposal for an Outage Measure Based on Customer Impact," Committee T1 Standards Contribution T1A1.2/94-003, January 1994.

9. "A Technical Report on Enhanced Analysis of FCC-Reportable Service Outage Data," Committee T1 Technical Report No. 42, August 1995.

10. "A Technical Report on Analysis of FCC-Reportable Service Outage Data," Committee T1 Technical Report No. 38, August 1994.

11. A. P. Oodan, K. E. Ward, and A. W. Mullee, *Quality of Service in Telecommunications*, London: The Institution of Electrical Engineers, 1997.

12. J. M. Bennett, "An Outage Index for the Wireless Industry Segment," Committee T1 Standards Contribution T1A1.2/97-012, January 20, 1997.

13. "A Technical Report on Network Survivability Performance (Supplement to Technical Report No. 24)," Committee T1 Technical Report No. 24A, August 1997.

14. B. Efron and R. J. Tibshirani, *An Introduction to the Bootstrap*, Chapman & Hall, New York, 1993.

15. P. McCullagh and J. A. Nelder, *Generalized Linear Models*, Chapman & Hall, New York, 1983.

10.5 A Knowledge-based Decision Support System for the Evaluation of a Terrestrial Cellular Network Investment

Themis T. Zambarakis, Nektarios N. Kokkinogenis, Harilaos G. Sandalidis
and Peter Stavroulakis

10.5.1 Introduction

The appearance of cellular networks and the rapid growth of them, has made the field of mobile, wireless communications accessible to most people, all around the world. The present challenge among communication researchers is to develop a worldwide *Personal Communication Network (universal PCN)* which can enable the reliable transfer of voice and data between two users traveling at any speed on land, sea, or in air, or located at any fixed point on earth. Also, an increasing demand of new services in this field is observed. To satisfy all these demands, new investments must be done for the deployment of large communication networks.

The evaluation of investments for cellular systems is necessary, because of the extremely high cost of them. This evaluation is usually done by expert designers with intensive knowledge and experience for large mobile cellular network design. The growing complexity of mobile communications means that designers have to possess a great deal of technical knowledge on the communications field. Such designers must be well informed about the most recent updates in the rapidly advancing field to be able to handle the available state-of-the-art technologies. Since designers of such caliber are difficult to be found and are usually very expensive, we propose the use of intelligent information systems to play their role and / or assist them in their task [1].

Basically, the efforts in supporting the whole decision-making process focused in the development of computer information systems providing the support needed. Two types of systems were developed: (1) *decision support systems (DSSs)* and (2) *expert systems (ESs)*. Although these two approaches were very promising, their implementation revealed several problems. To overcome these problems without missing the advantages of both ESs and DSSs, a new type of intelligent system called a *knowledge-based DSS (KBDSS)* has been proposed. The basic characteristic of this new approach is the integration of ES technology with models and techniques used in the decision support framework, such as mathematical programming methods, multicriteria-decision aid methods, and multivariate statistical methods [2].

Information systems have been successfully deployed in the various fields of communication networks such as for [3-7]:

- network and spectrum management.
- local area network configuration.
- monitoring and control of networks.
- fault detection, fault isolation, load balancing and diverting traffic from stressed resources.
- prediction of service degradation and traffic overload.
- project scheduling and resource allocation in network implementation.
- repair/expansion planning.
- telecommunication networks design.
- mobile communication planning.

The aim of this study is the development of a KBDSS that may be helpful to mobile operators for the design of the cellular networks. The system may decide if it is in such an operator's interest to install a cellular network by taking into account the existence of various standards, each of which has different technical specifications, different initial cost for deployment, variability in the offered services, market size e.t.c. Using this system, a service provider will be able to evaluate such an investment, by choosing a standard, performing technoeconomic estimations and computations, processing the results and developing criteria and conclusions. For an already installed cellular network the system may also be helpful for the analysis and evaluation of many situations which occur, when the economic parameters of this investment change while the system is in operation. (e.g. the cost of telecommunication equipment, the charge to the subscribers etc.) [8]. However it must be noted that the vital decisions are always made by the user, who may have other non-technical factors or preferences not considered by the system.

The structure of this study is as follows: Section 10.5.2 givesin brief the basic characteristics possibilities and perspectives offered by the KBDSS. A short description of the KBDSS architecture used together with a short explanation of the submodules of which it is composed is given in section 10.5.3. Section 10.5.4 presents more analytically the way the information system reaches the intermediate conclusions and the final decision. In parallel all the basic system actions and modules are described in more details. Section 10.5.5 presents the basic rules together with the relations among each other. Finally a typical cellular design application is given in section 10.5.6.

10.5.2 KBDSS Characteristics

The last few years quite a considerable number of researchers oriented their activities on the field of the *artificial intelligence (AI).* AI includes methods and techniques that provide the intelligent behaviour to the computers that is to say [9]:

- Learn and understand from the experience.
- Conclude in situation where fuzziness and uncertainty exist.
- Use logic in order to conclude.
- Use knowledge and experience to manipulate the environment.
- Think and reason.
- Understand and infer in ordinary, rational ways.
- Respond quickly and successfully to new situations.
- Recognise the relative importance of different elements in a situation.
- Make sense out of ambiguous or contradictory messages.

The contribution of the use of the AI was positive for the progress of many fields of the science and technology, utilising its advances (permanent storing of the knowledge, easy to transfer the knowledge, minimising the cost, operates into environment and situation where the man is not able to work). Even then there are also some disadvantages of the AI technology the main of which are focused on the facts that they are applied in a strictly small area, they are not constructive as the human one and they cannot conclude all the times taking into consideration a global knowledge.

Decision making in general has been significantly improved in the last two decades, through the rapid progress of information technology and computer science. The combination of decision theory with the new knowledge and the powerful tools offered by computer science and information technology, led to the development of new types of information systems able to support decision-makers and improve the decision-making process.

According to Klein and Methlie [10], ESs can be defined as computer programs that represent the knowledge and inference procedures of an expert to solve complicated problems providing possible solutions or recommendations. Their main characteristic and basic goal are their ability to simulate human logic and reasoning, to draw conclusions and to provide corresponding explanations concerning these conclusions. On the contrary a DSS is a computer information system that provides information in a specific problem domain using analytical decision models as well as techniques and access to databases, in order to support a decision maker in taking decisions effectively in complex structured problems. Thus the basic goal of a DSS is to provide the necessary information to the decision-maker, in order to help him get a better understanding of the decision environment and the alternatives he faces. The basic idea which characterizes and underlies the conceptual framework of a KBDSS is the combination of the capabilities provided by the classical DSS approach, such as access to data and information, and application of analytical decision models, with the reasoning and explanation capabilities provided by the ES technology. Therefore a KBDSS can be defined as a computer information system that provides information and methodological knowledge using analytical decision models, and providing access to data and knowledge bases to support decision makers in making decisions effectively in complex problem domains [1], [10].

The new possibilities and perspectives offered by KBDSSs, can be summarized by the following three main aspects [10], [11].

- *Understanding the operation and the results of the system*: incorporating a knowledge base in a DSS can help the decision-makers to understand the results of the mathematical models. The knowledge base provides the necessary explanations concerning the reasoning process of the system, its basic assumptions, and its operation. Thus, the methodology used by the system can be clearly presented to the decision-makers, helping them to learn the decision analysis process of the problem under consideration. At the same time, interpretation of the results is achieved, avoiding their misunderstanding and misinterpretation. The quantitative results are transformed to qualitative explanations, which are easier to understand, and are often of greater significance to the decision-makers.
- *The objectiveness and the completeness of the results are ensured*: the combination of the ES results with those derived from mathematical models and analytical techniques, increases the objectiveness of the final results. The subjective results of the ES (which depend on the expert's knowledge) are not the final estimations of the system. Apart from the expert's knowledge, represented by the ES results, the decision-maker is also provided with results derived from mathematical models from the fields of multicriteria analysis, multivariate statistical analysis and mathematical programming. The results of this combination correspond not only to the expert's opinion, but also to the preferences and the policy of the decision-maker.
- *The structuring of the proper decision analysis is achieved*: the usefulness and the suitability of decision analysis techniques often vary, depending on the available data of the examined decision problem. According to the existing data, the use of a certain analysis or a mathematical model may be unnecessary or even inappropriate and, consequently, the decision-maker could be directed to incorrect estimations and conclusions. The knowledge concerning the problem domain represented in a knowledge base can be used to face such problems, recommending the decision analysis process and the appropriate analytical models, which correspond to the available data. Thus, the structure of the decision analysis is achieved, ensuring the effectiveness, the applicability and the validity of the decision analysis models and techniques.

10.5.3 System Architecture

The information system that was developed follows the general structure of a KBDSS [8] (Fig. 10.5.1). The main components of this expert system shown in Fig. 10.5.1 are as follows:

- the expert system's knowledge base
- the inference engine
- the explanation component
- a mechanism for changing parameters
- a simple man-machine interface and a graphical user interface to display the simulation results.

The *knowledge base* contains all the data and knowledge, which are necessary for the successful deployment of a cellular network. The knowledge is represented in the form of *rules* (*predicates* in Prolog) or *facts*. Great attention has been paid for the knowledge base to be as complete as possible for the specific problem, but also to be stretchable for future modifications and improvements. For this reason, the rules in the knowledge base have the classic form:

If <condition> then <conclusion>.

For example, one of the rules containing in the knowledge base is the following:

"**If** *the cellular network is going to operate at 1800 MHz*
Then *you should use the DCS1800 standard.*"

This rule has been described in the knowledge base in the following form:

Use ("DCS1800"): -Frequency("1800")

When the user chooses the frequency of 1800 MHz, at the corresponding question, then the fact *Frequency ("1800 MHz")* is generated and imported to the knowledge base. After that the inference engine uses this fact and tries to find a rule which is "fired" by this fact. When the match with the rule has been made, the inference engine adds the conclusion of this rule to the knowledge base and the process of the questions and answers continues.

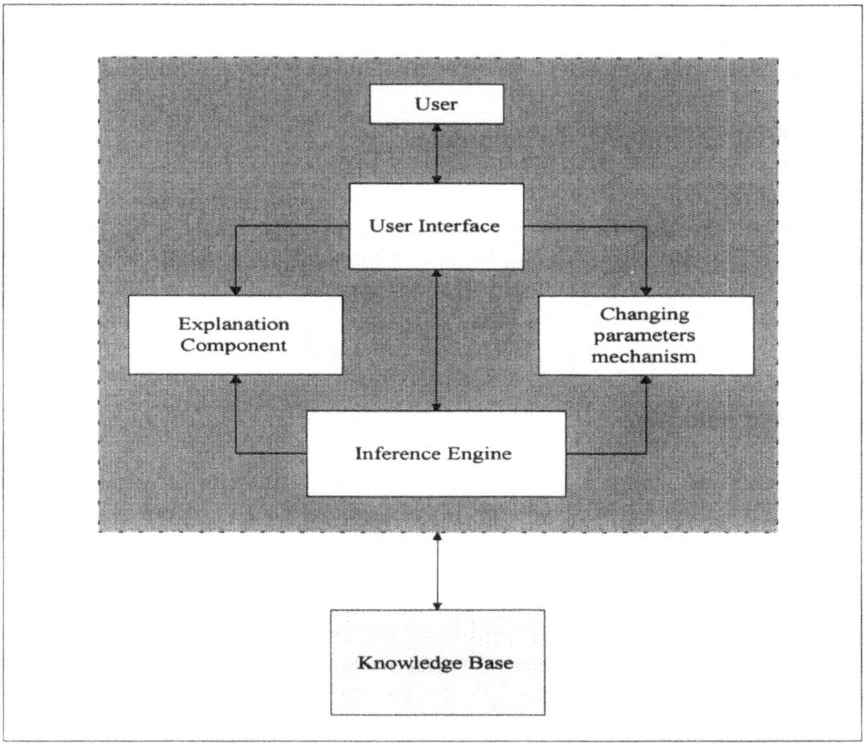

Fig. 10.5.1..System Architecture

There are two kinds of telecommunication rules included in the knowledge base. Rules are classified into two general groups: the technical ones (multiplexing, frequency assignment, available services etc.), which are presented in Table 10.5.1, and the general ones (about fading, co-channel interference, use of microcells etc.), which are generated after interviews with expert telecommunication engineers.

The system through the *interface utility* that is based on a windows based mechanism poses some questions whose answers will be used for the final decision. In each question the user has the possibility to ask for explanations. In that case the system gives a brief description about the specific query and the respective answer by means of the explanation component. The user has the capability to alter the data used in the system with the changing parameter module. This mechanism supports operations relating the entry and deletion of new rules in the knowledge base and the permanent and temporary parameter changes.

All the above modules cooperate with each other and the *inference engine* produces the final results concerning the choice of the standard that is proper for the specific cellular network that the user wants to install and the technoeconomic total expense and income analysis. Finally under the above circumstances the system finally decides if such an investment meets predetermined criteria [1], [8].

10.5.4 System Operation and Calculations

In general the operation of our KBDSS may be represented by the flow diagrams of Fig. 10.5.2a and 10.5.2b. In that figure the data flow is represented by directional arrows. These arrows are placed between operational modules or between operation modules and the knowledge base. A more analytical description of the most basic modules is presented in the following subsections

10.5.4.1 Standard selection

The selection of a proper communication standard, in virtue of which a cellular telecommunication network will be developed, forms an important determining factor that affects significantly the success of a prospective investment. The standard in question must satisfy the subscriber needs and demands as well the needs of the organization that will operate the network and supply the requisite services. The knowledge base contains data and information for various basic standards used in the field of terrestrial mobile communications (cellular and cordless) [12-14]. These standards are the GSM, DCS1800, IS-54, IS-95, JDC, DECT, CT-2, PHS, and PACS and their specifications (included in table 10.5.1) are briefly as follows:

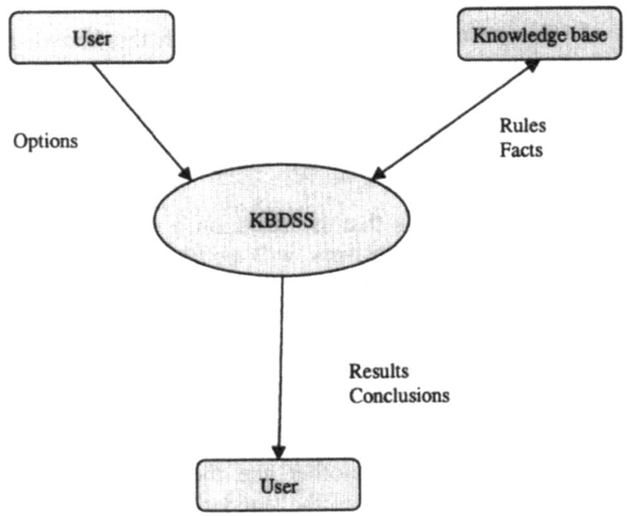

Fig. 10.5..2a. General flow diagram

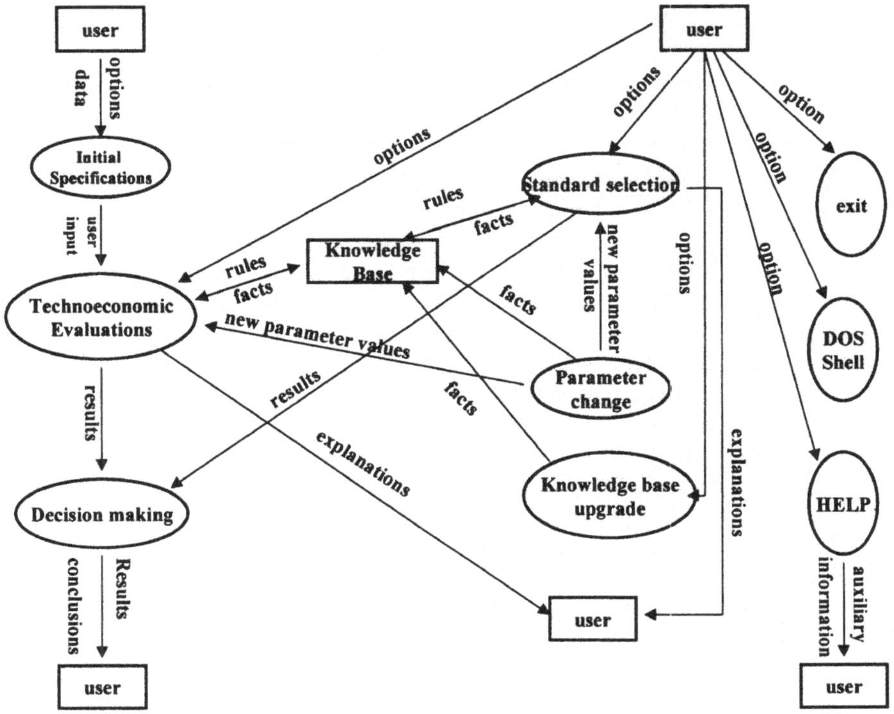

Fig. 10.5.2b. KBDSS flow diagram

Table 10.5.1. Technical characteristics of wireless systems described in the knowledge base

	DCS1800	GSM	IS-54	IS-95	JDC	DECT	CT-2	PHS	PACS	Un
Frequency range	1800	900	970	1900	880	1900	865	1900	1900	MH
Multiple access	TDMA	TDMA	TDMA	CDMA	TDMA	TDMA	FDMA	TDMA	TDMA	
Duplex	FDD	FDD	FDD	FDD	FDD	TDD	TDD	TDD	FDD	
Modulation	GMSK	GMSK	π/4 DQPSK	π/4 DQPSK	π/4 DQPSK	GMSK	GFSK	π/4 DQPSK	π/4 QPSK	
Frequency assignment	dynamic	dynamic	fixed	fixed	Fixed	dynamic	dynamic	dynamic	dynamic	
Channels/carrier	8 (full rate) 16 (half rate)	8 (full rate) 16 (half rate)	3 (full rate) 6 (half rate)	3 (full rate) 6 (half rate)	3 (full rate) 6 (half rate)	12	1	4	8	
Carrier spacing	200	200	30	30	25	1728	100	300	300	KH
Type	cellular	cellular	cellular	cellular	Cellular	cordless	cordless	cordless	cordless	
High Bit Rate Services	Yes	Yes	No	No	Yes	Yes	Yes	Yes	Yes	
Low Bit Rate Services	Yes	Yes	No	No	No	Yes	Yes	Yes	Yes	
Video	Yes	No	No	No	No	No	No	Yes	No	
Still images Transmission	Yes	No	No	No	No	No	No	Yes	No	

Abbreviations :

TDMA: Time Division Multiple Access
FDMA: Frequency Division Multiple Access
CDMA: Code Division Multiple Access
GMSK: Gaussian Minimum Shift Keying
GFSK: Gaussian filtered FSK
DQPSK: Differential Quaternary Phase Shift Keying
QPSK: Quaternary Phase Shift Keying

- The *Global System for Mobile Communications (GSM)* is a European digital cellular standard formerly known as Groupe Speciale Mobile. It is a TDMA digital system and has been extremely successful as a result of good technical design and the fact that it is an open standard allowing operators to multisource equipment both on the network and mobile side.
- Originally GSM was intended to be operated only in bands around 900 MHz. In 1989, the UK department of Trade and Industry started an initiative which finally led to the assignment of 150MHz near 1.8GHz for Personal Communications Networks (PCN) in Europe, and to the choice of GSM as a standard for that application. This system is called *DCS1800* for *Digital Cellular System 1800* and is an advanced version of GSM.
- The standard *IS-54* was developed in USA by the Electronic Industries Association (EIA) and the Telecommunications Industry Association (TIA) and is also based on TDMA.
- The EIA/TIA IS-95 standard is based on the CDMA system in which many users share a common channel for transmission. The standard has gained some success in the United States and Asia- Pacific regions and is now entering commercial deployment.
- The *JDC (Japan Digital Cellular)* was developed in 1991 under the auspices of the Ministry of Posts and Telecommunications and is based on TDMA method.
- Cordless systems are designed to replace wired phones and to provide mobility. *CT2* was initially developed in the United Kingdom as an alternative to analog cordless home phones.
- The *Digital European Cordless Telephone (DECT)* was standardized by ETSI during the early 1990s, and is designed as a flexible interface to provide cost-effective communication services to high user densities in picocells. The standard is intended for applications such as domestic cordless telephony, telepoint, cordless PBXs. Functionally DECT is closer to a cellular system than to a classical cordless telephone.
- The concept of the *Personal Handyphone System (PHS,* formerly *PHP)* was launched, on the basis of a digital cordless telephone and a digital network. The system objectives are to provide not only home and office usage, but also public access capability.
- *PACS (Personnal Access Communication Services)* was launched for industrial reasons and was designed to operate in the band of 2GHz

In order to find a suitable standard from the set of the aforementioned ones, the information system handles and processes a great number of information. The system queries for the user are designed to be as general as possible, assuming minimum user background. More specifically the user has to give answers to the following [1], [8]:

- *Queries about the proposed mobile services*: With these queries, the user defines some of the available mobile services (e.g. fax, video, still images) that are offered by the cellular network. It is obvious that a market research must be done first by the concerned organization, so that the demand for such services will be known.

- *Queries about the desired technical characteristics of the cellular network.* In particular they request information about the frequency range, the kind of multiple access, frequency assignment, duplexing and modulation, the carrier spacing and channels per carrier. Based on the answers of these queries (in combination with the previous queries), the appropriate mobile system for the development of the cellular network is specified.

When all the above queries have been answered by the operator, the system begins to process the acquired information. Initially it tries to find if the knowledge base has a specific standard that satisfies the user's demands. The system informs the operator if such a standard exist, otherwise it shows standards that may satisfy him. Moreover the user is informed about the characteristics that are common with the one he wants. In the latter case the user has to select one of the suggested standards.

10.5.4.2 Initial specifications

The information system requires some information in order to execute three of the following most basic operations:

- The specification of the telecommunication infrastructure.
- The estimation for the cash flow from the use of the network.
- The calculation of infrastructure, operational and future extension cost of the network.

In particular there are questions about, the surface of the area to be covered, the percentage of the area being urban, suburban and rural (for the estimation of the number of microcells, picocells and macrocells) and the morphology of the area (flat, hilly or mountainous terrain).

10.5.4.3 The determination of the telecommunication infrastructure

The first action concerning the determination of the infrastructure is the calculation of the cell number that is required for an area coverage that is in the user's interest, as well as the type of those cells. Namely,

- the densely populated regions (urban) will be covered by picocells (radius ~100m),
- the sparsely populated regions (suburban) by microcells (radius ~1000m), and
- the non-populated regions (rural) by macrocells (radius ~10km).

By using internal calculations the system is able to find how many square meters of rural, urban and suburban must be covered and the area of the macro/micro/picocell, and then calculates the approximate number of the required cells.

Afterwards the system proceeds to the calculation of the required base stations. In most of cases, a single base station exists. However, due to the fading phenomenon occurred by the existence of physical obstacles (buildings, hills, mountains) and also the particular characteristic of the current transmission media, there may be more economical and optimal to use more than one base stations. By properly processing the input data about the morphology of the structure and the characterization of a specific region, the system calculates the final number of stations capable to provide a good quality of service. That final number has to do with the initial infrastructure of the mobile network. In the course of time, due to the subscribers' population increase or decrease (leading to the traffic load increase/decrease) there must take place a gradual extension of the network with the addition/deletion of other stations. Such expansion starts after a time period and depends on the subscribers' number and the kind of services offered.

A typical terrestrial cellular system is supervised by the mobile telephone switching office (MTSO). Such an office controls a great number of base stations. In order to diminish overloading our KBDSS assigns one MTSO for every 500 base stations. This value may be changed of course under the appearance of new conditions.

10.5.4.4
Estimation of the infrastructure, operational and future extension costs

Infrastructure cost is the cost that will put a strain on the organization concerning the purchase and installation of the required telecommunication equipment. (Installation of base stations (BSs), connection of BSs with the mobile telephone switching office (MTSO)). That quantity is given as a parametrical value in order to follow market variations.

Operational costs contain basically the cost of manpower that is required to support a cellular network (engineers, technicians, managers e.t.c.). The system initially estimates the number of people need for the proper network operation. Personnel wages are calculated by using a parameter value that represents the mean annual value of an employee income. Operational expenses such as promotion, building hire e.t.c are also included as parameters in that cost.

The network extension and upgrade takes place gradually and starts when the initial infrastructure does not guarantee a satisfactory level of quality of service. Expansion happens when the total number of subscribers goes over 1% of the total initial population. An amount of 5% of the initial infrastructure cost will be appropriated then from the operator. The system can also calculate the specific time in months that the network becomes overloaded. From that time on the extension and upgrade expenditure is gradually added. The quantities of 1% and 5% are some typical figures used by some Greek carriers.

Obviously the total cost is the aggregate of the above three ones. The system has the capability to present that amount distribution per month where the evaluation takes place using proper graphical representations.

10.5.4.5
Income estimation

Before proceeding to income estimation, the information system makes a forecast for the initial subscribers and their increase rate by the course of time. That forecast is feasible after a proper market survey made by the concerned operator.

In more details, the system first calculates the number of subscribers per month. (e.g. there will be 5000 subscribers after the first month, 10000 after the second e.t.c.). Income is then calculated per each month which is the aggregate of the following three quantities:

- The payment of the basic fixed amount: All the subscribers should pay every month an amount that corresponds to the fixed sum of the basic mobile service (e.g. voice transmission).
- The payment of the mean unit usage charge: An indicative parameter value that corresponds to the mean unit usage charge has been given by the operator. All the subscribers are charged according to usage.
- The payment of the fixed amount for every additional service: For any additional mobile service, the operator defines two parameter values that have to do with the usage proportion and the fixed amount of every new service. For example 10% of subscribers will use the data transfer service charged a fixed amount of 5000 drs.

The KBDSS is capable to represent the monthly income distribution using graphical representations.

10.5.4.6
Decision making

After the technoeconomic computations have been made, the system is able to evaluate a possible investment for the development of a cellular network. The evaluation depends on the following criteria, which are generated after the above data have been inserted:

- The time needed for plant depreciation.
- The rate of income increases.

The above two criteria must be both taken into account for the evaluation. It is evident that for an investment evaluation regarding the time period for depreciation, the depreciation month (month when total depreciation occurs) must be calculated. In case that this calculation is difficult (depreciation does not occur in the specified time period) the system estimates the likely month with a reliability factor. Then the information system performs the evaluation for the real or the likely depreciation month.

A parallel evaluation concerning the income increase rate is also possible. Finally the operator-investor may compare the proceeds of his investment with other possibilities he might have and make an appropriate decision.

10.5.5 Relations between Basic Rules

This section contains the most basic rules that are used by the decision process. The relations among the rules i.e. which rules are preceded or followed by a specific rule and which rules are not permitted to be executed are described in table 10.5.2. Blanks in the table means that most of the rules are independent each other and are executed after the activation of the corresponding facts.

10.5.5.1 Rules

1. If natural obstacles in the geographical area of a cell exist (hills, mountains etc) then there is long term fading
2. If non-natural obstacles in the geographical area of a cell exist (small or large buildings etc) then there is short term fading
3. If some kind of fading exists then more base stations should be added
4. If the area to be covered has irregular shape then more base stations should be added
5. If co-channel interference or adjacent channel interference happens then we must use directional antennas
6. If we want a sufficient level for "grade of service" then we have to extend the network gradually
7. If an urban area must be covered then picocells should be used
8. If a suburban area must be covered then microcells should be used
9. If a rural area must be covered then macrocells should be used
10. If the number of subscribers grows up suddenly then traffic flow will also grow up
11. If traffic flow grows up then the "grade of service" decreases
12. If the fixed sum is reduced then the number of subscribers will grow up suddenly
13. If the fixed sum grows up then the number of subscribers will decrease suddenly
14. If we want high bit rate services then we can use one of the following: DCS1800,GSM, JDC, DECT, CT2, PHS, PACS
15. If we want low bit rate services then we can use one of the following: DCS1800,GSM, DECT, CT2, PHS, PACS
16. If we want video transmission then we can use one of the following: DCS1800, PHS
17. If we want transmission of image files then we can use one of the following: DCS1800, PHS
18. If we want the capability of further addition of new services then we can use one of the following: DCS1800, PHS, GSM, JDC, DECT
19. If the network must operate near 900 MHz then we can use GSM, IS-54, JDC or CT2
20. If the network must operate near 1800 MHz then we can use DCS1800

21. If the network must operate near 1900 MHz then we can use IS-95, DECT, PHS or PACS
22. If we want Time Division Multiple Access then we can use DCS1800, GSM, JDC, DECT, IS-54, PACS or PHS
23. If we want Frequency Division Multiple Access then we can use CT2
24. If we want Code Division Multiple Access then we can use IS-95
25. If we want Time Division Duplexing then we can use DECT, CT2 or PHS
26. If we want Frequency Division Duplexing then we can use DCS1800, GSM, JDC, IS-95, IS-54 or PACS
27. If we want GMSK modulation to be used then we can use DCS1800, GSM or DECT
28. If we want GFSK modulation to be used then we can use CT2
29. If we want $\pi/4$ QPSK modulation to be used then we can use PACS
30. If we want $\pi/4$ DQPSK modulation to be used then we can use IS-54, IS-95, JDC or PHS
31. If we want dynamic frequency assignment then we can use DCS1800, GSM, DECT, CT2, PHS or PACS
32. If we want fixed frequency assignment then we can use IS-54, IS-95 or JDC.
33. If we want the channels per carrier spacing to be equal to 1 then we can use CT2
34. If we want the channels per carrier spacing to be equal to 3 (full rate) or 6 (half rate) then we can use IS-54 or IS-95
35. If we want the channels per carrier spacing to be equal to 4 then we can use PHS
36. If we want the channels per carrier spacing to be equal to 8 then we can use PACS
37. If we want the channels per carrier spacing to be equal to 8 (full rate) or 16 (half rate) then we can use GSM or DCS1800
38. If we want the channels per carrier spacing to be equal to 12 then we can use DECT
39. If we want bandwidth to be equal to 25 KHz then we can use JDC
40. If we want bandwidth to be equal to 30 KHz then we can use IS-54 or IS-95
41. If we want bandwidth to be equal to 100 KHz then we can use CT2
42. If we want bandwidth to be equal to 200 KHz then we can use DCS1800 or GSM
43. If we want bandwidth to be equal to 300 KHz then we can use PHS or PACS
44. If we want bandwidth to be equal to 1728 KHz then we can use DECT
45. If the network is going to cover a large geographical area then we must use a cellular standard
46. If the network is going to cover a small geographical area then we must use a cordless standard
47. If we need a cellular standard then we can use GSM, DCS1800, IS-54, IS-95 or JDC
48. If we need a cordless standard then we can use DECT,CT2, PHS or PACS
49. If the network is going to take place in Europe then we can use DSC1800, GSM, DECT or CT2

50. If the network is going to take place in America then we can use IS-54, IS-95 or PACS
51. If the network is going to take place in Asia then we can use JDC or PHS.

Table 10.5.2. Rules interaction

Rule	Rules that are preceded or follow	Rules that are not permitted
1	3	-
2	3	-
3	1,2	-
4	-	-
5	-	-
6	10,11,12	-
7	-	8,9
8	-	7,9
9	-	7,8
10	6,11,12	-
11	6,10,12	-
12	10,11,6	13
13	-	12
14	-	-
15	-	-
16	14	-
17	14	-
18	-	-
19	-	20,21
20	-	19,21
21	-	19,20
22	-	23,24
23	-	22,24
24	-	22,23
25	-	26
26	-	25
27	-	28,29,30
28	-	27,29,30
29	-	27,28,30
30	-	27,28,29
31	-	32
32	-	31
33	-	34,35,36,37,38
34	-	33,35,36,37,38
35	-	33,34,36,37,38
36	-	33,34,35,37,38
37	-	33,34,35,36,38
38	-	33,34,35,36,37
39	-	40,41,42,43,44
40	-	39,41,42,43,44
41	-	39,40,42,43,44
42	-	39,40,41,43,44
43	-	39,40,41,42,44
44	-	39,40,41,42,43
45	-	46
46	-	45
47	-	48
48	-	47
49	-	50,51
50	-	49,51
51	-	49,50

10.5.6 Network Design Example

As an application of the methodology presented in this chapter, the following example illustrates the capability of this expert system in designing and evaluating a cellular network, which could cover the entire area of Greece. The queries and the answers for this specific example are as follows:

10.5.6.1 Phase 1: determine the ideal standard

Except voice transmission, a cellular network could offer a variety of services:

- *Are you interested for high bit rate services? [yes, no or don't care]*
  Choice: **Yes**
- *Are you interested for low bit rate services? [yes, no or don't care]*
  Choice: **Yes**
- *Are you interested for the transmission of still images? [yes, no or don't care]*
  Choice: **Don't care**
- *Are you interested for video transmission? [yes, no or don't care]*
  Choice: **Yes**
- *In what frequency range will the network operate?*
  1. *900 MHz*
  2. *1800 MHz*
  3. *1900 MHz*
  4. *Don't care*

  Choice: **2**
- *What kind of multiple access would you like?*
  1. *TDMA*
  2. *FDMA*
  3. *CDMA*
  4. *Don't care*

  Choice: **1**
- *What kind of duplexing would you like?*
  1. *TDD*
  2. *FDD*
  3. *Don't care*

  Choice: **2**
- *What kind of modulation would you like?*
  1. *GMSK*
  2. *GFSK*
  3. *QPSK*
  4. *DQPSK*

5. *Don't care*

Choice: **1**

- *What kind of frequency assignment would you like?*
  1. *Dynamic*
  2. *Fixed*
  3. *Don't care*

  Choice: **1**

- *How many channels per carrier would you like?*
  1. *1 channel*
  2. *3 (full rate) and 6 (half rate)*
  3. *4 channels*
  4. *8 channels*
  5. *8 (full rate) and 16 (half rate)*
  6. *12 channels*
  7. *Don't care*

  Choice: **7**

- *What carrier spacing would you like?*
  1. *25 kHz*
  2. *30 kHz*
  3. *100 kHz*
  4. *200 kHz*
  5. *300 kHz*
  6. *1728 kHz*
  7. *Don't care*

  Choice: **4**

The DCS1800 standard satisfies all your requirements.

The GSM standard satisfies many of your requirements, except frequency range.

Proposed standard: DCS1800.

10.5.6.2
Phase 2: technoeconomic evaluation

- *Enter the time period (in months) for evaluating the network:* ***72***
- *Enter the surface of the area (in* Km^2*) that will cover by the cellular network:* ***134000***
- *Enter the population of this area:* ***10000000***
- *What percentage of this area is urban?* ***0.3***
- *What percentage of this area is suburban?* ***3***
- *What is the terrain morphology?*
  1. *Flat terrain*

2. *Hilly terrain*
3. *Mountainous terrain*
4. *Combination of the above*

Choice: **4**

- *What is the density of various obstacles?*
  1. *High density*
  2. *Medium density*
  3. *Low density*

  Choice: **2**

The relation between the total expense and the cash flow from the subscribers is shown in the diagram of Fig. 10.5.3:

Time needed for plant depreciation: 60 months (Very Good)
Rate of income increasing: 0,28 (Good)
Investment: Encouraging.

10.5.7 Conclusions

The basic operations of a KBDSS to help mobile operators to install and evaluate a cellular system have been presented in this study. KBDSSs have proved to be very successful in formalizing the practical rules used by experts for cellular network design, formalizing the logic of solving such problems, and choosing the most suitable solutions for a certain networking requirement.

The above information system produced, helps perform technoeconomic evaluations and repetitive tasks, do complex mathematical calculations and keep track of large amounts of information, which would be much slower and more costly if this evaluation were made by expert staff hired by the operator. The first point that must be emphasized regarding the results of this chapter is that, for a specific implementation the results have been very encouraging. We had a reliable estimation (after comparison with real data) for the time needed for the operation to be profitable, assuming the implementation of a cellular network will cover the entire area of Greece. However, it must be pointed out that the vital decisions are always made by the operator, who may have other non-technoeconomic factors not considered by the system.

Our future work will include the development of a complimentary neural network decision tool to support the expert system for the exact designation of the locations of the base stations. An area that is open for further study is also the evaluation of decision rules utilized so that the tool proposed here become as versatile and include as many decision support situations as possible.

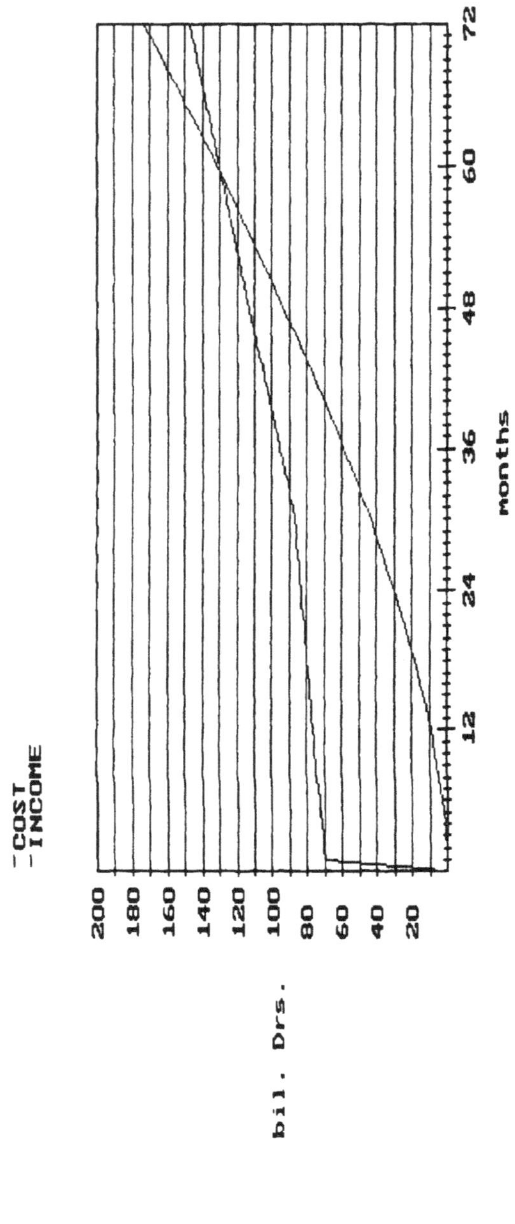

Fig. 10.5..3. Relation between income and expenditure with basic fixed amount 10.000 Drs. Region Coverage: Greece

References

1. Zambarakis T. and Stavroulakis P., "An Expert System for the Technoeconomic Evaluation of Mobile Cellular Network Implementation", *6th International Conference on Advances in Communications and Control (COMCON 6)*, 1997

2. Zopounidis C., Doumpos M. and Matsatsinis N.F., "On the use of knowledge-based decision support systems in financial management: A survey", *Decision Support Systems*, vol. 20, pp. 259-297, 1997

3. Liebowitz J., *Expert Systems Applications to Telecommunications*, John Wiley & Sons, 1988.

4. Covo A.A., Mouuzi T.M., and Peterson E.D., "Artificial Intelligence. Assisted Telecommunication Network Management", *Proc. IEEE GLOBECOM*, Dallas, TX, pp.487-491, 1989.

5. Muller C. et al., "Artificial Intelligence in Telecommunications", *Proc. IEEE GLOBECOM*, pp.883-887, 1993

6. Goyal S. and Worrest R., "Artificial Intelligence Directions in Telecommunications", *Proc. IEEE GLOBECOM*, pp.46.8.1-46.8.5, 1987.

7. Chan H.C. and Lee S.L., "Support for Mobile Communications Planning", *Decision Support Systems*, vol. 23, pp. 97-110, 1998.

8. Zamparakis T.T, Kokkinogenis N.N, Sandalidis H.G and Stavroulakis P, "Structure and Operation of a Knowledge-based Decision Support System for Terrestrial Cellular Networks", *International Workshop on Mobile Communications,* Chania, Greece pp. 375-382, June 1999

9. Siskos, Y. and Spyridakos A., "Intelligent Multicriteria Decision Support: Overview and Perspectives", *European Kournal of Operational Research* 113, pp. 236-246, 1999

10. Klein M. and Methlie L.B., *Knowledge-based Decision Support Systems with Applications in Business*, John Wiley, 1995.

11. Turban E., *Decision Support and Expert Systems: Management Support Systems*, McMillan New York, 1993

12. Wong P., *Mobile Data Communications Systems*, Artech House, 1995.

13. Webb W., *Introduction to Wireless Local Loop*, Artech House Inc., 1998.

14. Padgett J.E. et all "Overview of Wireless Personal Communications", *IEEE Communications Magazine*, pp. 28-41, 1995

The manufacturer's authorised representative in the EU is Springer Nature Customer Service Centre GmbH, Europaplatz 3, 69115 Heidelberg, Germany. If you have any concerns regarding our products, please contact ProductSafety@springernature.com

Printed and bound by CPI Group (UK) Ltd, Croydon, CR0 4YY

01/10/2026

02225968-0020